Y0-AAJ-150

ZINC

The Science and Technology of the Metal, Its Alloys and Compounds

Prepared with Cooperation of the American Zinc Institute under the Editorial Supervision of

C. H. MATHEWSON

Professor Emeritus of Metallurgy and Metallography
Yale University

with Chapters by Specialists

American Chemical Society
Monograph Series

REINHOLD PUBLISHING CORPORATION
NEW YORK

CHAPMAN & HALL, LTD., LONDON

THIRD PRINTING 1964

Library of Congress Catalog Card Number: 59-13266

Printed in the United States of America
THE GUINN CO., INC.
New York 14, N. Y.

CONTRIBUTORS

Edmund A. Anderson, *Chief of Products Application, Research Department, The New Jersey Zinc Company (of Pa.), Palmerton, Pa.*

G. W. Ashman, *Assistant to Manager of Pigment Division, The New Jersey Zinc Company, New York, N. Y.*

Alan M. Bateman, *Silliman Professor Emeritus of Geology, Yale University, New Haven, Conn.*

O. M. Bishop, *Zinc Specialist, U. S. Bureau of Mines, Washington, D. C.*

John H. Calbeck, *Director of Research, Pigment Division, American Zinc Oxide Company, Columbus, Ohio.*

H. D. Carus, *President, Matthiessen and Hegeler Zinc Company, LaSalle, Illinois.*

Nelson E. Cook, *General Superintendent of Galvanizing, Wheeling Steel Corporation, Wheeling, West Virginia.*

J. E. Crowley, *Division Head; Chemistry, Research Department, Bethlehem Steel Company, Bethlehem, Pa.*

C. T. Flachbarth, *Engineer, Walker Brothers, Conshohocken, Pa.*

J. C. Fox, *Die Casting Consultant, Toledo, Ohio.*

K. S. Frazier, *Chief Research Engineer, Fenestra Incorporated, Detroit, Michigan.*

H. H. Greene, *Metallurgist, Sheet and Strip Division, Republic Steel Corporation, Cleveland, Ohio.*

G. F. Halfacre, *General Manager, The New Jersey Zinc Company (of Pa.), Palmerton, Pa.*

R. E. Harr, *Chemical Engineer, Metal Finishing Development, Western Electric Company Incorporated, Chicago, Illinois.*

Ernest W. Horvick, *American Zinc Institute, New York, N. Y.*

W. T. Isbell, *Consultant, St. Joseph Lead Company, New York, N. Y.*

T. R. Janes, *Superior Zinc Corporation, Philadelphia, Pa.*

T. D. Jones, *Chief Lead Refinery Metallurgist, American Smelting and Refining Company, Barber, New Jersey.*

John L. Kimberley, *Executive Vice President, American Zinc Institute, New York, N. Y.*

Roland J. Lapee, *Metallurgist, Great Falls Department, The Anaconda Company, Great Falls, Montana.*

Walter L. Latshaw, *Director of Agricultural Department (Retired), United States Smelting, Refining and Mining Company, Salt Lake City, Utah.*

H. E. Lee, *Vice President, Research and Development, The Bunker Hill Company, San Francisco, California.*

T. J. Lennox, Jr., *Research Investigator, American Smelting and Refining Company, Central Research Laboratories, South Plainfield, New Jersey.*

G. T. Mahler, *Chief of Metallurgical Development, Research Department, The New Jersey Zinc Company (of Pa.), Palmerton, Pa.*

R. K. Martin, *Technical Director, Matthiessen and Hegeler Zinc Company, LaSalle, Illinois.*

C. H. Mathewson, *Professor Emeritus of Metallurgy and Metallography, Yale University, New Haven, Conn.*

K. D. McBean, *Metallurgical Engineer, The Consolidated Mining and Smelting Company of Canada Limited, Trail, B. C.*

Elmer L. Miller, *Assistant Superintendent, Amarillo Plant, American Smelting and Refining Company, Amarillo, Texas.*

S. W. K. Morgan, *Research Director, Imperial Smelting Corporation Limited, Avonmouth, England.*

C. W. Morrison, *Manager, Cherryvale Zinc Company Incorporated, affiliate of the National Zinc Company Incorporated, Bartlesville, Oklahoma.*

Richard H. Mote, *Chief, Branch of Base Metals, Division of Minerals, U. S. Bureau of Mines, Washington, D. C.*

K. Oganowski, *Associate Director, Research Laboratories, Armco Steel Corporation, Middletown, Ohio.*

Rollin A. Pallanch, *Consulting Mill Metallurgist, Salt Lake City, Utah.*

W. M. Peirce, *Assistant to Executive Vice President, The New Jersey Zinc Company, New York, N. Y.*

K. A. Phillips, *Assistant Chief Metallurgist, American Zinc, Lead and Smelting Company, East St. Louis, Illinois.*

J. N. Pomeroy, *Vice President, General Smelting Company, Philadelphia, Pa.*

James J. Rankin, *Production Engineer, St. Joseph Lead Company, Monaca, Pa.*

J. L. Roemer, Jr., *Superintendent, Butt Weld Tube Mills, The Youngstown Sheet and Tube Company, Youngstown, Ohio.*

Harvey W. Schadler, *Metallurgy and Ceramics Department, General Electric Research Laboratory, Schenectady, N. Y.*

R. Schuhmann, Jr., *Professor of Metallurgical Engineering, Purdue University, West Lafayette, Indiana.*

Lewis S. Somers, III, *Superintendent, Photoengraving Zinc Manufacturing, Imperial Type Metal Company, Philadelphia, Pa.*

W. H. Taylor, *Advertising Manager, Walker Brothers, Conshohocken, Pa.*

A. Paul Thompson, *Director of Research, The Eagle-Picher Company, Joplin, Missouri.*

J. Paul Tierney, *Assistant Superintendent; Rod, Wire and Conduit Departments, The Youngstown Sheet and Tube Company, Youngstown, Ohio*

Sam Tour, *Sam Tour and Company Incorporated, New York, N. Y.*

G. H. Turner, *Senior Research Engineer, The Consolidated Mining and Smelting Company of Canada Limited, Trail, B. C.*

Bert L. Vallee, *Assistant Professor of Medicine, Harvard Medical School and Scientific Director, The Biophysics Research Laboratory of the Department of Medicine, Harvard Medical School and Peter Bent Brigham Hospital, Boston, Mass.*

George F. Weaton, *Consultant, St. Joseph Lead Company, New York, N. Y.*

Floyd S. Weimer, *Manager, Great Falls Department, The Anaconda Company, Great Falls, Montana.*

T. H. Weldon, *General Superintendent, Metallurgical Division, The Consolidated Mining and Smelting Company Limited, Trail, B. C.*

Guy T. Wever, *Superintendent, Zinc Plant, Great Falls Department, The Anaconda Company, Great Falls, Montana.*

S. E. Woods, *Manager, Development Department, Imperial Smelting Corporation Limited, Avonmouth, England.*

GENERAL INTRODUCTION

American Chemical Society's Series of Chemical Monographs

By arrangement with the Interallied Conference of Pure and Applied Chemistry, which met in London and Brussels in July, 1919, the American Chemical Society was to undertake the production and publication of Scientific and Technologic Monographs on chemical subjects. At the same time it was agreed that the National Research Council, in cooperation with the American Chemical Society and the American Physical Society, should undertake the production and publication of Critical Tables of Chemical and Physical Constants. The American Chemical Society and the National Research Council mutually agreed to care for these two fields of chemical progress. The American Chemical Society named as Trustees, to make the necessary arrangements of the publication of the Monographs, Charles L. Parsons, at that time secretary of the Society, Washington, D. C.; the late John E. Teeple, then treasurer of the Society, New York; and the late Professor Gellert Alleman of Swarthmore College. The Trustees arranged for the publication of the ACS Series of (a) Scientific and (b) Technological Monographs by the Chemical Catalog Company, Inc. (Reinhold Publishing Corporation, successor) of New York.

The Council of the American Chemical Society, acting through its Committee on National Policy, appointed editors and associates (the present list of whom appears at the close of this sketch) to select authors of competent authority in their respective fields and to consider critically the manuscripts submitted. Since 1944 the Scientific and Technologic Monographs have been combined in the Series. The first Monograph appeared in 1921 and, up to 1957, 134 treatises have enriched the Series.

Owing to the prodigious expansion of knowledge in the broad fields of chemical science and technology, the Monographs are generally restricted to recognized areas of more or less limited specialization. In some cases, however, as with the present volume, the Monograph assumes a composite form, with individual chapters or sections contributed by selected authors and correlated by an editorial authority.

These Monographs are intended to serve two principal purposes: first, to make available to chemists a thorough treatment of a selected area in form usable by persons working in more or less unrelated fields to the end that they may correlate their own work with a larger area of physical science discipline; second, to stimulate further research in the specific field treated. To implement this purpose the authors of Monographs are expected to give extended references to the literature. Where the literature is of such volume that a complete bibliography is impracticable, the authors are expected to append a list of references critically selected on the basis of their relative importance and significance.

PREFACE

The purpose of this composite monograph is to offer a comprehensive treatise on the metallurgy of zinc with as much attention to economic, statistical, scientific, general technological and adaptive aspects of the metal as can be combined in a single volume.

The book was constructed on the basis of a planned outline submitted by the editor and critically reviewed by a committee of the American Zinc Institute composed of Messrs. C. R. Ince, Vice President, St. Joseph Lead Company; R. G. Kenly, Vice President, The New Jersey Zinc Company and S. D. Strauss, Vice President, American Smelting and Refining Company; aided by Mr. J. L. Kimberley, Executive Vice President of the American Zinc Institute.

Its substance is essentially a cooperative and composite contribution from the American, Canadian and English zinc industry, with additional coverage from academic or governmental sources in the United States. Obviously, there is some lack of coherence and uniformity of style, engendered by combining the writings of more than 50 authors on subjects which do impinge on one another in certain areas, but an effort has been made to reduce these nonconformities to a minimum. Where collateral subjects seemed to affect the work of the various authors, they have endeavored to establish priorities by consultation among themselves.

In general, the chapters are organized under broad titles with divisional coverage of the principal subordinate subjects by specialists. This is particularly the case with Chapter 9, The Processing and Uses of Metallic Zinc and Zinc-Base Alloys, in which a section on Zinc Coatings arranged by Mr. J. L. Kimberley makes use of a dozen contributors.

In the matter of corrosion, a separate chapter has not been included since the principal functions of zinc—as a sacrificial protective coating, and as a die cast or fabricated product often requiring a protective or decorative finish—have been considered in appropriate sections of the book. It may be noted that, among the more recent additions to corrosion literature, No. 129 of this Monograph Series is concerned with the technology of corrosion control.

By and large, the various chapters are written after the manner of technical papers, with appended references to representative work in the designated fields. In some cases, the reference material is closely associated with the text but in others it is introduced mainly for supplementary reading or study.

The individual authors have developed their own theses without any close attempt at supervision by the editor except in the matter of preliminary outlines or occasional advice.

With respect to zinc alloys extending beyond the scope of those based primarily on the metal itself, it cannot be presumed that this book occupies a dominant position by way of describing every function served by zinc in a general alloy technology. This aspect of the subject has not been neglected, however, and Chapter 10 contains material abstracted from various sources; notably, the chapter by V. P. Weaver of the American Brass Company on "Wrought Copper Alloys" in the ACS Monograph "Copper." It also contains a rather complete documentation of phase relationships of zinc in binary and ternary combination with other metals, based primarily on the diagrammatic material contained in Smithell's excellent "Metals Reference Book," kindly authorized by Butterworths (London) and Interscience Publishers, Incorporated (New York). There are, of course, many references to the Specifications of the American Society for Testing Materials and to published material issued by The American Institute of Mining, Metallurgical and Petroleum Engineers and The American Society for Metals (Metals Handbook), whose invaluable services to the American metal industry are so well recognized.

The editor's acknowledgment of cooperation and assistance must first of all be directed to the members of the aforementioned committee of the American Zinc Institute, whose knowledge and experience in the production and marketing of zinc products has insured expert treatment of the principal topics included in the book. Mr. J. L. Kimberley and his assistants in the New York office of the American Zinc Institute have given invaluable aid in furthering the editorial work and in reviewing the entire contents of the volume.

The many authors of sections and chapters have furnished the kind of cooperation which has brought much satisfaction and pleasure into the rather complex task of assembling and editing their contributions. They have in turn addressed a large number of acknowledgments to the many individuals and organizations that have furnished assistance in their compilations. From England we have had the valued cooperation of Lord David Kirkwood of the Imperial Smelting Corporation Limited.

C. H. Mathewson

Hammond Laboratory
Yale University
August 1959

CONTENTS

CHAPTER 1

HISTORICAL BACKGROUND

H. D. Carus

President, Matthiessen and Hegeler Zinc Company
LaSalle, Illinois

It is highly probable that zinc as a metal was known to the ancients, although its production and use were only occasionally encountered and the nature of the metal was very little understood. "False silver," (*pseudargyras* in Greek) was the first description given of zinc, by Strabos in the passage describing Andriera in Mysia. It is possible that this citation is too vague to give guidance for historical study, but it is interesting to note that near Balia, not far from the site of Andriera, there occur zinc deposits of blende, zinc sulfide, with iron pyrites and galena.

The oldest known piece of zinc extant is in the form of an idol found in the prehistoric Dacian settlement at Dordosch, Transylvania. The analysis of the idol showed zinc, 87.52 per cent, lead 11.41 per cent and iron, 1.07 per cent. There have been found in the ruins of Cameros destroyed 500 B.C., two bracelets filled with zinc, and in the ruins of Pompeii, destroyed 79 A.D., the front of a fountain, the upper part of which is covered with zinc.

The Romans, as early as 200 B.C., were well acquainted with brass, although the Greeks were not. Brass was prepared about the time of Augustus, 20 B.C. to 14 A.D., by a slow reducing fusion in a crucible of zincky wall accretions or oxide zinc ore and fragments of copper. In the operation, the zinc oxide was first reduced to metal, the zinc vapor permeated into the copper, and then the temperature was raised to fuse the charge. The medieval history of zinc is bound up with the history of alchemy in that period. The production of brass was well known to the alchemists and the knowledge gave rise to the belief that by the use of zinc and zinc bearing materials it might be possible to transmute copper into gold.

In Europe, the word "zinck" first appears to have been written in the fifteenth century in the work of Basilius Valentinius, but there is nothing to show that he meant metallic zinc. Actually the first writer to give the name "zinck" to the metallic form was Paracelsus (1490–1541) and his works show that he was acquainted with the metallic nature of zinc, in that he knew it to be fusible but not malleable in the ordinary sense, and was rather well informed regarding to the physical qualities of the metal,

In Asia, Kazwiui, called the "Pliny of the Orient," who died 630 A.D., stated that the Chinese knew how to render the metal malleable and used it to make small coins and mirrors. Zinc appears to have been known in India as early as 1000 to 1300 A.D. and was probably smelted commercially as early as the 14th century. Rana Laksh Singh, one of the Maharajas of Mewar State, was probably the first to work the mines, known as the Zawar deposits, around 1382. Mining and smelting were interrupted from time to time because of feudal wars, the great wars with invading Mogul emperors and the Maratha wars. The mines were abandoned by about 1830 and were not reopened until 1940.

Large heaps of lead- and zinc-bearing residues, slags and clay retorts of furnaces in the Zawar area bear testimony to an ancient smelting industry of impressive magnitude. There is no written record of it and inferences are based entirely on slags and other material left behind.

The unbroken retorts found in the Zawar ruins have walls $1/3$ to $1/2$ inch thick, of vesicular fused clay containing numerous fragments of phyllite quartz and quartzite, which suggest that the retorts were made from the local soils of the area. These retorts are doubly tapered, cylindrical vessels, closed at one end, but having a hole at the other end marked by a thin-walled tube of baked clay with an internal diameter of $1/2$ to $3/4$ inch. They clearly reveal construction in two parts, a cap with a tube being joined to the body, presumably having been added after the retort had been filled with ore and charcoal. All retorts bear evidence that they had been subjected to moderately high temperatures. Their interior surface is often partly glazed and blistered, and most of them are corrugated or wimpled as a result of fusion with subsequent partial collapse and flowage. The prolong end, which was apparently at a low temperature, does not show the same characteristics. The asymmetrical shape of the wrinkles or folds caused by flowage of the heat-softened clay walls indicates that the retorts had been in an inverted position when heated. They must have been closely spaced in the furnace, or in actual contact with each other, as many of them were found fused together.

The quantity of zinc residues in the Zawar area is estimated to be 130,000 to 170,000 tons. From this quantity of residue, a very large tonnage of zinc must have been produced.

During the seventeenth and eighteenth centuries large quantities of slab zinc, or the commercial spelter, were imported from the East. Various names were given to the metal, such as "Indian tin," "calaaem," "tutaneg," or "spiauter." In the early seventeenth century, about 1620, a Portuguese ship carrying spelter from the East Indies was seized by the Dutch. This metal was sold in Paris and other places under the name of "speauter" or "spialter." The name was latinized to "speltrum," from which comes spelter, the commercial designation of slab zinc. In 1745, a ship of the East Indies Company coming from Canton, China, was wrecked near Gothenburg, Sweden, with blocks of zinc aboard, whose per cent composition was: zinc, 98.990, iron, 0.765 and antimony, 0.245. Copper, nickel, silver, arsenic and lead were not present.

About the year 1730, the knowledge of smelting zinc was brought from China to England and in 1739 a patent was obtained for the process of distillation downward. At some time between 1740 and 1743, a smelter was erected at Bristol, England, which was the beginning of the manufacture of spelter. A production of 200 tons per year was mentioned. The process of distilling utilized a mixture of ore and carbon in a sealed clay pot, from the bottom of which a tube projected downward, letting the vapor condense in it and the metal drip into a basin.

Until 1758, oxides of zinc were used as ore, but in 1758 a patent was granted for making zinc from blende by roasting the ore, mixing the roasted ore with charcoal and smelting the mixture.

Toward the end of the eighteenth century, zinc distilling began in Silesia, and in 1798 a furnace was constructed at Wessola, utilizing the pots of a wood-fired glass furnace with zincky crust from an iron blast furnace as raw material. In 1805, the process was improved by using calamine as a raw material, which led to a change whereby the original glass pots assumed the forms of semicylindrical muffles. The gas and vapor emerged from the top of the charge instead of from the bottom as in the British prototype, but passed through a knee-shaped condenser, and the spelter dripped into a pan from which it had to be remelted.

Also near the end of the eighteenth century, a zinc furnace was built in Corinthia which exhibited some original ideas by employing small vertical pipes for distillation. These pipes were about 40 inches high and about 4 inches in diameter; their charge-capacity was small and the use of the Corinthian furnace never spread, nor survived very long. Although the process was incapable of commercial development, it introduced the idea of retorts set vertically and heated all around.

In the early nineteenth century, invention of the Belgian process of distilling zinc was accomplished by attaching a flower pot to a reverbratory furnace and condensing zinc in the pot. However, there is nothing here really suggestive of what subsequently became known as the "Belgian furnace."

In 1836, at Stolberg, a form of furnace was introduced for distilling zinc that was a compromise between the Belgian and Silesian furnaces, and was known as Belgian-Silesian or Silesian-Belgian, according to its predominant characteristics. The early composites of these furnaces were gradually molded into a distinct type that eventually became recognized as the

Rhenish. In their early composition, the retorts followed the muffle shapes of the Silesians, but they were less tall, and while the lower row sat on the hearth, an upper row bridged from shelf to shelf. The condensers extended from the retorts in a direct line as in the Belgian furnace, but they were enclosed in deep closets, after the Silesians. The retorts were banked in double tiers in an undivided combustion chamber and thus the Silesian ideas predominated. Gradually, three rows of retorts were provided, all of them bridged, and the retorts changed in cross section from muffle-shape to elliptical, the furnace then being more like the Belgian variety.

In the early days of smelting, the ore to be distilled was mostly calamine (zinc oxide ore), and the Silesians used relatively low-grade ore, pointing toward quantity and low treatment cost. In Belgium, high-grade ore prevailed and more consideration was given to percentage of metal extraction, while the Rhenish process utilized high-grade ore and aimed toward a compromise between cost and recovery.

In 1805 in Sheffield, England, it was discovered that the ordinarily brittle zinc could be rolled into sheets at a temperature ranging from 100 to 150°C, and the first sheet rolling mill was built at Liege, Belgium, in 1812. The first sheet zinc in the United States was made in Philadelphia in 1857, but the first industrial sheet zinc was rolled in the United States in 1866 by Matthiessen and Hegeler at LaSalle, Illinois.

In the United States, the first zinc was produced in 1835 at the Arsenal in Washington, D. C. The United States Government, seeking to establish definite standards of weights and measures, imported workmen from Belgium and built a small spelter furnace at Washington to make the zinc necessary to form the brass desired for the standard units of weight and measure. The ore smelted was zincite from the northern portion of New Jersey.

The first American attempt at the commercial production of zinc occurred at Newark, New Jersey in 1850. A Belgian zinc furnace was built for the smelting of franklinite (zinc-iron-manganese ore) but was not successful because of the high iron and manganese content of the ore. The high iron content caused excessive breakage of retorts, resulting in failure of the project. In 1856, a distilling furnace of the Silesian type was built at Friedensville, Pennsylvania, which also failed because of the ore used. In the western part of the United States, experiments in zinc smelting were carried on as early as 1855 at Potosi, Missouri, where a large furnace was erected in 1856, but not operated. In 1857, a small plant was built at Calamine, Arkansas, but this was not successful. However, in 1860, works were erected in LaSalle, Illinois, for the treatment of Wisconsin ores, and at South Bethlehem, Pennsylvania a Belgian furnace was erected, both plants achieving success in the distillation of zinc. These plants mark the beginning of the American zinc industry.

From 1840 to 1860, there was no noteworthy improvement in the distilling of zinc in Europe. The greatest concern at this time was the economical combustion of fuel, construction of furnaces that would stand up for a reasonable period and the manufacture of retorts that would be reasonably durable. With the enormous consumption of coal per ton of ore, these smelters established the principle that for the beneficiation of zinc ore it must be carried to the coal rather than the coal to the ore.

Between 1850 and 1860 in the United States, there was no scientific approach to the process of zinc distillation, except for one invention of importance. This was the grate of cast-iron plates containing small conical perforations with their small diameter at the top, adopted in 1851. Along with the muslin bag for the collection of fume, there was thus evolved what is known as the "Wetherill," or "American," process for the production of zinc oxide directly from ore.

The American process is the reduction of zinc oxide by carbon, distillation of the zinc and immediate burning of the vapor. The furnace is in effect a large retort in which, with the aid of an under-grate blast of air, coal is burned internally. Reduction of zinc oxide occurs with immediate combustion of the zinc, without the necessity of maintaining the reducing atmosphere that is essential in a closed retort. This process was put into use in 1854 and has been used since in many plants in the United States.

Between 1860 and 1880 in Europe, the smelters began thinking about gas-firing the furnaces and preheating the secondary air. Experiments in gas-firing a furnace were unsuccessful because refractory material was poor, heat-exchanging flues cracked, retort breakage was heavy, checker works clogged with zinc oxide and coal ash, and the gas producers were clumsy and inefficient.

The first regenerative furnace was introduced in the northern part of France, but proved a failure. In 1878, the same furnace was introduced at Peru, Illinois, where it was successful and became well known in America. The unappreciated concepts that the gas producer ought to be a part of the furnace and the sensible heat of the coal gas ought to be preserved held back the progress of distilling zinc. However, it was the invention and introduction of the mechanical gas producer that made gas-fired zinc distilling furnaces successful.

There was a line of experimentation and development at LaSalle, Illinois, that had a great effect upon zinc distilling, especially in the United States. In order to increase capacity, row after row of retorts were added and a long furnace like a tunnel was built so that the gas passed through it horizontally instead of from the bottom upward. The first large furnace of this character was built in 1872 and had 408 retorts (204 per side).

Much consideration was given to the imperfect condensation of zinc vapor and quite early in their practice, the Belgians extended their clay condensers and attached a cone or canister of sheet iron, which was termed a "prolong." The adoption of the prolong was not general in the United States because, with the more costly labor, the additional recovery in zinc did not pay.

In the 1880's, work was being done in the field of hydrometallurgy of zinc. Mixed sulfide ores were heap-roasted and sulfated so that the zinc could be leached out, giving a solution from which zinc sulfate could be crystallized, of sufficient purity to be used for lithopone manufacture.

Up to 1880, there was no scientific approach to blende roasting because the ore used in Europe was mostly calamine, and the calcination was done at the mine rather than the smelter. As the quantity of blende increased, it was usually sent to chemical works where sulfuric acid was produced and the calcines shipped to smelters.

In 1881, the roasting was greatly improved by introduction of a mechanically rabbled muffle furnace which was known as the "Hegeler furnace," and thus began the manufacture of sulfuric acid. In Europe, the development in blende roasting was perfection of the old muffle furnace which became known as the "Renania." The roasting furnaces were expected to produce a gas of high percentage in sulfur dioxide, desulfurization down to one per cent of sulfur, or less, and the minimum use of extraneous fuel.

The first experimental work on the electrolytic process was in 1881 when a solution of zinc sulfate, obtained by leaching roasted ore, was electrolyzed using insoluble anodes of lead and cathodes of zinc or copper. The solution was purified and the spent electrolyte treated with sulfuric acid for use in the digestion of more ore. The same principles are employed in the modern electrolytic process of extracting zinc. An electrolytic plant was built in the late 1890's, which was a failure, and the idea of electrolytic zinc extraction remained dormant for many years.

At about the same time the experimental work on the sulfate electrolytic process was being carried on, a process for the electrolysis of a zinc chloride solution was developed and put into practice. By this process, chlorine was liberated at the anodes, thereby yielding two commercial products, chlorine and zinc.

In 1895, natural gas was discovered in Kansas and then began the utilization of natural gas in zinc smelting. This led to the building of a number of smelters in the southwestern United States, which became one of the large areas of zinc production in this country.

At the beginning of the twentieth century, the commercial introduction of the froth flotation process on zinc ores practically revolutionized the zinc industry. It gave a new supply of high-grade sulfide ores, new ore supplies and ended the problem of mixed sulfide ores and the direct beneficiation of them. In the 1920's, the development and improvement of the process of preferential flotation, whereby a three-mineral separation instead of a two-mineral separation could be made, led to higher zinc concentrates for the smelters.

World War I had a great effect upon the zinc industry because, with the great demand for zinc, the price increased, causing more plants to be built, old ones to increase their capacity and those shut down to again start operation. It was at this time that the electrolytic extraction of zinc was again attempted, with solution of the many difficulties encountered earlier, so as to insure satisfactory continuation of the process in commercial form.

The idea of sintering calcines from blende roasting furnaces was first put into commercial application in 1917 and has grown throughout the zinc industry ever since. Sintering improves the physical condition of the ore, increases the capacity of the blende roasting furnaces, improves desulfurization and gives a higher percentage of zinc extraction.

The continuous distilling of zinc was considered as early as 1880 but was never practically realized until about 1925. The first vertical retorts were built in Germany and England, but were shut down after brief operation. This was then followed by the New Jersey Zinc process for continuous distilling, which has made good progress in the industry ever since.

The concept of distilling zinc in an internally, electrically heated retort dates back to the 1880's, and in 1901 a good deal of work was done in Sweden along this line. The first commercial plants were built in 1903, and production from the electrothermic smelting of zinc increased during World War I until 1931, when it ceased. At Josephtown, Pennsylvania, an electrothermic process was introduced for the production of zinc oxide, and was later adapted to the making of spelter. Today, the production of zinc by continuous distillation comes primarily from the New Jersey Zinc's vertical retort process aud the electrothermic process of the St. Joseph Lead Company.

At the present time, and ever since 1914, the United States has led the world in zinc production. Today, throughout the zinc industry there can be found many practices of producing zinc—by horizontal retort furnaces (both with natural gas and producer gas), electrolytic processes, electrothermic processes, vertical retort processes, and a blast furnace process—but the fundamental ideas are old and the industry has improved them by chemical and physical advances and by mechanization.

REFERENCES*

1. Bishop, O. M., and Mentch, R. L., "Mineral Facts and Problems—Zinc," Bull. 556, United States Department of the Interior, Washington, 1956.

2. Faloon, D. B., "Zinc Oxide," Princeton, D. Van Nostrand Co., Inc., 1925.
3. Georgius Agricola, "De Re Metallica," translated from the first Latin edition of 1556 by Herbert C. and Lois H. Hoover, London, *The Mining Magazine*, 1912.
4. Hofman, H. O., "Metallurgy of Zinc and Cadmium," New York, McGraw-Hill Book Co., Inc., 1922.
5. Ingalls, W., R., "The Metallurgy of Zinc and Cadmium," 2nd ed., New York, McGraw-Hill Book Co., 1906.
6. Mellor, J. W., "A Comprehensive Treatise on Inorganic and Theoretical Chemistry," Vol. 4, London, Longmans Green and Co., 1923. Good summary of history and early application of the metal.
7. "Metallurgy of Lead and Zinc," New York, American Institute of Mining and Metallurgical Engineers, *Trans. Am. Inst. Mining Met. Engrs.*, 121 (1936).

---

* Used generally, without citation, by the author and included for supplementary reading. (Ed.)

CHAPTER 2

ECONOMICS AND STATISTICS

RICHARD H. MOTE

Chief, Branch of Base Metals

and

O. M. BISHOP

Zinc Specialist, Division of Minerals, Bureau of Mines, U. S. Department of the Interior

Statistics of the zinc industry provide a numerical expression of its history and economic growth as well as a basis for predicting economic trends. But, as with all history, the early records in zinc are meager and often obscure. Data on mine production, for instance, are far less comprehensive than for smelter output. Moreover, such information as may be available is often difficult to interpret and correlate due to variations in methods of reporting mine output. In other statistical components, such as ore stocks at mines and smelters, metal consumption and secondary recovery, complete data are entirely lacking. In recent years, however, and particularly since World War II, comprehensive annual and monthly statistics have become regularly available in greater abundance. Gaps in continuity of information still persist and coverage is often inadequate, but despite these imperfections, reliable statistics on zinc are plentiful and a ready source of economic counsel.

In this chapter the statistics of the industry are reviewed and interpreted as to their economic significance. Figures on the closely related activities from mine production through smelting and refining and consumption of the metal are presented. A bibliography is provided for those readers who wish to examine the economics and statistics of the zinc industry in greater detail.

COMPOSITION OF DOMESTIC ZINC INDUSTRY

The zinc industry of the United States is composed of more than 500 firms which variously mine and concentrate domestic ores, import foreign ores, purchase zinc scrap, and smelt or process these materials to produce slab zinc, zinc pigments and salts, zinc dust and zinc alloys for sale. Some firms mine only, others process primary or secondary materials only, and yet others are integrated to combine all these functions.

TABLE 2-1. PRINCIPAL ZINC COMPANIES, MINES AND SMELTERS

Operating Company	Principal Zinc Mines			Zinc Smelters and Electrolytic Plants		
	Name and Location	Rank as of 1956	Important as a Lead Producer	Location	Capacity[1] (in short tons)	Type[2]
American Smelting & Refining Co.	Ground Hog unit, New Mexico	8	yes	Amarillo, Texas	54,400	H
	Page, Idaho	19	yes	Corpus Christi, Texas	100,000	E
	Van Stone, Washington	15	no	Beckemeyer, Illinois		S
				Los Angeles, California		S
				Sand Springs, Oklahoma		S
				Trenton, New Jersey		S
American Zinc, Lead & Smelting Co.	Grandview, Washington	24	no	Dumas, Texas	47,000	H
	Mascot #2, Tennessee	11	no	Fairmont City, Illinois[3]	12,000	H
	Young, Tennessee	25	no	Monsanto, Illinois	58,000	E
				Hillsboro, Illinois		S
The Anaconda Co.	Butte Mines, Montana	1	yes	Anaconda, Montana	86,500	E
	Darwin group, California	21	yes	Great Falls, Montana	165,000	E
Athletic Mining & Smelting Co.				Fort Smith, Arkansas	29,000	H
Blackwell Zinc Co.				Blackwell, Oklahoma	82,000	H
W. J. Bullock				Fairfield, Alabama		S
The Bunker Hill Co.	Bunker Hill, Idaho	14	yes	Kellogg, Idaho	57,800	E
	Star, Idaho	9	yes			
The Eagle-Picher Co.	Graham-Snyder-Spillane Feeham, Illinois	—	no	Henryetta, Oklahoma	46,000	H
	Shullsburg, Wisconsin	16	no			
General Smelting Co.				Bristol, Pennsylvania		S
Idarado Mining Co.	Treasury Tunnel—Black Bear—Smuggler Union, Colorado	12	yes			
Matthiessen & Hegeler Zinc Co.				La Salle, Illinois	31,000	H
				Meadowbrook, West Va.		V

Nassau Smelting & Refining Co.				Tottenville, New York		S
National Zinc Co.				Bartlesville, Oklahoma	39,000	H
New Jersey Zinc Co.	Austinville, Virginia	5	yes			
	Eagle, Colorado	4	yes			
	Hanover, New Mexico	13	no			
	Sterling Hill, New Jersey	—	no	Depue, Illinois		V
				Palmerton, Pennsylvania		V
New Mexico Consolidated Mining Co.	Kearney, New Mexico	22	no			
Ozark Mahoning Co.	Mahoning, Illinois	23	yes			
Pacific Smelting Co.				Torrance, California		S
Pend Oreille Mines & Metal Co.	Pend Oreille, Washington	20	yes			
St. Joseph Lead Co.	Balmat, New York	2	yes	Josephtown, Pennsylvania		V
	Edwards, New York	10	no			
Sandoval Zinc Co.				Sandoval, Illinois		S
Shattuck Denn Mining Co.	Iron King, Arizona	7	yes			
Superior Zinc Co.				Bristol, Pennsylvania		S
Tri-State Zinc Co.	Gray, Illinois	17	yes			
United Park City Mines Co.	United Park City Mines, Utah	18	yes			
United States Steel Corp., Tennessee Coal & Iron Div.	Davis-Bible group, Tennessee	6	no	Donora, Pennsylvania	51,000	H
United States Smelting, Refining & Mining Co.	United States & Lark, Utah	3	yes			
Wheeling Steel Corp.				Martins Ferry, Ohio		S
				Wheeling, West Va.[4]		S
					1,212,000	

[1] Plant capacity for electrolytic plants taken from 1956 Yearbook of the American Bureau of Metal Statistics; for horizontal retort plants estimated from daily plant capacities published in the 1956 Yearbook of the American Bureau of Metal Statistics. Working year for smelters estimated to be 340 days. Plant capacity for vertical retort plants estimated at 300,000 short tons based on data on plant capacities by distillation less horizontal plant capacity as published in the 1956 Yearbook of the American Bureau of Metal Statistics. Total capacity for secondary plants as taken from zinc preprint chapter of Bureau of Mines 1956 Minerals Yearbook was 53,400 short tons.

[2] H, Horizontal retort plant; E, Electrolytic plant; V, Vertical retort plant; and S, Secondary smelters.

[3] American Zinc, Lead & Smelting Company plant at Fairmont City, Illinois idle entire year.

[4] Wheeling Steel Company plant at Wheeling, West Virginia closed in March 1956.

The principal segments of the domestic industry are listed in Table 2-1. Major mines are ranked in order of production in 1956 and smelters are listed in terms of capacity and type.

Domestic mines range from small enterprises producing a few tons of crude ore a day to large units of many hundred tons. In 1956, about 520 mines in 19 states shipped domestic zinc ore or concentrate to smelters. The relative importance of many of these was small and 25 leading zinc-producing mines yielded about 70 per cent of the output, while the 3 leading mines accounted for 25 per cent of the zinc mine output. Nearly 60 per cent of zinc mine production is normally from mining districts in 9 western mining states. The more important western districts or regions are the Summit Valley (Butte) in Montana, the Cœur d'Alene in Idaho, the Big Bug in Arizona, the Central in New Mexico, the West Mountain (Bingham) in Utah, the Red Cliff in Colorado and the Pend Oreille and Metaline in Washington.

The remaining 40 per cent is chiefly from the Tri-State area of Missouri, Kansas, and Oklahoma; the upper Mississippi valley region of southwestern Wisconsin and northwestern Illinois; eastern Tennessee; Wythe County, Virginia; St. Lawrence County, New York and Sussex County, New Jersey.

Most of the mines produce mixed zinc-lead ores, but some—notably those in eastern Tennessee and Sussex County, New Jersey—contain no lead, and those in St. Lawrence County, New York, contain relatively little lead. Milling plants for the concentration of zinc ores are usually located at or near the mines and are owned and operated by the same company. Occasionally operations at custom mills, such as Eagle-Picher Company's Central mill (Cardin, Oklahoma), American Smelting and Refining Company's Deming mill (New Mexico) or United States Smelting, Refining and Mining Company's Bauer mill (Midvale, Utah), are based in part on purchased ore.

Market fluctuations, changes in the economy and mine consolidations have a marked effect on the number of mines in operation from year to year. Table 2-2 compares the variation in the number of producing zinc and lead mines and their output in groups of various sizes for the years 1944, and 1954 through 1956.

Zinc concentrates were smelted in 1956 in 17 plants, located in 8 states. Although some of these plants treated scrap metal for the recovery of zinc, most zinc scrap was smelted in 11 plants which process scrap exclusively. Primary zinc smelting operations are controlled by 12 companies, 7 of which are also large factors in mine production. Each of the 18 primary smelters purchase zinc concentrates domestically or abroad, and in some instances smelt concentrate on a fee or toll basis.

Industry control of mine production rests with relatively few com-

TABLE 2-2. COMBINED MINE PRODUCTION OF ZINC AND LEAD IN 1944 AND 1954–1956, SHOWING DISTRIBUTION OF MINES BY SIZE WITH PERCENTAGE OF TOTAL OUTPUT

Year	No. of mines and percentage of total zinc-lead output in mine groups having range of combined zinc-lead output in the indicated year									
	0–500 tons		500–2,000 tons		2,000–5,000 tons		5,000 tons & over		Total No. of mines	Combined total mine production of lead & zinc (in short tons)
	No. of mines	Percentage of total output	No. of mines	Percentage of total output	No. of mines	Percentage of total output	No. of mines	Percentage of total output		
1944	804	5.5	169	16.1	56	14.6	49	63.8	1,078	1,135,365
1954	524	3.3	61	8.3	27	11.3	33	77.1	645	798,890
1955	479	3.0	53	6.5	35	12.7	34	77.8	601	852,696
1956	557	3.0	56	6.8	46	16.8	37	73.4	696	895,169

panies. In 1956 about 70 per cent of zinc mine output was produced by 10 companies.

The integrated mining and smelting companies hold a commanding position in the zinc industry of the United States as well as the World. These companies operate the leading zinc-producing mines and substantial smelting operations which usually treat custom concentrates as well as company-owned mine output. The principal vertically integrated companies in the United States are those indicated in Table 2-1.

DOMESTIC PRODUCTION

The two major elements of domestic production are the domestic mines which produce ores and concentrates of zinc and the domestic zinc smelters and electrolytic and other plants which process domestic and foreign ores, or in some instances scrap, to produce zinc and zinc products.

Mine Production

Domestic mine production of recoverable zinc, including that recovered as pigments and salts directly from ores, fluctuates widely as economic conditions of the Nation and the industry vary. The greatest output occurred in 1926 when the total was 775,000 short tons, but only 6 years later (1932) output was but 285,000 tons. By 1941, domestic production had increased to 749,000 tons, and in 1942 reached 768,000 tons. Thereafter, despite the incentives of the Premium Price Plan during World War II and the stimulus of Government purchases for the strategic stockpile, annual mine production has generally trended downward (see Table 2-3).

The Tri-State district of Missouri, Kansas and Oklahoma has been the major producing district in the United States since before the turn of the century except in 1950, 1953 and 1956. For many years the district supplied more than half of all domestic mine output of zinc. This district has dropped from an output of 424,000 tons of recoverable zinc in 1926 to an average of 68,000 tons in 1952–1956. This decline which has not been wholly compensated by expanded production from other areas has been the principal reason for the general decline in United States production since the 1926 record year.

Mine output of zinc in the states east of the Mississippi River has increased in recent years and this trend appears likely to continue. Although production in the western states increased substantially following the depression years of the mid-thirties, output and percentage of total production has trended downward, particularly so since 1951.

Table 2-4 shows mine production by districts for the years 1952 through 1956. It is of interest that in the latter year the four principal districts—Summit Valley (Butte); St. Lawrence County, New York; Tri-State

TABLE 2-3. MINE PRODUCTION OF RECOVERABLE ZINC IN THE UNITED STATES, 1942–46 (average), 1947–51 (average) AND 1952–56, BY STATES, IN SHORT TONS

State	1942–46 (average)	1947–51 (average)	1952	1953	1954	1955	1956
Western States and Alaska:							
Alaska	—	11	—	—	—	—	—
Arizona	30,234	58,652	47,143	27,530	21,461	22,684	25,580
California	5,545	7,020	9,419	5,358	1,415	6,836	8,049
Colorado	37,637	46,620	53,203	37,809	35,150	35,350	40,246
Idaho	84,061	82,380	74,317	72,153	61,528	53,314	49,561
Montana	32,524	62,440	82,185	80,271	60,952	68,588	70,520
Nevada	17,730	19,350	15,357	5,812	1,035	2,670	7,488
New Mexico	46,622	37,927	50,975	13,373	6	15,277	35,010
Oregon	—	6	1	—	—	—	—
South Dakota	43	10	—	—	—	—	—
Texas	9	9	3	—	—	—	—
Utah	38,671	38,366	32,947	29,184	34,031	43,556	42,374
Washington	12,305	14,035	20,102	32,786	22,304	29,536	25,609
Total	305,381	366,826	385,652	304,276	237,882	277,811	304,437
West Central States:							
Arkansas	137	21	26	—	—	—	—
Kansas	54,524	32,517	25,482	15,515	19,110	27,611	28,665
Missouri	29,568	9,823	13,986	9,981	5,210	4,476	4,380
Oklahoma	98,179	47,821	54,916	33,413	43,171	41,543	27,515
Total	182,408	90,182	94,410	58,909	67,491	73,630	60,560
States east of the Mississippi River:							
Illinois	7,922	17,993	18,816	14,556	14,427	21,700	24,039
Kentucky	435	1,254	3,280	489	458	—	417
New Jersey	82,608	64,427	59,190	45,700	37,416	11,643	4,667
New York	36,968	37,005	32,636	51,529	53,199	53,016	59,111
Tennessee	37,001	32,898	38,020	38,465	30,326	40,216	46,023
Virginia	17,448	13,113	13,409	16,676	16,738	18,329	19,196
Wisconsin	13,840	9,372	20,588	16,830	15,534	18,326	23,890
Total	196,222	176,062	185,939	184,245	168,098	163,230	177,343
Grand total	684,011	633,070	666,001	547,430	473,471	514,671	542,340

(Joplin region) and the Cœur d'Alene contributed 226,000 tons or 42 per cent of the total domestic mine production. In 1926 the same districts contributed 527,000 tons or 68 per cent of all mine output of zinc.

Although United States mine production has been declining, world

TABLE 2-4. MINE PRODUCTION OF ZINC IN THE PRINCIPAL DISTRICTS[1] OF THE UNITED STATES, 1947–51 (average) AND 1952–56, IN TERMS OF RECOVERABLE ZINC, IN SHORT TONS

District	State	1947–51 (average)	1952	1953	1954	1955	1956
Summit Valley (Butte)	Montana	57,066	75,968	75,170	53,527	62,588	63,375
St. Lawrence County	New York	37,005	32,636	51,529	53,199	53,016	59,111
Tri-State (Joplin region)	Kansas, southwestern Missouri, Oklahoma	88,983	90,512	55,729	64,322	69,696	57,215
Cœur d'Alene	Idaho	79,703	70,316	68,650	58,736	50,527	46,738
Eastern Tennessee[2]	Tennessee	32,898	38,020	38,465	30,326	40,216	46,023
Upper Mississippi Valley	Northern Illinois, Iowa,[3] Wisconsin	21,436	34,716	26,286	25,441	31,411	38,498
Central	New Mexico	33,690	48,043	12,743	—	15,104	33,631
West Mountain (Bingham)	Utah	19,848	20,395	19,669	20,489	21,864	24,310
Red Cliff (Battle Mt.)	Colorado	20,067	26,000	16,850	18,604	21,322	19,766
Austinville	Virginia	13,113	13,409	16,676	16,738	18,329	19,156
Big Bug	Arizona	7,945	10,862	10,476	10,453	11,234	13,934
Park City region	Utah	9,454	7,746	4,848	6,650	12,295	10,983
Kentucky-Southern Illinois	Kentucky-Southern Illinois	7,183	7,968	5,589	4,978	8,615	9,848
New Jersey	New Jersey	64,427	59,190	45,700	37,416	11,643	4,667
Smelter (Lewis and Clark County)	Montana	2,083	2,807	2,924	5,301	4,077	4,361
Cochise	Arizona	2,409	4,266	3,893	3,566	3,295	2,795
Pima (Sierritas, Papago, Twin Buttes)	Arizona	5,776	3,472	11	—	1,310	2,786
California (Leadville)	Colorado	6,505	8,487	3,945	2,437	1,621	2,128
Flint Creek	Montana	131	1,084	[4]	1,290	1,400	2,046
Rush Valley and Smelter (Tooele County)	Utah	2,842	916	1,528	1,738	1,434	1,622
Yellow Pine (Goodsprings)	Nevada	716	1,464	—	—	716	1,603
Warm Springs	Idaho	1,813	2,142	3,026	2,584	1,833	1,388

Bayhorse	Idaho	405	217	264	[4]	790	1,203
Aravaipa	Arizona	845	1,315	1,732	1,366	1,670	1,185
Tintic	Utah	4,625	2,951	2,433	4,335	4,018	1,119
Magdalena	New Mexico	3,217	2,122	512	—	98	1,031
Creede	Colorado	506	1,024	858	1,111	745	927
Breckenridge	Colorado	521	620	1,200	1,186	615	830
Patagonia (Duquense)	Arizona	437	1,049	257	54	273	543
Cow Creek (Ingot)	California	[4]	[4]	—	—	4	20
Eureka (Bagdad)	Arizona	1,773	3,520	2,594	1,126	444	3
Old Hat (Oracle)	Arizona	4,121	3,368	—	1	—	1
Chelan Lake[5]	Washington	2,264	[4]	[4]	[4]	[4]	[4]
Coso[5]	California	3,824	5,479	[4]	[4]	[4]	[4]
Elk Mountain[5]	Colorado	65	303	—	—	[4]	[4]
Harshaw[5]	Arizona	3,219	3,924	4,186	4,193	[4]	[4]
Heddleston[5]	Montana	1,446	1,066	—	[4]	47	[4]
Metaline[5]	Washington	9,204	[4]	[4]	[4]	[4]	[4]
Northport[5]	Washington	2,454	[4]	[4]	[4]	[4]	[4]
Pioche[5]	Nevada	17,126	12,493	[4]	[4]	[4]	[4]
Pioneer (Rico)[5]	Colorado	2,372	2,734	2,634	2,896	[4]	[4]
Silver Bell[5]	Arizona	48	364	1,324	[4]	[4]	[4]
Smelter (Salt Lake County)[5]	Utah	25	—	—	—	3,148	[4]
Upper San Miguel[5]	Colorado	5,933	9,811	10,414	7,899	6,532	[4]
Pioneer (Superior)	Arizona	1,767	4,175	—	—	—	—
Verde (Jerome)	Arizona	4,553	4,360	959	—	—	—
Warren (Bisbee)	Arizona	24,165	4,791	1,182	—	—	—

[1] Districts producing 1,000 short tons or more in any year of the period 1952–56.
[2] Includes zinc recovered from copper-zinc-pyrite ore in Polk County.
[3] No production in Iowa since 1917.
[4] Quantity withheld to avoid disclosure of individual company confidential data.
[5] This district not listed in order of 1956 output.

TABLE 2-5. WORLD MINE PRODUCTION OF ZINC (content of ore),[1] BY COUNTRIES,[2] 1947–51 (average) AND 1952–56, IN SHORT TONS[3]
(Compiled by Augusta W. Jann, Division of Foreign Activities, Bureau of Mines)

Country[2]	1947–51 (average)	1952	1953	1954	1955	1956
North America:						
Canada	294,423	371,802	401,762	376,491	433,357	419,402
Cuba	—	—	—	—	1,134	3,276
Greenland	—	—	—	—	—	2,600
Guatemala	[4]3,765	9,000	6,700	4,400	10,400	12,000
Honduras[5]	[6]127	316	636	791	1,433	2,288
Mexico	210,947	250,638	249,715	246,441	296,961	274,351
United States[7]	633,070	666,001	547,430	473,471	514,671	542,340
Total	1,142,332	1,297,757	1,206,243	1,101,594	1,257,956	1,256,257
South America:						
Argentina	14,884	16,971	17,735	[8]22,000	23,260	26,100
Bolivia (exports)	22,819	39,263	26,427	22,403	23,509	18,818
Chile	[4]370	3,650	3,500	[8]1,650	3,200	[8]3,300
Peru	83,557	140,925	153,334	174,784	183,074	167,413
Total	121,630	200,809	200,996	[8]220,840	233,043	[8]215,630
Europe:						
Austria	3,082	5,496	4,826	5,140	5,787	5,868
Finland	2,712	7,700	3,500	5,000	23,300	43,000
France	10,339	16,100	14,600	12,500	11,400	13,800
Germany, West	58,988	88,956	100,506	103,867	101,558	101,803
Greece	3,694	8,000	8,300	7,900	13,500	22,300
Ireland	[4]917	1,892	1,819	1,719	2,769	2,127
Italy	86,731	124,466	117,102	129,707	131,891	134,912
Norway	6,554	6,160	5,661	5,917	7,411	7,055
Poland[8]	[9]108,000	110,000	130,000	129,000	139,000	138,000
Spain	60,800	95,000	92,000	97,000	102,000	96,000
Sweden	40,125	42,357	49,706	64,407	64,810	72,763
U.S.S.R.[8, 9]	138,000	214,000	241,000	258,000	300,000	351,000
United Kingdom	51	1,707	3,187	3,905	3,167	1,563
Yugoslavia	42,418	52,678	66,106	63,052	65,800	63,400
Total[2, 8]	577,000	799,000	869,000	931,000	1,016,000	1,103,000
Asia:						
Burma	—	2,400	4,300	6,400	9,100	8,000
India	[4]800	2,500	2,900	2,600	2,900	4,200
Iran[10]	[11]13,000	5,500	6,200	5,800	6,300	3,700
Japan	48,763	96,418	106,507	120,581	119,787	135,194
Korea, Republic of	74	550	22	—	—	440
Philippines	[11]165	1,770	830	—	—	1,050
Thailand (Siam)	[6]238	550	2,000	3,000	3,200	[8]2,200
Turkey[8]	820	990	4,400	6,100	770	670
Total[2, 8]	65,000	118,400	138,800	159,900	160,200	176,900

TABLE 2-5—*Continued*

Country[2]	1947–51 (average)	1952	1953	1954	1955	1956
Africa:						
Algeria	8,143	13,160	20,470	31,538	34,200	31,891
Angola	[11]386	50	110	—	—	3
Belgian Congo	67,594	109,071	138,661	94,015	74,700	126,235
Egypt	488	977	282	262	757	692
French Equatorial Africa	261	416	—	—	—	—
French Morocco	8,207	31,253	38,895	37,908	47,686	43,000
Nigeria	97	57	71	—	—	—
Rhodesia and Nyasaland, Fed. of Northern Rhodesia	[9]24,968	41,140	43,353	38,672	38,070	38,134
South-West Africa	12,078	[7]17,200	[7]17,400	[7]22,000	19,500	20,458
Tunisia	3,285	3,900	4,020	5,707	5,990	5,200
Total	125,507	217,224	263,262	230,102	220,903	265,613
Oceania: Australia	210,750	220,954	265,481	282,978	287,352	311,334
World total (estimate)[2]	2,240,000	2,850,000	2,940,000	2,930,000	3,180,000	3,330,000

[1] Data derived in part from the Yearbook of the American Bureau of Metal Statistics, the United Nations Statistical Yearbook, and the Statistical Summary of the Mineral Industry (Colonial Geological Surveys, London).

[2] In addition to countries listed, Bulgaria, Czechoslovakia, East Germany, Rumania, China, and North Korea also produce zinc, but production data are not available; esimates included in total.

[3] This table incorporates a number of revisions of data published in previous Zinc reports. Data do not add to total shown due to rounding where estimated figures are included in the detail.

[4] Average for 1950–51.

[5] United States imports.

[6] Average for 1948–51.

[7] Recoverable.

[8] Estimated.

[9] Smelter production.

[10] Year ended March 21 of year following that stated.

[11] Average for one year only, as 1951 was first year of commercial production.

mine production, exclusive of the United States, has increased since 1942 from 1,480,000 short tons in 1943 to 2,788,000 tons in 1956, and exploration and development programs indicate that further expansion of production is likely. The United States was the leading mine producer in 1956, followed in order by Canada, U.S.S.R., Australia, Mexico and Peru (see Table 2-5).

Smelter Production

Facilities for production of primary slab zinc consist of 12 distillation plants and 5 electrolytic plants. Three of the distillation plants have New Jersey Zinc Company's externally-gas-fired vertical retorts and a

TABLE 2-6. PRIMARY AND REDISTILLED SECONDARY SLAB ZINC PRODUCED IN THE UNITED STATES, 1947–51 (average) AND 1952–56, IN SHORT TONS

Year	Primary			Redistilled secondary	Total (excludes zinc recovered by remelting)
	From domestic ores	From foreign ores	Total		
1947–51 (average)	569,919	256,109	826,028	58,506	884,534
1952	575,828	[1]328,651	904,479	55,111	959,590
1953	[1]495,436	[1]420,669	916,105	52,875	968,980
1954	[1]380,312	[1]422,113	802,425	68,013	870,438
1955	582,913	[1]380,591	963,504	66,042	1,029,546
1956	[1]470,093	[1]513,517	983,610	72,127	1,055,737

[1] Includes a small tonnage of slab zinc further refined into High-Grade metal.

fourth is a St. Joseph Lead Company's electrothermic vertical retort plant. The remaining 8 retort plants are of the gas-fired horizontal type. These furnaces jointly have an estimated annual capacity of 694,000 short tons.

The electrolytic plants had a joint annual capacity of about 470,000 short tons at the end of 1956. They and the reflux refineries of The New Jersey Zinc Company were the sole producers of Special High Grade slab zinc.

In 1956 the zinc smelting industry of the United States produced a record 984,000 tons of primary slab zinc at the 17 primary plants discussed above. In the same period some few of the primary plants and 12 secondary smelters (see Table 2-1) also produced 72,000 tons of redistilled slab zinc.

Table 2-6 shows domestic production of primary and redistilled zinc for 1947 through 1956 and also shows the quantities of primary slab produced from domestic and foreign ores.

Table 2-7, which gives world smelter output of zinc by countries, shows that total smelter output of primary metal has increased from an average of 2,048,000 tons in 1947–51 to 3,110,000 tons in 1956. The greatest growth occurred in Europe, followed by North America, Asia, Africa, Australia and South America.

CONSUMPTION

Zinc is one of the essential major nonferrous metals required in the industrial economy. It is consumed primarily in the form of ingots, slabs and pigs of metal, but important quantities are consumed in the direct manufacture of pigments and salts, and yet other quantities are consumed as zinc-bearing scrap—metal to make alloys, zinc dust, chemicals and pigments. Major uses of zinc* are for galvanizing iron and steel products, in

* See especially Chapters 9, 10 and 12.

TABLE 2-7. WORLD SMELTER PRODUCTION OF ZINC BY COUNTRIES, 1947–51 (average) AND 1952–56, IN SHORT TONS[1, 2]

(Compiled by Augusta W. Jann, Division of Foreign Activities, Bureau of Mines)

Country	1947–51 (average)	1952	1953	1954	1955	1956
North America:						
Canada	200,766	222,200	250,961	253,365	256,542	255,601
Mexico	59,703	[3]55,542	[3]58,481	[3]60,477	[3]61,878	[3]62,136
United States	826,028	904,479	916,105	802,425	963,504	983,610
Total	1,086,497	1,182,221	1,225,547	1,116,267	1,281,924	1,301,347
South America:						
Argentina	5,521	11,023	12,787	[4]12,000	14,881	15,432
Peru	1,294	5,750	9,819	16,935	18,801	10,415
Total	6,815	16,773	22,606	[4]29,000	33,682	25,847
Europe:						
Austria	—	—	—	—	1,493	7,319
Belgium[5]	185,566	205,910	213,217	234,481	232,840	251,914
Bulgaria	—	—	—	—	1,497	6,435
Czechoslovakia	[4]2,480	[6]	[6]	[6]	[6]	[6]
France	66,919	88,255	89,219	122,249	123,624	124,106
Germany, West	90,924	162,278	163,430	184,804	197,026	204,964
Italy	35,534	60,463	66,214	74,356	77,761	81,086
Netherlands	17,877	28,555	27,780	28,702	31,347	31,980
Norway	44,461	43,248	42,767	49,010	50,176	53,171
Poland[4]	108,000	132,000	152,000	157,000	172,000	169,000
Rumania	[4]3,060	[6]	[6]	[6]	[6]	[6]
Spain	22,749	23,543	25,490	25,653	26,291	24,548
U.S.S.R.[4]	138,000	214,000	241,000	258,000	300,000	351,000
United Kingdom	77,145	76,984	81,433	90,989	91,108	91,247
Yugoslavia	10,394	15,943	16,038	15,040	15,176	15,436
Total[4]	803,000	1,059,000	1,127,000	1,249,000	1,329,000	1,420,000
Asia:						
China[4]	270	200	400	13,800	16,500	19,800
Japan	38,269	77,197	85,001	111,748	124,036	149,149
Total[4]	38,540	77,400	85,400	125,500	140,500	168,900
Africa:						
Belgian Congo	—	—	8,599	35,274	37,443	47,417
Rhodesia and Nyasaland, Fed. of Northern Rhodesia	24,968	25,636	28,370	29,736	31,248	32,396
Total	24,968	25,636	36,969	65,010	68,691	79,813

TABLE 2-7—*Continued*

Country	1947–51 (average)	1952	1953	1954	1955	1956
Oceania: Australia	87,887	97,930	100,999	117,066	113,220	117,592
World total (estimate)	2,048,000	2,460,000	2,600,000	2,700,000	2,970,000	3,110,000

[1] Data derived in part from the Yearbook of the American Bureau of Metal Statistics, the United Nations Monthly Bulletin and the Statistical Yearbook, and the Statistical Summary of the Mineral Industry (Colonial Geological Surveys, London).

[2] This table incorporates a number of revisions of data published in previous Zinc reports. Data do not add to totals shown due to rounding where estimated figures are included in the detail.

[3] In addition other zinc-bearing materials totaling 3,746 tons in 1952; 30,288 in 1953; 18,545 in 1954; 37,442 in 1955; and 39,554 in 1956.

[4] Estimate.

[5] Includes production from reclaimed scrap.

[6] Data not available.

the manufacture of the copper-zinc alloy, brass, and in the zinc-base alloy metals used in die castings by the automobile, electrical appliance, hardware and other manufacturing industries. In the form of rolled sheets, zinc is used extensively in the manufacture of dry-cell battery cans, photoengraving and lithographing plates, boiler plates, weather stripping, flashing and for numerous other purposes. Zinc oxide which is produced both directly from ore and from zinc metal, either primary or secondary, is consumed principally in the production of zinc pigments used in the manufacture of rubber, paints, ceramics, cosmetics and pharmaceuticals, coated fabrics, textiles, floor coverings and chemicals. Table 2-8 shows consumption of slab zinc in the United States in recent years by industries.

Detailed information on the consumption of slab zinc by industry and product, and by grades, according to geographic areas, beginning with 1940, has been published in the Minerals Yearbooks of the Bureau of Mines, United States Department of the Interior. Prior to 1940, detail on zinc consumption is available only on the basis of estimates and for some years only.

Except for the three war years, 1942–44, when increased quantities of slab zinc were required for cartridge and shell brass, the galvanizing industry has been the largest user of zinc. The percentage of total zinc consumed in galvanizing has varied somewhat through the years, but since World War II has averaged approximately 40 per cent of the total slab zinc consumed. Zinc coatings are recognized as the most economical provisions for protecting steel and iron products from atmospheric corrosion. Products galvanized include steel roofing sheets, wire, wire products, telegraph and telephone pole hardware, conduit, pipe and miscellaneous hardware. The recent widespread introduction of continuous hot-dip

Table 2-8. Consumption of Slab Zinc in the United States, 1947–51 (average) and 1952–56, by Industries, in Short Tons[1]

Industry and product	1947–51 (average)	1952	1953	1954	1955	1956
Galvanizing:[2]						
Sheet and strip	143,033	145,875	164,601	181,558	200,403	203,713
Wire and wire rope	47,594	48,645	44,100	44,882	48,171	42,937
Tubes and pipe	81,648	82,043	88,428	76,891	98,206	86,277
Fittings	14,625	10,366	10,330	10,513	10,586	10,652
Other	98,128	90,759	99,529	89,619	93,775	95,567
Total galvanizing	385,028	377,688	406,988	403,463	451,141	439,146
Brass products:						
Sheet, strip, and plate	56,347	71,706	94,826	52,284	67,550	56,207
Rod and wire	35,970	49,831	47,312	30,899	46,830	39,413
Tube	15,501	17,057	18,136	12,097	15,363	13,666
Castings and billets	4,254	7,262	8,145	5,499	7,518	6,337
Copper-base ingots	4,674	8,223	7,659	6,594	8,062	7,197
Other copper-base products	1,191	1,529	2,104	895	920	1,184
Total brass products	117,937	155,608	179,182	108,268	146,243	124,004
Zinc-base alloy:						
Die castings	241,742	225,877	297,280	279,676	417,333	349,200
Alloy dies and rod	4,612	9,235	7,140	8,857	11,754	9,322
Slush and sand castings	1,094	1,577	3,025	2,313	1,720	1,985
Total zinc-base alloy	247,448	236,689	307,445	290,846	430,807	360,507
Rolled zinc	67,016	51,318	54,649	47,486	51,589	47,359
Zinc oxide	16,147	17,205	20,675	18,701	22,433	19,160
Other uses:						
Wet batteries	1,493	1,396	1,417	1,264	1,420	1,345
Desilverizing lead	2,584	2,370	2,425	2,740	2,676	2,939
Light-metal alloys	1,456	3,266	5,939	3,526	3,484	5,830
Other[3]	4,299	7,243	8,207	8,005	10,019	8,500
Total other uses	9,832	14,275	17,988	15,535	17,599	18,614
Total consumption[4]	843,408	852,783	985,927	884,299	1,119,812	1,008,790

[1] Excludes some small consumers.

[2] Includes zinc used in electrogalvanizing and electroplating, but excludes sherardizing.

[3] Includes zinc used in making zinc dust, bronze powder, alloys, chemicals, castings, and miscellaneous uses not elsewhere mentioned.

[4] Includes 4,144 tons of remelt zinc in 1952, 3,710 tons in 1953, 3,589 tons in 1954, 2,997 tons in 1955 and 5,230 tons in 1956.

galvanizing lines has served both to improve the product and reduce costs, and has served to maintain the competitive position of galvanized steel versus aluminum sheets. Indeed in 1955, galvanized sheet shipments were at an all-time high and zinc consumed in galvanizing totaled 451,000 tons.

The use of zinc for brass production was the second largest use of slab zinc until 1946, but in recent years has been third. The all-time high demand for zinc in brass was in the war year, 1943, when 419,000 tons of slab zinc were consumed in the manufacture of brass products. This quantity alone represented over 50 per cent of the total zinc consumed in 1943. Since 1946, however, the quantity of slab zinc consumed for brass has been declining and has averaged less than 20 per cent of the total slab consumed. Developments in weapons and ammunition have reduced the military requirements for brass and many civilian applications have been replaced by other metals or alloys.

The second largest use of zinc is in zinc-base alloys for die- and other castings. Specific data on annual consumption of zinc for this use are not available prior to 1926 and from that period until 1940 the quantities used were comparatively small. In 1940 and 1941 the automotive industry increased its use of zinc die castings for such items as carburetors, fuel pumps, windshield-wiper motors, speedometer cases, door handles and other items of automobile hardware, increasing zinc consumption in this category substantially until automobile manufacture was curtailed at the beginning of World War II. After the war, consumption of zinc for die casting regained its second-place position and with few exceptions has been growing annually since. An all-time high of 417,333 tons was consumed for this use in 1955. This figure represents close to 40 per cent of the total zinc consumed in 1955 and is over 90 per cent of that used for galvanizing. Practically all the zinc used for die casting is Special High Grade.

The use of zinc in the manufacture of rolled zinc was at an all-time high of 97,585 tons in 1945. Since that year the use has trended downward and reached a 22-year low of 47,359 tons in 1956.

SUBSTITUTES

There are no suitable substitutes for zinc in its major uses, and materials that prove most promising are considerably higher in price or in too small supply. There is no satisfactory substitute for most galvanized iron products. Sheet aluminum is both higher in cost and structurally weaker than galvanized steel; and aluminized steel, because of high cost, is competitive in limited uses only. Cadmium is used in plating, but its high unit cost and limited availability make it uneconomical for most purposes where zinc is used. There are no substitutes for zinc in brass although other metals may be substituted in certain uses, such as builders' hardware. Aluminum and magnesium can be substitutes for zinc die castings in certain applications where weight limitations or reductions are important factors, or where an anodized finish is desired. The extent of substitution of these metals is influenced to a considerable degree by metal-price considerations and styling.

There are relatively few substitutes for zinc in its chemical applications. Aluminum and magnesium dusts and compounds could replace zinc as a reducing agent in some chemical reactions. Certainly titanium pigments have replaced lithopone to a marked degree, but in most paint formulations titanium oxide is used with zinc oxide. The principal use of lead-free zinc oxide in rubber has no competition. Because of the wide availability of zinc at moderate prices there appears little cause to seek substitutes.

STOCKS

Statistical data on zinc stocks are limited to information on slab zinc held by producers and consumers and to consumers' stocks of zinc-base scrap.

Both producers and consumers strive to adjust their slab-zinc stocks to conform to changes in industrial activity. During periods of high industrial activity producers usually consider about one month's output a desirable inventory. Consumers usually strive for a one-month to six-week supply on hand. Although these quantities are considered optimum inventories there have been times when both producers and consumers have misjudged the outlook for the industry and have found themselves with abnormally high or low inventories. The entry of the United States in World War II in 1941 caught primary zinc producers with slightly more than 24,000 tons of slab zinc on hand. This quantity was considered to be critically low. Similarly, the Korean War generated an unforeseen demand for zinc and as a result stocks of slab zinc on hand at smelters were reduced during 1950 to less than 8,000 tons on December 31 of that year. This was the smallest reserve at smelters since November 1925. During the period 1900 through 1956 the lowest beginning-of-the-year smelter inventory of slab zinc was 1,905 tons, in 1902. The highest quantity in inventory at the beginning of any year was 254,692 tons in 1946, but the monthly average for the 10 years, 1947 through 1956, was about 80,000 tons. Consumers' stocks of slab zinc were 105,000 tons at the end of 1956 and averaged 102,000 tons for the years 1952 through 1956.

MARKETING AND PRICES

The principal types of marketing transactions in the zinc industry include the purchase of zinc concentrates—and to a limited extent, ores—by custom smelters, the sale of slab zinc and the marketing of secondary zinc and zinc by-product scrap items. A factor common to each of these types of transactions is that the individual price determinations are almost always based on the quoted price per pound of Prime Western grade slab zinc, f.o.b. East St. Louis. In other respects, however, there is a notable lack of uniformity. Each type is uniquely different and involves dissimilar sales techniques and procedures.

Concentrates

Because the majority of zinc-mining companies do not have the smelters with which to treat their mine products they must sell their concentrates to companies which have such treatment facilities. The buying and treatment of ores and concentrates by companies operating smelters and refineries is known as custom smelting. Purchase of these materials by custom smelters is the result of negotiation and agreement between buyer and seller and the terms of the transaction are set forth in a contract called a "schedule." Custom smelters have standard or "open schedules" which are used to buy small shipments of ores and concentrates. The mining companies which are regular or large shippers to the smelters obtain special contracts which may provide more favorable terms than those obtained on the open schedule. Some contracts call for the treatment of concentrates on a toll basis with return of the metal to the original owner, or more commonly, with the smelters marketing the metal on a commission basis for the producers.

The prices paid for ores and concentrates under smelter purchase schedules vary with zinc content, market price of recoverable metals and nature and quantity of deleterious impurities. Special handling and sampling costs are charged against small shipments and all shipments must pay a base smelting charge.

The typical smelter schedule for the purchase for an individual shipment of zinc concentrates might read as follows: "Buyer to pay for 85 per cent of full zinc content at the East St. Louis price for Prime Western zinc, f.o.b. East St. Louis, as published in the *Engineering & Mining Journal*, for the calendar month following receipts of the shipment at Buyer's plant, less a deduction of 50 cents per 100 pounds of zinc accounted for. There shall also be deducted a treatment charge of $46.00 per dry short ton when the East St. Louis price is 10 cents a pound. For each cent increase or decrease in the settlement price there shall be a corresponding increase or decrease of $1.00 in the base treatment charge, with fractions in proportion." The smelters also normally charge the shipper for antimony and arsenic in excess of 1 per cent, iron and manganese in excess of 6 to 8 per cent and provide for changes in the base smelting charge depending on changes in the average hourly wages of the smelter workmen.

Numerous other factors covering sales of ores and concentrates are detailed in the smelter schedules. The purchase is based on assays of representative samples and it is customary for both buyer and seller to agree to settlement on the average of their individual assays if such assays are within close agreement. If they are at wide variance, however, a determinative "umpire" assay is made by a commercial assayer mutually acceptable to both parties. The quantities expected to be shipped under the terms of the contract and the period covered during the shipments are frequently stated. The basis for payment of zinc content, whether it be average price for month of shipment, price on date of arrival, or average of week of arrival, is always specifically stated. Provisions and conditions for advance payment are also frequently made.

Ores from the Tri-State district of southwestern Missouri, southeastern Kansas, and northeastern Oklahoma, unlike the complex sulfide ores and concentrates produced in the Rocky Mountain and far western states, are comparatively simple and remarkably constant mineralogically. Because of this, concentrates produced in this area are sold and purchased on the basis of a weekly quoted Joplin, Missouri, price in terms of dollars per short ton for 60 per cent zinc concentrates, f.o.b. at the mine. In 1957 when Prime Western slab zinc was quoted at 13.5 cents a pound f.o.b. East St. Louis, the Joplin market for 60 per cent zinc concentrate was $84.00 per short ton, and when the metal was quoted at 10 cents a pound, 60 per cent concentrate sold for $56.00 a ton.

Slab Zinc

Sales of slab zinc are usually made on the basis of individual contracts under terms which are subject to negotiation. Because the buyer and the seller of zinc are generally in disagreement as to what is considered the most favorable price the negotiated sale price of slab zinc may be more or less than the quoted East St. Louis price, plus necessary transportation and handling charges. When market conditions forecast a general increase in prices, the buyers of slab zinc usually attempt to negotiate firm prices in contracts for future delivery, whereas the sellers attempt to arrange terms for future deliveries made on the basis of quoted prices applicable at the time of delivery. On the other hand, when economic conditions suggest a forthcoming decline in prices, the sellers prefer contracts for future delivery on the basis of firm prices and the buyers attempt to purchase under contracts which specify that deliveries be made on the basis of quoted prices at the time of delivery.

Market prices for slab zinc are generally quoted in cents per pound, f.o.b. East St. Louis, Illinois, regardless of the point of production. The metal is also quoted on a New York basis as well as in London which is the market center for foreign transactions. The most widely published quotation is for Prime Western grade zinc which is the type of zinc generally used for galvanizing.

The quoted price of Prime Western grade zinc is usually the basis for determination of prices of other grades of slab zinc as well as zinc concentrates and scrap. It is also a factor in the determination of the prices for zinc pigments and zinc chemicals, but the quoted prices for these items are less susceptible to short-term market fluctuations than are the quotations on concentrates and metal.

The other five market grades of zinc are sold at a premium above the Prime Western quotation. Currently quoted premiums (November 1957) in the East St. Louis market are as follows: Selected, 0.10 cent; Brass Special, 0.25 cent; Intermediate, 0.5 cent; High Grade, 1.35 cents and Special High Grade, 1.75 cents. Both Special High Grade and High Grade zinc are sold on a delivered price basis, whereas the other three grades with smaller premiums are usually sold on the basis f.o.b. East St. Louis.

There has been an increasing tendency on the part of sellers of zinc to absorb freight charges in excess of 0.5 cent per pound and, since September 1953, price quotations on Prime Western, Selected, Brass Special and Intermediate have been on both f.o.b. East St. Louis and on a delivered basis. Thus the delivered price per pound for zinc at points where freight from East St. Louis exceeds 0.50 cent a pound is 0.50 cent per pound higher than the f.o.b. East St. Louis quotation.

There follows a tabulation of average annual prices of Prime Western slab zinc at East St. Louis and of zinc concentrate (60 per cent zinc content) at Joplin, Missouri (Table 2-9).

Secondary Zinc

The marketing of secondary zinc materials, such as dross, ash, residue, or skimmings, is done on a much more informal basis than the marketing of concentrates and slab zinc. Formal contracts seldom are written and most agreements are simple documents usually in the form of memoranda or letters. The quantities involved may be quoted precisely if it is a spot purchase or may be indicated by such terms as "sellers' output, but not exceeding (specified quantity) per month." Unlike ore purchase contracts, assays are seldom stated except in general terms such as "same as sample submitted" or "same at last shipment."

Except for sal ammoniac skimmings, all secondary materials are sold on the basis of the market quotations for Prime Western grade slab zinc price. Sal ammoniac skimmings are sold on a flat price per ton, either f.o.b. or delivered. Because the processor of the secondary zinc materials usually operates on a narrow profit margin, he must avoid purchasing materials with potentially low yields.

TABLE 2-9. AVERAGE ANNUAL PRICES[1] OF SLAB ZINC AND ZINC CONCENTRATE

1900–1956

	Prime Western Slab Zinc		Zinc Concentrate (60% zinc content)	
Year	Price in cents per pound at E. St. Louis	Price converted to 1954 dollars	Price in dollars per ton, Joplin, Mo.	Price converted to 1954 dollars
1956	13.49	13.02	86.18	83.16
1955	12.30	12.25	77.50	77.19
1954	10.69	10.69	65.72	65.72
1953	10.86	10.88	64.65	64.78
1952	16.21	16.02	116.10	114.71
1951	17.99	17.29	120.00	115.32
1950	13.88	14.85	87.39	93.51
1949	12.15	13.51	72.28	80.38
1948	13.58	14.34	86.37	91.21
1947	[2]10.50	12.01	66.20	75.72
1946	[2]8.73	12.24	51.12	71.67
1945	[2]8.25	13.22	55.28	88.61
1944	[2]8.25	13.46	55.28	90.22
1943	[2]8.25	13.58	55.28	90.99
1942	[2]8.25	14.17	55.28	94.97
1941	7.48	14.53	49.80	96.71
1940	6.34	13.68	41.87	90.36
1939	5.12	11.27	34.15	75.20
1938	4.61	9.95	27.83	60.06
1937	6.52	12.82	39.87	78.38
1936	4.90	10.29	31.95	67.13
1935	4.33	9.18	28.81	61.11
1934	4.16	9.42	27.14	61.47
1933	4.03	10.38	26.88	69.27
1932	2.88	7.55	17.83	46.71
1931	3.64	8.47	22.69	52.80
1930	4.56	8.96	31.97	62.85
1929	6.49	11.57	42.39	75.54
1928	6.03	10.58	38.66	67.81
1927	6.25	11.12	41.02	72.97
1926	7.37	12.51	48.94	83.05
1925	7.66	12.55	53.43	87.57
1924	6.35	10.98	42.54	73.55
1923	6.66	11.24	40.95	69.08
1922	5.74	10.07	35.01	61.41
1921	4.67	8.13	23.78	41.38
1920	7.77	8.54	46.07	50.63
1919	7.04	8.62	43.63	53.40
1918	8.04	10.40	51.68	66.87
1917	8.93	12.89	70.52	101.76
1916	13.57	26.92	83.26	165.19
1915	14.16	34.54	78.47	191.39

TABLE 2-9—*Continued*

Year	Prime Western Slab Zinc: Price in cents per pound at E. St. Louis	Prime Western Slab Zinc: Price converted to 1954 dollars	Zinc Concentrate (60% zinc content): Price in dollars per ton, Joplin, Mo.	Zinc Concentrate (60% zinc content): Price converted to 1954 dollars
1914	5.11	12.71	39.43	98.06
1913	5.61	13.62	42.62	103.44
1912	6.93	17.03	51.95	127.64
1911	5.70	14.88	39.90	104.18
1910	5.42	13.06	40.42	97.41
1909	5.39	13.54	41.08	103.19
1908	4.62	12.45	36.63	98.72
1907	5.90	15.36	43.68	113.74
1906	6.10	16.76	43.30	118.95
1905	5.79	16.36	44.88	126.78
1904	4.97	14.12	35.92	102.05
1903	5.25	14.96	33.72	96.07
1902	4.64	13.37	30.33	87.41
1901	3.89	11.97	24.41	75.11
1900	4.24	12.81	26.50	80.06

[1] Compiled from Metal Statistics 1957, published by American Metal Market.
[2] Excludes Government Premium payments to certain producers.

FOREIGN TRADE

The pattern of foreign trade of the United States as it relates to zinc has changed several times during the history of the domestic zinc industry. Until the latter part of the 1930's the United States was generally self-sufficient in zinc supplies and exports exceeded imports. A shift in the balance of zinc trade occurred in 1937 when imports increased markedly, initiating a trend of growing dependence on foreign sources of supply. Ores, concentrates, blocks, pigs, slabs and scrap have been the major components of foreign trade in zinc through the years, each of which at one time or another has predominated in exports or imports.

Imports

Imports did not become a significant factor in the domestic zinc supply until the late 1930's. Prior to that period, the small quantities imported consisted largely of ores and concentrates. Slab-zinc imports were unimportant but in 1937 they jumped to a then all-time high of 39,128 tons. Thereafter, total imports expanded rapidly as a result of the increased industrial and military requirements of World War II and the expanding demand for zinc metal which outpaced the productive ability of domestic mines.

In 1956 imports totaled 770,000 tons, nearly 60 per cent of the total domestic consumption (including secondary), and during the period 1940 through 1956 imports have averaged about 480,000 tons a year or about 40 per cent of consumption in all forms.

The all-time high for slab-zinc imports was reached in 1956 when nearly 245,000 tons were received from foreign sources. The peak year for ores and concentrates was in 1943 when 539,049 tons were received.

Canada and Mexico have been the principal sources of zinc imports although significant quantities have been received from Peru, Australia, and Argentina.

Exports

Changes in world trade patterns resulting from industrial developments and war have greatly affected the kind and quantity of zinc exports. Prior to World War I, most exports were in the form of ores or concentrates shipped to Europe, and in 1902 there were 55,733 tons of zinc in ore exported, compared to 3,237 tons of slab zinc. However, during the war, the zinc smelting districts of Belgium and France were occupied by Germany, new smelting facilities were built in the United States, export of ores trended sharply downward and export of metal trended upward to reach an all-time peak of 201,962 tons in 1917.

Export of zinc ore again became significant beginning in 1922 and extending through 1929. The all-time peak was reached in 1926 when 95,252 tons were exported principally to smelters in Belgium, Germany, and France. Export of slab zinc increased during the same years and reached a high of 76,351 tons in 1925.

Exports of zinc ores and concentrates have been negligible since 1929 and for most of the 1930's exports of slab zinc were small. With the advent of World War II exports of zinc metal again rose rapidly and reached a peak of 134,000 tons in 1942. Although exports of slab zinc since the war have varied from year to year, they have trended downward and have ceased to be a significant factor in the United States foreign trade in zinc.

Tariff

Imported zinc ores were first made subject to an import duty in 1846. Subsequently, a duty was imposed on slab zinc and zinc in ores was excluded from the tariff acts. From 1883, when the tariff on slab zinc was 1.5 cents a pound, to 1930, when the last Tariff Act was enacted, the tariff on slab zinc changed as follows: Act of 1890, 1.75 cents a pound; Act of 1894, 1.00 cents a pound; Act of 1897, 1.50 cents a pound; Act of 1909, 1.375 cents a pound; Act of 1913, 15 per cent ad valorem; Act of 1922, 1.75 cents a pound; and Act of 1930, 1.75 cents a pound.

In 1909 the duty was again imposed on zinc in ores and under the Act of

TABLE 2-10. TARIFF RATES SPECIFIED UNDER TARIFF ACT OF 1930 AND SUBSEQUENT MODIFICATIONS

Item	Tariff rate in—			
	1930	1945	1948	1951
	Cents per pound; percent ad valorem			
Par. 77:				
Zinc oxide and leaded zinc oxides containing not more than 25 per centum of lead:				
In any form of dry powder	1¾	(1) 1⅒	(2) 6/10	6/10
Ground in or mixed with oil or water	2¼	(1) 1½	(2) 1	1
Lithopone, and other combinations or mixtures of zinc sulfide and barium sulfate:				
Containing by weight less than 30 per centum of zinc sulfide	1¾	(3) 1½	(2) ⅞	⅞
Containing by weight 30 per centum or more of zinc sulfide	1¾ 15%	1¾ 15%	⅞ (2) 7½%	⅞ 7½%
	Cents per pound of zinc content			
Par. 393: Zinc-bearing ores of all kinds, except pyrites containing not more than 3 per cent zinc	1½	(1,4) ¾	(2) ¾	(5,6) 6/10
	Cents per pound			
Par. 394:				
Zinc blocks, pigs, or slabs	1¾	(1,4) ⅞	(2) ⅞	(5,6) 7/10
Old and worn-out zinc, fit only to be remanufactured, zinc dross, and zinc skimmings	1½	(1,7) ¾	(2,7) ¾	(7) ¾
Zinc dust[8]	1¾	(1,4) ⅞	(2) ⅞	(5) 7/10
Zinc sheets	2	(1) 1	(2) 1	1
Zinc sheets coated or plated with nickel or other metal (except gold, silver, or platinum), or solutions	2¼	(1) 1⅛	(2) 1⅛	1⅛

[1] Trade agreement with Mexico, effective Jan. 30, 1943, through Dec. 31, 1950.

[2] General Agreement on Tariffs and Trade (GATT) (Geneva), effective Jan. 1, 1948.

[3] Trade agreement with Netherlands, effective Feb. 1, 1936, through Dec. 31, 1947.

[4] Rate previously reduced in the trade agreement with Canada, effective Jan. 1, 1939, through Dec. 31, 1947, to 1⅕ cents per pound of zinc content on zinc-bearing ores and to 1⅖ cents per pound on zinc blocks, pigs, and slabs, and on zinc dust.

[5] GATT (Torquay), effective June 6, 1951.

[6] Duty suspended from Feb. 12, 1952, to July 23, 1952, inclusive (Public Law 258, 82d Cong).

[7] Duty on metal scrap suspended for practically the entire period from Mar. 14, 1942, to June 30, 1953, inclusive (Public Law 497, 77th Cong.; Public Laws 384 and 613, 80th Cong.; Public Law 869, 81st Cong.; and Public Laws 66 and 535, 82d Cong.).

[8] Since the enactment of Public Law 497 (77th Cong.), effective Mar. 14, 1942, and subsequent amendments (see note 7 above), providing for temporary suspension of duties on metal scrap, quantities of zinc dust have been entered free of duty under this law. No information is available as to the distinction between the zinc dust which has entered free of duty and that which has entered as dutiable.

1909 ores containing less than 10 per cent zinc were duty free; those containing 10 to 20 per cent zinc paid 0.25 cents a pound of zinc content; those containing 20 to 25 per cent paid 0.50 cents a pound and those containing 25 per cent or more paid 1.00 cents a pound.

The Act of 1913 provided a 10 per cent ad valorem tax. The Act of 1922 provided that ores containing less than 10 per cent zinc should be free; those containing 10 to 20 per cent zinc should be assessed 0.50 cent duty per pound of content; those with 20 to 25 per cent, 1.00 cent duty per pound; and those containing 25 per cent or more, 1.50 cents a pound. The Act of 1930 imposed a duty of 1.50 cents a pound on the zinc content of all imported zinc ores.

Subsequently the Tariff Act of 1930 was modified by the Canadian Trade Agreement (Canada, 1939), the Mexican Reciprocal Trade Agreement (Mexico, 1943), the Reciprocal Trade Agreements Act of 1945, the General Agreement on Tariffs and Trade (Geneva, 1948) and the Reciprocal Trade Agreement (Torquay, 1951). Duties established in 1930 and under these trade agreements are shown in Table 2-10.

As noted under footnotes 6 and 7 of Table 2-10, the duty on slab zinc and zinc in ore was suspended from February 12, 1952 to July 23, 1952 inclusive by Public Law 258 of the 82nd Congress, and duty on zinc metal scrap was suspended most of the time from March 14, 1942 to June 30, 1953. The tariff duties on zinc now in effect in 1957 were agreed to at Torquay, England and became effective in June 1951.

Since 1951 world supplies of zinc have been in excess of requirements. Much of the excess was shipped to the markets of the United States and added to the distress of the domestic mining industry. Domestic smelters processed both domestic and foreign ores at near capacity and were little affected except to the degree that they owned or operated marginal mines.

Spurred by its distress, the Lead-Zinc Industry established an Emergency Lead-Zinc Committee to seek relief through legislation, "escape clause" provisions of the Trade Agreements Extension Act of 1951, as amended, or other means. In 1954, the Tariff Commission investigated the industry under provisions of both the Trade Agreements Act and the Tariff Act of 1930 and recommended that the President invoke maximum duties under existing legislation. The recommendation was not followed. In November 1957 the Tariff Commission again made a similar investigation.

REFERENCES*

1. "Materials Survey—Zinc," National Security Resources Board, Washington, U. S. Bur. Mines and Geol. Survey, 1951, 651 pp.

* Used generally, without citation, by the authors and included for supplementary reading. (Ed.)

2. "Metal Statistics," New York, American Metal Market, 1957.
3. "Minerals Yearbook," Zinc chapter, Washington, Bur. Mines, 1956.
4. Pehrson, Elmer W., "Zinc," Chapter of summarized data on zinc production, in "Mineral Resources of the United States, Part I, Metals," pp. 675–727, Washington, U. S. Bur. Mines, 1929.
5. Ransome, Alfred L., "Consumption of Slab Zinc in the United States by Industries, Grades, and Geographic Division, 1940–45, Including a Summary of Consumption Since 1900," U. S. Bur. Mines, Inform. Circ., 7450, 1948, 30 pp.
6. U. S. Tariff Commission, "Lead and Zinc," Report on Investigation Conducted Under Section 332 of the Tariff Act of 1930 Pursuant to a Resolution by the Committee on Finance of the United States Senate, dated July 27, 1953, and a Resolution by the Committee on Ways and Means of the United States House of Representatives, dated July 9, 1953. Parts I, II, III, IV, V, April 1954, 5532 pp.
7. "Yearbook," New York, American Bureau of Metal Statistics, 1956.

CHAPTER 3

GEOLOGY OF ZINC DEPOSITS

ALAN M. BATEMAN

Silliman Professor Emeritus of Geology, Yale University
New Haven, Conn.

and

Editor, Journal of Economic Geology

Zinc was discovered as a metal in 1520, but bracelets filled with zinc were found in the ruins of Cameros, which was burned in 500 B.C. The Greeks and Romans unknowingly used it to make brass, as they found that copper melted with the mineral smithsonite resulted in a metal more yellow than bronze. In the 16th century zinc was imported into Europe from India and China, and mining of zinc in Europe began in 1740. In the Americas, lead ores carrying zinc were mined by the pre-Spaniards, and then by the Spaniards in Bolivia and Peru. The first production of zinc in the United States was at the Washington Arsenal in 1835 from New Jersey ores.

Zinc ores are commonly accompanied by lead ores and the two are mined together. They may also have associated copper or other metals.

GEOGRAPHIC DISTRIBUTION OF ZINC DEPOSITS

Zinc deposits are widely distributed throughout the world although about 60 per cent are located in only four countries, namely, United States, Mexico, Canada and Australia. Other important contributing countries are Italy, Belgian Congo, Peru, Spain, Germany, Japan, Poland and Russia. A dozen other countries contribute in a lesser degree to world supplies.

In the United States, zinc deposits, in order of their contribution, are distributed as follows: (1) Montana, (2) New York, (3) Idaho, (4) Tennessee, (5) Colorado, (6) Utah, (7) New Mexico, (8) Oklahoma, (9) Kansas (10) Washington, (11) Arizona, (12) Wisconsin, (13) Illinois and (14) Virginia. Four other states have smaller deposits.

Zinc Provinces

Zinc, like several other metals, tends to occur in delimited areas, called "metallogenetic provinces." For example, in North America there is an eastern metallogenetic province characterized by deposits of zinc ores, extending from Tennessee through Virginia, West Virginia, Pennsylvania,

New Jersey and New York, and with an intervening break, reappearing in New Brunswick, Gaspé and Newfoundland. A more isolated area occurs in northern Quebec. A Mississippi valley province extends across the corners of Oklahoma, Kansas and Missouri (Tri-State) and northward into Illinois and Wisconsin. A Rocky Mountain province includes parts of New Mexico, Arizona, Colorado, Utah, Montana, Idaho and British Columbia. Another province, which however is dominantly lead with lesser zinc, extends southward through central Mexico.

A North African province extends along the Middle Atlas mountains from Morocco through Algeria and Tunisia to Sicily. A minor Alpine province extends from France to Italy, branching into Yugoslavia. Other minor European provinces lie in south central Spain, in upper-Silesia, Germany, Poland, Great Britain and Scandinavia. An Andean province includes parts of Peru, Bolivia, Argentina and Chile. Still another, and an important province, occupies parts of central Australia. A distinct zinc-lead province lies along the Japanese arc.

Other isolated areas that can hardly be termed metallogenetic provinces occur in Greenland, Manitoba, Washington, southwestern Mexico, Belgian Congo, Northern Rhodesia, Burma, Austria and Czechoslovakia.

MINERALS OF ZINC DEPOSITS

The mineralogy of zinc ores is simple. Only seven minerals are commercial sources of zinc—two sulfide- and six oxygen-containing minerals. They are:

		Per Cent Zinc
Zinc blende or sphalerite	ZnS	67.0
Wurtzite	ZnS	67.0
Smithsonite	$ZnCO_3$	52.0
Hemimorphite (calamine)	$Zn_4Si_2O_7(OH)_2 \cdot H_2O$	54.2
Hydrozincite	$Zn_5(OH)_6(CO_3)_2$	56.0±
Zincite	ZnO	80.3
Willemite	Zn_2SiO_4	58.5
Franklinite	$(FeZnMn)(FeMn)_2O_4$	15–20

Zinc blende or sphalerite is the chief ore mineral. Smithsonite and hemimorphite are important ore minerals in many localities. They result from the oxidation of the sulfide during weathering. Zinc ores generally range from 2 to 12 per cent zinc metal.

Associated Metallic Minerals

In the Eastern United States Appalachian provinces, zinc minerals tend to occur by themselves, except for contained cadmium. In most other localities of the world, however, zinc minerals are generally associated with the minerals of some other ore. The most common and widespread associate is the lead mineral, galena, or its weathering products, cerrusite ($PbCO_3$)

and anglesite ($PbSO_4$). Zinc blende rarely occurs alone. Most of the cadmium produced in the world comes from zinc ores, in which it rarely exceeds 0.5 per cent, and is won as a by-product. Chalcopyrite ($CuFeS_2$) and pyrite (FeS_2) are common associates; silver minerals are likely to be present and add considerably to the value of the zinc ore; gold is present less commonly. Germanium, indium and thallium may also be by-products of zinc ores.

Associated Gangue Minerals

The most common gangue mineral in zinc deposits is calcite ($CaCO_3$), or dolomite ($CaMgCO_3$), which may have been introduced along with the zinc minerals, or may merely be the country rock in which the zinc minerals occur. Other common gangue minerals are fluorite (fluorspar, CaF_2), which may constitute a valuable by-product, rhodochrosite ($MnCO_3$) and rhodonite ($MnO.SiO_2$), which also may be valuable by-products, barite ($BaSO_4$), celestite ($SrSO_4$) and ubiquitous quartz (SiO_2).

ORIGIN OF ZINC DEPOSITS

Many kinds of processes operate within the earth to form various kinds of economic mineral deposits, but the processes that give rise to zinc deposits are relatively few. For the most part they originate through the action of mineral-bearing solutions that emanate from intrusive bodies of molten magma that consolidate deep below the earth's surface to form igneous rocks, such as granite; some hydrothermal solutions may be sweated out of older rocks.

Magmas may be considered as high-temperature mutual solutions of silica, silicates, metallic oxides, metals and always some volatile constituents such as much water, carbon dioxide, sulfur, arsenic, chlorine, fluorine and boron. They are not of the same composition as the igneous rocks that crystallize from them since most of the volatile substances are expelled when the magma finally consolidates. The volatile substances, although minor in amount, play an important role in decreasing viscosity, in lowering the melting point and in collecting and transporting metals. Since magmas are solutions and obey the laws of aqueous solutions, crystallization of the constituent minerals, which is dependent upon their solubility in the rest of the magma, will commence when the magma temperature falls below their individual saturation points.

Since the solution of one constituent in another tends to give a lower melting point than that of either constituent, a magma may remain fluid at a temperature below the melting point of all its constituents. Crystallization commences below this point. As the magma slowly cools, the minerals of highest crystallizing temperature solidify first, followed by those of lower temperature of crystallization. This is generally in the order of decreasing basicity. The water and other volatile constituents, with their dissolved metals, are largely excluded from these rock minerals and accumulate in the residual liquid fraction. Gradually the residual fraction becomes enriched in the volatiles, metals and silica, which characterize the mineralizing solutions.

Contact-Metasomatic Deposits

During the period of cooling of the magma, the heat given off from it may bake and metamorphose the country rocks adjacent to it. This is particularly the case if the country rock is a carbonate one, such as limestone or dolomite, where carbon dioxide is driven off and the metamorphic effects are most pronounced. This is called a "contact-metamorphic zone." Solutions carrying metals and other substances may stream off from the magma chamber into this altered contact zone and by reactions with the invaded rocks produce still greater alteration and deposit quantities of metallic and nonmetallic minerals. This gives rise to a contact-metasomatic ore deposit. The mineralizing solutions in this type of high-temperature deposit are generally considered to have been gases and vapors. Such deposits are made up of iron oxides, sulfides and other ore minerals, and a suite of diagnostic high-temperature nonmetallic minerals such as garnet, wollastonite and others. Vast quantities of materials are added to and subtracted from this altered contact zone. For example, Lindgren[6] showed that at Morenci, Arizona, the astonishing amounts of 2.8 tons of materials were added, and 1.8 tons removed, from each cubic meter of limestone. The deposits always lie quite close to the contact of the igneous rock, and all large deposits are developed only in limestone or dolomite.

Contact-metasomatic processes give rise to deposits of iron, copper, lead, zinc, tin, tungsten, molybdenum, gold and other materials, but those of zinc are not common; Hanover, New Mexico, and Kamioka, Japan, are examples.

Hydrothermal Deposits

With the continued freezing of the magma, the residual liquid will contain all the excess water of the magma and the metallic and other constituents in solution. Such solutions with their metallic load may be expelled or sucked off from the magma chamber into fractures, breccias, or other openings in either the frozen part of the magma or in the nearby country rocks. These constitute hydrothermal solutions that may give rise to one type of hydrothermal ore deposits; most zinc deposits fall under this group.

Hydrothermal solutions have also been thought to have originated through the heating up of ground water by magmatic contributions, either in the form of gases and vapors, or liquids. In the case of hot springs that deposit metals, the magmatic contributions are considered to be minor and the main body of the solutions is thought to be ground water. The metals are considered to have come from the magma or to have been dissolved from underlying rocks. Such solutions are thought to have given rise to some of the low-temperature (epithermal) zinc deposits. In the case of deep-seated medium-to-high-temperature (mesothermal to hypothermal) deposits that occur in nonsedimentary rocks, it seems probable that ground water formed little or no part of the hydrothermal solutions, since experience over the world has shown that in most deep mines in nonsedimentary rocks the ground water disappears at depths around 2500 feet; the lower levels of mines are dry and free from ground water except that which seeps down and through mine openings. Thus, such deep deposits probably were formed many thousands of feet below the zone of ground water.

Another type of hydrothermal solution has also been proposed, namely, water locked up in the rocks being sweated out under conditions of heavy loading or mountain building, being heated to the rock temperature, dissolving earlier formed minerals and rising to deposit their content in overlying rocks. To such deposits the term "alpine type" has been applied. This origin has been advanced for deposits where intrusive igneous rocks are unknown.

Hydrothermal solutions may give rise to two main groups of hydrothermal deposits—cavity-filling and replacement; and each of these in turn may form high-temperature (hypothermal), intermediate temperature (mesothermal) and low-temperature (epithermal) deposits, each characterized by more or less distinctive minerals. Most zinc deposits fall within the intermediate or low-temperature groups.

Cavity-Filling Deposits. There are many types of rock openings or cavities in the rocks of the earth's crust, and some of these may become filled with zinc and other minerals deposited from the hydrothermal solutions to form zinc ore deposits. The chief receptacles for zinc ores are fissures, breccia openings, shear zones, joints, bedding planes, stockworks, caves and other solution openings, and pore spaces. Fissure veins and breccia fillings are the most common types of cavity fillings carrying zinc ores.

The deposition of the ore and gangue minerals from solution is brought about by more than one process. Probably the most important are changes in temperature and pressure of the solutions. A drop in temperature of the solutions decreases solubility of the mineral substances, and deposition of some or all of them may then take place. Likewise, a drop in pressure promotes deposition of minerals from solution, and this may occur rapidly if the solutions enter large openings. Other factors that affect deposition are chemical reactions with the wall rocks or other vein minerals, changes of *p*H in the solutions, reactions with other solutions, and chemical complexes.

Cavity filling commonly proceeds by deposition of successive crusts of minerals from the walls toward the interior of the openings. Thus, in a fissure vein, mineral *a* may be deposited upon either wall with its crystal faces pointed toward the opening and mineral *b* may be deposited as a next crust or layer, followed by still others. The central zone may contain irregular openings called "vugs," which may contain one or more sequences of crystals perched on the walls, and are eagerly sought by mineral collectors because they often prove to be treasure houses of rare and beautiful crystals. The fragments of a breccia may be similarly crusted, giving rise to cockade ore. Such crustification is indicative of formation from solutions in open spaces.

Replacement Deposits. Replacement is a process of essentially simultaneous solution and deposition by which a new mineral is chemically substituted for a rock mineral or an earlier-formed vein mineral. The replacing minerals are carried in solution and the replaced substances are carried away by the solution; it is an open circuit. The petrification of wood is a replacement of wood by silica. In replacement, the substitution is not molecule for molecule, but an equal-volume substitution. For example, if in a brick wall, each brick were removed one by one, and a similar silver brick of each size substituted, the end result would be a wall of the same size and form except that it would be composed of silver instead of clay. The parts exchanged are infinitesimally small—of molecular or atomic size.

Replacement involves the necessity of continuing supplies of new material and removal of the dissolved material. Consequently, channels are necessary to permit the mineralizing solutions to be delivered to the places of attack. Such channels may consist of fissures, faults, shear zones, joints, bedding planes, shattered zones, pipes and solution openings, cracks in

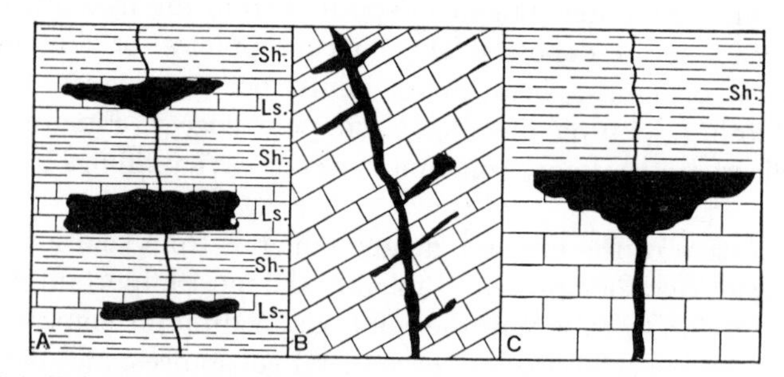

Figure 3-1. Relation of replacement ore bodies to features of sedimentary rocks. Black is ore. (A) Relation to beds of limestone lying between beds of shale. (B) Relation to bedding planes. (C) Relation to overlying impervious shale. (From Bateman, "Economic Mineral Deposits," Ref. 1, p. 148.)

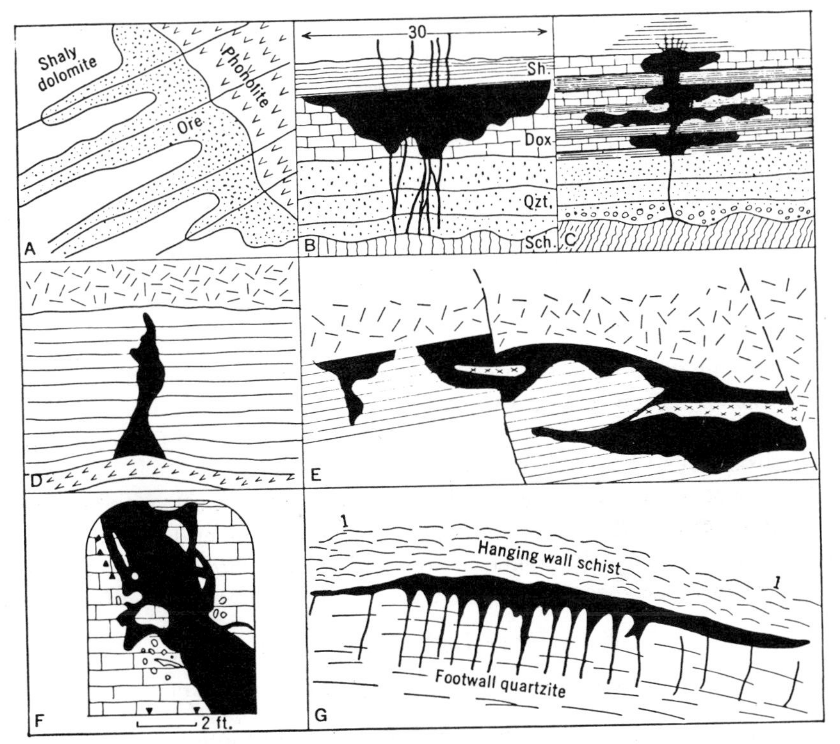

Figure 3-2. Forms assumed by replacement deposits.

A. Ore spreading out along fissures, Mineral Farm, Black Hills, S. Dakota. (After Irving.)

B. Ore replacing limestone outward from fissures but restricted above by impervious shale, Union Mine, S. Dakota. (After Irving.)

C. Ore spreading outward from a channel fissure along beds of dolomite, but not in intervening shale, Portland Mine, S. Dakota. (After Irving.)

D and E. Cross and longitudinal sections of Iron Hill, Leadville, Colo. (After Irving.)

F. Replacement vein in limestone showing irregularities of outlines of ore, Jumbo Mine, Kennecott, Alaska.

G. Relation of ore to fissures in quartzite and to overlying schist, Ferris Haggerty Mine, Encampment, Wyoming. (After Spenser.)

(From Bateman, "Economic Mineral Deposits," Ref. 1, p. 154.)

folded rocks, vesicular lavas and pore spaces in rocks. Most replacement deposits are spatially related to one or another of such channels. At the actual place of deposition, diffusion may aid the metal ions in reaching the site of replacement. Thus, structural features of the rocks largely localize the places of deposition. Figure 3-1 shows the relation of replacement bodies to sedimentary features and Figure 3-2 shows some of the forms assumed by replacement ore bodies.

Replacement bodies may form in any kind of rock, although carbonate rocks such as limestone or dolomite are more congenial host rocks than most, and shale is the least favorable.

Replacement deposits may be large, irregular, massive bodies largely made up of ore minerals, or they may be tabular bodies formed by the replacement of the walls of a fissure or shear zone, giving rise to replacement lodes, or they may consist of scattered grains of ore minerals in a host rock, forming disseminated deposits. All three types are represented by replacement deposits of zinc ores.

Replacement may take place by means of high-temperature gases and vapors, by hydrothermal solutions of high, intermediate or low temperature and by cold surface waters. Most zinc replacement bodies have been formed by low-temperature hydrothermal solutions.

Oxidation Deposits

Many zinc deposits have originated through processes of surficial weathering and oxidation, giving rise to bodies of zinc carbonate (smithsonite, $ZnCO_3$) and zinc silicate (hemimorphite or calamine, $Zn_4Si_2O_7(OH)_2 \cdot H_2O$), or hydrozincite ($Zn_5(OH)_6(CO_3)_2$).

Consider an ore body consisting of sphalerite (ZnS), pyrite (FeS_2) and some galena (PbS), with perhaps some gangue mineral. When exposed to weathering in which oxygen, water and carbon dioxide take part, the pyrite readily becomes oxidized and yields ferric sulfate and sulfuric acid:

$$FeS_2 + 7O + H_2O = FeSO_4 + H_2SO_4$$

$$2FeSO_4 + H_2SO_4 + O = Fe_2(SO_4)_3 + H_2O$$

Both ferric sulfate and sulfuric acid, particularly the former, are powerful solvents of most of the metal sulfides. The ferric sulfate attacks pyrite to yield ferrous sulfate, which in turn oxidizes to additional ferric sulfate. Thus, more solvent is continuously generated. The ferric sulfate will also attack the sphalerite and galena:

$$ZnS + 4Fe_2(SO_4)_3 + 4H_2O = ZnSO_4 + 8FeSO_4 + 4H_2SO_4$$

$$PbS + Fe_2(SO_4)_3 + H_2O + 3O = PbSO_4 + 2FeSO_4 + H_2SO_4$$

The zinc sulfate so formed is readily soluble in surface waters, giving rise to a solution of zinc sulfate. The lead sulfate, however, is not soluble in water and so precipitates. Some of the ferric sulfate hydrolyzes to ferric hydroxide, which in turn changes over to hydrous ferric oxide, or the mineral goethite ($Fe_2O_3 \cdot H_2O$), which makes up most of the substance commonly called "limonite." The ever-present limonite characterizes all oxidized zones and forms the universal brownish iron stain that colors most oxidized ore outcrops to form the iron cap or gossan.

The zinc sulfate solution may trickle downward or to the side of the oxidized zone. If it comes in contact with limestone, a chemical reaction takes place by which zinc carbonate and calcium sulfate are formed. The

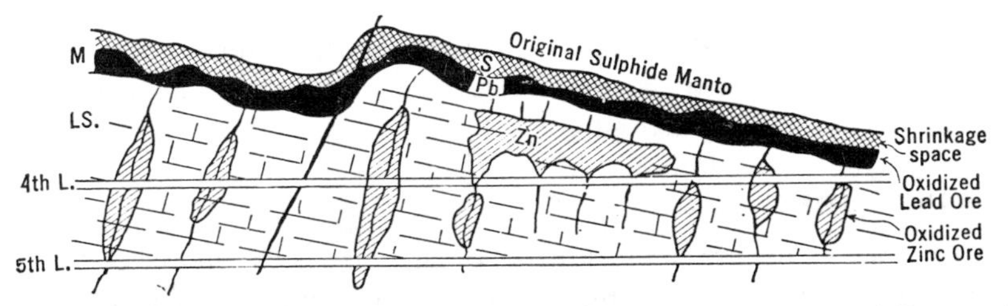

Figure 3-3. Sketch of longitudinal section of part of ore manto of Encantada Mine, Sierra Mojada, Mexico, illustrating separation of metals by oxidation. Original body (M) composed of zinc blende, galena, and pyrite. Oxidation yielded body of oxidized lead-silver ore (Pb) overlain by open space (S) caused by shrinkage due to removal of part of sulfides, and underlain by large bodies of zinc carbonate ore (Zn) aligned along fissures and joints in iron-stained dolomite. The zinc moved in solution out of ore body (M) downward along cracks and reacted with the carbonate rock to form zinc carbonate. (Sketch by author.)

soluble calcium sulfate goes away in solution and the zinc carbonate remains. Or, the zinc sulfate may come in contact with carbonated waters, and a similar deposition of zinc carbonate occurs. The lead sulfate may also be changed to lead carbonate by contact with carbonated waters. If the zinc sulfate comes in contact with soluble silica a different chemical reaction takes place, by which zinc silicate is formed. The silica is made available in large quantities by the breakdown of feldspars in igneous rocks during weathering. Large deposits of zinc carbonate and zinc silicate are formed by these simple processes.

The oxidation processes may also bring about a separation of mixed metals. In the case of a sphalerite-galena-pyrite deposit, the soluble zinc sulfate may travel downward and away from the original deposit, the lead sulfate may remain in its original site and much of the iron of the pyrite will be removed. This separation and the formation of zinc carbonate bodies illustrated in Figure 3-3.

The oxidation processes generally cease when the permanent water table is reached. If the original deposit extends beneath the present or a former water table, then zinc sulfide may be expected beneath the oxidized zinc ore. No underlying zinc sulfide ore may be expected beneath oxidized zinc deposits that have been transported, as in Figure 3-3.

LOCALIZATION OF ZINC DEPOSITS

Why ores are found in certain places and not in others has continued to be a challenge to geology. Although in many cases the answer is known, in many others the answer is still an unsolved problem. Much has been learned in recent years regarding structural control of ore deposition and there is a tendency to consider that this is the only control. However, orogenic elements, igneous instrusives, stratigraphic, chemical, physical, and other factors must also be considered. Generally more than one operate together to localize ore deposits.

Igneous Activity

A primary factor in the localization of zinc ores is the source of the zinc. This may be an igneous center, or underlying sediments. The very widespread association of ore deposits close to and around areas of intrusive igneous rocks, and their relation to intrusives in time, lends high probability to magmas being the source of zinc for most deposits. Consequently, centers of intrusive igneous activity in a large way may be considered favorable for broad localization of zinc ores. Where deposits do show spatial and time relations to a specific intrusive, it should not be inferred that the metals emanated from the exposed intrusive; more probably both the intrusive and the metals emanated from the same underlying magmatic reservoir.

However, many deposits occur where intrusive igneous rocks are unknown. The metals of such deposits may have come from an underlying hidden igneous source, or may have been dissolved out of underlying sediments. Both are conjecture and positive evidence for either is mostly lacking.

Metallogenetic Provinces and Shield Areas

Metallogenetic provinces of zinc, previously mentioned, are other broad localizers of ore, but just why such provinces exist is unknown. If, for example, one were to explore for new zinc deposits, the chances of discovery would be greater within a zinc metallogenetic province than outside it.

Shield areas are also broad localizers of ores. Here there have been sedimentation, volcanism, intrusive activity, mountain forming, folding and fracturing, and deep erosion to lay bare the depth zones where so much ore deposition is thought to have formed. The Canadian, Australian, African, Russian, Scandinavian and Brazilian shields have proved to be great storehouses of minerals, and new deposits are being discovered each year. A great many zinc deposits, however, have been localized outside shield areas.

Mountain Building and Folding

Mountain building is accompanied by folding of rock formations, extensive and deep faulting, lesser fracturing and igneous activity. Erosion following uplift has excavated great valleys and laid bare deep rock formations, which are the home of so many mineral deposits. Most of the zinc deposits, although not the greatest ones, occur in present or former mountain areas.

Structural Controls

Faulting and Fissuring. Large regional faults may serve as trunk channels through which mineralizing solutions move from the depths upward into subordinate connecting channels. For example, Singewald and Butler[11] consider the large London fault in Colorado to have been a major locus for the zinc-lead-silver deposits of that region.

Lesser faults, fissures and shear zones play an important role as immediate localizers of ore. They may themselves become filled with mineral substances or they may give access to solutions that bring about replacement of their walls to form tabular replacement deposits. They may also serve as channels to introduce solutions into beds of easily replaceable carbonate rocks, giving rise to bedded deposits.

Other Structural Openings. Smaller detailed structural features are for the most part the direct and immediate cause of ore being in one place rather than in another. They are features which can be readily observed and mapped, and owe their origin to structural processes. These include such features as stockworks in which a rock mass is traversed by small intersecting fissures, saddle reefs formed where hard competent rocks, such as sandstones interbedded with softer slates, are sharply folded giving rise to actual or potential openings at the crests of the folds, and ladder veins developed in hard dike rocks subjected to stress. Gentle slumping of beds gives rise to openings along the bedding ("flats") and connecting highly-inclined openings ("pitches"), such as are characteristic of the upper Mississippi valley zinc-lead deposits. Breccia openings, caused by shearing, intersections of fissures and slumping are very important localizers of zinc deposits in the Appalachian province.

Stratigraphic Controls

Stratigraphic features, both major and minor, commonly cause the localization of zinc and other ores.

Geosynclines. Geosynclines are large, down-sinking, troughlike areas of the earth's crust where thicknesses of many thousands of feet of sediments accumulate as the sinking progresses. Geosynclines are the forerunner of certain classes of folded mountains. The uplift and folding of the geosynclinal sediments is commonly accompanied by prominent faulting, overthrusting and intrusion of great masses of igneous rocks. Mineralizing solutions may accompany the granitic intrusives and deposit ores in the folded, fractured and faulted sediments. Thus in a very broad way geosynclinal areas bring about localization of ore deposits.

Plateau Areas. Positive plateau areas are commonly surrounded by belts of thick sedimentation, and as pointed out above, mineralization may be associated with belts of strong folding and faulting associated with basins of thick sedimentation. Butler[3] has shown that for much of pre-Tertiary time, the Colorado Plateau existed as a positive area, around which were marginal areas of extensive sedimentation. These areas were later strongly folded, faulted, intruded and mineralized. Butler plotted the many mining districts of Colorado, Utah, Arizona and New Mexico, and this brought out strikingly the absence of base-metal ore deposits within the plateau and their concentration around the margins of the plateau in the areas of thick sedimentation.

Unconformities. Stratigraphic unconformities indicate sites of erosion and weathering between underlying and overlying beds of sediments. During the process of weathering, pre-existent bodies of zinc sulfide may become oxidized and undergo concentration and residual accumulation. Also, the weathered zones beneath unconformities are generally more porous than contiguous rocks and thus may serve as channels for mineralizing solutions and sites for deposition of zinc and other ores.

Bedding Planes. Bedding planes facilitate the movement of underground fluids through sedimentary rocks and form favorable sites for ore replacement to take place (Figures 3-1 and 3-2). Ore deposits are commonly found extended along such bedding planes as, for example, in the Tri-State district.

Lenses. Lenses of carbonate rock intercalated between shales or sandstones are known to have localized replacement deposits of zinc and lead ores.

Impervious Covers. Beds of impervious shale commonly have acted as barriers to the upward flow of mineralizing solutions. Consequently, ore replacement has mushroomed out beneath the impervious beds, as may be seen in Figures 3-1 and 3-2. In Ouray, Colorado, for example, zinc deposits are localized beneath an impervious black shale bed.

Impervious Floors. Impervious base rocks have similarly impounded descending zinc-bearing solutions formed as a result of oxidation of sulfide deposits. As a result, concentrations of zinc carbonate have been formed resting upon the impervious base.

Physical and Chemical Controls

As has been mentioned, physical and chemical properties of host rocks exert an important influence upon ore deposition and thereby help to localize mineral bodies. However, the exact properties that aid or cause mineral deposition are not always evident; they are often surmised. In general they operate along with structural features. A fissure that traverses sedimentary beds, for example, acts as a channel for solutions, but a congenial host rock is also necessary to induce deposition (Figure 3-1). The formation of the ore only where limestone is the wall rock is an observable fact, but the cause of the deposition is uncertain.

Permeability. The mineralizing solutions that give rise to ore deposits require openings to reach the site of deposition. Fissures may serve as main channels, but connecting rock pores are necessary to permit the passage of the solutions or the solutes through the wall rock to the sites of deposition. In some deposits main channels are apparently absent and permeability seems to have been the main or only factor of ore localization.

Brittleness or Toughness. Brittle rocks tend to crackle readily under stress; such crackling has generally been observed in regions of metallization. The high permeability thus created makes favorable sites for ore deposition. Limestone, chert and quartzite are found in these regions. Tough or pliable rocks, such as shale, do not crackle and consequently are less permeable. The relative effects of the permeability resulting from these physical properties can also often be observed in secondary deposits of zinc ore.

Chemical Properties. The chemical character of the wall rock has for a long time been considered to play a dominant role in the localization of ore deposits and ore shoots. Deserving as it is of this role, it has also been overrated, occult attributes having been ascribed to it. Some rocks unquestionably are more congenial hosts to ore than others and in many cases for no observable reason.

Carbonate rocks are especially favorable rocks for ore. In district after district, fissures carry ore in limestone or dolomite and not in other rocks (Figure 3-2). This is a selective replacement due to chemical control. In many of the zinc deposits of northern Mexico, zinc deposits occur in dolomite but not in limestone. The reason for this is not clear since there is no observable difference in the two rocks; the porosity is about the same in both rocks. However, one contains magnesium and the other does not—a chemical difference. On the other hand, for no explainable reason, ore at the Potosi mine in Chihuahua, Mexico, occurs in the limestone but not in dolomite beds.

The chemical character of the country rock controls directly the location of secondary zinc carbonate or silicate deposits formed through weathering and oxidation of former sulfide bodies. If the zinc sulfate solutions released during oxidation come in contact with limestone or dolomite, reaction between them precipitates zinc carbonate and deposits of zinc carbonate may thus be formed. On the other hand, if the zinc sulfate solutions meet unreactive rocks, the zinc will not be precipitated and it will be dissipated. Similarly, if colloidal silica is released by rock-weathering, bodies of zinc silicate may result.

ZONAL DISTRIBUTION OF ZINC ORES

It has been observed in many mineral districts that the metals are zonally arranged with respect to some center—generally an igneous intrusive. The ores or minerals formed at higher temperatures lie closest to the center and those formed at lower temperatures lie farthest out. This zonal arrangement may be horizontal or vertical, or both, and it may involve ores of different metals or minerals of the same metal. Thus at Butte, Montana, Sales[9] has clearly demonstrated a pronounced zoning of metals and minerals both horizontally and vertically. In the central zone (hottest) are copper ores consisting chiefly of quartz, pyrite, enargite, bornite and digenite. In the next or intermediate zone, less enargite and chalcocite are found and chalcopyrite and tennantite are important. Toward the margin of this zone sphalerite is common and zinc and manganese carbonates are present. The peripheral zone is characterized by zinc, silver and manganese and the dominant minerals are sphalerite, galena, rhodochrosite and calcite.

Similarly at Bingham, Utah, copper-molybdenum-gold ores occur in and next to the porphyry pluton, but farther out in the sediments the ores consist of zinc, lead and silver. It has long been stated that in Cornwall, England, in some veins, tin lies below copper, and copper below zinc and lead. At Zeehan, Tasmania, tin lies in granite, tungsten and copper are next, copper, lead and zinc next, and farthest out are zinc-lead-silver-antimony ores.

Such a zonal arrangement of ores with zinc and lead on the outside is commonly presumed to be a function of the temperature of deposition of the minerals, the least soluble minerals being deposited near the centers and the more soluble ones farther out. However, it should be pointed out that factors other than temperature may have been involved in the deposition of minerals. Pressure, relative concentration, *p*H, Eh, reactions with wall rocks, chemical complexes and other factors may have participated.

It should be realized that this zonal arrangement is not a common one and it should not be inferred that copper ores are to be expected to lie beneath zinc ores. In some deposits there are actual reversals where zinc lies below copper and lead below zinc. Lindgren (Ref. 6, p.122) points out that in low-temperature (epithermal) veins there is generally no zoning except a common impoverishment in depth.

KINDS OF ZINC DEPOSITS

It was stated earlier that zinc deposits are formed by contact metasomatic, hydrothermal and oxidation processes. The resulting deposits are called by the same names. Some examples of each will be given in this section.

Contact-Metasomatic Deposits

Hanover, New Mexico. The Hanover-Fierro areas are a part of the Central Mining District of New Mexico, which has been the fourth largest zinc-producing region of the United States. The district also includes the large Chino porphyry copper deposit of the Kennecot Copper Corporation at Santa Rita. Around the porphyry copper deposit are a group of hydrothermal replacement and vein deposits of zinc ores. To the north are contact-metasomatic deposits centering around Hanover.

Underlying the region is a group of sedimentary formations ranging in age from Cambrian to Cretaceous. These are intruded by four stocks of granodionite and related sills and swarms of dikes. Each stock has a contact metamorphic halo around it and mineralization is associated with each. Parts of the area are covered by Tertiary volcanics and later alluvium. The region has undergone a complex history of uplift, folding, faulting, igneous intrusions, metamorphism, more faulting, lava eruptions, further faulting and deep erosion exposing the igneous bodies and older formations.

The Hanover-Fierro granodionite pluton is about 2½ miles long in a northerly direction by ½ mile wide. A contact-metamorphic halo surrounds much of it, consisting of an inner zone of garnet-pyroxine rock, an intermediate zone of marble and an outer actinolite-tremolite zone. The granodionite, porphyry dikes, quartz-dionite sills and altered sedimentary rocks are cut by quartz-latite dikes.

The ore bodies lie chiefly at the outer edge of the garnet-pyroxene zone and most of them have one marble wall. The ore minerals replace the silicates along stratigraphic horizons or along fissured zones that had been altered to contact-silicates. According to Lasky and Hoagland,[5] there are two main zinc zones in the contact metamorphic halo, one along the east side and the other on the west side toward the south end of the granodionite stock. The ore bodies have been formed by replacement of silicated limestones of Mississipian and Pennsylvanian age. Individual bodies may be 2500 feet long, 30 feet wide and 80 feet high. The Pewabic ore body is a gently dipping chimney 600 feet long by 30 feet in diameter; blanket-like fins extend 200 feet or more off from the chimney. The ore averages about 14 per cent zinc, 0.1 lead and 6 iron.

The ores are made up of early magnetite and specularite, and the chief silicates andradite, garnet, epidote, salite, ilvaite, bustamite and quartz. These were followed by sphalerite with quite minor specularite, pyrite, galena, chalcopyrite and quartz.

Generally similar zinc deposits occur at Copper Flat about 2 miles southwest of Hanover. Other contact-metasomatic zinc deposits occur at Magdalena, New Mexico; Success Mine, Idaho and Kamioka, Japan.

Hydrothermal Deposits—Cavity Fillings

Fissure Veins. *Butte, Montana.* Montana has become the largest zinc-producing state in the United States, the main source being fissure vein deposits at Butte. Here there are seven intersecting fissure systems, three of which carry copper ores; two of these also yield zinc ores. One is an east-west vein system and the other a northwest system of veins that fault and displace the east-west veins. The veins range in width from a few feet to a few tens of feet, in length up to a few thousand feet and in depth to over 3000 feet. The ores of the district are zoned both horizontally and vertically. A central zone carries copper ores, an intermediate zone contains copper and zinc ores and an outer zone is dominated by zinc blende along with a little galena, chalcopyrite and silver, in a gangue of calcite, rhodochrosite, rhodonite and quartz. These ores yield zinc, manganese and silver. They are typical hydrothermal fissure veins in granodionite of the Boulder batholith. The annual production is around 85,000 tons of zinc from ores that average 12 to 14 per cent.

Cœur d'Alene District, Idaho. This district has yielded ores worth well over a billion dollars in lead, zinc and silver. The mines occur within an area about 25 miles long and 15 miles wide. The country rocks consist of the Pre-Cambrian Belt Series which have been intruded by Cretaceous quartz, monzonite, and a few dikes of diabase and lamprophere. The Belt Series is over 20,000 feet in thickness and is made up of quartzites, argillites and calcareous rocks that have been intensely altered and bleached to sericite schists and slates and carbonates. The ores occur in the altered sediments and to a minor extent in monzonite. Minor contact-metasomatic ores occur adjacent to the monzonite intrusives.

The area is crossed by great barren faults several miles long; the Osborn fault has a vertical displacement of 15,000 feet or more and a horizontal displacement measured in miles; other faults have displacements exceeding 5000 feet. Between these great barren faults are minor faults and shear zones that have been filled and replaced to form the lodes. They attain 7,000 feet in length, average 9 feet in width and have been followed downward over 5,300 feet.

The ores consist of disseminated grains and masses of siderite, calcite and quartz along with galena, zinc blende, pyrite and argentiferous tetrahedrite. Some pyrrhotite, boulangerite, magnetite, arsenopyrite and copper minerals occur in minor amounts. The ores carry 3 to 12 per cent lead, 3 to 6 per cent zinc and 2 to 6 ounces of silver per ton. The annual production is around 70,000 tons of zinc, 60,000 tons of lead and 7 million ounces of silver. Oxidation is shallow and irregular.

Fresnillo, Zacatecas, Mexico. This famous mine was discovered in 1554 and has been worked almost continuously since then, having yielded about one-half billion dollars. The deposits consist of six main veins and 30 minor ones, which occur in slate, graywacke and altered volcanic rocks. A dike of granodionite porphyry is disclosed in the deepest workings. The veins average 1.4 meters and are being worked to a depth of 1000 meters and for a strike length of 2 kilometers. The ore reserves are 6 and ¼ million tons averaging 4.9 per cent zinc, 5.1 lead, 0.38 copper, 7.1 ounces silver and 0.017 ounce of gold per ton. The annual yield is about 57,000 tons of zinc concentrate and 32,000 tons of lead concentrate.

The ores consist of pyrite, sphalerite, marmatite, galena, chalcopyrite, pyrrhotite, arsenopyrite and ruby silver in a gangue of quartz and calcite. The veins are notably zoned. High silver and minor base metals occur in the upper levels and the reverse in the lower levels. Hot springs with a temperature of 57°C occur in the lower workings.

San Francisco del Oro, Chihuahua, Mexico. This important mine contains 10 chief veins and a few minor ones. The veins occur in shale of Cretaceous age which has undergone wall-rock silicification. Rhyolite and diabase dikes cut the shale and veins and are unconformably overlain by gravel and basalt flows. The veins are strong fault fissures of small displacement. They range in size up to 2 kilometers long, 800 meters deep and from 0.8 to 2 meters wide. They yield about 750,000 tons of ore annually, containing 80,000 tons of zinc concentrate and 57,000 tons of lead concentrate. The ore reserves exceed 4 and ½ million tons, containing 0.18 gram of gold and 154 grams of silver per ton of ore, and 7.75 per cent zinc, 5.65 lead and 0.57 copper. The ore consists of pyrite, sphalerite, galena, chalcopyrite, arsenopyrite and silver minerals in a gangue of quartz calcite and fluorite.

Other Vein Deposits. Other important vein deposits of zinc ores occur at Santa Barbara, Parral and Taxco in Mexico; Morococha, Casapalca and San Cristobal in Peru; Lake George in Australia; Ikuno and Hosokuro in Japan; Park City, Utah, San Juan and Red Cliff, Colorado and Pioche, Nevada, in the United States; the Noranda district in Quebec and Keno, Yukon, in Canada; Raibal, Trentino and Sardinia in Italy; Pulcayo and Matilda in Bolivia; Read-Roseberry and Lake George in Australia; Ridder-Sokolny in Kazakstan; Badwin in Burma; the Linares district in Spain; and Altenberg in Saxony.

Breccia Fillings. *Tri-State District.* This broad district, centering around Joplin, Missouri, is the greatest zinc district of the world. Its area is over 2,000 square miles and it has yielded about one billion dollars in metals—mostly zinc. The ores occur almost exclusively in the Boone formation of Mississipian age, which is made up of limestone, chert, limy chert and some shaly limestone. The beds are almost flat-lying but have suffered some gentle deformation and, due to their brittle character, some shattering, shearing and brecciation. The limestone members suffered some ground-water solution with resultant collapse of the shattered chert beds giving rise to chert breccias. Into such broken ground, the mineralizing solutions deposited sphalerite, a little galena, pyrite, marcasite, chalcopyrite and enargite, along with calcite, dolomite and jasperoid. The ores occur in runs and circles, and as sheet or blanket ore along the Grand Falls chert parallel to the bedding. The ore minerals fill in around the chert fragments, in part fill up other cavities and in part replace limestone. The runs range from 10 to 300 feet wide, up to 80 feet thick and extend up to 2 miles. The circles surround sinkholes and attain dimensions of 800 by 650 feet, with a thickness up to 30 feet. The sheet ore is generally 10 to 12 feet thick.

The ores formerly were thought to have been deposited by meteoric waters but are now considered to be hydrothermal in origin. No igneous rocks occur in the region.

East Tennesee Zinc District. The extensive zinc deposits of this district center around Mascot-Jefferson City in the Appalachian valley, which is a region of intense folding and thrust faulting. The zinc ore occurs in dolomites and dolomitized limestone of Lower Ordovician age. Intrusive rocks are absent.

The ore deposits are of two types: (1) filling and replacement of tectonic and collapse breccias of dolomite localized by faults and fault intersections along minor fold axes; and (2) breccia and fracture filling, accompanied by replacement along faults. Pre-ore flexing of the beds provided conditions favorable for the brecciation of the brittle dolomite forming large areas of "crackle" breccia. The openings in the breccias localized the deposition of the ores.

The ores consist of light yellow sphalerite, dolomite and calcite, with traces of galena, pyrite, barite, fluorite and quartz. The first event in the formation of the deposits was the fracturing of the limestone and dolomite, followed by dolomitization of the limestone and deposition of introduced dolomite. Pyrite was next deposited, followed by sphalerite and then by the sparse galena, fluorite, barite, a little quartz, and calcite. The individual ore bodies average from 15 to 100 feet in thickness, 125 to 150 feet in width and have been followed to depths of over 1000 feet.

Replacement Deposits. Most of the zinc production of the world is won from replacement deposits, chiefly in limestones and dolomites, but also in other rocks. Descriptions of some of the more outstanding deposits follow.

Sullivan Mine, British Columbia. This is the largest zinc mine in the Western Hemisphere, and perhaps in the world. Its annual production is around 185,000 tons of zinc; and 170,000 tons of lead, 6 million ounces of silver and a little tin.

The ore body is a sulfide replacement of argillaceous and silty beds of the Pre-Cambrian Aldridge formation. It is a large lens that occupies the crest and eastern flank of an anticlinal nose. The ore zone dips gently eastward and is underlain by footwall conglomerate and overlain by quartzite. Gabbro intrusives occur east and west of the mine while dikes cross the ore zone and are slightly mineralized. Some faults cross and dislocate the ore. Wall-rock alteration is pronounced; tourmalinization has taken place in the footwall, and chloritization and albitization in the hanging wall.

The ore consists of galena, sphalerite (marmatite), pyrrhotite, pyrite and magnetite; with minor chalcopyrite, arsenopyrite, boulangerite and some recoverable cassiterite: quartz, sericite, and metamorphic minerals constitute the gangue minerals. The galena is later than the pyrrhotite and sphalerite. The lower part of the main body is largely pyrrhotite. Tourmalinization extends 1,500 feet beneath the ore body.

Flin Flon Mine, Manitoba. This large mine of the Hudson Bay Mining and Smelting Company is an example of a sulfide replacement associated with quartz porphyry along a contact between pyroclastic volcanic rocks and andesite of Pre-Cambrian age. The ore bodies are a series of sulfide lenses localized along the northeast limbs of drag folds and the limbs of a major fold. The crests and troughs of the drag folds have been crumpled and the ore terminates against the crumpled portions.

The ores are of two types, massive and disseminated. The massive ore consists of solid pyrite containing sphalerite and chalcopyrite. This ore also carries silver, gold, arsenic, lead, selenium, tellurium, cadmium and cobalt. Other minerals present are magnetite, arsenopyrite, pyrrhotite, cobaltite, cubanite, tetrahedrite, enargite and tellurides. The ore averages 3.00 per cent copper, 4.24 zinc; 0.09 ounce of gold and 1.30 ounces of silver.

Nearby, generally similar deposits were formerly worked at Mandy and Sheritt-Gordon.

Bingham, Utah. If Bingham were less overshadowed by the colossal Utah Copper porphyry mine, it would be better known for its lead-zinc-silver deposits.

Surrounding the Utah Copper stock of quartz-monzonite porphyry, and in part around the Lost Chance stock, are several thousand feet of folded, faulted and thrusted Pennsylvanian sediments, chiefly quartzites and impure limestones. Within the Utah Copper stock is the great disseminated deposit of low-grade porphyry copper ore. In the limestones next to the porphyry ore are some contact-metasomatic copper ores. Farther out from the stock are replacement deposits and fissure veins of copper ore. Beyond this, zonally outward from the stock, are prolific replacement deposits of lead-zinc ores carrying silver and very minor copper. Thus, there is an ideal zonal arrangement of central porphyry copper ore, replacement and fissure vein copper ore, lead-zinc-silver ores, and farthest out, siliceous silver ores.

The replacement lead-zinc ores extend outward from the stock along the strike within each of several limestone horizons up to a distance of 8,000 feet and to a depth of over 3,000 feet. The successive ore zones are considered to have been formed by waves of hydrothermal solutions that emanated out of the monzonite reservoir. Complete replacement of limestone beds up to a stratigraphic thickness of 100 to 150 feet are known.

The minerals consist chiefly of galena, sphalerite, pyrite, and a little tetrahedrite, and chalcopyrite and bornite. Oxidation has been relatively minor.

Franklin Furnace, New Jersey. The unique zinc deposits of Franklin and Sterling Hill have no exact counterpart of any size in the world. The deposits have been known since the 17th century and actively mined since about 1850. The production of the two mines has exceeded 15 million tons.

The region is made up of Pre-Cambrian gneisses, "white" limestone (Franklin), and "blue" limestone (Kittatinny). Numerous pegmatites intrude the white limestone and locally there are dikes of basic trap rock (Triassic?). The zinc ores occur in the white limestone and outside of the two mines no other deposits have been found. Magnetite deposits have been mined from the base of the white limestone and from the gneiss.

The ore deposits are synclinal bodies, shaped like a canoe with one side lower than the other and the keel pitching 25° at Franklin and 50° at Sterling. The Franklin deposit is fish-hooked in shape but with depth the hook part lessens until it joins with the shank to form a single fat ore body (Figure 3-4).

The longer west limb had an outcrop length of about 2,600 feet and the east limb about 600 feet. The keel has been followed for a down-dip distance of 1,500 feet and to a depth exceeding 1,150 feet. At Sterling, the west limb has an outcrop length of about 600 feet and the east limb about 1,500 feet and the mine has been developed to a depth of 2,450 feet. A branch extends diagonally from the west limb nearly to the east limb.

The ore consists (Pinger[8]) of approximately: franklinite 40 per cent,

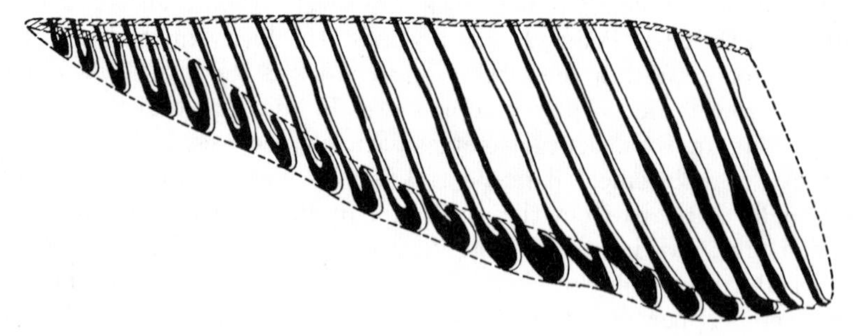

Figure 3-4. Stereogram of Franklin ore body. Black is ore; shaded area is outcrop. (After Spencer, U. S. Geol. Surv.[10])

willemite 23 per cent, zincite 1 per cent, gangue silicates 11 per cent and gangue carbonates 25 per cent. Franklinite and zincite do not occur in quantity elsewhere. Sulfides are essentially absent; sparse sulfides are known to be later than the zinc ore.[8] The silicates consist chiefly of garnets, pyroxenes and mica.

The origin of the ores has been variously attributed to contact metasomatism, sedimentary deposition, igneous injection, original sulfide masses oxided to zinc carbonate and silicate and later metamorphosed (Palache[7]), and to hydrothermal replacement of white limestone. The last hypothesis is the one most generally accepted.

Cerro de Pasco, Peru. The famed Cerro de Pasco district was for centuries a large producer of silver from oxidized ores. Modern operations began in 1901 and the district became a multi-metal producer, yielding copper, zinc, lead, silver, gold, bismuth, cadmium and indium. The chief mining centers are Cerro de Pasco, Morococha, Casapalca, San Cristobal and Vulcan. The deposits occur in an elongated belt at high elevations just east of the Andean divide.

The annual production of the Cerro group of mines is about 30,000 tons of copper, 65,000 of lead, 95,000 of zinc concentrates; 10 million ounces of silver and 45,000 of gold; and 700,000 pounds of bismuth.

The zinc-lead deposits occur in a variety of rocks and structures, but all in common are closely related to local Tertiary intrusives of intermediate composition and have been formed by hydrothermal replacement. In places the ores show a pronounced zoning from contact-metasomatic to low-temperature hydrothermal deposits. At Cerro, the largest producer of zinc-lead ores, the ore bodies occur as portions of a huge pyritic replacement body that is noted chiefly for its copper-rich sections. The crescentric pyritic body is formed along a large fault and has replaced volcanic breccia and quartz-monzonite porphyry on one side and phyllite and limestone on the other side. The zinc-lead ores occur in that portion of the pyritic body which has replaced limestone. At Cerro the pyritic mass is 6,000 feet long by 1,000 feet wide at the surface and is being mined below 2,500 feet.

At Morococha, sedimentary and volcanic rocks have been pierced by a stock of quartz monzonite. In and next to the stock are veins carrying copper ores; farther out are replacement veins and mantos in limestone carrying the zinc-lead ores. At San Cristobal, monzonite porphyry intrudes phyllites, and volcanics and accompanying fissure systems carry replacement bodies, mainly in phyllite, of copper-zinc-lead-silver ores. At Casapalca, again folded sediments and volcanics are cut by monzonite porphyry, near which is a zone of fractures carrying copper-silver nearest the porphyry, zinc-lead-silver farther out, and these ores pass upward and endwise into silver ores and stibnite and realgar mineralization.

The zinc ores consist chiefly of pyrite, pyrrhotite, sphalerite, galena, chalcopyrite, tetrahedrite, silver minerals and small quantities of other minerals including a little cassiterite. The copper averages 0.15 per cent. The ores are pronouncedly zoned and lead decreases whereas zinc increases downward.

These deposits are examples of replacement by hydrothermal solutions in diverse rocks, intimately associated with monzonite intrusives.

Broken Hill, Australia. The Broken Hill "Lode" is one of the great deposits of the world, extending nearly 4 miles in length, with a total metal production in excess of 1 and ½ billion dollars and an annual production of zinc in excess of 175,000 tons.

The lode occurs in Pre-Cambrian gneisses, schists and quartzites cut by numerous pegmatites and aplites. The beds were sheared and closely folded, forming major and minor folds that localized the ore and were replaced to form sheets, lenses and saddles of ore. Seventeen stratigraphic horizons have been recognized by Gustafson, of which two closely adjacent favorable horizons have been selectively replaced to form lead ores, and three stratigraphically higher horizons were selectively replaced to form zinc lodes, each with distinctive lead-zinc ratios. Oxidation extended down to 300 feet.

The ores consist of sulfides of lead and zinc, with minor values in silver, cadmium, gold and copper. The gangue minerals are either dominantly carbonate or silicate; the former are mainly calcite and quartz, the latter, quartz, manganese garnet, rhodonite and fluorspar; green feldspar is common in the pegmatites. The ores range from 11 to 13.7 per cent zinc, 13 to 17 lead and 4 to 10 ounces of silver per ton. Gustafson and associates believe the lode to be a high-temperature hydrothermal replacement deposit formed long after folding and metamorphism.

Mount Isa, Australia. The large Mount Isa mine yields ores that average about 10 per cent zinc, 9 lead and 7 ounces of silver per ton. The reserves are large. The deposit illustrates large-scale replacement. It occurs in a 2-mile thickness of crenulated Pre-Cambrian shale lying between quartzites, along a zone of folding and shearing parallel to the bedding, which, according to Blanchard and Hall,[2] resulted from an overthrust. No intrusives occur nearby. The commercial bodies occupy part of a mineralized zone that extends for 5 miles. The individual deposits are replacement bodies up to 2,000 feet long, 250 feet wide and exceed 1,500 feet in depth.

The ore consists of massive sulfide that exhibits remarkably fine banding inherited from the shale. Zinc, lead and iron sulfides are the chief metallic minerals, along with minor chalcopyrite, arsenopyrite, tetrahedrite, silver minerals, vallerite, pentlandite and considerable graphite. The mineralization started with the deposition of silica, followed by pyrite, and more silica, after which came other sulfides, followed by carbonate. Oxidation extends to a depth of 250 feet and in this zone are bodies of oxidized lead and zinc ores. Copper bodies underlie the zinc-lead ores.

The deposits illustrate noteworthy structural control of ore localization and remarkable replacement preservation of fine shale banding and contortion. They constitute another example of large tonnage replacement by solutions of intermediate temperature without evidence of an igneous source.

Santa Eulalia, Mexico. The chief mine of this district is the El Potosi, famed since 1703 as a rich silver-lead producer and later as a zinc producer. Its 275 miles of workings disclose mantos, chimneys, irregular replacement deposits and fissure veins, all in gently dipping Cretaceous limestone. The whole district is credited with a metal production of about one-half billion dollars from 500 miles of workings.

The limestones have been folded into a gentle anticline along which the ore bodies are localized. Rhyolite volcanics of premineral age cap the limestones and both have been intruded by dikes and sills of dolerite and diabase.

The deposits are massive replacements of limestone along fracture zones in several favorable beds. The ore bodies may be mantos along single beds or chimneys that cut several beds and connect with mantos. The mantos attain lengths of 1,000 or more meters and cross sections of from 100 to 5,000 square meters. The large "P" chimney is 310 meters high and 1,100 square meters in cross section. It throws off mantos into upper and lower beds. Other bodies are fissure replacements that extend as flanges into favorable beds.

The ore is of three classes: (1) high-grade sulfide ore, (2) low-grade sulfide-silicate ore, (3) oxidized ore (largely exhausted). The massive sulfide ore consists chiefly of galena, sphalerite, barite and silver minerals, carrying 9 to 10 per cent zinc and lead, with 9 to 10 ounces of silver. The silicate ore consists of skarn silicates, iron oxides, base-metal sulfides and silver minerals. The silicate ores grade into, overlap and also lie above the sulfide ores. Oxidation is complete to the 10th level, partial to the 13th and absent at lower levels.

These interesting deposits illustrate both intermediate hydrothermal and high-temperature conditions of formation combined in one mine.

Buchans, Newfoundland, Canada. The Buchans mine, discovered by geophysical prospecting, yields a complex ore carrying about 8 per cent lead, 17 zinc, 1.4 copper; 3 ounces of silver, 0.05 ounce of gold, and 30 per cent barite. It yields about 32,000 tons of zinc annually.

The deposit is a replacement of tuffs between strong arkose beds that have been folded into an anticline. These rocks are intruded by sills of quartz porphyry, which are offshoots from a nearby granite batholith.

Folding sheared the tuffs and quartz porphyry and provided access for the hydrothermal solutions.

The ore consists of pyrite, sphalerite, galena, chalcopyrite, tetrahedrite and tellurides, in a gangue of barite and quartz. There is no oxidized zone.

This is another example of replacement in rocks other than limestone.

Katanga, Belgian Congo. The Prince Leopold mine, Kipushi, in the midst of the great Katanga copper belt, is a sulfide pipelike replacement deposit in limestone, extending from the Pre-Cambrian Series de Mine beds up into Kundelungu beds. The rich ores consist of chalcocite, bornite, chalcopyrite, galena, sphalerite, pyrite, renierite, tennantite and silver minerals, quartz and barite. The ores yield some 150,000 tons of zinc concentrates annually.

The deposit is elongated along and parallel to a large fault between breccia and dolomite and calcareous shale and dolomite. The ore body in part leaves the fault and extends outward into dolomite, which it has replaced. The ore body dips 80°, is 20 to 40 meters wide and 400 to 600 meters long. Workings extend to 712 meters and the lowest working level is at about 500 meters. Oxidation extends to a depth of 100 meters. The chief mineralization is of copper and the zinc ores tend to occur about the margins of the deposit.

Upper Silesia, Poland. The chief zinc-lead district of Europe lies astride the pre-World War II boundary between Poland and Germany and, before confiscation, belonged to The Anaconda Copper Company. The deposits were first mined for silver and lead in the 13th Century. The first zinc works were erected in 1808, although oxidized zinc ores, now largely exhausted, had been shipped for making brass during the 17th and 18th centuries. The deposits yield zinc, lead, silver, cadmium and thallium; the annual zinc production is about 110,000 tons. Here, zinc-lead bedded deposits in slightly warped Triassic dolomites overlie Carboniferous coal seams—a fortunate combination. There are three ore zones in favorable stratigraphic horizons: one at 80 to 90 meters depth, at the base of the dolomite; another 12 to 15 meters above it; and a third bed 25 to 35 meters above the lowest, partly mineralized. The lowest deposit is 1.5 to 4 meters thick, is regular, extensive and nearly horizontal. The intermediate bed is 1 to 3 meters thick. A few pipes and irregular cavities occur. The upper bed is largely oxidized. The ores occur in synclines traversed by numerous faults.

The primary ore consists of sphalerite, wurtzite, galena, pyrite, marcasite and some rare arsenic and antimony compounds of lead. The oxidized ores consist of smithsonite, calamine and limonite.

The deposits are considered to be a combination of replacement and cavity filling by low-temperature hydrothermal solutions. The first event was dolomitization of the limestone beds, followed by the deposition succesively of sphalerite, wurtzite, galena, rare lead arsenates and antimonides, marcasite, pyrite, sphalerite and galena. The filling of cracks, crevices and small cavities was accompanied by replacement of the dolomite. Proof of the ascending origin of the solution is found in veinlets of similar ores in the underlying Carboniferous coal beds.

Other Replacement Deposits. Other important replacement deposits of zinc ores occur at Leadville, Colorado; Tintic, Utah; Bisbee, Arizona; Austinville, Virginia and at Balmat-Edwards, New York,—in the United States, where there are massive replacements in limestone; at Normetal, Waite-Amulet and New Calumet, in Quebec; Sierra Mojada, Mexico; Trepca in Yugoslavia; Aguilar in Argentina; Rammelsberg in Germany; Tsumeb in S. W. Africa and Broken-Hill in Rhodesia.

Oxidation Deposits

As shown in an earlier section, when zinc sulfide deposits containing pyrite are exposed to weathering, soluble zinc sulfate may be formed. If the zinc sulfate meets carbonate rocks (limestone or dolomite) or a bicarbonate solution, reactions take place by which zinc carbonate is deposited. Similarly, in contact with available silica, zinc silicate may be deposited. Thus, many zinc-containing primary deposits are capped by oxidized zinc ores composed of smithsonite (zinc carbonate) or hemimorphite (calamine or zinc silicate) or in places, hydrozincite. The zinc sulfate may migrate beyond the confines of the primary ore body and oxidized zinc ores may be deposited nearby. If the depth of oxidation is considerable, large oxidized deposits may result. Commonly, oxidized zinc ores give way to sulfide ores with depth. Consequently, there are many deposits that consist of oxidized zinc ores above and sulfide ores below. In a few places the entire deposit consists only of oxidized ores. During oxidation of mixed lead-zinc ores, the zinc is commonly separated from the lead. Most of the oxidized zinc ores are now largely exhausted.

Some Examples of Oxidized Zinc Deposits. The Polish deposits of Upper Silesia, described briefly under "Replacement Deposits," underwent oxidation at the outcropping edges of the synclines and in the upper bed, and deposits 20 to 30 meters in thickness were formed. The material is largely zinc carbonate with a little admixed zinc silicate, called "galman." The galman was discovered about the end of the 16th century (Zwierzycki[12]) and, with the lead ores, was the chief object of mining until the 19th century. The first metallic zinc extracted in Poland, in 1808, was from galman ore and for three-quarters of a century it was the chief source of metallic zinc. These ores are now largely exhausted.

At Leadville, Colorado, some time after the heyday of this great camp, a large body of zinc carbonate was discovered in the early 20th century, which had even been tunneled through and overlooked. This led to the discovery of other zinc carbonate bodies and to a rejuvenation of this famous mining area. The zinc carbonate was a replacement of limestone, adjacent to oxidized sulfide bodies.

In the Tintic district of Utah a large deposit of zinc carbonate was formed in limestone, quite remote from the sulfide deposits. It represented a migration of zinc sulfate to a site where limestone was encountered.

In Northern Mexico many lead-zinc sulfide deposits had associated zinc carbonate ore bodies either above or adjacent to the original sulfide deposits. At Sierra Mojada, Coahuilla, (Figure 3-3) several zinc carbonate bodies were found along fractures in limestone beneath a large manto deposit of sulfides that had become oxidized. Also at nearby Santa Elena large bodies of oxidized zinc ore were developed.

In Southern Rhodesia, at Broken Hill, much of the zinc production has come from oxidized zinc bodies. This is true also of much of the Tunisian zinc production and also of Sardinia.

WHERE ZINC DEPOSITS MAY BE EXPECTED

From the preceding discussions of the origin and localization of zinc deposits, and the descriptions of occurrences, it becomes evident that various geologic factors have determined the sites of zinc deposits. Chief among these are igneous activity to supply mineralizing solutions and structural and stratigraphic features. Other geologic features also play a part; some may operate together, others singly.

Field and laboratory evidence demonstrate that most zinc deposits the world over have been formed by hot mineralizing solutions, whatever their source. Consequently, the source of mineralizing solutions becomes of prime importance. In the case of contact-metasomatic deposits, the source of the mineralizing solution is obviously the magmatic reservoir that gave rise to the intrusive igneous body with which the deposits are in contact. Most zinc deposits, whether cavity filling or replacement, appear to have been formed by hydrothermal solutions. As may be seen from the descriptions of occurrences, most of the hydrothermal deposits show a direct field relationship to bodies of intrusive igneous rocks, indicating a connection between the igneous activity and the hydrothermal solutions. In a few instances, as in the Tri-State, Tennessee, Alpine and Kupferschiefer deposits (and a few others), no associated igneous activity is known; nor is it known that there may not be deep-seated hidden intrusives.

Areas of folded, faulted and sheared rocks that have undergone pronounced erosion are favorable structural sites for channeling mineralizing solutions and for loci of ore deposition. Such features are characteristic of mountain ranges, both younger and old. Where intrusives cut such rocks, a doubly favorable condition exists for the localization of ore deposits. The large faults and shears may serve as trunk channels for mineralizing solutions to reach sites of deposition such as fissures, joints, breccias, bedding planes and rocks congenial to replacement. Thus, favorable places to expect zinc deposits are folded mountains cut by intrusives and eroded deeply enough to expose the deep-seated portions. Again, eroded shield areas exposing the roots of old mountains and intrusives, such as the Pre-Cambrian shield of Canada, the Guayan shield of Venezuela-Brazil, the Siberian, Central African and Australian shields, are also likely places to expect zinc deposits.

As for deposits in relatively horizontal sediments where no intrusives are exposed, there is little geological evidence to indicate where they may be expected.

EXPLORATION FOR ZINC DEPOSITS

In highly developed countries, most zinc deposits that crop out at the surface have probably been all or mostly discovered. In underdeveloped regions, however, actual outcrops may still await discovery. They may be recognized by:

1. Zinc sulfide at the surface in unweathered regions (glaciated or rapidly eroding areas, or where runoff is rapid).
2. Oxidized zinc minerals such as smithsonite or hemimorphite, generally accompanied by limonite.
3. Sphalerite boxwork left where the zinc has been leached; this also is generally accompanied by limonite derived from pyrite associated with the original sphalerite.
4. Float, in the form of boulders, pebbles and grains of sulfide or oxidized zinc minerals, or sphalerite boxwork, that has moved down-slope from the original deposit, and which can be traced up-slope to the original deposit. Such float may be found in stream gravels, stream terraces, outwash plains, talus slopes or glacial deposits.

In exploration for hidden deposits, geochemical, geophysical, or photogeologic methods have to be employed.

Geochemical Prospecting. Zinc is taken into solution and migrates more readily than in the case of most metals. Hence zinc is fairly readily detected by geochemical means. Such tests may be made rapidly in the field by present-day wet techniques for testing trace amounts of zinc and other metals. Testing can be carried out on soil samples, swamp peat, the ash of twigs, leaves and grasses, stream waters, lake waters and underground water. Abnormal amounts of zinc in parts per million may indicate a nearby source of zinc in amounts greater than normal in the rocks, and zinc contours or anomalies may point to a center of supply. Such large-scale geochemical prospecting is being carried on extensively in many countries today, notably in the United States, Canada, Russia, Scandinavia, Australia and Rhodesia. The low cost of geochemical prospecting permits it to be carried on as a reconnaissance exploration over wide areas.

Geophysical Prospecting. Geophysical prospecting may be divided into two types, ground and aerial. The aerial is employed both for large area reconnaissance prospecting, and for pinpointing anomalies determined by geochemical or other means. The ground methods, because of their cost, are usually restricted to specific areas where other indications are present, such as geochemical or aerial anomalies, outcroppings of ore, extensions along structural features known to carry ores nearby or other favorable geologic features.

The ground methods employed in the search for mineral deposits are of several types. The simplest one is the magnetic needle, like a compass needle, mounted on a horizontal axis. In passing over a magnetic body the needle is pulled down. The large ilmenite deposit at Allard Lake was discovered by this means.

Another type of prospecting is the magnetometer method, whereby the magnetic susceptibilities of different rocks and ores are recorded. Certain rocks and minerals increase the earth's magnetic field; others decrease the strength of the earth's magnetic field from place to place. Both horizontal and vertical components are measured. This is a precise method but is rather slow. However, it is widely and effectively used for ore finding.

A third method commonly used is the self-potential, or natural current method by which the natural currents generated by an ore body itself can be measured and the presence of hidden ore bodies thereby detected. It requires no outside energizing force. The currents can be measured by a galvanometer and points of equipotential can be traced out on the ground. A modification of this method supplies some current to the ground. This method is also widely used in suitable climates.

Another electrical method is known as equal-potential, which has several variations. High-frequency current is passed into the ground from bare copper wires connected with a generator. Current passes between the two grounded electrodes, and if a conducting ore body lies between them, the lines of equal potential are distorted and staked out on the ground. It has been very successfully utilized, notably in Sweden and Canada.

The resistivity method measures the resistance offered to an electric current placed in the ground from stakes set at chosen intervals apart. Barren rocks offer high electrical resistance. The method is quick, of low cost to operate and has been widely used.

Electromagnetic methods are the most favored of electrical methods in the search for ore deposits. A loop of insulated cable is laid out on the ground through which a high-frequency current is passed. Along parallel traverses the primary field is measured, but if an ore body is present additional induced current is picked up and its outline indicates the horizontal and vertical depth of the deposit. An alternative is to use a long single cable grounded at both ends. Still other variations are used. The method is more precise than equal-potential methods.

The gravimeter is used to detect differences in gravity between wall rocks and ore bodies, the heavier ore body exerting a greater gravitative attraction than lighter rocks. This method is extensively used in petroleum exploration and has replaced the slower and more cumbersome torsion balance and pendulum for detecting differences in gravity. The gravity method is coming into greater use for ore exploration.

Aerial exploration is being increasingly used for ore exploration and for checking strong geochemical anomalies. There are three varieties: magnetometer, electromagnetic, and scintillometer. The last is commonly added to one of the former for the detection of radioactive minerals. In the aerial magnetometer method, the instrument is towed by a long cable underneath and behind the airplane so as to be at a distance from the disturbance of the metal parts of the plane. The height flown determines the sensitivity: the lower the height, the stronger the reception. Its use is limited in regions of strong relief. The method is fast and the cost is low for large areas. Many large areas in the United States have now been flown by private companies and by the United States Geological Survey, which has made several magnetometer traverses across the country. In Canada also, mining companies and the Government have flown vast areas. Much geological mapping is now preceded or accompanied by flying-magnetometer surveys. The flying-magnetometer surveys detect changes in rock formations, faults and other features in addition to mineral deposits. Several large hidden ore deposits have been found by this method. A strong anomaly disclosed by the flying magnetometer is generally checked by ground geophysical methods and then by diamond drilling. Many anomalies are disclosed that are not caused by ore deposits, such as graphite, carbonaceous slates, iron formations and intrusives of basic igneous rocks.

The flying-electromagnetic method was quite recently developed in Canada by the International Nickel Company and its early use was attended by spectacular discoveries in the Sudbury region of Ontario, in Northern Canada and by the discovery of zinc-lead-copper deposits in New Brunswick, Canada. In three years its use has spread widely in North America, particularly in the United States in 1956 and 1957; and several ore discoveries are credited to it. The principle is the same as for the ground-electromagnetic method. The current is supplied to the ground from cables in the frame of the airplane, the pickup from the ground is automatically received in the plane, and anomalies are plotted. Although expensive to operate, the cost is low per mile of traverse, and much ground can be covered rapidly. Like the flying magnetometer, it also records anomalies that are not always ore deposits, e.g., graphitic rocks.

Aerial geophysical methods will supplant much of the ground activity and areas the size of small countries can be covered in a few months. These are new devices for exploration for hidden ore deposits that should contribute greatly to increasing the world mineral resources, among them zinc deposits.

REFERENCES

1. Bateman, Alan M., "Economic Mineral Deposits," 2d ed., New York, John Wiley & Sons, Inc., 1950.
2. Blanchard, R., and Hall, G., "Mt. Isa Ore Deposition," *Econ. Geol.*, **32,** 1042 (1937).
3. Butler, B. S., "Ore Deposits of the Western United States," Chapt. 6, New York, Am. Inst. Mining Met. Engrs., 1933.
4. "Geology, Paragenesis and Reserves of the Ores of Lead and Zinc," London, Intern. Geol. Cong., 18th Session, Part 7, 1950.
5. Laskey, S. G., and Hoagland, A. D., "Central Mining District, New Mexico," West Texas Geol. Soc. Guidebook, Field trip No. 3, pp. 7–24, 1949.
6. Lindgren, Waldemar, "Mineral Deposits," 4th ed., New York, McGraw-Hill Book Co., Inc., 1933.
7. Palache, Charles, "The Minerals of Franklin Hill and Sterling, N. J.," U. S. Geol Surv., Prof. Paper, 180, Washington, D. C., 1935.
8. Pinger, A. W., "Geology of the Franklin-Sterling Area, Sussex, N. J.," In Ref. 4, pp. 77–83.
9. Sales, Reno H., "Ore Deposits at Butte, Montana," *Trans. Am. Inst. Mining Met. Engrs.*, **46,** 54–67 (1913).
10. Spencer, A. C., Kummel, H. B., Wolff, J. E., Salisbury, R. D., and Palache, C., U. S. Geol. Surv., Geologic Atlas, No. 161, p. 24, Washington, D. C., 1908.
11. Singewald, Quentin, and Butler, R., "Ore Deposits in the Vicinity of the London Fault, Colorado," U. S. Geol. Surv., Bull. 911, pp. 39–49, Washington, D. C., 1941.
12. Zwierzycki, J., "Lead and Zinc Ores in Poland," In Ref. 4, pp. 314–24.

CHAPTER 4

CHEMISTRY AND PHYSICS OF ZINC TECHNOLOGY

R. SCHUHMANN, JR.

Professor of Metallurgical Engineering, Purdue University
West Lafayette, Indiana

and

HARVEY W. SCHADLER

Metallurgy and Ceramics Department, General Electric Research Laboratory
Schenectady, New York

The arts of extracting zinc from its ores and of adapting zinc and its alloys and compounds to manifold engineering uses have developed of necessity with only fragmentary understanding of the physics and chemistry involved. However, in recent years the sciences of physical chemistry and chemical thermodynamics have proved increasingly valuable, first in analyzing, controlling and improving existing technology and then as a basis for developing new methods. The thermodynamic approach has been particularly practical because in many situations it yields clear-cut, direct and quantitative information without requiring detailed knowledge of structures and reaction mechanisms. In contrast, the newer sciences of quantum physics and solid-state physics focus attention on fine details of structures and atomistics, and deal with a field largely inaccessible to thermodynamics—the structure of matter and the relation of structure to behavior. The future development of zinc technology will no doubt involve both the gross thermodynamic approach and the detailed atomistic and structural approach. However, reflecting the present state of application of science to zinc technology, this chapter will present a brief summary of the structures and properties of zinc and its compounds and then will deal in more detail with the thermodynamics of zinc extraction processes.

STRUCTURES AND PROPERTIES OF ZINC AND ITS COMPOUNDS

Electronic Arrangements of Zinc Atoms and Ions

Zinc, cadmium and mercury constitute Group IIB in the periodic table. Atoms of these elements are characterized by having two *s*-orbit electrons in their outer shells, as shown in Table 4-1. For comparison, Table 4-1 also shows the ground-state electronic arrangements for the rare gases and for

TABLE 4-1. ELECTRONIC CONFIGURATIONS OF THE ELEMENTS: GROUPS O, II A, AND II B

Element	Atomic Number	Shell / Orbit: *K* *s*	*L* *s*	*L* *p*	*M* *s*	*M* *p*	*M* *d*	*N* *s*	*N* *p*	*N* *d*	*N* *f*	*O* *s*	*O* *p*	*O* *d*	*O* *f*	*P* *s*	*P* *p*	*P* *d*	*P* *f*	*Q* *s*	*Q* *p*	*Q* *d*	*Q* *f*
Group O (Inert Gases)																							
He	2	2																					
Ne	10	2	2	6																			
A	18	2	2	6	2	6																	
Kr	36	2	2	6	2	6	10	2	6														
Xe	54	2	2	6	2	6	10	2	6	10		2	6										
Rn	86	2	2	6	2	6	10	2	6	10	14	2	6	10		2	6						
Group II A																							
Be	4	2	2																				
Mg	12	2	2	6	2																		
Ca	20	2	2	6	2	6	2																
Sr	38	2	2	6	2	6	10	2	6			2											
Ba	56	2	2	6	2	6	10	2	6	10		2	6			2							
Ra	88	2	2	6	2	6	10	2	6	10	14	2	6	10		2	6			2			
Group II B																							
Zn	30	2	2	6	2	6	10	2															
Cd	48	2	2	6	2	6	10	2	6	10		2											
Hg	80	2	2	6	2	6	10	2	6	10	14	2	6	10		2							

Group IIA elements. The alkaline earth elements also have two outer *s*-orbit electrons. However, when the atoms are ionized by removal of the two outer electrons, Group IIA ions (Mg^{++}, Ca^{++}, etc.) are left with the highly stable electronic configurations of the inert gases, whereas Group IIB ions (Zn^{++}, Cd^{++}, and Hg^{++}) are left with outer shells containing the complete but less stable configuration of 18 electrons (s^2, p^6, d^{10}).

The electronic structures of free zinc atoms and free zinc ions are shown schematically in lines *A* and *B*, respectively, of Table 4-2. The dots, indicating individual electrons in the various orbitals, are grouped as pairs of electrons with opposite "spins," in accordance with the Pauli exclusion principle. In ordinary salts of zinc and cadmium, the two electrons in the outermost shell are regarded as the valence electrons, so that zinc, like the alkaline earth elements, exhibits the ionic valence of only +2.

TABLE 4-2. POSSIBLE ELECTRONIC CONFIGURATIONS OF ZINC ATOMS AND IONS

Description	Orbitals							
	1s	2s	2p	3s	3p	3d	4s	4p
		Stable argon core						
A. Free Zn atom, ground state	:	:	:::	:	:::	:::::	:	
B. Free Zn^{++} ion	:	:	:::	:	:::	:::::		
C. Neutral Zn atom with maximum unpaired electrons. Valence = 6	:	:	:::	:	:::	:::..	.	...
D. Neutral Zn atom, with four unpaired electrons and one empty orbital	:	:	:::	:	:::	::::.	.	..
E. Zn atom plus six electrons	:	:	:::	::	:::	:::::	: (shared)	::: (shared)
	Krypton structure							

When metallic bonding and electron-pair bonding are considered, a maximum valence of 6 may conceivably be achieved by making full use of all the available orbitals of approximately the same energy level outside the inert gas core (3*d*, 4*s*, and 4*p*). Thus structure *C* in Table 4-2 provides 6 unpaired electrons which are available for forming metallic bonds with other atoms. Pauling[1, 2] has suggested that the metallic bond in solid metals involves resonance of the valence bonds among the various available interatomic positions. Furthermore, resonance is best achieved when each atom has on the average one unoccupied orbital to receive an electron jumping from a neighboring atom. Such an arrangement is shown in structure *D* in Table 4-2. Following this approach, and also considering other data on neighboring elements in the periodic table, Pauling assigns a metallic valence of 4.5 to zinc, cadmium and mercury.

In crystals such as zinc sulfide, where the bonding is predominantly covalent, each metal atom is commonly bonded to four surrounding atoms in a tetrahedral arrangement. The zinc atom shares one pair of electrons with each of the four surrounding atoms, to make a total of eight shared electrons around each zinc atom. Two of these shared electrons are furnished by the zinc atom itself, but the other six are supplied by the electronegative atoms (e.g., sulfur). In this way, the zinc atoms may achieve something like the highly stable electronic configuration of krypton (see structure *E* in Table 4-2).

Crystal Structures

Much of the data on crystal structures of elements and compounds can be correlated and new structures can sometimes be predicted on the basis of a model in which the atoms or ions are considered to be hard spheres. According to this model, the crystal structure results from the stable and regular three-dimensional packing of these spheres under the influence of the bonds or attractive forces between the spheres. In many crystals the stable atomic arrangement corresponds simply to the most compact packing, so that the structures and lattice constants are predictable to a good approximation simply from a table of atomic or ionic radii. More generally, however, the atomic arrangements in crystals are also affected by the natures, magnitudes and directions of the interatomic forces or bonds.

Metallic zinc has a hexagonal close-packed structure (c = 4.937 Å), (a = 2.660 Å),[3] but the axial ratio (c/a = 1.856) is substantially larger than that theoretically obtained by close packing of spheres (for hcp spheres, c/a = 1.633), indicating extension in the direction of the hexagonal axis and closer packing in the basal plane. Thus each zinc atom has twelve near neighbors as in the regular hcp structure, but in zinc six of these neighbors are at one distance (2.660 Å) and six at another (2.907 Å),[3] resulting in an average metallic radius of 1.39 Å. The structure and axial ratio of cadmium are very similar to those of zinc, and both zinc and cadmium are in contrast to magnesium and other hcp metals in which the axial ratio is close to that for close-packed spheres. The conclusion that the bonds *in* the hexagonal layers of zinc atoms are stronger than those *between* layers is consistent with measured properties of the crystals, which show larger values of compressibility and coefficient of thermal expansion in the direction of the hexagonal axis.[1]

Zinc sulfide is dimorphic and the two crystalline forms, wurtzite (high-temperature form) and zinc blende (low-temperature form), are the type structures for a large number of *AB* compounds.[4] The difference between the wurtzite and blende structures can be described most easily in terms of the arrangements of the sulfur atoms, which in wurtzite correspond to a hexagonal close-packing of spheres and in blende correspond to a cubic close-packing of spheres. In typical *AB* compounds with these structures, the ratio of valence electrons to atoms is 4:1. For example, in zinc sulfide, each zinc atom contributes 2 electrons and each sulfur atom 6 electrons; in silicon carbide, each silicon and each carbon atom contributes 4 electrons; in aluminum nitride, each aluminum contributes 3 electrons and each nitrogen, 5 electrons; etc.). This electron-to-atom ratio favors the stability of the tetrahedrally coordinated wurtzite and zinc blende structures, in which each zinc atom is bonded by electron pairs to each of four sulfur atoms and each sulfur atom is in turn bonded to four neighboring zinc atoms. Thus

TABLE 4-3. TETRAHEDRAL COVALENT RADII (*Pauling and Huggins*)

I	II	III	IV	V	VI	VII
	Be	B	C	N	O	F
	1.06	0.88	0.77	0.70	0.66	0.64
	Mg	Al	Si	P	S	Cl
	1.40	1.26	1.17	1.10	1.04	0.99
Cu	Zn	Ga	Ge	As	Se	Br
1.35	1.31	1.26	1.22	1.18	1.14	1.11
Ag	Cd	In	Sn	Sb	Te	I
1.53	1.48	1.44	1.40	1.36	1.32	1.28
Au	Hg	Tl	Pb	Bi		
1.50	1.48	1.47	1.46	1.46		

the chemical bonding in wurtzite and zinc blende is largely covalent or homopolar in contrast with the primarily ionic or electrostatic bonding in crystals such as NaCl, MgO, CsCl, etc. All compounds of both zinc and cadmium with oxygen, sulfur, and the other Group VI elements, except for cadmium oxide, crystallize in either the wurtzite or the zinc blende structure (or in both). The interatomic distances in these crystals are quite well correlated in terms of the hard-sphere model, using the "tetrahedral" atomic radii of Pauling and Huggins[1] which are given in Table 4-3. The tetrahedral covalent radius of zinc is very close to the average radius of zinc atoms in zinc metal (1.39 Å), but is of course considerably larger than the radius assigned to Zn^{++} in ionic crystals (0.74 Å).[1]

The tetrahedral coordination of oxygen around zinc, which characterizes the wurtzite structure of zinc oxide ZnO, is also observed in willemite (zinc orthosilicate: Zn_2SiO_4). Thus, zinc orthosilicate is not isomorphous with such structures as Mg_2SiO_4 and Fe_2SiO_4 which involve sixfold coordination of $O^{=}$ around Mg^{++} or Fe^{++}. However, in still other oxy-salts of zinc, the Zn-O coordination is sixfold and the structures may be conveniently regarded as made up of ions rather than atoms. Examples are $ZnCO_3$, $ZnWO_4$, $ZnSO_4 \cdot 7H_2O$.[5] The ionic radius of Zn^{++} (0.74 Å) is close to those of Mg^{++} (0.65 Å), Fe^{++} (0.75 Å), Mn^{++} (0.80 Å), Co^{++} (0.72 Å) and Ni^{++} (0.69 Å) so that oxy-salts with six-coordinated zinc ions are commonly isomorphous with oxy-salts of magnesium, iron, manganase, cobalt and nickel.

Some of the interesting properties of zinc oxide, zinc sulfide, zinc silicate and other zinc compounds are attributed to the behavior of imperfections and defects in the crystals. Solid-state research has disclosed many types of defects, including interstitial metal atoms, missing atoms, free electrons, impurity atoms, electron holes, dislocations and others.[6, 7] Moreover, it has been shown that a large family of properties, the so-called "structure-sensitive properties," are closely related to the kinds, quantities and behaviors of these defect structures rather than to the regular crystal structure itself. These structure-sensitive properties include electrical conduction, semiconduction, luminescence, phosphorescence, photoconductivity, optical properties and others. Zinc compounds, especially the oxide and sulfide, have proved to be especially appropriate experimental materials for studies of many of these structure-sensitive properties. Moreover, a constantly growing number of commercial applications are coming from these studies, including such products as luminescent paints, television tubes, electron microscopes, cathode ray oscillographs and photocopying devices.[8]

Figure 4-1 shows schematically some of the kinds of imperfections which have been proposed to account for the structure-sensitive properties of zinc oxide. Wagner and co-workers[9] showed that the electronic conductivity of zinc oxide can be increased greatly by prolonged heating. This increase in conductivity is obtained more readily by heating in a reducing atmosphere. Accordingly, Wagner proposed that the semiconductor properties result from

(*a*) *Perfect zinc oxide crystal.*

```
O   Zn  O   Zn  O   Zn  O   Zn
Zn  O   Zn  O   Zn  O   Zn  O
O   Zn  O   Zn  O   Zn  O   Zn
Zn  O   Zn  O   Zn  O   Zn  O
O   Zn  O   Zn  O   Zn  O   Zn
Zn  O   Zn  O   Zn  O   Zn  O
```

(*b*) *Interstitial zinc atoms and ions with conduction electrons.*

```
O   Zn          O   Zn  O               Zn          O   Zn
Zn  O           Zn  O   Zn⁻             O           Zn  O
                                Zn⁺
O   Zn          O   Zn  O               Zn          O   Zn
Zn  O           Zn  O   Zn              O           Zn  O
         Zn
O   Zn          O   Zn  O               Zn          O   Zn
                                              Zn⁺
Zn  O           Zn  O   Zn⁻             O           Zn  O
```

(*c*) *Monovalent foreign atoms and anion vacancies.*

```
O   Zn  O   Zn  O   Zn  O   Zn
Zn  O   Zn  O   Zn  O   Zn  O
O   Zn  O   Li  O   Zn      Zn
Zn      Zn  O   Zn  O   Zn  O
O   Zn  O   Zn  O   Li  O   Zn
Zn  O   Zn  O   Zn  O   Zn  O
```

FIGURE 4-1. Possible defect structures of zinc oxide.

formation of interstitial zinc atoms, as shown in Figure 4-1(b). The interstitial atoms at moderate temperatures give up an electron which can move among the neighboring zinc atoms in the crystal. The over-all process of forming interstitial ions and conduction electrons by reduction can be represented by an equation, for example,

$$ZnO\ (\text{crystal}) + H_2\ (\text{gas}) = Zn^+\ (\text{interstitial}) + e^- + H_2O\ (\text{gas}).$$

Although the conductivity may be increased by an order of magnitude or more through removal of oxygen atoms and formation of interstitial metal atoms, the accompanying change in stoichiometric composition (e.g., Zn to O ratio) of the crystal may still be too small to measure by chemical analysis.

Small percentages of foreign oxides or other impurities in zinc oxide can also have marked effects on the electrical conductivity, catalytic activity, and on the other structure-sensitive properties, particularly if the foreign atoms differ in valence or in size from the zinc and oxygen atoms. One theory holds that additions of oxides of monovalent metals (e.g., lithium- or sodium oxide) tend to produce anion vacancies, Figure 4-1(c), while additions of oxides of trivalent metals (e.g., aluminum oxide) tend to reduce the number of anion vacancies.[10]

The above brief discussion of defect structures and properties of zinc oxide hardly does justice to the many significant experimental findings and interesting theories which in recent years have come from solid-state research on zinc oxide, zinc sulfide, and other similar materials. Certainly, however, it has become clear that solid-state physics promises to lead to important future developments in zinc technology.

Zinc oxide is a constituent of several spinel-type compounds of the general stoichiometric formula (AB_2O_4), including $ZnAl_2O_4$, $ZnFe_2O_4$, $SnZn_2O_4$, and $TiZn_2O_4$.[5] In the normal spinel structures, the $O^=$ ions are arranged in cubic close-packing with the *A* atoms occupying tetrahedral positions between four $O^=$ ions, and the *B* atoms occupying octahedral positions between six $O^=$ ions. $ZnAl_2O_4$ and $ZnFe_2O_4$ have this normal structure. In the inverse spinel structure the tetrahedral holes are occupied by half the *B* atoms while the octahedral holes contain all the *A* atoms and half the *B* atoms. Examples of these structures among the ferrites are given below:[6]

	Tetrahedral Sites	Octahedral Sites		
Normal	Zn^{++}	Fe^{+++}	Fe^{+++}	$(O^=)_4$
	Mn^{++}	Fe^{+++}	Fe^{+++}	$(O^=)_4$
	$Zn^{++}_{0.2}\ Mn^{++}_{0.8}$	Fe^{+++}	Fe^{+++}	$(O^=)_4$
Inverse	Fe^{+++}	Fe^{++}	Fe^{+++}	$(O^=)_4$
	Fe^{+++}	Ni^{++}	Fe^{+++}	$(O^=)_4$
Mixed	$Zn^{++}_{0.2}\ Fe^{+++}_{0.8}$	$Ni^{++}_{0.8}$	$Fe^{+++}_{1.2}$	$(O^=)_4$

All the above ferrites, with the exception of $ZnFe_2O_4$, are ferromagnetic, but, unlike iron and magnetic alloys, have very high electrical resistivities and thus low eddy current losses in electromagnetic applications. This combination of properties is especially useful in high-frequency apparatus. However, a surprising feature of the magnetic ferrites is that combinations of $ZnFe_2O_4$ with Fe_3O_4, $MnFe_2O_4$, $NiFe_2O_4$, etc., have substantially higher magnetic moments than the undiluted ferrites, even though the pure zinc ferrite has itself zero magnetic moment. This behavior has been explained in terms of antiferromagnetic interactions between the magnetic ions on the tetrahedral sites and those on the octahedral sites.[6, 7] These interactions are lessened by substitution of nonmagnetic Zn^{++} for the Mn^{++}, Fe^{+++}, or other magnetic ions in the tetrahedral sites. As a result of the discovery of these beneficial effects of substituting zinc in magnetic ferrites, the use of these materials in television and other electronic equipment has grown markedly since World War II.

Physical Properties of Zinc

Table 4-4 lists important properties of elemental zinc. Because zinc crystals are strongly anisotropic, values of some of the properties are different for polycrystalline zinc and for single crystals. As mentioned earlier, properties of single crystals measured in the direction of the *c* axis may be quite different from those measured in the direction of the *a* axis. Another factor which sometimes causes difficulties in accurate measurement of the properties is the effect of small, almost immeasurable percentages of impurities. Further data on the properties of the metal are given in Chapter 9.

Physical Properties of Zinc Compounds

Properties of a few zinc compounds of metallurgical interest are given in Table 4-5. Standard reference books and journals should be consulted for more complete data on the properties of zinc compounds. Chapters 12 and 13 provide further information about the properties of industrially and biologically important compounds.

PHYSICAL CHEMISTRY OF ZINC METALLURGY

Maier's publication[18] of "Zinc Smelting from a Chemical and Thermodynamic Viewpoint" in 1930 was an important early milestone in the development of metallurgical thermodynamics as an important and practically useful approach. In particular, Maier demonstrated how data on thermodynamic properties can be combined with stoichiometric and other data to predict the best operating conditions for existing processes and to appraise the feasibility of proposed new metallurgical processes. This same general approach will be used in the following sections to present the essential features of various chemical reactions of zinc roasting and smelting.

TABLE 4-4. PROPERTIES OF ZINC[1, 11, 12, 13, 14, 15]

Atomic number: 30
Atomic weight: 65.38
Stable isotopes, relative abundance:

Mass number	64	66	67	68	70
Per cent	48.89	27.81	4.07	18.61	0.62

Radioactive isotopes:

Mass number:	62	63	65	69	72	73
Half life:	9.5 hr	38.3 min	250 days	57 min	2.1 days	<2 min
Decay particles or process:*	K, β^+	β^+, K	K, β^+	β^-	β^-	β^-

Crystal structure and orientation:
  Hexagonal close-packed, a = 2.664Å, c = 4.9469Å, c/a = 1.856
  Glide Plane (0001); glide direction [11$\bar{2}$0]
  Twinning plane (10$\bar{1}$2)
Atomic sizes:
  Metallic radius (12-coordinated), 1.38Å
  Tetrahedral covalent radius, 1.31Å
  Ionic radius (Zn^{++}), 0.74Å
Density:
  solid at 25°C, 7.133 g/cm³
  solid at 419.5°C, 6.83 g/cm³
  liquid at 419.5°C, 6.62 g/cm³
  liquid at 800°C, 6.25 g/cm³
Melting point: 419.5°C (692.7°K)
Boiling point (1 atm): 907°C (1180°K)
Heat capacity:
  solid − Cp = 5.35 + 2.40 × 10^{-3} T (298 − 692.7°K.) cal/mol
  liquid − Cp = 7.50 cal/mol
  gas (monatomic) − Cp = 4.969 cal/mol
Heat of fusion: 1765 cal/mol at 419.5°C
Heat of vaporization: 27,430 cal/mol at 907°C
Linear coefficients of thermal expansion:
  polycrystalline (20 − 250°C), 39.7 × 10^{-6} per °C
  a-axis (20 − 100°C), 14.3 × 10^{-6} per °C
  c-axis (20 − 100°C), 60.8 × 10^{-6} per °C
Volume coefficient of thermal expansion (20 − 400°C): 8.9 × 10^{-5} per °C
Thermal conductivity:
  solid (18°C) 0.27 cal/sec cm°C
  solid (419.5°C) 0.23 cal/sec cm °C
  liquid (419.5°C) 0.145 cal/sec cm °C
  liquid (750°C) 0.135 cal/sec cm °C
Modulus of elasticity:
  10 to 20 × 10^6 psi (actually no region of strict proportionality between stress and strain in polycrystalline zinc)
Surface tension (liquid): γ = 758 − 0.09(t − 419.5°C) dynes/cm
Electrical resistivity:

polycrystalline (t = 0–100°C)	R = 5.46(1 + 0.0042t) microhms/cm³
along a-axis (20°C)	5.83 microhms/cm³
along c-axis (20°C)	6.16 microhms/cm³
liquid (423°C)	36.955 microhms/cm³

Magnetic susceptibility (diamagnetic):
  polycrystalline (20°C) − 0.139 × 10^{-6} cgs electromagnetic units
  along a − axis (20°C) − 0.124 × 10^{-6} cgs electromagnetic units
  along c − axis (20°C) − 0.169 × 10^{-6} cgs electromagnetic units

* β^+ refers to positrons; β^- to negative electrons; K to capture of orbital electron by the nucleus.

TABLE 4-5. PROPERTIES OF ZINC COMPOUNDS [13, 16, 17]

Formula	Mineral Name	Crystal System	Refractive Indexes	Specific Gravity	Melting Point	Boiling Point	Heat Capacity, Calories/mole °K
					(°C)	(°C)	
$ZnCO_3$	Smithsonite	trigonal	1.621 1.849	4.398	decomposes		$9.3 + 33.0 \times 10^{-3}T$ (298–780°K)
$ZnCl_2$		hexagonal	1.687	2.91	283	732	$15.0 + 10.85 \times 10^{-3}T$ (294–556°K)
ZnO	Zincite	hexagonal	2.029 2.013	5.68	1975		$11.71 + 1.22 \times 10^{-3}T - 2.18 \times 10^{5}T^{-2}$ (298–1600°K)
ZnS (II)	Sphalerite	isometric	2.37_{Na}	4.0	II → I, 1020°C		$12.16 + 1.24 \times 10^{-3}T - 1.36 \times 10^{5}T^{-2}$ (298–1200°K)
ZnS (I)	Wurtzite	hexagonal	2.378_{Na} 2.356_{Na}	3.98	sublimes, 1182°C		
$ZnSO_4$	Zinkosite	orthorhombic	1.658 1.670 1.669	3.7	decomposes		$17.07 + 20.8 \times 10^{-3}T$ (298–1000°K)
$ZnSO_4 \cdot 7H_2O$	Goslarite	orthorhombic	1.457 1.484 1.480	2.2	dehydrates, about 500°C		93.7 (282°K)
Zn_2SiO_4	Willemite	trigonal	1.719 1.691	3.9	1512		

Equilibrium Constants and Free Energy Data

Systematic thermodynamic studies of chemical reactions yield data on the energy requirements for the reactions and data on the conditions of equilibrium. Since chemical reactions tend to proceed toward equilibrium, given sufficient time, the knowledge of equilibrium conditions permits an evaluation of the maximum yield or extent of reaction which is thermodynamically possible for any given conditions of temperature, pressure and compositions of reacting materials.

For any reaction represented by

$$aA + bB \rightarrow cC + dD \tag{1}$$

the equilibrium conditions at a given temperature are conveniently given by the equilibrium constant, k, at the given temperature. This constant is defined by

$$k = \frac{(a_C)^c(a_D)^d}{(a_A)^a(a_B)^b} \tag{2}$$

in which a_A, a_B, a_C, a_D are the respective activities of the reactants and reaction products. For a pure solid or liquid a common procedure is to define the activity as unity. For gaseous reactants or reaction products, the activity is generally taken as equal to the partial pressure in atmospheres. The numerical value of k for a given reaction depends upon these choices of standard states in which the activities are equal to unity.

Equilibrium constants are readily calculated from free energy data by the relation

$$\Delta F° = -RT \ln k = -4.575\ T \log_{10} k \tag{3}$$

in which $\Delta F°$ is the standard free energy change (calories per mole) associated with the given reaction when all the reactants and reaction products are in their standard states of unit activity, R is the gas constant (cal/mol $-$°K) and T is the temperature (°K). Thus, equilibrium calculations for any reaction are readily made if thermodynamic data are available giving $\Delta F°$ for that reaction as a function of T. A further simplification in the handling of thermodynamic data is afforded by the fact that when chemical equations are algebraically added and subtracted, the corresponding $\Delta F°$'s can be added and subtracted in the same way. Therefore, in compiling the available thermodynamic data, it is not necessary to tabulate or plot values of $\Delta F°$ vs. T for all possible reactions, but it is necessary only to present the values for a small number of basic reactions which can easily be combined algebraically to obtain $\Delta F°$ for any possible reaction. In particular, the custom is to present the $\Delta F°$'s only for standard reactions of formation from the elements and to obtain the rest by simple algebraic computations.

Table 4-6 gives standard free energy changes as a function of absolute temperature for the reactions of formation of metallurgically important zinc compounds as well as other substances of interest in zinc metallurgy. The standard states of the reactants and reaction products are indicated by the parenthetical notations (s), (liq) and (g) after each formula in the chemical equations: (s) means pure solid, (liq) pure liquid and (g) gas at a partial pressure of one atmosphere. To facilitate computations, free energy data for the same reaction but with different standard state combinations are given on separate lines. Also, for convenience in dealing with zinc processes, free energy values at the melting point (692.7°K or 419.5°C) and boiling point (1180°K or 907°C) of zinc metal are included. Since $\Delta F°$ for a given reaction varies approximately linearly with temperature, free energy values at intermediate temperatures are found simply by arithmetic interpolation.

As an example of the use of the data in Table 4-6, consider the following important reaction:

$$ZnO(s) + CO(g) \rightarrow Zn(g) + CO_2(g) \text{ at } 1400°K\ (1127°C) \qquad (4)$$

The equilibrium constant is given by

$$k_4 = \frac{(p_{Zn})(p_{CO_2})}{p_{CO}} \qquad (5)$$

in which p_{Zn}, p_{CO_2} and p_{CO} are the respective partial pressures of Zn, CO_2 and CO in the gas phase in equilibrium with pure solid ZnO. From Table 4-6, values of $\Delta F°$ are found for each of the following reactions at 1400°K:

$$C(s) + O_2(g) \rightarrow CO_2(g) \qquad \Delta F° = -94{,}690 \qquad (6)$$

$$Zn(g) + \tfrac{1}{2}O_2(g) \rightarrow ZnO(s) \qquad \Delta F° = -43{,}700 \qquad (7)$$

$$C(s) + \tfrac{1}{2}O_2(g) \rightarrow CO(g) \qquad \Delta F° = -56{,}310 \qquad (8)$$

Adding Eq. (6) and subtracting Eqs. (7) and (8), and performing the corresponding algebraic combinations for the free energy changes, gives chemical Eq. (4) and

$$\Delta F°_4 = -94{,}690 + 43{,}700 + 56{,}310 = +5{,}320 \text{ cal/mole} \qquad (9)$$

Now, utilizing the relation $\Delta F° = -RT \ln k$, the equilibrium constant is calculated:

$$\log k_4 = \frac{-\ 5{,}320}{4.575 \times 1400} = -0.832 \qquad (10)$$

$$k_4 = 0.147 = \frac{(p_{Zn})(p_{CO_2})}{p_{CO}} \text{ at } 1400°K. \qquad (11)$$

Table 4-6. Standard Free Energies ($\Delta F°$) of Compounds Important in Zinc Metallurgy[13, 14, 15, 17, 19, 20]

$\Delta F°$, calories per mole

Reactions	298°K	400°K	600°K	692.7°K	800°K	1000°K	1180°K	1400°K	1600°K	1800°K	2000°K
$Zn(s) + \frac{1}{2}O_2(g) \rightarrow ZnO(s)$	−76,100	−73,650	−68,950	−66,800							
$Zn(liq) + \frac{1}{2}O_2(g) \rightarrow ZnO(s)$				−66,800	−64,050	−58,900	−54,350				
$Zn(g) + \frac{1}{2}O_2(g) \rightarrow ZnO(s)$				−78,430	−73,060	−63,120	−54,350	−43,700	−34,200	−24,800	−15,500
$Zn(s) + \frac{1}{2}S_2(g) \rightarrow ZnS(s)$	−57,000	−54,600	−50,100	−48,000							
$Zn(liq) + \frac{1}{2}S_2(g) \rightarrow ZnS(s)$				−48,000	−45,400	−40,500	−36,200				
$Zn(g) + \frac{1}{2}S_2(g) \rightarrow ZnS(s)$				−59,630	−54,410	−44,720	−36,200	−25,700	−16,300		
$Zn(s) + \frac{1}{2}S_2(g) + 2O_2(g) \rightarrow ZnSO_4(s)$	−217,900	−207,100	−186,200	−176,500							
$Zn(liq) + \frac{1}{2}S_2(g) + 2O_2(g) \rightarrow ZnSO_4(s)$				−176,500	−165,000	−144,000	−125,600				
$Zn(g) + \frac{1}{2}S_2(g) + 2O_2(g) \rightarrow ZnSO_4(s)$				−188,130	−174,010	−148,220	−125,600				
$Zn(s) + Cl_2(g) \rightarrow ZnCl_2(liq)$			−78,000	−76,000							
$Zn(liq) + Cl_2(g) \rightarrow ZnCl_2(liq)$				−76,000	−73,200	−68,600					
$Zn(g) + Cl_2(g) \rightarrow ZnCl_2(g)$				−76,800	−75,100	−71,900	−69,100	−65,600	−62,400	−59,300	−56,100
$H_2(g) + \frac{1}{2}O_2(g) \rightarrow H_2O(g)$	−54,635	−53,520	−51,150	−50,000	−48,640	−46,030	−43,630	−40,640	−37,880	−35,110	−32,310
$C(s) + \frac{1}{2}O_2(g) \rightarrow CO(g)$	−32,808	−35,010	−39,360	−41,370	−43,680	−47,940	−51,730	−56,310	−60,430	−64,480	−68,510
$C(s) + O_2(g) \rightarrow CO_2(g)$	−94,260	−94,320	−94,440	−94,500	−94,540	−94,610	−94,660	−94,690	−94,730	−94,720	−94,720
$C(s) + 2H_2(g) \rightarrow CH_4(g)$	−12,140	−10,050	−5,500	−3,210	−530	+4,610	+9,330	+15,450			
$\frac{1}{2}S_2(g) + O_2(g) \rightarrow SO_2(g)$	−81,310	−79,560	−76,080	−74,450	−72,570	−69,060	−65,930	−62,080	−58,590	−55,110	−51,630
$\frac{1}{2}S_2(g) + \frac{3}{2}O_2(g) \rightarrow SO_3(g)$	−98,100	−94,050	−86,000	−82,350	−78,100	−70,250	−63,250	−54,850			
$H_2(g) + \frac{1}{2}S_2(g) \rightarrow H_2S(g)$	−17,470	−16,400	−14,360	−13,330	−12,110	−9,790	−7,660	−5,090	−2,720	−330	
$Pb(liq) + \frac{1}{2}O_2(g) \rightarrow PbO(liq)$							−24,800	−20,550	−16,850		
$Pb(g) + \frac{1}{2}O_2(g) \rightarrow PbO(liq)$							−42,600	−33,600	−25,600		
$Cd(liq) + \frac{1}{2}O_2(g) \rightarrow CdO(s)$			−48,000	−45,600	−42,800	−37,700					
$Cd(g) + \frac{1}{2}O_2(g) \rightarrow CdO(s)$			−58,300	−53,700	−48,300	−38,600	−29,800	−19,200	−9,700	−300	+9,000
$Zn(s) \rightarrow Zn(g)$	+22,580	+19,710	+14,150	+11,630							
$Zn(liq) \rightarrow Zn(g)$				+11,630	+9,010	+4,220	0	−5,070	−9,600	−14,060	

Vapor Pressures

The relatively high volatility of metallic zinc is an important property which accounts for many of the essential features of zinc technology. Reliable experimental data are available for vapor pressures of both solid and liquid zinc, and Kelley[14] has carefully correlated these data with other thermodynamic data. The following free energy and vapor pressure equations are based on his recommendations:

$$Zn(s) \rightarrow Zn(g) \tag{12}$$

$$\Delta F^\circ = 31{,}392 + 0.64T \log T + 1.35 \times 10^{-3}T^2 - 31.17T \tag{13}$$

$$\log p_{Zn(s)} = 6.813 - \frac{6{,}862}{T} - 0.140 \log T - 0.295 \times 10^{-3}T \tag{14}$$

$$Zn(liq) \rightarrow Zn(g) \tag{15}$$

$$\Delta F^\circ = 30{,}902 + 6.03\, T \log T + 0.275 \times 10^{-3}T^2 - 45.03T \tag{16}$$

$$\log p_{Zn(liq)} = 9.843 - \frac{6{,}755}{T} - 1.32 \log T - 0.06 \times 10^{-3}T \tag{17}$$

In Eqs. (14) and (17), p_{Zn} is given in atmospheres. Although the vapor pressure of solid and liquid zinc is small at the melting point (2×10^{-4} atm or 0.15 mm Hg at 419.5°C), it is large enough to insure that wholesale evaporation or sublimation can be easily carried out in a modest vacuum at temperatures in the vicinity of the melting point. The vapor pressure of liquid zinc increases rapidly with temperature to reach one atmosphere at 907°C, the normal boiling point. In view of this low boiling temperature, a variety of distillation and retorting processes are used by the metallurgist. Examples are the retort processes and associated condensers for ore reduction, reflux stills for refining zinc, vacuum dezincing of liquid lead and the distillation of zinc from Parkes process crusts. In other processes, such as zinc oxide manufacture and slag fuming, production of zinc vapor is an essential intermediate step.

Alloys. The partial pressures of zinc vapor in equilibrium with various liquid and solid alloys have been measured experimentally, using a number of different techniques.[21] For still other alloys, the zinc partial pressures are easily calculable from available activity data on the alloys, since at a given temperature, $p_{Zn} = a_{Zn}p^\circ_{Zn} = \gamma_{Zn}x_{Zn}p^\circ_{Zn}$, where p_{Zn} is the equilibrium partial pressure of Zn over the alloy, a_{Zn} is the activity of zinc based on pure Zn as the standard state of unit activity, p°_{Zn} is the known vapor pressure of pure Zn, γ_{Zn} is the activity coefficient of Zn in the alloy and x_{Zn} is the mole fraction of zinc. In view of their relatively low melting points and other favorable properties, many binary and ternary zinc alloys

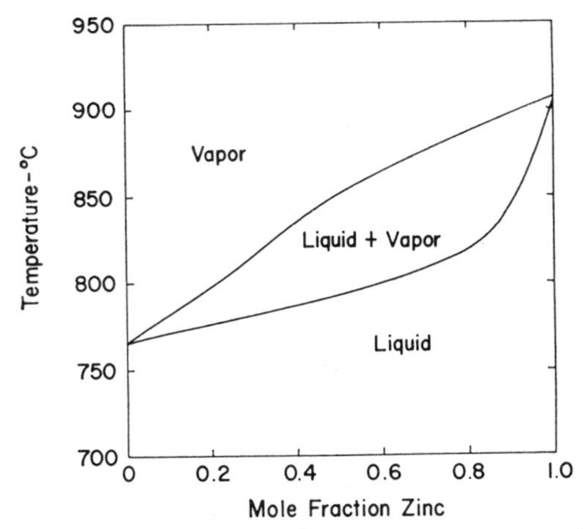

Figure 4-2. Liquid vapor equilibrium diagram for Zn-Cd system. (After Lumsden.[22])

in the liquid state have been studied thermodynamically in the laboratory so that good data are available for many systems.[21, 22]

The Cd-Zn system is of particular interest because both components are comparably volatile. Figure 4-2 gives the liquid-vapor equilibrium diagram for this system at one atmosphere total pressure.[22] This diagram is of the type very familiar to chemists and chemical engineers engaged in separating organic compounds by fractional distillation, and the principles of design and operation of fractionating distillation columns for separating zinc and cadmium are the same as those involved in distillation separations of organic liquids.

Condensation. The efficiency of a condenser in recovering liquid zinc from a mixture of zinc vapor with inert or nonreacting gases (e.g., CO) depends on the relative amount of nonreacting gases and on the condenser temperature. If the condenser is well designed and if no side reactions such as zinc reoxidation occur, the maximum condensation recovery can be estimated on the assumption that the gases leaving the condenser are at equilibrium with the liquid condensate at the condenser temperature. The theoretical percentage recovery (R) in a steady state, continuous condenser is then given by

$$R = 100\left(1 - \frac{yp_{\mathrm{Zn}}}{P - p_{\mathrm{Zn}}}\right) \tag{18}$$

in which y = moles inert gas per mole Zn in the incoming gases, p_{Zn} = partial pressure of Zn and P = total gas pressure at the condenser exit. Results of calculations based on this relation and on the vapor pressure data given previously are shown in Figure 4-3. The importance of minimiz-

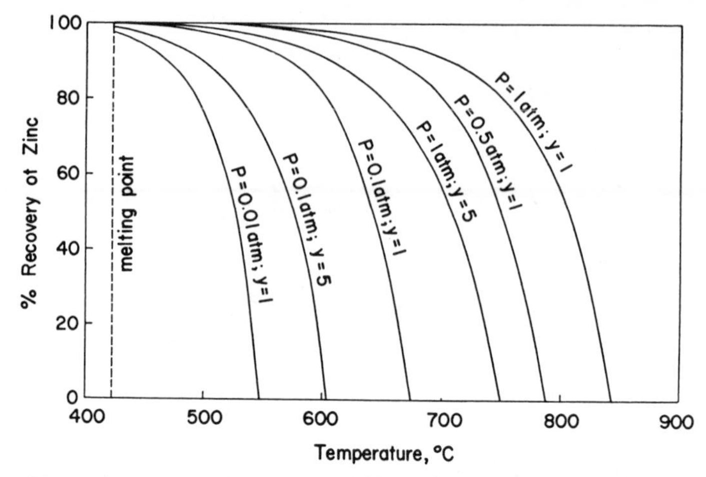

Figure 4-3. Theoretical efficiency of steady-state zinc condenser.

ing dilution of the incoming zinc vapor and the necessity for close control of condenser temperature are clearly demonstrated. A modest vacuum (0.5 atm) reduces the theoretical recovery slightly, but evacuating the condenser to $P = 0.1$ or 0.01 atm can result in substantial losses of uncondensed zinc, even if the temperature is held just above the melting point of zinc.

Physical Chemistry of Roasting

When ZnS is roasted in air, the principal reaction products are ZnO, $ZnSO_4$, SO_2 and SO_3. The stoichiometry of the process can be expressed in terms of the following chemical equations:

$$ZnS + 3/2O_2 \rightarrow ZnO + SO_2 \tag{19}$$

$$ZnO + SO_2 + 1/2O_2 \rightarrow ZnSO_4 \tag{20}$$

$$SO_2 + 1/2O_2 \rightarrow SO_3 \tag{21}$$

Eq. (19) represents the primary over-all reaction in most zinc roasting operations, and is a strongly exothermic reaction. Eq. (20) represents the sulfatizing reaction, which accounts for the presence of zinc sulfate in the calcine. Finally, Eq. (21), which will be recognized as the reaction involved in sulfuric acid manufacture, also proceeds to some degree in the roaster to account for the presence of SO_3 in the roaster gas. Some observers[18, 19] have considered the possible formation of basic zinc sulfates, such as $3ZnO \cdot 2SO_3$, but the evidence is not clear and therefore this possibility will be ignored in the subsequent discussion.

When roasting is completed in the presence of some excess air or oxygen, the zinc can be present either as oxide or as sulfate, depending on the temperature and on the atmosphere. Assuming that equilibrium is reached between gases and solids, the conditions for formation of oxide or sulfate can be predicted by a combination of equilibrium and stoichiometric calculations. The equilibrium constants for reactions (20) and (21) are

$$k_{20} = \frac{1}{(p_{SO_2})(p_{O_2})^{1/2}} \tag{22}$$

$$k_{21} = \frac{p_{SO_3}}{(p_{SO_2})(p_{O_2})^{1/2}} \tag{23}$$

One further condition can be specified by a total pressure equation

$$P = p_{O_2} + p_{SO_2} + p_{SO_3} \tag{24}$$

Actually, if the roaster is operated with air in the usual way at atmospheric pressure, P will be approximately 0.2 atm because the roaster gas will contain on the order of 80 per cent nitrogen. On the other hand, if oxygen is substituted for air and no nitrogen is present, P will be approximately one atmosphere. In principle, P can be varied over wide ranges also by varying the ambient pressure, which might be feasible for some kinds of roasting furnaces and laboratory operations.

At any given temperature, k_{20} and k_{21} are easily calculated from the data in Table 4-6. If P is also specified, Eqs. (22), (23) and (24) can be solved simultaneously to find p_{SO_2}, p_{O_2} and p_{SO_3} in the gas phase which would be at equilibrium with both ZnO and $ZnSO_4$. Gas mixtures departing from this equilibrium composition will be in equilibrium with ZnO if $(p_{SO_2})(p_{O_2})^{1/2}$ is actually less than the equilibrium product calculated from k_{20}, while $ZnSO_4$ will be in equilibrium if $(p_{SO_2})(p_{O_2})^{1/2}$ is greater than this equilibrium value. Results of these calculations are summarized in Figure 4-4. The curves in this figure represent the variation of $p_{O_2}/P \times 100$, or volume per cent O_2, with temperature for systems containing ZnO and $ZnSO_4$ at equilibrium. Each curve is calculated for a different value of total pressure P. For conditions to the left of each curve, $ZnSO_4$ will be the stable phase, and to the right, ZnO will be the stable phase. Study of this "phase diagram" brings out a number of factors which should be significant in controlling the formation of $ZnSO_4$. In the first place, the temperature required to produce ZnO is seen to be relatively insensitive to gas composition and to amount of excess oxygen over a large range (roughly from $p_{O_2}/P = 10$ per cent to $p_{O_2}/P = 80$ per cent). For calcines in equilibrium with ordinary roaster gases ($P = 0.2$ atm; p_{O_2} between 0.02 and 0.1 atm) these calculations point to a minimum sulfate-decomposition temperature of substantially 860°C. This decomposition temperature is raised by increasing P, which can be accomplished by substituting oxygen gas for the air

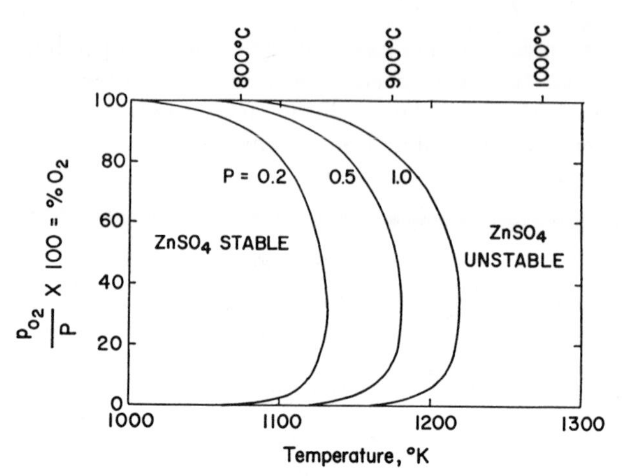

Figure 4-4. Effect of pressure, temperature, and gas composition on stability of zinc sulfate.

supplied to the roaster. Rapid decomposition against an external pressure of one atmosphere theoretically will require a temperature of 945°C. In view of the possibility of basic sulfate formation and in view of some uncertainties in the thermodynamic data for $ZnSO_4$, the calculated decomposition temperatures may be in error by as much as 50°C. However, these uncertainties should have little effect on the general conclusions regarding effects of the variables of pressure, temperature and gas composition.

In some parts of the roasting system (for example, deep in the bed on a hearth-type roaster), there may be a deficiency of O_2 and an excess of ZnS. Under these conditions, the gas phase will be chiefly SO_2 and N_2, with p_{SO_2} approximately equal to P. The question as to whether or not these conditions favor sulfate decomposition is readily answered by considering equilibrium data for the following reaction:

$$3ZnSO_4 + ZnS \rightarrow 4ZnO + 4SO_2 \tag{25}$$

$$k_{25} = (p_{SO_2})^4 \tag{26}$$

Calculations of the equilibrium p_{SO_2} for this reaction, using data in Table 4-6, show that p_{SO_2} theoretically reaches one atmosphere at a temperature of about 765°K (492°C) and that p_{SO_2} increases rapidly with increasing T. Thus, under conditions of excess ZnS and deficient O_2 and at all normal roasting temperatures (>500°C), ZnO should be formed in preference to $ZnSO_4$. Also, these calculations show that it is thermodynamically possible to convert $ZnSO_4$ to ZnO by moderate heating with an excess of ZnS, and that this conversion will occur at a considerably lower temperature than is required for the decomposition of $ZnSO_4$ in the presence of excess oxygen. The practical conditions for carrying out the sulfide-sulfate reaction presumably will be determined by kinetic rather than thermodynamic factors.

Another possibility to be examined for oxygen-deficient conditions is the reaction of oxide and sulfide to form liquid zinc or zinc vapor:

$$2ZnO + ZnS \rightarrow 3Zn + SO_2 \quad (27)$$

This type of reaction is of course well known both in lead metallurgy and in copper metallurgy. However, thermodynamic calculations based on data given in Table 4-6 show that this reaction cannot proceed to any significant extent, to produce either liquid or gaseous zinc, at temperatures below 1100°C. In fact, the available thermodynamic data indicate that temperatures somewhat in excess of 1300°C are required to make the oxide-sulfide reaction proceed spontaneously against a pressure of one atmosphere. Thus, it may be concluded that a mixture of ZnO and ZnS is chemically stable at all roasting temperatures in an SO_2 or SO_2-N_2 atmosphere deficient in O_2.

Roasting Behavior of Lead and Cadmium. The chemistry of roasting galena has been presented elsewhere,[23] and under oxidizing conditions the factors controlling the formation of lead oxide, sulfate and basic sulfates are similar to those involved in roasting zinc ores. However, when mixtures of lead and zinc sulfides and oxides are heated together under reducing conditions, such as within a bed of partially roasted ore, the possibility arises of forming metallic lead. One reaction to be considered is

$$ZnS(s) + 3PbO(s) \rightarrow 3Pb(liq) + ZnO(s) + SO_2(g) \quad (28)$$

Thermodynamic calculations indicate that this reaction might proceed at atmospheric pressure at temperatures as low as 750°C. Thus, although ZnS is not powerful enough as a reducing agent to reduce ZnO, it may nevertheless be powerful enough to reduce PbO within the temperature range of the roasting process.

The corresponding reaction for reduction of CdO is

$$ZnS(s) + 3CdO(s) \rightarrow 3Cd(liq\ or\ g) + ZnO(s) + SO_2(g) \quad (29)$$

Free energy calculations at various temperatures (based on data in Table 4-6) show that this reaction will not proceed to yield liquid cadmium, but that it will proceed spontaneously to yield cadmium vapor at temperatures above 900°C. Cadmium vaporized according to this reaction from within a bed of partially roasted ore would of course become oxidized to a CdO fume on contacting the oxidizing gases above the bed.

Formation of Zinc Ferrites. Zinc oxide reacts readily with ferric oxide at temperatures of 650°C and above to form zinc ferrite.[24] The metaferrite is $ZnO \cdot Fe_2O_3$, which has the normal spinel structure and is nonmagnetic, as discussed previously. However, the ferrites with spinel structures form solid solutions with each other. For example, $ZnO \cdot Fe_2O_3$ and $FeO \cdot Fe_2O_3$ (magnetite) may form a complete range of solid solutions.[25] This would account for the observed magnetic properties of ferrites formed in roasting.

Data on the thermodynamic properties of the ferrites are too scanty to justify quantitative calculations of their stabilities under roasting conditions. However, the results of various experimental studies of zinc ferrite preparation and of rates of reaction between ZnO and Fe_2O_3 indicate that zinc ferrite is quite stable under all ordinary zinc roasting conditions. Thus, the extent to which zinc occurs as zinc ferrite in roaster calcine will depend on the relative amounts of zinc and iron present and on the factors which determine the reaction rates between zinc oxide and iron oxides. Presumably, the ferrite forms by a solid-solid reaction and thus its formation is favored by intimate contact of zinc and iron oxides. Hopkins[24] showed that the reaction rate increases rapidly as the temperature is raised above 650°C.

Direct Reduction of Zinc Sulfide

In the quest for technological improvements in the extractive metallurgy of zinc, some efforts have been made to develop a one-step process for direct reduction of zinc sulfide to metal. These efforts have been based mainly on reduction with metallic iron, according to the reaction

$$ZnS(s) + Fe(s) \rightarrow FeS(s) + Zn(g) \tag{30}$$

Kelley[19] reported a detailed thermodynamic study of this reaction for the temperature range 1300 to 1800°K, together with the corresponding reaction for reducing PbS. Kelley's calculations indicated the technical feasibility of conducting the direct reduction at about 1600°K to produce zinc vapor at atmospheric pressure from a liquid phase containing ZnS, Fe and FeS. However, his calculations also pointed to the likelihood of blue-powder formation when the gases are cooled to condense zinc. In 1948, Gross and Warrington[26] showed that the direct reduction of ZnS with Fe could be carried out at lower temperatures (<1000°C) in a vacuum retort. Under these conditions, the ZnS, Fe and FeS remained as solids. More recently, Bethune and Pidgeon[27] measured the equilibrium vapor pressures of Zn obtained both in iron reduction and in copper reduction at temperatures from 850 to 1000°C, with the results summarized in Figure 4-5.

Reduction of Zinc Oxide

The reduction of zinc oxide to metal differs chemically from the reduction of other base-metal oxides both because of the relatively high stability of

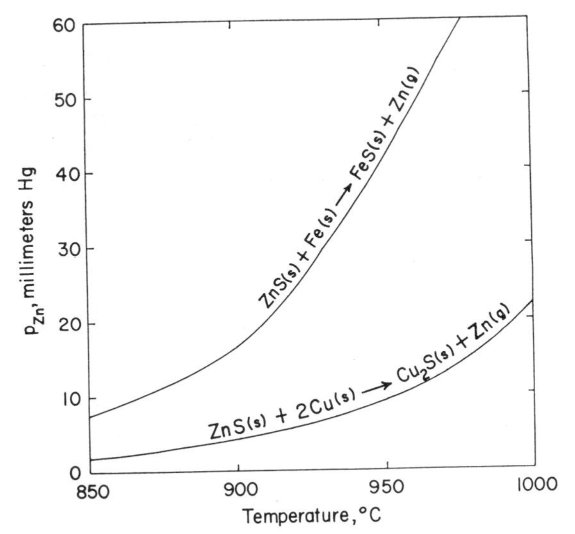

Figure 4-5. Vapor pressures of zinc from direct reduction of zinc sulfide. (After Bethune and Pidgeon.[27])

zinc oxide and because of the volatility of the metal. The combination of these two factors accounts for the fact that high-temperature reduction with solid carbon followed by condensation of the zinc has remained the only commercially practicable approach to the reduction of zinc oxide. The carbon reduction may be carried out in batch retorts, continuous retorts, electric furnaces or even the blast furnace. The retort processes and the electrothermal process discharge a solid residue, but in the Sterling process the zinc reduction is combined with a smelting process yielding liquid slag and pig iron. All these processes deliver Zn vapor with CO and small amounts of CO_2 to a condenser where the Zn is condensed in the liquid state. As will be shown below, the gases leaving the smelting furnace must be low in CO_2. Thus, the over-all reaction of carbon reduction is substantially

$$ZnO(s) + C(s) \rightarrow Zn(g) + CO(g) \tag{31}$$

A direct solid-solid reaction between ZnO and C is a most unlikely reaction mechanism. Instead, the carbon reduction may be regarded as the result of two simultaneous gas-solid reactions:

$$ZnO(s) + CO(g) \rightarrow Zn(g) + CO_2(g) \tag{32}$$

$$CO_2(g) + C(s) \rightarrow 2CO(g) \tag{33}$$

The equilibrium constants for these reactions are

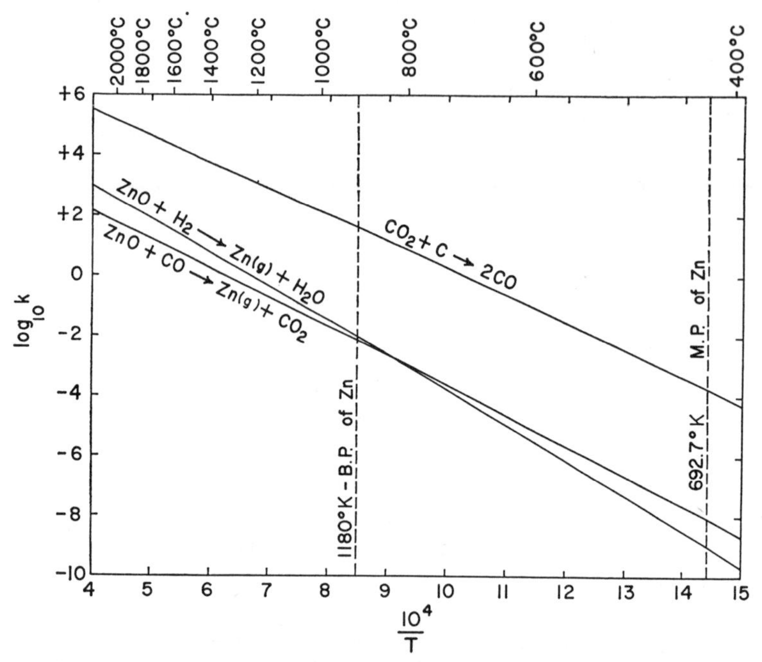

Figure 4-6. Equilibrium constants for zinc reduction processes.

$$k_{32} = \frac{(p_{Zn})(p_{CO_2})}{p_{CO}} \tag{34}$$

$$k_{33} = \frac{(p_{CO})^2}{p_{CO_2}} \tag{35}$$

These equilibria have been studied directly, and also the thermodynamic properties of the various reactants have been evaluated by other means. As a result, values of k_{32} and k_{33} are known with a good degree of confidence over the entire temperature range for zinc smelting. These data (based on Table 4-6) are conveniently summarized in Figure 4-6. For comparison, this also provides equilibrium data for the hydrogen reduction of zinc oxide:

$$ZnO(s) + H_2(g) \rightarrow Zn(g) + H_2O(g) \tag{36}$$

$$k_{36} = \frac{(p_{H_2O})(p_{Zn})}{p_{H_2}} \tag{37}$$

Comparison of the curves for H_2 reduction and CO reduction shows that insofar as the equilibria are concerned H_2 and CO are about equally effective as reducing agents, with the H_2 becoming somewhat more effective at higher temperatures. The primary equilibrium data given in Figure 4-6 can be applied to a wide variety of processes. Useful results are obtained

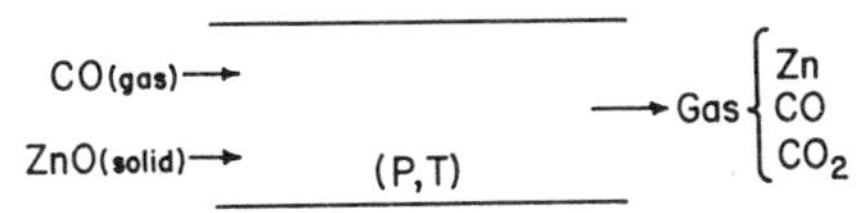

Figure 4-7. Continuous gaseous reduction of zinc oxide.

by combining the equilibrium constants with quantitative stoichiometric relationships specifically applicable to the process in question. This procedure will be applied to various systems under the subsequent headings.

Gaseous Reduction. As a first step in the study of zinc oxide reduction, the simple gaseous reduction system indicated in Figure 4-7 will be considered. This is a continuous steady-state process in which solid ZnO and gaseous CO are fed into a furnace which delivers a continuous gas stream containing Zn vapor, CO and CO_2. The extent of reaction and the zinc content of the effluent gas will depend on the rate of gas flow, the surface area of gas-solid contact and other kinetic factors. Under the most favorable kinetic conditions, however, the zinc content of the effluent gas will be a maximum and the effluent gases will be at equilibrium with solid ZnO. At equilibrium, the partial pressures of Zn, CO, and CO_2 in the effluent gas will satisfy the relation

$$k_{32} = \frac{(p_{Zn})(p_{CO_2})}{p_{CO}} \tag{38}$$

The stoichiometry of the process shows that one mole of CO_2 is produced for each mole of Zn. Therefore, in the effluent gas

$$p_{Zn} = p_{CO_2} \tag{39}$$

Also, the total gas pressure (P) is the sum of the partial pressures of the three gas constituents:

$$P = p_{Zn} + p_{CO} + p_{CO_2} \tag{40}$$

If P and T are given, k is also known from Figure 4-6 and the three equations can be solved simultaneously. The solutions are

$$p_{Zn} = p_{CO_2} = \sqrt{k_{32}(P + k_{32})} - k_{32} \tag{41}$$

$$p = P - 2(\sqrt{k_{32}(P + k_{32})} - k_{32}) \tag{42}$$

Results of these calculations are represented in Figure 4-8 where the zinc content of the effluent equilibrium gas (volume per cent Zn $= p_{Zn}/P \times 100$) is plotted against temperature for various values of total pressure. For a total pressure $P = 1$ atm, these results show that CO is a very inefficient reducing agent below about 1000°C, but gradually improves with increasing temperature to become relatively efficient above 1200°C. Also, comparisons of the curves for various values of P show that decreasing pressure

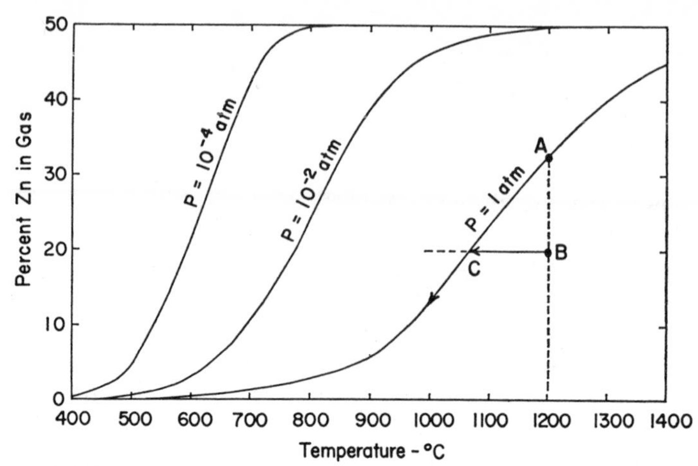

Figure 4-8. Effect of temperature and pressure on carbon monoxide reduction of zinc oxide.

markedly increases the zinc yield at a given temperature. It should be kept in mind that these curves all represent maximum or equilibrium values of the zinc content of the effluent gas at given P and T. If kinetic factors are unfavorable, the actual percentage of Zn obtained in the effluent gas will fall well below these curves.

Possible operation of a CO reduction process at some convenient temperature and pressure, say 1200°C and one atmosphere, may now be considered further. For these conditions, point A in Figure 4-8 represents an equilibrium gas percentage composition of 32.5 Zn, 32.5 CO_2 and 35 CO. Point B represents an arbitrary effluent-gas percentage composition of 20 Zn, 20 CO_2 and 60 CO which conceivably might be obtained by operating the system at a reasonable gas-flow rate with only moderately favorable kinetic factors. The cooling of this gas as it leaves the furnace is represented in Figure 4-8 by the horizontal arrow BC. Where BC intersects the equilibrium curve at 1070°C, the gas will be just at equilibrium with solid ZnO However, to the left of the curve, $(p_{Zn})(p_{CO_2})/p_{CO}$ will be greater than the equilibrium constant for the gaseous reduction reaction, and therefore the gas will react to precipitate ZnO (reversal of original reduction reaction). Then, as the gas is cooled further, more and more ZnO will be deposited while the gas composition continues to follow the equilibrium curve to lower temperatures. This reversal of the gaseous reduction reaction on cooling the reduction products accounts for the impracticality of a simple gaseous reduction process for making zinc. That is, a condenser placed after the furnace specified in Figure 4-7 would collect mainly ZnO rather than Zn metal.

Analogous calculations for H_2 reduction lead to curves very similar to

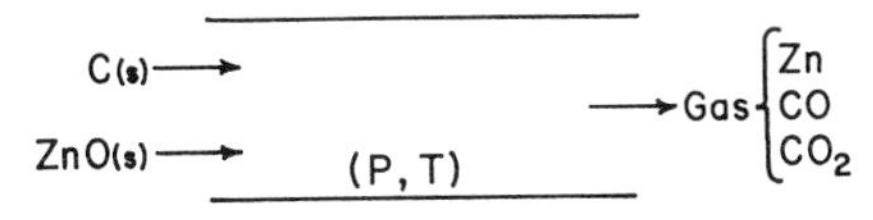

Figure 4-9. Carbon reduction of zinc oxide.

those in Figure 4-8, but with the curve for $P = 1$ atm rising somewhat more steeply to give better Zn yields in the effluent gas at higher temperatures.

Carbon-Reduction Equilibrium. The direct reduction of ZnO by C is indicated schematically in Figure 4-9, which can represent either a continuous or a batch process. The feed materials are solid ZnO and solid C, and the Zn vapor leaves the system in a mixture with CO and CO_2. As a starting point, the assumption can be made that the effluent gas is at equilibrium with both ZnO and C at the furnace temperature. That is,

$$\frac{(p_{Zn})(p_{CO_2})}{p_{CO}} = k_{32} \tag{43}$$

$$\frac{(p_{CO})^2}{p_{CO_2}} = k_{33} \tag{44}$$

Each mole of ZnO reduced gives one mole Zn(g) and 1 mole O as CO and CO_2. This stoichiometric relation can be expressed

$$n_{ZnO\ \text{reacted}} = n_{Zn(g)} = n_{CO} + 2n_{CO_2} \tag{45}$$

in which n refers to the number of moles of ZnO reacted and to the number of moles of Zn, CO and CO_2 produced. In view of Dalton's law of partial pressures, this relation leads to

$$p_{Zn} = p_{CO} + 2p_{CO_2} \tag{46}$$

The total pressure condition is given by

$$P = p_{Zn} + p_{CO} + p_{CO_2} \tag{47}$$

Thus, if equilibrium conditions are assumed at a given T, there are four equations and these are sufficient to determine P, p_{Zn}, p_{CO} and p_{CO_2}. P represents the maximum total gas pressure which can be generated by the reaction. If the retort is operated at a gas pressure or venting pressure less than the calculated equilibrium P, the carbon reduction reaction can proceed. On the other hand, if the retort pressure is raised above P, the reaction will be reversed and no reduction can occur. The dashed curve AB in Figure 4-10 gives values of P as a function of temperature, calculated by simultaneous solutions of the four equations given above. To the left of point A, liquid zinc appears, as will be shown later, and a different method of calculating P must be used. Curve AB can also be regarded as

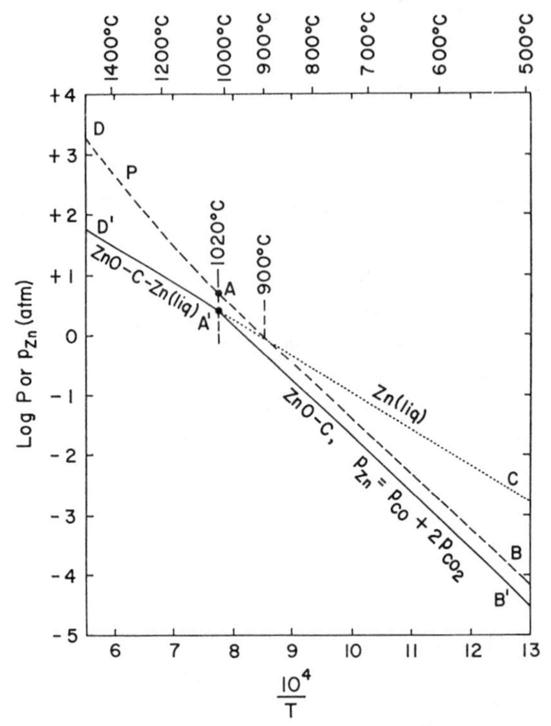

Figure 4-10. Equilibrium pressures *vs.* temperature in carbon reduction.

giving the minimum temperature at which carbon reduction can be carried out for a given venting pressure in the retort. At one atmosphere, for example, a temperature of 900°C or higher is required. Curve *AB* indicates furthermore that carbon reduction might be achieved at substantially lower temperatures in a vacuum retort (for example, at a temperature of about 575°C if $P = 10^{-3}$ atm). However, the possibility of commercially achieving low-temperature reduction in a vacuum retort is rather remote, owing to low rates of reaction at the lower temperatures.

Curve *A′B′* in Figure 4-10 shows p_{Zn} as a function of temperature, also obtained by simultaneous solution of the above equations. For comparison, the dotted curve *A′C* gives the vapor pressure of liquid zinc as a function of temperature. The partial pressure of zinc attainable by carbon reduction is seen to be below the vapor pressure of liquid zinc at all temperatures up to 1020°C (point *A*). Above 1020°C, the partial pressure of zinc calculated by simultaneous solution of Eqs. (43), (44), (46) and (47) is greater than the vapor pressure of liquid zinc, so that the calculation cannot be valid above 1020°C. Thus, curve *A′D′* giving the maximum partial pressure of zinc above 1020°C is simply the extension of *A′C*, the vapor pressure curve of liquid zinc. The total pressure P at temperatures above 1020°C (curve AD) is calculated by simultaneous solution of Eqs. (43), (44) and (47), taking p_{Zn} equal to the vapor pressure of liquid zinc. Thus curve AD represents the necessary conditions of P and T for the univariant equilibrium of the four phases: zinc oxide, carbon, liquid zinc and gas. These four phases cannot exist at equilibrium under any other conditions of total pressure and temperature than those represented along AD.

Curves AD and $A'D'$ are of particular interest in relation to the possibility of developing a smelting process yielding liquid zinc directly. The lowest temperature and lowest pressure at which liquid zinc can be formed by carbon reduction (point A) are 1020°C and 5 atm, respectively. Under these minimal conditions, however, the ratio of zinc to carbon monoxide in the gas phase is nearly 1 to 1 so that virtually all the zinc produced by carbon reduction will be carried out of the furnace with the carbon monoxide. At higher temperatures and pressures a greater proportion of the reduced zinc will be liquid. However, by combining the data in Figure 4-10 with straightforward stoichiometric calculations, it can be shown that a yield of 90 per cent of the zinc in liquid form would require a temperature above 1300°C and a pressure over 200 atmospheres. Hence, a smelting process using carbon reduction to yield liquid zinc directly (for example, a high-pressure electric furnace) does not seem commercially feasible. Before Maier's thermodynamic calculations on this problem were published in 1930, a number of elaborate and costly attempts to smelt zinc oxide to liquid in pressure blast furnaces had failed.[18] It might be pointed out that conditions in a blast furnace are even less favorable than those calculated previously, owing to the extra volume of gases involved in burning the coke at the tuyères.

Minimizing Oxide and Blue-Powder Formation. When the gaseous mixture of Zn, CO and CO_2 from the retort is cooled, either the CO_2 or the CO might oxidize the Zn back to ZnO

$$CO_2(g) + Zn(g) \rightarrow ZnO(s) + CO(g) \tag{48}$$

$$CO(g) + Zn(g) \rightarrow ZnO(s) + C(s) \tag{49}$$

Fortunately, the rate of the second reaction appears to be quite slow under ordinary operating conditions, so that formation of oxide and blue powder can be minimized by controlling the CO_2 content of the gases passing from the retort to the condenser. The basis for this control can be understood in terms of the graph of CO_2/CO ratio and temperature shown in Figure 4-11. Curve A gives the equilibrium CO_2/CO ratio for a gas mixture which is in equilibrium with solid ZnO and which also satisfies the total pressure and stoichiometric requirements of the carbon reduction process. That is,

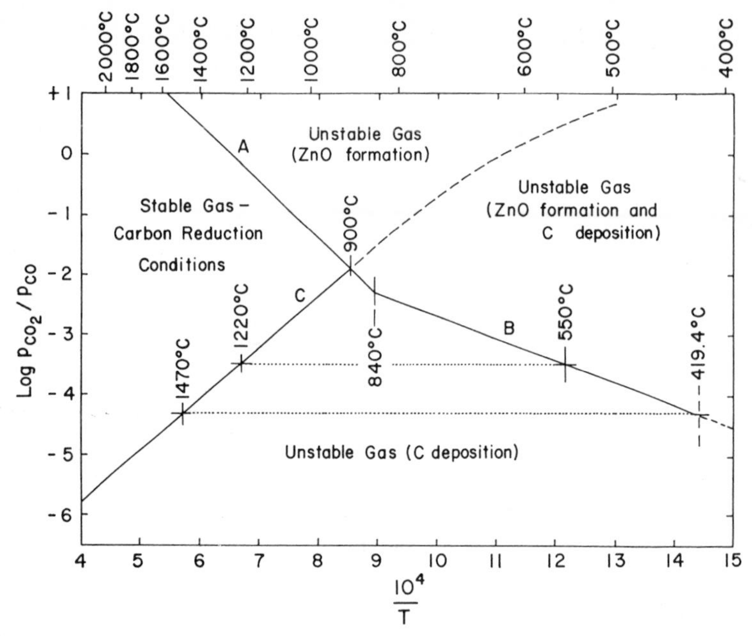

Figure 4-11. Equilibrium CO_2-CO ratios in carbon reduction and condensation at 1 atmosphere.

curve *A* represents the simultaneous solution of the following three equations as a function of temperature.

$$k_{32} = \frac{(p_{Zn})(p_{CO_2})}{p_{CO}} \tag{50}$$

$$P = 1 \text{ atm} = p_{CO} + p_{CO_2} + p_{Zn} \tag{51}$$

$$p_{Zn} = p_{CO} + 2p_{CO_2} \tag{52}$$

At a given temperature, the gas mixtures with CO_2/CO ratios below curve *A* are stable and will not form ZnO by reversal of the gaseous reduction reaction. Gas mixtures with CO_2/CO ratios above curve *A* will tend to deposit ZnO until the CO_2 content of the gas drops to the value given by the curve. Curve *B* represents the equilibrium CO_2/CO ratio for formation of ZnO from gases in equilibrium with liquid Zn—that is, in the condensing temperature range. Curve *B* is simply a plot of the equilibrium constant for the following reaction:

$$\text{ZnO(s)} + \text{CO(g)} \rightarrow \text{Zn(liq)} + \text{CO}_2\text{(g)} \tag{53}$$

$$k_{53} = \frac{p_{CO_2}}{p_{CO}} \tag{54}$$

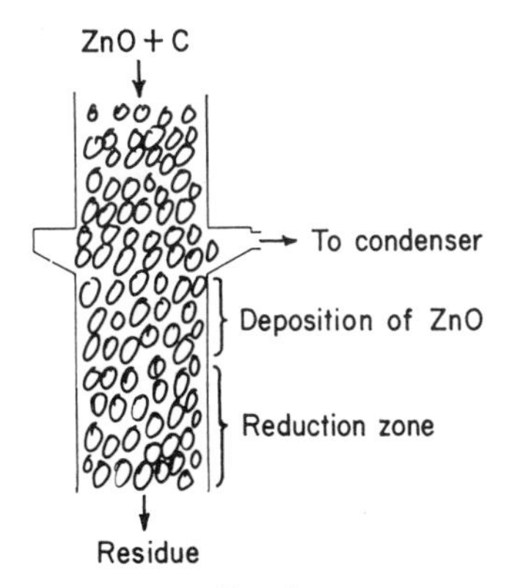

Figure 4-12. Continuous retort.

From curve *B*, it can be seen that the CO_2/CO ratio in the gases leaving the retort would have to be below 5×10^{-5} to avoid oxide formation at 419°C, the melting point of zinc. For a condenser temperature of 550°C, the CO_2/CO ratio should be below 3.2×10^{-4}. These ratios correspond to about 0.0025 per cent CO_2 and 0.016 per cent CO_2, respectively, in the gases leaving the retort.

One practical method of minimizing the CO_2 content of the effluent gases in continuous retorts is shown schematically in Figure 4-12. Before leaving the retort, the gases coming up from the high temperature zone of active reduction are cooled by the countercurrent downward stream of ZnO and C. As the gases cool, ZnO is formed and CO_2 is consumed so that the gas composition will tend to follow down curve *A* in Figure 4-11. The ZnO deposited on the cool charge is thus returned to the reduction zone as a circulating load. To minimize the CO_2 content of the effluent gas, the effluent gas temperature should be kept as low on curve *A* as possible. At the same time, to avoid condensing metallic zinc in the charge, the temperature should be above the temperature of intersection of curves *A* and *B* (840°C). If, for example, the gases leave the retort at equilibrium with ZnO at 900°C, Figure 4-11 shows that the CO_2/CO ratio will be about 0.012, corresponding to 0.6 per cent CO_2. In the connecting duct and condenser, this CO_2 probably will react to produce a small amount of zinc oxide (up to a maximum of 1.2 per cent of the total zinc in the gas stream).

Curve *C* in Figure 4-11 gives the equilibrium CO_2/CO ratios as a function of temperature for gas mixtures which are in equilibrium with solid carbon and which also satisfy the total pressure and stoichiometric requirements of the carbon reduction process. That is, curve *C* represents the simultaneous solutions of the following three equations as a function of temperature:

$$k_{33} = \frac{(p_{CO})^2}{p_{CO_2}} \tag{55}$$

$$P = 1 \text{ atm} = p_{Zn} + p_{CO} + p_{CO_2} \tag{56}$$

$$p_{Zn} = p_{CO} + 2p_{CO_2} \tag{57}$$

At a given temperature, gases with CO_2/CO ratios above curve C tend to react with solid carbon to reduce their CO_2 contents: $CO_2 + C \rightarrow 2CO$. In the absence of solid carbon, gases with CO_2/CO ratios above curve C are stable. On the other hand, gases with CO_2/CO ratios below the curve tend to deposit carbon and thus to increase in CO_2 by the carbon deposition reaction: $2CO \rightarrow CO_2 + C$.

Curves A, B and C of Figure 4-11 combined furnish a kind of phase diagram in which the necessary conditions for the various parts of the retort and condensing system can be conveniently delineated. In the first place, carbon reduction proceeds only under conditions in the region between curves A and C at the left. In this region, both ZnO and C react with the gas. If kinetic factors are more favorable to the reaction of the gas with solid ZnO, the retort-gas composition will be closer to curve A. On the other hand, if the kinetic factors favor the reaction of the gas with solid carbon, the retort-gas composition will be closer to curve C. The minimum reduction temperature at one atmosphere corresponds to the intersection of curves A and C (see also Figure 4-10). To the right of curve C and below curves A and B, carbon deposition should occur, but without simultaneous oxide deposition. Fortunately, the carbon-deposition reaction is quite slow, so that it does not constitute a practical problem. In fact, it is interesting to note that the retort-condenser process of making zinc would not be feasible at all if the carbon-deposition reaction were allowed to proceed to equilibrium as the retort gases are cooled for condensation of zinc.

Using the diagram of Figure 4-11, another practical procedure for controlling oxide deposition in batch retorts can now be understood. This procedure consists in providing an excess of carbon, or a hot carbon blanket, near the mouth of the retort. With this arrangement, the effluent-gas composition should approach the CO_2/CO ratio given by curve C for the temperature of the carbon blanket. That is, low CO_2 contents are favored by high blanket temperatures, and it is possible to produce an effluent gas which will have little tendency to deposit ZnO or blue powder on cooling in the condenser. Assuming no carbon deposition occurs, a blanket temperature of 1470°C theoretically would prevent oxide formation down to 419°C while a blanket temperature of 1220°C would prevent oxide formation down to 550°C (see Figure 4-11).

It should be kept in mind that the diagram in Figure 4-11 refers only to operation at one atmosphere total gas pressure. Also, the stoichiometric equation

$$p_{Zn} = p_{CO} + 2p_{CO_2}$$

involves the assumption that pure ZnO, without appreciable content of other reducible oxides such as those of iron, is fed to the retort. Still another assumption in calculating curve *A* was that ZnO is present in the retort as the pure solid oxide so that $a_{ZnO} = 1$. For different processes, such as vacuum retorting, high pressure retorting, reduction of mixtures of zinc and iron oxides, smelting with ZnO at less than unit activity in liquid slags, etc., the basic calculations represented in Figure 4-11 are easily repeated with the appropriate pressure, stoichiometry and activity data.

The successful development of blast furnace smelting (see Chapter 6) has been made possible by still another method of avoiding reoxidation of zinc during condensation. A typical blast furnace gas might analyze, in per cent, 5 zinc, 7–10 carbon dioxide, 25 carbon monoxide, balance, nitrogen. If such a gas were fed to a condenser operating nearly at equilibrium conditions, as discussed previously, one would expect to find more zinc oxide than metallic zinc in the condenser. However, the problem was solved satisfactorily by shock chilling the blast furnace gases with a spray of unsaturated solution of zinc in liquid lead at 550 to 600°C, thus condensing and collecting the zinc without allowing time for it to react with carbon dioxide to form zinc oxide.

Reaction Rates during Carbon Reduction. The retort reduction of roasted zinc ores requires treatment and retention times totalling many hours in order to accomplish reasonably complete reduction. In the first place, reduction does not even start until the feed materials are heated above the minimum reaction temperature of about 900°C. Since the over-all reaction is strongly endothermic, additional heat must be supplied to the interior of the charge to maintain the reaction. Thus, the rate of heat transfer to the charge is one possible rate-determining step.

As pointed out previously, the carbon-reduction process may be considered as the sum of two simultaneous gas-solid reactions: $ZnO + CO \rightarrow Zn(g) + CO_2$, and $CO_2 + C \rightarrow 2CO$. In a steady retort operation, these two reactions must proceed at approximately the same gross rate, since each reaction furnishes one necessary reactant for the other reaction. Accordingly, the rate of the over-all process will be determined by whichever is the slowest or rate-determining step. Also, the composition of the effluent gas from the retort will be affected by the relative rates of the two gas-solid reactions. If the ZnO-CO reaction is more rapid, the gas composition will tend to approach that corresponding to curve *A* in Figure 4-11.

If the CO_2-C reaction is more rapid, the retort gas will correspond more closely to curve C in this diagram.

Truesdale and Waring[28] have studied the relative reaction rates involved in ZnO reduction by C, CO and H_2 under laboratory conditions. Also, many studies have been made of the factors affecting the rate of the CO_2-C reaction.[29] These results indicate that at temperatures just above the minimum reduction temperature (just above 900°C), the CO_2-C reaction is inherently slow and is rate-determining. At the higher temperatures (1100 to 1300°C) necessary to obtain commercially acceptable reaction rates, it appears likely that both reactions tend to occur at about the same rate, and that this rate is determined primarily by rates of diffusion and rates of flow of CO and CO_2 between the ZnO reaction sites and the C reaction sites. Accordingly, such factors as particle size and porosity of the ore and carbon will affect the over-all reaction rate.

Methane Reduction. The process of gaseous reduction of zinc oxide with methane was studied both thermodynamically and experimentally by Maier,[18] Doerner[30] and co-workers in the United States Bureau of Mines in the late twenties and early thirties. The principal over-all reaction is

$$ZnO(s) + CH_4(g) \rightarrow Zn(g) + CO(g) + 2H_2(g) \tag{58}$$

This reaction proceeds rapidly and with a favorable equilibrium constant at temperatures of 950°C and above, and thus at first appears very promising as the basis for a reduction process. The primary reaction is accompanied by the additional gaseous reduction reactions:

$$ZnO(s) + CO(g) \rightarrow Zn(g) + CO_2(g) \tag{59}$$

$$ZnO(s) + H_2(g) \rightarrow Zn(g) + H_2O(g) \tag{60}$$

Thus the gas phase in the retort will contain the following six constituents: Zn, CH_4, CO, CO_2, H_2 and H_2O.

An understanding of the chemistry of the methane reduction can be gained by examining first the equilibrium process. If the total pressure P and temperature T are given, there remain six unknowns: p_{Zn}, p_{CH_4}, p_{CO}, p_{CO_2}, p_{H_2} and p_{H_2O}. Thus, six independent equations are required. Three of these can be the equilibrium constants for reactions (58), (59) and (60):

$$k_{58} = \frac{(p_{CO})(p_{H_2})^2(p_{Zn})}{p_{CH_4}} \tag{61}$$

$$k_{59} = \frac{(p_{Zn})(p_{CO_2})}{p_{CO}} \tag{62}$$

$$k_{60} = \frac{(p_{Zn})(p_{H_2O})}{p_{H_2}} \tag{63}$$

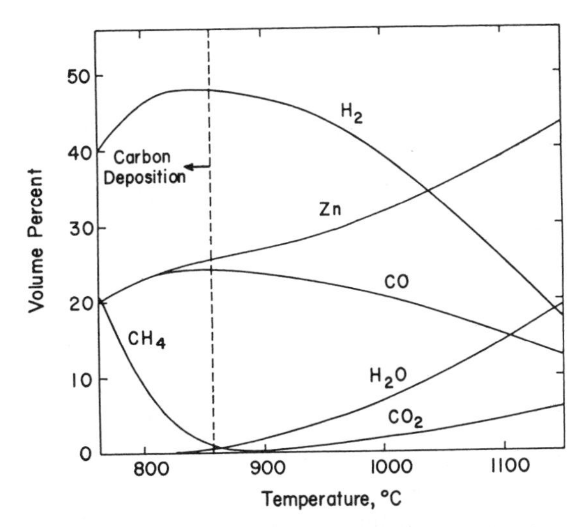

Figure 4-13. Equilibrium gas compositions for methane reduction of zinc oxide at 1 atmosphere.

A fourth relation is the total pressure equation:

$$P = p_{Zn} + p_{CH_4} + p_{CO} + p_{CO_2} + p_{H_2O} + p_{H_2} \tag{64}$$

The other two needed relations are based on the stoichiometry of the process and are readily derived from the element balances for the process, assuming ZnO and CH_4 as pure starting materials:

$$p_{Zn} = p_{CO} + 2p_{CO_2} + p_{H_2O} \tag{65}$$

$$p_{H_2} + p_{H_2O} = 2p_{CO} + 2p_{CO_2} \tag{66}$$

Results of simultaneous solution of these equations for temperatures from 800 to just over 1100°C at a total pressure P of one atmosphere are plotted in Figure 4-13. At temperatures below 860°C, the calculated equilibrium gas mixtures will tend to deposit carbon (that is, the equilibrium constant for the reaction $C + CO_2 \rightarrow 2CO$ is exceeded). Large scale methane reduction tests did show that carbon deposition could be troublesome under some operating conditions.[30] Above about 850°C, the calculations show that the methane reacts almost completely, as the methane contents of the equilibrium gases are under 1 per cent and too small to show on the diagram. Also, it can be seen that the Zn, CO_2 and H_2O contents of the gas all increase with increasing temperature owing to the increasing importance of reactions (59) and (60).

As is the case for the carbon reduction discussed previously, cooling of the gaseous reduction products from methane reduction results in reversal of some of the reduction reactions and formation of oxide. However, the gases produced by methane reduction of zinc oxide are lower in Zn vapor and in addition contain substantial percentages of H_2O. As a result, the extent of reoxidation of zinc on cooling is much greater than in the carbon-reduction process. The equilibrium data in Figure 4-13 afford the basis for estimates of the extent of oxide deposition between the active reduction zone and the condenser. For example, if the retort is operated at 1000°C to obtain reasonably rapid reduction, the corresponding equilibrium gas percentage composition given by Figure 4-13 is 31.5 Zn, 21 CO, 2 CO_2, 39.5 H_2 and 6.5 H_2O. If this gas mixture is cooled to 850°C, the final percentage equilibrium gas composition, ignoring possible carbon deposition, is substantially 25 Zn, 25 CO and 50 H_2, with negligible percentages of CO_2 and H_2O. A simple stoichiometric calculation shows that such a change in gas composition between reduction and condensation corresponds to oxide deposition of 27 per cent of the original zinc, leaving only 73 per cent to condense as liquid metal.

If the retort gases are withdrawn from the reduction zone and from contact with ZnO and then without cooling are reacted with additional methane, the CO_2 and H_2O contents can be reduced to the point where cooling and condensation of metal occur without oxide formation. This procedure corresponds roughly to the use of a carbon blanket in a standard batch retort. Unfortunately, however, the reactions of CH_4 with CO_2 and H_2O are relatively slow, and a catalyst is necessary to obtain usable rates of reaction. Carbon deposition may foul the catalyst under some conditions. The prevention of zinc reoxidation by catalytically reacting the hot retort gases with methane was successful in the small scale tests reported by Doerner.

Fuming Processes. A number of useful and industrially important processes are based on reduction of zinc oxide with carbon or with reducing gases to produce zinc vapor, followed by oxidation of the zinc vapor to produce a zinc oxide fume (see Figure 4-14). The oxide produced when zinc-bearing vapors are mixed with oxidizing gases is of very small particle size, like a smoke, and thus is carried out of the furnace with the flue gases for subsequent separation and recovery. These fuming processes furnish relatively simple and economical means of concentrating ZnO or separating ZnO from low-grade or impure materials such as residues, oxide ores and slags.

The basic physical chemistry of the reduction-oxidation reactions was discussed in preceding sections of this chapter. The relationships among gas composition, partial pressures, reduction temperatures and stoichiometric variables for the reduction steps of the various fuming processes are essentially the same as those for the processes of carbon reduction and gaseous reduction. That is, the compositions or possible composition ranges

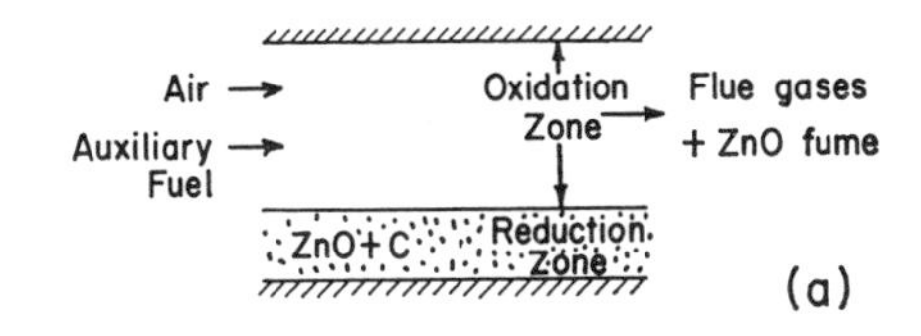

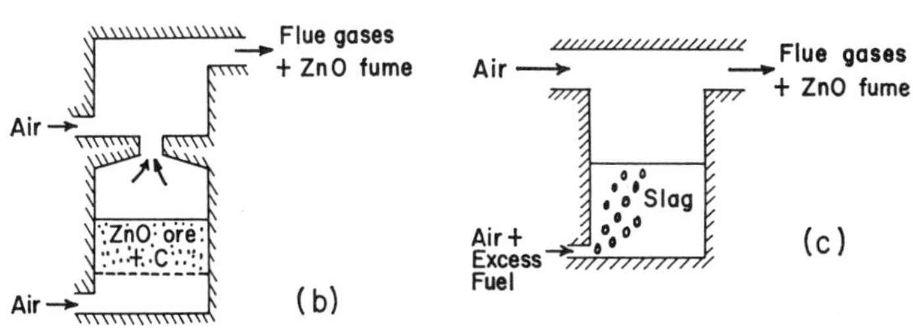

Figure 4-14. Fuming processes (schematic). (a) Waelz process (b) zinc oxide manufacture (direct process) (c) slag fuming.

for the gases leaving the reduction zone under various operating conditions are readily estimated by assuming that these gases leave the reduction zone at equilibrium with the ZnO and/or C, and then solving the appropriate equilibrium constants simultaneously with the total pressure equation and with other equations which represent the stoichiometry or materials balance for the process. When both fuel and air are fed into the reduction zone (Figures 4-14b and 4-14c), the gases leaving the reduction zinc will contain Zn, N_2, CO, CO_2, H_2 and H_2O in various proportions depending on the composition of the fuel and on the ratios of fuel and air to zinc oxide. In general, the zinc contents of these gases will be considerably lower than zinc contents of retort gases. However, since these fuming processes do not involve condensation of liquid zinc and since reoxidation of the zinc vapor is sought rather than avoided, the fuming processes are chemically more flexible than reduction processes designed to yield a metallic zinc product. Another characteristic of the fuming processes shown in Figure 4-14 is that they involve no external supply of heat. In effect, the fuels serve both as reducing agents and as sources of heat. Hence, the heat balance furnishes still another quantitative relation which must be satisfied in calculating or analyzing the operating conditions for these processes.

An excellent thermodynamic study of the slag-fuming process was published recently by Bell, Turner, and Peters.[31] Their calculations followed the general principles discussed above and demonstrated the value of calculations of this type in understanding and predicting the effects of such operating variables as fuel composition, oxygen enrichment of the combustion air, air preheating, blowing times and others. The reported

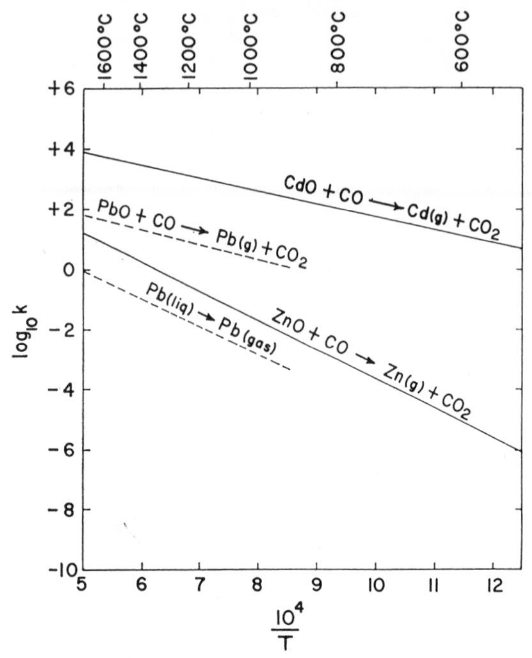

Figure 4-15. Gaseous reduction of cadium oxide, lead oxide and zinc oxide.

correlation between calculated and observed performance affords substantial support of the assumption underlying all these calculations that the gases leaving the reduction zone closely approach equilibrium with the condensed phases with which they are in contact.

Behavior of Lead and Cadmium in Reduction Processes. Since lead and cadmium are often present in zinc ores, the behavior of these elements in the various reduction processes is of considerable practical importance. Cadmium metal is more volatile while metallic lead is considerably less volatile than zinc. However, both CdO and PbO are much easier to reduce than ZnO. Figure 4-15 gives a comparison of the equilibrium constants for the reactions,

$$MeO(\text{s or liq}) + CO(g) \rightarrow Me(g) + CO_2(g) \tag{67}$$

where *Me* stands for Zn, Cd and Pb, respectively. In comparing these graphs, it is useful to regard the equilibrium constants as numerically equal to the partial pressure of the metal vapor in equilibrium with the oxide when the CO_2/CO ratio in the gas mixture is unity. The relative positions of the three curves for CdO reduction, PbO reduction and ZnO reduction, respectively, show that CdO should be reduced first, then PbO and finally ZnO. However, in processes involving both reduction and removal of the metals in gaseous form, the lead content of the gas cannot exceed that corresponding to the vapor pressure of liquid lead, which is also plotted in Figure 4-15.

The large spread between the CdO and ZnO reduction curves in Figure 4-15 indicates that it should be comparatively easy to separate zinc and cadmium by selective gaseous reduction of the mixed oxides.

REFERENCES

1. Pauling, L., "The Nature of the Chemical Bond," Ithaca, N. Y., Cornell University Press, 1948.
2. Pauling, L., "Atomic Radii and Interatomic Distances in Metals," *J. Am. Chem. Soc.*, **69**, 542 (1947).
3. Darken, L. S., and Gurry, R. W., "Physical Chemistry of Metals," New York, McGraw-Hill Book Co., Inc., 1953.
4. Evans, R. C., "An Introduction to Crystal Chemistry," Cambridge University Press, 1952.
5. Wells, A. F., "Structural Inorganic Chemistry," Oxford, Clarendon Press, 1952.
6. Goldman, J. E., (Ed.) "The Science of Engineering Materials," New York, John Wiley and Sons, Inc., 1957.
7. Dekker, A. J., "Solid State Physics," Englewood Cliffs, N. J., Prentice-Hall, 1957.
8. New Jersey Zinc Company, "Zinc Oxide Rediscovered," New York, 1957.
9. Wagner, C., and Schottky, W., "Theorie der geordneten Mischphasen," *Z. physik. Chem.*, **B11**, 163 (1930); von Baumbach, H. H., and Wagner, C., "Die elektrische Leitfahigkeit von Zinkoxyd und Cadmium-Oxyd," *ibid.*, **B22**, 199 (1933).
10. Weyl, W. A., "Atomic Interpretation of the Mechanism of Solid-State Reactions and of Sintering," *Ceramic Age*, **60**, 28 (No. 5, 1952).
11. Smithells, C. J., "Metals Reference Book," Vols. I and II, New York, Interscience Publishers, Inc., 1955.
12. Consolidated Mining and Smelting Company of Canada, "The Properties of Zinc," Trail, B. C., 1956.
13. Kelley, K. K., "High Temperature Heat Content, Heat Capacity, and Entropy Data for Inorganic Compounds," U. S. Bur. Mines, Bull. **476**, 1949.
14. Kelley, K. K., "The Free Energies of Vaporization and Vapor Pressures of Inorganic Substances," *ibid.*, Bull. **383**, 1935.
15. Coughlin, J. P., "Heats and Free Energies of Formation of Inorganic Oxides," *ibid.*, Bull. **542**, 1954.
16. Larsen, E. S., and Berman, H., "The Microscopic Determination of the Nonopaque Minerals," 2nd ed., U. S. Geol. Survey, Bull. **848**, 1934.
17. Rossini, F. D., and others, "Selected Values of Chemical Thermodynamic Properties," U. S. Bur. of Standards, Circular 500, 1952.
18. Maier, C. G., "Zinc Smelting from a Chemical and Thermodynamic Viewpoint," U. S. Bur. Mines, Bull. **324**, 1930.
19. Kelley, K. K., "The Thermodynamic Properties of Sulphur and Its Inorganic Compounds," *ibid.*, Bull. **406**, 1937.
20. Kellogg, H. H., "Thermodynamic Relations in Chlorine Metallurgy," *Trans. Am. Inst. Mining Met. Engrs.*, **188**, 862 (1950).
21. Wagner, C., "Thermodynamics of Alloys," Cambridge, Mass., Addison-Wesley Press, 1952.

22. Lumsden, J., "Thermodynamics of Alloys," London, Institute of Metals, 1952.
23. Wenner, R. R., "Thermochemical Calculations," New York, McGraw-Hill Book Co., Inc., 1941.
24. Hopkins, D. W., "The Formation of Zinc Ferrite from Zinc Oxide and Ferric Oxide," *J. Electrochem. Soc.*, **96**, 195 (1949).
25. Kato, Y., and Takei, T., "Studies on Zinc Ferrite. Its Formation, Composition, and Chemical and Magnetic Properties," *Trans. Am. Electrochem. Soc.*, **17**, 297 (1930).
26. Gross, P., and Warrington, M., "Physical Chemistry of Process Metallurgy," *Discussions Faraday Soc.* **No. 4**, 215 (1948).
27. Bethune, A. W., and Pidgeon, L. M., "Vapor Pressure of Zinc in the Reduction of ZnS by Cu and Fe," *Trans. Am. Inst. Mining Met. Engrs.*, **197**, 804 (1953).
28. Truesdale, E. C., and Waring, R. K., "Relative Rates of Reactions Involved in Reduction of Zinc Ores," *ibid.*, **159**, 97 (1944).
29. Lowry, H. H., "Chemistry of Coal Utilization," Committee on Chemical Utilization of Coal, National Research Council, Vols. I and II, New York, John Wiley and Sons, Inc., 1945.
30. Doerner, H. A., "Reduction of Zinc Ores by Natural Gas," *Trans. Am. Inst. Mining Met. Engrs.*, **121**, 636 (1936).
31. Bell, R. C., Turner, G. H., and Peters, E., "Fuming of Zinc from Lead Blast Furnace Slag," *ibid.*, **203**, 472 (1955).

CHAPTER 5

THE TREATMENT OF ORE OR CONCENTRATE FOR SUBSEQUENT PROCESSING BY LEACHING OR PYROMETALLURGICAL METHODS

The Concentration of Zinc Ores

ROLLIN A. PALLANCH*

Consulting Mill Metallurgist
Salt Lake City, Utah

More than 90 per cent of the present zinc production of the world comes from ores which contain their zinc mineral in the sulfide form. The pure zinc sulfide mineral sphalerite (ZnS) assays about 67 per cent zinc and 32 per cent sulfur. Sphalerites normally contain impurities such as cadmium, indium and gallium. Iron usually replaces some of the zinc isomorphously, ranging all the way from a trace to 15 per cent iron and even higher. Occasionally, manganese also replaces zinc isomorphously. Sphalerites range from white to black in color, depending upon the amount of iron in solid solution. The yellow variety (yellow jack) contains up to 1 per cent iron, the brown up to 3.5 per cent, the dark brown up to 8.0 per cent and the black as high as 15 per cent. Sphalerites such as the last two are usually referred to as "marmatite" (black jack). A single specimen of zinc ore may contain sphalerites of varying colors and compositions.

As a general rule it may be stated that the large bodies of zinc ore found in limestone and dolomite beds, such as in the Tennessee, Wisconsin and the Tri-State districts, contain the lighter-colored sphalerites. On the other hand, much of the zinc production of western Canada, western United States and Mexico comes from ore containing marmatite. There are of course exceptions.

Effect of Grade of Sphalerite on Beneficiation

When a particular ore contains sphalerite as ZnS it is naturally easier to produce a high grade of zinc concentrate, assaying in excess of 60 per cent zinc. At the other extreme, when the zinc mineral of an ore occurs as a dark-colored marmatite running more than 15 per cent iron the resultant grade of zinc concentrate will not assay as high as 50 per cent zinc. This has an

* Formerly Superintendent, Midvale Mill of the United States Smelting, Refining and Mining Company.

important bearing on concentrating economics, particularly in view of the fact that a zinc concentrate running as low as 40 per cent zinc is seldom marketable. It follows therefore that the greater the degree of marmatite the lower the zinc recovery economically possible, for greater sacrifice of zinc needs to be made to maintain an economic grade of zinc concentrate.

Effect of Association With Other Sulfide Minerals on Benefication

In any discussion of the milling of ores containing zinc minerals it is of course impossible to rule out other minerals with which they are associated. In the case of sulfide ores the sphalerite usually occurs in association with galena, chalcopyrite and other sulfides of copper; pyrite, marcasite, pyrrhotite, arsenopyrite, etc. Vein constituents may consist of quartz, calcite, dolomite, magnesite, feldspars, chlorites, micas and so-called "clay" minerals. Other gangue materials from the wall-rock hosts of the ore may be present: e.g., limestone, dolomite, chert, andesite, monzonite, granite, quartzite, slate and schist. Minerals are sometimes present which are desirable to beneficiate for their own sake, such as rhodochrosite, rhodonite, fluorspar, barite, magnetite and others.

Effect of Degree of Intermingling of Sulfides and Gangue on Beneficiation

Some ores are coarsely crystalline, making it possible to concentrate them at the coarser sizes by jigging, tabling, electrostatic and/or magnetic separation. There is gradation from the coarse-grained ores through greater and greater intimacy of crystallization to the extremely fine-grained massive sulfide ores that require very fine grinding, say all through 200 mesh, before concentration into products can even begin.

As a general rule the difficulty and complexity of milling an ore increases with the fineness of grinding required previous to concentration, other conditions being equal. Fine grinding promotes overgrinding, which is the reason that so many present mills use several stages of grinding and classification, dewatering, conditioning and the like in order to get the fineness of grinding desired and yet hold the adverse effects of oversliming to a minimum.

Effect of Oxidation on Zinc Sulfide Concentration

It is well known that nonsulfide minerals of zinc are not amenable to normal flotation methods. The same holds true for zinc sulfide minerals that have become sulfated or oxidized at their surfaces by weathering and the influence of adverse soluble salts. Both the grade of zinc concentrate and the zinc recovery are adversely affected. When the ore is a lead-zinc or copper-zinc sulfide ore that has become sulfated the difficulties increase. Then, a lead or copper concentrate will invariably contain excessive amounts of sphalerite which has become activated by soluble copper salts. Loss into the tailing of oxidized zinc mineral, together with the loss of activated sulfide zinc into the lead or copper concentrate, combine to reduce zinc recovery.

When an ore such as described is part of a blend of ores containing otherwise clean sulfide ore, the damage to zinc metallurgy may be out of proportion to the amount of adverse ore in the mixture being milled. It is like the rotten apple in a barrel of good apples.

Because of the wide variation of ores containing zinc, both as to the degree of association with other minerals and the degree of oxidation or sulfatization, the problem of beneficiating zinc is usually more complex, and consequently more difficult, than that of the other common nonferrous metals. Not only is the problem of physical separation of the zinc mineral more difficult, but the marketing demands are more stringent. The grade demand of zinc concentrate is much higher than that of either copper or lead concentrate. High treatment and freight charges combine to bring this about even when the smelter places no lower limit on zinc grade. A penalty for iron in excess of a certain figure, say 5 per cent, adds to the difficulty when the zinc mineral is marmatitic.

Under present conditions of technology and market, the somewhat more complex metallurgy, both in the milling and in the smelting and refining of zinc, results in a lower economic return to the miner from zinc as compared with returns from lead and copper.

HISTORICAL BACKGROUND

Zinc Concentration Before the Advent of Flotation

Prior to 1915 most of the zinc production of the United States was carried out in several eastern states, the Mississippi valley and Utah. The New Jersey Zinc Company was an important producer. Most of its production came from nonsulfide minerals of zinc such as calamine, willemite, zincite and franklinite, occurring in northwestern New Jersey, Allowing for the improvement in practice over the years, present methods described in the December, 1953 issue of *Mining Engineering* were already in vogue then. Wetherill high-intensity magnetic separators, developed by New Jersey Zinc, removed a franklinite product. Jigging and tabling followed, resulting in the recovery of willemite and zincite.[1]

Important zinc production of the United States also came from gravity mills in the Tri-State, Tennessee and Wisconsin zinc fields. In all of these, high-grade sphalerite zinc mineral existed in coarsely crystalline form in limestone and dolomite beds. It was possible to recover rich zinc concentrate (60 per cent zinc) by the use of jigs and tables. What galena there was could readily be separated from the sphalerite because of the wide divergence in the specific gravity of the two minerals (7.5, 4.0). The low pyrite content made it possible to produce high-grade coarse zinc concentrate. Much of the sphalerite found its way into the slime portion of the final tailing.

Of somewhat lesser importance from the standpoint of production, but of interest nevertheless, was the recovery of zinc mineral at the gravity mill of the United States Smelting Company at Midvale, Utah, which was treating company and custom lead-zinc-iron sulfide ores from the Bingham and other districts. This mill was unique in that it made a sphalerite-pyrite table concentrate in addition to the usual production of galena-pyrite concentrate by jigging and tabling. The sphalerite-pyrite concentrate, after dewatering, was dried in a rotary kiln dryer, following which it was screened dry into four sand sizes. Each of the four sizes was treated separately by Huff electrostatic machines. Pyrite was repelled by the electrodes, being thrown clear of the falling feed stream. The zinc sulfide particles fell vertically, being unaffected by the electrodes. Galena and copper sulfides tended to concentrate with the pyrite. Gangue minerals, being mostly unaffected by the electrodes, concentrated with the zinc product.[2]

Simple, Nonselective Sulfide Flotation

Simple flotation was of immediate help to many mills that handled sphalerite ores successfully by jig and table treatment. Flotation circuits were introduced into these mills to handle the slime portion of the ore (partly the result of previous comminution). They entailed a minimum of capital expenditure and added little to the operating cost. The flotation machines made a bulk float. Metal ratios in the product pretty much reflected that in the original ore. The higher the ratio of galena or pyrite to sphalerite in the original ore the lower the concentrate assayed in zinc, because of contained galena or pyrite.[3]

In the Rocky Mountain regions of North America, simple flotation proved of lesser benefit, for the resultant flotation product was largely a combined galena-pyrite-sphalerite product of low market value. The zinc assay usually was so high that a severe zinc penalty was exacted when the product was marketed as lead concentrate. It ran much too low in zinc content to be acceptable as zinc concentrate. The western concentrators that were handling lead-zinc sulfide ores continued making lead or lead-iron concentrate by gravity methods with respectable recoveries of gold, silver and lead. But methods were needed that would separate galena, sphalerite and pyrite from each other into concentrate products separately marketable.

Differential Flotation of Zinc Sulfide Ores

As usual, necessity became the mother of invention. The bulk flotation of the naturally floatable sulfide minerals, using coal tar and petroleum oil derivatives as collectors, and cresylic acid, hardwood creosote, turpentine and pine oils as frothing agents, gradually gave way to more selective methods. Sulfuric acid and acid salt-cake came into use to enhance the flotation of slow-floating sphalerites. This led to the discovery of the superior qualities of copper sulfate (blue vitriol) as an activator of sphalerite.

Copper Sulfate as a Sphalerite Activator. The fact that laboratory testing of Tennessee and other Mississippi valley sphalerite ores with sulfuric acid produced favorable results which could not be duplicated in the field led to the realization that the corrosive effect of the acid on brasses and bronzes used in the construction of the laboratory flotation cells was causing copper sulfate to form. The result was that copper sulfate came into wide use as a sphalerite activator in this country. The experience of mills in the Broken Hill field of Australia had brought about a similar result there. Copper sulfate is the most universally used reagent in zinc flotation. Its affinity for zinc sulfide faces makes it the ideal reagent in the preparation of sphalerite for flotation.

Application of Sulfur Dioxide and Alkaline Sulfites for Sphalerite Control. In Australia, Leslie Bradford had introduced his sulfur dioxide gas method for the selective flotation of galena from sphalerite. This happened before 1918. The patent application specified the maintenance of acid pulps so as to prevent the formation of sulfites.

Successful application of sodium sulfite to lead-zinc differential flotation was accomplished in the laboratory of the Midvale, Utah, gravity lead-zinc mill of the United States Smelting Refining and Mining Company, beginning August, 1919. A United States patent for the use of alkaline sulfites covered the idea.

About a year later the Midvale Minerals Leasing Company, working on slime from the current tailing of the Midvale gravity mill, achieved the separate concentration of lead and zinc sulfides by resorting to what was referred to as "The Bradford Process in an Alkaline Circuit." Undoubtedly calcium sulfite was formed in the lead circuit by the reaction of the sulfur dioxide with the lime carbonate in the flotation pulp. Sulfuric acid was used for the reactivation of sphalerite. This destroyed the zinc-depressing effect by the driving off of sulfur dioxide gas into the air.

Cyanide as a Depressant for Pyrite and Sphalerite. Successful use of cyanide in lead-zinc differential flotation was made at the Timber Butte operation in Butte, Montana, and patented by Sheridan and Griswold. The Timber Butte ore had a high ratio of zinc to lead. A bulk lead-zinc flotation product was made which was then retreated, using zinc sulfate and cyanide to discourage the flotation of sphalerite. The use of cyanide with or without zinc sulfate has become a common practice in the differential flotation of galena and sphalerite. In most cases these reagents are used at the start of the flotation operations in order to float a lead concentrate first, before zinc activation has begun.

By itself, cyanide is an excellent reagent for the depression of pyrite, with or without the addition of lime. If sufficient cyanide is used it can be made to depress chalcopyrite effectively as well.

When oxide copper minerals are present in an ore, cyanide, like ammonia, tends to put copper into solution, causing activation rather than retardation of sphalerite.

Thiocarbanalid and Xanthate in Lead-Zinc-Pyrite Flotation. The patents covering the use of thiocarbanalid were issued to Perkins in 1921. The reagent is used quite extensively today, serving principally as a collector of galena. It is used largely in the form of "Aerofloat 31" which is "Aerofloat 25" (phosphocresylic acid)* to which an additional 6 per cent of thiocarbanalid has been added.

Keller and Lewis of the Minerals Separation Company, developed the use of potassium ethyl xanthate as a collector for sulfide minerals in 1923. The first patents were not issued until 1925. Later patents have been issued covering the use of the higher xanthates of sodium and potassium. The various xanthates have come into extensive use.

Other Contributors to Flotation Technology. Many other valuable contributions to differential flotation technology were made between the years 1915 and 1923. Much research and testing was done by the staffs of operating companies. Testing and consulting companies, such as the General Engineering Company, also contributed improvements. The metallurgical staff of the Consolidated Mining and Smelting Company of Canada, Limited, did extensive research on the ore from the Sullivan mine in British Columbia, soon to become one of the foremost producers of lead and zinc in the world.[4]

Differential Flotation of Mixed Sulfide Ores. The advances in flotation technology led to tremendous expansion of the copper-lead-zinc industry by adding greatly to the productive ore reserves of the world. The zinc content of many reserves of mixed sulfide ores had now become an asset instead of a liability. The mines, mills and smelters of the operating companies were greatly stimulated. Many important new producers of copper-zinc and lead-zinc ores appeared. The shipping of many lead-

* Phosphocresylic acid, the first of the "Aerofloats" of the American Cyanamid Company, was developed and patented by F. T. Whitworth at the Utah Copper concentrators at Magna and Arthur, Utah.

zinc custom ores was encouraged by Salt Lake valley mills such as International's Tooele, United States Smelting's Midvale, and Combined Metals' Bauer plants (1924).

The Sunnyside Mill, Eureka, San Juan County, Colorado. The now-dismantled mill of The Sunnyside Mining and Milling Company, a subsidiary of the United States Smelting Refining and Mining Company, is of historic interest. Beginning operation in April, 1918, it was one of the first to produce separate lead and zinc concentrates from a single ore by flotation methods without previous gravity concentration. Built on a steep, hillside slope (elevation 10,000 ft), it received ore by aerial tram from the Sunnyside mine three miles away (elevation 12,000 ft). The ore analysis varied from 3.0 to 5.5 per cent lead, from 4.5 to 7.5 per cent zinc, with 0.3 to 0.5 per cent copper and about 3.0 per cent iron. It also carried some gold and silver. The base metals occurred as sulfide minerals, with some slight oxidation. A quartz-rhodonite gangue made the ore very hard. Separations could be made as coarse as 50 per cent through 200 mesh. The initial rating of 500 tons daily was gradually increased to 1000 tons, chiefly by coarsening the grind.

Tables were used in the Sunnyside flowsheet for pilot operation and to recover some slight additional values from the flotation tailing in a pyritic table concentrate. Minerals Separation flotation machines were used.

Up to 1923 the lead reagents were soda ash, sodium silicate, coal tar collecting oils, with pine oil and cresylic acid as frother. Acid salt-cake or sulfuric acid was used for zinc activation. Hardwood creosote and pine oil were used as zinc frothers. In 1923 sodium sulfite displaced sodium silicate. Potassium ethyl xanthate came into use in March, 1924, serving as a collector in both the lead and the zinc circuits. At the same time copper sulfate displaced sulfuric acid as the zinc activator.

CURRENT CONCENTRATION PRACTICE

Preconcentration by Heavy Media Separation of Sulfide Ores

Bull jigs had been used in Mississippi valley operations for many years to treat the coarse fractions of the originally crushed ore ($-\frac{3}{4}$ in.) for preconcentration purposes. The coarse jig tailing went directly to waste so as to eliminate further expensive crushing and grinding of very low-grade material. This led eventually to the development of heavy media separation, often referred to as "Sink-and-Float," as a substitute for coarse jigging. Plant tests of heavy media methods were conducted as early as 1935.

Heavy media separations rely upon the difference in specific gravity of the metallic and the gangue minerals within the ore treated. The medium usually consists of a thickened water suspension of either fine galena or ferrosilicon particles, the density of which is held at some predetermined figure, ranging from 2.6 to 3.0, depending largely upon the specific gravity of individual components of the ore. The tank containing the heavy medium suspension receives the sink-and-float feed, usually the plus $1/4$ to $5/16$ in. fraction of the ore after it has been crushed to $-1\frac{1}{2}$ in. size. All chunks of the ore averaging lower specific gravity than the top density of the heavy medium rise to the surface as float, to be removed for further disposal.

Washing of the float and the sink products has to be done in order to remove adhering galena or ferrosilicon, to be reclaimed later. The sink product and the original screen undersize go to further processing for the recovery of the concentrate products. The float reject is usually extremely low in grade. Uses are found for the material such as mine stope fill, concrete aggregate, road ballast and the like.

Because of the relatively high proportion of the original ore rejected as float, there is some sacrifice of recovery, in inverse proportion to the original grade of the ore. In all cases where the process is used, this sacrifice of values is justified by decreased grinding cost, increased mill capacity, market for the coarse reject, etc.

Important examples of heavy media preconcentration of zinc sulfide ores appear in the flow sheets of the Mascot concentrator of the American Zinc Company of Tennessee and the Central Mill of the Eagle-Picher operation near Miami, Oklahoma. Later installations of the process were made at the all-flotation operations of the West Mill of Bunker Hill and Sullivan at Kellogg, Idaho, and the Sullivan Concentrator of the Consolidated Mining and Smelting Company at Kimberley, British Columbia.

Ferrosilicon washings are reclaimed by the use of magnetic separators. Galena washings are recovered by flotation.

Grinding and Classification

Grinding sections of the many mills treating zinc sulfide ores vary greatly in methods and practice. It would be surprising if this were not so, because of the widely differing conditions both as regards tonnage and ore characteristics.

Single-Stage Grinding. Single-stage grinding may be defined as a wet, closed circuit grind in which a rod or a ball mill discharges to its classifier and the classifier returns its sand product to the feed of the mill. The classifier overflow is the finished grind. When crude ore is being ground, the previously-crushed ore is normally introduced directly into the mill, with dilution water being added to the mill and classifier in the optimum proportions. When material from a previous wet treatment within the plant is being ground it is sometimes introduced into the classifier, so as to avoid excessive dilution within the grinding mill.

Single-stage grinding is usually employed in small tonnage mills, e.g., up to 200 tons, where low capital expenditure and operating cost are paramount. The flowsheet is greatly simplified thereby. Single-stage grinding is also employed, even with larger tonnages, when it is possible to do satisfactory work at a relatively coarse size of flotation feed, e.g., 50 per cent −200 mesh and coarser. If the ore has been previously crushed to a relatively fine size, or if it is very easily ground, single-stage grinding often suffices. A few plants employing fine grinding get along with single-stage grinding by greatly diluting the classifier overflow. This of course requires high grinding capacity because of the large per cent of circulating sand return.

Two-Stage Grinding. When ore is ground by two mills in series it is referred to as "two-stage grinding." There are a number of variations of this method:

1. *Series grind with series classification.* The primary mill (or its classifier) receives the initial feed. The primary classifier receives the mill discharge. It returns sand to the mill and overflows the preliminarily-ground pulp at high solids, e.g., 50 to 60 per cent. This pulp is laundered, pumped or conveyed to the secondary classifier. The overflow of the secondary classifier is kept at lower solids, e.g., 30 to 40 per cent, to allow the coarse sand to drop and find its way as sand return to the secondary mill. The secondary mill discharge goes to the secondary classifier. The secondary classifier overflow is the finished grind.

2. *Series grind, with parallel classification.* The primary mill (or its classifier) receives the original feed. The primary mill discharge is divided between the feeds of the primary and secondary classifiers. The primary mill receives the sand return from the primary classifier and the secondary mill receives the sand return from the secondary classifier. The secondary mill discharge goes to the secondary classifier. Both classifier overflows are the finished grind, the dilution being adjusted to the fineness of grind desired. The fact that both classifiers overflow finished grind adds to classifier capacity. If the grinding mills have reserve capacity, this means greater grinding capacity as well, for the bottleneck of too little classifier capacity has been widened.

In this case it may be possible to substitute one larger classifier to take over the entire classifier function. The classifier sand return would then be divided between the two mills and the primary mill discharge would feed to the classifier. The overflow would be the finished grind.

3. *Open circuit primary grind, with secondary mill in closed circuit.* In this case the primary mill discharge goes to the classifier (or classifiers)

with no sand being returned to the primary mill. The secondary mill receives the total sand return and its discharge returns to the classifier. The classifier overflow is the finished grind. In an exceptional case the discharge of the primary open circuit mill may go directly to the feed of the secondary mill rather than to the classifier.

The above three procedures may be, and often are, varied or used in combination. For instance, some of the original feed may be introduced into the secondary mill. In case 2 this idea may be stretched to the point of having two parallel single-stage grinding units, with classifier overflows joining as flotation feed. It then becomes possible to grind two ores separately that require different kinds of grinding and still mix the ores for subsequent flotation. Reagent treatment in the grinding section can then be varied to suit the two conditions.

Rod Mills as Primary Mills with Ball Mills as Secondary Mills. The practice of using a rod mill for the initial grinding, with a ball mill to do the final grinding, is finding favor in the milling of some lead-zinc and copper-zinc sulfide ores. The line contact of the rods effectively reduces the whole of the feed to a sand size with a minimum of sliming. Rod mill discharges resemble the undersize of an 8 or 10 mesh screen, indicating the screening effect of the mass of parallel rods within the mill. The line contact protects the ore particles from being overslimed. This is an advantage in lead-copper-zinc flotation metallurgy where it is desirable to avoid the deleterious effect of overgrinding. Rod mills are essentially open-circuit in their nature because of the minimum of oversize in the discharge. When in closed circuit with a classifier, the percentage of circulating sand return is considerably less than that of a ball mill circuit on similar duty.

The very fact that rod mills deliver a sandy discharge with a minimum of slime works to disadvantage in the case of ore particles that require a fine grind. For this reason ball mills usually serve better as the secondary mills of a grinding circuit. The material that requires the impact, point-contact type of grind for proper attrition then drops in the secondary classifier to return as feed to the secondary ball mill. The oversize normally present in the discharge of primary ball mills has been eliminated by the substitution of the primary rod mill grind. This greatly reduces the amount of excessively coarse material going to the secondary ball mill circuit. The ball mill is then better able to devote itself to the grinding down of the more tough and resistant sandy particles, which succumb to the hitting effect of the balls within the mill.

Sink-and-Float Effect of Classifiers. Mechanical classifiers in closed circuit with grinding mills have a sink-and-float effect. This is particularly true when using rod mills for the primary grind, the discharges of which contain little material coarser than 10 mesh. Much of the light, relatively barren gangue material then escapes over the finished classifier overflow without ever re-entering a mill. As it is fine enough and light enough to pass over the classifier weir it also is suitable as flotation feed. Thus it finds its way to the flotation tailing at a relatively coarse size, usually much coarser than the sulfide minerals present in the flotation feed.

The heavier the mineral the greater is its tendency to return to the mill, even at a relatively fine size. Galena grinds down very easily. Excessive grinding of galena is usually averted by its flakiness. The introduction of collecting reagents such as naphthalene (in coal tar derivatives), thiocarbanalid, xanthate, "Minerec," etc., reduces the sliming effect upon galena by causing it to float at the classifier pool, to be carried off with the overflow. Pyrite, which has a specific gravity of 5.0 and is usually quite resistant to grinding, tends to make up the major portion of a classifier return when working on lead-zinc-pyrite ore. Sphalerite, having a specific gravity of 4.0 and being only moderately resistant to grinding, usually is relatively minor in the classifier return. The latter mineral tends to pass on to the flotation feed in a somewhat coarser condition than do the other sulfides. This is an advantage in lead-zinc and copper-zinc differential flotation because the zinc mineral is then easier to keep down in lead or copper flotation. Usually, however, the coarser sphalerite particles make a scavenger zinc middling float necessary, with the incorporation of a regrinding step before returning the scavenger zinc middling to the main zinc circuit. Many mills have resorted to zinc middling regrind practice.

Differential Grinding. The term "differential grinding" is here applied to those cases in which a flotation step is placed between the primary and the secondary grind. The overflow of the primary classifier passes to the rougher lead or copper circuit at the coarser size. After a portion of the rougher concentrate has been removed the tailing returns to the classifier of the secondary grinding circuit for additional grinding before passing on to further flotation. In a sense this could be called a regrinding operation, but it is preferable to refer to it as differential grinding in order not to confuse the term with that of the regrinding of materials within the flotation circuits.

Many mills employ a differential grind. The Flin Flon, Manitoba, concentrator of the Hudson Bay Mining and Smelting Company early adopted the practice of floating at a preliminary size of approximately 60 per cent −200 mesh to remove a talc concentrate, followed by a portion of the rougher copper concentrate. The preliminary flotation tailing then went on to further grinding to about 90 per cent −200 mesh for the final copper rougher flotation, followed by the zinc rougher flotation. Another mill employing differential grinding is the East Mill of the Pend Oreille Mines and Metals Company at Metaline Falls, Washington. The classifier of the primary grind overflows at about 50 per cent −200 mesh, which passes on to the lead flotation roughing circuit. The lead rougher tailing goes to the classifier of the secondary grinding circuit, the overflow of which is about 74 per cent −200 mesh. This overflow then passes on to the zinc rougher flotation circuit.[5]

"Unit Cell" practice, which introduces a flotation step between a grinding mill and its classifier, is another illustration of a differential grind.

Multiple-Stage Grinding. By multiple-stage grinding is meant the grinding in more than two stages, with or without a flotation step between any of the stages. Some mills employ tertiary and even quaternary grinding. The grinding flowsheet of the Sullivan Mill of the Consolidated Mining and Smelting Company provides a good example. Besides the original rod mill open circuit grind this plant employs primary, secondary and tertiary closed circuit ball mill grinding. Actually there are four grinding stages, although the rod mill job is classed as "fine crushing" at Sullivan.[6]

Blending and Segregation. When mills handle ores from many different sources, such as is the case where many custom and lease ores are treated, attention has to be given to proper blending of ores that work well together and proper segregation of ores that tend to "fight" each other. Blending can be very valuable in the averaging-up of ores of widely varying grades, resulting in a much improved feed, with the hills and valleys of metal grades well ironed out. Very rich sulfide ores often cause poor conditions at the classifiers, because of the unduly high specific gravity. Solids control at the classifier overflow is then apt to be difficult. Low-grade ore, with clean gangue, mixed with an ore of this kind makes for improved grinding practice.

Segregation is of equal importance in the case of ores that do not work well together. It is not wise to mix two ores of widely-differing grain size. Grinding that suits the one cannot possibly be right for the other. Flotation rates also make it wise to keep some ores separate. An ore that has an easy-floating sphalerite which is difficult to depress should not be mixed with one that has slow-floating galena, requiring a hard pull at the lead machine for proper recovery. Unfortunately, the same ore will sometimes have fast-floating sphalerite and slow-floating galena, as in the case of some copper-containing lead-zinc sulfide ores.

Probably the best reason for keeping two ores separate is in the case of one having interfering soluble salts, which not only vitiate the ore itself but harm any clean ore with which it may be mixed.

Flotation of Zinc-Containing Sulfide ores

Much of the success in the flotation of ores hinges upon previous proper grinding and classification. Following this, the wise use of conditioning intervals, maintenance of the best specific gravity of pulp by proper thickening or dilution, suitable flotation equipment with correct flotation intervals and flow, judicious use of regrinding equipment when helpful, and effective reagent introduction and control all combine to promote metallurgical efficiency. Some point must be reached that economically balances metallurgical efficiency with the expense of operation.

Practice at the various mills handling zinc sulfide ores varies greatly, for each plant has problems which are peculiarly its own. In the case of operations handling clean zinc sulfide ores which contain insignificant amounts of other sulfides, there is really only one basic job to be done, that is, to recover the sphalerite in as high a grade of concentrate as practicable. In the case of operations handling mixed lead-zinc or copper-zinc sulfide ores, two basic things have to be done to the sphalerite. First, the mineral has to be kept from reporting in lead or copper concentrate. Second, the mineral has to be recovered in as high a grade of concentrate as practicable. This, of course, is where differential flotation comes in.

As Gaudin so clearly points out in a recent contribution,[7] most substances are naturally water-avid, or readily wetted. This is true of particles of ground metallic ore. None of the particles will attach themselves to a water bubble except in the presence of some natural floater, such as a hydrocarbon. However, when such a hydrocarbon is introduced in very small amounts in a water slurry of the ground ore, it tends to attach itself to galena, chalcopyrite and other sulfide minerals, causing such particles to become water repellant. Introduced air causes bubbles to form to which the water-repellant particles cling and are thereby floated to the surface of the receptacle. Nonactivated sphalerite is an exception. The hydrocarbon does not attach itself to this mineral, so it remains water-avid. However, the presence of a very minute amount of soluble copper causes sphalerite activation by the attraction to it of the copper ion. It then behaves like the galena and chalcopyrite, floating with them when a hydrocarbon has been introduced. The sphalerite then needs to be rendered water-avid again if it is to be depressed.

Zinc Ion Responsible for Sphalerite Depression. As stated before, sphalerite does not respond to hydrocarbon collectors unless previously activated by the copper ion. For this reason the zinc mineral of many sulfide ores is naturally depressed. This is particularly true in the case of ores that have very little or no copper content. What sphalerite does come up is very apt to have been brought up mechanically. Examples of this are the high-grade sphalerite at the Central Mill of Eagle-Picher and the much lower-grade marmatite of the Broken Hill operations in Australia.

There exists such strong affinity between zinc and copper ions that the presence of only a very small amount of oxide or soluble copper in a lead-zinc or copper-zinc sulfide ore may readily bring about excessive activation of sphalerite. Zinc salts then are needed, in addition to the amount naturally present in the ore, to counteract such activation. Zinc sulfate is the normal vehicle for introducing the zinc ion, although zinc chloride and other soluble compounds of zinc have been effectively used. In the case of highly-pyritic ores zinc sulfate, whether natural or introduced, has a tendency to encourage pyrite to float into both the lead concentrate and the succeeding zinc concentrate. The introduction of cyanide or sulfite relieves the condition. The latter reagents are often used in the same circuit. Each can do things that the other cannot do as well. If an ore contains calcium sulfâte and oxidized copper mineral it is better to introduce sulfite into the initial grinding circuit so as to convert the calcium sulfate to calcium sulfite. This counteracts the tendency to form copper sulfate with the oxide copper, thus lessening the danger of sphalerite activation. Cyanide can then be introduced later to help in depressing the pyrite. Reaction of the sulfite and the cyanide with zinc sulfate presumably results in the formation of zinc sulfite and zinc cyanide, both depressing sphalerite.

Copper Ion Responsible for Sphalerite Activation. The use of copper sulfate for sphalerite activation is universal in sphalerite flotation. The tendency of copper sulfate to activate pyrite makes it necessary to use pyrite depressants, such as hydrated lime, cyanide and/or sulfite, in the same circuit when considerable pyrite is present in the ore. A conditioning period of from one to thirty minutes, as the case may be, is usually required ahead of zinc flotation. The copper ion is attracted to the surfaces of the sphalerite particles, making them susceptible to later flotation by xanthate or other collector. Presumably, some of the copper sulfate converts to the hydrate, cyanide or sulfite, depending upon the reagent used for pyrite depression. Very effective zinc conditioning can be done when copper sulfate and sodium sulfite solutions are introduced together, allowing copper sulfite precipitate to form before entering the pulp stream. But it is usually better to allow the two reagents to search each other within the pulp after each reagent has been introduced separately.

A seeming paradox is the fact that so-called "sphalerite depressants" may also act as sphalerite conditioners. This is not surprising, for the fact that a sphalerite particle has been depressed may mean that it has some zinc compound at its surface which could attract the copper ion when copper sulfate is present or introduced.

The question of whether a pyrite concentrate should be floated following zinc flotation has nothing to do with zinc metallurgy, except as it might add to the possible over-all profit of the operation. This is also true of the question of whether the copper and lead minerals of a lead concentrate should be separated into their component products by a retreatment operation. Pyrite flotation following zinc flotation may be advisable from the standpoint of additional gold, silver and lead recovery or from the standpoint of possible sulfuric acid production and iron beneficiation. Lead-copper separation may be advisable if the separation is easy to make, or if the copper content of the lead concentrate is too great to ignore. Only test work and experimentation can determine this.

In a few mills handling lead-copper-zinc ores, a copper flotation circuit precedes lead and zinc flotation. This is the case at the Lake George operation at Captain's Flat, New South Wales, Australia.[8]

Other mills, such as the concentrators of the Hudson Bay Mining and Smelting Company at Flin Flon, Manitoba, and the Quemont Mining Corporation at Noranda, Quebec, are strictly copper-zinc operations, with no lead to beneficiate.

Flotation Reagents

Many reagents, other than those already mentioned, are used in the treatment of zinc-containing sulfide ores. Hydrated lime is employed in many operations, either to neutralize acidity or to maintain high alkalinity. Where lime is not suitable, sodium carbonate is often used. Sodium bichromate, sodium bicarbonate, sodium sulfide and sodium silicate are resorted to for specific effects. Numerous frothing alcohols, such as methyl isobutyl carbinol, pentasol alcohol, "B23" and "Dowfroth" have come into extensive use. The various "Aerofloats" of the American Cyanamid Company and the several "Minerec" compounds of the Minerec Corporation are frequently employed.

The ore itself may introduce chemical compounds into the flotation pulps. The mill water used may have compounds in solution which have a bearing upon the problem. High lime and magnesium bicarbonates cause high hardness that may need to be counteracted. Acid waters from mine workings may require neutralization.

It is a mistake to regard reagents as competitive; rather they should be looked upon as tools. No one reagent can effectively do the whole job when the problem is at all complex. There is some overlapping of functions, but each type of reagent has its own characteristics and effects.

Very few chemical reagents retain their identity after they come into contact with the pulp stream. This, of course, does not refer to conditions where a high excess of a compound is required to furnish the proper "climate" for good flotation, such as in high lime circuits where a high *p*H is desirable.

Use of Xanthates. It is well known that soluble xanthate compounds form metallic xanthate precipitates when they come into contact with metallic salts. Zinc xanthate probably forms in most lead circuits and copper xanthate in most zinc circuits where xanthate is the collector used in both circuits. It is possible to very effectively accomplish good galena flotation in the lead circuit with the use of zinc xanthate precipitate at the classifier overflows, provided that sufficient time interval exists to allow the zinc xanthate to pervade the flotation feed. On the other hand, copper xanthate precipitate is ineffective as a zinc flotation collector, as the xanthate must be attracted to the copper ion that exists on the face of a conditioned particle of sphalerite. If the xanthate is already present as copper xanthate there is little such attraction.

Use of Bichromates. Sodium and potassium bichromate have long been known as very effective depressants of galena. Were it not for the fact that galena is naturally much more floatable than sphalerite, making it "against nature" to float sphalerite ahead of galena, bichromate might well have become more prominent in differential zinc-lead flotation work. Another reason for floating the lead concentrate first is that the first concentrate to be floated tends to carry the greater part of the precious metal values. Lead and copper smelter schedules pay better for gold and silver than do zinc smelter schedules.

The galena depression by bichromate might be attributed to oxidation, except that other oxidizing reagents do not have a like effect. The forming of a film of lead chromate on the galena face undoubtedly has much to do with the success of the reagent as a galena depressant. Chromium compounds seem to have a quickening effect upon sphalerite flotation as well, adding to the differential between sphalerite flotation and galena depression.

Regrinding of Flotation Middlings

In order to avoid excessive grinding of the entire pulp of a flotation circuit, it is often possible to accomplish successful flotation at a more moderate grind by depending upon a scavenging pull to make the final recovery. If such a scavenged middling is returned directly to the circuit there is apt to be a build-up of coarse circulating material. Subjecting the scavenged middling to a regrinding operation before it is returned to the main circuit avoids this difficulty.

It is quite a common practice to regrind scavenged lead or copper middling, either by sending it to the classifier or cyclone of a separate ball mill regrind circuit, or to one of the final classifiers of the original grinding circuit. Sometimes the lead or copper cleaner tailing contains enough coarse material for like treatment.

Scavenged zinc middling from the zinc circuit of an operation is usually pumped to the classifier or cyclone of a separate zinc-regrinding circuit. The classifier sand or the cyclone underflow goes to a zinc-regrind ball mill whose discharge returns to the classifier, or joins the classifier or cyclone overflow returning to the main zinc circuit.

Retreatment of Flotation Concentrates

The term "retreatment" as here used means only those cases in which an otherwise final flotation product is retreated for separation into two final products, with no return of middling or of cleaner tailing to the main flotation circuits. The retreatment concentrate is one product; the retreatment tailing is the other.

When treating the more refractory types of complex lead-zinc or copper-zinc sulfide ores an impasse is often reached, making it necessary to compromise flotation results. A certain amount of easy-floating sphalerite may have to be allowed to escape into the lead or copper concentrate to effect anything like a satisfactory recovery of lead or copper. On the other hand it is equally imperative that this loss of zinc be kept within bounds by permitting some of the slower-floating galena or copper mineral to report in the succeeding zinc concentrate. The proper balance is achieved when the highest economy results.

In cases such as this the situation can often be relieved by resorting to separate two-way separation of otherwise finished concentrate by retreatment flotation. As no material is returned to the main flotation circuits, there is no risk of endangering previous flotation work. Decoppering, deleading and dezincing retreatment circuits, when used to advantage, raise metallurgical efficiency.

Bichromate is the reagent usually employed when galena depression is desired. Hydrated lime and cyanide are used to depress pyrite. Cyanide is used to depress chalcopyrite and sulfurous acid to activate it. Zinc sulfate and cyanide are used to depress sphalerite.

Retreatment of lead concentrate to produce a copper concentrate is done at the Buchans Mining Company operation at Buchans, Newfoundland and at the mill of the Idarado Mining Company near Ouray, Colorado. In the first case a copper concentrate is floated, using sulfur dioxide gas to promote chalcopyrite flotation and sodium bichromate to depress galena. In the second case sodium cyanide is used to depress the chalcopyrite, permitting the galena to float.[9] This is also done at Buchans. In this case the retreatment tailing is the copper concentrate.

At the Sullivan mill of the Consolidated Mining and Smelting Company the lead concentrate is dezinced in a retreatment operation that employs sodium bichromate to depress the galena, copper sulfate to activate the zinc and lime to depress pyrite and pyrrhotite. In this operation it is considered important to hold the temperature at about 40°C and the dilution at about 30 per cent solids.

At the Midvale mill of the United States Smelting Refining and Mining Company zinc concentrate has been subjected to separate retreatment since 1942, for deleading of the zinc concentrate from the more difficult ore. Sodium bichromate is used for the purpose, the tailing of the operation joining the lead concentrate of the mill. At the Parral and Santa Barbara mills of the American Smelting and Refining Company in Chihuahua, Mexico, deleading of zinc concentrate is accomplished by retreatment with zinc sulfate and sodium cyanide to depress the zinc. The floated material is sent to lead concentrate.[10]

Concentration of Oxidized Zinc Ores

The nonsulfide minerals of zinc usually listed are:

Smithsonite, $ZnCO_3$	52.1%	zinc
Calamine and Hemimorphite, $Zn_2(OH)_2SiO_3$*	54.2%	"
Willemite, Zn_2SiO_4	58.0%	"
Hydrozincite, $ZnCO_3Zn(OH)_2$*	60.5%	"
Franklinite, $(Zn, Mn)Fe_2O_4$		

* Other formulas may be assigned to these compounds. See for example, Chapter 3 (Ed.).

Gravity and Magnetic Concentration. In the concentration of zinc ores whose zinc minerals are nonsulfide, the practice of New Jersey Zinc Company at its Franklin and Sterling operations is of interest. An excellent account by Pellett, with accompanying flowsheets, appears in the New Jersey Zinc issue of *Mining Engineering*, December 1953, from which we quote:

"Necessity for removal of the franklinite and various still less magnetic silicate minerals led to the invention and development of the Wetherill high intensity magnetic separators, and in 1896 to construction of a 150-ton mill at Franklin. Shortly thereafter, jiggling and tabling equipment was installed for discarding the calcite gangue. This was followed by the production of willemite-zincite concentrate suitable for metal production. With many improvements, the same separating principles have been employed on the present mill site at Franklin since 1901, and at Ogdensburg for Sterling ore since 1915."

Flotation. The opening paragraph of an informative paper by Rey *et al.*[11] reads as follows: "Concentration of oxidized copper and lead ores by flotation has been practiced for 30 years, but flotation of oxidized zinc ores has remained unsolved until a few years ago."

Smithsonite responds to fatty acid flotation but the reagent is not selective enough, floating gangue constituents as well. This mineral also can be floated with the higher xanthates or the higher mercaptans. When such ore is sulfidized, treated with copper sulfate and conditioned, it sometimes responds to flotation by ordinary xanthate.

Quoting again from the paper, they state: "Bunge, Fine and Legsdin published in 1946 some interesting results obtained on Missouri oxidized zinc ores. With sodium oleate as a collector and a combination of sodium hydroxide, sodium silicate and citric acid as a limestone depressor, they secured, after four cleaning operations, a good grade of concentrate. With many ores this reagent lacks adequate selectivity."

In 1939, M. Rey began research work on oxidized zinc ore which resulted finally in the use of fatty amines as collector. It was necessary to sulfidize in order to float zinc carbonate and zinc silicate. The result of this work was that two milling plants were put in operation in Sardinia in 1950. Ferruginous clay slime and calcium sulfate were found to be very deleterious. It proved easier to successfully treat the lighter-colored ores.

Lead sulfide and oxide minerals are floated first, followed by sphalerite if present, using the customary reagents. The tailing from this operation goes on to the new process, which comprises conditioning, sulfidizing, and addition of collectors and frothers. Rey *et al.* found that in amine flotation of the zinc minerals, an excess of sodium sulfide had no depressing effect and conditioning was not necessary. Addition of an alkali sulfide was found to be indispensable. The authors had not succeeded in obtaining a selective float of the zinc minerals without its use.

The San Giovanni and the Buggerru mills of the Societa Mineraria e' Metallurgica di Pertusola, in southern Sardinia, are discussed in the paper. A simplified flowsheet of the Buggerru mill is included.

The Buggerru ore contains zinc carbonate with a small amount of lead carbonate. It is crushed to −2 in. and deslimed. The plus ⅛ in. material is treated in a sink-and-float plant. Daily capacity of this plant is 240 tons. With the density of the heavy medium at 2.7 the float assays about 2 per cent zinc. From 50 to 60 per cent of the weight of original ore is discarded as float. The sink and the original fines are combined and are ground for later flotation. Conditioning is done with sodium silicate, sodium carbonate and sodium sulfide, followed by lead flotation. After further conditioning of the lead tailing with sodium carbonate and sodium silicate, the pulp goes on to oxidized zinc flotation, using the acetate of an aliphatic amine as collector, along with sodium sulfide. Positive-flow Minerals Separation machines are used for flotation. The zinc concentrate is cleaned twice. It is shipped to the electrolytic zinc plant at Crotone, Italy.

The paper lists the following results for the Buggerru operation in September 1953:

	Per Cent Lead	Per Cent Zinc
Heads (presumably flotation feed)	1.11	13.07
Lead carbonate concentrate	57.62	10.97
Zinc carbonate concentrate	1.31	44.44
Tailings	0.02	1.51
Recovery	81.7	90.0

The reagent consumptions are given as follows for the same month:

		Lead Circuit	Zinc Circuit
Sodium carbonate, lb/ton		2.00	3.80
Sodium silicate	"	2.00	3.60
Organic colloids	"		0.20
Sodium sulfide	"	2.38	6.00
Sodium cyanide	"	0.03	
Amyl xanthate	"	0.12	
Amine acetate	"		0.23
Pine oil	"	0.20	0.20

It is to be noted that no copper sulfate is used for oxidized zinc flotation. This contrasts with the practice of the mill of Gorno Mines, Gorno, Italy, where copper sulfate is used for both sulfide and oxidized zinc flotation.[12]

Influence of Marketing on Milling

Freight and smelter schedules have an important bearing on the practice at mineral dressing plants. Nowhere is this more true than in the case of mills treating complex sulfide ores. The problem is simple when such ores are clean, fairly coarse-grained and amenable to sharp separations, permitting the marketing of lead, copper and zinc concentrates of good grades and with high recoveries. But such cases are exceptional. Ordinarily, compromises are necessary. It is then that a particular smelter schedule may have a strong influence on practice.

Penalty for zinc in lead or copper concentrate has become the exception rather than the rule. Differential flotation has had a lot to do with this, for it has greatly decreased the amount of zinc going to lead and copper smelters, making the slagging-off of zinc much less troublesome. Some lead and copper smelters now have slag-fuming adjuncts to permit recovery of zinc as zinc oxide from slag. In such cases the sending of some zinc into the lead or copper concentrate may be welcomed by the smelter. The mill staff still needs to exercise care, however, as zinc should only be permitted to go into such a concentrate to the degree that the over-all efficiency of the milling operation is benefited by the sacrifice. However, knowing that the zinc which reports in lead or copper concentrate is later beneficiated rather than lost, the mill operator may at times elect such zinc sacrifice in favor of a higher recovery of other values in the lead or copper concentrate.

Zinc smelter schedules also have a bearing upon mill practice. If the market is an electrolytic zinc plant, the pay is better for residue values such as gold, silver, lead and cadmium, than if the market is a zinc retort smelter. Such values reporting in the zinc concentrate may then be tolerated to permit a better grade of lead concentrate ahead of zinc flotation. If the zinc smelter schedule pays little or nothing for such values, the tendency is to sacrifice zinc in the lead concentrate in order to increase gold, silver and lead recoveries.

Increasingly high freight rates and treatment charges tend to encourage the shipment of higher grades of concentrate, at the sacrifice of some recovery, in order to offset the higher cost of marketing.

Economic Recovery and Metallurgical Efficiency

The term "economic recovery" as here used means the ratio of the net combined return to the mill from the sale of all products, less freight and treatment charges, to the total gross value of the metals present in the original ore.

The term "metallurgical efficiency" as here used means the ratio of the net combined return, less freight and treatment, to what the return would have been had theoretically perfect products been made, with 100 per cent recoveries of gold, silver, copper and lead in the lead concentrate and 100 per cent recovery of zinc in the zinc concentrate. Such perfection is of course impossible to attain, but it does provide a realistic metallurgical yardstick for a particular operation.

It is assumed in the above that not enough copper mineral is present in the lead concentrate to make a copper-lead separation advisable. The shipment of a perfect chalcopyrite product on the lead smelter schedule determines the base for copper. The shipment of a perfect galena concentrate on the lead smelter schedule determines the base for lead. Freight and treatments are deducted as though such products had been shipped. Gold and silver base is that of the pay in the lead smelter schedule.

Perfect zinc concentrate grade in the case of a highly marmatitic sphalerite may be as low as 55 per cent zinc, whereas in the case of pure sphalerite it would be 67 per cent zinc. Usually 60 per cent zinc grade makes a convenient figure for the average condition. The shipment of perfect zinc concentrate on the zinc smelter schedule determines the base for zinc.[13]

Some Present Operations Treating Zinc Ores

Published descriptions of the milling practices of many of the interesting and important operations treating zinc-containing ores have appeared in the literature. Because of the ever-changing picture and limited space, allusion to present practices is confined to a few articles appearing during the past five years.

Cominco Operations. Repeated reference has been made to the operations of the 11,000 ton Sullivan Concentrator of the Consolidated Mining and Smelting Company of Canada Limited. The May, 1954 issue of the *Canadian Mining Journal*[14] is devoted entirely to the operations of this company. Part Three, on Ore Dressing, gives an excellent description of the Sullivan plant, along with flowsheet and reagent listing. The fact that tin is recovered from the flotation tailing adds to the interest. The chapters on the Company's Blue Bell, Tulsequah, and H. V. concentrators, all of which produce zinc concentrate, are also very informative.

European Operations. European operations are described in the volume "Mining and Dressing of Low-Grade Ores in Europe."[15] Brief descriptions of the practices at various concentrators in the United Kingdom, Sweden, Norway, Germany, Italy and France are included, along with flowsheets. Among the plants producing zinc concentrate are: the Boliden Mining Company, Sweden; the Erzbergwerk Grund and the Erzbergwerk Rammelsberg operations in Germany; the Gewerkschaft Auguste-Victoria operation in the Ruhr coalfield in Germany; the S.A.P.E.Z. Gorno Mines operation in Italy; and the Roros Kobbervery and the Sulitjelma in Norway.

Of particular interest in the Erzbergwerk Rammelsberg operations is the fact that a barite concentrate is floated following the flotation of lead, zinc and pyrite. Zinc sulfate, sodium bisulfite and sodium cyanide are used in lead work, along with xanthate and frothers. Lime, copper sulfate and sodium cyanide are used for zinc work, along with xanthate and frothers. Sulfuric acid, xanthate and frother are used for pyrite flotation. Sulfuric acid, sodium silicate and oleic acid are used for barite flotation. There are two milling operations. The 650-metric-ton Rammelsberg mill handles the richer ore. The 400-metric-ton Bollrich mill handles the lower grade ore.

The 1200-metric-ton Auguste-Victoria operation, in the Ruhr coalfield, does nice work on clean lead-zinc ore from a deposit (Seam Stein V) in the coal measures. The operation uses a common shaft with the coal mining operations. A coal product is floated before lead-zinc flotation, going to waste. "Flotigol C. S." is used for coal flotation. Zinc sulfate, sodium cyanide, amyl xanthate and frother are used for lead flotation. Copper sulfate, isopropyl xanthate and "Sapinol" frother are used for zinc flotation. A 77 per cent lead product is made with high lead recovery, followed by a 60 per cent zinc product with high zinc recovery.

The S.A.P.E.Z. Gorno operation in Italy is another illustration of the flotation of oxidized zinc mineral. A sulfide lead flotation is first made, followed by a sulfide zinc flotation. Both of these circuits are heated to 40°C. The zinc sulfide tailing is conditioned in series in six conditioners, the total conditioning time being about 25 minutes, at 40°C. Amyl xanthate and sodium sulfide are used at the conditioner feed, with copper sulfate and collecting and frothing reagents being introduced in the later conditioning. The conditioned pulp goes to a calamine flotation circuit, where the oxidized zinc minerals are floated. These consist principally of calamine and smithsonite.

Australian Operations. In a consideration of lead-zinc ores, Nixon[16] includes very clear and informative discussions of the milling of both coarse-grained and fine-grained ores. Two plants handling coarse-grained ores, North Broken Hill and Broken Hill South Limited, recover lead concentrate by jigging, tabling and flotation. Only flotation is used to recover zinc. Two other plants handling coarse-grained Broken Hill ores belong to Zinc Corporation Limited. Lead and zinc products are recovered by flotation. No zinc depressants are listed as used in the Broken Hill operations, although Broken Hill South uses sodium sulfite.

In a discussion devoted to fine-grained ores, Nixon describes the practices of Mount Isa Limited, Queensland; the Lake George Mines Proprietary Limited, Captain's Flat, New South Wales; and the Electrolytic Zinc Company of Australasia Limited, Rosebery, Tasmania. Grinding in all three cases is extremely fine, about 75 per cent −325 mesh. Sulfur dioxide gas is used in a flotation tower at Mount Isa before lead flotation with lime, cyanide and zinc sulfate. At Lake George, sulfurous acid is used to make an initial copper concentrate float before lead flotation. At Electrolytic Zinc Company's Rosebery mill, sodium sulfite is used in the flotation of an initial copper-lead concentrate, the pulp being heated to 24°C. Sodium cyanide is used in the succeeding lead circuit. In all cases zinc flotation follows lead flotation, with the customary reagents.

Two American Operations. The unique underground Eagle Mill of the New Jersey Zinc Company at Gilman, Colorado, is described by Craig.[17] This 1000-ton operation uses "Agitair" flotation machines for roughing and Denver machines for cleaning. Lime, sodium cyanide, xanthate and pine oil are used for zinc work. The ore runs about 1.7 per cent lead and 11.0 zinc. The zinc mineral is marmatite. The ore is washed before grinding to remove injurious soluble salts. Grinding is two-stage to −65 mesh.

The Minerva Oil Company operation at Cave-in-Rock, Illinois, described by Anderson[18] as a 250-ton operation, is especially interesting because of the flotation of a fluorspar product after the sphalerite has been floated. The ore is lead-free. The sphalerite contains very little isomorphous iron. Multiple-stage grinding is carried out to produce a fine grind (76 per cent −200 mesh) without overgrinding. The ground ore is conditioned with sodium cyanide and copper sulfate. Flotation of sphalerite is with sodium "Aerofloat," pine oil, "B 23," "Reagents 311 and 633." A 96-per cent recovery of zinc is made in a 63.5-per cent product. Denver machines are used. The succeeding fluorspar concentrate contains over 90 per cent of the fluorite in the ore, in three grades ranging from 87 to 96 per cent fluorite. The original ore assays 3.8 per cent zinc, 29.6 CaF_2, 35.7 $CaCO_3$ and 21.5 per cent SiO_2.

The Tsumeb Mine, S. W. Africa. The mill of the Tsumeb mine is described by Ratledge, Ong and Boyce.[19] The operation includes a 1500-ton sulfide unit and a 350-ton oxide unit. "Agitair" flotation machines are used in both.

The sulfide unit, working on ore with a percentage composition of 2.82 copper, 13.78 lead and 6.36 zinc, makes a preliminary two-stage grind to 60 per cent −200 mesh. There are three copper-lead roughing circuits, with differential grinding between each. The tertiary roughing feed is 81.4 per cent −200 mesh. The tertiary tailing goes on to thickening and conditioning for zinc flotation. Sodium cyanide, zinc sulfate and sodium carbonate are used for lead flotation. Lime, copper sulfate, ethyl xanthate, isopropyl xanthate, cresylic acid and frother are used for zinc flotation. A germanium concentrate product is recovered from a fraction of the copper-lead concentrate.

The oxide unit, working on dump ore containing 3.67 per cent copper, 5.26 lead and 2.94 per cent zinc, grinds two-stage. A bulk lead-zinc sulfide float is made. The cleaned bulk concentrate is conditioned with bichromate and "Reagent 610." Retreatment of the bulk concentrate floats the sphalerite in a final zinc product and depresses the galena in a product which goes to copper-lead concentrate. The tailing from the bulk float goes to oxide flotation, using sulfidizing reagents. The oxide lead product goes to copper-lead concentrate. The oxide flotation tailing goes to waste.

Complete descriptions of the various concentrators visited by the author in 1957 are beyond the scope of the present work, but the practice observed in two of them is summarized in some detail below.

Central Mill, Eagle-Picher Mining Company. The Central Mill of the Eagle-Picher Mining Company at Picher, Oklahoma, began operations in 1932. It was the outgrowth of a great need of centralization of ore milling in the Tri-State field. Before 1939, when preconcentration by heavy media separation was introduced, the capacity of the mill was between 6000 and 7000 tons daily. Heavy media separation more than doubled this to as much as 15,000 tons daily. Operations by February, 1957 were at the 12,000-ton rate for continuous operation.

The average grade of the ore is low, hovering around 0.7 per cent lead and 2.3 per cent zinc. The lead mineral is galena. The zinc mineral is a high-grade, light-colored sphalerite, with very little isomorphous iron. The ore at times contains bituminous, pitch-like material which tends to blind screens and interfere with the making of good flotation froths. The gangue consists of chert, dolomite, limestone, calcite, jasperoid and "cotton rock." The latter is partially silicified limestone. The specific gravities for these gangue materials are:

Dolomite, both gray and pink	2.85
Calcite	2.70
Jasperoid	2.67 to 2.70
Chert	2.65
Limestone	2.45
Cotton rock	2.02 to 2.40

The gangue which is contemporary with the sulfide minerals is apt to be mineralized to a degree that would require fine grinding for the release of sphalerite. Much of the jasperoid is of this category. As it tends to be part of the float reject, it is responsible for the greater proportion of the zinc loss of the mill. Both the jasperoid and the chert are very resistant to grinding. The fact that much of this goes into the coarse float reject avoids otherwise costly grinding.

Five hoppers receive the run-of-mine ore, four supplied by railroad cars and one by truck. Provision is made for the separate crushing and sampling of individual lots of custom ore and ore from various leases, so as to permit accurate accounting with the shippers.

The ore is crushed in two Webb City jaw crushers (18- by 36-inch). The crushed grizzly oversize plus the grizzly undersize is conveyed to a Butchard sampler, which takes a 7.5 per cent cut for the sampling plant. The final sample of each lot going to the assay office weighs about 40 pounds. All sampler rejects are returned to the main ore stream going from the Butchard sampler to a 6000-ton primary storage bin. Five 42-inch belt feeders take ore from the bin to a conveyor which supplies a secondary storage bin at the head of the secondary crushing department. Five 42-inch belt feeders take ore from the secondary storage bin to five 5 by 10-foot Tyler vibrating screens, each supplied with 1½-inch square mesh. The oversize from these five screens is conveyed to two 7-foot Symons S.H. cone crushers, the discharges of which are conveyed back to the secondary storage bin.

Concentrating operations begin with the undersize of the 1½-inch screens. The combined undersize is divided between five Robbins 5- by 10-foot vibrating screens. The decks of the two outer screens have ½- by 3⁄16-inch woven wire mesh. The middle three decks have ⅜-inch square woven wire mesh. The oversize is about 73 per cent of the original weight of ore. It is conveyed to the heavy media cone plant.

The float product from the cone operation, after washing and drainage, is conveyed to the chat storage pile. The sink product, after washing and drainage, is conveyed to a concentrate storage bin. The medium employed is −100 mesh ferrosilicon. This replaced powdered galena in 1945. Crabtree[20] discusses the conversion from galena to ferrosilicon, concluding with this summary:

> "The results of the comparison described herein indicate that the use of ferrosilicon has been worthwhile, particularly considering the results attained during the past few months. The respective operating costs using the two media and the grades of tailings produced are approximately equal. With ferrosilicon the grade of cone concentrate is materially higher; the cost of fine-grinding the cone concentrate for flotation treatment is less; the lead content of the flotation zinc concentrate has been reduced; better treatment of muddy ores is possible because of the greater ease of cleaning the ferrosilicon medium."

The specific gravity of heavy medium at the Central Mill is held at 2.67 at the top and 2.77 at the bottom. About two-thirds of the weight of original ore goes into the float product. Jeffrey counterflow magnetic separators are used in the reclamation of ferrosilicon from the washings. One-third of the reclaimed material is densified; two-thirds is filtered on disc filters. The densified and filtered materials are blended and sent to storage for re-use at the cone plant, along with the originally-introduced ground ferrosilicon.

The undersize of the five Robbins vibrating screens goes to double drags for desliming. Up to this point the distribution in terms of original ore approximates:

	Weight (%)	Assay Pb (%)	Assay Zn (%)
Ore	100.0	0.68	2.4
Oversize of Robbins screens	73.0	0.20	1.7
Cone float	66.4	0.02	0.4
Sink	6.6	5.0	15.0
Double drag sand	20.0	1.0	3.0
Overflow	7.0	2.0	8.0

The accompanying flowsheet, Figure 5-1, depicts the concentrator operations. All jigs are of the Bendelari type. Esperanza-type drag classifiers are used extensively within the jigging circuits for desliming of the successive stages. All flotation machines are 66-inch Fagergrens. No concentrating tables are used, except for a small size pilot table working on a split from the final tailing stream. Two grades of lead concentrate are shipped separately: a jig lead concentrate and a flotation lead concentrate. The zinc concentrate is entirely a flotation product. The rejected chats, consisting of the coarse float from the sink-and-float operation and the final tailing from the rougher and rechat jigs, are stacked in huge storage piles. Shipments of this material are made, the principal market being for concrete aggregate and road ballast.

Grinding is single-stage, with each of the four 8- by 6-foot Marcy ball mills in closed circuit with a Morse Brothers classifier. The overflows of the four classifiers go to a distributor which divides the material between two 8-cell lead rougher flotation machines. The dilution of the overflows is carried at about 50 per cent solids, resulting in a relatively coarse grind of about 6 per cent on 35 mesh and 70 per cent through 100 mesh. No regrinding of flotation middling is done separately in either the lead or the zinc flotation circuits. The lead cleaner tailing passes to a flume which returns it to drag classifiers. Thus the sandier portion has a chance of getting back into the original ball mill grinding circuit.

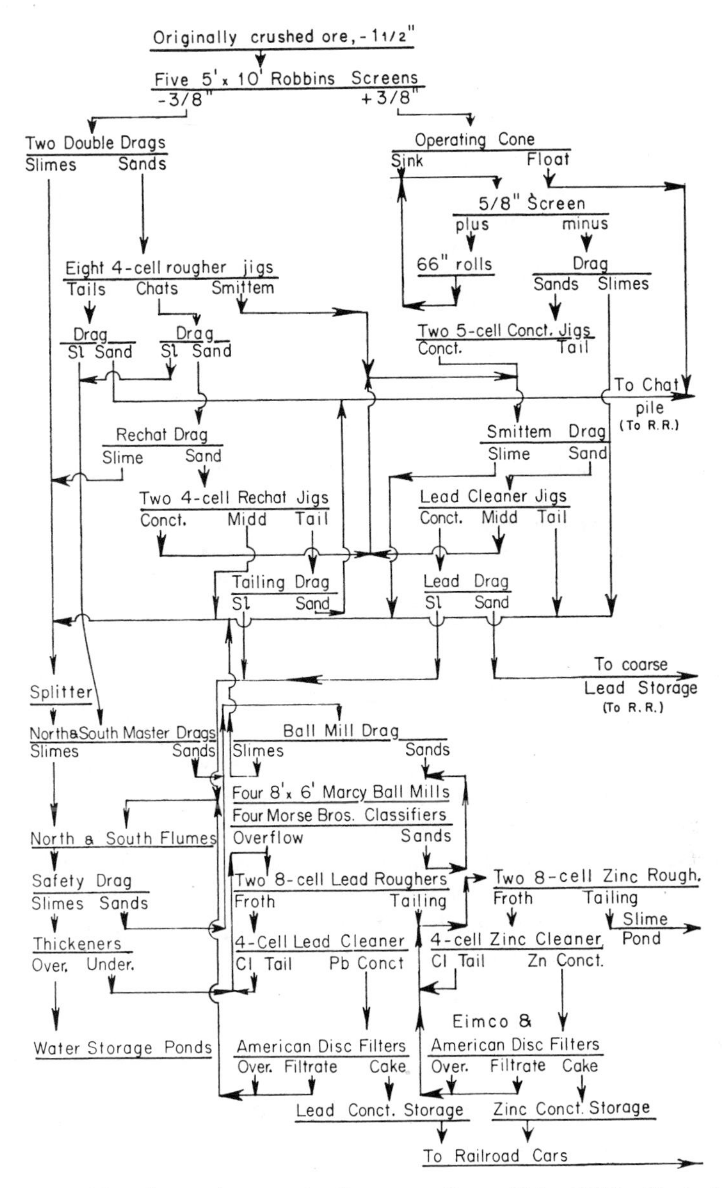

Figure 5-1. Flowsheet of concentrating operations (Feb. 1957), Central Mill. (*Courtesy Eagle-Picher Mining Company*)

No zinc depressant is used in the lead circuit, for the sphalerite normally does not float until conditioned with copper sulfate. The reagents used are:

Circuit	Material	Consumption, lb/ton of flotation feed
Grinding	None	
Lead	"Reagent 343" (isopropyl xanthate)	0.05
	Methylisobutyl carbinol frother	
	Sodium silicate used occasionally	
Zinc	Copper sulfate crystals	1.0
	Sodium "Aerofloat"	0.2
	Methylisobutyl carbinol frother	
	Soda ash used in nonwinter periods	

Mill metallurgical results reported for December, 1956, follow:

Product	Assay	
	% Pb	% Zn
Jig lead concentrate	83.72	0.97
Flot. lead concentrate	65.57	5.30
Flot. zinc concentrate	0.74	60.71
Tailing (by diff.)	0.088	0.407
Original ore	0.79	2.26

The zinc loss in the tailing is proportioned about as follows:

In float of cone operation	63%
In rejected jig tailing	18%
In final flotation tailing	19%

The jig lead concentrate is shipped to the company's Galena, Kansas, operation. Here it is rejigged and the concentrate ground and floated. The result is a product assaying over 86 per cent lead, or almost pure galena. This product is used to make lead sulfate and lead oxide. The tailing from the jigging and flotation operation assays about 70 per cent lead. It is shipped to the St. Louis smelter of the St. Joseph Lead Company. This smelter also gets the flotation lead concentrate from the Central Mill. The zinc concentrate is shipped from the mill to the retort smelter at Henryetta, Oklahoma.

Lead concentrate is penalized if it runs under 58 per cent lead, or if it runs over 9 per cent zinc. This is based upon monthly shipment averages.

The Buchans Mining Company Mill, Buchans, Newfoundland. The concentrator of the Buchans Mining Company Limited, a subsidiary of the American Smelting and Refining Company, first started operations in 1928. The initial capacity rating was 500 tons per day. In 1931 the capacity was increased to 1300 tons daily. Until 1936 concentrating tables were included in the flowsheet, working on primary cleaner tailing from both the lead and the zinc circuits before their return to their respective flotation circuits. Galena, pyrite and precious metals were thus short-circuited directly to finished concentrate. This practice was discontinued after 1936. Other improvements after this time were:[21]

Adoption of a very fine grind, now 86 per cent −325 mesh.
Use of sulfur dioxide gas along with bichromate to float chalcopyrite from lead concentrate in a separate treatment. (Decoppering.)
Use of cyanide to float galena from copper concentrate in a separate retreatment. (Deleading, only done occasionally.)
Use of corduroy mats in middling circuits for precious metal recovery.
Increase in the cleaning cycle in lead flotation from two to five cycles.

A pilot plant of 25 tons daily capacity was erected in 1951 to help out in the difficult problem of translating test laboratory findings to commercial practice. It is a complete flotation metallurgical operation by itself, following the main mill practice very closely. Starting with a small fraction of the regular mill feed, it goes through the successive stages of grinding and classification, lead rougher and zinc rougher operations. Lead cleaning is five-cycle, zinc cleaning is two-cycle, both as in the main mill. No copper separation is made. The final products join the corresponding products of the main mill previous to filtering. The tailing joins the main mill tailing. With the pilot plant holding close to mill conditions as a control, it is possible to vary practice at some point or points and note the effect of such variation. It is felt that the expense of the operation is many times justified by the information obtained.

The Buchans mill feed is a very complex and fine-grained sulfide ore. In the order of their amounts, the base metal minerals are sphalerite, galena, pyrite, chalcopyrite, tetrahedrite, tennantite and covellite. For the complete liberation of minerals it is thought that grinding would have to be all through 25 microns. The principal gangue minerals are barite, silicates and calcite.

Crushing of the ore is done in open-circuit, two-stage operation. As the ore comes quite wet from the mine, water sprays are used to keep the sticky materials from hanging up in the crushers and chutes. The sieve analysis of the crushed ore runs about as follows:

Plus 1-in	1.7%
Minus 1-in plus ¾-in	17.9%
Minus ¾-in plus ½-in	23.4%
Minus ½-in	57.0%

The crushed ore is divided between two 1000-ton bins. Two feeds are taken from one bin for the single-stage initial grinding unit. One feed is taken from the other bin for the two-stage initial grinding unit.

The accompanying flowsheet (Figure 5-2) illustrates the grinding flow. In the unit employing single-stage initial grinding each of two 8-foot by 72-inch conical ball mills receives its feed from one of the bins. Each of these mills is in closed circuit with a rake classifier. The two rake classifier overflows are reclassified in a 12- by 30-foot bowl classifier, the return sand of which is divided between the feeds of the two 8-foot by 72-inch ball mills. In the unit employing two-stage initial grinding the single feed from the other bin goes to an 8-foot by 60-inch conical ball mill in closed circuit with a rake classifier. The classifier overflow passes to a 12- by 30-foot bowl classifier which is in closed circuit with another 8-foot by 60-inch conical ball mill. The overflows of the bowl classifiers of the two units join and are pumped to three 12-inch cyclones. The cyclone underflow is ground in a 5- by 8-foot ball mill, the discharge of which is returned to the cyclones. The cyclone overflow passes on to flotation. Its dilution is carried at about 23 per cent solids, resulting in a grind of about 86 per cent −325 mesh.

Both grinding circuits described give about the same grind. Test work indicates equivalent metallurgy. The two-stage grind has the advantage of lower cost, but the single-stage grind has the advantage of greater flexibility. The present flowsheet was adopted to retain both advantages.

Ball consumption for the mill is about 1.55 pounds per ton of original ore. The conical ball mills on single-stage grinding receive a rationed charge of 2- and 3-inch balls. The initial mill of the two-stage grinding circuit receives 3-inch balls. The secondary mill of the circuit receives 1¼-inch balls. The 5- by 8-foot ball mill, working on cyclone underflow, uses ¾-inch balls.

The cyclone overflow passes to a 4- by 18-foot wooden absorption tower, where the pulp descends against an ascending column of SO_2 gas, generated by the burning of elemental sulfur in a Buchans burner. (This idea was initiated at Buchans.) From here the pulp passes to a 12- by 10-foot conditioning tank. The conditioned pulp passes on to the lead-copper rougher flotation circuit. The accompanying flowsheet shows the subsequent flow. It will be noted that the final lead-copper concentrate goes on to another SO_2 tower for conditioning with the gas ahead of the separate copper-flotation, lead-depression step. The percentage of solids at this point is about 30. Sulfur dioxide activates the chalcopyrite while sodium bichromate depresses the galena. Both the lead-copper and the copper flotation are done in acid circuits (*p*H 6.2 and 5.2). When the copper concentrate contains too much lead, it is deleaded by the use of cyanide to depress the chalcopyrite in a separate flotation operation. This is not indicated on the accompanying flowsheet. The practice is only occasional.

Five cleanings of the lead-copper concentrate are needed because of the ease with which the sphalerite floats in the lead-copper concentrate.

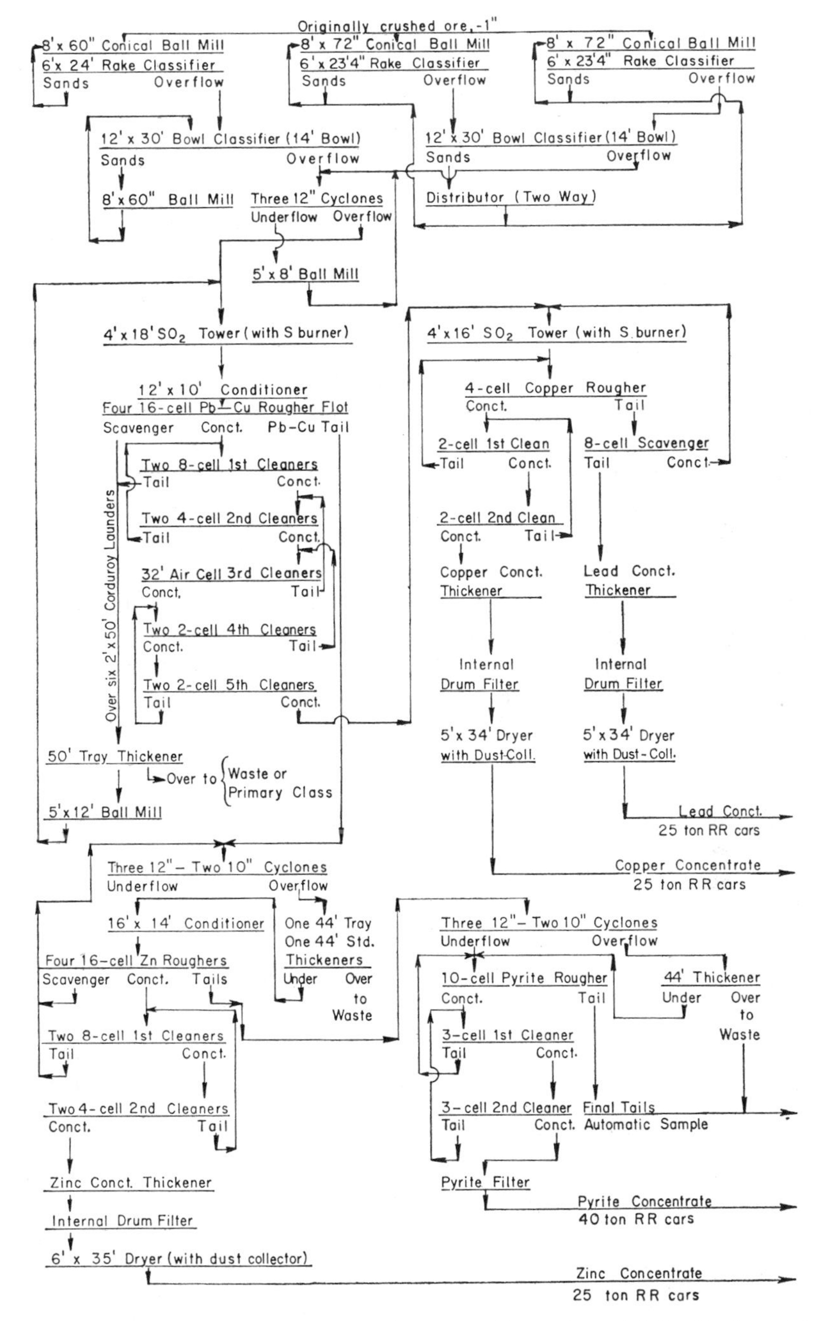

Figure 5-2. Flowsheet of concentrating operations (Feb. 1957), Buchans Mill. (*Courtesy Buchans Mining Company Limited*)

At the same time the galena tends to be sluggish. Regrinding of the lead-copper scavenger middling is essential before returning it to the SO_2 tower.

The tailing from the lead-copper rougher circuit is thickened to about 25 per cent solids before conditioning for zinc work. Hydrated lime is introduced into the zinc circuit to bring the *p*H up to about 10.2. Copper sulfate is used to reactivate the sphalerite.

The flowsheet (Figure 5-2) shows the zinc rougher tailing as going to cyclones for rethickening ahead of pyrite flotation. However, the pyrite flotation step is omitted at times. The zinc rougher tailing then passes directly to waste. All roughing flotation operations are performed in "Sub-A" type machines, with 42- by 42-inch cells.

An interesting feature of the Buchans mill is the fact that all discarded materials, such as thickener overflows, filtrates, etc. join the tailing stream to be included in the automatic sample of the material leaving the mill. There is no tailing disposal problem, as the tailing stream flows directly into the Buchans River, which flows into one of the large lakes of the island.

The copper, lead and zinc concentrates are all filtered in internal drum filters. The filter cakes are dried, with provision for dust collection. The railroad is 42-inch gage. Concentrate cars of 25 ton capacity are used.

Reagent consumptions during 1956 were as follows (pounds per ton ore):

	Pb-Cu Section	Decoppering Section	Deleading Section	Zn Section	Pyrite Section
Sulfur	1.47	0.35			
Sodium cyanide	0.12		0.32	0.03	
Zinc sulfate	1.70				
"Reagent 301"	0.12				0.28
Thiocarbanalid	0.04				
Hydrated lime			0.26	1.67	
"Dowfroth"				0.06	0.023
Cresylic acid	0.06				0.036
Copper sulfate				0.88	
Isopropyl xanthate			0.002	0.24	
Sodium bichromate		0.85			
"Reagent 242"			0.004		
"Separan 2610"					0.008
Soda ash					0.74

Metallurgy for 1956 was as follows:

	Assays					
	Oz per Ton		Per Cent			
	Au	Ag	Cu	Pb	Zn	Fe
Copper concentrate	0.20	30.0	27.5	5.9	6.2	25.2
Lead concentrate	0.11	15.4	2.2	60.8	12.1	3.6
Zinc concentrate	0.04	4.0	0.6	2.3	59.8	2.0
Gravity concentrate	23.15	225.1	2.1	21.5	30.8	9.9
Pyrite concentrate	0.07	3.9	0.6	1.4	5.5	35.8
Tailings	0.01	0.5	0.08	0.31	0.85	2.8

The mill tailings for December, 1956, when no pyrite was made, assayed:

Au	Ag	Cu	Pb	Zn	Fe	S	$BaSO_4$	SiO_2	CaO	MgO
0.014	0.74	0.08	0.31	1.20	3.8	3.2	46.4	26.6	2.1	1.7

For further information the reader is referred to an interesting article on the Buchans operations by the company staff.[22]

ACKNOWLEDGMENT

The courtesy and generous treatment the author received from the managements and metallurgical staffs of the various companies and milling plants whose operations were visited during the preparation of this section of Chapter 5 is hereby acknowledged. These plants include the Central Mill of the Eagle-Picher Company at Cardin, Oklahoma; the Mascot Mill of the American Zinc Company of Tennessee at Mascot, Tennessee; the Buchans Mill of the Buchans Mining Company Limited, subsidiary to the American Smelting and Refining Company, at Buchans, Newfoundland; the Quemont Mill of the Quemont Mining Corporation Limited, Noranda, Quebec; and the Midvale Mill of the United States Smelting, Refining and Mining Company, Midvale, Utah.

The fact that due to space limitation it has been impossible to include here the rather detailed descriptions made of all these plants by the author, and approved for publication by the managements concerned, in no way reduces the appreciation of and obligation to the managements and staffs of these operations.

Thanks are also due to the various publications studied, references to which have been included in the text; also to the several able readers who made valuable and helpful criticisms and suggestions.

REFERENCES

1. Pellett, J. S., "Milling at Franklin and Sterling," *Mining Eng.*, **5**, 1211–13 (1953).
2. Lemke, C. A., "Milling Practice at Midvale," *Trans. Am. Inst. Mining Met. Engrs.*, **73**, 342–53 (1926).
3. Burris, S. J., Anderson, C. O., Illidge, R. E., Howes, W., and Harbaugh, M. D., "Milling Practice in the Tri-State Zinc-Lead Mining District of Oklahoma, Kansas and Missouri," *Trans. Am. Inst. Mining Met. Engrs.*, **112**, 861–924 (1934).
4. Diamond, R. W., "Ore Concentration Practice of Consolidated Mining and Smelting Company of Canada Ltd.," in "Flotation Practice," pp. 95–106, New York, Am. Inst. Mining Met. Engrs., 1928.
5. Crampton, J. C., "Pend Oreille Mines and Metals Company," *Mining Eng.*, **7**, 846–9 (1955).
6. Staff, Canadian Mining Journal, "The Sullivan Concentrator," *Can. Mining J.*, **75**, 211–223 (May, 1954).
7. Gaudin, A. M., "Separating Solids with Bubbles," *Scientific American*, **195**, 99–110 (1956).
8. Nixon, J. C., "Treatment of Lead-Zinc Ores," in "Ore Dressing Methods in Australia," Vol. 3, Chapter 2, Melbourne, Australasian Inst. Mining and Met., 1953.
9. McQuiston, F. W. Jr., "Milling Practice at Idarado Mining Company," *Trans. Am. Inst. Mining Met. Engrs.*, **183**, 95 (1949).
10. Boeke, C. L., and Gunther, G. G., "Deleading Zinc Concentrate at the Parral and Santa Barbara Mills," *Mining Eng.*, **4**, 495–8 (1952).

11. Rey, M., Sitia, G., Raffinot, P., and Formanek, V., "Flotation of Oxidized Zinc Ores," *Trans. Am. Inst. Mining Met. Engrs.*, **199**, 416–20 (1954).
12. Billi, Marcello, "How Gorno Recovers Oxidized Zinc," *Eng. Mining J.*, **158**, 82–6 (April, 1957).
13. Pallanch, R. A., "Metallurgical Efficiency—A Yardstick in Lead-Zinc Flotation Metallurgy," *Trans. Am. Inst. Mining Met. Engrs.*, **183**, 283–8 (1949).
14. Staff, Canadian Mining Journal, "Cominco, A Canadian Enterprise," *Can. Mining J.*, **75**, May, 131–393 (1954).
15. "Mining and Dressing of Low Grade Ores in Europe," Organization for European Co-operation, Paris, June, 1955.
16. Nixon, J. C., in "Ore Dressing Methods in Australia and Adjacent Territories," Vol. 3, Chapter 2, Melbourne, 1953.
17. Craig, J. G., "The Underground Mill," *Mining Eng.*, **5**, 1225–7 (1953).
18. Anderson, O. E., "Minerva Oil Company," *Deco Trefoil*, **17**, 7–14 (March–April, 1953).
19. Ratledge, J. P., Ong, J. N., and Boyce, J. H., "Development of Metallurgical Practice at Tsumeb," *Trans. Am. Inst. Mining Met. Engrs.*, **202**, 374–83 (1955).
20. Crabtree, E. H. Jr., "Comparison of Galena and Ferrosilicon in Heavy Media Separation," *ibid.*, **183**, 252–6 (1949).
21. Hellstrand, G. A., and George, P. W., "Milling Practice at Buchans Mine, Newfoundland," *ibid.*, **112**, 841–60 (1934).
22. Staff, Canadian Mining and Metallurgical Bulletin, "Buchans Operation, Newfoundland," *Can. Mining J.*, **48**, 349–53 (1955).

Roasting, Sintering, Calcining and Briquetting

K. D. McBean

Metallurgical Engineer
The Consolidated Mining and Smelting Company of Canada Limited
Trail, British Columbia

The processes of roasting, sintering, calcining and briquetting have been included in one paper because they are all basically of a preparatory nature and each one fills an essential part in the preliminary stages of some current method of recovering zinc from its concentrates.

Because all of these processes are used widely outside the zinc industry, and as a complete survey would be beyond the scope of this paper, descriptions and discussions have been confined to methods and apparatus as used in the treatment of zinc-bearing materials.

The paper has been divided into sections, with a section devoted to each one of the processes. In each section has been given a brief outline of the history and development of the process, followed by a more complete description of modern equipment and operation, with a final summary.

ROASTING

The main purpose of roasting zinc ores or concentrates is to convert the zinc sulfide into zinc oxide and to expel the sulfur dioxide gas. In practice this objective is never attained completely because some zinc sulfide remains unoxidized and some of the zinc oxide is converted to zinc sulfate.

The roasting reaction is usually expressed by the following equation:

$$2ZnS + 3O_2 = 2ZnO + 2SO_2$$

This reaction is exothermic and about 1200 kilogram-calories per kilogram of concentrates oxidized are released. Theoretically this is ample heat energy to sustain the reaction and, provided that the correct conditions are maintained, no extraneous fuel, other than that to start the reaction, should be required.

The more important factors in roasting are temperature, time and air or oxygen supply, and, in order to gain the optimum effect from each of them, suitable physical conditions must necessarily be provided.

In efforts to improve these physical conditions, roasting methods have progressed from the primitive heap roasting to the hand-rabbled reverberatories, such as the Deplace, and later to mechanically-rabbled reverberatories, of which the Hegeler kiln and the Ropp furnace were the most important and have survived the longest.

A significant advance began with the development of the circular multiple-hearth roaster, the prototype of this design being the McDougall, out of which evolved the Wedge, Zelewski, Haas, Ord, Skinner and Herreshoff roasters and several others of the same type. Contemporaneously with these developments, the de Spirlet and Barrier revolving-hearth furnaces were successful variations of the circular type of roaster.

During the last two or three decades, progress in roasting has greatly accelerated following the successful development of the suspension and later the fluidizing techniques.

Except for the mechanically-rabbled multiple-hearth roasters, which have partially met the challenge by continual improvement, the old type of roasting operation has virtually yielded to these modern systems. The suspension roasters, in company with the variety of fluid-bed reactors, all capable of high unit capacity, autogenous roasting, waste-heat recovery and the production of rich sulfur dioxide exhaust gases, have about captured the field.

In each classification of roasting systems, there are many individual types which differ from each other in certain details but which operate on the same principle. For this reason, the discussion that follows has been limited to the description of typical installations for each class of furnace

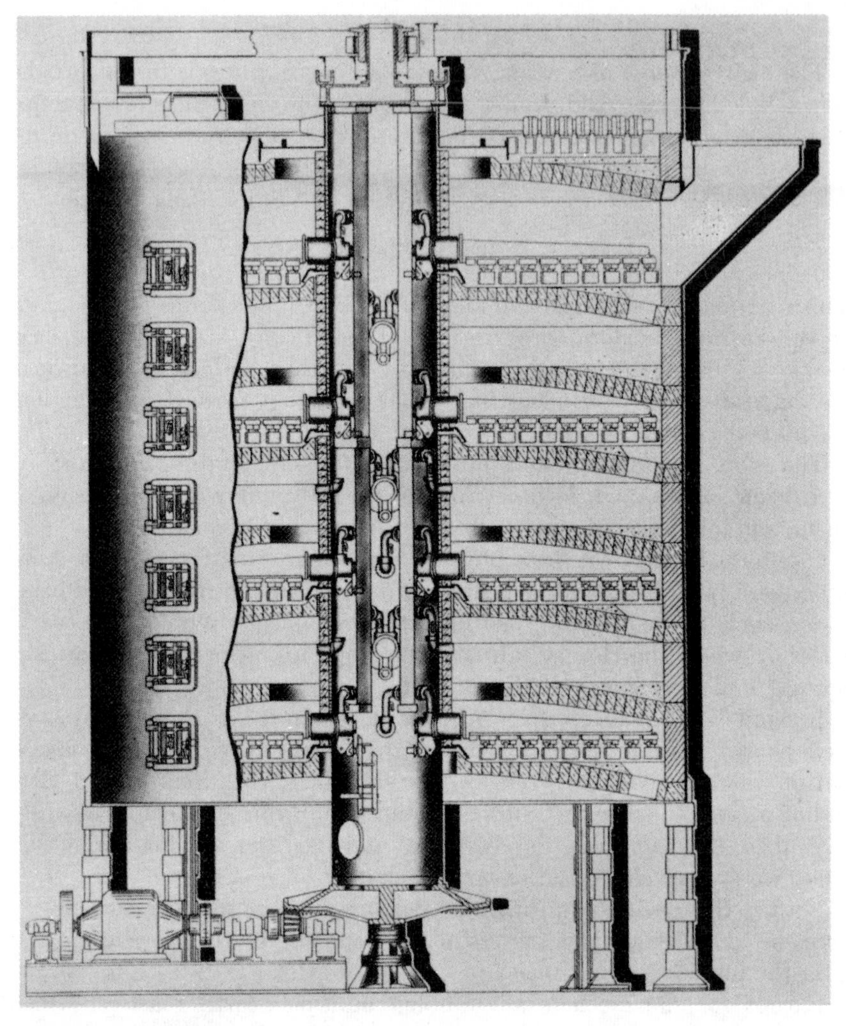

Figure 5-3. Seven-hearth Wedge type roaster. (*Courtesy Bethlehem Foundry and Machine Company*)

except where the operating results have been sufficiently interesting to warrant further remarks.

Conventional Multiple-Hearth Roasters

With the exception of the Split Draft roaster, which is used only by the Hudson Bay Mining and Smelting Company, the various types of multiple-hearth roasters in zinc metallurgy are designed and operated on the same general principle, with modifications depending upon the quality of calcine required for subsequent processing.

In general, this type of roaster consists of a brick-lined steel shell which supports a number of brick arches or hearths, usually seven, but up to sixteen in some cases. These hearths are spaced between two and three feet apart, according to the particular design and diameter of the furnace.

Each hearth has two to four air-cooled arms equipped with rabbles which are carried by a central shaft passing vertically through the middle of the furnace. By means of the rotation of this shaft, the arms and rabbles turn over and advance the roasting concentrate across each hearth in succession until it is finally discharged as finished product from the bottom hearth. Usually some heat from burning oil, gas or coal is added on the lower hearths, and the products of combustion, mixed with the sulfur dioxide gases, are exhausted from the roaster to flues at the top.

Oxidizing air for the concentrate is normally supplied by the excess air used in burning the extraneous fuel supplemented by heated air from the arms and by air admitted through doors and other openings.

Operation. The concentrates, usually at 8 per cent moisture, are fed to the roaster, generally by means of a weightometer belt feeder, and are dropped upon the so-called "dry hearth" where the ploughs move them to the inside seal and feeding device. By this means about two per cent of the moisture is eliminated and the concentrates are more evenly distributed over the first roasting hearth. Also, by having the feed exposed on the dry hearth, before dropping into the roasting zone, accretions and tramp material are more easily removed.

Ignition normally takes place about half way across the first hearth and roasting continues autogenously from hearth to hearth until the sulfide sulfur is decreased to 3.5 per cent. At this stage, hot gases from the burning of the extraneous fuel maintain the temperature and roasting continues until generally only a few tenths of sulfide sulfur remain in the finished product.

The exhaust gases leave the roaster through the top hearth flues at about 500°C and carry, as suspended matter, about 12 to 15 per cent of the total product. The sulfur dioxide strength in the gases may be as low as 1.5 per cent if the gases are wasted to atmosphere, but may reach up to 6 per cent if for acid manufacture or when roasting is followed by sintering and sulfur elimination from the calcine is not so essential.

In some plants, the gases pass through cyclones where about 80 per cent of the dust burden is precipitated. The remainder of the dust burden is collected in electrostatic precipitators with a recovery of about 98 per cent. The cleaned gases are usually treated in an acid plant but, in some plants, they are discharged directly to a stack and atmosphere.

The dust which is collected in the electrostatic precipitator is either treated separately for lead and cadmium recovery or returned by screw conveyor or other means to the lower hearths of the roaster, along with the cyclone and flue dusts.

Data on Multiple-Hearth Roaster Installations. The Donora Zinc Works, of the American Steel and Wire Company, have two Nichols-Herreshoff roasters, each 20 feet in diameter, with sixteen hearths, which are used as preroasters for sintering machines. Each roaster treats about 55 dry tons of concentrate per day and produces calcine with an average of one per cent sulfide sulfur. The exhaust gases run over 6 per cent sulfur dioxide and carry a dust burden of only 4 tons of dust per roaster per day, or about 7.3 per cent of the feed rate. This low dust burden is almost completely precipitated in a Cottrell before the gases pass to a chamber acid plant. The dust is not returned to the roaster but is sent directly to the sintering operation.

The furnaces are extraneously heated with fuel gas on the twelfth, fourteenth, and sixteenth hearths at the rate of 1,000 cubic feet per ton of ore roasted. Temperatures vary from a maximum of 920°C on the fourth hearth to an average of about 800°C on the bottom hearths.

A fair elimination of lead and cadmium is obtained. The calcine assays 0.02 per cent lead and 0.2 per cent cadmium from a concentrate content of 0.35 per cent lead and 0.3 per cent cadmium. These metals are concentrated in the Cottrell dust which assays up to 5.0 per cent lead and 1.7 per cent cadmium.

The furnaces at Donora produce satisfactory calcine and sulfur dioxide gases but the high temperature of operation is very destructive to rabbles, which have a short life of about three months on the upper hearths and not more than six months on the lower hearths. This high maintenance has detracted in some degree from the advantages usually associated with large-capacity furnaces having many hearths.

The Anaconda Company also uses multiple-hearth roasters, of the Wedge type, for roasting zinc concentrate. They have twelve roasters, each 25 feet in shell diameter, with seven roasting hearths plus a drying hearth.

The furnaces roast concentrate for the electrolytic zinc process and the calcines are leached directly without sintering or other processing. Since solubility of the zinc in the calcine is the main objective and as there is no acid plant in conjunction with the roasters, the strength of sulfur dioxide in the exhaust gases is of minor importance compared to the degree of sulfur elimination, or to the prevention of high temperatures on the hearths to minimize ferrite formation. In order to obtain these optimum conditions, large volumes of oxidizing air are used and, as a consequence, weak exhaust gases at 1.5 per cent sulfur dioxide and carrying up to 20 per cent of the feed as dust are produced. However, by not exceeding 680°C on any of the hearths, the life of arms, rabbles and brickwork is greatly prolonged and a calcine low in sulfide sulfur and with high zinc solubility is obtained.

The Mexican Zinc Company, Rosita, Mexico, operates four Skinner, twelve-hearth roasters, each 20 feet in inside diameter, as preroasters for sintering machines.

The concentrates are predried in a rotary kiln and are fed by gas-tight screw conveyors to the first roasting hearths. An average of 93 dry tons of concentrate per furnace day is treated. Acid is produced from one roaster only and this furnace is operated at higher temperatures with less air than the other furnaces in order to produce a rich gas of 6.2 per cent sulfur dioxide. The flue and cyclone dusts from all the roasters are returned to the seventh hearth of this particular roaster; the calcine contains about 5.4 per cent sulfide sulfur as compared to under 0.1 per cent for the other furnaces. The composite calcine of all the furnaces assays about 4.3 per cent total sulfur, of which about 62 per cent is sulfide sulfur.

Auxiliary heat is supplied by burning coke oven gases at several openings on the bottom hearths. Temperatures of 890°C are maintained on the upper hearths and 615°C on the cooler hearths on the furnace supplying gases to the acid plant, with temperatures about 50°C lower on the other furnaces.

The dust carry-over averages 13 per cent of the concentrates and separate flue systems for the acid plant gas and the waste gases are used. A system of dust chambers, cooling tubes, cyclones and Cottrells cools and cleans the gases. The dusts are returned to the acid plant roaster, except the Cottrell dust, which is delivered directly to the sintering plant.

The calcine from all of the roasters is mixed and screened through a 5-mesh screen in closed circuit with rolls before delivery to the sintering plant.

The zinc concentrate roasters at the Hudson Bay Mining and Smelting Company, Flin Flon, Manitoba, are an unique adaptation of draft control to multiple-hearth roasters. They operate on the so-called "split draft principle," whereby the gases are split at a neutral area between the third and fourth hearths. The gases from the first three hearths leave the roaster at the top and the gases from the fourth and lower hearths leave the roaster at the bottom. The volumes of gases are controlled by dampers in both outlets and are usually divided with about 25 per cent flowing upward and the remainder downward. By this means, no extraneous fuel is required, except occasionally for delays or starting up, and excessive temperatures on any hearth are avoided. The gases which travel upward dry and ignite the incoming concentrate. The hot gases which flow downward maintain the temperatures on the lower hearths in spite of the gradually diminishing sulfur in the roasting material.

Although this system is more sensitive to control and has less unit capacity than the standard method, these disadvantages are favorably balanced by the production of a calcine low in sulfide sulfur and having a high zinc solubility, with the minimum formation of zinc ferrites. Also, the process requires no carbonaceous fuel and has almost negligible maintenance for arms, rabbles and brickwork.

Another efficient application of the multiple-hearth principle is obtained by the de Spirlet roaster and a modification called the "Barrier roaster," at one time used extensively and still used in Europe and Africa. The special feature of this furnace is the use of revolving hearths. There are normally six superimposed hearths, three stationary and three revolving. All hearths, except the bottom hearth, have replaceable rabbles on the underside, affixed to steel blocks which are incorporated in the brickwork.

This design of roaster permits the hearths to be placed very close together which thus forces the maximum of contact between the oxidizing gases and the burning concentrates. As a result, the roasting rate per unit of hearth area is above normal, the sulfur dioxide strength in the gases is high, and generally no auxiliary fuel is required. The structural design, however, limits the capacity of individual furnaces to about 20 tons per day and the complicated construction creates high maintenance charges.

Although increasing labor costs are gradually forcing out small unit operations in favor of the large capacity roasters, the de Spirlet furnaces are still being used at a number of plants where the small tonnages roasted or the favorable labor conditions do not justify a change.

At Sogechim, an associated company of the Union Minière du Haut Katanga, in the Belgian Congo, there are 18 de Spirlet roasters, each 18 feet 6 inches in shell diameter, with six roasting hearths plus one cover hearth and one bottom cooling hearth. Each roaster treats between 14 and 20 metric tons per day of zinc concentrate, containing 57 per cent zinc, 3.0 per cent iron and 32 per cent sulfur, without auxiliary fuel. The calcine assays under 0.3 per cent sulfide sulfur and 1.4 per cent sulfate sulfur, and a gas running about 6.0 per cent sulfur dioxide is delivered to the contact acid plant.

These excellent results are due partly to the design of the furnace and partly to the practice of recirculating over 40 per cent of the calcine. The entire furnace discharge is screened through a 6-mesh screen and the oversize, which assays over 5 per cent sulfide sulfur, is crushed, then mixed with the Cottrell dust and fresh concentrates and returned as feed to the furnace. Only the minus 6-mesh material is finished calcine. It is moistened in a revolving drum with 4 per cent water and shipped to another plant for the electrolytic recovery of zinc.

The Suspension Roasting Process

The evolution of the suspension roasting process for zinc concentrates followed naturally upon the production of enormous tonnages of finely ground flotation concentrates which were ideally suited for speedy oxidation. Before this period, experiments along flash-roasting principles had not progressed to the commercial stage, mainly because the available concentrates were coarse, and satisfactory methods of avoiding accretions were not discovered.

The process which was developed at the Trail plant of The Consolidated Mining and Smelting Company of Canada Limited was the first, on a commercial scale, for the treatment of zinc concentrates. The eventual success and efficiency of this process can be gaged by the fact that nearly all the electrolytic zinc plants and many of the retort zinc plants in the world have adopted it. Other suspension roasting systems have also been developed, which have been highly successful in the treatment of iron sulfides, but have not apparently been considered suitable for zinc concentrates.

The latest type of "Cominco" furnace can be built for capacities exceeding 500 tons per day and there are roaster units in operation at Palmerton, Pennsylvania, and at Galena, Missouri, which are each capable of treating over 300 tons of zinc concentrates daily.

Some plants have Cominco suspension roasters with unit capacity of only 70 tons per day, but the majority of the installations are in the capacity range of 140 tons per day.

The design and arrangement of the latest type of furnace and its auxiliaries are shown in Figures 5-4 and 5-5. The furnace proper consists of a brick-lined cylindrical steel shell, which encloses a relatively large combustion chamber, having one or two collecting and desulfating hearths at the base of the chamber and one or two drying hearths under the calcine hearths. Each hearth has four air-cooled arms equipped with very large rabbles. The arms are attached to a truncated shaft which extends a very short distance into the combustion chamber. The volume of the combustion chamber determines to a large extent the capacity of the furnace, so height and diameter are not fixed relationships but may be altered to suit various design requirements.

The auxiliary equipment usually includes a ball mill, with the size depending upon the screen analysis of the raw concentrate and the sulfur elimination desired.

The calcine-conveying apparatus normally consists of water-cooled screw conveyors but other types, such as "Baker" coolers, "Holo-flite" conveyors, push-pull and chain drag conveyors, are used at some plants. The gas-handling system generally includes waste-heat boilers, cyclones and

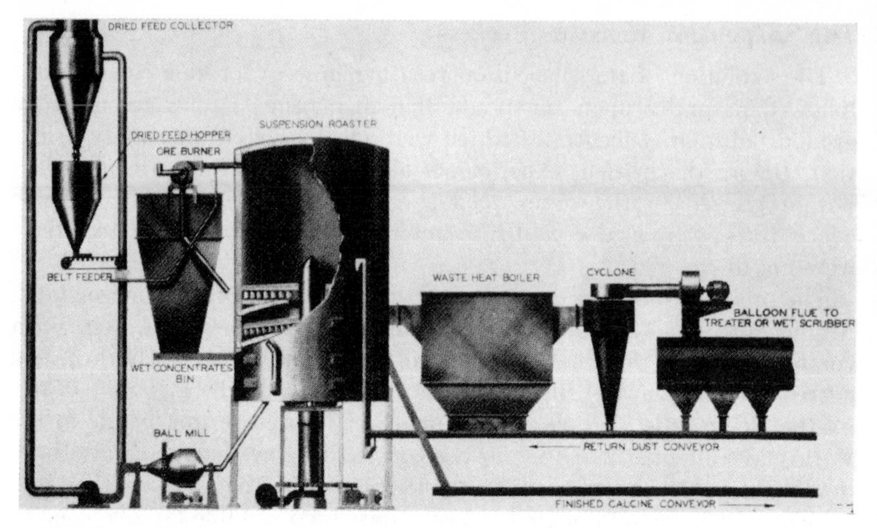

Figure 5-4. "Cominco" suspension roasting system. (*Courtesy Bethlehem Foundry and Machine Company*)

electrostatic precipitators, although, in some installations, cooling pipes and heat exchangers are substituted for boilers. Various types of boilers have been found satisfactory. At present, the three-drum type with horizontal gas flow is the most favored, but the forced circulation boilers of the Lamont type are gaining in popularity in Europe.

Operation. The operation of the furnace is essentially straight-forward and follows the general practices worked out for burning pulverized coal. As in coal firing, it was soon discovered that finer particle sizes not only improved the combustion efficiency but greatly increased the capacity of single units, a principle that applies generally to gas-solid reactions.

In the majority of suspension-roasting installations, the wet concentrates are fed directly to the furnace without any preliminary treatment. Drying is carried out on the drying hearths by the recirculation of a regulated volume of hot-chamber gases. This system of drying is virtually automatic and problems connected with dust catching, auxiliary heating, or accretions, are nonexistent.

The dried concentrates are discharged to grinding apparatus, generally a ball mill with air classification, in order to break up the agglomerations and to grind further if necessary. The ground material is collected in a feed bin, from which it is fed at a controlled rate into the combustion air stream, which carries the dried concentrate, along with the drying gases, to two diametrically opposed burners at the top of the combustion chamber.

The concentrates burn rapidly in suspension and the sulfur elimination is 95 per cent complete by the time the particles settle on the hearth or

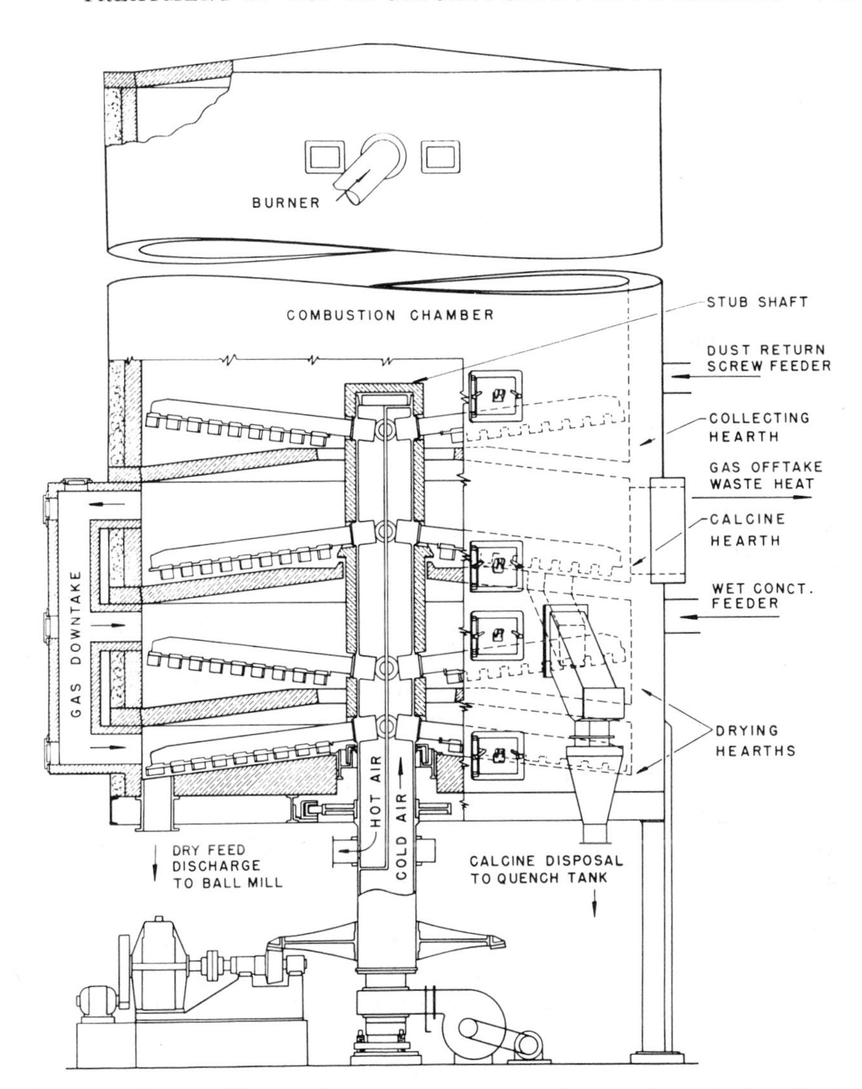

Figure 5-5. Skinner "Cominco" suspension roaster. (*Courtesy Colorado Iron Works*)

leave the furnace with the gas stream. The dusts which are precipitated in the boiler and cyclones are normally returned to the hearth at the base of the combustion chamber in order to complete the desulfurization and to improve the settling and filtering qualities. The product usually assays under 0.1 per cent sulfide sulfur and the minimum of sulfate sulfur but, in some installations, sulfur elimination is controlled to give a product having 3 to 4 per cent sulfur as a preroasted feed for sintering or nodulizing.

The gases, with 9 to 10 per cent sulfur dioxide and under 0.1 per cent sulfur trioxide, leave the furnace at 1800°F and are cooled in the boiler to 550°F. Steam is generated generally above 350 psig to keep tube temperatures above the dew point of the gases. With typical concentrates, about one ton of steam per ton of concentrate roasted, is recovered. In some plants, this steam is used for process heating or electric power generation, in others the steam is used directly in individual turbines to drive the ball mill, fans and blowers and other equipment at the roaster and acid plant.

In addition to the basic simplicity of the suspension roasting process, the outstanding feature is the tremendous range in unit capacity that is available. Units have been constructed for capacities as low as 10 tons of concentrates per day and as high as 350 tons of concentrates per day, while units up to 500 tons daily capacity are feasible. These high unit capacities connote low labor, maintenance and installation costs, and the flexibility in design and simple operation lend themselves favorably to instrumental control.

Fluidization Processes

The fluidization processes have come into prominence in the zinc industry only within the last decade, yet, in that short time, at least five variations of the process have been developed.

Dorr-Oliver, Inc. has been in the forefront in the general exploitation of the system in the zinc and other industries while other companies, including the St. Joseph Lead Company, the New Jersey Zinc Company and the Société Overpelt-Lommel, have all adapted the principle of fluidization for more specified purposes. Also the Badische-Anilin and Soda-Fabrik, AG, in Germany, have recently adapted their turbulent-layer process to the treatment of zinc concentrates, which was originally developed for iron sulfides.

The application of the fluidization principle to the roasting of metallic sulfides followed the successful use of the process in the oil industry. Using this technique, a very high roasting rate per unit of chamber space could theoretically be obtained. However, in large scale practice, due to the limitations imposed by temperatures, velocity of gases, variations in fineness or composition of the feed plus normal operating difficulties, all of the theoretical advantages have not been realized. Actually some of the fluid-bed processes for zinc concentrates have evolved into a combination of fluidizing and flash roasting, with some of the advantages and disadvantages of both systems. This departure from true fluidization is evident from the comparatively large size of chamber, the low density of the gas-concentrate mixture in the freeboard section, and from the large percentage of dust carry-over, which are all of the same order of magnitude found in flash roasting systems.

A more detailed description of the several fluidizing systems emphasizes some of these points.

Dorr-Oliver FluoSolids Roasting Process. The process was originally developed for the roasting of gold-bearing pyrites but was soon extended to treat zinc concentrates and various other metallurgical materials. The large-scale pioneer work on zinc concentrates was made at the Arvida plant of the Aluminum Company of Canada, where a roaster treating about 150 tons of zinc concentrates per day has been operating for more than four years. This furnace is 22 feet in inside diameter and has a chamber 22 feet from the bed plate to the roof. The combustion air is supplied by a 9,000 cfm turboblower at 6 psi and fluidization of the bed is maintained by passing the combustion air through a dispersion plate having 1,400 quarter-inch holes uniformly spaced at six-inch centers.

The concentrate storage and feeding system includes a thaw shed and storage pit for the fresh concentrates besides a pug mill, a slurry holding tank, two slurry feed tanks and two special pumps for feeding the reactor.

The gas cooling and cleaning system is the same as in suspension-roaster installations of equal capacity and service, and consists of a three-drum waste heat boiler, standard type cyclones and an electrostatic precipitator. The gases are humidified and purified in a "Peabody scrubber" in addition to mist Cottrells in the contact acid plant.

Operation. The concentrate is fed as a slurry of 78 per cent solids at two points above the fluidized bed by means of Moyno pumps. When operating at a rate of 150 tons of concentrates per day, about 7,500 cfm of air at 4 to 5 psi are used. With normal concentrate of 58 per cent zinc and 31 per cent sulfur, a gas averaging 10 per cent sulfur dioxide and 0.3 per cent sulfur

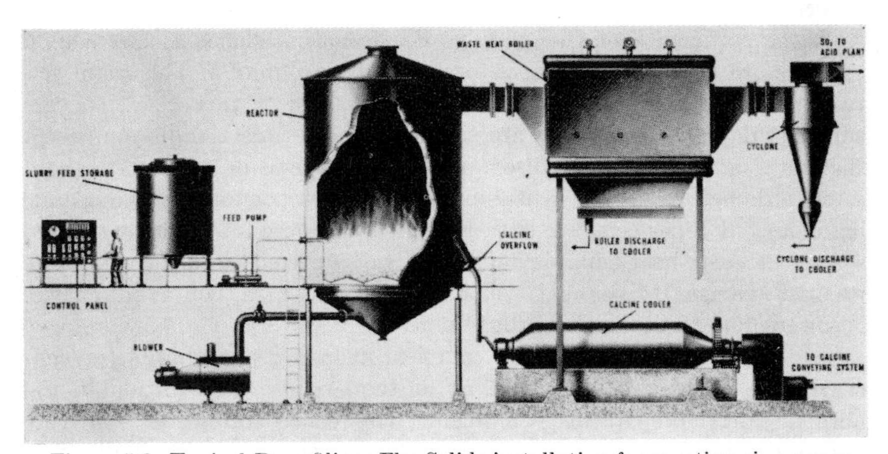

Figure 5-6. Typical Dorr-Oliver FluoSolids installation for roasting zinc concentrates. (*Courtesy Dorr-Oliver-Long Company*)

trioxide is produced. The temperature in the bed is controlled automatically by water sprays and is maintained at 1,640°F. The temperature in the freeboard or upper section of the chamber is also the same but occasionally may rise about 50 degrees higher than the bed. At normal rates of feed, about 30 per cent of the total product remains in the bed and finally overflows the discharge port in the wall of the reactor as finished calcine. The remaining 70 per cent is carried out with the gas stream and is progressively precipitated; 23 per cent of the total product as boiler dust, 44 per cent as cyclone dust and 3 per cent as Cottrell dust. The dust products are not returned to the furnace but are delivered via screw conveyors and Fuller-Huron air slides to a cooler and thence by Fuller-Kinyon pneumatic pumps to the calcine storage bins.

The gases are cooled in the boiler to about 465°F and an average of 0.94 net ton of steam per ton of concentrate roasted is recovered.

The total power consumed, including all uses from the receipt of concentrates to the delivery of calcine, varies between 125 kwh and 145 kwh per ton of concentrate treated.

The composite calcine which includes the reactor discharge, boiler, cyclone and Cottrell dusts all mixed together has a typical analysis of 0.54 per cent sulfide sulfur and 2.53 per cent sulfate sulfur and is mostly minus 200 mesh. This final product has a good zinc solubility and is a very desirable feed for the electrolytic zinc plant to which it is shipped.

The Dorr-Oliver roasting process is also used at several small electrolytic zinc plants in Japan at rates between 20 and 140 tons of zinc concentrates per day, and recently for rates of 140 tons daily at the Anaconda plant in the United States, with results and procedures comparable to the Arvida experience.

At the National Zinc Company in Bartlesville, Oklahoma, Dorr-Oliver reactors are used as preroasters for the sintering process. This plant has two 18-foot inside diameter furnaces, each treating 95 tons of zinc concentrate per day. These reactors are similar to the Arvida installation except that no waste heat boiler is used. The general operating results are about same, although the power consumption at 60 kwh per ton and the operating delays at 5 per cent are both less, probably accounted for by the absence of a waste-heat boiler installation. The sulfide sulfur in the composite product averages 0.7 per cent, which is abnormally high, but, as the calcine is sintered later, it is not a serious fault.

The feed handling is typically complex, including screening of the concentrate and swing-hammer milling of the oversize, in addition to the normal slurry preparation treatment. The concentrates are relatively coarse, a typical analysis showing only 60 per cent minus 200 mesh. This could account for the poor sulfur elimination and the small dust carryover which amounts to only 50 per cent of the total product.

The dust carry-over is precipitated very efficiently in a primary Ducon cyclone, in series with twin Buell secondary cyclones, and only about 1.2 per cent of the weight of feed, equivalent to two grains per cubic foot of gases at standard conditions, passes as dust to the acid-plant Cottrells.

General Remarks. From the foregoing descriptions, it is evident that the Dorr-Oliver FluoSolids process has many excellent features but is not entirely free from defects. The process normally requires no auxiliary fuel and the exhaust gases are rich in sulfur dioxide. It operates at an easily regulated and constant temperature and has the flexibility to produce various types of calcine. Also, having no moving parts inside the reactor, draft interruptions are minimized and, at the tonnage rates carried to date, fine grinding of the concentrate is unnecessary. A possible defect seems associated with the method of temperature control. The addition of water reduces the steam recovery and increases the hazards of corrosion in the boiler and gas-handling system.

The extreme care required in preparing and feeding the concentrate is also an undesirable feature and the lack of a method of sulfate sulfur control in the calcine might be a handicap under certain conditions.

St. Joseph Lead Company Fluid-Bed Process. This is a combined multiple-hearth and fluidizing roasting process specifically designed to produce a final calcine low in lead and cadmium and to concentrate these metals in a small bulk of Cottrell dust.

The zinc concentrates are first given a partial roast in the multiple-hearth furnace with insufficient oxidizing air, while a temperature of 1,650°F is maintained by extraneous heat supplied from oil or gas burners.

The semiroasted product at 23 per cent sulfur is screened through a one-inch trommel screen in closed circuit with a swing hammer mill and is then fed to a St. Joseph Lead Company type fluid-bed reactor by specially designed twin screw conveyors.

The reaction chamber is operated under a pressure of 5 inches water gage and at a steady temperature of 1650°F. Control is by sprays of zinc sulfate solution or water and by variations in the rate of feed. About 50 per cent of the product leaves the furnace as dust, of which 90 per cent is precipitated in two brick-lined cyclones, one at each gas outlet. These cyclones are operated under pressure, with a pressure drop of 4 inches water gage. The gases from the cyclones pass to a waste heat boiler, under a negative pressure of 0.1 inch at the inlet and about 2 inches at the outlet, and then flow to a Cottrell and acid plant.

The products, which are discharged from the reactor bed and from the cyclones, are at 1,650°F and are practically denuded of lead and cadmium. They are kept separate from the boiler and Cottrell dusts, which are united with the preroaster Cottrell dust, for lead and cadmium recovery.

A gas concentration of about 10 per cent sulfur dioxide is normally ob-

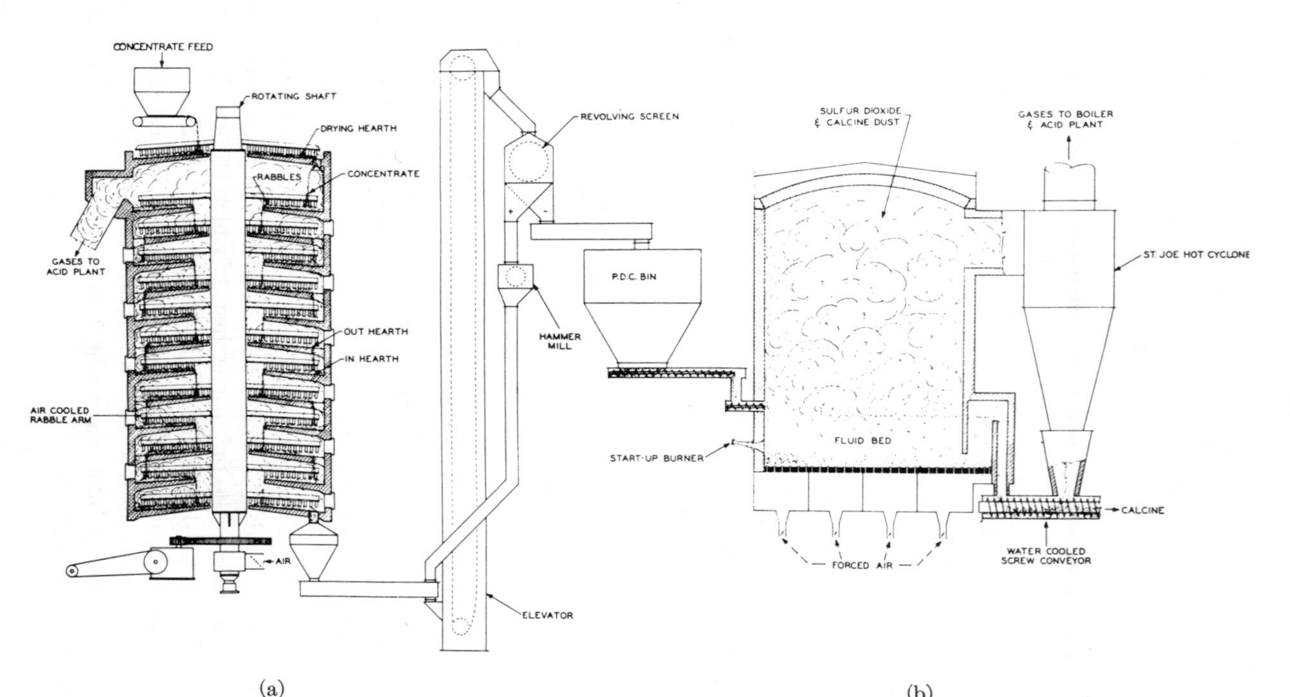

Figure 5-7. Two-stage fluid-bed roasting process. (a) Hearth roaster for first stage partial roasting (deleading and partial desulfurization). (b) Reactor for second stage fluid-bed roasting (desulfurization). (*Courtesy St. Joseph Lead Company*)

tained in the outlet gases and this is kept reasonably constant by automatic regulation of the combustion air supply on the basis of the percentage oxygen in the boiler inlet gases.

A combination of two 25-foot diameter Wedge roasters and one fluid-bed reactor treats up to 250 tons of concentrates per day. The final calcine normally assays under 0.5 per cent sulfide sulfur and the minimum of sulfate sulfur. Most of the lead is concentrated in the Cottrell dust with from 92 to 97 per cent eliminated from the calcine. The cadmium is also concentrated in the Cottrell dust but to a lesser extent.

The outstanding features of the process are the efficient elimination of lead from the calcine and the effective use of cyclones ahead of the boiler, in addition to the carefree operation of the fluid bed.

The New Jersey Zinc Company Fluid-Column Process. While this is a new process which has not yet undergone much commercial experience, it has several features which could prove to be attractive for certain applications.

The furnace is rectangular in plan, long and narrow at the base and wider in the upper section, similar to a blast furnace. Combustion or fluidizing air is admitted progressively in three stages; some through a grate at the base and the remainder through tuyères along both sides, in two zones, one above the other.

The feed to the furnace consists of dried pellets between 4 and 20 mesh, composed generally of a mixture of zinc concentrates and process dust, plus a little sulfuric acid for binder. The pellets are charged to the furnace through side ports above the top of the fluidized bed of material and the roasted product is discharged through a chute, with a valve control, which is regulated to adjust the depth of bed in the furnace.

A temperature of 1,050°C is normally set and is controlled by the rate of feed and by water additions to the bed, with lances through the roof of the furnace.

Air is supplied by a blower at 5 psig and a gas of about 12 per cent sulfur dioxide is produced at the furnace outlet. The dust burden which leaves the furnace does not exceed 20 per cent of the total product and is precipitated in the boiler, cyclone and Cottrell. This dust is returned to the furnace after pelletizing with the concentrates.

The calcine is also in the form of pellets and assays about one per cent total sulfur, with 0.25 per cent as sulfide sulfur.

The boiler is operated at 450 psig and 0.685 pound of steam per pound of concentrate roasted is recovered.

The largest fluid-column reactor which has been installed to date has operated at the La Oroya, Peru, plant of the Cerro de Pasco Company on a rate of about 100 tons per day, but it is understood that a much larger furnace is now under construction. What the maximum size of this design of furnace might be has, of course, not been determined but it is believed that capacities exceeding 200 tons of concentrate per day should be attainable. The limiting factors would appear to be the rate at which the sulfur could be oxidized at the center of the pellets or the amount of water which could safely be used for cooling.

The attractive features of the process over other fluid-bed processes are claimed to be the low percentage of dust carry-over and the potentially high capacity of a reactor without loss of fluidization from incipient fusion in the bed, plus the elimination of short-circuiting of unroasted material to the discharge port.

However, some disadvantages remain, including the necessity first to pelletize the feed and later to grind the calcine when treated by the electrolytic zinc recovery process. Also, the ferriting ratio between iron and zinc oxides must be high because of the intimate contact within the pellets at high temperatures.

Other Fluidization Systems. The Badische-Anilin and Soda-Fabrik, AG, have had a turbulent-layer process operating successfully on pyrites for some years, which has recently been modified for zinc concentrates.

The furnace is divided into two sections to prevent short circuiting of the feed to the discharge port and the temperature control is obtained principally by means of steam generating elements in the fluidized bed. The concentrates are pelletized to 6 millimeters, with varying proportions of dust according to the screen size and calorific value of the concentrates, and are fed to the furnace by an enclosed table feeder, sealed to prevent gas leakage.

The auxiliary equipment is of the conventional type, normally including a boiler, cyclones and an electrostatic precipitator.

The dust carry-over is said to be about 8 per cent of the concentrates treated and the steam produced to be between 0.75 and 1.0 tons per ton of blende treated. Other operating results are comparable to those with fluidizing furnaces generally.

There is also an Overpelt-Lommel fluid-bed process developed in order to eliminate the sintering step in the preparation of zinc calcine for the horizontal retorts, and some information gathered from various sources has indicated that the results have been very satisfactory.

It is understood that the furnace is rectangular in plan, approximately 4 feet wide, 23 feet long and some 20 feet high above the grate or dispersion plate, and has a capacity to treat about 180 metric tons of concentrates per day.

The feed to the furnace is in the form of practically dry pellets composed of zinc concentrate mixed with the furnace-dust products and some acid for binder. The pellets are fed above the bed, at two or three locations on each side and near one end, and the roasted product overflows a weir at the opposite end.

Air for fluidizing and oxidation is supplied by blowers at pressures varying between 1 and 2.5 psi. The temperature is maintained just below the sintering point—approximately 950 to 1,050°C—and is controlled by adjusting the rate of feed at the several feed points, according to conditions in that particular section.

The gases at 7 to 10 per cent sulfur dioxide and at about 1,000°C carry a dust load varying between 10 and 20 per cent of the feed, depending on the fineness and composition of the concentrates.

The gases are cooled in a waste heat boiler and about 15,000 pounds of steam per hour at 600 psi and 400°C are produced. The steam is consumed in turbines and some 2,000 to 2,500 kilowatts of electric energy are generated.

The final granular product from the furnace normally assays under 0.1 per cent sulfide sulfur, with the minimum possible of sulfate sulfur, and is fed directly to the horizontal retorts.

SUMMARY

The foregoing descriptions of processes have indicated that there are several types of efficient roasters available but that, for certain applications, some one system may have advantage over the others. Before a decision could be satisfactorily made, certain factors should be carefully considered and evaluated. These factors are: firstly, the metallurgical requirements, including the capacity in tons per day, the physical and chemical qualities of the product, the means of sulfur dioxide recovery and the versatility desirable in tonnage rates and quality of calcine; with, secondly, an estimation of operating costs, including labor, power, fuel and maintenance; and, finally, the comparison of capital costs which are mainly affected by the number, size and complexity of the various parts of the installation.

In this connection, a brief outline of the significant features of each system of roasting might be of some value.

Multiple-Hearth Roasters

The most attractive feature of this system of roasting is the general simplicity of the process. The furnaces can treat an unlimited variety of concentrate without elaborate feed preparation and the calcine obtained is normally very constant in quality and, at a moderate temperature, not difficult to handle. The operating controls are few and easy to adjust, and the percentage of operating lost time is very low. Also, means of eliminating lead and cadmium from the zinc calcine are readily available.

Against these advantages, however, should be placed the low capacity of the units, indicating high capital and labor costs per ton treated, the consumption of fuel, except under special conditions, and the production of low sulfur dioxide gases, generally contaminated with carbon dioxide. Moreover, steam recovery is seldom worthwhile because of the large volumes of low-temperature gases, and maintenance costs for brickwork and rabbling are generally substantial.

Fluidization or Turbulent-Bed Roasters

The outstanding credits of the several types of fluidization roasters include the production of rich sulfur dioxide gases, no requirement for mechanical equipment operating under high temperatures, no auxiliary fuel, a fair steam recovery, a reasonably high unit capacity, good control of the roasting conditions and the production of the desired quality of calcine.

These advantages are modified to some extent by the exacting methods required to feed the reactors. The preparation of pellets, slurries, or of partially roasted concentrates needs considerable equipment and operating attention, which adds to the costs. Other controversial disadvantages seem to be the methods of cooling since injection of water reduces the steam production and is limited in the range of application, whereas cooling by steam-producing elements in the bed has not been fully proved for zinc concentrate roasting.

Suspension Roasters

This system of roasting has many of the advantages of both multiple-hearth and fluid-bed processes, and some of their faults.

The outstanding features are the versatility and high capacity of the units. Almost every type of zinc concentrate can be treated directly without special preparatory treatment, and furnaces can be operated within the range of half to full capacity and can be built for capacities from 25 to 500 tons per day. The operation of a furnace can be quickly changed to produce dead-roasted or high-sulfide sulfur calcines as desired. Sulfate sulfurs can also be controlled within the range of 0.5 per cent up to 3.0 per cent. Rich sulfur dioxide gases are obtained, no auxiliary fuel is required and the steam recovery normally exceeds one ton per ton of concentrate roasted. The ferritization ratio in the calcine is at a minimum and high zinc solubilities are obtained.

The process is specially adaptable to oxygen enrichment of the combustion air because of the efficient method of restricting temperatures by recycling cooled gases. Also, through this system of cooling, the process can be switched to iron sulfide roasting with only minor adjustments.

The disadvantages, which include grinding of the dried concentrates and mechanical rabbling of the calcine, are more apparent than real. Grinding after drying is not an absolute necessity with the average zinc concentrate but, if the percentage of sulfur remaining in the calcine is the measure or limiting factor of the capacity of a unit, then grinding will raise the capacity or, alternatively, it will lower the sulfur in the product.

Mechanical rabbling provides a simple and efficient method of decomposing sulfates in the dust products or of adding other materials to make a single homogeneous product. With modern alloys and methods of design, difficulties of rabbling have become of minor importance.

SINTERING

The practice of sintering zinc ores was adopted, in a fairly well developed state, from the lead industry, although many important modifications were later necessary to suit the special needs of the thermic reduction processes for zinc-bearing materials.

The first recorded use of the sintering process for zinc concentrates was made at Port Pirie, Australia, in 1917. Here, some green ore and partially roasted concentrate, averaging 9 per cent sulfur, were mixed and sintered by a single pass on a Dwight-Lloyd machine. This new procedure was remarkably successful and it was soon followed by other zinc plants, notably at the Baelen plant of the Société Anonyme des Mines et Fonderies de Zinc de la Vieille Montagne, although at this plant the process was modified to treat dead-roasted calcines mixed with 6 per cent of fine coke.

During the same period the original Port Pirie process was installed at Bartlesville, Oklahoma, by the National Zinc Company, but it was later changed to a method of their own, using calcines with only 2 to 3 per cent of sulfur, supplemented with 3 per cent of coke breeze and, sometimes, anthracite coal.

Since these early experiments in the sintering of zinc ores and concentrates, numerous procedures have been tried, covering many types of sinter feed, ranging from green concentrates, and partially roasted concentrates alone, to several qualities of calcine mixed with different proportions of carbonaceous fuel. However, currently established practices have fairly well settled down to a choice of four methods which could be briefly summarized as follows:

1. The practice of sintering dead-roasted, or low-sulfur calcine, with the addition of carbonaceous fuel.
2. The two-stage sintering of dead-roasted calcine mixed with carbonaceous fuel.
3. The direct sintering of unroasted zinc concentrate mixed with a high proportion of sinter returns.

4. The sintering of partially roasted concentrate, 7 to 9 per cent sulfur in the calcine, without carbonaceous fuel.

Although recent installations have indicated a trend to the direct sintering of fresh concentrates, the process of sintering dead-roasted calcines with fuel has been the more common practice, especially in the United States. In this system, the preroasting has generally been accomplished in Hegeler, Ropp, and circular multiple-hearth roasters, but gradually suspension roasters and fluid-bed reactors are supplanting them.

The preference for this combined system of dead-roasting and sintering has been largely influenced by a number of factors; in some cases by the fact that roasters and acid plant were already in existence and, in others, by the requirement for an especially uniform and robust quality of sinter. Also, the economics of good sulfur recovery, increased sintering capacity and the availability of low-cost fuel were important considerations. Some of the interesting plants using this system will be described.

Sintering Dead-Roasted and Low-Sulfur Calcines with Fuel

The New Jersey Zinc Company, at their Palmerton and Depue plants, sinter dead-roasted calcines from suspension roasters on Dwight-Lloyd machines of American Ore Reclamation Company design. The larger sintering machine at Palmerton, which has an hourly capacity of 20 tons of finished sinter, is 105 feet long by 6 feet wide. There are twelve windboxes each 6 feet square.

The feed preparation equipment consists of a 4- by 12-foot double shaft, paddle mixer, and a pelletizing drum 10 feet in diameter by 14 feet long.

The sinter, after discharge from the machine, is crushed in slugger rolls, set at 1½-inch gap, and is then ground in a 6-foot diameter rod mill in closed circuit with a Hummer screen.

The exhaust gases are handled by one fan and are delivered, after humidification and dust settling in an acid-proof brick-lined chamber, to an electrostatic precipitator and then to a 300-foot stack.

The feed to the machine, in the form of pellets ranging from fines up to ¼ inch in diameter, is composed of calcine, ground sinter, and anthracite screenings, mixed in the proportions of 100 calcine, 70 sinter and 8 coal. The pellets contain under 0.5 per cent sulfide sulfur and under 2.0 per cent sulfate sulfur. They are fed to the grates by a swinging spout to form a bed about 6 inches deep. The bed is ignited by an oil-fired muffle, consuming about 2.5 gallons per ton of sinter produced. A windbox draft varying between 10 and 16 inches of water gage is maintained. The gas temperature at the fan inlet varies between 125°C and 175°C. The sinter, as discharged from the machine, could be described in general terms as hard, porous and moderately friable. No mechanical means are provided for shaving off or separating any portion richer in lead and cadmium and the entire machine discharge is crushed and ground to minus 14 mesh. About 45 per cent of the total is returned to the feed preparation system and the remainder is the finished product for subsequent processing by the vertical retort process.

This combined suspension roasting and sintering procedure has given very satisfactory results and was chosen because the installation and operating costs were estimated to be less than with other methods. Also, it was reasonably certain that a high sulfur recovery from the rich roaster gases would be obtained and that the sintering operation itself would be much simplified when unrestricted by acid-plant requirements.

The practice at Palmerton is generally typical of that followed by several other companies in America, although many plants have instituted numerous changes in design and operation to suit their particular conditions.

The National Zinc Company at Bartlesville, Oklahoma, sinters calcine containing about 3 per cent sulfur, preroasted in a Dorr FluoSolids reactor. The calcine is mixed and pelletized in a pugmill with about 60 to 70 per cent of the final sinter produced, along with flue dust, and 4 per cent of an equal mixture of fine coal and coke breeze. Moisture is supplied by waste liquor from the cadmium recovery plant. The pellets range in size from ½ inch to minus 300 mesh, with over 60 per cent in the plus 20-mesh fraction. They are fed by a swinging spout and ignited by gas-fired muffles. Desirable high temperatures in the bed are attained by good porosity. High temperatures greatly improve the lead and cadmium elimination, which is also helped by some chlorine in the cadmium plant waste liquor.

The sinter as produced for the horizontal retort process is under one per cent sulfur, weighs about 118 pounds per cubic foot and is quite porous and easily crushed.

The current practice at Bartlesville has definite advantages over the former method of sintering high-sulfur calcines. Although coke and coal are now consumed and some sulfur dioxide wasted, the capacity of the sintering machines has shown a 30 per cent increase. Also, the sinter has been lower in sulfur, more uniform in quality, and the lead and cadmium elimination has been substantially increased.

Another retort zinc producer, the Donora zinc plant of the American Steel and Wire Division of the United States Steel Corporation, sinters multiple-hearth calcine, containing under 2 per cent sulfur, mixed with 30 per cent recirculated sinter and 5 per cent crushed coke. In order to increase the lead and cadmium elimination, the feed is moistened with water-soluble chlorides and the chlorine maintained at 0.4 per cent in the mixed feed. In addition, an auxiliary gas burner is fired over the second windbox section.

About 240 tons per day of sinter are produced. This is crushed in rolls to about 58 per cent through an 8-mesh screen and is treated in horizontal retorts.

The Matthiessen and Hegeler Zinc Company's sintering practice is very similar in method and results to that of the Donora plant, with the exception that no chlorides are added to the feed and satisfactory lead and cadmium elimination is obtained by higher temperatures in the sinter bed.

At the Amarillo plant of the American Smelting and Refining Company, calcine from Ropp furnaces at about 5 per cent sulfur is mixed on a revolving table with 20 per cent return sinter and 15 per cent crushed coal or coke, and the whole moistened with 13 per cent water. This mixture is fed directly to the Dwight and Lloyd sintering machine without further pelletizing or nodulizing.

The gases, at about one per cent sulfur dioxide, are cleaned in a conventional baghouse, preceded by polyclones, and are discharged to atmosphere through a 400-foot stack.

The sinter, which is described as hard and dense with medium porosity, is treated without crushing or grinding in horizontal retorts.

Two-Stage Sintering of Low-Sulfur Calcines

An interesting variation of the preroast sintering method is practiced at the Rosita plant of the Mexican Zinc Company, whereby the concentrates after roasting in Skinner roasters to between 3.5 and 4 per cent sulfur are then sintered in two steps on a pair of Dwight and Lloyd machines in series. The charge to the first machine is a pelletized mixture of calcine, sized undercut from the second machine, Cottrell dust from the roasters, and coke breeze wetted down in a rotary drum to 12 per cent moisture. The first machine discharge is crushed and screened through ½-inch mesh, then pelletized in a rotary mixer with ¼-inch coke and 8 per cent moisture, and fed to the second machine by a swinging spout.

The second-over machine has a Bruderlin cutter, which removes the top 9 to 10 inches as finished sinter. The depth of cut can be varied to regulate the amount of lead and cadmium in the final product. The undercut, normally about 20 per cent of the total discharge, is crushed and sized and constitutes the return sinter for the first-over machine.

The gases from both machines are cleaned in a nine-sectioned standard baghouse, using some woolen and some Orlon filter bags. The fans on the second-over machine only are preceded by polyclones to remove the abrasive particles.

Since changing from single-pass to double-over sintering, definite advantages have been gained. The sinter is denser and more uniform in quality and the elimination of the lead, cadmium and sulfur has been improved.

The St. Joseph Lead Company's method of sintering dead-roasted calcines is, in principle and apparatus, very similar to the foregoing practices, but has been modified in many ways to meet the special needs of the electrothermic reduction process. This process requires an exceptionally hard and strong sinter, crushed and screened to a relatively uniform size. High porosity and density, though desirable, are less important.

In order to obtain these qualities, double sintering with silica flux added to the charge of the second-stage machines is practiced. In addition, greater than normal amounts of coke breeze and sinter returns are required on both machines. Zinc sulfate liquor, a by-product of the cadmium plant, is added to improve the pelletizing and, incidentally, to recover the zinc in the liquor.

Although this sintering procedure might seem to be very complex, several improvements in the machines, including a new design of grates and mechanical rappers, drop bar seals, roller bearing pallet wheels and the elimination of the sprocket on the discharge end, have contributed to an efficient operation.

Sintering of Unroasted Concentrates

The Robson or Green-Ore Sintering Process. This sintering process was originated and successfully developed by the National Smelting Company at Avonmouth, England, in 1930, and has been adopted in principle by the Blackwell Zinc Company in America and by Unterharzer Berg- und Hüttenwerke at Goslar, Germany. Another German company, "Berzelius," Metallhütten Gesellschaft at Duisberg is now in process of installing a new plant using the Robson system, which will replace the existing Deplace roasters and the Dwight and Lloyd round-table sintering machine.

Briefly, the Robson process is essentially a roasting and sintering operation performed entirely on a sintering machine unit. This double function is made possible through the expedient of diluting the sulfur content of the concentrate with a high ratio of sinter returns. By this means, excessive temperatures are avoided and good porosity through the bed is maintained. The sintering conditions are further aided by the recirculation of windbox gases, which not only improves the combustion control but enriches the sulfur dioxide gases for the acid plant.

The advantages of this system are counterbalanced to some extent by the necessity for considerable handling and conveying of dusty materials and by a relatively complicated and sensitive operating procedure.

At the National Smelting Company's works, the sinter feed is in the proportion of five of returns to one of zinc concentrate. The returns are first broken down to $\frac{5}{32}$ inch in a Sturtevant rotary cone crusher and then are mixed with the green concentrates and water in a double-shafted paddle mixer similar to a pugmill. The mixture is discharged as pellets, with 7 per cent moisture and 6 per cent sulfur, varying in size from $\frac{3}{16}$ inch to minus 200 mesh. The pellets are fed to the sintering machine by an apron feeder discharging to a distribution tray set at 60 degrees. Ignition of the charge is by means of an oil-fired suspended brick firebox.

Among the important features developed at Avonmouth and also practiced at Goslar, Germany, has been the system of recirculating the windbox gases from the last half of the machine to a hood covering the first half of the bed. By this means, the total exhaust gases have been enriched to 6.5 per cent sulfur dioxide and a better control of bed temperatures has been obtained.

The many progressive features of the operation have contributed to the production of a soft, porous and friable sinter, very desirable for the horizontal and vertical retort processes.

Since the adoption of this sintering method in 1930, operating costs have been decreased and a better quality of sinter, with increased production, has been obtained.

The process at Goslar, Germany, is virtually the same, except in details of equipment. A Lurgi Dwight and Lloyd type machine is used and, in addition to the same ratio of sinter returns, calcium sulfate is added to the pelletizing drum to aid in temperature and porosity control.

The Blackwell Zinc Company at Blackwell, Oklahoma, which has adopted a modified form of the Robson process, is also of special interest because of the magnitude of the machine and the manner in which the concomitant problems have been solved.

The machine (Figure 5-8) has a hearth width of 12 feet, a windbox length covering 168 feet, and has the capacity to produce up to 550 tons of finished sinter per day from zinc concentrates running 31 per cent sulfur. The principal features in the design, as described by Urban,[34] included a method of minimizing the corner wear on pallets, the prevention of serious damage to the machine by out-of-step pellets, the moderation of impact forces at the discharge end, and the provision for longitudinal expansion. Failure to have provided these features would undoubtedly have caused heavy maintenance and other troubles on such a large machine.

The Blackwell modification of the Robson process relates chiefly to the changes in the recirculation of the gases. At Blackwell, the gases are wasted to atmosphere and a strong sulfur dioxide is not desired. However, in order to improve the elimination of lead and cadmium from the sinter and to reduce the volumes and dust loads handled by the baghouse, the gases are recirculated from the first ten windboxes to a hood covering the bed above the last nine windboxes.

Normally, as at Avonmouth, no carbonaceous fuel is used, but in certain circumstances when a denser sinter is desired, a separate campaign is run on the sintering machine and the crushed first-run sinter is mixed with about 3 per cent of fine coke and resintered.

Figure 5-8. Aorco 12 x 168-ft sintering machine at Blackwell Zinc Company, Blackwell, Oklahoma. (*Courtesy Koppers Company*)

Figure 5–9. Discharge end of 12-ft sintering machine at Blackwell Zinc Company, Blackwell, Oklahoma. (*Courtesy Koppers Company*)

The operating results of this large machine since 1951, as compared to previous results obtained from the sintering of Ropp roaster calcine on small machines, have shown an improved sulfur and lead elimination, a more uniform sinter, less operating labor per ton sintered, and an over-all reduction in operating costs.

Sintering Partially Roasted Concentrates

This was the original sintering process, developed by Gilbert Rigg in Australia, for the sintering of zinc calcines containing 9 per cent sulfur, but there remain only a few followers of this practice at the present time. The Overpelt-Lommel Company in Belgium formerly used this method in conjunction with a special process for feed preparation, whereby spaghetti-like pieces of feed were obtained by extruding a paste of calcine and acid through a perforated plate with reciprocating rollers. A small amount of coke breeze was added to compensate for heat losses at the sides and surface of the bed, and ingenious methods to accomplish this were used to distribute the feed and coke on the pallets. The gases were concentrated to over 6 per cent sulfur dioxide by recirculation of the discharge-end windbox gases to a hood over the bed. The final exhaust gases were mixed with the preroaster gases for conversion to acid.

The sinter was low in sulfur, of good density and porosity, and very suitable for speedy reduction in the horizontal retorts. However, in recent years, this combined roasting and sintering process has been superseded by a turbulent-bed roasting process, which produces a very satisfactory feed for the retorts without sintering and with less labor and total costs.

The Société Anonyme des Mines et Fonderies de Zinc de la Vieille Montagne at Baelen, Belgium, and also the Berzelius Company in Germany, pending the installation of the new plant, sinter partially roasted calcines. At both plants, green concentrates are added to the preroasted calcine, which, along with the return sinter, brings the charge to about 7 per cent sulfur. A small amount of low-grade coal is also added at Baelen in order to improve the density of the sinter. No recirculation of windbox gases is practiced but a sulfur dioxide concentration of 4.5 per cent is nevertheless obtained, which is converted to acid. The sinter is low in sulfur and has excellent physical characteristics for the horizontal retort process.

CONCLUSION

The foregoing has been chiefly confined to a general description of current sintering procedures with very little discussion about the merits or demerits of any particular system. However, because no process has yet attained perfection, some remarks concerning the potential field for experiment and progress might be appropriate.

Since the rate of growth in the size and capacity of sintering machines has been exceedingly great within recent years, some slowing down in that direction might be expected, at least until more data on large machines have been accumulated and their value fully assessed. In the meantime, schemes for improving other phases of the process should bear investigation.

Possible methods of eliminating the consumption of ignition fuel should be studied because, theoretically, there are ample calories available for completely autogenous sintering. The same problem baffled roaster metallurgists for many years but was finally solved by various techniques.

Also, methods of reducing the volumes of exhaust gases should be developed more generally. The potential gains here are appreciable both in the installation and the operation. Reduced volumes of gases connote smaller flues, fans and dust-catching apparatus and lower handling costs. The recirculation of windbox gases has been a step in the right direction, but improved methods of combustion air and temperature controls and better protection against false air infiltration would be advantageous. In this connection, the redesign of sintering machines along updrafting principles would simplify this problem.

No attempt to recover waste heat from the gases in sintering operations has been recorded. If some means of temperature control could be obtained without the necessity for excessive volumes of air, high moistures in the charge and the tremendous circulating loads, waste heat boiler installations would be practical and a worthwhile production of steam could be obtained.

The biggest item in sintering costs is connected with the handling of materials due to the recycling of huge tonnages of sinter. These large circulating loads are inevitable under existing designs and practices in order to obtain good sulfur elimination and the desired quality of final sinter. However, the potential operating savings that are possible should stimulate the search for better methods and apparatus to obtain these results.

The use of oxygen-enriched air for the sintering process does not seem to have been investigated or very seriously considered. The operating advantages from the successful use of oxygen enrichment would be impressive and could include, among other advantages, greatly increased unit capacity without any increase in size, relatively small volumes of exhaust gases with high concentration of sulfur dioxide, improved elimination of lead and cadmium, completely autogenous sintering and a potentially valuable heat recovery as steam from the exhaust gases.

Many of the foregoing possibilities for improving the sintering process are somewhat visionary at the present time, but it is believed that some design or scheme of combining suspension roasting and sintering in one contained unit might offer a solution to all these problems and should provide considerable opportunity for investigation.

CALCINING

The term "calcining" may be defined as the process of applying heat to ores and other metallurgical materials, without oxidation of the minerals, for the purpose of expelling the water of hydration, and the carbon dioxide and other gaseous dissociation products which would otherwise interfere with the subsequent treatment.

The process is similar to a drying operation but differs from it in the higher temperatures employed. The zinc minerals must be heated to 400°C to give up their carbon dioxide and water of hydration, and the gangue minerals require temperatures up to 800°C in some cases. During calcination, carbonate ores lose up to 30 per cent of their original weight.

In the zinc industry, calcination processes formerly were used almost exclusively for the preparation of carbonate ores for the recovery of zinc by the horizontal retort process.

The calcination of carbonate ores was originally carried out in brick kilns. Alternate layers of ore and fuel were charged at the top of the kiln—a circular brick structure about 10 feet in diameter and 20 feet high—and combustion air was admitted through tuyères around the base. About 40 tons of ore, with from 4 to 10 per cent of fuel, were treated per day. In another type of kiln, the ore was charged alone and the charge was heated from external fireboxes. Although more fuel was used, this method was generally preferred because the ore did not become contaminated with ashes from the fuel. Later, the brick kiln and shaft furnaces were superseded by rotary kilns of the White-Howell type. The kilns were inclined and were charged at the upper end with counter-current flow of the heating gases from coal fires at the discharge end.

Fuel consumption varied between 8 and 14 per cent of the weight of ore. The dust carry-over was high and the methods of recovering the dust were usually inadequate, and losses were excessive. However, as the mines which produce carbonate ores are becoming depleted and as other methods of treating oxidized ores are being employed, calcination practices in the zinc industry are now practically limited to the treatment of secondary materials for the removal of chlorine, fluorine, carbon and some other elements deleterious to the recovery process.

At the present time the few companies that utilize calcination methods on zinc materials have established that the multiple-hearth rabbling furnaces are the most satisfactory for the purpose—the dust carry-over is low, fuel consumption is not excessive and maintenance costs at the low temperatures involved are negligible. This, of course, does not include the treatment of zinc oxides from fuming operations where deleading is practiced at temperatures much above the generally accepted figure for calcination.

Under these conditions, brick-lined rotary kilns, some with coolers included, are almost universally used.

Zinc materials that currently require calcining as distinguished from de-leading are mainly zinc oxides from fuming operations which are to be treated by the electrolytic process.

The Société des Mines et Fonderies de Zinc de la Vieille Montagne, at their Baelen plant, treat zinc oxides from slag reduction furnaces to remove carbon and to complete the oxidation of the constituent elements. A Wedge type six-hearth furnace, 20 feet in diameter, is used for calcination. The feed contains 50 per cent zinc, 15 per cent lead, 4 per cent sulfur, 12 per cent moisture and about 5 per cent carbon. Normally, 20 tons per day are fed to the furnace. No extraneous fuel is required because sufficient carbon is present in the feed to support the combustion, and temperatures of 700 to 800°C are obtained on the third hearth, which is the hottest hearth. The dust carry-over is less than 2 per cent and it is practically all recovered in a standard baghouse.

The fume, after calcination, has become densified to an apparent density of 1 to 1.5 from less than 1.0 in the feed. The carbon is all removed and the total sulfur, which is all sulfate sulfur, is reduced to 2 per cent, while the lead and zinc contents are practically unchanged. The product is a desirable feed to the electrolytic zinc plant for the production of metallic zinc.

The Hudson Bay Mining and Smelting Company at Flin Flon, Manitoba, also treats zinc oxide fume from the slag-fuming plant in order to expel the chlorine, fluorine and occluded sulfur dioxide, and to oxidize the arsenic before leaching in the electrolytic zinc process.

Two seven-hearth Wedge roasters which have a maximum capacity of 90 tons each per day are used, although only about 130 tons of fume are cal-cined per day. A temperature of 650°C on the fourth hearth and between 550 and 650°C on the seventh hearth is maintained by two fuel oil burners on the fifth hearth plus three burners on the seventh hearth. The oil is burned in Dutch oven combustion chambers attached to the shell, and about 6 gallons of Bunker C oil per ton of fume treated are consumed. The fume tends to fuse on the fourth hearth and becomes difficult to rabble. This is minimized by means of double spacing the rabbles on this hearth and by avoiding temperatures in excess of 650°C. The calcined product is discharged to water-cooled screw conveyors and thence to a Redler elevator and a $\frac{1}{4}$-inch vibrating screen in closed circuit with an Allis-Chalmers pulverizer.

The calcined product, from which 99 per cent of the fluorine and sulfur dioxide have been eliminated, and with the troublesome impurities, arsenic and antimony, rendered less soluble in the sulfuric acid solution, is delivered to the oxide leaching plant.

CONCLUSION

Due to the scarcity of oxidized zinc ores and the development of new processes for zinc recovery, the need for calcination methods has gradually become of minor importance in the zinc industry. On the other hand, calcination methods are widely used in the cement and other branches of the chemical industry. Modern practice in these industries has predominantly favored the rotary kiln, but fluidization processes are gradually gaining ground and may eventually equal the use of kilns.

BRIQUETTING

Briquetting is a process for compressing fine materials into small blocks, or agglomerates, of sufficient mechanical strength to resist reasonable handling and weathering without deterioration.

The main purpose of briquetting in the zinc industry is to form strong briquettes composed of a homogeneous mixture of zinc-bearing material

Figure 5-10. Single pocket briquette press. (*Courtesy Komarek & Greaves*)

and reducing agent, which will not impede the flow of gases or seriously disintegrate during the reducing operation.

Many of the basic problems were first solved in the briquetting of coals, particularly in Belgium and Germany, where the art flourished in the latter part of the last century. The first tangential, or roll press, was invented by a Belgian monk in 1865 and this was the prototype of the various designs most generally used today.

Several types of machines were developed, including many varieties of plunger presses, with pressure applied to one side only, or to both sides, extrusion presses and tamping presses. The roll press is now used almost exclusively for the briquetting of zinc-bearing materials, although plunger presses are used for some purposes.

The successful production of briquettes from zinc-bearing material depends upon the characteristics of the material being briquetted, as well as the proper selection of reagents and the careful functioning of each stage of the operation.

Although each material may require certain variations in procedure, the

Figure 5-11. Heavy duty pug mill. (*Courtesy Komarek & Greaves*)

briquetting process for zinc-bearing materials could, in a general way, be divided into five principal steps as follows:

1. *Preparation of the Zinc-Bearing Materials.* This step usually entails some drying and grinding because the material must be dry and fine enough to mix uniformly with the other ingredients.

2. *Preparation of the Coal or Coke.* The reducing material should also be reasonably dry and uniform in size, generally in the 6- to 12-mesh range.

3. *Mixing and Plasticizing.* This step is usually accomplished in two stages; firstly, the mixing of the material and binder in a rotary drum or paddle-mixer; and, secondly, the plasticizing in Chilean mills (edge runners).

4. *Pressing of the Plastic Mixture in the Briquetting Machine.* Tangential, or roll-type presses, are in most general use today. Pressure is applied by a pair of horizontal rolls revolving in opposite directions. The faces of the rolls are exchangeable shells, provided with molds, preferably in staggered rows. Plunger-type presses are also in use in some plants in Europe, especially where larger-sized briquettes are desired. Extrusion presses have also been used for special applications. Pressures applied range from 1,200 up to 5,000 psi, depending on the machine and materials being briquetted.

5. *The Drying, Screening and Storage of the Finished Briquettes.* The briquettes should be handled carefully on pan or belt conveyors to avoid breakages. In some cases they pass through dryers or coking tunnels and over grizzlies. Normally they are conditioned in storage for 24 hours or more before use.

A description of briquetting, as developed for the vertical retort process in America, has been most ably described in Chapter 6, The Vertical Retort Process, but some account of briquetting methods, as currently practiced in Europe, might also be of interest.

At the Baelen, Belgium, plant of the Société de la Vieille Montagne, briquetting of slags has been practiced for a number of years in order to produce zinc oxide fume for the electrolytic zinc process.

The granulated slag, 14 per cent zinc, 23 per cent iron, 2 per cent lead and from 3 to 5 per cent moisture, is mixed with fine coal in a horizontal paddle-mixer along with 4 per cent of dried sulfite liquor. The mixture passes directly to the briquette press through a special type of feeder consisting of an agitator tank, with a patented distribution box having three regulating gates.

The briquette press is the pocketed-roll type made by the Sahut Conreur Company in France. It produces 7 tons per hour, equivalent to a rate of 30,000 briquettes per hour.

The briquettes are discharged to an apron conveyor which passes through a coking tunnel held at 200°C. Coking normally requires about three hours.

The briquettes are egg-shaped, 75 millimeters long and weigh about 220

Figure 5-12. Various types of briquettes. (*Courtesy Komarek & Greaves*)

grams apiece. They contain 12 per cent zinc, 11 per cent carbon and 25 per cent ferrous oxide and have a high degree of compressive strength. The per cent recycled, due to degradation in handling, amounts to less than 10 per cent of the total.

The briquette machines have proved very satisfactory, major repairs being normally limited to the replacement of the pocketed wheels after the production of about 7,000 tons of briquettes.

The Sahut Conreur briquetting press is also used by the Société de la Vieille Montagne for briquetting zinc calcines at another plant in France.

The calcine, containing 69.7 per cent zinc, 3.7 per cent iron, about 0.8 per cent lead, and 1.5 per cent sulfur, is mixed with bituminous coal in the ratio of 600 calcine, 400 coal and 120 recycled briquettes. An equal mixture of pitch and tar, to the extent of from 5 to 6 per cent of the total weight of the mixture, is added as a binder. The materials are mixed in a steam heated paddle-mixer and then plasticized in a Chilean mill. The temperature at the start is about 75°C and drops to about 45°C after discharge to the bin which feeds the press.

The press produces 7 tons per hour, equivalent to 70,000 briquettes of 100 grams each. They are somewhat egg-shaped, 54 millimeters long by 36 millimeters thick and 47 millimeters high, and are composed of about 36 per cent zinc, 28 per cent carbon and 1.6 per cent iron. The briquettes are conveyed by a belt conveyor to storage and remain there at least 24 hours for seasoning. These briquettes are reasonably strong, not friable, and have resistance to crushing between plates, equal to between 70 and 120 kilograms per square centimeter. The percentage recycled, because of breakage and other reasons, amounts to 5 at the press and about 2 in handling and storage.

Because of the shape and small size of the briquettes, the wear on the alveolar wheels of the machine is above normal, though not excessive.

In England, the National Smelting Company produces briquettes from zinc sinter containing 60 to 63 per cent zinc, 8.5 to 10 per cent iron, 1.0 to 1.7 per cent lead, and 0.3 to 1.0 per cent sulfur, mixed with about 1 per cent of high-grade oxidized zinc material.

The sinter and other materials are crushed in a 2-foot Symons cone crusher, in closed circuit with an 8- by 14-mesh Sherwin vibrating screen. The product, which ranges between 50 per cent minus 36 mesh and 20 per cent minus 240 mesh, is mixed with fine bituminous coal, along with dried ball-clay binder, some recycled coked fines and coke breeze. These raw materials, in the proportions of 50 to 55 per cent sinter, 24 to 25 per cent bituminous coal, 4 per cent ball clay, 6 to 10 per cent recycled coked fines, and about 10 per cent coke breeze, are mixed in a 32-foot by 4-foot diameter rotary mixer. Sulfite liquor, equivalent to 1 per cent, and water up to 5.5 per cent of the mixture is added. After the rotary mixer, mixing and plasticizing are completed in three Chilean mills in series, with water additions made to each mill to give a final mix of 6.5 per cent moisture.

The mixed product is fed directly to the briquetting stage which consists of a densifying press and a briquetting press, both of the two-roll pocketed type designed by National Smelting Company engineers. The briquette press has three rows of staggered pockets and produces 15 to 17 tons per hour, equivalent to 28,080 briquettes per hour.

The briquettes are pillow-shaped, size 4⅛ inches by 2⅞ inches by 2 inches, and weigh approximately 1.3 pounds. Typical assays are 37.0 to 37.5 per cent zinc, 17 to 20 per cent carbon, and 5.5 to 6.0 per cent moisture. They are passed over a static bar grizzly and are slowly conveyed by a basket conveyor, during which time they lose about 1 per cent moisture, to the storage pile for aging and more drying.

The briquette machine is said to be very satisfactory but suffers the usual amount of abrasion, especially to the pockets, which occasionally hinders the release of the briquettes.

Before World War II, the briquetting of leaching plant residues from the electrolytic zinc process was practiced at several plants in Europe for the production of zinc oxide by the Philipon, or ash fusion, gas-producer process.

The mixture of partially dried residue and coke was mixed with 7 per cent moisture in paddle-mixers. No binder was required. The mixture was fed to Brück rotary presses of the plunger type and disk-shaped briquettes, 130 millimeters in diameter and 80 millimeters thick and weighing about 1,600 grams each, were produced. A press produced at the rate of 40 briquettes per minute. After pressing, the briquettes were air-dried in the open for a few days and were then ready for the cubilot treatment.

Within recent years, other means of treating leaching residues have been successfully developed and the furnace treatment of briquetted lixiviation residues has been gradually superseded by these processes mainly due to the increasing cost of coke. However, briquetting for many special purposes such as the continuous distillation process will continue to occupy an important place in zinc metallurgy in spite of new developments in sintering, roasting and nodulizing processes.

ACKNOWLEDGMENT

The author is indebted to the following individuals and organizations for valued assistance or cooperation in the preparation of this section of Chapter 5.

W. W. Armstrong, Aluminum Company of Canada, Ltd., Arvida, Quebec
A. Bellier, Société Anonyme des Mines et Fonderies de Zinc de la Vieille Montagne, Paris, France
Th. Boving, Société Anonyme des Mines et Fonderies de Zinc de la Vieille Montagne, Baelen, Belgium
B. F. Buff, National Zinc Co., Inc, Bartlesville, Oklahoma
H. D. Carus, Matthiessen and Hegeler Zinc Co., La Salle, Illinois
E. J. Flynn, The New Jersey Zinc Co. of Pa., Palmerton, Pennsylvania
M. K. Foster, Mexican Zinc Co., Rosita, Coah., Mexico
Paul D. Haigh, Komarek-Greaves and Co., Chicago, Illinois
H. E. Hornickel, American Steel and Wire Co., Donora, Pennsylvania
Messrs. Huser and Horedt, Unterharzer Berg- und Hüttenwerke, Goslar, Germany
M. Johannsen, Badische Anilin Soda Fabrik, A. G., Ludwigshafen a. Rhein, Germany
A. E. Lee, Blackwell Zinc Co., Blackwell, Oklahoma
G. M. Meisel, Colorado Iron Works Co., Denver, Colorado
R. G. McElhanney, Dorr-Oliver-Long, Ltd., Orillia, Ontario
W. M. Peirce, The New Jersey Zinc Co., New York
Robert E. Powers, Freyn Department, Koppers Co., Inc., Pittsburgh, Pennsylvania
E. Preib, Berzelius Metallhütten Gesellschaft, Duisburg-Wanheim, Germany
J. Quets, Union Minière du Haut Katanga, Belgian Congo
P. R. Rose, American Smelting and Refining Co., Amarillo, Texas
Robert Salmonsen, F. L. Smidth and Co., New York
H. J. Sanford, Mexican Zinc Co., S. A., Rosita, Mexico
V. W. Uhl, Bethlehem Foundry and Machine Co., Bethlehem, Pennsylvania
John G. Wehn, St. Joseph Lead Co., Josephtown, Pennsylvania

REFERENCES

For supplementary reading in the designated phases of zinc metallurgy.

Roasting

1. "Air Jets Control Fluid Bed," Feature News, *Chem. Eng.*, pp. 134-6 (August, 1956).

2. Anderson, T. T., and Bolduc, R., "Fluosolids Roasting of Zinc Concentrates," *Chem. Eng. Progr.*, **49,** 527–32 (1953).
3. Austin, E., and McFadden, W. E., "The Electrolytic Zinc Plant, Hudson Bay Mining and Smelting Company," *Trans. Can. Inst. Mining Met.*, **59,** 208–23 (1956).
4. Bray, John L., "Non-Ferrous Production Metallurgy," pp. 134–49, 2d ed., New York, John Wiley & Sons., Inc., 1947.
5. Carr, John D., and Reikie, M. K. T., "The Flin Flon Zinc Plant," *Trans. Can. Inst. Mining Met.*, **38,** 287–310 (1935).
6. Coates, W. H., (British Titan Products Co.), "Flash Roasting," *Proc. Fertilizer Soc.*, **No. 26,** pp. 23–48 disc. pp. 49–54, (1954).
7. Cyr, H. M., Siller, C. W., and Steele, T. F., "Roasting Metallic Sulphides in a Fluid Column," *Trans. Am. Inst. Mining Met. Engrs.*, **200,** 900–904 (1954).
8. Duncan, W. E., "Zinc Metallurgy," *Eng. Mining J.* **140,** 81–2 (1939).
9. Freeman, H., "The Use of Pyrites Concentrates in Sulphite Pulp Manufacture," *Trans. Am. Inst. Chem. Engrs.*, **33,** 180–8 (1937).
10. Hopkins, J., "Zinc Concentrate Roasting," *Chem. Eng. Mining Rev.*, **27,** 5–11, 52–55, 95–101, 134–138, 179–182 (1934–5).
11. Ingalls, W. R., "History of the Metallurgy of Zinc," *Trans. Am. Inst. Mining Met. Engrs.*, **121,** 339–73 (1936).
12. Ingalls, W. R., "Roasting Furnaces," Chapter in "Metallurgy of Zinc and Cadmium," pp. 67–121, 1st ed., New York, McGraw-Hill Book Company, Inc., 1922.
13. Jessup, A. W., "How Dowa's Plant Extracts Cu-Zn from a Single Electrolyte," *Eng. Mining J.*, **155,** Jan., 73–4 (1954).
14. Kurushima, H., and Tsunoda, S., "Hydrometallurgy of Copper-Zinc Concentrates," *J. Metals*, **7,** 634–8 (1955).
15. McBean, K. D., "The Cominco Suspension Roasting Process," *Mining Congr. J.*, **38,** 36–39 (1952).
16. Mulliken, F. R., "Flash Roasting and Its Applications," *Mining and Metallurgy*, **18,** 279–82 (1937).
17. Nakanishi, K., "How Mitsubishi's New Akita Plant Makes 99.997% Electrolytic Zinc," *Mining World*, **17,** 56–59 (October, 1955).
18. Rigg, G., and McBride, W. J., "The Roasting of Zinc Blende Ores," *Mining Mag.*, **26,** 139–145 (1922).
19. Skogmo, S., (Det Norske Zinkkompani), "New Methods for the Roasting of Sulphide Ores," *Tidsskr. Kjemi, Bergvesen Met.*, **10,** 198–207 (1949).
20. Snow, W. C., "Electrolytic Zinc of Risdon," *Trans. Am. Inst. Mining Met Engrs.*, **121,** 482–526 (1936).
21. Stimmel, B. A., Hannay, W. H., and McBean, K. D., "The Electrolytic Zinc Plant of The Consolidated Mining and Smelting Company of Canada, Limited," *ibid.*, pp. 450–72.
22. Stephens, F. M., "The Fluidized-Bed Sulfate Roasting of Non-Ferrous Metals," *Chem. Eng. Prog.*, **49,** No. 9 (Sept. 1953.)
23. U. S. Patent 2,665,899 and Canadian Patent 522,223, Cyr, H. M., *et al.* (to The New Jersey Zinc Company).
24. U. S. Patent 2,665,899, Fassotte, A. D. H. L. (to Cie. des Metaux d'Overpelt-Lommel).
25. Canadian Patent 528,170, Quintin, A. A. J. (to Cie. des Metaux d'Overpelt-Lommel).

Sintering

26. Dwight, A. S., "Metallurgical Treatment of Flotation Concentrates," *Trans. Am. Inst. Mining Met. Engrs.*, **76**, 527–36 (1928).
27. Gyles, T. B., "Horizontal Retort Practice of the National Smelting Company," *ibid.*, **121**, 418–26 (1936).
28. Ingalls, W. R., "Intermittent Zinc Distilling from Ore," *ibid.*, pp. 612–14.
29. Lee, A. E. Jr., "Sintering Zinc Concentrates on the Blackwell 12 x 168 Ft Machine," *J. Metals*, **5**, 1631–33 (1953).
30. MacMichael, H. R., "Sintering Zinc Ore at Rosita, Mexico," *Trans. Am. Inst. Mining Met. Engrs.*, **102**, 278–9 (1932).
31. Najarian, H. K., Peterson, K. F., and Lund, R. E., "Sintering Practice at Josephtown Smelter," *ibid.*, **191**, 116–19 (1951).
32. Seiffert, R., Koch, E., and Matthies, H., "Berzelius Zinc Smelters," *Eng. Mining World*, **2**, 485–9 (1931).
33. Stehli, H. J., "Sintering Zinc Ores," *Trans. Am. Inst. Mining Met. Engrs.*, **121**, 374–86 (1936).
34. Urban, W. J., "The World's Largest Sintering Machine," *Blast Furnace Steel Plant*, **39**, 339–42 (1951).

Calcining

35. Austin, E., and McFadden, W. E., "The Electrolytic Zinc Plant of the Hudson Bay Mining and Smelting Company, Limited," *Can. Mining Met. Bull.*, **49**, 344–59 (1956).
36. Bray, John L., "Non-Ferrous Production Metallurgy," pp. 478–85, 2d ed., New York, John Wiley & Sons, Inc., 1947.
37. Ingalls, W. R., "The Metallurgy of Zinc and Cadmium," pp. 63–67, 1st ed., New York, McGraw-Hill Book Company, Inc., 1922.
38. Mast, R. E., and Kent, G. H., "The Operation of the Flin Flon Smelter and Fuming Plant," *Trans. Can. Inst. Mining Met.*, **57**, 208–24 (1954).

Briquetting

39. "Fuel Briquetting," Canadian Dept. of Mines and Resources, No. 775, Mines and Geology Branch, Bureau of Mines, Ottawa, Ontario.
40. Liddell, D. M., "Handbook of Nonferrous Metallurgy," Vol. 1, 2d ed., pp. 409–28, New York, McGraw-Hill Book Company, Inc., 1945.
41. Johnson, W. M., "Fine-grinding and Porous-briquetting of the Zinc Charge," *Trans. Am. Inst. Mining Met. Engrs.*, **59**, 156–61 (1918).
42. Van Oirbeek, J., "Treatment of Residues from Electrolysis of Zinc and of Lead Furnace Slags in Ash Fusion Gas Producers," *Trans. Am. Inst. Mining Met. Engrs.*, **121**, 693–701 (1936).
43. Ronge, Wilhelm, "High Pressure Briquetting," *Iron and Coal Trades Rev.*, **159**, 1281–82, (1949).

CHAPTER 6

METALLURGICAL EXTRACTION

Electrolytic Zinc Processes

FLOYD S. WEIMER, *Manager*

GUY T. WEVER, *Superintendent of Zinc Plant*

ROLAND J. LAPEE, *Metallurgist*

The Anaconda Company
Great Falls, Montana

NATURE AND IMPORTANCE OF THE ELECTROLYTIC PROCESSES

History

The commercial production of zinc by hydrometallurgy, and the recovery of zinc by electrolysis, was proposed and patented by Léon Létrange of France in 1881. The method proposed by Létrange contained many of the basic elements in commercial use today. Several other experimenters worked on various methods, using sulfate, chloride and alkaline solutions for the solvent and electrolyte. At least two small commercial plants were built, one in England and another in Cockle Creek, New South Wales, but neither were operated with any degree of success.

The First World War gave an added impetus to the developing of the electrolytic zinc process. The metallurgists of The Anaconda Copper Mining Company, and of the Consolidated Mining and Smelting Company, began intensive work on the problem. Both groups developed processes that were commercially feasible and economically sound. The first successful electrolytic zinc plant began production at Anaconda, Montana, late in 1915. Although the plant at Anaconda, Montana, reached a capacity of 25 tons per day, it was actually a pilot plant for developing the process. A zinc plant having a capacity of 100 tons per day was built by The Anaconda Copper Mining Company at Great Falls, Montana, in 1916. The Consolidated Plant at Trail, B. C., began production early in 1916. These were followed by the Electrolytic Zinc Company of Australasia with a plant at Risdon, Tasmania, in 1918. These three companies are regarded as the pioneers of the modern electrolytic zinc process. Several other electrolytic zinc plants followed, all based on the same fundamentals, but differing widely in their individual application of the process. Even the

Figure 6-1. Electrolytic zinc plant of Det Norske Zinkkompani A/S, Eitrheim, Odda, Norway.

Figure 6-2. Electrolytic zinc plant of The Consolidated Mining and Smelting Company of Canada Ltd., Trail, B. C., Canada.

Figure 6-3. Electrolytic zinc plant of The American Smelting and Refining Company, Corpus Christi, Texas.

Figure 6-4. Electrolytic zinc plant of the Anaconda Company, Great Falls, Montana.

pioneer companies, starting with practically the same methods, have become highly individualistic in applying the same basic fundamentals. While economic conditions, markets and engineering developments all contribute, the most important reason for differing methods is the large variation in the impurity content of the source material from which zinc is recovered.

Basic Steps in the Process

The dissolving of zinc-bearing materials in dilute sulfuric acid to form zinc sulfate solution, the purification of the zinc sulfate solution to obtain

Figure 6-5. Zinc plant of the Electrolytic Zinc Company of Australasia Ltd., Risdon, Tasmania.

Figure 6-6. Electrolytic zinc plant of The Bunker Hill Company, Kellogg, Idaho.

a satisfactory electrolyte and the recovery of metallic zinc from the electrolyte as a high-purity product are the basic steps in the hydrometallurgy of the electrolytic zinc process. These steps are not always clearly defined in actual practice. Leaching and purification steps often take place together, and purification and electrolysis may occur simultaneously.

The chemistry of the process seems rather simple until it is realized that each of the many variable factors encountered has its influence on the reactions involved. The hydrometallurgy of the process becomes complex due to the very narrow margin by which it is possible to deposit zinc from a solution by electrolysis. The comparatively low market value of zinc adds to the problem, causing the economic necessity of producing zinc at a low cost and a high recovery.

Source of Materials

The main source of zinc for the production of electrolytic zinc is from calcine produced by the roasting of zinc sulfide concentrates. A secondary source is from zinc fumes obtained by the retreatment of zinc-bearing residues.

Zinc calcine will contain from 50 to 65 per cent zinc. The bulk of the zinc is in the form of zinc oxide. A small portion, up to 4 per cent or so, may be present as zinc sulfate, and a portion, up to several per cent, may be present as a complex zinc-iron compound commonly called zinc ferrite.

Zinc oxide is soluble in dilute sulfuric acid and zinc sulfate is soluble in water. Zinc ferrite is practically insoluble under the ordinary methods of leaching with dilute sulfuric acid. The amount of zinc sulfate is regulated during the roasting process to offset the amount of sulfate ions lost in the residue. The control of the amount of zinc sulfate is quite important unless a source of cheap sulfuric acid is available. The amount of zinc ferrite formed during roasting is a function of the amount of iron in the concentrates and the control of the roasting process. The amount of zinc ferrite in the calcine has a considerable effect on the recovery possible.

Zinc calcine or zinc fume will also contain minor amounts of other elements, many of which are also soluble in dilute sulfuric acid. The control of these impurities constitutes one of the major problems of the process.

THE PREPARATION OF A SUITABLE ELECTROLYTE

Dissolving of Zinc-Bearing Materials

The primary purpose of the first step in the production of zinc by the hydrometallurgical process is to dissolve or "leach" as much as possible of the soluble zinc in the source material.

Dilute sulfuric acid is the solvent used in all successful commercial operations. Zinc oxide is soluble in dilute sulfuric acid according to the reaction:

$$ZnO + H_2SO_4 = ZnSO_4 + H_2O$$

Zinc sulfate, which may or may not be present, is soluble in water.

The sulfuric acid used, commonly called "return acid," is the acid regenerated during the electrolyzing step:

$$ZnSO_4 + H_2O + e = Zn + H_2SO_4 + O$$

Sulfuric acid from outside sources can be used to make up for losses encountered in the system, but the regenerated acid must be used if the process is to be economical. The electrolyzing step becomes very costly if more than about two-thirds of the zinc contained in the electrolyte is removed and the zinc remaining in the depleted electrolyte must be recovered. Also, the acid regenerated is formed in any case and is available at no cost.

Recycling continuously can result in several disadvantages, most of which can be attributed to the build-up of the elements present in addition to zinc. Detrimental impurities, such as the metals electronegative to zinc, and elements like chlorine and fluorine which are difficult to remove, are again introduced into the system and may build up to the point where they are difficult to control. Other elements, such as sodium, potassium and magnesium are also found in the regenerated acid. These elements are not normally considered directly harmful to the electrolysis of zinc, but they are cumulative and if allowed to build up to large amounts will become detrimental. Aluminum and calcium also appear in the regenerated acid, but are not accumulative.

Each commercial plant has its own method of avoiding the build-up of cumulative elements or compounds. In some cases the process is practically automatic. In other cases, a periodic increase in the amount of zinc left in the final residue, with a corresponding elimination of the other elements is sufficient to keep the system in balance. Some plants strip the zinc from a portion of the electrolyte and discard depleted acid. Others bleed off a portion of the neutral solution and use it for the production of crystalline zinc sulfate.

Equipment. Standard types of equipment for leaching, settling, filtering and clarifying have been adapted to the needs of the process. The initial dissolving of calcine and other reaction steps are commonly carried out in either air-operated Pachuca tanks or in mechanically agitated tanks. Standard thickeners of various sizes are often employed for thickening the pulp. Vacuum filters of either the batch or continuous types are used for thickening the pulp or filtering thickened pulp. Standard pressure filters are normally used for clarifying solution. The pumping equipment may be of many different types. Centrifugal pumps are used almost universally at the present time. Some of the larger plants have developed their own design of centrifugal pump especially adapted to this type of operation.

Corrosion and abrasion are the main factors determining choice of materials of construction for the various items of equipment used. Tanks in contact with acid are usually of wood lined with lead. Some wood or iron tanks lined with lead and with a secondary bottom or a complete lining of acid-proof brick are also used where both corrosion and abrasion are encountered. Tanks in contact with solutions that are neutral or very slightly acid may be of wood or, in a few cases, concrete.

Lead, wood, stainless steel, bronze, aluminum, rubber, glass and various forms of plastics are all employed for the equipment used for filtering and clarifying. The applications are so many and varied that each must be studied to determine the best material for that particular purpose. The same is true of pumps and pipelines.

The recent developments in plastics is of great interest to the electrolytic zinc industry. Plastics, within their range of temperature and pressure, have replaced several of the older materials and can be expected to come into greater use as the plastics industry improves on their products. Plastic filter media are also replacing older materials to some extent.

Rubber is also coming into wider use as better types and methods are developed. Rubber pipelines can be arranged to be practically self-cleaning where the flow of solution is intermittent and the lines are subjected to considerable flexing.

Stainless steel has a wide application and various types are used according to the type of corrosion encountered.

Wood, lead and bronze are the old standbys of the electrolytic zinc industry and are still widely employed. Wood is strong, relatively cheap, and if used properly, has a long life. Lead and bronze have a high resist-

Figure 6-7. Burt filters, American Smelting and Refining Company, Corpus Christi, Texas.

Figure 6-8. Moore filter, The Consolidated Mining and Smelting Company Ltd., Trail, B. C., Canada.

Figure 6-9. Fifty-foot Dorr thickener, The Consolidated Mining and Smelting Company Ltd., Trail, B. C., Canada.

Figure 6-10. Shriver press section, Hudson Bay Mining and Smelting Company Ltd., Flin Flon, Manitoba, Canada.

Figure 6-11. Rotating disk filter, Electrolytic Zinc Company of Australasia Ltd., Risdon, Tasmania.

ance to certain types of corrosion and have the advantage of a high scrap value.

Methods of Leaching. The dissolving of zinc oxide in dilute sulfuric acid is, in itself, a simple process. It is complicated, however, by the simultaneous dissolving of unwanted elements and compounds that affect the electrolyzing step or influence the recovery of zinc. Impurities such as arsenic, antimony, copper, iron, cadmium, cobalt, nickel, tin, germanium, selenium, tellurium, silica, alumina and others are often present in varying amounts. Many of these can be reduced in amount or eliminated by neutralizing zinc sulfate solution with zinc oxide, with the formation and precipitation of ferric hydroxide. The leaching methods employed by the various plants are influenced more by the necessity of minimizing the amount of impurities in the solution than by any other factor.

The formation and precipitation of ferric hydroxide, commonly called an "iron purification," is carried out by first oxidizing any ferrous iron present to the ferric state and precipitating the ferric sulfate by adding a neutralizing agent until a hydrogen-ion concentration of *p*H 5.6 to 6.0 is reached. Good elimination of impurities is obtained if the amount of iron present is at least ten times the amount of impurities to be removed. A solution of ferric sulfate may be added if there is not enough soluble iron present. Iron solution can be obtained by dissolving scrap iron in sulfuric acid and oxidizing to the ferric state by a suitable reagent, or by dissolving and oxidizing any other suitable iron compound that is available.

Ferric hydroxide can be precipitated with zinc oxide, limerock, milk of lime or any other suitable neutralizing agent. Zinc calcine, being mainly zinc oxide, is the cheapest and most readily available material. When zinc calcine is used to neutralize, an excess is necessary to complete the reaction. With zinc oxide or calcine the reaction is:

$$Fe_2(SO_4)_3 + 3ZnO + 3H_2O = 2Fe(OH)_3 + 3ZnSO_4$$

The precipitation of ferric hydroxide assists in the removal of certain other impurities such as arsenic, antimony, germanium, silica and the like. The part played by iron in removing other impurities may be due to the formation of basic salts with such impurities, an adsorption process, or a combination of the two.

The formation of ferric hydroxide is closely coupled with the hydrogen-ion concentration of the solution. The theoretic relationship between the concentration of iron in moles per liter and *p*H is: $p\text{H} = 2.014 - \frac{1}{3} \log \text{Fe}$.

Ferric hydroxide does not begin to precipitate until *p*H 2.6 is reached. The precipitation then follows the hydrogen-ion concentration closely until *p*H 5.7 to 6.0 is reached, where zinc sulfate solution becomes buffered and further additions of zinc oxide will not affect the *p*H value. Ferric hydroxide, precipitated between *p*H 2.6 to 4.0, is in the form of a colloid which is practically impossible to settle or filter unless the solution is boiled to coagulate the solids. Foaming also occurs in this range. Coagulation of the ferric hydroxide begins at *p*H 4.0 if zinc oxide is used. Zinc calcine, for some unexplained reason, will cause coagulation to take place when *p*H 3.7 is reached. The coagulation point is easily determined by a visual inspection of a sample of the solution. Time and temperature have an effect on coagulation but to a lesser degree than the hydrogen-ion concentration. Excessive riding time tends to cause the coagulated solids to break down into slimes but is of considerable benefit in reducing the germanium content.

Some impurities begin to precipitate while the hydrogen-ion concentration is quite high, but most will not precipitate to any extent until after the ferric hydroxide begins to coagulate. Good elimination requires an excess of zinc oxide and a hydrogen-ion concentration of at least *p*H 5.0.

The methods used to dissolve zinc-bearing material in dilute sulfuric acid with good extraction, and to eliminate as much of the impurity as possible while doing so, are as varied as the number of commercial plants in operation. The methods can be divided broadly into types such as "single" and "double" leaching or into batch and continuous leaching. It should be understood that no commercial operation follows exactly any one such type.

Single leaching combines the dissolving and purifying steps into a single operation. This type of leaching can be carried out as either a batch or a continuous operation, although the batch system is the more easily controlled. If followed exactly, it is impossible to get a good recovery of zinc by this method if zinc oxide is used to neutralize. Limerock can be substituted to precipitate the iron and coagulate the pulp. Limerock will precipitate zinc from a neutral zinc sulfate solution, so a great deal of care must be taken to avoid excessive losses of zinc. The sulfate ion used up by neutralizing with limerock must be replaced by new acid or by zinc sulfate from the roasting process. The residue from single leaching will be bulky and high in moisture content, which will tend to increase the losses of zinc. The greatest advantage of single leaching is the lower initial cost of equipment.

The double leaching system differs from single leaching in that no attempt is made to complete all operations in a single step. This allows an excess of the neutralizing agent to be used without impairing the recovery of zinc. Zinc calcine, which needs an excess to complete the reaction, is universally used in the double system. Double leaching can be done in either batch or continuous operations but it is usual to employ double leaching in the continuous systems where large volumes are to be handled.

The batch system of leaching is ideal for plants that operate at high current density in the electrolyzing cells and a high zinc content in the electrolyte. The control of each reaction can be within closer limits than with the continuous system. The batch system is also advantageous when treating diversified feed as each batch can be held for testing. The size of the plant and the volume of material handled are the determining factors in the batch system. As the size of a plant increases, the time lost in filling and discharging tanks becomes important to the economics of the process.

The degree of control possible with the batch system has resulted in most plants, regardless of size, operating some portions of their process by this method.

The continuous leaching system is at its best advantage where large volumes are handled and uniform feeds are available. A variable feed requires that a continuous system be operated at all times under the conditions required by the worst material introduced. The low-density plants employ continuous leaching because fewer tanks are needed for an equal volume of material than with the batch system. As the continuous system sacrifices a certain amount of control, it is usually applied only to those portions of the system where reactions are rather rapid and minor fluctuations are relatively unimportant. The steps requiring close control or long reaction times are carried out in batches. Coagulating agents are frequently used to aid settlement or filtration in the leaching steps.

The American Smelting and Refining Company at Corpus Christi, Texas, uses a batch system of single leaching. The system used is actually a two-stage treatment as the initial leaching step is followed by a careful neutralization step. Large mechanically agitated tanks are used. Roasted concentrates are delivered to the tanks by screw conveyors. Acid from the electrolyzing cells diluted with other solutions or with water is used. Manganese dioxide is added for oxidizing purposes. The leaching is carried out very carefully over a 3- to 5-hour period. Calcine is added slowly and the hydrogen-ion concentration checked frequently. The leaches are discharged to Burt filters. The filtered solution is delivered to the purification step of the process. The residue is settled in Dorr thickeners, the overflow being returned to the leaching step. The underflow from the thickeners is filtered through Oliver filters, dried and shipped to a lead smelter for further treatment. The leaching of zinc oxide fume is carried out in much the same manner.

The Anaconda Company operates a zinc plant at Great Falls, Montana with a rated capacity of 13,500 short tons of slab zinc per month, and a plant at Anaconda, Montana, with a rated capacity of 7,500 short tons per month. The combined rated capacity of 21,000 short tons per month makes the Anaconda Company the largest producer of electrolytic zinc.

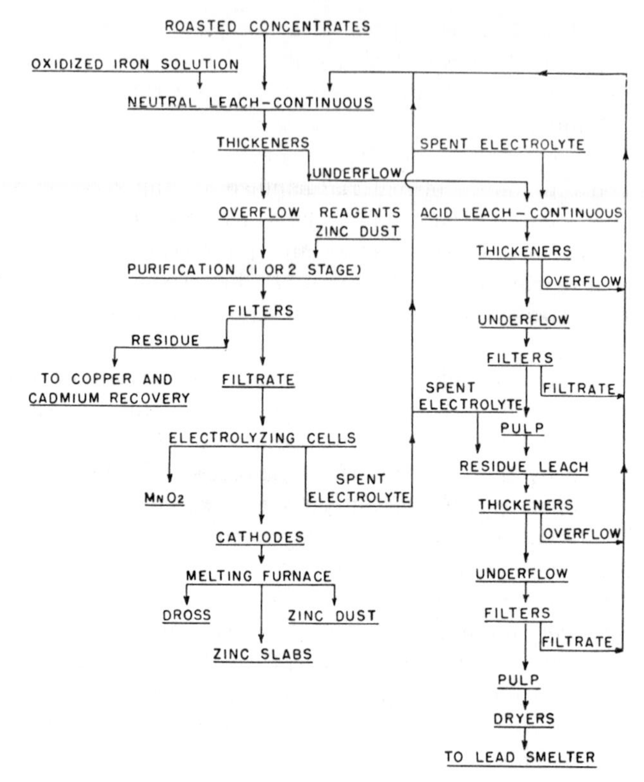

Figure 6-12. Simplified flowsheet of typical low-density plant.

Both of these plants electrolyze at low current density. Leaching is carried out by a continuous, double-leaching system.

Calcine is added to an acid solution consisting of a mixture of return acid and leach solutions from the acid leach and other cycles in the plant. The reaction is carried out in a series of Pachuca tanks. An excess of calcine is added for complete neutralization of acid, removal of iron and some other impurities, and coagulation of the pulp. No attempt is made to leach more than about one-half of the zinc in the calcine in this step. While some leaching takes place, this step is more of a purification than a leach. The discharge from this step is sent to thickeners for separation of solids and solution. The solution overflowing the thickeners is further purified with zinc dust and is used as feed solution to the electrolyzing cells. The solids settled out in the thickeners are leached in a series of Pachuca tanks with sufficient return acid to dissolve all the zinc oxide present and to leave a slight excess of acid in solution. The discharge from this second, or acid, leach is sent to thickeners for separation of solution and solids. The solution overflowing the thickeners is returned to the first, or neutral, leach and mixed with spent electrolyte to form the solution used in the first leach. The solids from the thickeners are filtered and further leached by a batch process for additional recovery of zinc.

Det Norske Zinkkompani at Eitrheim, near Odda, Norway, has a rated capacity of 4,500 tons of slab zinc per month. The leaching methods used in the Norway plant are very similar to those used at Great Falls and at Anaconda.

The Consolidated Mining and Smelting Company of Canada Limited, operates an electrolytic zinc plant at Trail, B. C. This zinc plant is the largest single zinc plant in the world, with a rated capacity of 16,000 short tons of slab zinc per month. A continuous double-leaching system is used in which excess return acid is used in the first leach. The discharge from the first leach is sent to thickeners for the separation of solutions and solids. The solution is treated with excess calcine to remove iron and neutral insoluble impurities, again settled in thickeners, and the resulting solution further purified. The residue from the first leach is filtered and eventually, with or without further treatment to extract zinc, finds its way out of the plant as a final residue. The solids from the neutralizing step are returned to the first leach. Current practice is to treat about 80 per cent of the calcine directly in the first, or acid, leach, and 20 per cent in the neutralizing step.

The Trail Plant also operates a leaching plant for the treatment of zinc oxide fumes obtained from the fuming of slag from lead blast furnaces. Part of the lead plant treatment is zinc plant residue which contains zinc, and some zinc is always present in lead concentrates and ores. Due to the fact that the zinc plant and lead plant by-products are intercycled, there is a concentration of impurities in the oxide fume. This concentration, plus the amount of fluorine introduced from lead circuit, makes a separate treatment highly desirable. The method of leaching used is to dissolve the oxide fume with return acid, adding enough ferric sulfate to precipitate the impurities, and neutralizing the excess acid with either calcine or oxide fume. The pulp is settled and the residue is filtered, washed and returned to the lead smelter. The solution is also filtered and further purified at high temperature.

The Bunker Hill Company at Kellogg, Idaho, employs a preliminary treatment step on zinc concentrates containing dolomitic gangues before roasting. The concentrates are leached with a weak sulfuric acid solution. The leach solution is removed by decantation and the pulp is washed several times by the same method. This treatment removes much of the calcium and magnesium contained in the concentrates. The final slurry is filtered and the pulp is sent to the concentrate storage. The leach and

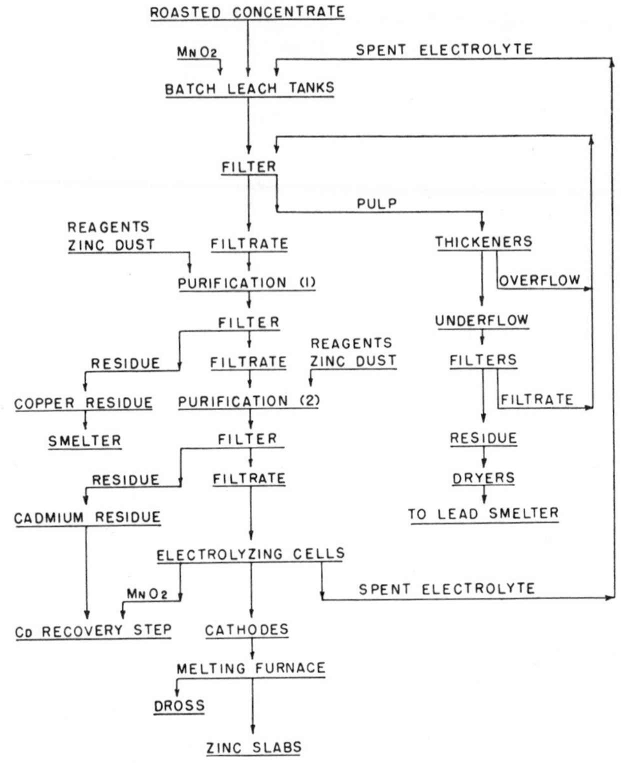

Figure 6-13. Simplified flowsheet of typical high-density plant.

wash solutions are treated in flotation cells for recovery of suspended particles of concentrates.

The Kellogg plant has long been characterized by the use of a high-acid, single-leaching system. Recently, a low-acid single-leaching system has been developed. A summary of the high-acid system is given. Details of the low-acid system are not available.

Leaches in the high-acid system were made by adding a known amount of return acid to a mechanically agitated leach tank. The acid was heated to about 60°C and a measured amount of calcine added to neutralize approximately two-thirds of the acid. Manganese dioxide or roasted manganese concentrates were added if necessary to bring the manganese content up to about 4 grams per liter.

The temperature of the leach rose to the boiling point due to the heat of reaction. Agitation was continued for 3 to 5 hours at which time the acid strength was between 40 to 60 grams per liter of sulfuric acid. The remaining acid was then carefully neutralized by the careful addition of calcine until the iron was precipitated. Reacidification of the leach to about 20 grams per liter acid and again carefully neutralizing was of great benefit in improving the filtering rate. Enough excess calcine was added to precipitate the iron. The addition of 0.25 pound of glue for each ton of pulp also was an aid to the filtering step.

The leaches were discharged to Burt filters for clarification. The residue was dewatered in Oliver filters, dried, and sent to the lead smelter for further treatment. The filtrate was sent to the purification step.

The Electrolytic Zinc Company of Australasia, at Risdon, Tasmania, uses a single-stage batch process for the leaching of calcine. The leach is made with return acid from the electrolyzing cells in a series of Pachuca tanks. The leach is maintained acid. In the later stages of the leaching cycle, the amount of acid is reduced to a low amount by the very careful addition of calcine. The leaching step requires about 3½ hours to complete. The finished leach is discharged through classifiers to thickeners for settlement.

The material removed by the classifiers is ground and treated in flotation cells for removal of the sulfides present which are returned to the roasting step. The remainder is further leached to dissolve the soluble zinc oxide present and sent to thickeners for settlement.

The thickener underflow is agitated with limerock which neutralizes the acid, precipitates iron, silica, etc., and brings the pulp to a free filtering condition. The pulp is then filtered in a Moore filter, dried and sent to a lead smelter.

The thickener overflow is agitated with manganese dioxide and fine calcine in a series of both mechanically- and air-agitated tanks to precipitate the iron. The ideal condition is almost instantaneous neutralization. A certain amount of silica in solution also aids in removing impurities. The pulp is filtered in a Moore filter, washed by repulping and again filtered. The filtrate is sent to the purification step for further treatment.

A portion of the pulp from the Moore filter is recirculated to bring excess calcine back to the neutralizing stage of the oxidizing step. A cake is first formed on the filter and the remaining pulp is pumped back to the second tank of the oxidizing step. In addition to this, the cake formed on the filter is occasionally blown off and added to the recirculated pulp.

The Hudson Bay Mining and Smelting Company at Flin Flon, Manitoba, uses a two-stage continuous leaching process. The first stage, or neutral leach, is a combined leach and purification. The amount of zinc calcine added is in excess of the amount needed to neutralize the acid present. Approximately 40 per cent of the soluble zinc present in the calcine is dissolved in this stage. The excess of zinc oxide in the calcine precipitates practically all the iron, arsenic and antimony from solution. The overflow solution from the oxide leaching plant-neutral thickeners is also added to this stage. The discharge from the first stage is settled in thickeners for separation of solution and solids. The clarified solution from the thickeners is sent to the zinc dust purification step.

The solids from the neutral thickeners are treated in the second, or acid, leach stage, with an excess of acid for the recovery of zinc. The discharge from the second stage is settled in thickeners. The solution overflowing the thickeners is recycled to the head of the first stage. The solids from the thickeners are filtered, washed with water, dewatered, and sent to a copper smelter for further recovery of metal values.

The leaching of zinc oxide fume at Hudson Bay is carried out in a separate two-stage, continuous, countercurrent leach. In the first, or neutral stage, an excess of oxide fume is leached with return acid from the electrolyzing cells to produce a neutral slurry. This slurry is settled in thickeners to produce a clarified solution and an underflow containing undissolved zinc oxide and precipitated impurities. The clarified solution is added to the first step of the sulfide circuit for subsequent purification and electrolysis.

The second, or acid leach, stage treats the thickener underflow in mechanically agitated leach tanks with a slight excess of return acid for zinc recovery. The discharge from this stage is settled in thickeners. The underflow from these thickeners is filtered and this filtrate, together with the overflow from the thickeners, is returned to the first stage. The filtered residue is stockpiled pending treatment at a future date.

Residue Treatment. The residue from the leaching steps will contain unleached zinc oxide, some entrained zinc sulfate solution, zinc as zinc ferrite and all the elements either insoluble or precipitated with the iron purification. Some plants carry this material through another cycle in which the residue after being filtered is again leached with hot cell acid in order to recover as much of the zinc oxide as possible and dissolve a portion of the zinc ferrite.

A major problem in the treatment of residue from the zinc process is avoiding the loss of zinc as entrained solution during the filtering steps. The residue contains gelatinous precipitates of iron alumina and silica which are difficult to wash. Washing by dilution, by replacement or a combination of the two, are the only satisfactory methods. Repulping a residue containing a low percentage of sandy material will cause the material to form a frothy, sticky mass that is practically impossible to filter or dewater. Pulp of this nature is usually washed by replacement using equipment such as a Moore filter. This step is followed by a second filtration to dewater the solids as much as possible.

The economics of the process determines the length to which further leaching and washing can be carried out. When the cost of more leaching or washing to recover zinc exceeds the value of the zinc recovered, the operation is no longer economically sound. Each plant must determine this point individually because of the many factors involved.

Classes of Residue. The residue from the leaching steps can be divided into four main classes. The first class is one in which there are not sufficient values to justify further treatment. Such a residue may be considered a waste product and disposed of by the cheapest possible method.

Residue of the second class will contain only lead in amounts that will pay for the cost of recovery. Several methods may be employed in treating such a residue. The Waeltz process or various wet methods may be used or the material can be used as a feed to a lead smelting operation.

The third class of residue is one in which lead, gold, silver and copper may all be present in recoverable amounts, but in which the value of the lead exceeds the value of the copper. This is the most common class and is usually treated by a lead smelting operation. Treatment in a reverberatory furnace will produce a copper matte containing most of the gold and silver and a fume product containing most of the lead. The same methods used for recovery of a class two residue can be then used for the recovery of lead.

A class four residue is much the same as a class three residue except that the value of the copper present exceeds the value of the lead. Such a residue may be treated in a reverberatory furnace as mentioned in the recovery of a class three residue, or in a copper blast furnace or copper reverberatory furnace.

The fume from zinc residue treated by a Waeltz process will contain zinc and other volatile metals which can be recovered. The slag from a lead smelting operation treating the residue from a zinc plant will contain zinc which can be recovered as an oxide by a slag-fuming process and used for further recovery of zinc.

Roasting any class of residue with sulfuric acid will convert much of the insoluble zinc into sulfates which can then be recovered by a water leach.

By-products. The available markets usually determine what by-products can be produced profitably from a zinc plant. Cadmium is a common by-product of the electrolytic plants as cadmium accompanies zinc in most of its ores and must be removed during the purification process to produce a high-grade zinc. A large portion of the cost of producing cadmium is borne by the zinc process. Starting with a high-grade cadmium metal sponge, the leaching of cadmium in dilute sulfuric acid and the purification of the resulting cadmium sulfate solution is a relatively simple procedure. Cadmium metal is recovered from the purified solution by electrolysis, using equipment and methods almost identical to those used for zinc.

Sulfuric acid can be recovered from roaster flue gases and is a desirable by-product under suitable market conditions. Zinc dust, zinc dross and zinc sulfate can all be produced if desired. Lead, gold, silver and copper can be considered by-products, although these metals are usually sold to other producers. Indium, gallium and germanium often accompany zinc in its ores and the recovery of these metals is profitable if present in sufficient amounts.

Purification of Neutral Solution

Zinc-Dust Purification. The further purification of neutral solution is necessary because the solution will contain copper and cadmium as major impurities and also small amounts of arsenic, antimony, germanium, cobalt, etc., if these impurities were present in the feed material to any extent. Special methods are often used for certain impurities when present in more than the normal amounts or if they have not been reduced to low values in other cycles. The main purification step at this point is with zinc dust.

The basic reaction is that of electrolytic reduction of those metals whose positions in the electromotive series for sulfates are below that of zinc. There is a close relationship between the purification with zinc dust and the electrolysis of zinc. Those metals which are injurious to the electrolysis of zinc will also cause a more active reduction in the zinc dust purification, and for the same reason. Zinc metal tends to protect itself against corrosion by the formation of an inert oxide film. Any factor which tends to dissolve the oxide film or lower the hydrogen overvoltage on zinc will cause the zinc to become more active.

Metallic zinc in contact with another metal, such as copper, will form a couple which is much more effective in removing other impurities than zinc alone. Copper is often added to the solution in order to remove other impurities more difficult to precipitate.

The presence of copper or other elements which cause an increased activation of the zinc dust also increases the tendency of cadmium to redissolve. Metallic cadmium is oxidized and the oxide reacts with zinc sulfate to form soluble cadmium sulfate. Air agitation or aeration of the solution increases the tendency of cadmium to oxidize and should be avoided. The presence of germanium also increases the tendency of cadmium to redissolve.

The purification with zinc dust is carried out in one or more steps depending on the amount and kind of impurities present and the grade of zinc desired. One step is often sufficient if High Grade zinc is to be produced, but the production of Special High Grade zinc usually requires at least two stages to insure extreme purity of the solution.

Copper and cadmium are usually the only recoverable elements present to any extent. The purifications may be carried out in such a manner that a partial separation of these elements is obtained, or they may be precipitated together and separated in another recovery cycle.

A two-stage purification is carried out by the American Zinc Company of Illinois in their plant at Monsanto, Illinois. Enough copper sulfate is added to the first stage to raise the copper content of the solution to 0.4 to 0.5 gram per liter. Arsenic trioxide is also added at the rate of 0.5 pound per ton of slab zinc produced. Copper sulfate is added as crystals and the arsenic trioxide is added as a powder while the tank of solution to be purified is being filled. The solution is heated, zinc dust is added, and a very strong mechanical agitation is used in this step. The discharge from the first stage is filtered through Shriver presses. The filtrate is sent to the second stage for further purification and the residue is mixed with the regular zinc plant residue for additional treatment.

The second-stage purification is made with the addition of a small amount of copper sulfate and agitation with zinc dust. The discharge is filtered through Shriver presses and the filtrate is used as feed for the electrolyzing cells. The residue is leached for the recovery of cadmium and zinc.

The Corpus Christi electrolytic zinc plant of the American Smelting and Refining Company is divided into two separate operations. One operation produces zinc from sulfide concentrates and the other produces zinc from densified zinc oxide fume. The methods of purification are similar in both operations.

Purification in the oxide plant is performed in two stages, each of which is carried out in a large single tank mechanically agitated at about 6.5 rpm. The first stage is primarily for the reduction of the germanium content and the second stage for removal of the other impurities present.

The first stage treats the filtrate from Burt filters. The solution is heated to about 90°C and the acid content is adjusted to about 2.0 grams per liter of sulfuric acid. Part of the press cake from previous charges, and all the press cake from the second stage, is added during the initial period of operation. The remainder of the operation, usually a four-hour period, consists of adding zinc dust at a gradual rate with a belt feeder. A reasonable assurance of satisfactory germanium removal can be obtained by maintaining the solution during this part of the operation at *p*H 4.0 to 5.0. The discharge is filtered through Shriver presses. The filtrate is sent to the second-stage purification. A portion of the residue is reused in the first stage. The unused portion of the residue is sent to a copper smelter for further treatment.

The second stage usually requires about four hours for completion. The purification is started by adjusting the solution to *p*H 4.0 and adding copper sulfate and a small amount of arsenious oxide. Zinc dust is added at a uniform rate and additional copper sulfate as required. Cell acid is added to maintain the solution between *p*H 4.0 to 5.0, and the temperature of the solution is permitted to decrease from about 80 to 70°C. The purification discharge is filtered through Shriver presses. The filtrate is cooled and used as the feed to the electrolyzing cells. The residue is added to the first stage.

The purification of solution at the Great Falls plant of The Anaconda Company is normally carried out in a single-stage batch operation. The overflow solution from the neutral Dorr thickeners is pumped to a series of mechanically agitated tanks. The tanks are hooded and equipped with forced-draft ventilation. At the start of each purification a small amount of potassium antimony tartrate is added. Zinc dust is added and the solution is agitated for about 120 minutes. The tanks are filled and discharged in rotation so that a continuous flow to the filtering section is obtained.

The amount of copper in the solution to be purified varies from 0.30 to 0.75 gram per liter depending on the amount of soluble copper in the material treated. The amount of zinc dust used is on a sliding scale depending on the amount of copper and cadmium present in the solution. Copper or copper-bearing materials are added to the roaster feed during periods when the normal feed is low in soluble copper.

The purification tank discharge is pumped directly to Shriver presses for filtration. Aeration is avoided as much as possible to prevent re-solution of impurities. The filtered solution is used for the electrolysis of zinc. The residue is removed from the presses and transported by belt and screw conveyors to railroad cars. The residue is further treated for the recovery of zinc, cadmium and copper.

A two-stage purification is sometimes used. In this case, the clarified solution from the first stage is pumped continuously through mechanically agitated reaction tanks to which is added continuously a stream of zinc dust. The discharge is filtered through Shriver presses. The solution is the feed to the electrolyzing cells. The filtered residue is added to the first stage where is replaces a portion of the regular zinc dust.

The Anaconda plant of The Anaconda Company carries out the purification of solution in a manner similar to that used at Great Falls.

The Consolidated Mining and Smelting Company of Canada Limited, operates both a calcine treating plant and an oxide fume treating plant. The purification of solution in the calcine treating plant is carried out in two steps, both of which are continuous. The solution to be purified is added to a constant-level pump box to which is also added a closely controlled stream of extremely fine zinc dust. The amount normally added is 1.28 pounds per volume-ton of solution. The discharge from the pump box is passed continuously through two contact tanks, a settler and a surge tank. Aeration in the system must be avoided to prevent re-solution of impurities. The precipitation of impurities proceeds at a rapid rate due to the fineness of the zinc dust and the intimate mixing of the material passing through the pumps.

The discharge from the contact tanks is settled in a thickener to settle out the heavy metallics which are delivered to the cadmium recovery section. The overflow from the thickener passes to a surge tank and from there to pressure filters for clarification. The residue is treated for cadmium recovery. The filtrate is again treated with fine zinc dust and filtered through pressure filters. This clarified solution is the finished electrolyte. The residue is treated for the recovery of cadmium.

The purification section of the zinc oxide plant also operates in two stages. The solution is clarified and heated before being purified. The first stage is used to rough-out the cadmium contained in the solution. This process is carried out continuously in a tank to which is added only enough zinc dust to precipitate the bulk of the cadmium. The pulp is filtered through a pressure filter. The residue is used for the recovery of cadmium. The filtrate is sent to a series of batch-operated purification tanks. The solution is heated to 70°C. Copper sulfate and about 3 to 4 grams per liter zinc dust is added and agitation is carried out for 90 minutes. The batches are discharged to a thickener where the bulk of the solids settles out. The overflow from the thickener is pumped to pressure filters for final clarification. Extra zinc dust is added to the intake of the pump supplying the filters to counteract any tendency for cadmium to redissolve. The clarified solution is used for the electrolysis of zinc. The residue is further treated for the recovery of zinc and cadmium.

The purification of solution with zinc dust is carried out in a single stage at Risdon. Mechanical agitation is used and the required zinc dust is added in two stages, 90 per cent at the beginning of agitation and 10 per cent shortly before filtration begins. As a precaution against the re-solution of cadmium, the tanks are not completely discharged.

Filtration is carried out in Dehne filter presses. The filter plates are covered with woven wire or chain-wire screens to prevent the filter cloth from pressing too closely to the plates and reducing the drainage space. Two filter cloths are normally used on each plate. The dirty cloth is removed each time for cleaning. The second cloth is moved to the outer position and a clean cloth placed next to the plate. The residue from the filtering step is sent to the cadmium circuit for further treatment.

A portion of the clarified solution is further treated with nitroso-beta-naphthol for the removal of cobalt. The amount treated may range from 40 to 60 per cent of the total depending on the amount of cobalt present. After filtration of the material from the cobalt purification, the two solutions are blended and used as the feed to the electrolyzing cells.

A small portion of the total purified solution is treated for the elimination of chlorine. This purification is carried out by batches in mechanically agitated tanks. Return acid is added in sufficient quantity to raise the acidity slightly, and a calculated amount of silver sulfate added to remove all but a small amount of the chlorides present. An excess of silver sulfate is avoided. The purification is settled and the clear solution decanted through filter presses. The clarified solution joins the main portion of the purified solution where the basic zinc sulfates present are sufficient to neutralize the acid present.

The silver chloride residue remaining in the tanks after the decanting process is allowed to accumulate through several charges. It is then collected in a small Pachuca tank where it is agitated with water acidified with sulfuric acid. Zinc dust is added to form zinc chloride and metallic silver. The material is filtered, the zinc chloride is discarded and the metallic silver is treated with sulfuric acid and heat to regenerate silver sulfate. The excess acid is fumed off and the silver sulfate is ground in a ball mill.

Det Norske Zinkkompani, Eitrheim, Odda, Norway, uses a two-stage purification. The overflow solution from the neutral thickeners is heated to about 60°C by stainless steel heat exchangers before purification. The first stage purifications are made in batches in mechanically agitated tanks. Copper and cadmium are precipitated in the first stage. The purification discharge is filtered through Shriver presses. The filtrate is sent to the second purification stage where a continuous flow of solution is treated with a continuous addition of zinc dust. The discharge is again filtered through Shriver presses. The clarified solution is used in the electrolyzing cells. The residue from both purification stages is further treated for the recovery of zinc, cadmium and copper.

The zinc plant of the Hudson Bay Mining and Smelting Company, at Flin Flon, Manitoba, uses a two-stage purification system with both stages being operated on a continuous basis. Each stage consists of five mechanically agitated tanks in series. The solution to be purified enters the primary stage in the first tank in the series. Each tank overflows into the top of the next tank in series. The underflow from each tank is pumped to the top of the next tank in series in order to prevent a build-up of pulp in the tanks. Zinc dust is added continuously to each tank in the series by vibrating feeders. Copper sulfate solution is also added to the feed launder when additional copper in solution is desired. The discharge from the last tank is pumped to Shriver filter presses. The filtrate is discharged to a surge tank from which it is drawn as feed to the second stage.

The second stage of the purification step is carried out in an operation very similar to that used in the first stage. A small amount of copper sulfate is added continuously to the first tank. Zinc dust is added by vibrating feeders as in the first stage. The discharge from the second stage is filtered through Shriver presses. The filtrate is either used directly as a feed to the electrolyzing cells or further purified for cobalt.

The residue from both the first and second stages in the purification step is leached with return acid from the electrolyzing cells for further recovery of zinc, copper and cadmium.

Approximately 50 per cent of the solution from the zinc dust purification step is further purified for the removal of cobalt. This purification is operated on a batch basis. The solution is adjusted to a *p*H of about 3.0 and reagents consisting of beta-naphthol, sodium nitrite and caustic soda, previously dissolved and mixed, are added. The discharge from this purification is filtered through Shriver presses. The filtrate is added to the feed to the electrolyzing cells. The residue is added to the residue from the acid leach step in the sulfide circuit.

An intensive batch purification, consisting of either three or four steps, is carried out at the Kellogg plant. All steps are made in mechanically agitated tanks with relatively slow agitation over a long period of time.

The solution to be purified is heated to 70 to 80°C and sufficient zinc dust added to precipitate practically all the copper. The purification is then filtered through Shriver presses. The filtrate is sent to the second step for cobalt removal. The removal of cobalt is aided by a small amount of copper in solution. A small amount of solution from which no copper has been removed is added to the main bulk of solution and the whole heated to about 75°C. Zinc dust is added to precipitate cobalt and nickel. The purification is again filtered.

Antimony and arsenic are reduced in a third step by again heating the solution, adding enough copper sulfate to bring the copper content to about 0.07 gram per liter, and again adding zinc dust. The purification is filtered. Steps 2 and 3 can be carried out simultaneously if desired and this practice is sometimes followed.

The residue from the first three steps is sent to a copper smelter. The clarified solution is sent to the fourth step for removal of cadmium. Cadmium is removed by adding about 0.07 gram per liter copper sulfate and sufficient zinc dust to precipitate all the copper and cadmium. Filtration at this step produces a solution suitable for electrolysis containing about 220 grams per liter zinc as zinc sulfate. The residue from the final step is treated in the plant for the recovery of cadmium.

Purification Residue. The residue, or residues, from the purification of neutral solution contains enough zinc, copper and cadmium to make the recovery of these metals desirable.

Zinc and cadmium are dissolved from the purification residue with return acid from the electrolyzing cells. The residue can be leached in either the oxidized or unoxidized condition. The residue can be oxidized in a furnace or by heap roasting. Oxidation increases the solubility of zinc and cadmium slightly and the solubility of copper to an extent determined by the amount of oxidation.

The leaching of unoxidized residue or partially oxidized residue should be carried out with extreme caution as the presence of arsenic in the residue is likely to cause the formation of arsine gas.

Copper and cadmium can be precipitated from the clarified solution, obtained by settlement or filtration, by the use of zinc dust. Copper is precipitated first by the careful addition of dust in order to remove all the copper possible without precipitating any cadmium. The copper removed from solution can be separated by filtering and used as the feed for a copper reverberatory furnace or can be recovered by some other method. With the proper precautions to insure a pure product, this copper residue can also be used as a source of copper for use in purification steps in the zinc plant.

Cadmium is removed from solution by agitation with zinc dust or contact with metallic zinc. Careful control will result in a high-grade cadmium sponge suitable for use in the recovery of cadmium. The sponge produced is separated from the solution by filtering, or settling and filtering.

The solution remaining after the removal of copper and cadmium contains a considerable amount of zinc and may also contain impurities such as cobalt and nickel. This solution is sent back to the main portion of the plant where it is commonly introduced along with the head acid or at some point where it will receive the full benefit of all purification steps.

The solution is sometimes given a strenuous purification before being reintroduced into the main flow of the plant. A purification with nitroso-beta-naphthol to remove cobalt, or a hot copper-arsenic-zinc dust purification to remove both cobalt and nickel, is desirable.

Arsine. The presence of arsenic in the materials used for the extraction of zinc makes precautionary measures imperative. Arsenic hydride gas, commonly called arsine, is an extremely toxic poison. Small amounts breathed into the human system will cause serious illness or even death. Only two conditions are necessary for the formation of arsine: the presence of arsenic and hydrogen. No other condition, such as the presence of acid or metallics, is absolutely necessary, although these may also be contributing factors. Arsine has been known to be formed in highly caustic solutions under the proper conditions.

Arsine is heavier than air and it is natural to assume that it would be most likely found in the low places in the plant. This is true of accumulated arsine and should be taken into account. It should also be remembered that the formation of arsine is accompanied by the formation of hydrogen, often in copious amounts, and that the arsine-hydrogen mixture will be lighter than air and will rise. Positive ventilation and an adequate arsine detection and warning system are the best insurance against this hazard in operations where arsine may be encountered.

Antimony and germanium also form hydrides although the formation with these elements is not as likely as with arsenic. The protective measures taken for arsenic will usually cover the danger from the hydrides of these elements.

THE ELECTROLYSIS OF ZINC SULFATE SOLUTION

Polarization and Overvoltage

The recovery of zinc by electrolysis is accomplished by the application of an electrical current through insoluble electrodes, causing a decomposition of the aqueous zinc sulfate electrolyte, and the deposition of metallic zinc at the cathode. Oxygen is released at the anode and sulfuric acid is formed by the combination of hydrogen and sulfate ions. The operation of a zinc electrolyzing cell is affected by several factors which must be closely controlled.

Polarization and overvoltage are not synonymous terms, but the effects of polarization are in large part caused by overvoltage. Polarization may be due to counterforces set up in the cell to offset a portion of the applied current, to the ion concentration of the electrolyte, or to the accumulation of gases at the electrodes.

The theoretical voltage to deposit zinc from a zinc sulfate solution is about twice the voltage necessary to decompose water. It would appear that the electrolysis of zinc sulfate solution would result simply in the decomposition of water with the production of hydrogen at the cathode. This actually happens when an electrode of platinum black is used. On zinc, however, considerable resistance (overvoltage) to the production of hydrogen is encountered.

Comparing a platinum cathode and a zinc cathode in an electrolyte of sulfuric acid shows that hydrogen will be produced on the platinum electrode at a potential of about 1.7 volts, while hydrogen will not appear on the zinc electrode until a potential of 2.4 volts is reached. This difference can be called the hydrogen overvoltage of a zinc electrode as compared to a platinum electrode. The potential required to deposit zinc on a zinc surface from an electrolyte of zinc sulfate and sulfuric acid is 2.35 volts, while the potential required to deposit hydrogen under the same conditions is 2.4 volts. This difference of 0.05 volt is the narrow margin which spells the success or failure of the production of zinc by electrolysis.

There are many theories that have been advanced to explain polarization or overvoltage. Regardless of the many explanations offered, it can simply be said that the difference between the electrode potential required for the passage of current and the equilibrium value is the overvoltage of the electrode. Overvoltage is an electromotive force that acts counter to the applied electromotive force during electrolysis, and represents the excess of energy required to form a substance over that given off by the resolution of the product to its original state. The hydrogen overvoltage of an electrode is the difference between its potential when hydrogen is liberated during electrolysis and the potential of the reversible hydrogen electrode, both potentials being referred to the same electrolyte.

The consideration of overvoltage effects is usually limited to hydrogen and oxygen, although other gases and the metals also have definite overvoltages. The hydrogen overvoltage is an important factor in the deposition of zinc.

Any controllable factor that will cause a lowering of the hydrogen overvoltage on zinc should be avoided if the elimination of such a factor is also consistent with the over-all economics of the system. In a like manner, all factors that tend to increase the hydrogen overvoltage on zinc should be utilized where possible.

Some of the known factors that affect the hydrogen overvoltage on zinc follow:

Electrode Surface. The overvoltage is lower on a rough than on a smooth surface. The effect is accumulative as deposition progresses and accounts in part for the extreme sensitivity of a zinc cell. The original roughness may be produced by some other factor but, when once formed, is in itself a contributor to a lowering of overvoltage.

Concentration of Ions. A lack of available ions in the electrolyte produces an effect at least similar to a lowering of the hydrogen overvoltage. This effect, sometimes called "concentration polarization," can be diminished by agitation of the electrolyte and by the use of grid anodes to increase the circulation. An increase in current density increases the necessity for improved circulation.

Purity of Electrode Metal. A detrimental impurity existing even in trace amounts may seriously lower the hydrogen overvoltage at this point. "Local-cell" action may be set up which will prevent the plating of zinc in the immediate area and will also contribute to the re-solution of zinc.

Time of Deposit. The maximum hydrogen overvoltage may not be reached immediately and may also vary considerably as deposition progresses. The deposition time that can be tolerated in the economics of the system is dependent on most of the other factors that effect the plating of zinc. As the other factors vary, the deposition time can be varied from 8 to 72 hours on a commercial basis. Under laboratory conditions, the deposition time has been extended to 30 days or more with excellent results.

Temperature. The temperature at which electrolysis is carried out may appreciably affect the overvoltage. An increase in temperature in the electrolysis of zinc usually lowers the overvoltage and also increases the detrimental effects of certain other factors.

Presence of Colloids. The hydrogen overvoltage of a zinc electrolyzing cell is normally increased by the presence of colloids. The generally accepted theory is that the colloidial particles migrate under the influence of the applied electromotive force, by an action known as electrophoresis, due to charges on the surfaces of the particles. These particles deposit in much the same manner as electrolytic ions. Points of crystal growth, or any other projection on the surface of the cathode, will be of higher current density than that of the normal surface and the colloidial particles will deposit in clusters on these points. The result is partial insulating action which causes the ions of zinc to deposit preferentially on the normal surface. This tends to smooth out the deposit.

Presence of Organic Substances. The presence of organic substances, other than certain colloids, usually tends to lower the hydrogen overvoltage.

Current Density. An increase in current density tends to increase the overvoltage. Other factors, such as an increase in temperature, an increased activity of impurities and a reduction in the ion concentration, may accompany the increase in current density and offset any improvement gained by this factor.

Impurities in the Electrolyte. This is probably the most important of all factors affecting the hydrogen overvoltage and also one of the least understood. The effect of single impurities has been studied in great detail and their effects under standard conditions can be quite accurately predicted. The effect of two or more impurities in combination is a much more complex problem. The large number of possible combinations and their varied effect on the hydrogen overvoltage makes this factor a never-ending problem.

Commercially "pure" zinc sulfate solution is a solution which, when used as an electrolyte, will give a good deposit of zinc, with a high current efficiency. All commercial solutions are only relatively pure as all will contain other elements to some extent.

A really pure zinc sulfate electrolyte is almost impossible to obtain. The zinc cell is so sensitive to impurities that even C. P. reagents must undergo purification before they can be used. Even the vessels used to prepare such a solution may cause contamination. The electrodes used are a further source of impurities. Zinc produced from a pure zinc sulfate electrolyte possesses properties not commonly associated with the commercial metal. It does not tarnish in air, it is practically insoluble in sulfuric acid, and is so ductile that it can be beaten like gold leaf.

Voltage

The theoretical decomposition voltage of a zinc sulfate solution can be calculated from the heat of reaction and is 2.35 volts. This theoretical decomposition voltage can be obtained under laboratory conditions and with platinum electrodes. The commercial, working value for the decomposition voltage is about 2.67 volts. Other factors of resistance within the electrolyzing cell bring the actual voltage applied to about 3.41 volts.

The voltage necessary to maintain a given current density is proportional to the total resistance of the cell. The resistance of the electrolyte, of the electrodes, and at the surface of the electrodes must be taken into account. The resistance of the solution is by far the most important due to several factors. The electrode spacing, the concentration of zinc sulfate and of sulfuric acid, and the temperature of the solution are some of the factors involved. An increase in the temperature decreases the resistance of solution and, if corrosion did not also increase with temperature, the use of hot solutions would be desirable. The resistance of the electrodes is usually quite small except when the accumulation of lead peroxide or manganese dioxide at the anode reaches the point where it forms a barrier to the passage of current. Gas accumulation at either electrode will also increase the electrode resistance.

Current Density

The current density at the cathode is an important factor in the operation of a zinc cell. Current density is usually expressed in amperes per square foot of cathode area. A current density of 25 to 40 amperes per square foot is considered a low density in the electrolysis of zinc. In this range, the control of temperature of the cell usually can be carried out by cooling coils in the cell. External cooling must be used at higher current densities. Because of the external cooling required, it is about as practical to operate in the range of 80 to 100 amperes per square foot as it is to operate in the range of 50 to 60 amperes per square foot.

The cathode sheets in a zinc cell are usually larger than the anode sheets in order to reduce the effect of the sharp edges setting up points of high current density. Some type of edge strip of rubber, plastic or wood is often used to reduce this effect and also to prevent the deposited zinc from bridging around the edges of the sheet.

A high current density tends to increase the detrimental action of any impurities present and extreme measures must be taken to provide a very pure electrolyte. High current density requires a higher zinc ion concentration in the electrolyte and a higher rate of circulation of the electrolyte within the cell to avoid areas of low ion concentration. Grid-type anodes are often used to improve circulation. Gas evolution at the anode increases with the current density and increases the circulation.

The length of the deposition period for a given amount of zinc is shortened with an increase in the current density. A shorter deposition period will tend to offset the detrimental effects of high current density.

The current density employed has a considerable influence on the energy efficiency of a zinc cell. One gram of zinc should be deposited with each 1.927 watt-hours of energy. There is an optimum range of acidity in which zinc can be deposited at the greatest energy efficiency for each current density employed. This range is quite narrow and is affected by other factors present. In general, the higher the current, the higher the acidity required for best results.

Temperature

The operating temperature of a zinc cell can have a considerable effect on the current efficiency, the amount of lead in the deposited zinc, and the amount and physical form of manganese dioxide depleted from the electrolyte.

A temperature range between 30 and 40°C has been found to be the best when all factors are considered. The detrimental effect of impurities is increased with an increase in temperature, although a smooth deposit will show fewer ill effects than a rough deposit. The electrolysis of zinc is complicated by the formation of 1.505 grams of sulfuric acid for each gram of zinc deposited. Corrosion of the deposited zinc is a function of the amount and kind of impurities present and the concentration of acid in the electrolyte. Corrosion, being a chemical reaction, is more rapid as the temperature increases.

The amount of lead deposited with zinc at the cathode is a function of the cell temperature. The optimum temperature is 35°C for the production of low-lead zinc.

The physical character of the deposited zinc varies with the operating temperature. At 20°C, the deposit of zinc is a tangled fibrous mass resembling steel wool when viewed under a microscope. At 30°C, some of the fibrous character is still present but the surface begins to show the familiar "warty" deposit. At 40°C, the zinc surface has lost practically all of its fibrous character and is almost totally "warty" in appearance.

The manganese dioxide deposited at the anodes of a zinc cell operating at 20°C is finely divided and does not adhere to the anodes to any extent. The amount of manganese depleted from the solution is relatively large and some of the manganese is present as permanganic acid. At 30°C, the manganese dioxide formed is a mixture of finely divided particles and scale. Both are loosely adherent to the anode. The amount of manganese depleted from a given solution, at this temperature, is less than the amount depleted at 20°C. At 40°C, the manganese dioxide scale is mainly of the flake form. A larger portion of the deposit adheres to the anode but is fairly easy to remove. The amount of manganese dioxide depleted from solution again increases at temperatures above 40°C. The scale formed adheres tightly to to the anode and is very difficult to remove.

Purity of Electrolyte

The most important factor in the successful electrolysis of zinc sulfate solution is the purity of the electrolyte. The solution must be free of elements electronegative to zinc. If the solution is not free of mechanically held impurities, the sulfuric acid generated in the cell will dissolve any such impurities that are soluble. With an extremely pure solution, other factors may vary over a considerable range without disasterous results. While the effects of certain impurities can be minimized by changes in operating conditions or by the use of addition agents, the only guarantee of satisfactory results is purity of electolyte.

Purity is only relative and no electrolyte will be completely free of impurities. The zinc cell is so sensitive to certain impurities that the reaction of the cell to these impurities is often a better indication of their presence than ordinary methods of analysis. Ideally, the solution to be used as an electrolyte should be thoroughly tested before actually being fed to the cells. This is not always economically feasible and each commercial plant must decide whether to follow this practice.

The purity of the electrodes, tank linings and other auxiliary equipment that may come into contact with the electrolyte or may contribute to the contamination of the electrolyte is also of prime importance. Anodes should be of chemically pure lead or, if lead-alloy anodes are used, the lead and alloying agents should be as free as possible of unwanted elements. The aluminum used for the cathodes should be of the highest purity obtainable. Tank linings, if of lead, should be of the purest grade. A great deal of care should be taken that no impurities accidentally find their way into the tanks. Bus bars should be so designed and located that they are not a source of contamination.

Addition Agents

Addition agents such as glue, goulac and various other organic substances are often used to increase the ampere efficiency of a zinc cell and to improve the physical character of the deposit. Metal salts have a limited use for the same purpose.

One theory on the use of addition agents is that certain substances added to the electrolyte will combine with the unwanted impurities present and cause them to precipitate with the anode sludge. Compounds containing oxidizing radicals are thought to cause the precipitation of certain elements as oxides. Cyanamide, tannic acid and the like are in this class of addition agents.

A colloidal solution or suspensoid, such as glue, is formed of neutral substances which are influenced by an applied electromotive force to migrate to one or the other electrode in an electrolyzing cell. This migration is called "electrophoresis," and any suspended solid, liquid or gaseous colloidal particle exhibits this effect. The particles migrate in a definite direction and at a definite velocity. The migration is due to charges appearing on the surface of the suspended particles. Electrophoresis explains why colloidal addition agents appear at the electrodes in proportions greater than can be accounted for if migration did not take place. It may also explain why a certain amount of lead is found in the zinc deposit. Water molecules also migrate to the cathode by electrophoresis. This action is accelerated in dilute solutions and decreases the concentration of metal ions in the vicinity of the cathode.

Colloidal particles migrate and deposit at the electrodes at much the same velocity and manner as do electrolytic ions. Points of crystal growth or any other projection on the electrode will be of higher current density than the normal surface and the colloidal particles deposit in clusters on these points and, by an insulating action, reduce the amount of current to these points. This action tends to smooth out the deposit. Only the colloidal particles that migrate to the cathode are of particular interest in the electrolysis of zinc.

Power Equipment

The electrolysis of zinc requires d.c. power which, in most cases, must be converted from a.c. The amount of power required is large enough that a small difference in conversion efficiency between different types of conversion equipment becomes an appreciable cost item.

Motor-generator sets, synchronous rotary converters, and mercury-arc rectifiers are the main types of equipment used. Motor-generator sets usually operate in a conversion efficiency range of between 85 and 88 per cent. Synchronous rotary converters operate in an efficiency range of 92 to 94 per cent. Mercury-arc rectifiers will operate at an efficiency of 94 per cent or slightly higher, depending on the voltage.

Mercury-arc rectifiers are the most efficient of the types in use and are superior in the amount of operating and maintenance personnel required. A mercury-arc rectifier station can be made completely automatic in operation. Power surges, voltage fluctuations or other power-line variations may cause a loss of current to the conversion station. In the case of mercury-arc rectifiers, power can be restored immediately and automatically as soon as the trouble is cleared up. With motor-generator sets and synchronous rotary converters, considerable time and manual labor are necessary to restore normal operation.

A loss of power to the electrolyzing cells is serious as zinc begins to redissolve as soon as the voltage drops below that necessary to plate zinc. The back emf set up is large enough that, with proper equipment, this battery action can be utilized to operate emergency equipment in case of a complete power failure.

Mercury-arc rectifiers suitable for use in the plating of zinc are relatively new. Many plants continue to operate older types of equipment because the advantage in operating efficiency would not justify the cost as long as the present equipment is in good operating condition.

Electrodes

The cathode plates are rolled aluminum sheets of high purity. The thickness of the sheets varies from $\frac{3}{32}$ to $\frac{3}{16}$ inch and the submerged area varies from about 8 to 12 square feet. The sheets are either riveted or welded to support, or header, bars of copper or aluminum.

Copper header bars may be of rolled or extruded stock while aluminum bars may be of rolled, cast or extruded stock. Copper header bars require smaller cross section for the same current capacity than do aluminum bars, and electrical contacts are easier to maintain. Copper bars have the serious disadvantage of being soluble in the acid spray from the cells. The copper sulfate formed is a source of contamination of the electrolyte and the presence of copper sulfate on the cathode sheet, especially in the area between the solution line and the header bar, causes an increased corrosion rate of the aluminum.

Aluminum header bars require copper contacts which are either cast into the aluminum or welded on at the point where contact is made with the bus bar. Welded contacts result in a low resistance junction between copper and aluminum. The same results may be obtained with copper cast into the aluminum if the copper insert is first coated with an intermediate metal such as cadmium or nickel.

The use of aluminum header bars usually requires that the sheets be welded to the header bars as a riveted contact of aluminum to aluminum is not particularly good. Shielded-arc welding is the most satisfactory method.

Several months of service can be expected from an aluminum cathode sheet. Corrosion of the sheet is most severe in the area between the solution level and the header bar and the worst point is at the solution line. Several methods have been devised to reduce the amount of corrosion in this area. Various paints and coatings have been used but none have been completely satisfactory. Increasing the thickness of the sheet in this area is of benefit, but increases the amount of aluminum in service and the cost of fabrication is high. Changing the solution level periodically by adjusting the overflow from the cell will increase the life of the sheet at this point and is a distinct advantage is some cases. Sheet failure may also be due to the body of the sheets becoming thin, especially on the edges. The presence of either chlorine or fluorine compounds in the electrolyte increases the amount of corrosion encountered. Fluorine is particularly detrimental and the presence of an unusual amount of fluorine can result in the zinc deposit sticking to the aluminum to such a degree that the deposit cannot be removed except by dissolving in acid.

Aluminum sheets that have failed at the solution line can be sheared into smaller sizes and the remaining good aluminum welded to form a full-sized sheet. The weld should be well below the solution line for best results.

Cathode edge strips are usually used when electrolyzing at the higher current densities. The use of edge strips prevents zinc from being plated around the edges and also avoids points of high current density and the growth of "trees" due to the sharp edges of the sheets. Several types of edge strips are used, the usual materials being wood, plastic and rubber. The cost of wood strips made from pine is low, but the life is only a few days. The cost of either plastic or rubber is such that they must last for several months to be economical. There is some thought that the use of pine strips contributes to better efficiency through supplying wood products which act as addition agents in much the same manner as glue.

The anodes used by commercial producers of electrolytic zinc are of pure lead or of lead-silver alloy. They may be cast or rolled and either solid sheets or of grid construction. The thickness of the sheets varies from 3/16 to 3/8 inch. The anodes are normally slightly smaller than the cathodes in order to minimize the effect of high current density.

Cast anodes have the advantage of being cheaper than rolled anodes and the lead from scrap anodes can be used to cast new anodes. The header bars used to support the anode sheet and provide electrical contact between the anode sheet and the bus bars are usually copper bars covered with lead except at the point of contact with the bus bar. The anode headers can be cast as an integral part of the anode or the headers may be cast separately and welded to either cast or rolled sheets. Header bars consisting of an iron bar entirely covered with lead may be used, but in this case it is necessary to provide electrical contact by burning the lead to a lead-covered bus bar. When copper bars are used, an improved electrical contact between the copper and the lead can be obtained by heating and fluxing the copper bar in zinc chloride solution immediately prior to casting. Used copper bars should be well cleaned of lead and copper salts. Tumbling in a horizontal barrel will clean and straighten the bars without the use of grinding media.

Anode sheets of grid construction are sometimes used to improve the circulation within the cell and to reduce the weight of lead required. The low-density and medium-density plants usually use one more anode per cell than the number of cathodes, but a high-density plant may use two anodes for each cathode. Grid anodes are cast separate from the header bar and welded on. A construction which will produce results similar to cast grid anodes can be obtained by punching holes in rolled sheets.

Rolled anode sheets are somewhat more resistant to corrosion than cast anodes and may be used if the cost is not prohibitive. The cost of joining either cast or rolled sheets to the header bars is a considerable item of expense as compared to the cost of an anode sheet and header cast as a complete unit.

Numerous materials and many alloys of lead with other metals have been tried as anodes. The most suitable are either pure lead or a lead-silver alloy. Lead-silver alloy anodes containing from 0.5 to 1.0 per cent silver are more resistant to corrosion than pure lead, resulting in a longer life and a smaller amount of lead deposited with the zinc. Lead-silver anodes may cause a slight decrease in the ampere efficiency of a zinc cell, but the advantages offset losses from this cause. Copper is one of the impurities that may accumulate in anode lead through reuse. The build-up of copper can be avoided by either melting the header bars separately or by purifying the lead by scorification.

Anodes should be straight and the spacing between the anodes and cathodes should be as uniform as possible. Spacers or guides are used to obtain the desired result. Some plants use one or more porcelain or plastic buttons inserted in the lower portion of the anode as spacers. Pine edge strips serve the same purpose and have the advantage of preventing some of the warping that may occur. Anodes may also be fitted into guides fastened to the inside walls of the cell. Cast anodes should be cooled and then straightened before use to prevent excessive warping.

Electrolyzing Cells

The cells or tanks used in electrolyzing zinc are constructed of wood, concrete or pitch concrete. Wood and concrete cells must be lined with an acid resisting material while pitch concrete can be used unlined. Many materials have been tried in an effort to find the ideal material, without outstanding success. Wood or concrete lined with lead is the most common construction. Concrete cells lined with rubber, sulfur sand, or acid-proof brick are often used and in some cases have proved to be satisfactory. Pitch concrete cells were used extensively at one time as they were apparently the best approach to the ideal tank. Pitch concrete is subject to cracking through thermal or mechanical shock and many plants have discontinued its use for these reasons.

The ideal tank should be impervious to the action of dilute sulfuric acid in the presence of highly oxidizing conditions. It should be a nonconductor, reasonably low in cost and give a long period of service life. The ideal tank should be capable of supporting not only the weight of solution and accumulated sludge, but also the weight of bus bars, electrodes and the deposited zinc.

The size of tank depends on the number and size of the electrodes used and also upon the spacing between electrodes. The formation of manganese dioxide during electrolysis requires that some space for accumulated sludge be provided below the electrodes.

The tanks are insulated from their foundations by the use of nonabsorbent materials such as glass or porcelain blocks. Bus bars must be well insulated from the tanks and supports to avoid current leakage. Solution feed lines and depleted acid lines or launders must be of nonconducting material or well insulated. Metallic pipelines in the electrolyzing building are often fitted with short sections of rubber hose to avoid current leakage.

Zinc electrolyzing cells are placed in series electrically according to the power equipment available. The voltage of the system is often the determining factor. The maximum voltage available from the equipment divided by 3.75, is a conservative figure to use in determining the number of cells that can be carried in one unit.

Several different methods may be employed in the solution flow system to and from the cells. Most plants arrange the cells in a cascade system in order to economize in the space required. Six to nine cells may be in cascade with the discharge flow from each cell passing through the cells below it in each cascade. Two such cascades may be placed side by side to form a double cascade of cells with a working aisle between each double cascade. When arranged in this manner, the adjacent tanks in each cascade may even be built as a double tank with a common wall.

Cells may also be arranged in parallel instead of series. In this system, each cell discharges directly to the return-acid launders. This system requires either that more space be used or that the men employed in stripping walk across the tops of the tanks. The latter arrangement has been used and is satisfactory.

The purified solution to be electrolyzed is fed by gravity to individual cells in most of the low-density and medium-density plants. Launders or pipelines of iron, plastic or porcelain are used. The flow is accumulative when the cells are cascaded, but it is necessary to feed each cell individually and to maintain an even acidity throughout the cascade for best results. High-density plants usually use the parallel feed system. The electrolyte is circulated quite rapidly through each cell and the discharge from the cells is carried through a separate cooling section. In this system it is customary to feed the new solution in batches as required and to remove a portion of depleted electrolyte in a like manner. Depleted electrolyte is removed from the electrolyzing cells through plastic pipes, plastic launders or launders lined with lead.

A considerable amount of heat must be dissipated if the cells are to operate within the optimum range of temperature. The amount of heat to be dissipated increases with an increase in current density. Up to a certain point, the cells can be cooled sufficiently by means of cooling water passed through coils in the ends of the cells. Beyond this point, external cooling systems are used and the electrolyte is recirculated quite rapidly through the cells and sets of cooling cells equipped with cooling coils.

The cooling coils are of lead, aluminum or stainless steel. Lead is commonly used as it has a long service life. Lead coils are comparatively poor in heat transfer. Aluminum coils are in extensive use where the amounts of chlorine and fluorine in the electrolyte are low enough to allow a reasonable service life. Stainless steel coils are in limited use at the present time but show considerable promise as they can be thin enough to give a good heat transfer and are resistant to chemical attack.

The control of acidity within each cell is important. For each current density employed, there is a narrow range of acidity in which the zinc is most economically deposited. Frequent testing and regulation of the flow to the cells is necessary. The tests for acidity can be made by titration of the acid with carbonate solution, by the use of special hydrometers or by the use of electrical devices to measure resistance.

Periodic cleaning of the cells is necessary in order to remove the accumulated manganese sludge, to clean, straighten and cull the anodes, to renew cell insulation and to inspect the tanks for leaks. Most plants are equipped with spare tanks that can be cut into the operating circuit so that the cleaning operation can proceed without loss of production.

The removal of manganese sludge can be done by hand, by a vacuum system or by means of mechanical digging machines. During periods where full production is not required, it has been found advantageous to shut down and clean an entire unit in one operation. By this method, the deposited zinc is not removed from the cathodes prior to the shutdown. The cells are decreased to a very low acidity by flooding with neutral electrolyte. The deposited zinc protects the aluminum sheets from any remaining acid, which is soon completely neutralized by the re-solution of zinc. The electrodes are then removed and the cells emptied and cleaned.

The above described method of shutting down a unit for cleaning can also be employed for shutting down the entire operation. In this case, the cells are not emptied and cleaned. Electrolyzing cells shut down in this manner can be started-up with very little loss in ampere efficiency and with no damage to equipment due to the shutdown.

The Bunker Hill Company electrolyzes in 620 lead-lined wood tanks arranged into 4 units. Each cell is individually fed from an overhead launder system through polyethylene pipes. Each cell is divided into two sections by the electrode and feed arrangement. There are 7 cathodes and 14 anodes in each section of the cell with the overhead feed pipe serving as the dividing point. The overflow from the cells is carried through polyethylene pipes to collection launders beneath the floor. Cells are placed side by side. The anode bars are bolted to the bus bar and the cathode bars fit into clips on the bus bar on the adjacent cell. The bus bar between each adjacent cell acts as an anode bus for one cell and the cathode bus for the next.

The cathode current density is approximately 100 amperes per square foot. In each cell are two lead-covered wooden frames, to which resin-treated wood cathode guides are attached to hold the electrodes in proper relation to each other. The cathodes slide up and down in a groove in the center of the guide. Polyethylene anode spacers are used on the cell frame.

The anodes are of cast lead alloyed with 1.0 per cent silver. They are $3/16$ inch thick and are of grid construction in which 40 per cent of the total area is taken up by the holes through the material.

The cathodes are of $1/8$-inch aluminum plates welded to aluminum header bars. The high current employed requires that very good contact be maintained to avoid heating. A clip for each cathode is soldered and bolted to the bus bar, and the contact area of the header bar that fits into the clip is plated with copper.

Electrolyte is circulated through the cells at a rate of about 10 gallons per minute per cell. The electrolyte also passed through cooling tanks at a higher elevation than the electrolyzing cells so that the flow back to the cells is by gravity. The cooling tanks are fitted with plate-type lead coolers.

Fresh electrolyte is added to the recirculating system in batches rather than continuously. Depleted electrolyte is removed in the same manner. About one-fifth of the total volume of electrolyte is removed when the acidity has reached the optimum point, and an equivalent amount of purified zinc sulfate solution is added.

Cathode deposits are removed at the end of either an 8- or 12-hour deposition period. The aluminum cathodes carrying the deposited zinc are removed one at a time and a clean aluminum cathode is immediately dropped into place. The cathodes carrying the zinc deposits are placed on a battery-powered truck and removed to a separate room for washing and removal of the zinc.

A froth is maintained on the top of the cells by the addition of a mixture of cresylic acid, sodium silicate and gum arabic and effectively keeps down the mist from the cells.

The Consolidated Mining and Smelting Company zinc plant operates electrolyzing cells in units of 144 cells. Each unit is made up of 4 groups of 36 cells arranged in four cascades of 9 cells each. The cells are usually concrete, rubber lined, and are cast in pairs with a common center wall. The average current density is 46 amperes per square foot.

The anodes are of rolled lead-silver sheets $\frac{3}{16}$ inch thick, containing 0.75 per cent silver. There are 23 anodes in each cell, each with a submerged area of 9.2 square feet, set at 3-inch spacing. The anodes are burned to cast lead header bars which are in turn burned to a lead-encased copper bus bar. Two maple or laminated spacer-sticks along the bottom of each set of anodes maintain the electrode spacing.

The cathodes are of No. 8 B and S gage, high-grade aluminum sheet. A copper connector is fastened with rivets to one side of one end of the sheet and an iron support bar is fastened to the other side to serve as a header bar.

Electrolyte is fed individually to the top eight cells of a nine-cell cascade, the overflow being accumulative. The addition of electrolyte is governed by the acidity as determined by hydrometer readings. From 50 to 55 per cent of the zinc is plated out in each cycle of solution through the cells. Glue is added at the rate of 0.3 pound per ton of cathode zinc. Lead cooling coils are used to control the temperature to an optimum of 30°C. Cathodes are stripped each 24 hours.

The humid climate at the Corpus Christi plant requires special designs for control of stray currents in operating a 600 volt d.c. circuit. Brick-lined concrete cells are used and plastic launders carry the electrolyte to and from the cells. The feed launders are located about 6 feet above each bank of cells and the return launders are from 3 to 4 feet below the outlet of the cells. Removable porcelain pipes are used for directing electrolyte from the feed launders to each of the cells and rubber pipe to convey the cell discharge to the return launders.

The electrolyzing cells are arranged in parallel rows of 10 cells each in the sulfide division and of 14 cells each in the oxide division. The current density is about 70 amperes per square foot.

The anodes are of cast grid construction, $\frac{5}{16}$ inch in thickness, and contain 1.0 per cent silver. The main body of the anode is burned to a header bar of antimonial lead. Each anode is fitted with a large screw-in porcelain insulator at the lower center. Spacing is maintained at 3-inch centers.

The cathodes are of $\frac{3}{16}$-inch aluminum sheet. The sheets are welded to silicon-aluminum alloy header bars. Each cathode header bar is fitted with a cone-shaped copper contact. The anode header contacts are shaped so that each side of the anode contact fits around approximately one-half of each adjacent cathode contact.

Electrolyte is fed individually to each cell. A closed system is used for continuous flow to each cell. The circulated solution is cooled in evaporative-type cooling towers. A frothing agent of cresylic acid, sodium silicate and gum arabic is added to the feed to cells. A slurry of strontium carbonate is also added for control of the lead content of the cathode zinc.

The cells are stripped after either a 16-hour or a 24-hour period of deposition. Alternate cathodes are removed at one time for stripping.

Electrolysis at the Risdon plant is carried out in units of 144 cells. The average current density is about 28 amperes per square foot. A normal load of 18,000 amperes is used per unit although this may be increased to 20,000 amperes during periods of off-peak power.

Figure 6-14. Tank room, Electrolytic Zinc Company of Australasia Ltd., Risdon, Tasmania.

The anodes used are $\frac{3}{8}$-inch cast lead-silver alloy containing 0.11 per cent of silver as a stiffener only. The cathodes are $\frac{3}{16}$-inch rolled aluminum sheets spot-welded to rolled aluminum header bars. Copper contacts and steel lifting lugs are rivetted to the header bars. There are 44 cathodes and 45 anodes in each cell. Spacing varies from $3\frac{3}{4}$ inches to 4 inches.

Electrolyte is fed to the cells individually in each cascade, the overflows being accumulative. An addition agent, consisting of glue, beta-naphthol, and a small amount of antimony, is used to promote better efficiency. The operating temperature of the cells is maintained at an optimum of 35°C by the use of three or four cooling coils in each cell.

The zinc deposition period is 72 hours. This is an outstanding accomplishment of the Risdon plant that is not approached by any other commercial producer of electrolytic zinc.

A small separate section of the electrolyzing division uses anodes of 1.0 per cent lead-silver alloy and an addition agent of strontium carbonate to produce low-lead zinc.

The zinc plant at Great Falls is divided into eight electrolyzing units of 156 cells each. Each unit is made up of 12 cascades of 6 cells and 12 cascades of 7 cells. The cells used are of concrete, lined with sheet lead.

The anodes are of cast lead containing 0.5 per cent silver. An anode and its header support are cast as one piece with a copper support and contact bar cast into the header. The anode sheet is tapered from $\frac{5}{16}$ inch at the top to $\frac{3}{16}$ inch at the bottom, except for a narrow strip on each side which is $\frac{3}{16}$ inch for the full length. Wood strips on each edge aid in preventing warping and maintain electrode spacing. Thirty-three anodes are used in each cell.

The cathodes are high-purity aluminum sheets, $\frac{3}{16}$ inch thick. The sheets are welded to cast aluminum header bars. The header bars are equipped with cast aluminum lifters and a short copper contact bar cast in one end. Edge strips of plastic are used to prevent points of high current density at the edges of the sheet and to aid stripping by preventing the zinc from bridging around the edges. The submerged area of each sheet is 12.2 square feet and the current density is 34 amperes per square foot of cathode area.

Electrolyte is introduced into each cell individually but the overflows

Figure 6-15. Stripping zinc, The Anaconda Company, Great Falls, Montana.

in each cascade are accumulative. A portable electrical measuring device is connected to recording meters and indicating lights to measure acidity, and the flow is regulated accordingly. Sixty per cent of the zinc is removed from the solution in each cycle through the cells. Glue is added to improve zinc deposition. Lead, aluminum and stainless steel cooling coils are used to cool the solution. The zinc deposit is stripped from the cathodes each 24 hours.

The Anaconda, Montana, zinc plant is divided into six electrolyzing units of 156 cells each. The cells are grouped in cascades of 6 cells each. The cells are of wood, lined with sheet lead.

The anodes are of cast lead, 3/16 inch thick, and contain 0.5 per cent silver. The anodes and anode header bars are cast together with a copper support and contact bar cast in the header. Porcelain buttons in the lower portion of each anode serve as spacers.

The cathodes are high-purity aluminum sheets, 3/16 inch thick. The sheets are welded to extruded aluminum header bars. Contact between the aluminum header and the bus bar is through a copper clip welded to one end of the header bar.

Electrolyte is introduced individually to each cell of the six-cell cascade, the overflows being accumulative. The addition of electrolyte is regulated by means of special hydrometers mounted in wells in the bottom of each cell. Aluminum coils are used to maintain the temperature between 30 and 35°C. The cathodes are stripped at the end of a 24-hour deposition period. Polyethylene strips are used on the sides and the bottoms of the cathodes.

The electrolyzing units at the Hudson Bay plant are made up of groups of 160 cells. The cells are of unlined prodor pitch construction. They are placed end to end in cascades of 10 cells each. The cathodes used are of No. 8 gage high-purity aluminum sheets, 40 inches long and 24 inches wide. The sheets are welded to cast aluminum header bars in which is cast a short, nickel-plated, copper insert. Grooved plastic strips are fitted to the vertical edges of the cathodes to prevent the zinc from being plated around the edges.

The anodes are 3/16 inch thick, 23 inches wide, and 39½ inches long. The anode sheet is cast of silver-lead alloy containing about 1.0 per cent silver. The sheets are welded to pure lead header bars. The lead header bar is cast with a copper insert to insure good electrical contact with the bus bar. Conical plastic insulators are screwed onto studs inserted into the anodes near each bottom edge to serve as spacers.

The units are normally operated at about 560 volts with a current flow of 6000 amperes. With 17 cathodes per cell, this is equivalent to a current density of 34 amperes per square foot of cathode area. During certain periods, the current has been increased to 8000 amperes per unit or about 46 amperes per square foot.

Cooling of the electrolyzing cells is by the use of one or two aluminum cooling coils per cell. The optimum operating temperature of the cells is 32°C. Glue is used to inhibit the action of impurities in the solution and to promote the formation of a smooth cathode deposit.

The zinc deposited on the aluminum cathodes is stripped each 24 hours and sent to the casting division for melting and casting into shapes for market.

The Norway plant operates lead-lined concrete electrolyzing cells, each equipped with 29 lead anodes containing either 0.5 or 1.0 per cent silver. The 28 aluminum cathodes per cell are welded to aluminum header bars. Two lead-tube coils are used in each cell for cooling the electrolyte. The cells are arranged in double cascades of 12 cells each. Solution is fed to the cells individually and the overflow from cell to cell down each cascade is accumulative. The current density is about 38 amperes per square foot of cathode area. Strontium carbonate is added to the cells to control the amount of lead in the cathode zinc. The control of acidity in each cell is by periodically checking each cell with a portable electronic conductivity meter. The zinc deposit is stripped at the end of either a 24- or 48-hour deposition period.

The American Zinc Company of Illinois at Monsanto (East St. Louis), carries out the electrolyzing step in lead-lined concrete cells which are 8 feet long, 2 feet, 10 inches wide and 4 feet, 10 inches deep, inside measurements. Brick-lined cells are also being tested. Each cell contains 27 cathodes and 28 anodes. Each unit of cells is operated at 14,400 amperes or at approximately 46 amperes per square foot of cathode area. The cathodes are ¼-inch rolled aluminum sheets to which are welded ⅝-inch by 2 and ⅞-inch extruded aluminum header bars. A copper clip welded to the contact end of each bar serves as a contact with the bus bar. Rubber cathode edge strips are used.

The anodes are rolled sheets of lead-silver alloy containing 1.0 per cent silver. The sheets are punched with ⅜-inch square holes in a grid pattern. The anode sheets are burned to soft lead header bars cast around a copper contact and support bar. The anodes are fitted with porcelain buttons which serve as separators and prevent contact with the cathodes.

The lead cell linings are maintained cathodic, at 2.5 volts, by a connection through a resistor. Improved life of the linings and a low amount of lead in the cathode is obtained in this manner.

Purity of Cathode Zinc

The three metals that normally deposit with zinc are copper, cadmium and lead. The purity of the electrolyte is the main factor that influences the amount of either copper or cadmium in the deposit. Both are fairly easy to remove in the zinc dust purification. Copper from bus bars or cathode header bars can also be a source of contamination. The amount of lead present in the solution as lead sulfate is very minute and contamination from this source is negligible. The main source of lead in the deposit is from the lead anodes or lead cell linings and is caused by corrosion. Any factor that will prevent lead corrosion or will remove the products of corrosion will aid in reducing the amount of lead in the deposit.

The presence of cobalt is quite effective in suppressing the amount of lead in the deposit, and is the cheapest method if it can be used. The cobalt content can be carried as high as 0.01 gram per liter in the absence of germanium.

Low cell temperatures are beneficial to ampere efficiency and consequently to lower lead in the deposit. A temperature of 35°C has been found to be the optimum for obtaining the lowest lead.

A high ampere efficiency will result in a lower lead content of the cathode zinc than a lower efficiency. This and the preceding factor are tied in with the impurity content of the electrolyte.

The use of silver-lead anodes containing from 0.5 to 1.0 per cent silver is beneficial in reducing anode corrosion and consequently the amount of lead in the zinc. This method is costly to initiate but has the advantage of being permanent. There may be a slight lowering of ampere efficiency due to the use of alloy anodes. New anodes containing silver will deplete more manganese from solution than new pure lead anodes. This effect is reduced after the anodes have been used for some time.

The frequent cleaning of electrolyzing cells and remelting of the anodes are actually two separate factors which are closely tied together. Frequent cleaning will benefit ampere efficiency by removing the anode sludge and scale. Anode scale and sludge may become "poisoned" by the accumulation of impurities and cause the cell to operate at a lower than normal efficiency. New anodes, especially those of new lead and of low copper content, are beneficial in lowering the lead content of the deposit. New silver-lead anodes, after a short conditioning period, will give the same results.

The use of either strontium carbonate or barium hydroxide as addition agents is very effective in lowering the lead content of the zinc deposit if the temperature of the cells is held below 40°C. The use of strontium carbonate is a patented process developed by the American Smelting and Refining Company and the use of barium hydroxide is a patented process developed by the Cerro de Pasco Corporation.

The Effect of Impurities

The zinc electrolyzing cell will react to impurities in amounts too small to be determined by ordinary means. Even the most rigorous purification steps in the leaching and purification portion of the process will not entirely eliminate the impurities present. The impurities are controlled and reduced only to amounts that normally allow electrolysis to be carried out at a reasonable ampere efficiency.

The effect of two or more harmful impurities in combination is much more detrimental to electrolysis than the effect of single impurities. An example of this is the effect of cobalt and germanium in combination. In electrolytes containing little or no germanium, cobalt can be tolerated in amounts up to 0.01 gram per liter without appreciable effect. In fact, cobalt is sometimes deliberately carried in rather high amounts as cobalt in the electrolyte will tend to suppress the amount of lead in the deposited zinc. In electrolytes containing even a small amount of germanium, however, the amount of cobalt that can be tolerated without affecting the efficiency of electrolysis is in the order of 0.001 gram per liter or less.

The metallic ions which act as cathodic impurities in zinc electrolysis may be grouped into classes according to their effects. Class I impurities are those metals whose sulfates decompose at a voltage above that of zinc sulfate (2.35 volts). The common members of this class are sodium, potassium, magnesium, aluminum and manganese. These impurities have no effect on electrolysis as far as cathodic reactions are concerned. Their presence may affect the conductance of the electrolyte.

Class II impurities are those metals whose hydrogen overvoltage is above 0.65 volt, but whose decomposition voltage is less than that of zinc. The members of this class affecting the electrolysis of zinc are cadmium and lead. These metals, when present, will deposit with the zinc, causing an impure product.

Class III impurities are those metals whose hydrogen overvoltage is below 0.65 volt, and whose decomposition voltage is above that of sulfuric acid. The common members of this class are iron, cobalt and nickel. These metals are injurious to the electrolysis of zinc, but are not deposited with zinc to any appreciable extent. They are soluble in sulfuric acid due to their low hydrogen overvoltage. Their action in the zinc cell is probably one of being alternately deposited on the zinc surface and dissolved by the sulfuric acid present. The presence of these impurities can cause localized lowering of the hydrogen overvoltage and spot burning of the deposited zinc. Their action also consumes power that would otherwise be used to deposit zinc.

Class IV impurities are those metals whose sulfates decompose at a voltage below that of the decomposition voltage of sulfuric acid, and whose hydrogen overvoltage is less than 0.65 volt. The members usually considered as belonging to this class are copper, arsenic, antimony, germanium and tellurium. The members of this class, if present, are deposited with zinc, causing points of low hydrogen overvoltage and the evolution of hydrogen. The action of these impurities is very detrimental to the electrolysis of zinc.

Further groups or classes can be drawn up, but these four cover the cathodic impurities normally encountered in the electrolysis of zinc.

Sodium, Potassium, Magnesium, Aluminum. These elements and their compounds are regarded as harmless impurities unless they build up sufficiently to displace zinc in solution. The increase in specific gravity due to excessive amounts of any of these elements can hinder settlement and filtration and may even increase the resistance of the electrolyte.

Manganese. The compounds of manganese are normally considered harmless, or even beneficial, to electrolysis. Soluble manganese compounds are almost always present in the electrolyte. Manganese is usually present in the calcine treated and manganese dioxide or other manganese compounds are commonly used as oxidizing agents in the leaching step. Manganese dioxide is deposited at the anode during electrolysis and a certain amount of the more highly oxidized compounds of manganese may also be formed. Only a portion of the manganese in the electrolyte is deposited and the amount deposited depends upon other cell conditions rather than the amount of manganese present. Some of the deposit adheres to the anode, some falls to the bottom of the cell as a sludge, and a smaller portion is carried out of the cell with the depleted electrolyte. The amount deposited and the amount that adheres to the anode is influenced by temperature, current density, acidity, anode age, anode composition and the presence of certain impurities such as cobalt or iron.

The amount of manganese dioxide adhering to the anodes and the amount falling to the bottom of the cells as a sludge are the main factors in determining the rate at which cells must be removed from service for cleaning.

Excessive amounts of manganese dioxide adhering to the anodes can cause partial insulation and a high anode current density on the remainder of the surface. Manganese in solution has little effect on the voltage of a zinc cell, but deposits on the anode will raise the voltage.

The presence of manganese in the electrolyte is of some benefit when other impurities such as arsenic, antimony and cobalt are present. The presence of manganese dioxide in the cell and its effect on anode corrosion is somewhat open to question. Under certain conditions, the presence of manganese dioxide will reduce the amount of anode corrosion and aid in lowering the amount of lead deposited with the cathode zinc. Under other conditions, the reverse will be found. New anodes will precipitate more manganese from solution than old anodes, and the rate at which cleaning is required is influenced by this factor.

The amount of manganese that can be tolerated in a zinc cell will vary according to the other cell conditions. A concentration of 6 to 7 grams per liter usually has no noticeable effect on electrolysis, but more than this amount may be detrimental.

Cadmium. The presence of cadmium up to 0.15 gram per liter in the electrolyte has no detrimental effect on ampere efficiency. Cadmium will deposit preferentially to zinc but at a lower voltage. Cadmium is one of the major impurities of either High Grade or Special High Grade zinc, and the close control of cadmium is necessary for this reason. The amount of cadmium in per cent appearing in the slab zinc will be roughly double the amount of cadmium in grams per liter in the electrolyte. Cadmium is usually well controlled in the zinc dust purification.

Lead. The presence of lead in the electrolyte has little effect on the electrolysis of zinc. Lead is a major impurity of slab zinc and the amount in the cathode deposit must be closely controlled if a pure product is to be obtained.

Iron. Considerable amounts of iron may be present in the electrolyte without noticeable effects on electrolysis. Ferrous iron is oxidized at the anode and is in turn reduced at the cathode. This action consumes power that would otherwise be used to deposit zinc. The presence of iron in the electrolyte, while not particularly harmful in itself, is a good indication that adequate purification has not been carried out and that other, more detrimental, impurities are likely to be present.

There is some evidence that the use of iron as an addition agent will reduce the amount of lead in the zinc deposit. The action of iron is probably that of a weak depolarizer which decreases the formation of soluble lead compounds and reduces the corrosion of the lead anodes.

Cobalt. Controlled amounts of cobalt in the electrolyte are beneficial in reducing the amount of lead deposited with the zinc. The amount of cobalt that can be tolerated without detrimental effect on the ampere efficiency is mainly dependent on the presence of other elements such as germanium, and also on the acid concentration used in the cell. The presence of cobalt in the electrolyte can be observed by inspecting the deposited zinc. Cobalt causes re-solution of the deposit and produces characteristic holes, more or less round, on the back of the sheet. These holes may extend through the deposit, but are usually larger on the back than on the face, or exposed, side of the sheet.

Cobalt is somewhat difficult to control in the purification steps of the process and special purification methods are usually employed. In theory, the zinc dust purification should remove cobalt but in actual practice zinc sulfate appears to suppress the ionization of cobalt sulfate. The addition of potassium antimony tartrate in small amounts to the regular zinc dust purification will eliminate as much as 0.002 gram per liter from the solution. The use of potassium permanganate also aids in eliminating small amounts of cobalt. Larger amounts can be removed from solutions free of iron and copper by the use of nitroso-beta-naphthol which reacts with cobalt to form insoluble cobalti-nitroso-beta-naphthol. The presence of either iron or copper will reduce the efficiency of the operation as compounds with these elements are formed in preference to cobalt.

Sodium ethyl xanthate will remove cobalt from solution if used in conjunction with an oxidizing agent such as potassium permangante or copper sulfate. Zinc xanthate is first formed which, when copper is used, reacts to form cuprous xanthogenate, and the oxidizing process is transferred to the cobaltous salt to form insoluble cobaltic xanthogenate.

A purification made with copper, arsenic and zinc dust is effective in removing cobalt and the efficiency of the reaction is increased with elevated temperatures. At 80°C or higher, this purification will also remove nickel from solution. The copper used may be in the form of copper sulfate or metallic copper. The arsenic is usually in the form of the trioxide dissolved in a caustic solution.

Nickel. The zinc cell will tolerate as much as 0.0001 gram per liter nickel without ill effects unless cobalt is also present. Nickel is usually eliminated by the normal process but may build up in the system over a long period of time. The only economical commercial method of eliminating nickel from solution is the hot copper-arsenic-zinc dust purification used for cobalt. The presence of nickel in the electrolyte causes large holes to be burned in the zinc deposit.

Copper. The presence of copper in the electrolyte causes a serious loss in ampere efficiency by lowering the hydrogen overvoltage. Copper is quite easy to remove with the regular zinc dust purification and no copper should ever be found in the electrolyte from this source. Copper is easily dissolved from bus bars and electrode contacts and care should be taken not to introduce copper into the cell in this manner.

Arsenic. Arsenic is less toxic than antimony and amounts up to 0.004 gram per liter will have no serious effects unless other impurities, such as cobalt, are also present. Arsenic causes beading on the surface of the zinc deposit if present in the electrolyte. Arsenic is readily removed in the iron purification. The presence of arsenic, while not extremely serious in itself, is a good indication that adequate purification has not been carried out and that other, more injurious, impurities are probably also present.

Antimony. Antimony has a very detrimental effect on electrolysis and amounts in excess of 0.00002 gram per liter are harmful, especially if either cobalt or germanium is also present. Antimony produces a zinc deposit with a very beady surface and causes severe re-solution due to lowering the hydrogen overvoltage. The condition of the anode surface has a considerable bearing on the effect of antimony. New lead anodes cause a heavy deposit of manganese of which a portion is in the form of a fine slime. A large portion of any arsenic, antimony or cobalt present is occluded with the manganese precipitate and forms corrosion centers upon contact with the cathodes.

Antimony may be present in either the trivalent or the pentavalent form. Trivalent antimony is fairly easy to control and is normally removed by the iron purification. Pentavalent antimony is not removed by ordinary methods. The source is usually zinc oxide fume or similar materials. The only control is treating such materials at a high furnace temperature during formation or in a separate roast.

There is some evidence that a small amount of antimony in the electrolyte is beneficial to electrolysis. If the presence of antimony is desirable, it should be introduced as an addition agent rather than allowing it to appear in the electrolyte because of insufficient purification.

Germanium. One part of germanium in 10 million parts of electrolyte may cause a serious loss of ampere efficiency. The combination of germanium with either cobalt or antimony is especially serious. Germanium will cause the deposit to form with numerous tiny holes which tend to overlap and produce larger holes. Only a portion of the germanium entering the system is soluble. Germanium hydrolyzes quite readily and a leach made with a high acid content will dissolve less germanium than a leach made with a low acid content.

Germanium is difficult to remove and special steps must be taken if more than small amounts are present. Some germanium is precipitated in the iron purification but its removal is less complete than that of antimony or arsenic. Small amounts of germanium can be removed in the zinc dust purification in the presence of copper and arsenic. Germanium causes a lowering of the hydrogen overvoltage in both the purification with zinc dust and in electrolysis. When precipitated as a metal on either metallic zinc or cadmium, hydrogen is liberated at the point of contact, and these metals will go into solution. Germanium in the zinc dust purification will cause the re-solution of cadmium and the formation of the basic sulfates of both zinc and cadmium. Germanium is removed in the zinc dust purification at a slow rate, hence, good agitation for a long time is beneficial to the reaction.

Tellurium. Tellurium is highly toxic in a zinc cell. Ampere efficiency is affected by the presence of 0.000005 gram per liter of tellurium, especially if cobalt is also present. The deposit is uneven and fairly large areas are raised above the normal cathode face accompanied by corresponding depressions on the back of the cathode. The back usually shows a black discoloration over large areas. Larger amounts of tellurium will cause large areas to be redissolved.

Tellurium is quite effectively removed with the iron purification but may be introduced into the zinc dust purification through slimes carried over by the clarified solution. The zinc dust purification has little effect on tellurium removal unless made at elevated temperatures. Tellurium is often found in concentrates that also carry a high copper content.

Selenium. Selenium will usually be eliminated in the normal methods of purification. The presence of 0.0001 gram per liter in the electrolyte produces a zinc deposit with heavy vertical ridges and raised spots on the face of the cathode. The raised spots are accompanied by depressions on the reverse side which may be filled with solution. Large areas of the cathode are often completely eaten away.

Thallium. Ampere efficiency is not affected by the presence of thallium but thallium in the electrolyte will tend to raise the amount of lead deposited with the zinc. Thallium will also deposit with zinc. Trivalent thallium precipitates along with iron in the iron purification but the monovalent form will not be precipitated. The state of oxidation of thallium can be controlled by the same control used for iron.

Tin. The presence of unusual amounts of tin in a concentrate is unfavorable. Tin should be eliminated by the ordinary methods of purification. The presence of tin in the electrolyte in very small amounts will cause a loss in ampere efficiency. The presence of tin in the leaching cycle will result in a zinc deposit with a shiny surface and a different crystalline structure even when the ampere efficiency is not affected and no tin can be found in the deposit.

Bismuth, Tungsten, Indium and Gallium. These impurities are not likely to be found in the electrolyte as they are either practically insoluble or easily removed.

Calcium. Calcium is usually present and does not cause trouble chemically. Calcium sulfate will crystallize out in launders, pipelines, filters and tanks and will cause considerable trouble mechanically.

Chlorine. The corrosion of lead in the zinc cell is due mainly to the presence of chlorine. The presence of manganese tends to temper the effect of chlorine. Chlorides are oxidized at the anode into perchloric acid. The amount formed is dependent on the oxidation potential at the anode. New anodes have a higher oxidation potential than old anodes and pure lead anodes have a higher oxidation potential than silver-lead anodes.

Chlorine is automatically eliminated in the leaching process if enough silver is present to combine with the chlorine to form silver chloride. Silver sulfate can be added to eliminate chlorine and the silver-bearing sludge can be regenerated by zinc dust and acid to recover the silver.

Fluorine. The presence of fluorine is the main cause of the zinc deposit sticking to the aluminum cathode. The presence of large amounts of fluorine may result in the zinc deposit being impossible to remove. Fluorine attacks the thin film of aluminum sulfate on the surface of the aluminum. The effect becomes noticeable at a fluorine concentration of 0.002 gram per liter or over. The only known way of combating the effects of fluorine is by conditioning the cathodes by immersing them in the electrolyte for a short period before electrical contact is made. This restores the film of aluminum sulfate and materially reduces sticking.

ACKNOWLEDGMENT

Grateful acknowledgment is made to the management of the following companies for supplying information on the operation of their electrolytic zinc plants and for their permission to quote from various papers concerning their methods of operation. The addresses given are the locations of their electrolytic zinc plants:

The American Smelting and Refining Company, Corpus Christi, Texas

The American Zinc Company of Illinois, Monsanto, Illinois (East St. Louis)

The Bunker Hill Company, Kellogg, Idaho

The Consolidated Mining and Smelting Company, Trail, British Columbia

The Electrolytic Zinc Company of Australasia, Risdon, Tasmania

The Hudson Bay Mining and Smelting Company, Flin Flon, Manitoba

Det Norske Zinkkompani, A. S., Eitrheim, Norway (near Odda, Norway).

REFERENCES*

1. Austin, E., and McFadden, W. E., "The Electrolytic Zinc Plant of the Hudson Bay Mining and Smelting Company Ltd.," *Trans. Can. Inst. Mining Met.*, **59**, 208 (1956).
2. Cunningham, G. H., and Jephson, A. C., "Electrolytic Zinc at Corpus Christi, Texas," *Trans. Am. Inst. Mining Met. Engrs.*, **159**, 194 (1944).
3. Davidson, L. P., Carpenter, R. K., and Tschirner, H. J., "Reverse Leaching of Zinc Calcine," *ibid.*, **191**, 134 (1951).
4. Ellsworth, John T., "The Effect of Single Impurities on the Electrodeposition of Zinc from Sulphate Solutions," *Trans. Electrochem. Soc.*, **42**, 63 (1922).
5. Hanley, H. R., Clayton, C. Y., and Walsh, D. F., "Investigation of Anodes for Production of Electrolytic Zinc," *Trans. Am. Inst. Mining Met. Engrs.*, **91**, 275 (1930); **96**, 142 (1931).
6. Moore, T. I., and Painter, L. A., "Electrolytic Zinc Plant at Monsanto, Illinois," *J. Metals*, **4**, 1149 (1952).

---

* Used generally, without citation, by the authors and included for supplementary reading. (Ed.)

7. Snow, W. C., "Electrolytic Zinc Plant at Risdon, Tasmania," *Trans. Am. Inst. Mining Met Engrs.*, **121**, 482 (1936).
8. Stimmel, B. A., Hannay, W. H., and McBean, K. D., "Electrolytic Zinc Plant of the Consolidated Mining and Smelting Company of Canada Ltd.," *ibid.*, p. 540.
9. Tainton, U. C., "Hydrogen Overvoltage and Current Density in the Electrodeposition of Zinc," *Trans. Electrochem. Soc.*, **41**, 389 (1922).
10. Tainton, U. C., and Leyson, L. T., "Electrolytic Zinc From Complex Ores," *Trans. Am. Inst. Mining Met. Engrs.*, **60**, 486 (1924).
11. Watts, O. P., and Shape, A. C., "Addition Agents in the Deposition of Zinc From Zinc Sulphate Solution," *Trans. Electrochem. Soc.*, **25**, 291 (1914).
12. Wolff, W. G., and Crutcher, E. R., "Electrolytic Zinc Plant of the Sullivan Mining Company," *Trans. Am. Inst. Mining Met. Engrs.*, **121**, 527 (1936).

Horizontal Retort Practice

K. A. Phillips

Assistant Chief Metallurgist
American Zinc, Lead and Smelting Company
East St. Louis, Illinois

FURNACES, RETORTS AND CONDENSERS

The first unalloyed metallic zinc was undoubtedly produced in small batch-type retorts. A natural oxidized zinc mineral, such as the carbonate or silicate, was mixed with anthracite coal or charcoal and heated in a simple closed vessel to yield zinc vapor. Today, roasted zinc sulfide concentrate is the ore most commonly charged, but the basic reactions are still the same:

$$ZnCO_3{}^* + \text{heat} \rightarrow ZnO + CO_2$$

$$ZnO + CO \rightarrow Zn_2 + CO_2$$

$$CO_2 + C \rightarrow 2CO$$

The chemistry of zinc roasting and smelting is discussed in Chapter 4.

The basic unit of antiquity—a small retort with a daily output of less than 100 pounds of metal—is still very important as a source of zinc. Many improvements have, of course, been made in the retorts and furnaces, and many of the operations have been partially or completely mechanized. Although several more-impressive methods of zinc production have been developed since 1910, about 40 per cent of the output of slab zinc in the United States still comes from horizontal retort furnaces, essentially adaptations of the earlier predominant and well-documented Belgian

* Present in natural ores, not in roasted concentrates.

retort furnaces. The total tonnage produced in these units in 1956 was greater than that for any previous year.

Almost all the early retort furnaces were fired with coal or with producer gas. Those having the highest thermal efficiency—and the highest construction cost—employed regenerative preheating of the combustion air in checkerwork preheaters similar to those used in open-hearth steel furnaces. The last American regenerative furnaces were abandoned in 1953, but descriptions oı these units may be found in the older books on zinc metallurgy.[7, 9] The regenerative furnaces are difficult to fire uniformly, and they can be operated competitively only where an ample supply of skilled, conscientious workmen is available at low or moderate wage rates.

Hegeler Furnaces

The furnaces best suited for mechanization and instrumental control are those using a modified Hegeler design and either producer gas or natural gas for fuel. They have a reasonably good over-all thermal efficiency only if the heat in the exit combustion gases is recovered in waste heat boilers.

Figure 6-16 shows a front view and two sections of a Hegeler-type furnace fired with producer gas. The two "sides" of the "block" are fired independently. The producer gas for one side enters at one end and is burned progressively as it passes along the furnace. The necessary combustion air is injected at the division lines between sections (of 16 retorts) or half sections, the amount added at each point being regulated to maintain the desired temperature in the downstream section. The gas mixture retains its reducing character until it passes the last retorts. The gases from the two sides are finally combined, burned with a small excess of air, and (usually) passed through a boiler.

The furnace of Figure 6-16 has four horizontal rows of retorts. Furnaces using five or six rows are also in use, but they are less popular because the top rows cannot be charged by a man standing on the floor. A furnace side contains 304 to 462 retorts.

Conventional refractories have an acceptable life in retort furnaces. Silica brick is ordinarily used for the center wall and shelves; first-quality fire-clay brick is required for the pillars and arches. Fine silica sand is used for the "bottoms" to catch the slag and retort fragments which fall to the floor and weld together into a tough mass.

Southwestern Furnaces

These are a modification of the Hegeler furnace described above, and they are fired with natural gas. The principal difference concerns the method of supplying the gas. Both natural gas and air are supplied at every half section, with inlets on either two or three levels. The combustion mixture

Figure 6-16. Hegeler-type furnace fired by producer gas. Upper left, partial view of one side of furnace. Stack, or uptake to boiler, is at left. Combustion air is injected in small holes located between sections (group of 16 retorts each). Upper right, longitudinal section of furnace. Hot producer gas enters the furnace from right. Lower center, transverse section of furnace. The center wall divides the block into two independent "sides." The chutes at each side are designed for the quick transfer of hot residues to the cellar below. Except when spent residues are being pulled, the mouths of these chutes are covered with a steel blue powder pan. (*Courtesy Harbison-Walker Refractories Company*)

in the furnace is always slightly reducing, but it does not change significantly in composition as it flows along the furnace. The furnace interior is kept under a small positive pressure, which varies from 0.02 or 0.04 inch of water at the beginning of the firing cycle to 0.20 or even 0.40 inch at the end. This favors a more uniform distribution of the hot gases along the entire length of the retorts.

The exit gases are usually discarded through a short stack; the steam from a waste heat boiler is of little net value in an area supplied with relatively cheap natural gas. At one plant, at least, the combustion gases are used to preheat retorts placed in a rectangular kiln located between the end of the furnace and the base of the stack.

Each furnace block fired with natural gas is usually divided into a front "stove" and a back stove. A stove is divided by its center wall into two "corners," and the two corners in line constitute a furnace "side" of from 400 to 416 retorts.

Retort Dimensions

The retorts used at most American smelters are round in cross section, with inside diameters (after firing) of 8¼ to 9¾ inches and inside lengths of from 53 to 63 inches. The volume varies from 1.75 to 2.50 cubic feet, but the size generally preferred at the present time is about 2.25 cubic feet. The butt thickness is usually 2 inches; the wall thickness varies from 1 to 1⅛ inches, according to the size of the retort and the corrosiveness of the residues.

Many foreign plants and one American plant use retorts having an oval cross section. The strength and thermal characteristics of such a retort are a little better than for one of round cross section, but these small gains may be more than offset by higher costs in making and handling the oval retorts and condensers.

Manufacture of Retorts

Retorts must be impervious to zinc vapor, they must be resistant to corrosion by ore residues, they must have a reasonably good tensile strength and resistance to plastic flow at temperatures up to 2300°F, and the cost of manufacture should not be excessive. Typical retorts have a lifetime of less than 60 days. A sound retort made from fire clay will meet most of these conditions but will fail in plastic flow under the combination of temperature and unsupported span which is encountered in modern furnaces. To meet this problem the "silica retort" was developed by the American Zinc Company in 1919.[16, 17] These retorts are now used everywhere, although their composition varies from plant to plant, as shown below.

	Parts by Volume	Per Cent by Weight
Plastic fire clay	44 to 50	48 to 54
Grog*	25 to 31	26 to 32
Silica flour	25	20

* Crushed firebrick or calcined clay.

The silica flour is very fine as received; the clay and grog must be crushed to at least ¼ inch and preferably to ⅛ inch before mixing.

The quality of a clay cannot be determined by a chemical analysis alone; physical properties are of equal importance.[18] However, the analyses of most of the plastic clays now being used in America will fall into the range shown below.

Compound	Per Cent
Al_2O_3	30–35
SiO_2	50–55
Fe_2O_3	1.0–2.5
CaO	0.1–0.9
MgO	0.1–0.5
Ignition Loss	10–14

The clay, grog and silica flour are first mixed with 10 to 12 per cent of water in a single-shaft or double-shaft pugmill. The product leaving the die is an endless rectangular ballot having a cross-sectional area of 50 to 80 square inches. This is cut into convenient lengths and repugged to give a more uniform mix. The second-pass ballot is cut into 18-inch or 24-inch lengths, covered with moist burlap, and stored in a tempering room for about ten days. During this period the mixture becomes more plastic because of the uniform distribution of the moisture and the natural action of the colloidal materials in good-quality clay.

The tempered ballots are then remixed in a single-shaft or double-shaft vacuum-extrusion machine for the production of deaired cylindrical ballots.[12] (See Figure 6-17.) In a double-shaft machine the burden of shredding the mix prior to effective deairing is placed upon a shaft with outboard bearings on each end. The lower shaft, with a bearing only on the driven end, merely forces the deaired clay through the final round die.

Figure 6-17. Vacuum extrusion machine for producing deaired ballots. The square ballots came from the tempering room. Each cylindrical ballot contains just enough mixture for one retort. (*Courtesy American Zinc, Lead and Smelting Company*)

Figure 6-18. Retort ready for removal from Wettengel press. Most of the press is located below floor level. The butt of the retort is up. The retort will be supported by the sheet-steel cradle, which protects the retort while it is lowered, trimmed, and trucked to a drying room. (*Courtesy American Zinc, Lead and Smelting Company*)

The deaired clay is less plastic than the original mix because of the removal of the gases and a little of the water; it is best handled in a hydraulic press of the Wettengel type (Figure 6-18). The retort is extruded to a length somewhat in excess of that actually required. It is then cut to size with a cylindrical template, and the rough edges of the mouth are rounded. The retort is wheeled to the drying room, where the retorts may be stacked in tight rows for slow drying or in open order for rapid drying. In the latter case there are usually some losses caused by bending prior to the initial set. Closely packed retorts are dried at room temperature (60 to 90°F) for from 10 to 30 days. Then the temperature is raised to 130 or 140°F, and the drying continues for another 30 days. Retorts in open formation are dried in from 4 to 30 days, usually with circulated air of closely controlled humidity and temperature.[4, 15]

Losses in drying and handling vary from 2 to 6 per cent. Causes of failure include local differences in drying rates, excessive bending and ordinary mechanical damage.

Figure 6-19. Hot retort being removed from firing kiln. (*Courtesy American Zinc, Lead and Smelting Company*)

The dried retorts are transferred to firing kilns for a preheating period of 16 to 24 hours. Spalling will occur if the retorts contain more than 0.3 per cent moisture and if heat is applied rapidly. The kiln temperature is raised gradually to a maximum of from 1800 to 2100°F. The transfer of the retorts from kiln to furnace is accomplished at as high a temperature as is practicable. If the temperature falls below one of the troublesome inversion points for silica (1600°F), strains will be set up in the retort walls when it is quickly reheated inside the furnace. The handling of red-hot retorts weighing over 150 pounds is not pleasant, but it can be done, as shown in Figures 6-19 and 6-20.

Visible losses in preheating and installing vary from 3 to 7 per cent. If hot retorts are not used on the intended cycle they should be stored at or above 1600°F. If they are cooled and then reheated they will usually fail in the kiln or after a few cycles in the furnace.

The life of an individual retort will vary from one day to over a hundred; it will tend to be shortened by the following factors:

1. Use of poor raw materials.
2. Use of faulty manufacturing and handling methods.
3. Short smelting cycles, with frequent thermal shocks.
4. Use of charges yielding corrosive residues.
5. High furnace temperatures.

Figure 6-20. Hot retort being slid into place in furnace. The butt must be placed on a shelf in the center wall. The open space around the mouth of the retort is then filled with ½ to 2 inches of plastic clay, stamped into place with a hot tool. (*Courtesy American Zinc, Lead and Smelting Company*)

With ore yielding noncorrosive residues, and with a 48-hour smelting cycle, well-made silica retorts should have an average life of 90 days. The average for all smelters is not much above 45 days.

Silicon Carbide Retorts

Silicon carbide has a better thermal conductivity and a better resistance to iron-rich slags at high temperatures than does fire clay. To achieve these advantages, two smelters in North America used for many years a retort mix consisting of about two parts of silicon carbide grains to one part of plastic clay.[1] The average life of retorts made with new silicon carbide grains was about twice as great as for silica retorts, but the over-all economic picture eventually led to the abandonment of the silicon carbide retort.

Manufacture of Condensers

Condensers are subject to much different service conditions than retorts; resistance to physical and thermal shocks is much more important than corrosion resistance. Because the life of a condenser is so short, its cost must be reduced to the lowest possible figure. The clay used may be about the same as that used for retorts, but the grog is always the cheapest obtainable. At most plants, used retorts are freed of the larger pieces of adhering slag and are then crushed for condenser grog.

Figure 6-21. Condenser machine. Balls of mud are moved from the stack at left to the buckets of the machine. The condensers are then formed by the spinning mandrel (enclosed by shield). At this plant, each wet condenser is crimped when it is dumped from the bucket. (*Courtesy American Zinc, Lead and Smelting Company*)

The raw materials are crushed at least to ¼ inch and then mixed once, using more water than can be tolerated in a retort mix. Six parts of plastic clay to four parts of grog are commonly used. No silica flour is ever used. The condensers made at most smelters have a simple conical form and are made in semiautomatic machines similar to the one shown in Figure 6-21. Most condensers have a length of from 18 to 24 inches, and the circular mouth has a diameter of from 2½ to 3¼ inches. The wall thickness is usually about ¾ inch.

After a preliminary drying period of about 48 hours, the condensers are stacked in a drying room and kept there for from one to two weeks. At a very few plants they are used without firing, but it is more common to fire them at 2000°F or higher to give greater mechanical strength. The final weight varies from 18 to 24 pounds, and the average life of a fired condenser is 5 to 8 days.

RETORT FURNACE OPERATION

The principal charge to the retorts at any given primary zinc smelter is either sinter or nodulized calcines. From 5 to 10 per cent of the charge (national average) consists of the densified oxide from the slag-fuming furnaces of lead smelters. A few primary smelters charge small quantities of secondary materials, such as ashes from galvanizing plants, brass mill fume, and certain copper converter fumes rich in zinc, lead and tin. Naturally oxidized ores, the so-called calamines, are suitable for treatment if the zinc content is sufficiently high. In any case most of the zinc charged should be in the metallic or the oxide form, and certain impurities, such as sulfur and chlorine, should not be present in excessive amounts.

There are a few secondary retort plants operating small blocks—often equipped with retorts of unusual dimensions. The charge consists of ashes, drosses and fumes, and the chlorine content is often high enough to introduce special problems in furnace firing and metal drawing. The retort residues may contain enough lead, tin or copper to require their retreatment at a suitable smelter.

Reduction Fuel

The most common reducing agents are anthracite coal, semianthracite coal and coke breeze. Mixtures may be employed to obtain the advantages inherent in two types of fuels. Fuel low in ash is always desirable, but the delivered cost of a pound of carbon is also a very important factor. The following data are for the reasonably priced fuels being used during 1956.

	Volatile Matter (%)	Fixed Carbon (%)	Ash (%)	Iron (%)	Sulfur (%)
Anthracite coal	3–7	80–90	7–11	1–2	1–2
Semianthracite coal	10–12	80–83	6–9	1–3	1–2
Coke breeze	2–4	81–86	10–16	1–2	0.5–1.5

At most plants the charging of excess ash in the reduction and loaming fuels must be balanced by the charging of less ore. The weight of the fuel is only 23 to 32 per cent of the ore weight (on a dry basis), but the bulk density of the fuel is so low that the latter contributes about 40 per cent of the total volume of the mixed charge.

Mixing

A typical batch of charge contains ore, "blue powder," reduction fuel, condenser concentrates, salt and enough water to bring the over-all moisture content to from 4.5 to 7 per cent. Most of these materials are minus ⅜ inch in size, so that a reasonably satisfactory mixture can be obtained in a device similar to a large concrete mixer. The salt dissolves in the water, so that corrosion of all handling equipment becomes a problem.

Densification

A cubic foot of wet mixed charge weighs from 88 to 93 pounds. If a charge containing sinter as the principal ore is passed through a rod mill, the bulk density will be increased by from 7 to 10 per cent. This results from the development of a more favorable range of grain sizes and by the filling of the pores of the sinter and coke with fine particles of charge. An excellent mixture of the several components is obtained, and this is not apt to be disturbed by segregation in bins and cars. The screen analyses of feed and products are tabulated below.

Screen Size, Mesh	Per Cent Retained Before Densification	Per Cent Retained After Densification
+3	3–8	0
−3 +6	15–20	2–5
−6 +10	25–30	20–35
−10 +20	20–25	25–30
−20 +35	10–13	15–20
−35 +65	5–8	8–12
−65	7–12	15–20

The bulk density of charges consisting principally of noduled calcine and densified fume increases 3 per cent or less during densification. If the ore is charged is quite fine there may actually be a decrease in bulk density.

The Charging Maneuver

The maneuver requires from two to five hours per corner or from three to six hours per side (depending upon the system employed and the number of retorts to be changed, as well as upon the skill and energy of the crew). Where a 48-hour cycle is used, the following steps are usually performed in the order listed.

1. The drawing of the last zinc from the condensers.
2. The breaking-down of the condensers.
3. The removal and recovery of the zinc-saturated loam from the mouth of the retort.
4. The removal and recovery of the rich crust of residue near the mouth of the retort.
5. The removal and disposal of the remaining residues.
6. The emplacement of the new mixed charge.
7. The setting and loaming of the condensers, about two-thirds of them having been cleaned up for re-use during the other operations of the maneuver.
8. The replacement of unsound retorts.

Methods of drawing metal will be discussed later. The condensers are left reasonably free of molten metal; they can be loosened with suitable bars and moved to a position 7 or 8 feet from the furnace face for cleaning.

A heavy "bumping" bar is used to break the loam from the mouth of each retort so that the condenser can be fitted properly on the next cycle. Very resistant accretions are dislodged with pneumatic guns at intervals. Both the loam and the rich residues (called the "sample" in America and the "front" in England) are raked onto a steel pan, and they are promptly transferred to the blue-powder pile for recharging. Practice varies widely, so that from 2 to 15 per cent of the spent residues may be included in the sample (or front) at a given plant. According to American terminology, blue powder includes all by-products of the furnace proper (such as ladle skimmings, loam and rich residues) which must be returned to the furnace. From 20 to 35 per cent of all metal produced may come from these secondary materials.

After a 48-hour distillation cycle the lean residues usually contain only 3 to 5 per cent of the new zinc charged. Those from a 24-hour cycle may contain 7 to 12 per cent of the new zinc. The zinc assays of the residues from a rich charge will tend to be higher than those from a charge which is high in iron, but the percentage of the new zinc will usually be lower.

The residues from charges low in iron may be blown from the retort with compressed air or steam; the latter is more economical, and it is generated by spraying the hot retort walls with fine jets of water. Residues containing more slag and gum must ordinarily be removed by hand with shovels and drags. At the present time the use of cleaning machines is becoming more common.[11] The hot residues fall into a chute leading to the cellar below the furnace. After they have cooled to an average temperature of 500°F or less, they are removed with scrapers or small power shovels. Residues are very dirty and dusty, but they are seldom moistened. The steam generated is as objectionable as the dust. Also, moist residues cannot be concentrated effectively in magnetic separators, in use at several smelters.

Ordinary cleaning methods tend to leave iron-rich slag or "retort gum" lying long the bottom element of the retort. This slag tends to ooze to the mouth of the retort, where it solidifies as the temperature falls slightly. This gum detracts from the usable volume of the retort and interferes with charging operations.

The lean residues which fall to the cellars weigh from 25 to 45 per cent as much as the new zinc ore charged. High iron assays or short distillation cycles tend to cause high residue weights. Typical assays are shown in Table 6-1.

Hand Charging

Except for numerous mechanical-charging experiments of relatively short duration, all retorts were charged by hand until 1950. Many charging machines had been built, and some of the basic principles tested have been proved sound today, but the necessary combination of metallurgical, mechanical and economic factors did not exist to justify the permanent adoption of machines at any plant. One factor was the liberal supply of strong and skillful workmen prior to World War II. A simple but effective design of the charge cars minimized the burden upon the chargers, especially at four-row Hegeler-type furnaces.

TABLE 6-1. ANALYSES OF RETORT CHARGES AND RESIDUES

Assay Per cent	Low-iron Charge*			High-iron Charge		
	Sinter Charged	Residues		Sinter Charged	Residues	
		24-hour	48-hour		24-hour	48-hour
Zn	72.1	18.7	13.9	61.8	13.5	6.6
Pb	0.12	0.08	0.04	0.32	0.31	0.07
Cd	0.032	0.01	0.008	0.05	0.01	0.005
Fe	2.0	5.1	7.0	10.4	26.0	28.7
Cu	0.10	0.20	0.23	0.78	1.92	2.08
As	0.005	0.01	0.01	0.05	0.12	0.14
Sb	0.002	0.004	0.004	0.02	0.04	0.05
Bi	—	—	—	0.003	0.007	0.008
In	—	—	—	0.01	0.005	0.005
S	0.4	1.4	1.5	0.2	1.0	1.3
C	0.0	37.2	28.6	0.0	32.2	24.7
CaO	0.3	1.1	1.2	0.2	0.6	0.7
MgO	0.2	0.5	0.5	0.1	0.3	0.4
Al_2O_3	1.4	6.2	6.7	1.4	4.6	5.7
SiO_2	4.7	21.3	25.6	3.4	16.6	22.6
Oz Au†	Tr	Tr	Tr	0.010	0.02	0.033
Oz Ag†	0.1	0.2	0.3	2.44	4.78	7.11

* Residue assays will also depend upon the kind of fuel used.
† Troy ounces per avoirdupois ton.

Hand charging has several inherent disadvantages, even with reasonably good workmen. The amount of charge which is placed in a given retort will vary widely. Retorts in the top row and those which are filled after the charger is tired will almost always receive less than their share. Retorts charged with blue powder or clean-up from the floor also will be lightly charged. Such irregularities adversely affect the average production per retort-day, the furnace recovery and the retort life. A lightly charged retort heats up more rapidly than its fellows, overloads its condenser early in the cycle and suffers from slag corrosion late in the cycle.

At some plants today (and at all plants a few years ago) the blue powder is moistened, mixed and charged by hand into the retorts of a few downstream sections of each corner or side. From 23 to 35 per cent of the total retorts will be used for this purpose, and the blue-powder sections—with their burden of metallic scrap—must be fired differently than the others. The zinc drawn from the blue-powder sections must be sampled and handled separately; this metal will tend to be a little higher in lead, iron and most other impurities than the metal from the sections charged with new ore.

At a growing number of plants the blue powder from one charge is returned to the mix room, where coarse pieces are removed and the fines are mixed with new charge. All charge is delivered to the furnace in cars for easier charging and every retort on the furnace receives the same type of charge. This method of operation is referred to as the "uniform charge" system. Excess fuel and chlorides in the blue powder are used to help smelt new ore. Also, all metal drawn from a side can be sampled and lotted as a unit.

Charging Machines

Several advantages can be obtained by the use of the uniform charging system and a charging machine of good design:

1. A heavier load can be placed in each retort.
2. The physical work of charging is largely eliminated; older workmen of high seniority automatically become the machine operators.
3. Each retort contains the same amount of the same kind of charge.

Some of the reasons why this third advantage is so important have already been discussed, indirectly, in the previous section on hand charging. Today, a new aspect has appeared. More furnaces are being fired by instrument rather than by eye, and it becomes more important to fire all sections of the furnace at the same temperature at any given point in the distillation cycle.

Under typical conditions the use of the above system at a Southwestern furnace will yield a 25 per cent increase in the output of slab zinc with little or no increase in the furnace labor requirement. The gain in furnace recovery which would be obtained as a result of uniform charging will, however, tend to be offset by the heavier fume losses caused by heavier loadings of all the condensers.

Pneumatic Charging Machine. The first unit to be employed on a plant-wide basis (since 1950) uses compressed air to project the charge into the retort. Fine, moist charge is delivered to the machine by an overhead conveyor belt which parallels the face of the furnace. The charge falls into the boot of a small bucket elevator, which supplies charge at a uniform rate to four blow guns. The particles of ore and fuel are given enough initial velocity to carry them to the back of the retort, although some segregation should theoretically occur en route. Under ideal conditions of air pressure and charge moisture, the retort loading will be reasonably high and uniform. The dust loss is much lower than might be anticipated with pneumatic projection. The cost of the machine itself is low, but the cost of the entire system, including conveyors and air compressors, is about average for the several systems.

Figure 6-22. Projector opened for inspection. The angle of the blades is very important in the control of erosion. The car in the background delivers charge to a bucket elevator, which in turn delivers the charge to the projector. (*Courtesy Athletic Mining and Smelting Company*)

Projector-Type Machines. Charge can be thrown mechanically into the retorts with more efficiency, more uniformity and with less probability of segregation. One unit is similar to a sand slinger and uses the projector shown in Figure 6-22. Figure 6-23 shows the assembled unit in operation. Very high retort loadings are obtained with this machine, and over-all capital cost is lower than for other systems currently in use. Projector maintenance cost tends to be high, and this problem requires good design and a wise selection of materials of construction.

Another type of machine uses a moving-belt projector, similar in general principle to a box car loader. A standard unit does not impart sufficient velocity to the particles of charge, so that a successful projector must hold the charge against the belt by centrifugal force during the period of ac-

Figure 6-23. Charging machine using a centrifugal slinger. The long spiess-rod must be inserted before the retort is charged and be pulled out after charging. The emplaced charge is too tight to be spiessed by conventional means. (*Courtesy Athletic Mining and Smelting Company*)

celeration. The results attained with the most recent units are quite satisfactory, although construction costs are relatively high.

Auger-Type Machines. A new machine of this type is in very successful use in at least two American plants. The augur is first advanced to the back of the retort, with the charge emplaced being limited to the "left-overs" from the previous retort. Then a heavy stream of new charge is delivered to the feed hopper and this is transferred to the back of the retort by a turning screw. As the retort fills with charge, the auger is gradually forced backward and out of the retort against a constant (and adjustable) pressure exerted by electrical brakes. Such a machine is sound in principle and has several obvious advantages. It is, of course, moderately expensive to build, and reliable operation requires a good program of inspection and preventive maintenance.

Special Problems. Most machines emplace the charge so firmly that a speiss rod must be inserted at the beginning of the operation and withdrawn later. Steam from the moist charge escapes through the small opening thus provided; otherwise, it is trapped at the back of the retort and tends to blow out the charge with some violence.

Any practicable machine can charge the retorts more rapidly than the crew ahead of the machine can clean them; however, if a breakdown occurs the crew must be kept from cleaning retorts which cannot be charged promptly. Retorts left open at full furnace temperature tend to be corroded by the slag which often lines them.

Condenser Setting

After the retort has been charged, a cleaned or new condenser is placed with the butt centered in the retort mouth and with the nose supported by an adjustable metal device. The bottom element of the cone should be level; this establishes the optimum conditions for good condensation, volumetric capacity for liquid zinc and easy drawing. The condenser is then luted to the retort with a wet, finely-crushed mixture called "loam." The loam must stick to the refractories when thrown into place, but it must not become as strong as the condenser after becoming saturated with zinc metal and zinc oxide. The base material is always a fuel similar to that used for reduction. The bonding action is strengthened by the addition of bentonite, clay or various furnace by-products.

To avoid the dislodging of the fresh loam, most workmen "stuff" the condenser mouth prior to loaming. The stuffing mixture is always similar to loam and it may be exactly the same. It should cling to refractory surfaces, it should not shrink excessively during the period of water evaporation, it should retain liquid zinc perfectly while allowing gases to escape freely, and it should be easily broken by the metal drawer. In practice, the property most often compromised in an effort to achieve the others is that of good gas filtration. Before the stuffing dries, a small hole is pierced along the top element of the condenser. The gases pass through this hole until the stuffing dries and achieves maximum porosity. Thereafter, if gas pressure accumulates in the condenser, the hole is reopened with a long, sharpened steel "spiessing" rod. Table 6-2 shows the screen analyses of the loam and stuffing mixtures used at various plants.

Changing Retorts

Leaking retorts may be replaced at either the beginning or the end of the maneuver of a 24-hour cycle; they are always changed at the end of the maneuver of a 48-hour cycle. The clay front or face is knocked out to free the mouth of the retort, and the damaged retort is slid out of its place.

TABLE 6-2. SCREEN ANALYSES OF LOAM AND STUFFING MIXTURES

Screen Size Mesh		Loam Only	Used for Both Loam & Stuffing			Stuffing Only
			A	B*	C	
	+6	1	2	1	12	7
−6	+10	8	20	7	15	19
−10	+20	25	21	12	18	24
−20	+35	21	20	30	20	19
−35	+65	15	16	23	15	13
−65		30	21	27	20	18

* Straight deslimed anthracite fines; would otherwise be too fine for stuffing.

It is replaced with a new, preheated retort, and a new clay front is "stamped" into place.

The retorts discarded contain about 2 per cent zinc by assay or about 0.2 per cent of the new zinc charged to the furnaces.

Organization of Crew on Charge Shift

Retort furnace blocks may be charged by sides or by corners. Where charging machines are used the machine operator and one or more extra men may work on both corners of the side. Such men report after the first few sections of the front corner are ready for charging and leave before the crew of the back corner change retorts and clean up. The following scheme shows the job assignments used for hand charging and machine charging at one plant.

	Corner System		Combination System		
	Front	Back	Front	Back	Both
Machine charger	—	—	—	—	1
Head charger	1	1	—	—	—
Second charger	1	1	1	1	—
Bumper	1	1	1	1	—
Shovelers	2	2	2	2	—
Stamper	1	1	1	1	—
Connie boy	1	1	1	1	—
Extra men	—	—	—	—	1
Total per side	14		14		

At other plants a single crew, with two men replacing each man of a corner crew, charges straight through one side. The total labor requirement with any system is usually from 15 to 17 men per side charged. Firemen, plumbers, loam chiselers, water boys and retort kiln tenders are examples of additional men employed at most plants but not assigned to a specific corner or side; they are not shown in the above table.

The Firing Cycle

All American plants now use either a 24-hour or a 48-hour cycle. Since World War II labor has insisted upon a regular working schedule, with the charging maneuver being performed in the cool morning hours. Cycles of intermediate length offer theoretical gains,[13] but they are no longer in use.

24-Hour Cycle. Only two American primary smelters used this cycle during 1956. A reasonable direct recovery can be achieved only by shortening the charging maneuver and by using the highest practicable furnace temperature at all times. There is little or no advantage in using large retorts, charge densification or machine charging, because the extra zinc cannot be distilled off in the time available. Either three or four metal draws (including that at the end of the maneuver) may be made, and the slab zinc production is usually 40 to 50 pounds per retort-cycle. Direct furnace recoveries are often only 80 to 85 per cent, but from half to two-thirds of the unrecovered zinc can be found in the residues. Most of this zinc can be recovered by Waelzing at a cost of about three cents per pound of zinc present in the densified oxide produced.

48-Hour Cycle. A 48-hour cycle offers a much lower labor cost per pound of zinc, usually at a small decrease in the output per retort-day. Because the product of one long cycle must bear the cost of 48 hours of firing and of four to six metal draws, it is essential to use the heaviest practicable retort loading. Large retorts, low reduction fuel ratios, charge densification and mechanical charging are among the tools employed to increase the amount of charge introduced into each retort. With good planning and good ore, the slab zinc production is from 80 to 95 pounds per retort-cycle, or just under the output per day of a furnace charged with similar ore and fired on a 24-hour cycle. The initial firing is usually more gentle than for a short cycle, but the distillation period is long enough that only 3 to 6 per cent of the new zinc is left in the residues. The zinc assay of the latter is so low that they can be discarded or be treated for the recovery of a fraction rich in gold, silver and copper.

Oldright[14] has recorded the most serious study of firing cycles and related problems, and the basic principles which he developed should be carefully considered by anyone interested in retort smelting problems. A relatively pure sinter can be started more slowly and then heated to a higher temperature than can a sinter high in iron. In the latter case a high initial temperature is absolutely essential to good condensation at the beginning of the cycle, and a reasonably low finishing temperature minimizes the corrosion of the retort by the spent residues. Figure 6-24 shows actual examples from practice at two Southwestern smelters. The temperatures shown are those of the hot gases surrounding the retort, as measured by thermocouples projecting through the arch. The charge inside the retort is cooled by the endothermic reaction involved in the reduction and volatilization of zinc and by some radiant heat loss to the condenser. Toward the

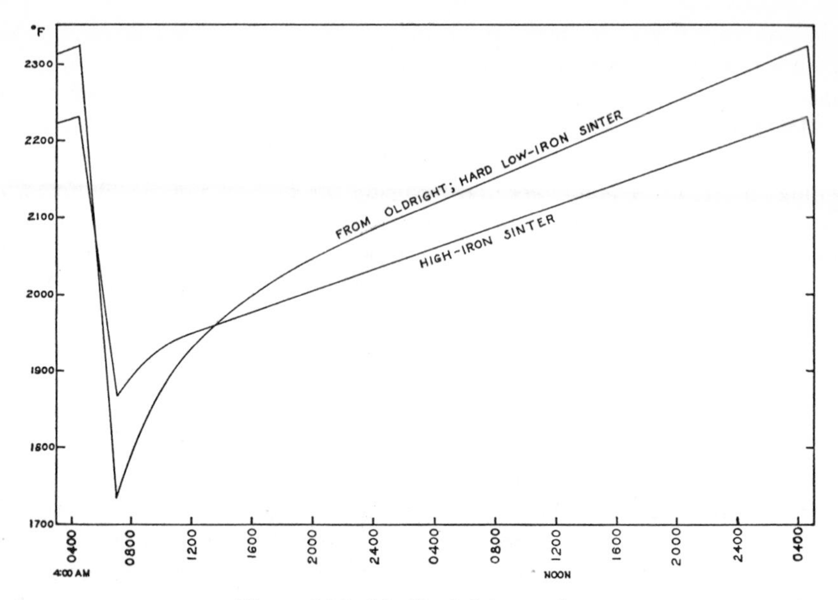

Figure 6-24. Idealized firing cycles.

end of the cycle the residues lying on the retort wall will tend to reach a temperature reasonably close to that of the surrounding gases.

At the beginning of distillation from any type of charge, the temperature should be raised quickly to a point at which the carbon dioxide produced in the primary reactions is promptly reduced to carbon monoxide. The percentage of carbon dioxide in the gas-vapor mixture will, of course, be higher where iron oxides are being reduced simultaneously with the most reactive particles of zinc oxide. Where the carbon dioxide content in the condenser is as high as one or two per cent, part or all of the zinc vapors will be condensed as blue powder or lost as fume. It should also be recognized that the retort temperature and the ambient temperature at the furnace face must be so balanced as to hold the condenser temperature at a level where the condensed zinc will remain in a molten condition. Some assistance is obtained from the heat of vaporization given up by the zinc vapor as it condenses. If the charge contains considerable metallic zinc, the gas-vapor mixture will tend to be rich in zinc, and the condenser will tend to be heated more rapidly by the condensing zinc.

It has already been mentioned that salt is added to the mixed charge at all plants operating under normal conditions. Zinc chloride forms and distills over, the rate being highest at the beginning of the cycle. In effect, the zinc chloride dissolves the oxide film from many of the freshly condensed zinc droplets and permits them to coalesce into a pool of liquid zinc.

Where no salt is employed, the amount of blue powder formed in the condenser during the first draw is at least twice as voluminous as it is when salt is employed. In the absence of salt the efficiency of condensation can be improved somewhat by using an ore charge containing at least 0.40 per cent lead. A reasonable amount of lead lowers the surface tension of the globules of condensed zinc and encourages the coalescense of adjacent droplets. Where salt is used, a maximum condensation efficiency can be achieved until the assay of the charge falls below about 0.25 per cent lead.

Obviously, there is also a maximum practicable starting temperature for each type of charge; zinc vapor must not be produced so rapidly that it cannot be condensed efficiently. Even with good operation the fume losses from the condenser during the first few hours of the cycle may constitute half of the total fume loss.

Fume losses will be increased by an unnecessarily hard firing cycle or by "checking" a furnace in mid-cycle. If the firing is not pyrometrically controlled, any furnace will be overfired at some time during an occasional cycle, and the fireman will note the sharp increase in the fume losses. The furnace temperature will then be held constant—or even reduced—until the fume losses appear to again be normal. In the meantime, the delicate balance between charge temperature and condenser temperature is destroyed, and the furnace will show a poor recovery for the cycle.

Distillation Rates

Under ideal conditions the rate of zinc vapor production would be constant throughout the cycle, falling suddenly to zero at the time of the charge maneuver. Actually, a reasonably uniform rate can be maintained only for the first two-thirds of the cycle, as is shown in Figure 6-25. Even this requires a good control of both the charge composition and the drop in furnace temperature during the charging maneuver.

Metal Drawing

The final draw is always made immediately prior to the charging maneuver, and the remaining draws are so spaced as to avoid overloading of the condensers during any distillation period. From 5 to 30 pounds of zinc is drawn from each condenser every time that it is opened. The stuffing is broken, and the bulk of the molten zinc flows by gravity into the ladle. The stuffing, together with some zinc, zinc dust and zinc oxide, is then also drawn into the ladle, usually with enough hot charge from the retort mouth to melt scrap zinc picked up from the floor. The bulk of the residues and blue powder floats harmlessly on the zinc, but some iron and other impurities are dissolved by the zinc before the skimmings are finally removed.

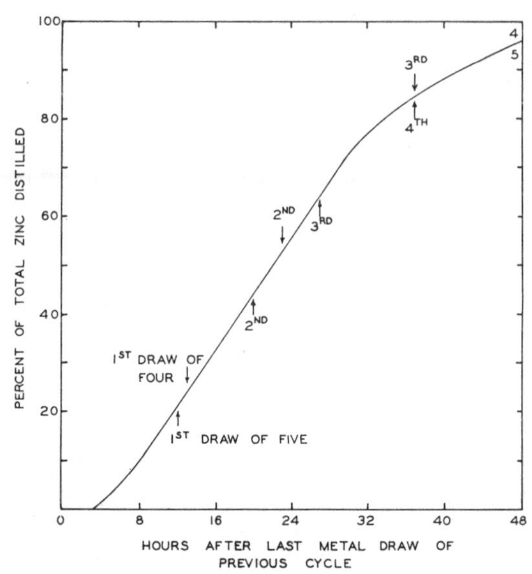

Figure 6-25. Graph showing cumulative production of zinc from retort charge.

Figure 6-26. Casting zinc into slabs. A ladle holds enough zinc for nine or ten slabs weighing 60 to 62 pounds each. (*Courtesy American Zinc, Lead and Smelting Company*)

A typical ladle has a capacity of 500 to 600 pounds of zinc and is carried in a track-mounted car or is suspended from the monorail which also carries the shields used to protect the charging crew. As shown in Figure 6-26, from eight to ten slabs of zinc are cast from each ladle. Most of the slabs are shipped as originally cast, so the molten zinc must be carefully skimmed after it has been poured into the mold. The use of a silicone-type release agent simplifies the dumping of the solidified plates from the molds.

Analysis of Metal

A sample is taken from each ladle, and a composite is prepared from all ladle samples from one side during one draw. The slab zinc will include metals which have a significant vapor pressure at retort temperatures, such as lead and cadmium, as well as several metals which are present principally because they have been dissolved by the molten zinc lapping against the hot charge in the front of the retort. Iron is also contributed by furnace tools and by the skimmings floating in the ladle.

Cadmium has a lower boiling temperature than does zinc; it tends to concentrate in the metal from the first draw. Figure 6-27 shows the cad-

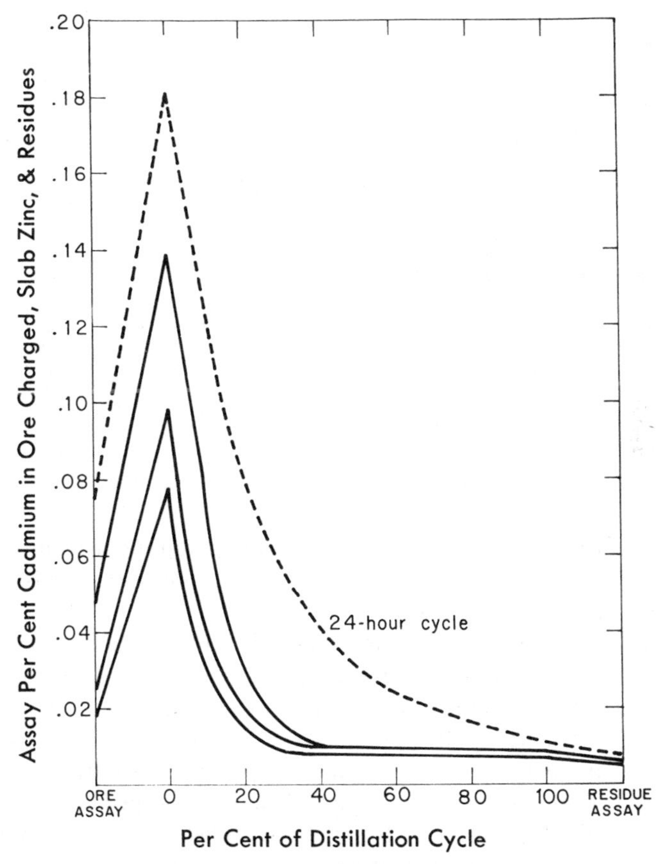

Figure 6-27. Graph showing cadmium assays of ore charged, of zinc being produced at any moment in the distillation cycle, and of retort residues produced.

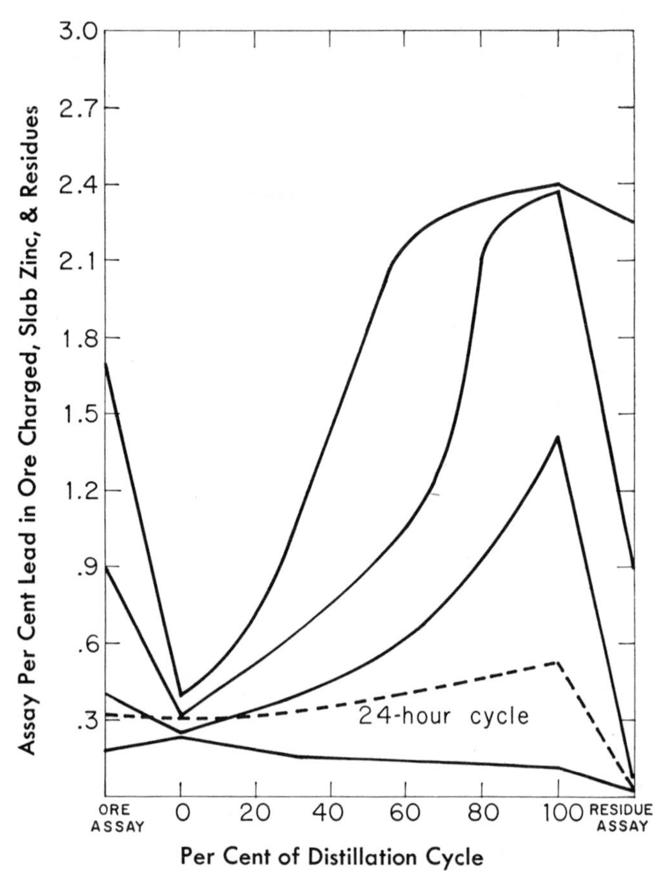

Figure 6-28. Graph showing lead assays of ore charged, of zinc being produced at any moment in the distillation cycle, and of retort residues produced.

mium assay of the vapor being distilled at any given moment in the cycle. Note that the assay of the first vapor is roughly proportional to the cadmium assay of the ore charged.

Lead has a high boiling point, which is never reached in a retort furnace. Lead does have an appreciable vapor pressure at operating temperatures, and Figure 6-28 shows that even the first-draw metal usually has a significant content. It will, however, be noted that the lead assay of the ore charged has much more effect upon the late-draw metal than upon the first-draw metal. If the assay of the ore charged is less than 0.15 per cent lead, the lead assay of the metal distilled will actually fall as the cycle progresses. This is because there is insufficient lead in the charge to keep the zinc vapor saturated at the existing temperature.

TABLE 6-3

Assay Per cent	High	Low	Average
Pb	0.53	0.12	0.30
Fe	0.034	0.007	0.025
Cd	0.08	0.01	0.03
Cu	0.005	0.001	0.003
Sn	0.005	0.001	0.002
Al	0.001	0.000	0.000
Mg	0.000	0.000	0.000
In	0.09	0.000	0.002*
Ga	0.000	0.000	0.000
Ge	0.001	0.000	0.000
As	0.0004	0.0002	0.0003
Sb	0.00005	0.00002	0.00003
Bi	0.001	0.000	0.000
Ag	0.0002	0.00005	0.0001

* One high assay not included.

Tin, if present at all, tends to concentrate in the metal distilled late in the cycle. Bismuth behaves in a similar fashion. The indium assay is influenced by two factors. One favors an early distillation as indium chloride, followed by a vapor phase reduction by zinc. The other factor favors a gradual and moderate increase in the indium assays, caused by the decrease in the volume of zinc vapor rather than by an increase in the rate at which the indium is distilled.

The iron and copper contents of the zinc vapor are both very low; these metals are dissolved from the charge. The assays tend to rise late in the cycle, when the spent charge becomes both hotter and richer in all nonvolatile impurities. The curves for antimony and arsenic are similar in general shape to those for iron and copper.

Table 6-3 shows the range of assays found in the metal from the first and second draws at four Southwestern smelters during one month of 1956.

Slabs having a poor physical appearance or chemical analysis are remelted, refined if necessary and carefully recast for shipment. Methods of refining are discussed in Chapter 7.

Condenser Treatment

Rejected condensers are heavy with crusts of zinc and zinc oxide containing 1.5 to 2.5 per cent of the new zinc charged. They are usually crushed and concentrated, using jigs and tables. From 60 to 80 per cent of the zinc in the condensers is recovered and returned to the mix room in the form of concentrates containing from 60 to 70 per cent zinc. Used condensers can also be treated in a Waelz kiln.

TABLE 6-4. DATA FOR RETORT RESIDUES

Element	Range of Assays (%)	Feed Left in Residues (%)
Zn*	5–22	3–12
Pb	0.04–3	5–50
Fe	7–29	95–99
Cd	0.001–0.015	1–10
Cu	0.05–3	93–96
Sn	0.00–0.03	5–50
As	0.000–0.02	90–95
Sb	0.004–0.40	90–95
Bi	0.000–0.02	90–95
C*	20–40	15–40
Au†	0.00–0.10	85–95
Ag†	0.10–20	80–90

* Highest assays are for a 24-hour cycle.
† Troy ounces per avoirdupois ton.

Treatment of Retort Residues

Retort residues contain varying amounts of valuable constituents, the assays for which depend principally upon those of the ores charged to the retorts. Typical assays and typical percentages of each metal left in the residues are shown in Table 6-4.

Residues low in iron are treated most economically in a Waelz kiln. Residues rich in iron are seldom valuable as a direct source of zinc. However, if an iron-rich fraction is removed by magnetic concentration, the nonmagnetic fraction will be a reasonably valuable source of both zinc and carbon. The magnetic concentrate may have value for its contents of copper, gold and silver. Heavy concentrates produced from residues by wet concentration methods will contain zinc as well as iron and the economic metals of a magnetic concentrate.

Over-all Recoveries at Horizontal Retort Smelters

From two to four per cent of the zinc received in the form of concentrates is lost in unloading, roasting and sintering. Of the zinc actually charged to the retorts, the over-all recovery of a horizontal retort furnace and its auxiliaries (condenser concentrator, Waelz kiln, magnetic separator, etc.) will range from 88 to 95 per cent. Metallurgical skill will have a very worthwhile bearing upon the outcome, but equal or larger changes in recoveries can result from variations in the skill and interest of the furnace workmen. Mechanization and instrumentation have permitted improvements in outputs and uniformity, but human factors still control the operations of condenser setting and metal drawing. Reasonable variations in the quality of workmanship in these two steps can change the furnace recovery by as much as 5 per cent.

REFERENCES*

1. Bruderlein, E. J., "Manufacture of Silicon Carbide Retorts," *Trans. Am. Inst. Mining Met. Engrs.*, **121**, 441–4 (1936).
2. Bruderlein, E. C., "Process for Charging Horizontal Zinc Retorts," U. S. Patent 2,633,257 (Mar. 31, 1953).
3. Burgess, J. W., "A Survey of Mechanical Charging Practices of Horizontal Retort Zinc Smelters." Paper presented before AIME meeting in New York, Feb., 1956.
4. Furlong, R. R., and Wertz, D. H., "Controlled Drying of Retorts," *Trans. Am. Inst. Mining Met. Engrs.*, **185**, 393–4 and 852–3 (1949).
5. Gyles, T. B., "Horizontal Retort Practice at the National Smelting Company, Limited, Avonmouth, England," *Trans. Am. Inst. Mining Met. Engrs.*, **121**, 418–26 (1936).
6. Harbordt, C. G., and Buff, B. F., "Retort Charging Machine," U. S. Patent 2,701,067 (Feb. 1, 1955).
7. Hofman, H. O., "Metallurgy of Zinc and Cadmium," pp. 198–562, New York, McGraw-Hill Book Company, Inc., 1922.
8. Ingalls, W. R., "Intermittent Zinc Distilling From Ore," *Trans. Am. Inst. Mining Met. Engrs.*, **121**, 610–35 (1936).
9. Ingalls, W. R., "Metallurgy of Zinc and Cadmium," pp. 198–562, New York, McGraw-Hill Book Company, Inc., 1906.
10. Loskutow, F. M., "Die Metallurgie des Zinks," pp. 87–158, Halle, Germany, Verlag von Wilhelm Knapp, 1950.
11. Morrison, T. P., "Retort Cleanout Machine," U. S. Patent 2,285,298 (June 2, 1942).
12. Neale, M. M., "Deaeration in Manufacture of Zinc Retorts," *Trans. Am. Inst. Mining Met. Engrs.*, **159**, 127–30 (1944).
13. O'Harra, B. M., and McCutcheon, F. G., "Effect of Length of Cycle on the Economics of Retort Zinc Smelting," *Trans. Am. Inst. Mining Met. Engrs.*, **182**, 213–25 (1949).
14. Oldright, G. L., "Zinc Smelting in the Horizontal Retort Fired with Natural Gas," U. S. Bureau of Mines Reports of Investigations, 4333, 4334, 4335, and 4336, 1948.
15. Page, H. R. and Lee, A., E. Jr., "Development of the Modern Zinc Retort in the U. S.," *Trans. Am. Inst. Mining Met. Engrs.*, **185**, 73–7 and 850–2 (1949).
16. Rossman, W. F., "Making of Zinc Retorts and Other Refractory Shapes," U. S. Patent 1,424,120 (July 25, 1922).
17. Spencer, G. L., Jr., "High-silica Retorts at the Rose Lake Smelter," Technical Publication, No. 378, AIME; *Trans. Am. Inst. Mining Met. Engrs.*, **96**, 119–24 (1931).
18. Wheeler, E. S., and Kuechler, A. H., "The Properties of Refractories in the Metallurgy of Zinc," *J. Am. Ceram. Soc.*, **10(2)**, 109–31 (1927).

* Including additions for collateral reading.

The Vertical Retort Process

G. F. Halfacre

General Manager
The New Jersey Zinc Company (of Pennsylvania)
Palmerton, Pa.

and

W. M. Peirce

Assistant to the Executive Vice-President
The New Jersey Zinc Company
New York, N. Y.

The extraction of metallic zinc from its oxide, carbonate, silicate, or from roasted sulfide ores, requires either the electrochemical reduction of the sulfate to metal, or chemical reduction by high-temperature reaction with a reducing agent. While considerable experimental work has been done with the use of natural gas (methane) as a reducing agent, coal or coke are the only reducing agents in use commercially.

The reduction of zinc oxide by carbon requires a high temperature and a high heat input. This heat can be supplied either by electrical resistance or arc heating, or by the combustion of a suitable fuel.

The vertical retort process discussed in this chapter is of this latter type; namely, a reduction of oxidized zinc materials with carbon in the form of coal or coke, with heat supplied by the combustion of fuel.

Basic Requirements of a Carbon Reduction Process

The reduction of zinc oxide by carbon may, for practical purposes, be considered to take place by the following reaction:

$$ZnO + C = Zn + CO$$

The heat requirement to support this endothermic reaction is approximately 1,565 Btu per pound. The sensible and latent heat content of the products of reaction—namely, zinc vapor, carbon monoxide gas, and the residue from an average coal and an average zinc concentrate—at an assumed final reaction temperature of 1150°C is approximately 1,500 Btu per pound of zinc. The first basic requirement of the process therefore is to supply approximately 3,100 Btu of heat to the coal-ore mixture for each pound of zinc produced.

The reaction begins at a temperature of around 900°C but a temperature in excess of 1000°C is necessary for a practical rate of reaction. This is well above the boiling point of zinc which therefore is evolved as a vapor. Furthermore, the complex equilibrium between the carbon monoxide and carbon dioxide in the heterogeneous system—C, ZnO, Zn, CO, CO_2—is such that it is desirable to reach a temperature in the range of 1100 to 1200°C in order to have a minimum carbon dioxide content in the zinc vapor-gas mixture going to the condenser. Even with the most rapid chilling, any carbon dioxide present reoxidizes zinc vapor.

Innumerable efforts have been made to develop a practical reduction process in which the required heat is developed by adding the necessary excess of carbon to the ore-coal mixture (above that required as a reducing agent), to generate the necessary heat when burned by the admission of air or oxygen to the reaction chamber. This blast furnace type of operation has presented formidable difficulties in the efficient condensation of the dilute zinc vapor-gas mixture. As a consequence, the methods which have gained wide commercial acceptance to date are retort processes in which the heat is generated by the combustion of coal or gas in firing chambers surrounding the retorts. The heat required for the reaction is transmitted through the walls of the retort to the mixture of coal and ore. (It may be noted here that a commercial carbon reduction process developed by St. Joseph Lead Company generates the necessary heat electrically directly in the charge, by the resistance of the coke constituent of the charge itself.)

The products of the reaction consist of zinc vapor and an equivalent volume of carbon monoxide plus that resulting from the reduction of other metallic oxides in the ore. The total volume of gas and vapor that must escape from the charge at the high temperature of reaction is very large. Retort design therefore must take into account the resulting pressures and maintain a proper relationship between length and charge porosity.

In the oldest and simplest of the retort processes—the Belgian retort process, in which the charge is a loose mixture of fine coal and ore—the practical limit of size is a retort approximately 8 inches in diameter and 5 to 6 feet in length. The diameter is limited by the rate of heat transfer from the inner surface of the retort walls through the poorly conducting charge, to bring about complete reaction at the center in a reasonable length of time. The length of the retort is for all practical purposes limited by the hot strength of the refractories economically available as well as by the problems of charging and cleaning long retorts of small diameter.

The mixture of zinc vapor and gas produced must be cooled under such conditions that the zinc vapor will not condense as a fine mist. If such is the case, the particles will fail to coalesce and will thus form a finely divided powder known as "blue powder."

The Belgian retort process is one requiring skilled and arduous labor un-

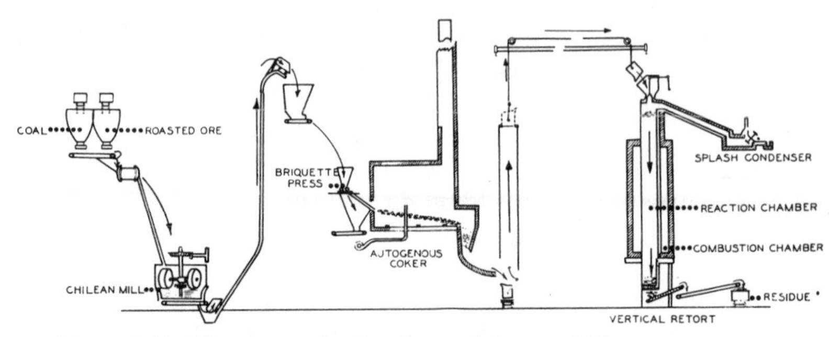

Figure 6-29. Diagrammatic flowsheet of the vertical retort process.

der conditions of extreme exposure to heat. The fuel economy is poor and the percentage recovery of zinc is low—generally in the high 80's.

The goal of zinc metallurgists from the beginning of the present century or earlier has therefore been to develop a continuous mechanized process which would improve working conditions, reduce the man-hours of labor per ton, and reduce the fuel consumption.

Vertical Retort Process[1]

Figure 6-29 is a diagrammatic flowsheet of the vertical retort process, a description of which follows.

A continuous retort process for the reduction of zinc concentrates by coal requires that the retort be a vertical one through which the charge can move by gravity. It requires that the charge be agglomerated for three reasons: (1) to insure a free movement of the charge through the retort, (2) to permit heat transfer to the center of a retort of commercially practical cross section, and (3) to provide sufficient porosity for the escape of the gases from a retort of commercial height.

The production of an agglomerated charge by briquetting a mixture of suitable zinc-bearing materials and coal proved to be a formidable problem. It is necessary that the briquettes possess sufficient strength to be handled by commercially practical methods up to the time they enter the retort, and that they retain their integrity as briquettes throughout their passage through the retort.

In the vertical retort process, these requirements have been met commercially. In order to give the briquettes the necessary mechanical strength, bituminous coal of carefully chosen coking characteristics in the amount of about 25 per cent of the mix, together with about 10 per cent of anthracite or coke, about 10 per cent of recirculated fines from the briquette coking step, and the balance of about 55 per cent of roasted zinc concentrates or other zinc-bearing materials, all carefully sized, are mixed with suitable binders.

Figure 6-30. Interior of rotary drum mixer.

These ingredients are well mixed—for example, in a rotary drum mixer—and then undergo further mixing and densification in chasers (otherwise known as "Chilean mills" or "pan mills"). Figure 6-30 shows the interior of such a mill. The moisture content of the mix is suitably adjusted and the mix is then formed into briquettes of appropriate size and shape in roll presses often referred to as "Belgian roll presses." The fines are removed by passing the briquettes over a grizzly; and in the most modern practice, the briquettes feed directly into an autogenous coking furnace of step grate design.[2] As the briquettes move through this furnace, they are coked by heat generated by burning the evolved volatile constitutents of the bituminous coal. The temperature is carefully controlled by regulating the admission of air for the combustion.

According to an earlier method of operation, the briquettes are coked by passing them through a vertical shaft through which the waste firing gases from the retort settings are passed transversely to supply the needed heat.

The autogenous method of coking first described is greatly to be preferred since it permits the complete separation of the operation of the vertical retorts from that of the coker, thus simplifying the control of both operations. Furthermore, since the exit gases from the firing of the retorts need not be high enough to coke the briquettes, it becomes possible to use countercurrent rather than concurrent recuperators and thus achieve substantially better fuel economy in the process. A further advantage of the autogenous coker is that the volatiles are burned in the coker and the small residual solids content of the gas, consisting of dry metallic oxides, can be recovered after passing the gases through a waste-heat boiler to recover heat and reduce their temperature.

The briquettes discharged from the step grate fall into large holding hoppers. In these, the final stages of coking are completed, as the briquettes pass through and are withdrawn from the bottom for charging to the retorts (see Figure 6-31).

The briquettes leaving the coking furnace at a temperature of about 700°C have attained a very strong coke structure, and their carbon content is sufficiently in excess of that needed for the reduction of the zinc and other metallic oxides present to leave a coke structure in the residue briquette which prevents its disintegration.

The greater part of the heat which must be transmitted from the interior walls of the retort to and through the briquetted charge is transferred by radiation. The lesser part is transferred by convection. It is therefore neces-

Figure 6-31. Discharge of briquettes from autogenous coker.

Figure 6-32. Preferred "loaf" shape of briquettes.

sary that the size and shape of the briquettes so be related to the distance from the heated retort wall to the center of the charge that paths exist for radiation. It has been found that a rectangular retort having a dimension of 12 inches between the two parallel heated walls requires a briquette weighing approximately 18 ounces. The shape of this briquette is controlled in large measure by the shape which it is practical to make on a roll briquetting press. The preferred shape is the so-called "loaf" briquette shown in Figure 6-32.

The choice of bituminous coal is of the greatest importance, and it does not follow that a coal suitable for the production of metallurgical coke will be suitable for this process. A variety of binders may be used. Combinations of sulfite liquor (which in many areas is available as a by-product from paper mills) and of bentonite or a suitable plastic clay may be used. In other instances, combinations of oil and clay may be used, and in still others pitch may be used as a binder. The choice of binders will be dictated by considerations of cost and of compatibility with the available coals. Careful testing is necessary in every case to develop a satisfactory mix with the concentrates, coals and binders available in a particular location.

Another important factor in the production of good briquettes is the character of the roasted zinc concentrate. If the very fine calcine from flash roasting is the zinc-bearing material, at least half the quantity used should be sintered, for example, by Dwight-Lloyd sintering. The proper correlation of screen size of the coals used and of the zinc calcine is essential to good briquette quality. Hammer mills or ball or rod mills may be used to grind the materials to suitable size.

Retort Construction. The development of a practical vertical retort requires that it be constructed of a refractory material with the best possible heat conductivity, together with suitable strength at high temperature

Figure 6-33. Cross section of retort structure.

and resistance to thermal shock, to give it a useful life. It is also necessary to employ such a refractory in a structure so designed as to minimize all stresses other than compressive stresses and to compensate for the stresses resulting from thermal changes.

In the vertical retort that was developed by The New Jersey Zinc Company, the retort walls are constructed of silicon carbide brick. The retort structure, a cross section of which is illustrated in Figure 6-33, consists of end walls which are not heated but which provide vertical slots into which the heated side walls are recessed. The walls are free to move in these slots to adjust for exapansion and contraction with changes in temperature.

The early vertical retort plants were constructed with retorts having a heated height of about 25 feet, an internal width of 12 inches, and an internal length of 5 or 6 feet. With improved construction and methods of operation, the height and length of the retorts have been increased and the most recently installed plants have retorts 35 feet in height and 7 feet in length. The productive capacity of the retorts is a direct function of the heated surface area, and modern retorts have an average production of 8 net tons per calendar day. This includes outages for maintenance and rebuilding.

The retort setting or furnace provides a combustion chamber outside each of the heated walls of the retort. Gas is admitted through ports at the top of this combustion chamber, and preheated combustion air is supplied at selected levels to control the rate of combustion and maintain a uniform temperature from top to bottom of the retort.

It is vital, in order to obtain good life from the retorts, not only that the temperature in the combustion chambers be uniform throughout, but also that the firing temperature be maintained at a uniform level from day to day. It is essential to have gas of adequate Btu content to achieve good firing control. In practice, this may be attained advantageously by the use of natural gas with its high calorific value. Where natural gas is not available, clean producer gas must be used. If anthracite coal or coke breeze is available, there is no difficulty in obtaining a clean gas. If bituminous coal must be used, a gas cleaning system is necessary. The Btu value of producer gas using anthracite will run between 140 and 150 per cubic foot. This gas is reinforced by mixing with it the cleaned carbon monoxide gas from the retorts, after it has passed through the condensers and scrubbing system. This gas not only supplies a large fraction of the total firing fuel requirement but also, as previously indicated, in the case of producer gas it raises the calorific value of the gas mixture to a point where the requisite combustion chamber temperatures can be uniformly maintained.

The combustion chamber temperature is held in the neighborhood of 1300°C and the waste gas has a high sensible heat content which is used to preheat the combustion air by passing it through countercurrent recuperators of special design. The waste gas leaving the recuperators at about 600°C goes to a stack.

Operating Procedures. Hot coked briquettes are transferred directly from the coker to the retorts which are charged on a regular periodic cycle (Figure 6-34). A discharge mechanism of special design discharges spent residue continuously, with the consequence that the level of the top of the charge subsides gradually between charging intervals. Above the furnace proper is an extension of the retort, called the "charge column," which is heavily insulated, but not heated, and is of sufficient height so that the charge level never drops below the level of the top of the combustion chamber.

The residue extraction mechanism is water-sealed, the spent briquettes being dropped into a quenching compartment from which they are continuously extracted (Figure 6-35). In order to prevent downward diffusion of zinc vapor into the extractor mechanism and the extension of the retort below the combustion chamber which connects the retort to the extractor mechanism, a small amount of displacement air is admitted. This, coming in contact with the hot residue containing a high percentage of carbon, is

Figure 6-34. Charging hot coked briquettes into vertical retort.

Figure 6-35. Retort residue extraction mechanism.

converted to carbon monoxide which, rising through the residue into the retort, prevents any back diffusion.

At intervals of some weeks, charging of the retort is stopped for a number of hours, and after the evolution of zinc from the charge in the retort has ceased, it is opened. Any accretions of zinc oxide (often called "rock oxide"), which have formed at the roof level and in the charge column by back reaction of carbon dioxide and zinc vapor, are then removed, after which operation is resumed. At the same time, the condenser system is thoroughly cleaned.

The normal life of the retort is three years or more, but some cracks ordinarily develop during its life, and these are usually patched during the shutdown for the cleaning out of zinc oxide accretions.

Condensation. A major problem, the successful solution of which has contributed vitally to the success of the vertical retort process, is the efficient condensation of the zinc vapor leaving the retort to molten zinc. The basic principle involved in efficient condensation of zinc vapor from the retort gases lies in causing the condensation to take place on a liquid zinc surface, the condensed zinc being thereby incorporated in a body of molten zinc. The formation of blue powder is thus avoided.

A simple and practical means of achieving this has been developed in the so-called "splash condenser."[3,4] (See Figure 6-36.) A bath of molten zinc is maintained at a thermostatically controlled temperature in the sump of the condenser. A motor-driven graphite rotor, its shaft extending through the wall of the condenser, dips into the bath of molten zinc and fills the condenser chamber with a rain of coarse drops of molten zinc and bathes the condenser walls with molten zinc. The zinc vapor condenses on these drops, which fall back into the bath, and on the zinc flowing down the walls into

Figure 6-36. General view of "splash condensers."

the bath. The augmented volume of zinc in the bath is continuously overflowed through an external well and laundered to a collecting pot. Heat is removed from the bath by water-cooled coils immersed in the outer well of the sump.

This device condenses about 95 per cent of the zinc vapor leaving the retort to liquid zinc. The 3 or 4 per cent of blue powder formed is periodically removed from the condenser when the retort undergoes its periodic removal of oxide accretions. The remaining 1 or 2 per cent of uncondensed vapor and entrained blue powder passing out of the condenser with the residual gases is scrubbed out by water sprays and the zinc is recovered from suitable settling equipment. One element of this scrubber system is a Venturi ejector which serves the dual purpose of scrubbing the gas and pumping the clean carbon monoxide gas into the producer gas main or directly into the setting.

Thermal Efficiency. The thermal efficiency of the process may be judged by the fact that in producer-fired plants the heat input comes from producer coal with a heat value of 5,000 Btu per pound of zinc produced, plus the recovered carbon monoxide gas from the condenser. The Btu value of the latter is something less than half that of the producer coal. It was stated at the outset that the heat required to support the reaction and to bring the reaction products to temperature is approximately 3,100 Btu per pound of zinc.

Waste-Heat Recovery. The exit gas temperature from the countercurrent recuperator used in modern plants having autogenous cokers is too low to make the use of waste-heat boilers at this point profitable. There is, however, another source of waste heat which, under certain conditions, can be profitably recovered. This is the off-gas from the cokers consisting of the products of combustion of the evolved volatiles which is at a temperature of over 1000°C. The use of waste-heat boilers at this point also serves to cool the gas from the cokers to a temperature where the small amount of zinc oxide fume, carrying also as much as 30 per cent of the cadmium and 2 to 4 per cent of the lead values, can be recovered.

Treatment of Residue. The spent residue briquettes contain 30 to 40 per cent carbon and 2 to 4 per cent zinc. The residue also contains some lead and whatever copper and precious metals may have been present in the roasted zinc concentrate. If the combined value of the steam which can be generated and the metallic values which can be recovered are sufficiently high, the short blast furnace method developed by the Unterharzer company of Oker, Germany[5] may be profitably used to treat the residues.

This process is applicable to residues that carry significant values in precious metals and sufficient copper to yield a matte. Formation of the matte normally requires some addition of sulfur-bearing material. The vertical retort residue, together with any required additions, is charged to a short-shaft blast furnace from which slag and matte are continuously tapped. The zinc and lead contents are volatilized and recovered as a crude oxide product in a settling chamber, boiler and bag room or Cottrell. The matte is usually too low grade to stand the cost of shipment any considerable distance and may require upgrading in a small converter.

Operating Results. In many years of practical operation, a wide range of ores have been treated successfully by the vertical retort process, as summarized in the following table.

TABLE 6-5. RANGE OF TYPICAL ORES TREATED BY VERTICAL RETORT

Type	Source	Analysis (per cent)				
		Zn	Pb	Cd	Fe	Cu
Silicate	New Jersey	45.7	—	—	2.2	—
Sulfide	Newfoundland	53	3.6	0.19	4.0	0.77
"	Canada	51.9	0.05	0.15	11.4	0.63
"	Spain	57.1	1.2	0.15	4.2	0.33
"	Mexico	57.5	1.5	0.25	5.9	0.09
"	South America	46	1.8	0.2	12.7	0.11
"	Virginia	59.4	0.9	0.17	2.4	0.3
"	Colorado	49	0.7	0.25	13.	0.7

High-iron ores present no difficulty up to 10 per cent iron, and ores of even higher iron content can be successfully treated if necessary. Lead contents below 1½ per cent are preferable, but ores of much higher lead content have been successfully treated. Other ordinary impurities occasion no difficulty. In the horizontal retort process, sulfur ties up zinc as zinc sulfide which remains with the residue. This effect has never been observed in vertical retort practice, although excessive sulfur in the calcine is undesirable.

The vertical retort process has successfully achieved the objectives which have been sought by metallurgists for so long. In this improved pyrometallurgical process for the extraction of zinc, the man-hours of labor have been drastically reduced as compared with the old horizontal retort process, and working conditions greatly improved; metallurgical recoveries have been increased and firing fuel consumption substantially reduced. In the most recent installations, with retorts of 8 tons a day output and equipped with countercurrent recuperators, the consumption of anthracite producer coal for firing fuel is less than 0.45 ton per ton of slab zinc produced, and direct labor requirements have been reduced to less than 7 man-hours per ton of metal. Zinc recovery is approximately 94 per cent from the delivery of roasted concentrate to the retort plant through to slab zinc.

The largest installation of vertical retorts is at the Palmerton, Pennsylvania plant of The New Jersey Zinc Company where the process was de-

Figure 6-37. Palmerton, Pa. plant of the New Jersey Zinc Company (of Pa.) where the largest installation of vertical retorts is located.

veloped (Figure 6-37). There are 43 retorts at that plant. The New Jersey Zinc Company also has a 16-retort plant at Depue, Illinois; Meadowbrook Corporation has an 18-retort plant at Meadowbrook, West Virginia; National Smelting Company, Limited has an 18-retort plant at Avonmouth, England; Unterharzer has a 32-retort plant at Oker, Germany; and Asturienne has a 17-retort plant at Auby, France.

REFERENCES

1. Bunce, E. H., and Handwerk, E. C., "New Jersey Zinc Company Vertical Retort Process," *Trans. Am. Inst. Mining Met. Engrs.*, **121**, 427–40 (1936).
2. U. S. Patent 2,536,365.
3. U. S. Patents 2,457,544 to −552, inclusive.
4. Bunce, E. H., and Peirce, W. M., "New Jersey Zinc Develops a New Condenser," *Eng. Mining J.*, **150**, 56–62 (March, 1949).
5. U. S. Patent 2,613,137.

The Sterling Process

G. T. Mahler

Chief of Metallurgical Development, Research Department
The New Jersey Zinc Company (of Pa.)
Palmerton, Pa.

and

W. M. Peirce

Assistant to the Executive Vice-President
The New Jersey Zinc Company
New York, N. Y.

In the Sterling Process, zinc and certain accompanying metals are reduced from their oxides by carbon*, the heat being supplied by open electric arcs. An attempt—extending over a period of some forty years—to develop this type of process was made during the early part of this century at Trollhättan, Sweden, and Sarpsborg, Norway,[1] but was abandoned. Major difficulties were the failure to control the operation of the furnace so as to produce a vapor condensable to liquid zinc, and inability economically to convert the blue powder formed to liquid and finally to slab zinc.

Unsuccessful attempts to develop an open arc process were made by others also. Another difficulty in these attempts, in addition to condensation troubles, was due to reliance on heat transfer from the slag bath to the unreduced charge, rather than on direct radiation to the charge. Excessive slag bath temperatures and consequent refractory problems result from this method of operation. Furthermore, any zinc that enters the slag by fusion of the ore prior to reduction is reduced at a very slow rate.

Figure 6-38 is a flowsheet of the Sterling Process.[2, 3] Roasted zinc concentrates or other oxidic zinc-bearing materials are mixed with just enough carbonaceous reduction agent to reduce all of the zinc, cadmium, lead and copper present, and all or as much of the iron as desired. Unless local conditions give the iron a value as a by-product, it is desirable to expend as little fuel and power as possible on reduction of the iron content of the ore. Furthermore, retention of an iron content of about 5 per cent in the slag is desirable to provide a buffer to prevent the unwanted reduction of other more difficultly reducible oxides. An iron content in the slag higher than about 5 per cent promotes decarburization of the iron bath, thus necessitating undesirably high temperatures for proper tapping fluidity. Where

* See Chapter 4 for theory of zinc reduction and preceding section for other basic requirements.

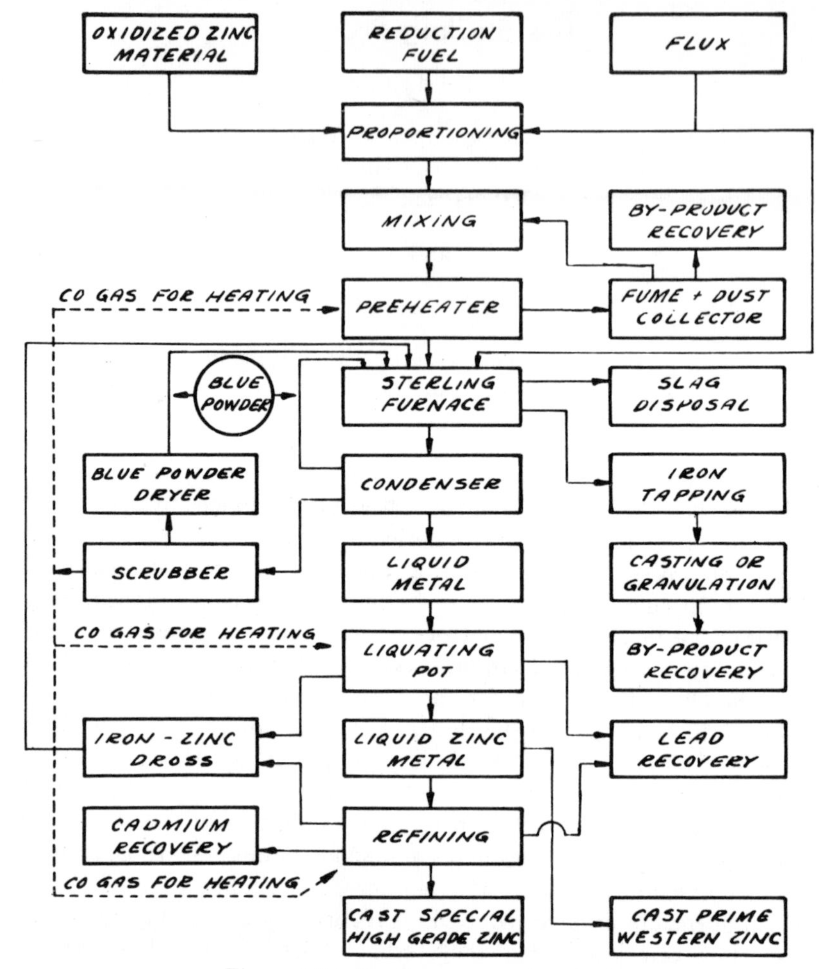

Figure 6-38. Sterling process flowsheet.

the iron has no value as a by-product and the iron content of the ore, if unreduced, would give an excessive iron oxide content in the slag, the matte smelting variation of the process mentioned later may be employed.

Either lime or silicious flux is added to maintain a lime-silica ratio of approximately 0.8:1 to 1.4:1 in the slag-forming residue.

The several charge materials are properly sized to avoid serious segregation during the preheating and charging steps. The mixed charge is preheated in a rotary kiln which may be fired to a temperature of between 850 and 900°C with the carbon monoxide gas recovered from the zinc condensers, or with other fuel. At this temperature, most of the volatiles in the coal are driven off and some reduction of iron and copper to the lower oxides takes place. Also at this temperature, a small amount of zinc, together with some lead and cadmium, is reduced and volatilized. These fumes, together with the dust from the kiln, are collected in a cyclone and bag-room system for recirculation.

The Sterling furnace itself is of reverberatory-type construction with three electrodes in line introduced through the arch. Charging ports are located in the furnace roof around the periphery of the furnace. This and the electrode arrangement are shown in Figure 6-39.

The furnace is constructed of high-grade firebrick. Special refractories are not required since the arch temperature is not over 1300°C. The slag and iron bath temperatures do not exceed 1450°C and the bottom and side walls are well protected by frozen skulls and the charge banks.

Also shown in Figure 6-39 are two of the condensers, the required number of which are located on two sides and one end of the furnace. At the other end of the furnace are two tapholes, one at a higher level for tapping slag, and one at hearth level for tapping iron.

In operation, the furnace carries a bath of molten iron covered by a bath of slag. Charge banks protect the two side and two end walls. The toes of

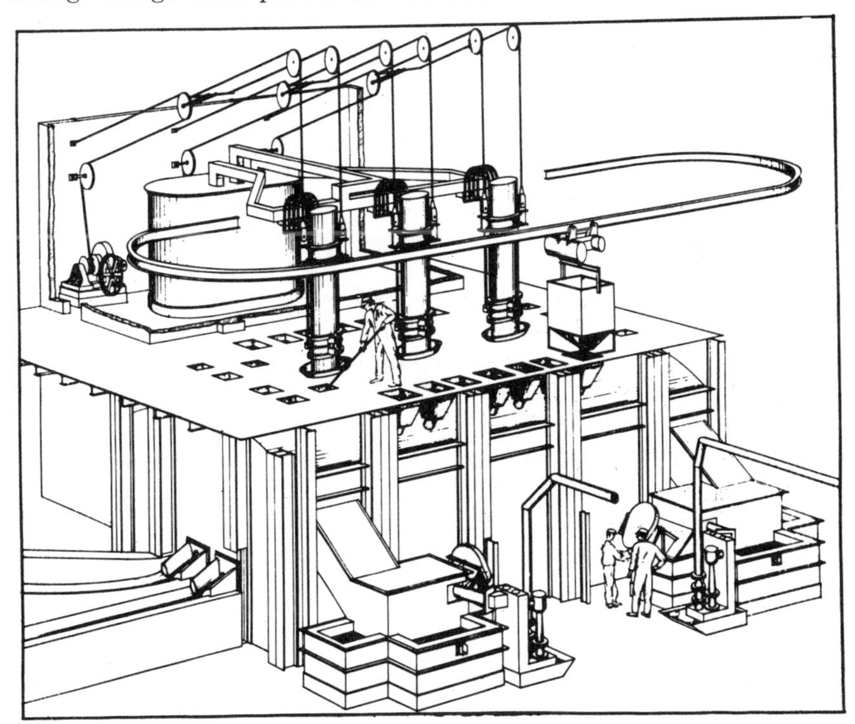

Figure 6-39. Outline of Sterling furnace construction showing charging ports, electrode arrangement, condensers and tap holes.

these sloping banks leave an exposed area of slag at the center of the furnace beneath the electrodes. The heat generated by the arc between the electrodes and the slag bath is rapidly absorbed, being transmitted mainly by direct radiation to the charge banks and, to a lesser degree, by reflected radiation from the furnace roof. Some heat, of course, is generated within the slag bath itself by its resistance to the current passing between the arcs striking its surface. The objective, however, is to maintain such a relation between the voltage and the current at which the furnace is operated that only sufficient heat is generated in the slag bath to maintain the slag and iron baths at the necessary temperatures for good fluidity, but to leave frozen skulls of iron and slag to protect the hearth and side walls. The heat of reduction, as previously stated, is transmitted directly to the charge banks where the bulk of the reduction reaction takes place.

As the reduction reactions are completed, the temperature of the charge rises to a point where the residue fuses and runs into the slag bath. Any reduced iron is melted and descends to the furnace hearth. New charge introduced at the top of the banks in frequent small increments is reduced as it works down the bank. By maintaining a clear slag bath surface to which arcing takes place preferentially, arcing to the charge banks is largely avoided, and there is a minimum of dusting and of volatilization of silica and other oxides which form a fume detrimental to good condensation. The elimination of zinc, lead and cadmium from the charge is virtually complete. Any copper and gold values enter the iron bath, and silver values are distributed between the iron and the volatilized metals. The carbon monoxide gas and metallic vapors leave the furnace through ports in the side walls and enter condensers located as close as is practicable to the furnace itself.

These condensers (described in the previous section of this chapter) are of the splash type commercialized earlier for use with the vertical retort process. A motor-driven graphite impeller throws a shower of zinc which scrubs the gas stream. The droplets of zinc serve to cool the gas and at the same time provide liquid zinc surfaces on which condensation of zinc takes place. The molten zinc bath in the condenser sump is held at 500°C by water-cooled coils. This method of condensation keeps the formation of blue powder to a minimum.

Venturi scrubbers following the condensers condense the last traces of zinc, control furnace pressure, and pump the clean carbon monoxide gas to points of use.

As in any process where zinc is reduced by carbon, it is important to keep as low as possible the carbon dioxide content of the gas entering the condenser which requires that it should leave the reaction zone at as high a temperature as possible. This is one of the reasons why preheating of the charge is essential, since if the gas stream leaving the furnace passed over cold charge, its carbon dioxide content would be increased. The presence of carbonates or the higher oxides of iron and copper in the charge if it were not preheated would also lead to high carbon dioxide content in the gas to the condenser.

The condensed zinc is liquated to remove iron, and, if derived from zinc concentrates high in lead, to remove lead. The dross from the liquation step is recirculated to the furnace, as is all blue powder from the condensers and scrubbers. Depending on the composition of the concentrates in the charge, a refining step may be required to produce salable metal. The vertical refiner described in Chapter 7 may be advantageously used.

Because of the virtually complete elimination of zinc from the charge, the process yields high recoveries even with low-grade concentrates and is of particular interest for the treatment of such concentrates and residue materials (e.g., the leach residues from the electrolytic process).

Zinc concentrates containing sufficient copper to make its recovery worthwhile may be treated by a modification of the process in which sufficient sulfur is introduced into the charge by incomplete roasting of the concentrates, or by the addition of green concentrates, to result in the formation of a matte rather than pig iron. The matte will carry virtually all of the copper and gold, and most of the silver. Part of the silver which volatilizes with the zinc will recirculate to the furnace with the dross and eventually appear in the matte. This method is, as stated earlier, useful with high-iron ores where iron recovery is not economical. Matte smelting in such cases can yield economies in fuel, power and roasting costs.

When treating high-grade concentrates, the process requires about 2800 kilowatt-hours per ton of combined nonferrous metals and iron reduced, and about 180 kilowatt-hours of auxiliary power. The electrode consumption using graphite electrodes is about 16 pounds per ton of metals reduced. As previously stated, the amount of reduction fuel is substantially the amount predicted by theory.

A furnace of full commercial size having an inside hearth area of 16 by 32 feet was operated at Palmerton, Pennsylvania, on a wide variety of zinc concentrates and other zinc-bearing materials. The furnace had a nominal rating of 6000 kva and was operated at inputs up to 4500 kva without reaching the limit of the furnace itself. Additional condensing capacity would have permitted higher outputs. When treating high-grade concentrates, the production of this furnace was over 35 tons of slab zinc per operating day with a recovery of 95 per cent. The power factor is 95 per cent or better, and the operating time (power on the electrodes) can be maintained between 90 and 95 per cent. Two furnaces of similar size and design have been built at Oroya, Peru, by Cerro de Pasco Corporation.

REFERENCES

1. Landis, W. S., "Trollhättan Electrothermic Zinc Process," *Trans. Am. Inst. Mining Met. Engrs.*, **121**, 573–98 (1936).
2. U. S. Patents 2,598,741, E. C. Handwerk and G. T. Mahler; 2,598,742, E. C. Handwerk and G. T. Mahler; 2,598,743, R. K. Waring, L. D. Fetterolf and T. L. Hurst; 2,598,744, E. C. Handwerk and G. T. Mahler, The New Jersey Zinc Company.
3. Handwerk, E. C., Mahler, G. T., and Fetterolf, L. D., "Electric Furnace Smelting of Zinc Ores—The Sterling Process," *J. Metals*, **4**, 581–6 (1952).

The Electrothermic Process of the St. Joseph Lead Company

George F. Weaton

Consultant
St. Joseph Lead Company
New York, N. Y.

The electrothermic process development of the St. Joseph Lead Company involved not only the electrothermic furnace itself but also the dependent and cooperating processes. These include the development or improvement of roasting, sintering, leaching, the nature of reduction fuel, and the furnace reject separation of coke, values and residues. Zinc oxide collection had to be improved, and, very important, the jet condenser for recovering zinc metal had to be developed. Figure 6-40 illustrates diagrammatically the integration of the plant processes.

The present plant at Josephtown, Pennsylvania, is equipped to handle up to 300,000 tons of zinc concentrates per annum and the corresponding fuel, power and supplies necessary to their production. It can store 125,000 tons of concentrates, 100,000 tons of coke and other necessary material. Shipment or receipt of material is by Ohio River, rail and truck. The plant has two diesel locomotives, numerous lift trucks—both engine and battery motor driven—cranes, whirlers, etc. There is a concentrate transfer building with overhead crane, blende hoppers, and storage capacity; there are Cottrell electrostatic precipitators of 450,000 cfm. serving the sinter plant and of 78,000 cfm. serving the roasting plant. A baghouse of 250,000 cfm. capacity serves the zinc oxide department. The plant has its own water system; analytical and physical laboratories for complete chemical analysis with polarographs, routine and research spectrographs, microscopes—dark and light field—electron microscopes, electronic control testing equipment; fully equipped rubber, paint, and ceramic laboratories; cafeteria, change houses, shops, auditorium with bowling alleys, basketball courts, stage, etc. Under construction is a 100,000 kilowatt coal-burning power

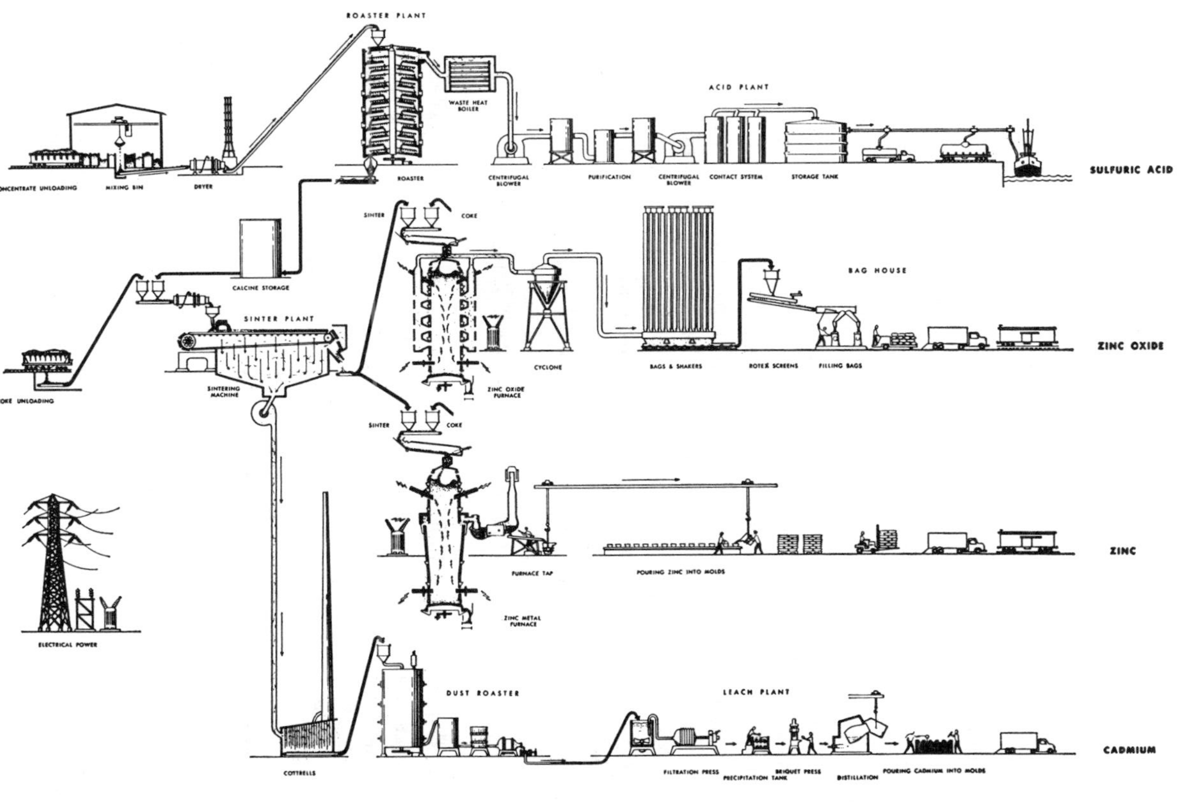

Figure 6-40. Diagrammatic flowsheet of the zinc smelting division, St. Joseph Lead Company.

plant to operate at 1,800 psi steam pressure, 1,000°F initial, and 1,000°F reheat. The smelter is a fully integrated unit.

Prior Investigations

The first investigation into the possibility of the reduction of zinc ores by means of an electric furnace was conducted by the Cowles brothers who designed and patented their process in 1885.[1] This was a direct-charge resistor furnace (submerged arc) utilizing the form of the ordinary Belgian retort, the end being closed with carbon plates which acted as the electrodes.

Following Cowles, many patents were taken out and considerable literature developed. However, there was no development of great importance until about 1898 when Gustave De Laval, at Trollhättan in Sweden, developed an open arc furnace. A company was formed to exploit De Laval's patents and good spelter with less than 0.1 per cent total impurities appeared on the European market in 1904. The development proceeded, and in 1911[2] there were 11 smelting furnaces of the old type (open arc). The capacity of each furnace was 2.8 metric tons of ore per 24 hours. These furnaces were slag-type. The ore mixture was 300 kilograms of Broken Hill slimes, 10 of calamine and 75 kilograms of coke, and the furnace production was principally blue powder which was worked up in a separate furnace. A later and more detailed account of the Trollhättan operation is given by Landis.[3]

The Trollhättan developments created much interest, and experimental work was undertaken by many. It may be conceded that the work of Johnson,[1] from 1903 until 1915, contained many ideas for continuous processes. The several furnaces of Johnson were of the resistor or submerged arc type. Condensation difficulties were not overcome.

The most important work was undertaken by Snyder and the Canada Zinc Company, and a 10-ton per day plant was erected at Nelson, B.C. in 1908. This was a vertical shaft furnace with water-jacketed shaft and vertical electrodes projecting through the charge into the slag. The furnace operated for about one year.[1]

In 1914, Fulton[1] proposed "an indestructible briquette" as a means of smelting zinc by electrothermic means; his briquette was to consist of 100 parts ore, 60 parts coke and 18 to 20 parts pitch, formed under pressures of 500 to 1,000 psi and baked at temperatures of 450 to 500°C. The briquettes were 9.25 inches in diameter and 21 inches long. They weighed about 90 pounds. The furnace was a cylinder and contained 12 or more columns of briquettes, stacked 3-high. Furnace power consumption was about 2,600 kilowatts per ton of zinc. One ton of zinc from 60 per cent concentrates required 1.11 tons of coke and 0.37 ton of pitch. Here also, much trouble was experienced with the condensation system. It is also notable that this was a resistance furnace and probably had the lowest power consumption per ton of metal produced of any electric furnace prior to the advent of St. Joseph Lead Company's development.

The major causes of failure might be attributed to two factors: (a) furnaces were generally too small to permit effective diagnosis and remedy of defects in the system or construction, and (b) there was no condensation system capable of condensing in a single unit several tons of zinc vapor per day. The open-arc furnace vapor was particularly difficult to condense due to the dust blown to the condenser by the arc. This iron and dust made it necessary to dross the condenser and re-treat the dross.

Rather comprehensive literature indicates the amount of money which has gone into these developments, yet today there are only five electrothermic zinc smelting plants in the world: one in Peru; one in Argentina; one in Japan, of which little is known; the electrothermic slag furnace at Herculaneum, recovering zinc, lead and copper matte from lead blast furnace slag and producing slab zinc by directly condensing the vapors; and the large Josephtown plant of St. Joseph Lead Company, the two latter and the Argentina plant being developments of the St. Joseph Lead Company.

The development work of St. Joseph Lead Company's electrothermic system was undertaken in 1926 when the late E. C. Gaskill introduced his idea of how electrothermic zinc might be produced commercially. The first furnace was 18 inches in inside diameter and about 18 feet high, too small and unsuccessful. The second furnace was 24 inches in diameter by 18 feet high, failing for the same reason. The third furnace was 48 inches in diameter and 16 feet between the top and bottom electrodes, with an over-all height of about 34 feet. This was later modified and the diameter increased to 51 inches. These furnaces produced zinc oxide only and had rather extensive runs.

The sinter for this furnace was produced from concentrates from the Edwards, New York, mine, which were desulfurized on the sinter machines. Here silica was added to the sinter charge to harden the sinter to permit crushing and proper sizing as required by the furnace. This work was carried on at Herculaneum, Missouri, and was concluded in 1929, when the decision was made to build a commercial plant. Much credit is due to Clinton H. Crane, then President of St. Joseph Lead Company, for the decision, inasmuch as the commercial success of the process was not proved, and its success would definitely depend upon future development. Ground was broken May 9, 1930, and the first production of zinc oxide was in January 1931. The location chosen was on the south side of the Ohio River about 30 miles from Pittsburg at what is now Josephtown, three miles down river from Monaca, Pennsylvania, and at the junction of the Beaver River in Beaver County.

The Roasting Process

The original equipment consisted of two 21-foot 6-inch diameter 12-hearth roasters, later modified, and ultimately there were five of these roasters. Purification was a major problem when using concentrates with a wide range of lead, cadmium, etc. It was found that by proper control of roasting temperature, recirculation and stage roasting, a high percentage of these impurities could be separated out and their values recovered. The original patent, U.S. No. 1,940,912, was granted in 1933.

Two suspension roasters were installed in 1947, and development work was started on the fluid types in 1948 on a one-meter diameter unit which later developed into 7-foot diameter and 15-foot diameter experimental units. As a result, one 22-foot 9-inch suspension unit was converted to a fluid system, and a 20-foot diameter unit was constructed, bringing the total roasting capacity of the plant to approximately 800 tons of 50 to 57 per cent zinc concentrates per day. The purification consisting of pre-roasting and lead elimination, with partial cadmium elimination on the hearth roaster, is followed by finishing the desulfurizing in the fluid roaster.

Roaster gases are cooled through waste-heat boilers, then pass through Cottrells and a wet scrubber and gas purification system. Finally sulfur is recovered as sulfuric acid in five Monsanto-type contact acid units. Recovery of sulfur from concentrates as sulfuric acid is 86 to 87 per cent.

Improvement in the acid plant was found necessary also, and in particular, the development of efficient coolers and means of recovering cadmium values from the scrubber liquors. This acts as a closed-system scrubber to a special type settler from which the pregnant liquor and the lead sulfate is withdrawn and sent to the leach plant for value recovery. The clear overflow liquor then is returned to the scrub system through the coolers. This circuit operates under limited acid build-up as a high acid content will not remove the arsenic and fluorine necessary for the protection of the contact mass. Acid is shipped by river barge, rail and truck, and is used by the steel, pigment, chemical, and synthetic fiber industries.

The Sinter Plant

Modern zinc sintering developed out of lead sintering practice,[4] current in 1905 when the basic principle of downdraft sintering was first demonstrated by Dwight and Lloyd. Rigg at Port Pirie, Australia,[5] in 1917, applied the downdraft principle to the sintering of zinc calcine.[4] Later, Vieille Montagne at Baelen, Belgium, sintered dead-roasted calcine with 6 per cent coke. During the late 1920's many of the American smelters installed sintering machines in order that the yield of zinc per retort and recoveries could be raised. Here, it may be interesting to note that early attempts at sintering were by the updraft principle lately revived at Port Pirie.

The Huntington-Heberlein pot method, patented in 1897–98, made satisfactory lead sinter by the updraft method. The chief difficulty of this system was that it was a batch system, and breaking the 10-ton lead sinter cakes was a difficult and dirty process and very unhealthy for the workmen.

Downdraft versus updraft has been discussed by Dwight,[4] who states:

"They found that with an updraft blast, the product while desulfurized was very irregular in physical character and contained a large proportion of fines. Careful study of the procedure when using the shallow grates and also deep updraft pots brought the conclusion that the fines resulted principally from the agitation of the particles by the air blast, and that if the particles could be held in quiescence in contact with one another along the surface of the exit of the gases, they could be sintered into a coherent cake. Various expedients were tried to effect this, but the simplest and the one that finally did the trick was to reverse the air blast."

The Herculaneum lead smelter had installed sintering machines in 1917 and when, in 1926, the experimental electrothermic zinc furnace was erected there, two of the machines were assigned to charge preparation for the electrothermic zinc furnace. In this case, there was no preroasting, the sulfide ore was a part of the mix and, excepting the ignition fuel, the sulfur was the fuel. The scalp of finished sinter was approximately 20 per cent of the sinter and the desulfurization was good, notwithstanding the silica added to balance the iron in the ore and to produce a hard sinter capable of being sized to the exacting requirements of the electric furnace. When the Josephtown plant was built, a complete study indicated considerable economy in preroasting followed by coke sintering, and such is the practice today.

This sinter development started with three 42-inch by 44-foot machines; these were later rebuilt to 60-inch by 44-feet, the limitation on width resulting from the size of the very hard sinter cake which invariably leaves the machine as a solid pallet-size cake difficult to convey and crush effectively in the required sizing limits. The sizing recovery runs about 30 to 35 per cent and the large circulating load on the machines poses a mixing problem, there being an excess of nuclei in proportion to the calcine fines available for proper balling of the mix. This may be corrected by grinding sinter returns, which, however, proves uneconomical and is only practical on a portion of the soft sinter produced as the first stage of a two-stage purification circuit. From this circuit, zinc of above 99.95 per cent purity is made. A high percentage of the mix is "sandy." While this appears to be a disadvantage, it also follows that unless the mix pellets could be size-graded to compare with the crushed size, net product capacity might be reduced rather than increased with complete pelletization of the mix. Figure 6-41 shows the character of the mix as it leaves the ignition muffle.

The charge bed is generally held at a depth of 10 inches; the ignition fuel is carbon monoxide gas recovered from the electric furnace zinc condensers. The sinter charge is made up on calcines, sinter fines and return fines from the furnace after screening and magnetic separation. Other additions are coke, high silica sand, oxides of zinc from various sources, and other residues. The resulting smelter grade is 54 to 56 per cent zinc.

Approximately 50 per cent of the concentrates are the produce of company-owned mines, the balance being purchased on world markets, and therefore subject to considerable grade variation. It is quite necessary to maintain the strictest laboratory control over all the elements entering into the sinter mix in order that grade, fluxes, etc. may be in proper relationship. If any of these elements are out of control, crushing, furnace efficiency and product quality are affected.

The sintering process also is the final step in purification before the furnaces; a portion of the lead left by the preroasting lead eliminating system is removed, as is a major portion of the cadmium contained in the concentrates. The gas fumes and carry-over dusts are passed through a conditioning chamber where fog nozzles increase the humidity of the gases and where much of the larger particulate matter is precipitated. These

Figure 6-41. Character of the sintered mix as it leaves the ignition muffle. (*Photography by Clyde Hare.*)

gases and fumes then pass through Cottrell precipitators where the fume, which is high in cadmium and with considerable lead, is precipitated; the conditioner precipitate is returned to the sinter mix. The Cottrell fume is sulfate-roasted according to U.S. Patent 2,777,752 and is sent to the leach plant for recovery of contained values.

After passing the Cottrells, the gases pass through huge cyclone washers where a considerable recovery of cadmium is made prior to their discharge through a 400-foot chimney.

The sinter discharged from the sinter machines is crushed first in a roll, with a series of pyramids on its face, to minus 3 inches, then in a closed-circuit screen and roll circuit until all sinter is minus 0.75 inch and plus 0.25 inch; 50 per cent must be plus 0.375 inch. This is the furnace feed sinter and may contain from 60 to 90 per cent of the zinc input to the furnace depending upon the burden of other feed in the circuit.

The sintering department also prepares the coke for the furnaces as well as for its own use in the sinter mix. The sinter-mix coke is crushed to minus 20 mesh. The furnace coke must be sized to approximately the same size and exacting requirement as the sinter; total plant coke approximates 0.8 ton per ton of zinc produced.

The Furnace Plant

The production of electrothermic zinc oxide is described in Chapter 8 of this volume; attention is directed presently to the metal furnaces exclusively.

Although the preparation system is common, there are four metal furnaces of 5-foot 9-inch diameter, each equipped with 8 electrodes; 4 at the top of the furnace, with single-phase circuits extending to the 4 bottom electrodes, the electrode spacing top to bottom being 24 feet. There are seven 8-foot diameter furnaces each having 16 electrodes; 8 at the top and 8 at the bottom, the spacing being 32 feet. There are 8 single phase circuits. The capacity of the 5-foot 9-inch furnaces is approximately 20 tons of zinc per day.

The plant capacity in slab zinc is 400 tons daily. The maximum one-day production of any furnace-condenser unit has been 70 tons. The average production of the 8-foot furnaces is 50 tons and is rising as improvements are made.

Each furnace has independent coke, sinter and other feed bins, each bin being equipped with charge weighing belt scales to accurately proportion the charge constituents. The charge is conveyed to the preheater drums where it is heated to 750°C by carbon monoxide gas recovered from the zinc condensers. The charge is then discharged into the furnace by means of a rotary distributor whose function is to build the resistor by proper charge segregation. The electric current will spread evenly over the furnace area and the smelting temperature is therefore uniform, and a minimum pinch effect is obtained.

As the furnace is a resistance furnace, the charge of coke and sinter being the electrical resistor, the temperature within the resistor is inverse to the resistance. The ultimate current effect would be to pinch. This does occur to some extent in practice, the highest temperature area being at or near the center of the furnace.

The rotary distributor is designed purposely to segregate the charge in order to control the pinch effect. Checking the furnace temperature dis-

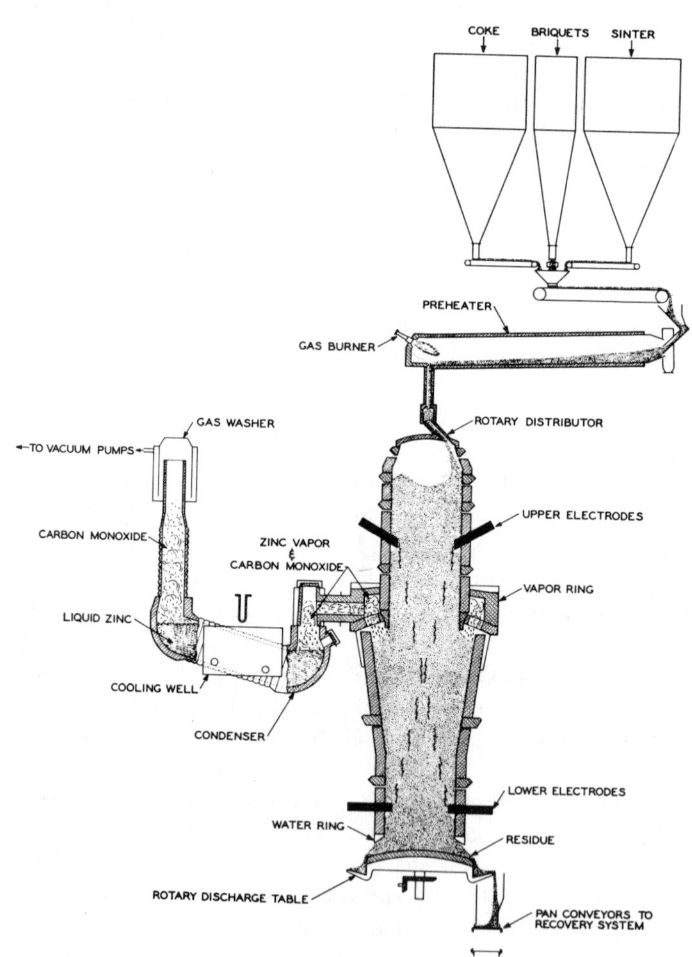

Figure 6-42. St. Joseph Lead Company electrothermic zinc metal furnace.

Figure 6-43. Action of open variable-speed table at bottom of electrothermic zinc furnace. (*Photography by Clyde Hare.*)

tribution is done with a steel bar inserted into and across the furnace at 120° spacings at various levels for a definite time. The temperature pattern is read from the bar temperature and recorded. The temperature at the center of the furnaces is 1300 to 1400°C.

Furnace power is automatically controlled by electronic devices developed in our own laboratories. The power input per electrode circuit is 600 to 700 kilowatts. The voltage is regulatable over a wide range; normally it is of the order of 230 to 240 volts. Twelve-inch round graphite electrodes are used.

The total height when the furnace is smelting is approximately 50 feet. The charge height when the furnace is smelting is about 46 feet. The feed into the furnace is continuous (Figure 6-42). Figure 6-43 shows a furnace with a variable-speed table at the bottom; the worked-off charge is shown leaving the furnace and spilling on the table apron, where it is scraped off on a pan conveyor. Also shown is the specially designed poker truck used to free the table of clinkers in order that discharge may be continuous. The table speed is adjusted to control the height of the charge at the furnace top. Charge settling also occurs from the shrinkage of the coke and the sinter charge, due to smelting-off the zinc. The residues from the table are sent to the treatment plant for separation and reclaim.

At about one-third of the distance between the top and bottom electrodes (Figure 6-42) the furnace widens out and the charge spilling out from the 8-foot diameter to the greater diameter gives a large vapor-disengaging surface; above this point there is a double arch annulus, the lower arch having ports whereby the vapors pass into the annulus and then around to the single vapor offtake to the condensers. At this point, the vapor under standard temperature and pressure conditions occupies 5.47 cubic feet per pound and the temperature is about 875°C. For each pound of zinc there is a slightly greater volume of carbon monoxide gas than the volume of the zinc vapor. The furnace smelting temperature is about 1300°C.

The furnace construction is sectional, there being usually seven sections, each supported independently on a skew ring which is hung on insulation. Those parts which are steel-enclosed—such as the vapor annulus, a part of the barrel below the annulus, and a portion of the upper furnace—all must be insulated as are the condenser connecting joints. The bottom section sits on a water-cooled ring which is also insulated. The rotary discharge table can be raised or lowered to maintain a proper discharge aperture between the water ring and the table. The preheater discharge gases pass to a baghouse for the ultimate recovery of fume and oxide which escape into this system through the rotary distributor orifice. The furnace system is the result of many years of improvement. It is safe to predict that this improvement will continue.

The Condensation Problem. The early difficulties of condensation of zinc vapors are well indicated by the Trollhättan experience as indicated by Landis,[3] Ingalls[5] and Doerner.[6] This problem was recognized by the engineers of St. Joseph Lead Company, and due to their experience in the condensation of steam in power generation, a comparison study was started in late 1928. It became apparent early in 1929 that using a surface type of condenser was the wrong approach, due to the fundamental requirement that energy must be applied if the powder or zinc dust formed by the chilling action of condensation is to be coalesced into liquid metal. However, the 48-inch furnace at Herculaneum was operated as a metal furnace from February 23 until May 13, 1929, being equipped with a multitude of somewhat larger than usual truncated cone condensers, like the standard horizontal retort condenser. The operation was continuous, if not too salubrious. However, the metal production was only two to three tons per day and again pointed up the necessity of condenser development to fit the furnace. The ideas returned again to steam condensation and it appeared that if materials of construction could be found, the so-called "jet type" condenser (steam) would fulfill the conditions required. Table 6-6 indicates the similarity of the thermal data between steam from water and "steam" made from zinc. These data indicated to us in 1929 the feasibility of large capacity jet or separate type condensers.

The Herculaneum furnace had been shut down after the May 1929 runs to make an appraisal of the results obtained and to evaluate the future possibilities of the program. It was decided to abandon any further work at Herculaneum and to build the Josephtown plant. The study of condensa-

Table 6-6. Comparative Thermal Data

	Water	Zinc
Pressure	760 mm Hg	760 mm Hg
Boiling point	100°C	920°C
Latent heat of vaporization	970 Btu/lb	802 Btu/lb
Total heat of vapor	1150.4 Btu/lb	1090 Btu/lb
Vapor volume per lb	26.83 cu ft	29 cu ft
Heat in liquid at 38°C per lb	68 Btu	—
Heat in liquid at 500°C per lb	—	148 Btu
Heat given up during condensation	1092.4 Btu	942 Btu
Heat given up to condenser by 26.4 cu ft CO gas	—	216 Btu

tion, and of course many of the furnace problems, was therefore suspended until the Josephtown plant became an operating unit in January 1931.

It may be noted that the great depression of the thirties was then under way. However, this did not stop development of the condenser and other plant equipment necessary to the progress and economy of the plant. Lack of money may have limited, and extended the time for completion of, the necessary developments and there may also be a question whether the capital available was sufficient to implement our ideas as they developed.

The first commercial metal unit went into production in August 1936. It was equipped with a St. Joseph Lead Company jet-type condenser. The time between January 1931 and 1936 had been devoted to its development. Many different condensation experiments were performed and the design characteristics were widely varied. As the work progressed, however, it became more apparent that any condenser not having a considerable energy component would not serve as a single unit for large scale vapor condensation, and that multiple units decreased the condenser's efficiency as well as complicated the maintenance problem.

The work was then directed to the single purpose of developing the St. Joseph Lead Company condenser. Several models were built with glass inserts. These were filled with mercury or other liquids to observe the behavior of steam or other condensable gases as they passed through the condensing medium. Bubble size, condensation time, angle of the unit (which has influence on bubble size), cooling rate, etc., were noted. From these small-scale tests a unit was built rated at 5 tons per day and put on one furnace in August 1933. Najarian[7] describes this operation as follows:

> "This experimental condenser was attached to one of the tewels of an oxide furnace, with a damper between the furnace and the inlet leg of the condenser. After preliminary heating of the condensers and priming with molten metal, the condenser was opened to the furnace and vacuum applied to the exit leg. We could hear the gas bubbling through the metal and soon the temperature began to rise and the condenser became excessively hot as we had no means of cooling the shell except by natural

Figure 6-44. Original 5-ton experimental condenser.

air circulation. After some six hours of operation, bubbling slowed down and soon stopped completely. After the metal was tapped out and the condenser allowed to cool, examination disclosed that the exit pipe (8 inches diameter) and part of the piping to the vacuum pump were almost plugged with metal splash from the condenser and with blue powder."

This condenser had a number of runs of short duration from which we learned much. Among the most difficult problems was to develop a proper washer to scrub the exit gases from the condensers. The gases passed upward along the top of the inclined plane and were exhausted through the 8-inch opening facing to the rear (Figure 6-44). Here was the major source of trouble. It finally became apparent that the exit gases should be above the melting point of zinc and that this cooling must be sudden if trouble were to be avoided. This was accomplished, and no change since has been made in the washer from the first unit installed in 1936. The research of C. C. Long* developed a valuable adjunct in the form of a high-velocity spray and impinger, resulting in a gas-cleaning efficiency of above 99 per cent.

Figure 6-44 is a view of the original 5-ton experimental condenser. In the lower foreground is a divided chamber part of which was the tap well, midway to the vapor inlet, and in the lower background is a well connecting

* Director of Research, St. Joseph Lead Company.

to the main condenser, intended as a cooling well. Cooling was to be effected by means of tungsten-coated water pipes. This was not effective, as the tungsten, which was welded to the pipe by means of an "atomic torch," could not be made into a continuous coat. The iron would bleed through and the zinc would cause failure in a short time. In later years the tube method of cooling was developed and is now in use. In the meantime, cooling was by means of water sprays directed on the shell of the condenser.

The condenser is a pressed-steel corrugated ovate section 5 feet high and 40 inches wide at the ovate crown. It contains 50 tons of liquid zinc, and is lined with carborundum, silica carbide brick and mica, next to the steel shell. It operates under 11 to 12 inches of mercury vacuum and has a condensation capacity of above 50 tons of liquid zinc per day, when operating on a carbon monoxide-zinc vapor gas of, for example, 40 per cent zinc vapor by volume, with an efficiency above 98 per cent plus. Figures 6-45, 6-46 and 6-47 are views of the present day condenser and cooling well.

Residue Reclaim and Retreatment. The residue discharged from the rotating table at the furnace consists of slag, low-grade sinter, reduced iron and coke. This is discharged to a pan conveyor, through a jaw crusher set to crush to about 1¼ inch, and ultimately is discharged into a trommel screen with ¾-inch openings. The oversize is low grade and classed as slag

Figure 6-45. Partial view of modern St. Joseph Lead Company condenser.

Figure 6-46. Partial view of condenser and cooling well.

Figure 6-47. Drawing metal from the condenser cooling well.

for rejection. The nonmagnetic is subject to air separation, wherein the coke is separated from the values and the plus ¼-inch coke returned to the furnace. The balance, fine coke and "poor" sinter, is ground to about minus 20 mesh and incorporated into the sinter mix. Iron balance of the

TABLE 6-7. ECONOMIC DATA FOR 1956 AND 1957

	Unit	
Year—1956		
Concentrate	tons	216,989
" zinc content	"	124,063
Other zinc content	"	20,083
Total zinc intake	"	144,147
Output commercial slab zinc	"	109,929
" " zinc oxide, zinc content	"	24,225
" total	"	134,154
Sulfuric acid, 100%		190,004
Power, per ton zinc produced smelting	kwh	2,449
" " " " " plant	"	721
" " " " " total plant	"	3,170
Coke " " " " sized	tons	0.687
" " " " " braize	"	0.098
Cadmium produced	"	156
Bonded refractories and cements per ton zinc	lb	75
Carbon " " " " " "	"	1.67
Graphite furnace electrodes per ton zinc	"	2.17
Man hours, total, sal. & P.R. per ton zinc	hr	18.02
Annual plant capacity as of June 1, 1957:		
Total zinc intake 56% conc.	tons	161,000
" " " other zinc content	"	20,000
" " "	"	181,000
" commercial slab zinc output	"	146,000
" " zinc content of zinc oxide	"	24,100
" " output	"	170,100
" sulfuric acid, 100%	"	246,000

entire furnace sinter system is maintained by a variation of the magnetic flux as the residues pass over the magnetic rolls. The rejected slag and high magnetic material account for about 3 per cent of the plant zinc intake.

The results for the year 1956, which are set forth in Table 6-7, indicate the principal economic data of interest in connection with the Josephtown zinc smelter.

ACKNOWLEDGMENT

It is a privilege to acknowledge the resources, ideas, engineering ability and common sense of my many associates in this development, in particular, J. J. Rankin, Louis Trudo (now deceased), H. J. Najarian, W. B. MacBride, C. C. Long, and the many who contributed their share; also to those whose continued support and confidence made the growth of the Josephtown plant possible: Clinton H. Crane, Andrew Fletcher, and C. Merrill Chapin.

REFERENCES

1. *Bull. U. S. Bur. Mines*, Department of the Interior, No. 208, "The Electrothermic Metallurgy of Zinc," 106 pages, Washington, D. C., 1922.

2. Harboard, F. W., "Zinc Smelting at Trollhättan," *Eng. Mining J.-Press*, **93**, 314–5 (1912).
3. Landis, W. S., "The Trollhättan Electrothermic Zinc Process," *Trans. Am. Inst. Mining Met. Engrs.*, **121**, 573–98 (1936).
4. Dwight, A. S., "A Brief History of Blast-furnace Lead Smelting in America," ("Preparation of the Ore"), *ibid.*, pp. 33–36.
5. Ingalls, W. R., "History of the Metallurgy of Zinc," *ibid.*, p. 362; Ingalls, W. R., "Intermittent Zinc Distilling from Ore," ("Reduction and Condensation"), *ibid.*, pp. 624–6.
6. Doerner, H. A., "Reduction of Zinc Ores by Natural Gas," ("Condensation"), *ibid.*, pp. 659–61.
7. Najarian, H. K., "Weaton-Najarian Vacuum Condenser," *Trans. Am. Inst. Mining Met. Engrs.*, **159**, 165 (1944).

The Production of Zinc in a Blast Furnace*

S. W. K. Morgan, *Research Director*

and

S. E. Woods, *Manager, Development Department*

Imperial Smelting Corporation Limited
Avonmouth, England

Metallic zinc has been produced on a commercial scale from two furnaces at the Avonmouth Works of Imperial Smelting Corporation Limited. These two furnaces were first operated in 1952, and since that date have produced over 85,000 tons of zinc. A considerable amount of investigational work was carried out on a small experimental furnace prior to their design, but a considerable degree of modification was found necessary as experience was gained on the larger scale. This has been carried out, and the units now form an appreciable part of the normal productive capacity of the company.

While the early hopes that the process would form an economical method of treating normal high-grade zinc concentrates have been realized, several other features which were only vaguely anticipated at the beginning of the project have been found to have considerable importance. Thus it is now apparent that the blast furnace can treat lower-grade concentrates more successfully than can the established processes. Both the existing retort processes and the electrolytic process become increasingly handicapped if the iron content of the concentrates treated rises above 10 per cent. With blast furnace operation, ores of considerably higher iron content can be tol-

* A description of this process by one of the authors has also appeared in a recent volume of the *Transactions of the Institution of Mining and Metallurgy*.[1]

erated, and in certain circumstances may be preferred. Again it has been found that lead (and silver) can be recovered at high efficiency as a bullion from the bottom of the furnace. Due to the fact that the condensers fitted to the furnaces can deal successfully with gas of high carbon dioxide content, within broad limits the lead recovered does not impose additional demands upon the carbon consumption. The lead can, therefore, be considered to have a "free ride" and the capacity of the furnaces to produce zinc simultaneously with lead is not reduced. If present in relatively low proportions, copper can be recovered at high efficiency from the bullion produced. There has not yet been an opportunity to treat charges containing higher proportions of copper, but there seems no reason why with these charges the additional copper cannot be recovered economically as a relatively high-grade matte.

The blast furnace process is a newcomer in the field and it is too early to define with precision its position relative to the established methods for zinc production. The largest furnace yet built at Avonmouth is one burning 55 tons of carbon per day with relatively high-grade concentrates. This is obviously not the optimum size. As the next stage in development, furnaces with a carbon-burning capacity of 105 tons per day are projected. What may be the ultimate limit is not clear and will depend upon local conditions, but one of the obvious potentialities of the process—that of large unit operation—will not be fully exploited except with large furnaces.

Previous Attempts at Blast Furnace Production

Some, or all, of the potential advantages of blast furnace production of zinc have been in the minds of metallurgists for a considerable number of years, and many expensive attempts in a number of countries are recorded. One of the earliest dates from 1860, when Müller and Lencauchez patented a number of features which are considered essential today.

One of the most refractory of the problems of zinc metallurgy arises from the fact that the reaction

$$ZnO + CO \rightleftharpoons Zn + CO_2$$

is readily reversible. Due to this, it is not easy to condense zinc efficiently, even from retort gases, which may contain as much as 40 per cent zinc vapor and 2 per cent carbon dioxide. For many years, the problem of condensing zinc from the much more dilute gases arising from a blast furnace was considered to be insuperable, and attempts at blast furnace production were aimed at producing liquid zinc from the bottom of the furnace. It was realized that for there to be any hope of doing this, it would be necessary to work the furnace at elevated pressures. Hofman refers to a trial[2] made at Warren, New Hampshire, in 1906, in a furnace which was airtight at 80 pounds pressure per square inch, but which was unsuccessful. It was not until 1930 that Maier[3] published a thermodynamic examination of the problem, and showed that success was possible only at temperatures and pressures far beyond the reach of any conceivable practical process.

For these reasons, we decided at Avonmouth that the only hope of success would be to operate the furnace at normal pressures, allow the zinc to volatilize, and to tackle the formidable problems of condensing zinc in the metallic state from the vapors outside the furnace.

Development of the Imperial Smelting Process

The gas from a practicable blast furnace operation is necessarily much lower in zinc content than that typical of retort smelting, and contains a substantial proportion of carbon dioxide, about equal to the zinc content. Typically, the zinc may have to be condensed from a gas containing 5 per cent zinc, 7 to 10 per cent carbon dioxide, 25 per cent carbon monoxide, and remainder, nitrogen. The necessity for the presence of so much carbon dioxide may not be immediately obvious, but with charges in which the zinc oxide is in a reasonably reactive form, the competition within the charge of the reactions

$$ZnO + CO = Zn + CO_2$$

and

$$CO_2 + C = 2CO$$

one of which leads to the formation and the other to the elimination of carbon dioxide, results in practice in a final gas containing substantial quantities of carbon dioxide. Furthermore, the presence of carbon dioxide is welcomed, since if the carbon were burned only to carbon monoxide in the furnace, the heat balance would necessitate an excessive fuel requirement per unit of zinc evolved.

Compared with retort smelting, therefore, two circumstances which militate against condensation to liquid metal become much more serious:

1. Due to greater dilution, the tendency to form fogs in the condenser is increased. This leads to "physical" blue-powder formation.

2. Sufficient carbon dioxide is present, completely to reoxidize the zinc vapor. It is necessary to suppress the incidence of the reoxidation reaction.

The solution of this formidable condensation problem was found in the development of a lead splash condenser.[4] The zinc-bearing gases are scrubbed in countercurrent stages with liquid lead copiously sprayed through the gas. The hottest lead, with which the gases leaving the furnace are scrubbed, is at a temperature of 550 to 600°C; the shock chilling thus achieved allows so little time for the reoxidation reaction that very little zinc oxide is, in fact, re-formed. The fact that fog formation in the condenser is reduced to quite manageable proportions must be attributed in part to the circumstance that the gases are scrubbed with an unsaturated solution of zinc in lead; the physical action of the violent rain of liquid lead in the condenser is also probably important.

The lead splash condenser was, in fact, developed at Avonmouth to condense satisfactorily the zinc content of blast furnace gases. The zinc leaves the condenser in the form of a solution in liquid lead at about 550°C. To complete a continuous process, this lead stream is cooled to yield zinc, equal in quantity to that removed in the condenser, as a second liquid layer prior to return to the condenser.[5] The lead thus serves as a solvent for zinc and as a heat transfer medium, with the very convenient property, as regards recovery of the zinc, of yielding a two-liquid system on cooling. By using liquid lead it is also possible to use steel for the manufacture of the splashing units and circulating pumps.

In the furnace charge the equilibrium conditions for the reaction

$$CO + ZnO \rightleftharpoons CO_2 + Zn$$

are such that zinc oxide can only be reduced at temperatures above about 1000°C. Furthermore, if the gas resulting from blast furnace smelting passes through a region in which it is cooled below the equilibrium temperature, reoxidation of zinc vapor must occur. Accordingly, the type of furnace first developed at Avonmouth provided for introduction of air blast both at top and bottom of a furnace shaft, with removal of gases at a middle level.[6] A successful design of this type was evolved and large scale furnaces were operated in this way. However, further development of the process showed that it was feasible to operate a furnace blown with bottom tuyères only. The gas leaving the top of the charge has to supply the deficiency of heat corresponding to the difference between charge temperature and the minimum reduction temperature. By using preheated charge, it is possible to reduce the deficiency to moderate proportions; the gas does, however, leave the furnace charge below the reoxidation temperature. This does not prevent operation of the process and is advantageous in some respects. Special measures are necessary in order to convey the gas to the condenser.[7]

The development of the blast furnace process at Avonmouth has entailed the evolution not only of the condensing system and the furnace arrangement, but also of satisfactory types of charge. The fuel used has, in the main, been metallurgical coke. The metalliferous materials are charged as sinter and development has been directed toward a single sinter containing all fluxes, recycle material, etc. It will be appreciated that the requirement of well-sized dust-free sinter, typical of blast furnace operation, is of special importance in the zinc blast furnace, which is characterized by a hot top and by being linked at its gas exit to a condenser in which the metal value is recovered.

The development of a process on the above lines at Avonmouth, directed solely at the production of zinc from roasted concentrates, already involved the handling of leady materials in so far as the condenser used liquid lead as a scrubbing medium. The dross and blue powder obtained in the condensing operation contained lead as well as zinc, and the return of this material to the furnace (either directly or in sinter) involved enhancing the lead content of the furnace charge. Thus, it was a logical extension of the process to consider the treatment of materials containing both zinc and lead in order to recover both values. Theoretical study indicates the peculiar suitability of the blast furnace process for this joint smelting treatment. In practice, the treatment of lead-rich charges in the zinc blast furnace presented a number of difficulties in relation to zinc condensation; the presence of large amounts of lead with the residual sulfur from a sinter-roasting can under some conditions give rise to excessive volatilization of lead sulfide. The difficulties were overcome, however, with the result that metalliferous materials have been successfully smelted in blast furnaces to yield their zinc content as metal from the condenser and their lead and silver contents as bullion tapped from the furnace.

Present Operating Units at Avonmouth

After a number of initial operating difficulties with the experimental furnace had been overcome, the process began to show promise and shed some of its complications. At this stage, it was decided to build the first pilot production furnace with a nominal zinc capacity of 20 tons per day. Before construction of this first unit was complete, a decision was taken to build a second slightly larger furnace, with a nominal capacity of 25 tons per day. As development has proceeded, these units have been modified, but their basic dimensions are unchanged. They have, now, capacities of 35 and 45 tons of zinc per day, respectively, and although they still suffer, as production units, from their somewhat premature origin, they have between them produced over 85,000 tons of zinc. Representative views of the plant and No. 2 furnace are shown in Figures 6-48 through 6-51.

In many respects the furnaces resemble an orthodox lead blast furnace with a water-jacketed bottom section, water-cooled tuyères and a brick-lined upper shaft. The smaller furnace is fitted with 16 tuyères and has a hearth area of 55 square feet; the larger, with 20 tuyères and an area of 69 square feet. The furnaces are tapped into external settlers for collection of matte and bullion, and from which the slag overflows into a granulating launder.

The top of the furnaces differs from that of a standard lead furnace in that it must be totally enclosed to prevent any ingress of air. A double-

Figure 6-48. General view of blast furnace plant.

Figure 6-49. General exterior view of blast furnace No. 2 showing slag granulation pit. (*Courtesy The Institution of Mining and Metallurgy*)

charging bell system is fitted for this purpose. The preheated, weighed charge of sinter and coke is fed to these bells by means of skip hoists.

The gas from each furnace passes to two condensers, one on each side of the furnace. Each condenser is divided into three stages with a continuous

Figure 6-50. No. 2 blast furnace-zinc tapping floor. (*Courtesy The Institution of Mining and Metallurgy*)

pool of flowing lead in the bottom. In each stage, there is a rotor fitted to a vertical shaft and electrically driven.[8] The rotor generates a dense shower of lead droplets which rapidly cools the gas to a temperature approximating that of the lead. Almost all the zinc vapor is condensed to droplets which dissolve in the lead and can be removed from the condenser. The uncondensed zinc in the gas leaving the condenser is reduced to less than 5 per cent of that entering it.

The lead flows through the condenser in a direction countercurrent to that of the gas. The hot lead containing the condensed zinc is removed from the gas inlet end and is pumped through a water-cooled launder which reduces its temperature from 600 to 450°C. As the lead cools, the excess zinc separates from it and is collected in a separator bath as a pool on top of the lead flowing below. From this pool the excess zinc overflows continuously into a heated bath where it is treated with sodium for arsenic removal, and its temperature raised for casting into slabs. The lead, now cooled to 450°C, flows from underneath the zinc in the separator bath back into the condenser.

In order to cope with the very considerable quantities of heat which must be removed from the condensers, a high rate of lead circulation is required through the condenser and the external cooling circuit. In practice, this rate of circulation through each condenser is 300 to 400 tons of lead per hour.

The zinc produced is of Prime Western grade and assays: lead, 1.1 to 1.3 per cent; iron, 0.02 to 0.03 per cent; arsenic, 0.001 per cent. The cad-

Figure 6-51. No. 2 blast furnace-slagging operation. One-ton lead pig is shown in left foreground. (*Courtesy The Institution of Mining and Metallurgy*)

mium content of the metal depends upon the amount present in the charge. Most of the cadmium content of the sinter is collected in the output zinc.

The gas leaving the off-takes of the condensers is scrubbed with water in a spray tower, and then passes to Theissen-type disintegrators where it is further cleaned. The cleaned gas (calorific value 70 to 80 Btu per cubic foot) is burned on the plant to preheat the air and the charge. A surplus is available which is burned in boilers. The liquor leaving the spray towers and disintegrators is passed through a thickener which collects the solids as a low-grade blue powder. This is filtered, dried and returned to the sinter machine feed.

Operations at the furnace bottom are similar to those of orthodox lead practice. The furnace is tapped periodically as required, into an external settler. Matte and bullion are retained in this settler and the slag overflows into a granulating system. The slag composition varies somewhat with the type of gangue in the concentrates treated. A slag of low zinc content can be obtained if the lime content of the slag is adjusted to a value higher than that usual in lead blast furnace slags. The zinc content of the slag may range between 1 and 5 per cent, but with close control of charge composition, and coke of a reasonably consistent ash composition, slags containing 2 to 3 per cent zinc can be steadily maintained. Accurate proportioning of coke and sinter in the charge is essential. Conditions in the hearth are more reducing than in normal lead furnace practice; with an efficient settler the lead content of the slag is below 0.5 per cent.

The high proportion of coke in the charge results in a lower and more consistent back pressure of the charge than that characteristic of the other nonferrous blast furnaces. This makes it possible, and steady operation of the condenser makes it desirable, to operate the furnace at constant blowing rate rather than at constant bustle-pipe pressure. The coke consumption rate is, therefore, approximately constant. The proportion of sinter to coke in the charge is varied in order to correct variations in zinc content of the slag; e.g., the sinter proportion is reduced if the per cent zinc in slag rises, or is increased if the per cent zinc in slag falls below target, or if there are signs of heavy iron reduction. A more immediate corrective to over-reducing conditions is available by lowering the blast preheat temperature; this is equivalent to reducing the proportion of fuel in the charge. The preheat is restored to target when the effect of a change in charge composition becomes apparent in the slag.

Vertical-shaft preheaters are used to preheat the charge. Furnace gas is used for this purpose. This is burned in a combustion chamber with automatically controlled additions of air to insure a neutral atmosphere. The burned gas then passes up the shaft of the preheater and heats the charge. A proportion of the gas leaving the top of the preheater is recycled in order to prevent overheating at the base. This is again controlled automatically. Alloy tube recuperators also fired with furnace gas are used to preheat the air blast to a temperature of 550°C.

All the main operations of the process can be, and are, controlled automatically. Reference has already been made to the firing of the charge and air preheaters. The furnace is operated at a constant controlled blowing rate, and the gas leaving the furnace is split between the two condensers by automatic damper control.

Theory of the Process

In the blast furnace the solid charge consisting of a mixture of coke and metalliferous sinter reacts in countercurrent with the gas stream, and passes through the following stages:

1. A preheating stage, possibly very short, in which hot gases raise the input charge to the temperature required for zinc oxide reduction. In this stage, some reoxidation of zinc may occur, and reduction of lead oxide can take place.

2. A dry reduction stage, in which the temperature is high enough for zinc oxide reduction to take place, but below the slag melting point. Desirably, a high proportion of the zinc oxide reduction occurs here.

3. A melting and "wet" reduction stage in which the reacting system consists predominantly of coke with molten slag trickling downward into the slag pool. In this zone the final zinc reduction is effected. Clearly, the

behavior in this zone is critical for the attainment of the required zinc elimination, for good separation of lead bullion, for formation of matte, etc.

The over-all behavior in the furnace conforms to a heat balance with heat input comprising heat of combustion of carbon to a mixture of carbon monoxide and dioxide, and the sensible heat introduced in solid charge and air blast. The heat is consumed in heat losses, in slag melting, in heating the gases to a temperature not greatly below the minimum temperature for zinc oxide reduction, and in the desired reduction of zinc oxide. The quantity of zinc which can be produced per unit of carbon is affected to an important degree by the heat input in air blast and by minimizing the heat loss, mainly in the water-jacketed furnace bottom; clearly also, the greater the carbon dioxide percentage in the furnace gases, the more favorable is the heat balance for zinc production. The proportion of carbon dioxide actually obtained depends upon the extent and temperature of the different zones in the furnace, the amount of zinc oxide fed per unit of carbon, and the whole complex of conditions which can be termed the "working" of the furnace. Typically, the gas leaving the charge contains 5 to 6 per cent zinc, with 8 to 10 per cent carbon dioxide, corresponding to a reoxidation temperature of 960 to 1000°C.

To reduce zinc oxide charges with the desired fuel efficiency and at the same time to yield a slag of low zinc content, the relevant factors are:

1. To maximize the reduction in the dry zone. This requires a reactive, porous sinter, i.e., the avoidance of too dense slagging of sinter.
2. Suitable slag composition, involving considerably higher lime content than is generally used in lead smelting. Desirably, the slag melting point should not be too low.

When the conditions are such as to obtain slags of low zinc content, they are necessarily close to those for reduction of iron oxide in the charge. Excessive iron reduction can be troublesome; on the other hand the removal of arsenic in an iron-rich speiss and of sulfur in an iron-rich matte can be advantageous. The best combination of factors is obtained with slags melting above 1250°C, and such that the activity coefficients of both zinc oxide and iron oxide are high. Under such conditions only a small proportion of the iron content of the charge is reduced, so that it is not necessary, as in a retort process, to supply the considerable quantity of heat which would be needed for complete iron reduction.

Successful condensation requires avoidance of fog formation and of the reoxidation reaction: $CO_2 + Zn = CO + ZnO$. By shock chilling with a lead shower at about 550°C, followed rapidly by scrubbing in presence of lead down to 450°C, the reoxidation reaction is "frozen." Rapid cooling is normally a condition promoting fog formation. In the present instance, the tendency to form a fog is lessened by the high superheat of the inlet gases which have a dew point of about 650°C and leave the furnace some 300°C above this dew point, and by scrubbing with unsaturated lead. However, it is probable that droplets of zinc are, in fact, formed but are themselves scrubbed and dissolved in the intense rain of liquid lead.

The countercurrent scrubbing of the furnace gas containing about 5 per cent zinc is carried out in the condenser under virtually adiabatic conditions; the heat is removed entirely by external cooling of the lead stream. Under these conditions the entering lead stream increases both in temperature and in zinc content, but becomes progressively more unsaturated. A typical situation is:

Cold lead—440°C, containing 2.02 per cent zinc (saturated)
Hot gas—5 per cent zinc, 1000°C, to be cooled to 450°C
Hot lead—to leave condenser at 570°C.

The lead leaves with its zinc content raised by only 0.24 per cent, containing 2.26 per cent zinc compared with a saturation value of 4.9 per cent. The lead circulation in this case would be $100/0.24 = 420$ times the rate of zinc supply to the lead circuit, in order to maintain this situation.

This type of condenser successfully treats the gas from a zinc blast furnace. Under favorable conditions 89 per cent of the zinc vapor entering the condenser is recovered as molten metal. The remaining 11 per cent is recovered in drosses which settle out from the lead leaving the condenser (or in blue powder washed out of the gases leaving the condenser). Both the dross and blue powder contain lead. Typical analyses are: blue powder, 32 per cent zinc and 45 per cent lead; and dross, 43 per cent zinc and 36 per cent lead.

To obtain good condensation efficiency the furnace conditions must be such as to avoid excessive carry-over of dust and fume. Prevention of dust carry-over involves charging dust-free sinter of suitable strength; with a furnace top at about 1000°C the problem is greater than with the normal cold-top furnace. Fume is principally a matter of sulfides. Lead sulfide is highly volatile in the hotter regions of the furnace, but can react with oxides of iron or calcium in the furnace charge, or with zinc vapor. To obtain satisfactory condensation with lead in the charge involves establishing conditions in which the lead sulfide vapor is prevented by these reactions from leaving the furnace. It is possible to attain these conditions, so that a mixed lead-zinc charge is smelted with just as high an efficiency of condensation as a pure zinc charge.

The reduction of lead oxide by carbon monoxide is slightly exothermic. The presence of lead in the burden thus does not increase the thermal load on the furnace provided that the actual reduction corresponds to that by carbon monoxide. This prediction, that the lead content of the charge is a "passenger" requiring no fuel allocation for its smelting, has been realized in practice. When mixed lead-zinc charges are treated in the blast furnace, the proportion of fuel required is calculated from the amount of slag to be melted and the zinc to be eliminated. The simultaneous reduction of lead oxide results in the gas containing more carbon dioxide than would otherwise be present, with some loss of calorific value.

Application of the Process

The blast furnace process for zinc production has only recently attained commercial status and it is still too early to see clearly what part it will play in zinc metallurgy of the future. With large units, the capital cost and labor and maintenance charges will be lower than for the established processes. Using high-grade zinc concentrates, the process will, therefore, be competitive with its rivals at least in areas where furnace coke can be readily obtained. Such furnaces have, however, a large reserve of reduction capacity, and can simultaneously produce considerable quantities of lead, without affecting the carbon consumption or the quantity of zinc which can be produced. Again, while the process can treat high-grade concentrates, these are not essential. In fact, in certain cases a lower grade of material may be more profitable to smelt; the criterion is the self-fluxing characterization of the gangue.

In addition to the patents referred to in the text and enumerated in the references, there are other patents and patent applications covering various aspects of the process and modifications of these. Corresponding patents and patent applications exist in numerous foreign countries.

REFERENCES

1. Morgan, S. W. K., "The Production of Zinc in a Blast Furnace," *Trans. Inst. Mining Met.*, **66**, Part II, 553–65 (1956–57).
2. Hofman, H. O., "Metallurgy of Zinc and Cadmium," New York, McGraw-Hill Book Co., Inc., 1922; also p. 54 of Ref. 3.
3. Maier, C. G., "Zinc Smelting from a Chemical and Thermodynamic Viewpoint," *Bull. U. S. Bur. Mines*, No. **324**, 1930.
4. U. S. Patent 2,464,262.
5. U. S. Patent 2,671,725.
6. U. S. Patent 2,684,899.
7. U. S. Patent 2,682,462.
8. U. S. Patent 2,583,668.

The Waelz Process

C. W. Morrison

Manager
*Cherryvale Zinc Company, Incorporated**
Cherryvale, Kansas

Application of the Waelz Process

In the business of producing metals from ores and secondaries, the application of the Waelz Process is to produce commodities containing commercially important concentrations of certain elements—specifically, zinc, cadmium, lead and germanium—from source materials consisting of low-grade ores or secondaries. In the concentrates produced, the respective elements occur chiefly as oxides accompanied by minor quantities of sulfur compounds and sometimes chlorides of the respective elements. Thus essentially the Waelz Process is used as a method of beneficiating ores or secondaries.

Certain processes, such as deleading, desulfurizing, densifying and sintering zinc oxide fume or zinc sulfide concentrates, as well as the manufacturing of lime and Portland cement, make use of a rotary kiln and fume collection apparatus of the same types as are used in the Waelz Process. Occasionally such processes have been confused with the Waelz Process. However, because the methods of such processes are quite different from those of the Waelz Process, the term "Waelzing" is not descriptive of, or applicable to, them.

The fume produced by the Waelz Process is used for the pyrometallurgical production of slab zinc, after removal of much of the lead, cadmium and sulfur, which the crude product usually contains, by a heat treatment such as sintering on a Dwight-Lloyd machine or in a rotary kiln. The fume is also used for the electrolytic production of slab zinc after removal by wet methods of all or nearly all of the lead, cadmium, and other impurities. Waelz fume is also frequently used for making commercial grades of refined zinc oxide, zinc sulfate and lithopone.

The Waelz Process has been used recently for beneficiating horizontal zinc furnace residue containing germanium, to produce from the residue a fume containing commercially important concentrations of germanium. Frequently the germanium-containing Waelz fume is resmelted in horizontal retorts to produce slab zinc, and the retort residue resulting therefrom is Waelzed to produce a fume still higher in germanium tenor than the primary Waelz fume produced in this type of operation.

Very early in the evolution and development of the Waelz Process an

* Affiliate of the National Zinc Company Incorporated, Bartlesville, Oklahoma.

attempt was made in Germany to beneficiate low-grade South American tin ore by the application of this process. The attempt was only partially successful commercially because the Waelz Process was competing with conventional wet gravity concentration methods for attaining the same end.

Essential Parts of the Plant

A Waelz Process plant consists essentially of the following parts:

1. An installation for reducing the particle size of the various charge components to less than 3/8 inch thickness, with a minimum of fines production.
2. A battery of overhead cone or hopper-bottom feed bins equipped with proportioners for feeding continuously and separately, the mix components in measured quantities.
3. A device for blending the charge components.
4. A rotary kiln for heating and reacting the mix.
5. A chamber for settling out the fly ash carried over from the kiln and for burning the combustible constituents of the gas and fume evolved during the operation.
6. A flue for conducting and cooling the fume-laden gases from the combustion chamber (item 5) to a fume-collecting apparatus (item 8).
7. A fan, exhauster type, for impelling the fume-laden gases through the system.
8. A fume-collecting apparatus for removing from the gas stream the solid particles in the form of fume, such as zinc oxide, lead sulfate, germanium oxide and gangue material.
9. A slag-handling apparatus for transporting the Waelz slag from the kiln discharge to the dumping ground.

Metallurgy of the Process

Like many other extractive metallurgical methods, the Waelz Process method is simple in principle but complex in successful practice. In operation, zinc compounds are converted to elemental zinc, which is gaseous at temperatures maintained in the Waelz kiln. For example, the ore or secondary material entering the kiln may contain one or more of the following zinc compounds; oxide, silicate, carbonate, sulfide and sulfate. These zinc compounds then react with carbon and oxygen as exemplified by the following equations:

$$ZnO + C = Zn + CO$$

$$ZnSiO_3 + C = Zn + SiO_2 + CO$$

$$ZnCO_3 + 2C = Zn + 3CO$$

$$ZnS + 3O + C = Zn + SO_2 + CO$$

$$ZnSO_4 + 2C = Zn + SO_2 + 2CO$$

The bed of material formed by the reacting charge-mix progresses continuously through the operating kiln, impelled by the action of gravity supplemented by the action of the rotating motion of the inclined kiln. The charge-mix progresses first through a preheating region, where the moisture (water) it usually contains is evaporated, and the temperature of the bed is increased to its reacting temperature, approximately 2200°F. Increasing the temperature of the bed is effected by heat transfer from gases passing countercurrently over the traveling bed. The aggregate of particles constituting the bed then progresses through the reacting region of the kiln, where both reduction and oxidation of zinc and associated elements proceed. The reducing reactions occur beneath the surface of the traveling bed of material and the oxidation reactions occur on and above the surface of the bed, which is rolling continuously in the operating kiln, alternately submerging each of the particles constituting the bed and then causing it to cascade down the steeply inclined surface of the bed. Thus heat is alternately dissipated from, and imparted to, each particle of the charge-mix constituting the bed. When submerged, the heat is dissipated from the incandescent particle by conduction to, and eventually radiation from, the outer surface of the operating kiln, and by the effect of endothermic reactions.

The end products of the endothermic reactions between components of the charge of an operating kiln are gaseous zinc and cadmium; also, presumably, gaseous stannous oxide, stannous sulfide, germanium monoxide and carbon monoxide. There are also various solid or liquid metals in the elemental state; notably iron, lead, tin and germanium. Each time a particle of zinc oxide in the bed is submerged, a portion of it is reduced and the zinc in it volatilized, the remaining portion of the particle not being reduced because of the cooling effect of the heat losses previously described. Heat is imparted to the particle of charge when it is in motion on the surface of the bed of charge after emerging from below the surface of the bed. While on the surface of the bed the oxidizable particles, carbon, iron, lead, etc. are oxidized, the exothermic reactions generating heat. The heat imparted in this manner to the particle is supplemented by the heat transmitted from the preheated air entering the reaction region of the operating kiln at its discharge end, as well as by the burning of the gases generated beneath the surface of the bed as previously described.

The volatile elemental zinc and cadmium, as well as any stannous oxide, stannous sulfide and germanium monoxide which may have been formed, are oxidized in part on reaching the zone in the kiln above the bed, to the extent that available oxygen exists in that region. The products of oxidation are zinc oxide, cadmium oxide and the dioxides of tin and germanium. These oxidized materials form a fume consisting of very small solid particles which travel through the kiln suspended in the gas stream, ultimately flowing out through the opening in its upper or feeding end. In the upper reacting region of the operating kiln, available oxygen in the gas stream passing over the bed is insufficient for oxidizing the gases generated below the surface of the bed. These unoxidized gases, some of which may be condensed to fume particles flowing out of the kiln, are ultimately burned in the combustion chamber. This chamber is maintained at a temperature above the ignition point of the unoxidized gases and of fume entering it from the kiln. Air is deliberately admitted to the combustion chamber, preferably through the annular opening between the shell of the kiln and the wall of the combustion chamber.

Nonvolatile elements, such as elemental iron, lead, tin and germanium, are not oxidized in the upper reacting region because of insufficient oxygen. However, considerable oxidation of all these components can occur both below and above the surface of the bed in the downstream portion of the reacting region of the operating kiln, because of the presence there of a large supply of free oxygen. As a result of these oxidizing reactions in that portion of the reacting region, much of the elemental iron formed in the upstream region is oxidized to ferrous oxide, which in the presence of free silica combines with that component to form iron silicate. When the iron tenor of the charge is very high, a portion of the iron frequently escapes oxidation and is discharged from the kiln in the form of "cast iron balls." Also resulting from these oxidizing reactions, elemental lead, tin and germanium previously formed in the upstream reacting region are oxidized to their respective monoxides, which are gaseous at the operating temperature in this downstream portion of the reacting region of the operating kiln. Subsequently, on the surface of the bed and in the zone above the bed in this downstream region, certain gaseous components, formed as described, are oxidized to a fume consisting of solid particles of zinc oxide, cadmium oxide and dioxides of tin and germanium. The lead monoxide remains gaseous until it is condensed after reaching a zone in the system where the temperature is below its dew point.

General Specifications of Typical Equipment

The crushing equipment frequently consists of an apron feeder, a conventional jaw crusher, rolls, and a screen in closed circuit. The capacity is 150 to 200 tons per 8 hours.

Four or five overhead feed bins and feed proportioners are usually provided: one for the crushed source material; one for crushed coke breeze or low combustible volatile matter coal; one for material to be recycled—dust from the combustion chamber and occasional spill from the feed end of the kiln; and one or more for conditioning reagents such as silica in the form

Figure 6-52. Waelz kiln, with feed elevator and tanks.

of sand or lime in the form of limestone, which may be needed for altering the angle of repose of the bed composed of the charge-mix in the operating kiln, as noted in a following paragraph.

Material-handling equipment for the charge-mix usually consists of belt conveyors, belt or chain bucket elavators and an inclined water-jacketed feed spout, with an internal diameter of about 10 inches, inclined about 30° from the vertical and provided with an expendable steel liner on the bottom, which is subjected to the abrasive effect of the descending charge-mix.

The type of rotary kiln used is identical to the kilns used in the manufacture of Portland cement. While the diameter of these kilns may be as much as 11 feet, and the length as much as 150 feet, the size most commonly used is: length, 125 feet; shell diameter, 8 feet. Inclination from horizontal varies in different installations from about 3 to 4 per cent. The kiln is usually provided with a 9-inch thick lining made of refractory brick such as high-duty fire clay, silica, etc. In order to prolong the life of the refractory lining, the kiln shell is often cooled by water sprays, and in at least one case was completely water-jacketed. Both accomplish the desired result at the cost of lowered heat efficiency.

The opening in the feed end of the kiln is typically about 5 feet in diameter for a kiln of the indicated dimensions. An excessively large opening makes it necessary to operate with a pressure in the upper portion of the combustion chamber resulting in (1) escape of fume and gas to the ambient air through tbe upper portion of the annular opening between the wall of the kiln and the wall of the combustion chamber, where the upper end of the kiln projects into it; and (2) overheating of the steel shell of the end of the kiln due to heat transmitted from the escaping hot gases. An excessively small opening obviously restricts the draft too much for optimum operating results.

The combustion chamber for a Waelz kiln consists of a fire-clay brick structure, with the following inside dimensions: about 18 feet high, 7 feet wide, and 20 feet long. This chamber is frequently provided with a curtain wall made of high-duty fire-clay brick, the function of which is to protect the end of the kiln from the effect of the heat generated in the combustion chamber. This wall also functions to separate the spilled material, when spillage of the charge occurs, from the fly ash dropping out in the combustion chamber. The curtain wall is situated adjacent to the end of the kiln, a distance of from 6 to about 10 inches. Structurally, the curtain wall is a portion of the same wall through which the feed end of the kiln projects. The axis of the opening in the curtain wall is roughly in line with the axis of the kiln, the diameter of the curtain wall opening being a little larger than the inside diameter of the kiln.

The flue for conducting and cooling the gases flowing out of the combustion chamber is usually a steel tube 60 to 45 inches in inside diameter for a kiln of the indicated dimensions. The flue is provided with a fire-clay brick lining from 2½ to 4½ inches thick in the 150-foot portion nearest the combustion chamber. The remainder of the flue and return bend coolers are unlined and constructed of 10- or 12-gage steel sheet. Even in hot weather, an area of approximately 15,000 square feet usually provides sufficient surface for cooling the gas from a temperature of about 1500°F, at which it enters the flue, to a temperature about 250°F, at which it enters the bag filters. Most of the cooling surface is usually provided in the form of return bends equipped with dampers for by-passing the coolers to aid in temperature control.

The exhauster-type fan for impelling the gas is preferably situated so that it will discharge into the bag filters. Capacity for the kiln of the indicated dimensions is about 25,000 cubic feet per minute at 5 inches water gage static head. A 50-hp motor is required to drive an exhauster of this capacity handling fume-laden gases at a maximum temperature of 300°F. The maximum fume burden of the gas at the outlet of the exhauster at the maximum temperature is approximately 0.0025 pound or 17.5 grains per cubic foot at the stated temperature.

A bag-filter installation is preferably used for filtering the fume out of the gas. The capacities of bag filters vary from one to two cubic feet through-put of gas per square foot of bag area, for economical operation considering the stipulated fume burden. The well designed mechanical,

Figure 6-53. Fume cooling trail and "hairpin" coolers. Baghouse in background Shown during construction.

automatic types of bag-cleaning equipment have the most capacity per unit of filtering area. Gas temperature at inlet of bags is usually maintained above 220°F and below 300°F.

The incandescent material discharging from the kiln consists of a porous, semifused mass of particles varying in size from fractions of an inch to several inches, after quenching. Its composition depends on the composition of the gangue in the source material and the composition of the additives, coke, and conditioning reagents. An economical method of handling the Waelz slag is to provide for discharging it from the kiln into a hopper-bottom quenching pit from which it is removed continuously in operation by a chain-drag conveyor. The chain-drag conveyor then delivers the wet slag to an inclined belt conveyor delivering to a heap. From the heap the slag then is transported to the dump by various means: by a dump truck, a LeTourneau carry-all, or Sauerman-type drag line scraper, for example.

Operating Practice

Usually the charge-mix in operating practice for the purpose of making a zinc concentrate consists of: (1) ore or secondaries, the zinc tenor of which may vary from 8 to 35 per cent, and (2) sufficient coke or low-combustible volatile-matter coal to make the fixed carbon tenor of the mix about 20 per cent minimum. About 65 per cent of this carbon is oxidized in operation, the remainder forming a constituent of the residue or so-called "Waelz slag," which, in good operation, contains less than 2 per cent zinc. A Waelzing installation, consisting of a rotary kiln (8 feet in shell diameter by 125 feet in length) and its auxiliary equipment, has sufficient capacity to produce about 700 tons of Waelz fume per month, with zinc tenor 60 to 70 per cent, depending on the composition of the raw material. In addition to the reducing agents, coke or coal, it is frequently necessary to add one or more conditioners to the mix. One purpose of the conditioners is to make the mix, at the operating temperature, sufficiently sticky to maintain a slope of the tumbling heap of the mix in the operating kiln at approximately 40° from the horizontal. Another purpose of an additive is to make the slag less corrosive with respect to the kiln lining and thus reduce the cost of kiln-lining maintenance.

Silica or one or both of the two bases (lime or iron) may be used as a conditioner. If at the optimum operating temperature the mix melts, becoming so fluid that the angle of repose is too flat (the circumferential slope of the heap being less than 40° from the horizontal), then an additive, to make the mix less fluid, would be needed. If, on the other hand, the mix is so refractory at the operating temperature that it does not become somewhat sticky through incipient fusion, it would flow like fluidized dry powder, the angle of repose becoming excessively flat and the slope less than the desired 40° from the horizontal. In this latter case an additive would be required to promote stickiness through incipient fusion. Through elimination of the zinc and other volatile constituents, the composition of the mix traveling through the kiln continuously changes, causing a continuous change in its physical properties. The elimination of the zinc causes the mix to become progressively less refractory.

Accretions invariably form on the wall in the upper portion of the reacting zone of an operating kiln. These accretions are frequently very hard and may grow to such a thickness that the kiln must be stopped, cooled, and barred out manually. However, it is the usual practice to adjust the draft in the operating kiln so that an oxidizing zone is established in the region where the troublesome accretions occur. The result of this action is to ignite the particles of carbon incorporated in the accretions. This burning of the carbon particles so reduces the structural strength of the accretions that they usually break and fall out. Accretions seldom form in the zone at the discharge end of the operating kiln, because in that zone, due to the presence of a large supply of free oxygen for burning fixed carbon, the temperature on the wall of the kiln is at or near the maximum, relative to temperature of the atmosphere radially adjacent to it. On the other hand, the wall-temperature in the highly reducing zone of the kiln is at the minimum relative to the temperature of the atmosphere radially adjacent to it. Therefore, in the latter case, the particles in the charge-mix may partially fuse when, as they tumble in the heap or bed of the operating kiln, they are most remote from the wall; they may then freeze to the relatively cool wall or to the already formed mass of accretions as they come in contact with it.

It is, of course, the practice to supply heat at the lower or discharge end of the operating kiln, in addition to the heat generated by the carbon burning in the charge-mix. In operating a kiln of the indicated dimensions, about 200,000 cubic feet per 24 hours of 1,000-Btu-per-cubic-foot natural gas, or the equivalent in calorific power of fuel oil or pulverized coal, are usually required as fuel for this purpose.

It is the usual practice to rotate the kiln at a peripheral speed of about 18 feet per minute. A 30-hp motor is usually required to handle this load in operation.

The items of direct cost in Waelz kiln operation in the order of magnitude are labor, labor supervision, maintenance and operating supplies, fuel, and power.

ACKNOWLEDGEMENT

Acknowledgment is given to W. H. Leverett, President of the Cherryvale Zinc Company and National Zinc Company for helpful suggestions, and to E. M. Villareal, Jr., President, Zinc National, S. A., Monterrey, Mexico, for permission to use the illustrations herein.

REFERENCES*

1. Dedolph, Eduard, U. S. Patent 959,924 (May 31, 1910). (Extracting lead or zinc as oxide in the form of fume.)
2. Harris, W. E., "The Waelz Process," *Trans. Am. Inst. Mining Met. Engrs.*, **121**, 702–720 (1936).
3. Hayward, Carle R., "An Outline of Metallurgical Practice," pp. 263–4, Princeton, D. Van Nostrand Co., Inc., 1952.
4. Hoffmann, R., "The Waelz Process," *Trans. Am. Inst. Mining Met. Engrs.*, **76**, 537–53 (1928).
5. Jensen, C. W., "The Waelz Process," *Mining Mag.*, **92**, 73–9 (1955).
6. Johannsen, Friedrich (to Krupp Grusonwerk Aktiengesellschaft), U. S. Patent 1,618,204 (February 22, 1927). (Recovery of volatilizable metal.)
7. Liddell, Donald M., "Handbook of Nonferrous Metallurgy," Vol. 2, pp. 441–2, 464, New York, McGraw-Hill Book Co., Inc., 1945.
8. Samans, Carl H., "Engineering Metals and Their Alloys," p. 151, New York, The Macmillan Co., 1949.
9. Stephani, Hermann, (to Krupp Grusonwerk Aktiengesellschaft), U. S. Patent 1,755,712 (April 22, 1930). (Recovery of volatilizable metals.)

* Used generally, without citation, by the author and included for supplementary reading.

Extraction from Slag

HAROLD E. LEE

Vice President, Research and Development
The Bunker Hill Company
San Francisco, California

and

W. T. ISBELL

Consultant
St. Joseph Lead Company
New York, N. Y.

Processes for the extraction of zinc from metallurgical slags utilize the high vapor pressure of metallic zinc at elevated temperatures. Reduction of the zinc oxide component of the slag is usually accomplished by a dual reaction with secondary carbon monoxide and residual solid carbon resulting from controlled fuel combustion. While the exact reduction mechanism is a subject of speculation, it has been demonstrated that, under a given set of conditions, maximum fuming rate performance is related to the maintenance of a certain optimum carbon to carbon monoxide ratio.

The greatest bulk by far of the zinc recovered from metallurgical slags is produced from coal-fired systems wherein pulverized coal is air-blown into molten slag and the distilled zinc reoxidized and collected in a baghouse. A minor tonnage is recovered in the metallic state as in the electrothermic process developed by the St. Joseph Lead Company[8] at Herculaneum, Missouri.

COAL-FIRED SLAG FUMING PROCESS

General Practice

While having undergone many subsequent operational and facility refinements, the coal-fired slag-fuming process is basically unchanged from its first commercial application by The Anaconda Company (at East Helena, Montana) in 1927. A coal-air mixture is introduced, via multiple double inlet tuyères,[6] into a molten slag bath contained in a rectangular, completely water-jacketed furnace. The tuyères are equispaced, along the longitudinal sides, in a plane just above the furnace bottom. The coal-air flow is continuous, and slag charging and tapping cyclic. On an average, about a two hour "blowing time" between "charging" and "tapping" operations is required for adequate zinc elimination from the slag bath.

Between cycles, the molten feed slag is stored either in a holding furnace or in 5 to 6-ton pots.

Coal in the slag fuming process is utilized both for reduction and for the generation of sufficient heat to maintain the slag bath in a fluid state. Roughly, the coal requirement approximates 20 per cent of the slag weight. With a given type coal and a given flow rate, the corresponding weight of air employed approaches that required for the combustion of the contained carbon to carbon monoxide. While departures from this optimum coal-air ratio are practiced for limited time intervals to attain special effects,* in general, lesser air flows tend toward over-reduction and a viscous slag bath; conversely, greater air flows tend toward under-reduction and excessive fuel consumption.

Plant Types

Coal-fired slag fuming plants fall into two categories; those employing "indirect" coal feeding systems and those using "direct" coal feed. Indirect installations use a separate drying and grinding plant for preparing coal feed. The dry, pulverized coal is stored in intermediate bins for subsequent introduction into the furnace feeder lines. This practice is typified by the operations of The Anaconda Company at East Helena, Montana, The Consolidated Mining and Smelting Company of Canada Limited at Trail, B. C.[2, 3] and at the International Smelting and Refining Company plant at Tooele, Utah.

More recent installations, such as the Asarco plants[14] at El Paso, Texas; Selby, California;[15] and Chihuahua, Mexico; and the Hudson Bay operation at Flin Flon, Manitoba,[18] have adopted the direct coal-feeding system developed by the Bunker Hill Smelter[7, 9] at Kellogg, Idaho, in 1943. In these latter operations, combined drying and grinding of coal in Babcock and Wilcox pulverizers for direct entry into furnace feed lines is practiced. Raw coal is proportioned into the pressurized pulverizers which are directly connected to the furnace feed lines. Drying and transport of the pulverized coal are accomplished by a flow of preheated "primary" air which sweeps through the grinding mills from peripheral pulverizer wind boxes.

Reduction Practice

In all coal-fired plants, tuyère air is supplied from an adequate source in two flows, one termed "primary" and the other "secondary." Primary and secondary lines lead to corresponding separate, distributing headers which span each furnace side longitudinally. The primary air headers

* Some plants, particularly those charging appreciable tonnages of cold slag and/or pot shell, vary the coal-air ratio during each furnace cycle. A greater air flow than that defined above is used early in the cycle to develop extra heat for the melting of the solid slag charged. Somewhat lesser volumes are then later employed to effect the over-all degree of reduction sought.

connect to tuyère inlets nearest the furnace wall and the secondary headers to the outer tuyère inlets. Coal is continuously proportioned into the primary air flow, upstream from the distributing manifolds, and its subsequent introduction into the slag bath is facilitated by the sweeping action of a greater secondary air volume flowing from behind.

At the slag temperature necessary for sustaining economic fuming rates (2100 to 2200°F) and under conditions of limited oxygen supply, the indicated reactions within the slag bath are:

1. $C + O = CO$

2. $ZnO + CO = CO_2 + Zn$

3. $CO_2 + C = CO$

Product Handling Practice

As already noted, zinc from coal-fired slag-fuming operations is recovered as zinc oxide. The rising, zinc vapor-carbon monoxide mixture is oxidized above the slag bath by a "tertiary" air supply, either blown into the upper furnace chamber or drawn into the system through the charge-hole and/or other available openings by the provision of sufficient draft.

In the high-temperature range, immediately after the furnace, waste-heat boilers are employed where steam is of utility. Otherwise gas cooling is effected by high-pressure water sprays. At intermediate temperatures, cooling towers are used, with final baghouse temperature control effected by air quenching through automatically operated intake louvers.

Except at the Tooele, Utah, plant, gas filtration is accomplished in automatic baghouse units, usually employing Orlon bags. Continuous mechanical discharge and collection of precipitated zinc oxide dust are also generally practiced over the entire gas-transport and baghouse systems.

Slag tailing, containing on an average about 1 to 1.5 per cent residual zinc, is tapped at the close of each charge cycle either into pots or into water granulation systems for subsequent disposal to waste or to dump storage.

Zinc Oxide Disposal

Most metallurgical slags treated originate from lead smelting operations and the assay of the recovered zinc oxide ranges between 67 to 70 per cent along with 7 to 10 per cent lead. While the bulk of zinc oxide production from slag-fuming operations is disposed directly to either electrolytic zinc plants for recovery as slab zinc, or to pigment market outlets, an appreciable tonnage is shipped to zinc retort plants.

Both economic factors and impurity-density limitations require the removal of the bulk of the contained lead from fume shipments destined for zinc retorting. Deleading and densification of slag fume is effected by

Table 6-8. Metal Elimination Rate During Typical Furnace Cycle

Period	Time Elapsed (min.)		Slag Charged (lb)	Slag Assay (per cent)							Zinc (lb)				Residual Slag (lb)	Rate of Elimination of Zinc (lb per min.)	
											Charged	Fumed		Residual			
				Pb	PbO	Zn	ZnO	FeO	CaO	SiO_4		Interval	Accum.			Interval	Accumulative
Charging	Pot 1	0															
	Pot 2	6															
	Pot 3	11															
	Pot 4	16	69,200	2.1	2.26	15.7	19.66	35.0	13.4	22.5	10,910						
	Pot 5	20															
	Pot 6	24															
	Pot 7	28															
	Pot 8	32															
Blowing		33		0.6	0.65	10.5	13.70					4,334	4,334	6,576	62,626	131.3	131.3
		44		0.3	0.32	8.9	11.08					1,148	5,482	5,428	60,986	104.4	124.6
		54		tr.		7.5	9.34					958	6,440	4,470	59,595	95.8	119.3
		64		tr.		5.5	6.85					1,279	7,719	3,191	58,017	127.9	120.6
		79				4.0	4.98					916	8,635	2,275	56,862	61.1	109.3
		94				3.2	3.98					474	9,109	1,801	56,273	31.6	96.9
		119				2.3	2.86					522	9,631	1,279	55,623	34.8	80.9
Tapping	Start	134				1.4	1.74					509	10,140	770	54,986	33.9	75.7
	Finish	144		tr.		1.4	1.74	44.8	15.9	28.0		0	10,140	770	54,986	0	70.3

heat treatment in kilns at 2200 to 2300°F. In the passage through the kiln, a mixture of coke (1 to 2 per cent by weight) and zinc oxide dust, with a bulk density of about 40 pounds per cubic foot and a lead content of 7 to 10 per cent, is converted to a discharge product at 185 pounds per cubic foot and about 1 per cent contained lead. Natural gas or oil is used to supply the extraneous heat requirement and the lead-rich kiln fume is filtered in an auxiliary baghouse and recycled to the lead smelting system.

Operational Data

As might be expected, the fuming rate varies for a maximum at the time of charging to a minimum at the time of tapping. And the accumulative fuming rate, assuming given operational conditions, is directly related to the weight of zinc contained in the charge. Pertinent operational data are presented by Tables 6-8 and 6-9, and Figure 6-54. Table 6-8 illustrates

TABLE 6-9. PLANT CONDITIONS EXISTING DURING TYPICAL FURNACE CYCLE

Slag charge, tons	34.6
Cycle, min. {Charging, 32; Blowing, 102; Tapping, 10}	144
Treatment rate, 10 cycles per day, 346 tons hot slag per day	
Zinc:	
Content of slag heads, lb	10,910
Content of slag tails, lb	770
Eliminated during cycle, lb	10,140
Eliminated per minute, lb	70.3
Quantity eliminated, per cent	92.8
Coal consumed, lb:	
During cycle	14,500
Per lb zinc eliminated	1.45
Steam generated, lb:	
Per hour @ 240-lb gage	36,300
Per pound coal	6.0
Average air volume to tuyères, cu ft per min.:	
Primary	3,850
Secondary	4,900
Total	8,750
Average gas analysis at economizer outlet, per cent:	
CO_2	10.4
O_2	8.0
CO	0.0
N_2	81.6
Baghouse drafts cond., in. water:	
Inlet	2.5
Outlet	3.5
Differential	1.0

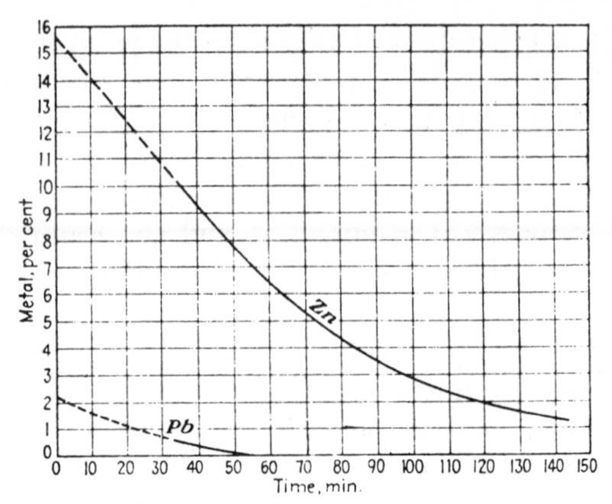

Figure 6-54. Metal elimination tests, run 221-7. Furnace charge 455.

operational procedure and the progressive fuming rates experienced during a typical furnace cycle. Table 6-9 presents a recapitulation of pertinent data accumulated during the same operational cycle. Metallurgical performance during this cycle is depicted by the metal-elimination curves of Figure 6-54.

ELECTROTHERMIC SLAG FUMING

After several years of experimentation, mostly concerning furnace-wall corrosion problems, a completely water-jacketed (except for arch) electrothermic slag-fuming unit,* 37 feet 9 inches by 22 feet 2 inches by 18 feet 8 inches, was placed in commercial operation[11] in 1955 at the Herculaneum, Missouri, smelter of the St. Joseph Lead Company.

Here, as in the case of coal-fired operations, batch charges are treated, the molten slag heads being charged at one end of the furnace and the slag tailing tapped from the opposite end. Three 30-inch diameter electrodes on 9-foot centers, operate through heavy, water-cooled cast copper aperture jackets which are symmetrically placed in a 15-inch thick firebrick sprung arch. The aperture castings also serve as support for electrode sealing devices. Four additional water-cooled jackets are provided along the crown of the arch to facilitate openings for the charging of coke.

Power for the maintenance of the slag bath in a fluid state is supplied to the electrodes through a multitap 5000-kilowatt transformer. The electrodes are supported by multishoe clamps and desired power input and electrode distribution are automatically maintained by balancing voltage and current relays. The furnace operates as a resistance unit at a high power factor.

Reduction is provided by coke charged to the bath surface and the

* U. S. Patents 2070101, 2766114 and 2766034.

resultant gas-zinc vapor is vented through a 25-ton per day capacity Weaton-Najarian zinc condenser, fitted with an external cooling well. Condenser zinc is tapped at above 500°C directly into a 90-ton capacity liquating furnace where it is cooled to about 450°C for the separation of excess lead and iron contaminant. Prime Western grade zinc is tapped from the liquation furnace and cast into 60-pound slabs for marketing.

Gas venting the condenser passes through a baffled washer and impingement scrubber before discharge through a vacuum pump system (at 80 per cent carbon monoxide) to a combustion flare. Blue powder collected in the scrubber and auxiliary settling ponds, accounting for about 45 per cent of the lead contained in the feed slag, is circulated back to the lead sinter plant.

Another 50 per cent of the slag-lead content is recovered from the furnace as metallic lead. Copper in the slag reports as a low-grade matte which is tapped through a water-cooled graphite block located in the furnace side. The metallic lead is sent directly to the lead refinery and the copper matte to the lead blast furnace.

Pertinent Herculaneum operational data are listed below:

Material	Tons per Day	Per Cent Analysis								
		Pb	Cu	Zn	CuO	MgO	SiO_2	S	FeO	Al_2O_3
Slag heads	150	2.29	0.63	13.8	8.2	4.3	24.9	3.0	34.6	4.8
Slag tailing	98	0.38	0.5	4.75	10.5	7.8	26.5	2.3	38.5	9.0
Matte	23	2.14	7.2	5.1	3.4	1.8	9.6	13.3	52.4	3.2
Blue powder	—	8.9	0.1	46.6	1.4	—	8.4	9.1	3.3 Fe	—
Slab zinc	12	1.25	—	—	—	—	—	—	0.033 Fe	—

Coke = 4.6 tons per day Power = 4.7 kwh per lb Zn Zn Recovery = 72%

REFERENCES*

1. Courtney, Guy, "The Recovery of Zinc from Lead Blast Furnace Slags," *Proc. Australasian Inst. Mining & Met.* (N. S.), **No. 38,** (1920).
2. Murray, G. E., "The Recovery of Zinc and Lead from Blast Furnace Slags at Trail, B. C.," *Trans. Can. Inst. Mining Met.*, **36,** 75–103 and 562–5 (1933).
3. McNaughton, R. R., "Slag Treatment for the Recovery of Lead and Zinc at Trail, B. C.," *Trans. Am. Inst. Mining Met. Engrs.*, **121,** 721–36 (1936).
4. Sokolov, V. S., "Reverberatory Melting of Slags," *Tsvetnaya Met.*, **14** (No. 7), 88–92 (1939).
5. Brenthel, F., "Utilization of the Zinc, Lead and Iron Content of Lead Slags," *Metall u. Erz*, **36,** 355–61 (1939).
6. Laist, A., and Baldwin, E. M., "Furnace with Tuyères, Suitable for Zinc Recovery from Molten Lead Smelter Slags," (to Anaconda Copper Mining Company), U. S. Patent 2,334,878 (Nov. 23, 1943).

* Cited in part by the authors and arranged for supplementary reading.

7. Feddersen, P. C., Schuettenhelm, J. B., Lee, H. E., Gittinger, D. R., and Dunn, G. W., "A Modern Practice of Removing Zinc from Slag," *Mining Congr. J.*, **29,** 22–31 (Oct. 1943).
8. Isbell, W. T., and Long, Carlton C., "Direct Production of Metallic Zinc from Lead Blast Furnace Slag," *Trans. Am. Inst. Mining Met. Engrs.*, **159,** 176–81 (1944).
9. Feddersen, P. C., Schuettenhelm, J. B., Lee, H. E., Gittinger, D. R., and Dunn, G. W., "Design and Operation of the Bunker Hill Slag-treatment Plant," *ibid.*, **159,** 110–26 (1944).
10. Hodgins, B. H., "Automatic Control Features Zinc Fuming Plant," *Eng. Mining J.*, **145,** 88–90 (Apr. 1944).
11. Weaton, G. F., and Isbell, Wm. T., (to St. Joseph Lead Co.), British Patent 600,434 (Apr. 8, 1948).
12. Dennis, W. H., "Zinc Volatilization," *Mining Mag.* (London), **80,** 80–5 (1949).
13. Editorial, "Fuming Plant Treats 12% Slag," *Eng. Mining J.*, **150,** 115 (July, 1949)
14. Editorial, "Asarco, Smelter Expansion," *ibid.*, **154,** 135 (July 1953).
15. Editorial, "Slag Fuming at Selby," *ibid.*, **154,** 95–7 (Dec., 1953).
16. Bell, R. C., Turner, G. H., and Peters, E., "Fuming of Zinc from Lead Blast Furnace Slags, a Thermodynamic Study," *Trans. Am. Inst. Mining Met. Engrs.*, **203,** 472–9 (1955).
17. Ellerman, Roy, "Processing of Zinc Oxide Fume at Flin Flon, Manitoba," *ibid.*, **203,** 813–22 (1955).
18. Mast, R. E., and Kent, G. H., "Slag Fuming Furnaces Recover Zinc and Lead from Copper Slag," *J. Metals* (Am. Inst. Mining Met. Engrs.), **7,** 877–84 (1955).

Sources and Recovery of Scrap Zinc

J. N. Pomeroy

Vice President, General Smelting Company
Philadelphia, Pennsylvania

and

J. E. Crowley

Division Head, Chemistry, Research Department
Bethlehem Steel Company
Bethlehem, Pennsylvania

Zinc is recovered from various types of scrap and the resultant products are slab zinc, zinc dust, zinc alloys, zinc oxide, zinc sulfate, zinc chloride, zinc ammonium chloride and lithopone. The slab zinc falls into three categories, namely, redistilled zinc, remelt zinc and remelt die-cast slabs.

The zinc scrap used in the various products is derived from galvanizers' dross, die castings, skimmings or ashes, sal ammoniac skimmings, chemical residues, flue dust, die-cast skimmings, engravers' plates, new clippings

TABLE 6-10. COMPOSITION OF SCRAP ZINC MATERIALS

Type of Material	% Composition Range							Notes
	Zinc	Lead	Tin	Copper	Iron	Chlorine	Aluminum	
Galvanizers' dross	93/96	0.3/2.5	Nil/0.2	Nil	3/5	Nil	Nil/1.5	
Die castings	90/95	Nil/0.5	Nil/0.3	Nil/2.5	Nil/2.5	Nil	3.5/8.0	(a)
Skimmings or ashes	60/85	0.3/2.5	Nil	Nil	0.2/1.5	2/12	Nil/0.3	(b)
Sal skimmings	35/60	0.3/1.0	Nil	Nil	0.2/1.5	15/35	Nil	(c)
Chemical residues	72/76	Nil/0.2	Nil	Nil	Nil	Nil/0.1	Nil	(d)
Flue dusts	35/55	0.3/9.0	0.3/4.0	0.3/1.5	0.5/1.5	0.1/0.7	Nil/0.2	(e)
Diecast skimmings	45/65	Nil	Nil/0.2	0.2/2.5	0.5/2.5	1.0/3.0	1.0/5.0	(f)
Engravers plates, new clippings, old zinc	Plates and clippings correspond to rolled zinc compositions and old zinc may be anything from Prime Western to die-cast.							

[a] Also contains up to 0.6% Ni.
b " " " " 0.5% NH_3 and from 1% to 50% metallics.
c " " " " 3% to 7% NH_3 and from 1% to 15% metallics.
d " " " " 0.6% combined sulfur and 1% water soluble salts. When in the form of carbonate it contains 55% Zn and 23% CO_2.
[e] Also contains up to 5.0% metallics.
f " " " " 0.3% Ni and from 3% to 10% metallics.

and old zinc. The composition, source and end use are shown in Tables 6-10 and 6-11.

Galvanizers' Dross

This material is a metallic alloy of iron and zinc plus some unalloyed zinc. The iron is derived from the steel being treated, the galvanizing pot and from flux which may be on the work. The alloy has a greater specific gravity than zinc and liquates to the bottom of the galvanizing pot. As it settles, free zinc is entrained. Also, as the dross rests on a layer of molten lead, some of this element is also entrained.

In the continuous process for galvanizing sheets aluminum is used as a minor addition to the zinc. The resulting zinc-aluminum-iron alloy has a lesser specific gravity than zinc and the dross rises to the top of the zinc bath. Such dross is also contaminated with oxides from slab zinc which naturally rise to the top of the pot, as well as with oxides formed by contact with air at the metal surface. A small portion of this dross settles to the bottom of the pot.

Dross is removed from the galvanizing pots manually with spoons and is cast into slabs. These slabs are used directly in the secondary smelting processes.

While the rate of dross formation is less in the continuous process, the tonnage of dross produced from all sources has not decreased. However, the quality of dross has deteriorated which results in lowered yields in the secondary recovery processes.

TABLE 6-11. DISTRIBUTION OF 1954 ZINC SCRAP CONSUMPTION BY TYPE OF PRODUCT, GROSS WEIGHT IN SHORT TONS OF SCRAP CONSUMED*

Type of Scrap	Product in which scrap was used									
	Redistilled Slab	Remelt Spelter & Rolled Zinc	Remelt Die-cast Slab	Misc. Zinc Alloys	Zinc Dust	Zinc Oxide Lead-free	Zinc Chloride[a]	Misc. Zinc Chemicals	Brass and Bronze	Total Zinc, Scrap Consumption
Galvanizers' dross	17,221	397	—	20	29,028	8,930	—	—	—	55,596
Die cast and rod and die scrap	9,566	—	12,825[c]	113	—	5,964	—	42	307	28,817
Skimmings and ashes	31,918	478	—	136	—	—	4,665[b]	—	—	37,197
Sal skimmings	449	—	—	—	—	—	20,066	111	—	20,626
Chemical residues	—	—	—	142	—	16,228[d]	—	350	—	16,720
Flue dust	16,083	—	—	—	—	—	—	1,343	—	17,426
Die cast skimmings	4,469	—	3,203	1,779	—	—	—	—	—	9,451
Engravers' plates	768	465	58	228	—	—	—	134	272	1,925
New clippings	638	4,337	320	78	—	—	—	34	815	6,222
Old zinc	1,244	1,841	329	—	—	207	—	22	740	4,383

* Table published by permission of United States Department of the Interior—Bureau of Mines.
[a] Includes zinc ammonium chloride.
[b] Used in production of zinc chloride, zinc ammonium chloride, zinc sulfate and lithopone.
[c] Includes 1,283 tons used in making die castings or dies.
[d] Used in production of zinc oxide and zinc sulfate.

Figure 6-55. Tilting type of furnace and condenser for recovery of zinc from dross and die-cast scrap.

A certain type of continuous galvanizing process produces a by-product which is similar to zinc skimmings. The treatment of this material is the same as described under Skimmings or Ashes.

There are three types of furnaces in general use in the production of slab zinc and zinc dust from zinc dross and die castings; namely, stationary, horizontal and tilting type. The stationary furnace is built of firebrick with inside dimensions of 51 inches long, 41 inches wide and 66 inches high. These furnaces are fired with either gas or oil through a port at the bottom of the furnace and the exhaust gas passes out of a vent in the crown. The horizontal furnace is of similar construction and size. The tilting type (Figure 6-55) consists of a large cylindrical steel casing, which is firebrick-lined and mounted on trunnions which permit it to be tipped at an angle. The inside diameter is 48 inches and the length 72 inches.

In each type of furnace a retort is used in which the charge is placed. The flame passes around the retort and the heat is transmitted through it to the metal, which is brought to the boiling point. The fumes pass through the neck of the retort into a condenser where they are cooled to form either metal or dust, depending on the type of condenser.

The retorts are of two designs. The same retorts are used in the stationary and tilting-type furnaces. They are shaped like bottles and are capable of taking a charge of 4000 pounds of either zinc dross or die-cast slabs. These retorts can be charged with either solid or molten metal. They are placed on a pedestal in the furnace at an angle usually of 30 to 45 degrees. After the heat is completed, which takes approximately 24 hours, a residue of iron and other foreign materials remains in the retort; this must be chipped and raked out by hand. It is at this point that the tilting-type furnace is advantageous. It can be revolved in a 45-degree arc which permits the residue to run out or be chipped and raked quite easily. In the case of the stationary furnace, the residue must be chipped and raked at the angle at which the retort is placed in the furnace. This requires considerably more skill. The life of the retort varies widely, but the average is 40 heats when charged with 4000 pounds.

The retort in a horizontal furnace is a cylindrical tube, 24½ inches in diameter, 58 inches long, and holding a 4000 pound charge. It is installed in the furnace at a slight angle, and the fumes pass out of the neck into the condenser. At the rear of the retort there is a raised section which is sealed, but may be opened to charge the retort. This construction permits the continuous charging of molten metal and does not entail removal of the condenser when the retort requires cleaning.

The condenser for producing zinc is cylindrical and is normally 37 inches in diameter and 8 feet 6 inches long. The shell is steel, lined with either firebrick or a castable refractory. When the vapors from the retort come in contact with the cool sides they are condensed and fall in the form of metal. The condenser is tapped at intervals through a sealed hole at the bottom of the extreme end. The first tap is normally zinc of Intermediate grade and the second may be either Brass Special or Prime Western, depending on the composition of the metal charged. By the use of charge of carefully selected die castings, which are produced from Special High Grade zinc by additions of copper and aluminum, it is possible to produce High Grade zinc.

The condenser for the production of zinc dust is constructed of steel and is not lined. Those in use vary in size, but the one most generally in operation is 8 feet long, 68 inches wide and 13 feet 9 inches high. Due to the large dimensions, the vapor from the retorts floats in the interior and eventually condenses and falls as a metal powder. This is screened later on vibrating screens into a final product containing 96 per cent metallic zinc, 96 per cent of it passing through a 325-mesh screen, and a lead content not in excess of 0.26 per cent.

Zinc oxide can be produced in the furnaces described above, but usually a muffle-type furnace is used.

Die Castings

Die castings constitute the second largest source of zinc scrap and within a decade may be the largest source. The composition is dependent upon source, and when the castings are derived from salvage operations there is contamination by iron.

Clean scrap is used in making galvanizing aluminum-additives or it may be slabbed for use as remelt zinc, inferior die castings or slush castings.

This is a melting operation as described under Miscellaneous Metallic Scrap. Clean scrap can also be used to make slab zinc as described under Galvanizers' Dross.

Die-cast scrap is converted to metal or oxide by the use of special types of retort or hearth furnaces. For this application it is necessary to remove unmeltables by a premelting operation. The clean metal is then distilled and condensed to be cast into slabs. Alternatively, zinc vapor may be mixed with air and the resultant oxide collected in bag filters. Under certain conditions, die-cast scrap can be charged to primary smelters' retorts.

Skimmings or Ashes

These materials are formed by oxidation on the surface of galvanizing baths when no flux blanket is used. Turbulence created by the withdrawal of work from the galvanizing bath governs the rate of skimmings formation and the tenor of metallic zinc inclusions.

The skimmings are removed manually with spoons and are handled as is any pulverulent material. It is important in handling skimmings that water is not inadvertently added, as the resulting heat generation could be sufficient to ignite containers made of wood.

Zinc skimmings may be charged directly for retort smelting. In this instance, the zinc recovery is comparatively lower because of the evolution of zinc chloride. This loss is compensated for by the penalties levied by the smelter on the chloride content. However, the vapors evolved constitute an operating nuisance and may, under certain climatic conditions, establish an unhealthy environment.

The usual procedure is to demetalize the skimmings using a ball mill with a trommel screen. The metallic materials are contaminated with oxide and are melted down as described under Miscellaneous Metallic Scrap. The fines are calcined in a rotary kiln. Flue dusts are recovered. Agglomerates from the kiln may be scalped with a coarse screen and put through a ball mill.

The product is charged directly for zinc smelting. Formerly, extensive usage of the roasted zinc ash was made in lithopone manufacture. This is being carried on to a lesser extent now in a manner similar to that described under Flue Dust.

Sal Skimmings

Skimmings also come from the surface of the galvanizing bath when flux is used. The flux, composed of zinc ammonium chloride, absorbs oxides and is removed manually with spoons as it loses activity. A variable tenor of metallics is present, dependent upon the turbulence associated with the galvanizing process. Molten flux is cast into large blocks in molds and barrels. These blocks are broken down manually before shipment, or at the secondary recovery plants by weathering or by jaw crushers.

Sal skimmings constitute the raw material used to make the original galvanizing fluxes. In the secondary recovery process the skimmings are leached with muriatic acid followed by filtration of the insolubles. Heavy metals are precipitated with zinc dust and are removed by filtration. Sulfates, when present, may be removed by precipitation with barium carbonate.

The liquor resulting from this treatment contains 46 per cent zinc chloride and 4 per cent ammonium chloride. Such liquor is used directly as tinning flux. When less than 1 per cent ammonium chloride is required, it is necessary to crystallize out some zinc ammonium chloride. Also, ammonia may be added and the double salts of zinc ammonium chloride are then formed by crystallization.

Sal skimmings are treated to a minor extent by first removing the water-soluble chlorides and then leaching the residue separately. Sal skimmings may also be converted to chloride, and subsequent reduction by electrolysis to zinc metal can be done in an amalgam cell; this is not an economical form of treatment, however.

Chemical Residues

These residues are formed in the manufacture of sodium hydrosulfite. Zinc dust is used in an intermediary step and is converted to impure basic zinc carbonate. The carbonate is removed by filtration and is shipped as a sludge. Recovery is made by two methods:

1. The sludge is calcined in a rotary kiln to eliminate organic matter and sulfites. The product is washed to remove soluble alkali salts and is classified with oversize through a ball mill. The finely divided product is dried in a rotary kiln with flue dust being caught in a bag filter. The product is zinc oxide.
2. Hydrosulfite residue is leached with sulfuric acid, filtered if necessary, and zinc sulfate crystals are produced in a conventional manner.

Flue Dust

This material results from the furnacing of copper-zinc alloys in the form of scrap. Zinc is oxidized and passes into the flues whence recovery is effected by bag filters, cyclone, Cottrell precipitator, wet scrubber or combinations thereof. Both copper smelters and ingot makers produce this material and the composition differs according to the process used, scrap employed and the product being made. Elements other than zinc result from contamination of the scrap by solder and iron. The dry flue dusts are very finely divided and are handled in bulk.

Usage of flue dust in the manufacture of lithopone was formerly extensive but is now on a small scale. In this process, the material is leached with sulfuric acid followed by filter pressing with or without prior thickening. The residue containing lead or tin-lead compounds is sold to nonferrous smelters in this country or abroad. The resulting solution is treated with zinc dust to precipitate heavy metals, which are removed by filtration. Copper cake is sold to copper smelters. A treatment with hypochlorite to remove heavy metals and organic matter may also be made. The purified zinc sulfate solution is used to manufacture lithopone.

The flue dusts are also smelted directly to yield slab zinc.

Die-Cast Skimmings

Die-cast skimmings are generated in the melting pots used to supply metal to the die-casting machines. These are essentially composed of oxidized material containing a small amount of entrained metallic zinc. Such material contains impurities which add to the cost of making zinc sulfate for lithopone. It is used extensively for direct charging to zinc smelting operations.

Miscellaneous Metallic Scrap

Engravers' plates are derived from the printing trade which manufactured them from rolled zinc. New clippings result from the forming of rolled zinc. Old zinc is a mixture of various metallic zinc compositions some of which are contaminated with iron.

These forms of scrap are used to make remelt zinc or are alloyed with die-cast scrap. The former product is used for galvanizing and the latter is used as a source of aluminum-additive for galvanizing.

Remelt zinc is produced by melting metallic scrap in a cast iron pot or crucible. These pots are of a stationary type, mounted in a brick setting and fired by gas or oil. When an iron pot is used very low temperatures and consequent low melting rates are likewise necessary to minimize contamination of the zinc. Dirt and unmeltables are raked out of the bath. The zinc is cast into slabs by hand. When remelt slabs are too contaminated to be sold directly, they are used as a raw material for making redistilled zinc or zinc oxide.

Melting and Casting of Market Shapes

T. H. Weldon

General Superintendent, Metallurgical Division
The Consolidated Mining and Smelting Company of Canada Limited
Trail, B. C.

The final step in the zinc recovery process is to cast the metal into a shape readily accepted by the market. The basic metallurgy of the process determines to a large extent the melting and casting technique used to produce the metal in salable form.

Primary zinc is marketed in six basic grades, and also in special grades made to customer specification. The six basic grades are covered by specifications limiting the maximum content of certain impurities as set forth in the various countries, such as the American Society for Testing Materials Standard Specifications in the United States of America (see Chapter 9, Table 9-2), the British Standards Institution Specifications in the United Kingdom, etc. These grades are: Special High Grade, High Grade, Intermediate, Brass Special, Selected, and Prime Western.

In recent years, more and more special grades are being required by industry for special purposes, and producers are adjusting the content of certain elements to rather close limits to meet the customer's requirements. This assures the customer of a uniform supply of zinc without the necessity of providing a mixing plant and the extra labor involved in procuring the necessary elements to make up the desired grade.

There are about a hundred brands of slab zinc in various shapes, sizes and weights available in the world market. The bulk of primary metal tonnagewise is sold in the characteristic shallow rectangular-shaped slabs. The length of the slab is roughly twice the width, the depth about 2 inches, and the weight is approximately 50 pounds. There are exceptions, notably a British brand that is cast in a somewhat deeper, narrower mold to give a slab more in the form of a pig. There has also been a small interest in jumbo slabs weighing up to one ton. Very minor quantities of zinc have been sold in the form of balls, slugs, rods and billets.

CASTING IN THE ELECTROLYTIC PROCESS

The electrolytic process produces high purity zinc in the form of cathode sheets, which are approximately 24 inches by 30 inches and ⅛ inch thick, somewhat rough on one surface and sharp on the edges. Until recent years, reducing these cathodes to a more marketable form was invariably accomplished by melting in reverberatory furnaces and hand casting on stationary mold racks. In the past few years, induction melting has shown considerable promise, and mechanical casting has proved not only feasible but much more economical.

As the "Cominco" plant[1, 2] situated at Trail, B. C., is the largest electrolytic plant in the world—where both reverberatory and induction melting of cathodes, and hand and machine casting of slabs are being carried out—a detailed review of the practice there will be given. Known variations in other plant practices will be discussed in appropriate places.

While the bulk of the cathode zinc produced meets Special High Grade specifications, the grade of metal cast from the furnaces is adjusted from time to time to meet market requirements. Cathode zinc is sampled daily from individual units in the cell rooms according to a pattern set up after statistical analysis. The selected sheets are saw-sampled and the sawings thoroughly mixed and pressed into disks for assay. Rapid determinations of lead, cadmium and copper are made on a direct reading spectrograph or "Quantometer." These assay results are available by the time the cathode sheets are stripped from the cells and loaded onto narrow gauge cars for delivery to the melting plant. The best quality cathode sheets, as determined by analysis, are charged to the furnaces producing Special High Grade slabs, while the balance is distributed to furnaces producing High Grade, Prime Western or other grades. All dross metal, skimmings and scrap are charged to the Prime Western grade furnace.

In some plants, the cathodes are washed and dried prior to melting, to remove any entrained electrolyte or adhering anode slimes, and to prevent "bumps" caused by moisture addition to the molten bath. Some lowering of dross production is probably accomplished as well.

Reverberatory furnaces used for melting cathode zinc are usually of overall dimensions, 30 feet by 15 feet, with a bottom sloping away from the firing end to an inverted siphon leading to a casting well. Bath depth is approximately 30 inches. Furnaces may be oil-, gas- or coal-fired, depending on local fuel costs. Tests with pulverized-coal fuel were not very successful due to the deposition of a layer of ash on the bath. Cominco furnaces were originally fired by hand, using run of mine coal, but were converted some years ago to oil-firing, using a Bunker "C" fuel which requires steam heating and line tracing for proper handling and ignition.

Cathodes are weighed and then charged through a water-cooled ring in the roof of the furnace. Normally there are four charges in an 8-hour shift. In starting a charge, a few sheets at a time are thrown into the furnace, then bundles are increased in size until approximately one ton at a time is dumped into the furnace. This practice is followed to prevent damage to the furnace bottom, and, in the case of undried cathodes, to prevent possible explosions caused by carrying moisture into the bath of molten metal.

The bath temperature is maintained at about 900°F (480°C) by two oil burners. The oil consumption varies with the rate of melting and the amount of dross on the bath, which in turn depends to some extent on the physical condition of the cathodes and the amount of scrap being charged. The oil consumption varies from 7 to 8 Imperial gallons per ton of zinc cast.

Hand Casting According to Earlier Cominco Plant Practice

Until recently, each reverberatory furnace was equipped with two rows of stationary mold racks firmly anchored to concrete piers. These were served by monorails carrying dippers or ladles of 300-pound capacity suspended from crawls (Figure 6-56).

The molten metal was ladled out of the dipping well, located in the end wall of the furnace but separated from the main bath by a bridge wall, and transported by the above-mentioned monorail along the row of molds. The metal was poured into 56-pound slabs and skimmed to make a clean, salable slab. The hot slabs were cooled from the top by air blown on them from bottom-drilled air pipes 30 inches above the mold racks, and, once the surface had set, the cooling was completed by water sprays turned on the bottom of the molds. The cooled slabs were dumped by hand and stacked in bundles of 50 in piling stands, where they were picked up by lift truck. The bundles were transported to the scales for weighing and hammer marking with an identifying lot number on each slab. From here the bundles were

Figure 6-56. Hand-casting crew showing dipper carried by crawl and monorail. Man in foreground is skimming surface of slabs as cast.

passed through a cooler and transported into boxcars for shipment, or placed in metal storage.

The molds used for hand casting by Cominco were individual molds producing a slab with over-all dimensions of 17 inch by 8½ inch by 1⅞ inch, with the grade of zinc and the "Tadanac" trademark incorporated in the mold so that the lettering appeared on the bottom of each slab. Indentations or dividers also cast into the slab made it reasonably easy to break a slab in three or four sections. The slab weighed 56 pounds.

The molds were made locally of cast iron containing 3.4 to 3.5 per cent carbon, 2.0 per cent silicon, 0.08 per cent sulfur, 0.1 per cent phosphorus, and 0.6 to 0.8 per cent manganese. This analysis appeared to give the best compromise between better stripping, corrosion resistance, thermal shock resistance and ease of machining. The corrosion rate of iron rises very rapidly with increasing temperature of molten zinc, and it is important—not only to mold life, but also to prevent the pickup of iron in the metal—to cast at as low a temperature as is feasible.

Variations in the practice of other plants include the tapping of the furnace into large mechanical casting pots and the use of racks with hinged molds to facilitate dumping of the slabs. Some plants also produce a type of interlocking slab and pallet slab to form a more stable bundle when stacked.

Sampling. As the molten metal was ladled out of the furnace, a sample dipper of metal was taken out of each ladle and emptied into a sample mold. Several of these sample bars were taken for each lot of metal cast, and marked with the lot number. Each of these bars was saw-cut according to a pattern, and the sawings were thoroughly mixed to make up a representative sample for the Assay Office. The sample bars were held as a reserve for three months, in case a recut was necessary. At the Assay Office, the sawings were compressed into disks for spectrographic analysis, or weighed out for wet analysis.

Dross and Flue Dust Treatment

Drossing of the furnaces is generally carried out once per 24 hours on night shift. If required, the through-put of a furnace can be increased and a slight saving in fuel accomplished by drossing on each of the three shifts at some extra labor cost.

At drossing time, the temperature of the furnace is raised and the dross on the surface of the bath is rabbled with ammonium chloride to free metallic zinc from the zinc oxide. The dross is then raked out of the furnace and transferred to drossing drums where further tumbling with ammonium chloride frees more metallic zinc, which may be up to 50 per cent of the weight removed from the furnace. Metallic zinc drains out of the drums to a collecting pot and is transferred back to the Prime Western furnace. The drums are then reversed and the dry dross dumped into cars for disposal. Dry dross contains 3.5 to 4.0 per cent of the cathode weight.

In the case of the Cominco process, this dross is converted into a very fine metallic zinc dust for use in the purification section of the leaching plants. The dross is first roasted at about 1,400°F (760°C) to drive off the chlorine, then mixed with coke and charged to an electric furnace. The zinc is reduced and driven off as a vapor which passes through a condenser where it is precipitated as a very fine dust.

Some plants have a market for their dross and it is shipped without further treatment. Others re-treat their dross through the head of the plant with incoming concentrates either as is, or with a prior wet treatment to wash out chlorine and make a further recovery of metallic prills.

The melting furnaces, drossing drums and spot ventilation are vented to a large stack through a system of flues under forced draft, and an electrostatic treater. The furnace gases enter the system at about 1,000°F (540°C) and are cooled by radiation and air infiltration down to 120°F (50°C) at the treater. The dust load from the furnaces amounts to 1.1 per cent of the zinc charged. This, incidentally, was found to increase with the use of oil firing compared to the old coal-fired grates. The quality of the dust produced by oil firing, moreover, is such that it does not lend itself as readily to electrostatic collection, having a very low bulk density and low conductivity. Special conditioning of the gas is necessary for an efficient catch. The dust catch contains about 2 per cent chlorine arising from the drossing operation, and requires a roasting step before return to the leaching plant for zinc recovery.

Low-Frequency Induction Furnace

Induction melting of zinc has been considered for some time, particularly where low-cost electrical energy is available, because of such practical advantages as: (1) improved operational control, (2) labor savings, (3) elimination of large volumes of combustion gases, and (4) better working conditions. By inductive heating, the heat is produced where required in the metal itself. Thus the metal is not subjected to exceptionally high combustion gas temperatures which tend to oxidize it and not only cause losses but also produce an insulating layer on the bath. There are no combustion gases to be handled requiring extensive flue systems and dust catching equipment, with attendant losses and operating expense.

In 1953, Cominco installed an induction furnace constructed by the Societa Italiana Construzioni Elettrotermiche of Milan, Italy (Figure 6-57). This was the first furnace of this type used in America, although two considerably smaller ones had been in operation in Europe.

Figure 6-57. Societa Italiana Construzioni Elettrotermiche induction melting furnace with control panel to the right and casting machine in the foreground.

The furnace is composed of three interconnected chambers. The over-all steel body, approximately 25 feet by 21 feet and 7 feet high, is lined with insulating brick, then firebrick, and in the lower half by a carefully rammed mix. There are four 750 kva inductors operating at 600 volts, two in each of the spaces between the three chambers. The secondaries of these transformers are channels connecting the chambers at the bottom, filled with molten zinc. Power is supplied through a step-down transformer and regulating transformers, which have push-button controlled tap changers. A battery of condensers is installed to correct the power factor to better than 0.9.

A control panel carries instruments to indicate or record voltage, amperage, power factor and power consumption, also tap-changing buttons for each transformer.

There are also transformer temperature alarms and a knockout switch to cut off the power in case of a cooling-air fan shutdown. An auxiliary gasoline driven generator, capable of driving emergency cooling fans, is installed in case of power failures.

In operation, cathodes are charged in 1,100-pound bundles from the top of the furnace through vertical shafts at two places in each chamber. One sidewall of each shaft opens up to form a platform. The bundles of cathode zinc are lifted off the cars by tongs handled by crane and placed on these platforms, which tilt and discharge the cathodes vertically down the shaft. Thus, there is a column of cathodes on edge down from the charging platform through the bath to the bottom of the furnace. The shafts and chambers are ventilated so that not only are fumes from ammonium chloride drawn off, but some hot air is drawn through the bundles carrying off residual moisture. Melting takes place through the contact of the molten bath with the cathodes. There is a fair amount of stirring in the bath due to electrodynamic movement which increases the melting rate and aids in the drossing action. Dross production is considerably less from this furnace, averaging 2 to 2.5 per cent, and of a dry character, obviating the necessity of the usual drum treatment. The furnace is drossed frequently, depending on the production rate. The maximum rate of melting to date has been 496 tons through the furnace in 24 hours. Power consumption is 105 kwh per ton of zinc cast. The small amount of ventilation gas pulled through the furnace can be treated in a small baghouse. The dust burden is about 400 pounds per 24 hours.

Mechanical Casting as Practiced in the Cominco Plant

All slabs are now cast mechanically except for very minor tonnages of special metal. For many years, it was known that any movement or vibration to zinc during the freezing period resulted in slabs of poor appearance, and considerable pains were taken to anchor the casting racks to solid ground to overcome this. After the success of their lead-casting wheel, Cominco experimented with a semicommercial size casting wheel for zinc, and proved that mechanical casting could be applied to zinc. Due to the physical layout of the plant, it was decided that a straight-line machine would be preferable, and, since Sheppard and Sons in Great Britain had already constructed such a machine for casting zinc alloys, tests were arranged on one of their machines. Results were satisfactory, and one machine was purchased for installation in conjunction with the induction furnace (Figure 6-58). The operation of this casting machine was so successful that two more were subsequently purchased and are in operation, two on the induction furnace and one on a reverberatory furnace.

The over-all length of the single-mold strand casting machine is 60 feet. Each machine is equipped with 140 molds mounted on endless conveyor chains, driven by a tractor-type booster unit that engages the return chain. This in turn is driven by a variable-speed drive, which in one case is a fluid drive, and in the other two cases a dynamically-controlled slip coupling. A very smooth, steady motion is achieved. The molds travel in an inclined plane set at 0.7 inch per foot to gain elevation at the discharge end. Molds are attached to the chain through slotted pins and wedges and can be readily changed when a change in grade is to be scheduled. The mold design

Figure 6-58. Sheppard casting machine.

gives exactly the same type of slab as was obtained on the hand casting lines.

In operation, this machine has a capacity of 15 tons per hour. The empty molds pass under a ladle mounted over the lower or trailing end of the machine, and are filled with zinc pumped from the furnace to the ladle. At present, this ladle is manually controlled, but design is nearly perfected to make this automatic. The slabs are skimmed immediately after casting. As the molds travel up the incline, they are cooled by air to set the surface and by water, on the bottom, to solidify the zinc. The slabs are discharged at the head end onto a short roller-conveyor, where they are picked up and stacked by hand. The bundles are then handled by lift truck (Figure 6-59).

The molds are treated daily with a brushed-on application of Dow Corning "Silicone No. 8" emulsion, diluted to a 3½ per cent solution with water. This treatment has improved the appearance of the mold face by preventing the build-up of zinc oxide in the lettering on the mold.

Sampling. Sampling has been modified to suit a recently installed direct-recording spectrograph. The stream of zinc flowing from the ladle is cut at frequent intervals, and from each cut a specimen is cast for direct arcing. One-half the samples are submitted for analysis, and the balance is retained as a reserve. Thus, the assay of a shipping lot is the average of a number of assays of samples taken throughout the cast.

Strapping. A fairly substantial tonnage of metal is now shipped in strapped bundles weighing approximately 3,000 pounds. The bundles for strapping should be cooled for 24 hours to insure tight bundles, otherwise the shrinkage is sufficient to slake up the straps. Three steel bands 1.25 by

Figure 6-59. Lift truck loading loose bundles of zinc into a boxcar.

Figure 6-60. Strapped bundle of 50 slabs, total weight 2,800 pounds.

0.035 inch are used and are tightened by a tool and crimped with seals (Figure 6-60).

Other Plants[3, 4]

While several plants are now using casting methods as outlined above, another variation is employed at one plant. Cathodes are melted in a gas-fired reverberatory furnace featuring automatic control. Furnace draft is induced by a fume-recovery unit and is controlled at about 0.05 inch water.

Molten zinc is laundered to a 35-kw "Ajax" holding furnace, and from the holding furnace to a casting wheel. The wheel is driven through a fluid clutch. A lug on the wheel trips a switch disengaging the drive, and the wheel decelerates. The same lug trips another switch, which actuates another valve, allowing air to enter the holding furnace and force molten zinc to flow from under the surface of the bath to three pouring spouts. In the meantime, the lug contacts another switch which brakes the wheel to a standstill. A time switch regulates the length of pouring time and the air pressure is released. At the same time, the drive is engaged and a solenoid-operated valve opens, allowing the holding furnace to build up the zinc bath to a predetermined level.

The casting wheel contains 36 water-jacketed molds. One revolution produces one stack of slabs with four "foot" or "pallet" slabs. The slabs are removed from the molds by vacuum suction cups connected to an ingenious linkage. The suction lifter takes three slabs off the wheel, back over a slat conveyor, where the slabs are dropped and then stacked. The molds are dressed periodically with a liquid graphite. Two types of slabs are produced, a usual slab weighing 60 pounds, and an interlocking slab of the same size.

Zinc oxide in the furnace gas is recovered by water scrubbing. A 30-inch brick-lined flue connects the furnace with a vertical-spray cooling tower of the recovery unit. A 24-inch and a 30-inch water-jet fume scrubber are in series with the spray tower. Clean gas leaves the last chamber. Oxide pulp is withdrawn from the lower part of a cone-bottomed steel tank forming the scrubber unit, is filtered, washed, and the cake reprocessed.

The melting furnace is drossed on each shift and the dross is treated through drums, then roasted and retreated in the leaching plant.

CASTING IN THE DISTILLATION PROCESSES

Zinc recovered by distillation is in the molten state as a fundamental part of the metallurgy of the process. The subsequent handling then depends on whether further processing is practiced before casting into marketable shapes.

In horizontal retort operations, up to 35 per cent of the zinc distilled may be condensed as zinc dust or blue powder, and, if not marketed as such, is reworked in the retort charge. The condensers collecting the molten metal are tapped at desired intervals during the distillation runs and the metal is hand cast into slab zinc (see Figure 6-26). The collection pot and slab racks are usually mounted on a mobile car.

In Germany, a furnace was developed to recover slab zinc directly from blue powder. Blue powder and miscellaneous zinc-bearing drosses are charged to a slowly rotating, externally fired, refractory-lined horizontal furnace. After rotating for 6 hours at 12 rpm, the zinc is run out into a liquating pot and the residue removed from the furnace.

The vertical-retort process has made possible the condensation of zinc vapor in large condensers. The more recent of these is of a splash type, where molten zinc from the bath is splashed mechanically to give more intimate contact with the incoming zinc vapor, thus aiding in the cooling and condensation of the vapor.

Molten zinc is tapped from the condenser sump into ladles and cast into slabs. The zinc dust from the condenser may be marketed or recharged to the retort.

Zinc, as produced above, is uaually of Prime Western grade. The retort zinc may be further refined by redistillation. In this case, the zinc is transported by pot or laundered to rectification columns and refined to give a product running as much above 99.99 per cent pure zinc as the requirements of use may necessitate. The Special High Grade metal is collected from the base of the cadmium column in refined-metal holding pots, and from here is cast into slabs in the usual manner.

CASTING IN THE ELECTROTHERMIC PROCESSES

In the electrothermic processes, zinc metal is again recovered in the molten state, similarly to the vertical retort process, through condensation of the vapor in special condensers. The bath of molten metal, at approximately 500°C, is kept at a relatively constant level by continuous overflow, or by periodic tapping into a ladle. This metal may be transferred to a blending and liquating pot, where impurities are removed to the extent that the zinc will meet the requirements of the Prime Western specification, or the molten liquated metal may be refined to produce Special High Grade zinc. In some cases, the zinc vapor from the process is oxidized and the zinc oxide bagged for shipment.

SPECIAL GRADES AND SHAPES

Zinc is marketed in special grades and shapes to customer specification. For example, the recent development of continuous strip galvanizing plants requires a special zinc containing about 0.4 per cent lead and 0.25 per cent aluminum. This special zinc is made in Cominco's operations on a continuous basis. Molten zinc is pumped from the melting furnace at a constant rate to a graphite mixing pot of 1,000-pound capacity.

The predetermined amount of lead is added at frequent intervals in short lengths of an extruded round section. Molten aluminum is added by a small measuring dipper. The pot is mechanically agitated with a graphite impeller driven by a ½ hp geared "Lightnin" mixer. The pot overflows continuously to the casting machine pouring ladle.

A similar setup is used to make other grades of special zinc, and also to debase the high-purity electrolytic zinc with lead to meet the Prime Western specification.

Metallic zinc is also sold in odd shapes, such as bars, balls, pencils and many other cast forms. At Cominco, many shapes, including bar-stock pencils and specially shaped anodes for electrogalvanizing, are made by extrusion from cast cylindrical billets. Other cast forms are made for sale, such as billets, pipeline anodes, having a steel core, and nodules made by splashing a thin stream of molten zinc in water. The nodules so produced are dried and shipped in jute bags.

Metallic zinc dust made by atomizing a thin stream of molten metal by high pressure air or steam, or by shock cooling zinc vapor, is also produced for the market and shipped in metal containers.

REFERENCES

1. Staff of the *Canadian Mining Journal*, "Cominco, A Canadian Enterprise," *Can. Mining J.*, **75**, 131–393 (May, 1954).
2. Nicholson, J. H., "Cominco's Zinc Melting and Casting Facilities," *J. Metals*, **9**, 1125–8 (1957).
3. Cunningham, G. H., and Jephson, A. C., "Electrolytic Zinc Plant at Corpus Christi, Texas, "*Trans. Am. Inst. Mining Met. Engrs.*, **159**, 194–209 (1944).
4. Moore, T. I., and Painter, L. A., "Electrolytic Zinc Plant at Monsanto, Illinois," *ibid*, **194**, 1149–59 (1952).

CHAPTER 7

THE REFINING OF ZINC

ELMER L. MILLER

Assistant Superintendent, Amarillo Plant
American Smelting and Refining Company
Amarillo, Texas

The various distillation processes for winning zinc will at the same time reduce other metals present in the charge. Some of these metals are carried with the zinc vapors and are found as impurities in the slab zinc. Lead, cadmium and iron are the most common and troublesome elements appearing in zinc; some others—especially copper, tin and arsenic—may give occasional concern.

All impurities in zinc are considered detrimental for certain industrial uses. The commercial grade of zinc which a manufacturer can use is determined by the amount of impurities, individually or in total, that can be tolerated in the finished product. The large use of zinc in die-casting alloys has resulted from an understanding of the effects of various impurities and from the development of Special High Grade zinc.

Zinc producers strive to produce the desired quality of marketable zinc either by careful selection and blending of ores and concentrates or by employing special techniques in the roasting and sintering operations of their smelting process. Some slab zinc smelted from the lower, more complex zinc ores and concentrates may still have one or more contaminants in such quantities as to be off-grade for any commercial classification, or the metal may not be pure enough for an especially strong Special High Grade market. Such metal can be made salable by a refining process. Only a primary zinc produced by a distillation process will require refining since an electrolytic zinc can be produced as Special High Grade in one operation.

Two practical methods are available for refining zinc: liquation and redistillation. *Liquation* is used to remove lead and iron to a limited degree from a molten crude zinc, and is based on differences in solubility and specific gravity. *Redistillation* is used to upgrade a spelter to high purity by utilizing the difference in boiling points of zinc and its impurities.

While electrolytic refining of zinc is possible, it is seldom used for the purification of zinc metal, although an off-grade zinc is often used to produce the zinc dust with which the zinc sulfate solution is purified. The phrase "electrolytic refining" is generally applied to the primary electrolytic production of a High Grade or Special High Grade zinc.

Refining by Liquation

Western horizontal-retort smelters have long practiced liquation refining to secure a Prime Western grade of zinc from an unmarketable off-grade zinc, produced chiefly in the last draws of the distillation. In general, the percentage of lead carried by the zinc vapors into the horizontal-retort condensers will increase with the amount of lead in the charge and the retort temperature attained in the final drive for zinc recovery. Much of the iron enters the hot molten zinc by solution, from coming in contact with the charge and the iron tools used during drawing and casting. A small part of the iron is entrained in the zinc vapors entering the condenser.

In the Sterling Process (Chapter 6), zinc is reduced and distilled in the intense heat of an electric arc. Superheated zinc vapors carry a relatively greater quantity of impurities by supersaturation and entrainment into the condenser. Liquation is then employed to recover a marketable Prime Western zinc or is used as an intermediate step before redistillation to Special High Grade.

Robson and Derham[1] describe a new application of the principles of liquation in a patent assigned to the National Smelting Company of England, wherein a small amount of zinc is recovered from a large amount of molten lead used as a shock cooling and condensing medium in a process for smelting zinc ores in a blast furnace. The zinc-enriched lead is cooled to separate and recover a part of the zinc in solution, and the resulting zinc-lean lead is reused to condense more zinc vapor.

If a crude zinc containing lead and iron in amounts too large to meet Prime Western grade specifications—i.e., lead 1.60 per cent and iron 0.080 per cent—is melted and held in a quiescent state for a period of time at a temperature just above the melting point, the molten metal separates into three strata. The lighter zinc with some lead and iron still in solution will form the top stratum. The amount of lead or iron retained will depend on their solubilities in zinc at the holding temperature. Waring, Anderson, Springer and Wilcox[2] studied the mutual solubility of lead and zinc and their data are presented in Table 7-1. Truesdale, Wilcox and Rodda[3] published data on the solubility of iron in the molten zinc, some of which are

TABLE 7-1. MUTUAL SOLUBILITY OF LEAD AND ZINC[2]

Temp. (°C)	Pb in Zn (%)	Zn in Pb (%)	Temp. (°C)	Pb in Zn (%)	Zn in Pb (%)
417.8	0.7	2.0	650	9	8
450	1.4	3.2	700	15	12
500	2.3	3	750	24	19
550	4.0	4	775	32	26
600	5.9	6	790	Complete miscibility	

TABLE 7-2. SOLUBILITY OF IRON IN MOLTEN ZINC[3]

Temp. (°C)	Iron (%)	Temp. (°C)	Iron (%)
419.4	0.018	600	0.70
425	0.02	700	3.85
450	0.03	800	7.70
475	0.06	(900)	(9.75)
500	0.10		

given in Table 7-2. Lead not remaining in solution sinks because of its higher specific gravity to form the bottom stratum. Zinc is carried into the lead zone, according to its solubility, as also shown in Table 7-1. Excess iron subsides as a mushy iron-zinc alloy mixture of solution and crystals to form a stratum over the lead. The supernatant layer of molten zinc is readily drawn or dipped off from the lower layers of impurities. Only lead and iron are materially altered by liquation. If the lead and other impurities in zinc except iron are within the specification for Intermediate, Brass Special or Selected Grades, this element may be brought within limits by careful liquation.

The usual practice is to carry out the liquation of a crude zinc in a firebrick reverberatory furnace. The design, shape and size of the furnace will vary greatly to meet the user's needs. Limited commercial operations have been performed in cast iron kettles, crucibles, ladles or other convenient container that can be heated.

A natural-gas-fired reverberatory of simple design as used at a horizontal-retort smelter, is a firebrick chamber 20 feet long, 8 feet wide and 8 feet high, built in a 4-foot deep welded sheet-steel pan. The bottom is formed by an inverted firebrick arch. The 13 and ½-inch side walls rise 2 feet above the pan to provide for the necessary openings and to support a low covering arch. Natural-gas burners are placed along the sides at an angle so the flames impinge on the arch and heat the metal by radiation. The furnace is filled to near the height of the pan and has a capacity of 75 tons of molten zinc. Crude zinc is charged through doors placed in one side toward a corner. At the diagonal corner a channel permits the liquated zinc to be drawn from the surface of the bath. A vertical slot in one side is packed with fire-clay mud when the furnace is filled; this is cut out when necessary to provide an opening for emptying.

The reverberatory furnace is placed in service by gradually raising the temperature to a point well above the melting point of zinc. Plates of crude zinc are charged as rapidly as they will melt to prevent any marked overheating of the metal bath. Floating oxides are skimmed off as soon as the filling is completed. The molten zinc is cooled to a surface temperature of

425°C over a period of 16 to 24 hours, during which time the freed lead and iron-zinc alloy precipitate from the upper region of the bath. Operations are started by furnishing additional heat to melt the feed and to raise the surface-metal temperature sufficiently for the metal to remain molten while being drawn and cast. Off-grade zinc is fed into the furnace at the same rate as the liquated zinc is drawn—from 27 to 45 tons per day. Impure zinc slabs sink to the iron-zinc alloy layer to melt. The amount of lead and iron remaining dissolved in the liquated zinc depends on the temperature at which it rises and leaves the impurity zone; thus, a finer zinc is produced than would be indicated.

A liquated zinc containing 0.025 per cent iron and 1.2 per cent lead is produced in normal operation. However, a slightly purer product may be obtained by drawing off a cooler zinc and reheating it in a receptacle outside the furnace before molding. To maintain a given zinc analysis as the level of impurities rises, it is necessary to reduce the surface temperature gradually until finally the furnace is unworkable and must be emptied. The condition of the furnace and the thickness of each layer of metal are easily determined by means of an iron bar, which when thrust downward into the bath passes freely through the upper zinc zone, meets a definite resistance in the heavy, granular iron-zinc alloy and has a soft feel again on entering the lead layer.

Before emptying, the accumulated impurities in the furnace are heated close to the burning point or to approximately 750°C for two or three days to effect some solution, by diffusion, of the heavier iron-zinc compound. To prevent the reprecipitation of iron-zinc crystals, the dross alloy is drawn off through the side slot at as high a temperature as it can be handled, about 600°C, and either cast for use in a manufacturing process or granulated for recirculation in the smelting operation. Upon reaching the lead level, the furnace is cooled to freeze the remaining iron-zinc dross and to reduce the percentage of zinc dissolved in the lead. The cooled, molten lead containing approximately 2.0 per cent zinc is drawn and cast for future refining in another reverberatory furnace. The last of the iron-zinc dross is remelted and drawn off at high heat after which the slot is mudded up for another run.

The percentage of zinc recovered in a marketable form depends on the amount of lead and iron present in the crude zinc and the extent to which they are reduced in the refining operation.

Loss of zinc in the surface skimmings is undesirable and with good practice is kept under 1.5 per cent of the feed. This is accomplished (1) by preventing the entry of more air over the molten metal than is required for combustion, (2) by not agitating the liquid zinc with unnecessary skimming or rough charging, and (3) by adding a small amount of ammonium chloride to the skimmings to free entrapped metal from the oxides. A reverberatory furnace designed for continuous liquation is described later as a part of the continuous redistillation process.

Refining by Redistillation

During World War I, the specifications for ordnance brasses and matériel demanded a finer zinc, in larger quantities than the horizontal-retort smelters could produce by careful practice. To meet this market, some plants diverted some of their retort furnaces from primary production to redistillation, by which was produced a High Grade zinc, near 99.9 per cent pure.

The furnaces were adapted for redistilling zinc by first removing the bottom row of retorts. The butt end of each upper row of retorts was then placed on the next lower shelf, so that the inclined retort was capable of holding molten zinc. A dam or constricted opening at the large end of the condenser prevented the redistilled zinc from flowing back into the retort. Regular-production zinc from draws low in cadmium content was cast into small bars for refining. The furnace was charged by placing a feed bar in a sheet-iron trough and ramming it with a wood stick through the condenser, past the dam into the retort. Redistillation was carried on at a lower furnace temperature than that used for primary distillation. Condensers were drawn at 4 to 6 hour intervals and the metal was held in an equalizing reverberatory furnace for casting.

Condensers were removed from some of the retorts each day and the accumulated lead, iron-zinc dross and oxide residues were cleaned out with iron scrapers. High Grade quality zinc was extremely hard to maintain. As lead and iron concentrated in the retort, more of these impurities were showered by the violent boiling action among the zinc vapors to be carried by supersaturation and entrainment into the condenser; also, the easily distilled cadmium was reduced only by a loss in condensation.

Special redistillation retort furnaces, designed to facilitate the charging of impure zinc and the withdrawal of the residue of impurities, were built at some smelters.

Interest in the horizontal-retort redistillation process faded with the market in the postwar years. The New Jersey Zinc Company, however, in a diligent research program continued to explore this method of refining, and through a series of patents[4] gradually evolved a continuous, vertical redistillation process capable of refining zinc to a high purity: 99.995 per cent plus. Spectrographically pure zinc has been produced by repeating the redistillation process.

Components and constructional details of the fractional distillation apparatus used in the refining are shown in Figure 7-1. Impure zinc is charged

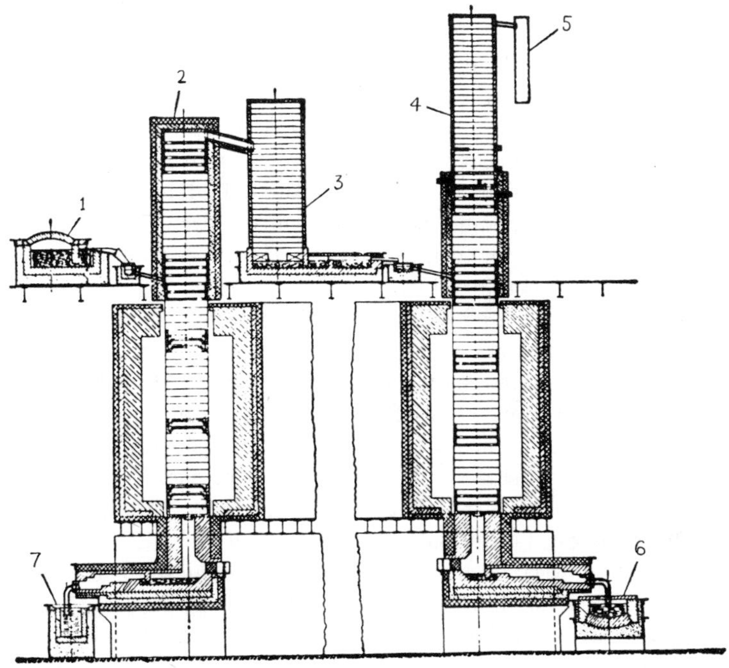

Figure 7-1. Apparatus for refining zinc by fractional distillation. (From "Outline of Metallurgical Practice," 3rd edition, C. R. Hayward, Copyright 1952, D. Van Nostrand Company, Inc., Princeton, New Jersey.)

1—Melting furnace
2—Lead column
3—Condenser
4—Cadmium column
5—Cadmium canister
6—Refined-metal pot
7—Liquating pot

and melted in a furnace (1) so as to give a continuous metal flow into the lead column (2). Here a large part of the zinc is vaporized and the vapors rectified to remove the relatively high-boiling-point impurities such as lead and iron. The resulting vapors are cooled in the condenser (3) to give a liquid metal which is partially distilled and rectified in the cadmium column (4) to eliminate cadmium vapors into the canister (5). Virtually pure molten zinc flows from the bottom of the column into the refined-metal pot (6). Residue lead and iron are removed from the lead column into the liquating pot (7).

Each rectifying column consists of a series of superposed, rectangular trays made of a bonded silicon carbide refractory. Silicon carbide possesses the desirable qualities of high thermal conductivity, high structural strength and unusual resistance to thermal shock.

Each tray is a molded and fired monolithic piece. With experience and improved manufacturing techniques, the tray has progressively increased in size to the present 48 by 24 by 7⅞ inches, with a wall thickness of 1½ inches. Trays are assembled in a column so that an opening through the floor near one end of each tray is alternated to give a baffled flow for descending liquid metal and ascending vapors. Top and bottom edges of the side walls are beveled so that the stacked trays will fit firmly in a vertically aligned position. The intimate, beveled surfaces of adjacent trays are sandblasted to remove a glaze and expose the sharp silicon carbide grain for a firmly anchored zinc vapor sealing joint made with a refractory cement.

The lead column is composed of two types of trays. A section containing boiler trays is within a fired chamber and supports a section of refluxing trays above the furnace.

Boiler trays have a cross section similar to a letter W that has been elongated at its midpoint, thus V‾V. The V legs form a continuous perimetrical trough that holds the impure molten zinc in contact with a large proportion of the externally heated wall surface. This provides favorable conditions for the transfer of heat and a high yield of zinc vapors. Around the upper edge of the opening in the uplifted floor of the tray is a raised rib for holding a shallow layer of molten zinc. Contaminated zinc vapors rise from the heated V trough into the raised center of the superior boiler tray, where they are surrounded and protected from any superheating, either by metal or a portion of the silicon carbide tray containing metal at the boiling point of zinc. More molten zinc must be fed to the boiler section than is required to establish and maintain metal in the trays. From 20 to 33 per cent of the feed zinc is not distilled, but serves to wash the concentrating lead and iron from the V troughs downward to the supporting sump and out into the liquating pot. Ebullition and the solubilities of impurities in the boiling zinc help the runoff zinc to clean the troughs.

The silicon carbide grain of the lower boiler trays is attacked and destroyed by any undue concentration of iron caused by too high an iron in the feed metal and insufficient runoff zinc. Continued iron corrosion leads to the eventual failure of the tray and loss of the column. To prevent iron contact with the silicon carbide grain of the lower trays, the inside wall of the V, next to the fire chamber, is sandblasted and painted or troweled with a $\frac{1}{16}$-inch coat of mullite cement. There is some question as to the effectiveness of this procedure, for when the mullite coat fails the attack may be even more rapid on the exposed, unglazed grains.

At each tray opening, rising zinc vapors pass through a shower of descending molten metal which removes mechanically entrained and superheated vapor impurities. The zinc vapors are purified until they tend to approach equilibrium with the partial pressures of the impurities in the feed metal entering the boiler section. While relatively pure vapors reach the top of the boiler section, they are still further refined by rectification in the section of refluxing trays.

Refluxing trays are flat-bottomed with several transverse ribs on the top surface that are alternately notched to cause the liquid metal to flow back and forth as it travels the length of the tray. One or two silicon carbide shapes, termed "tray dividers," placed longitudinally in the upper refluxing trays, reduce vapor channeling and furnish added surfaces for the fall-out of vapor contaminates. Ascending zinc vapors continue to be purified from the high-boiling-temperature metals by contact with refluxing metal in the pools, on the walls and at the tray openings, until they reach the top tray and cross over into the condenser. Internal heat is conserved within the refluxing section by a 9-inch wall of insulation brick.

The lead-column condenser is a simple boxlike structure made of silicon carbide brick. The bricks are laid, with an air-setting cement, on the long edge so as to present the large face and form a wall 2½ inches thick. Silicon carbide plates cover the top. The firebrick work of a sump for the condensed metal, contained in a welded sheet-steel pan, supports the condenser. Suitable cleanout openings are provided for cleaning the inside walls of the condenser and the metal bath in the sump.

The rate of heat dissipation and zinc condensation is regulated by adjusting insulating curtains about the condenser. By careful control, the boiling and condensation rates are balanced in the united lead column and condenser to eliminate fluctuations in the internal vapor pressure. Blue powder or zinc dust is not produced during uniform operating conditions.

Wherever molten zinc enters or leaves the apparatus in the presence of zinc vapors, a liquid metal seal prevents the loss of vapor or the entry of air. The seal is made by a curtain wall partly submerged in a pool of molten zinc retained by a dam across the canal. The metal flows under the curtain and over the dam.

The cadmium column is composed entirely of the refluxing-type trays. A combustion chamber surrounds the lower part of the column while the upper part rises outside the furnace and is wrapped with 9 inches of insulation brick. A cadmium condenser similar in construction to the lead column condenser surmounts the column of trays.

Partially purified zinc, entering the cadmium column, is heated sufficiently to eliminate the cadmium in its downward flow through the trays, which will at the same time distill a part of the zinc. The mixed vapors rise and undergo rectification in the refluxing trays until a cadmium-enriched zinc vapor reaches the cadmium condenser. A purified zinc flows from the bottom of the column into the refined-metal pot.

The cadmium condenser furnishes the desired liquid metal for refluxing, the amount being controlled by adjusting the insulation. Cadmium-enriched zinc vapors leave the condenser through a refractory tube in one side near the top and enter the sheet-steel canister. The vapors condense and are collected as a partially oxidized metallic dust.

A cadmium column built with trays the same size as used in the lead column is capable of handling the metal produced by two lead boilers, or one lead column can be served by a smaller cadmium column.

A supporting sump serves as a foundation for each column and as a liquid-sealed passageway for molten metal to escape from the trays into a receiving pot. The lead-column supporting sump is built of a double-fired firebrick for its reduced thermal conductivity, as the runoff metal must not cool enough to permit the precipitation and freezing of an iron-zinc alloy in the channel. The cadmium-column sump is constructed of silicon carbide brick to help dissipate a part of the heat contained in the larger volume of refined metal before it is discharged into the refined-metal pot. The masonry work throughout the entire apparatus requires the highest quality of craftsmanship.

Refractory-walled combustion chambers, 6 feet 6 inches by 5 feet 4 inches by 11 feet 4 inches high, enclose the lower part of each column. They are natural-gas fired by pipes entering ports in the furnace roof on each side of the column. Part of the combustion air is aspirated through the burner opening, but the length and temperature of the flame are controlled by the air admitted at each of three levels of air flues and inlets in the furnace walls. Combustion products are removed by two channels in the side walls at the bottom of the furnace and discharge directly into the stack without heat recuperation. Furnace temperatures are observed by six thermocouples projecting into each laboratory.

The melting pot is a natural-gas fired reverberatory furnace with a surface area of 36 square feet and a capacity 12 tons of zinc. Metal is heated by radiation and the pot is capable of melting and heating 32 tons of metal to 650°C per day. Since the molten metal in the pot is hot enough on the surface to burn, if agitated, the feed metal is charged into a chute to enter the bath under the metal level. Slab zinc is fed at regular intervals to give a uniform flow of molten metal from the pot into the lead column.

The liquating pot, designed for continuous operation, is 22½ inches wide by 9½ feet long. The floor of the pot slopes from the shallow end, where the runoff zinc enters, to the opposite end, where another slope at a right angle leads to the lowest point, and under a partition into a lead well outside the pot. On the opposite side is a larger well to collect the supernatant liquated zinc. The runoff metal from the lead column is rapidly cooled to

Table 7-3. Typical Percentage Analyses of Redistillation Products

	Fe	Pb	Cd	Zn
Feed zinc	0.027	0.62	0.05	
Special High Grade zinc	0.0003	0.0022	0.0003	
Liquated zinc	0.032	1.22	Tr	
Lead				1.8
Cadmium-zinc dust			17.0	80.0

450°C or less. As the lead falls out of solution it flows down the slope and under the partition into the lead well. Collected lead is dipped from the well with an iron ladle and cast for refining. Mushy iron-zinc dross is scooped out with perforated shovels and granulated for recirculation in the smelting process. Liquated zinc entering the large side well is drawn and cast in salable Prime Western slabs.

The refined-metal pot is a covered holding pot. The hot pure zinc is cooled to near 500°C before casting into 63-pound slabs for shipment.

First- and second-draw zinc from the horizontal retorts is used as feed, because it is lowest in iron and lead and highest in cadmium. It is considered advisable to use a feed zinc with less than 0.30 per cent iron to reduce the possibility of damaging the lower trays and for longer column life. Lead content in the feed is a limiting factor in the distillation rate of the boiler as high lead is not effectively removed with an excessively high vapor velocity in the refluxing trays. Cadmium is easily eliminated and is a valuable by-product.

A lead boiler with 21 by 42-inch trays distilled 24 tons of Special High Grade zinc per day with 22 per cent of the feed as runoff. Typical analyses of the zinc and by-products are given in Table 7-3. Very little metal is lost in the process as 99.7 per cent of the total metals in the feed can be accounted for.

REFERENCES

1. Robson, Stanley and Derham, Leslie Jack (to National Smelting Co., Ltd., London, England) U. S. Patent 2,671,725.
2. Waring, R. K., Anderson, E. A., Springer, R. D., and Wilcox, R. L., *Metals Technology*, Technical Publication 570-E, September 1934; *Trans. Am. Inst. Mining Met. Engrs.*, **111**, 254–63 (1934).
3. Truesdale, E. C., Wilcox, R. L., and Rodda, J. L., *Metals Technology*, Technical Publication 651-E, Am. Inst. Mining Met. Engrs., October 1935.
4. Holstein, Leon S., and Ginder, Phillip M. (to The New Jersey Zinc Company) U. S. Patents 1,994,345, 1,994,346 and 1,994,358.
   Ginder, Phillip M. (to The New Jersey Zinc Company) U. S. Patents 1,994,347 and 1,994,348.
   Ginder, Phillip M., Peirce, Willis M. and Waring, Robert K. (to The New Jersey Zinc Company) U. S. Patents 1,994,349, 1,994,350, 1,994,351, 1,994,352, 1,994,353 and 1,994,357.
   Ginder, Phillip M. and Hixon, Harold G. (to The New Jersey Zinc Company) U. S. Patent 1,994,354.
   Peirce, Willis M. and Waring, Robert K. (to The New Jersey Zinc Company) U. S. Patents 1,994,355 and 1,994,356.
   Ginder, Phillip M., Mahler, George T. and Cyr, Howard M. (to The New Jersey Zinc Company) U. S. Patent 1,980,480.

CHAPTER 8

THE MANUFACTURE OF ZINC OXIDE

Grate Processes

JOHN H. CALBECK

Director of Research, Pigment Division
American Zinc Oxide Company
Columbus, Ohio

Grate processes as discussed here will include not only the processes normally employing Wetherill grate furnaces, but also other direct methods using fuel to supply the heat reduction. The *direct* methods of manufacturing zinc oxide are those that use zinc ores as a source of zinc, and the product is commonly referred to as "American process" zinc oxide. The *indirect* methods use zinc metal as a starting material and the product is known as "French process" zinc oxide.

All direct processes use oxidized zinciferous material, such as calcines or sinter; these are reduced in a furnace to metallic zinc vapors which are subsequently oxidized to zinc oxide. The indirect process described in the ensuing section of this chapter employs metallic zinc which is vaporized in a retort or other chamber, and the pure metallic zinc vapors, uncontaminated by the products of reduction and combustion, are then oxidized to zinc oxide.

An important direct process for making zinc oxide is a modification of the electrothermic process for producing metal described in Chapter 6. A complete description of this process is introduced later on in the present chapter. Indirect methods using scrap metal and waste zinciferous materials are described in the last section of this chapter.

Zinc oxide is a very important pigment and industrial chemical. In 1956, there were used in the United States about 182,000 tons of zinc oxide, lead-free and leaded, with a metallic zinc content of over 138,000 tons representing 12 per cent of the total zinc consumed in the country. Tables 8-1 and 8-2 show the more important industrial uses of zinc oxide covering the period from 1946 to 1956 inclusive. It is important to note that during this period 73.4 per cent of the lead-free production was consumed in the paint and rubber industries.

The metallurgy and engineering involved in the production of American process zinc oxide is relatively simple. The starting material must be an oxidized zinciferous material, such as zinc calcines or sinter. To this is

TABLE 8-1. NATIONAL SHIPMENTS OF AMERICAN AND FRENCH PROCESS LEAD-FREE ZINC OXIDE IN SHORT TONS, 1946–1956*

	Rubber	Paint	Ceramics	Textiles†	Other	Total
1946	83,776	34,785	9,056	10,022	20,212	157,851
1947	82,248	32,867	11,350	9,100	25,206	160,771
1948	82,895	26,779	12,327	9,474	19,483	150,958
1949	58,496	26,205	6,982	5,200	13,249	110,132
1950	82,944	39,699	12,679	6,303	19,204	160,829
1951	71,507	32,934	10,324	7,265	25,686	147,716
1952	72,774	31,424	7,760	6,262	23,990	142,210
1953	78,439	31,920	8,862	8,718	20,688	148,627
1954	71,058	31,157	8,990	6,322	22,758	140,285
1955	86,677	33,932	10,617	11,263	26,052	168,541
1956	80,459	32,485	10,160	8,447	23,404	154,955

* Source: Bureau of Mines, United States Department of the Interior.
† Includes rayon, coated fabrics and other textile uses.

TABLE 8-2. NATIONAL SHIPMENTS OF LEADED ZINC OXIDE IN SHORT TONS, 1946–1956*

Year	Total Shipments (Primarily Paint)
1946	67,971
1947	81,459
1948	67,441
1949	36,722
1950	63,973
1951	44,341
1952	37,892
1953	39,712
1954	33,972
1955	32,661
1956	27,159

* Source: Bureau of Mines, United States Department of the Interior.

added a solid fuel, such as coke or anthracite coal, in an amount sufficient (1) to provide heat to raise and maintain the mixture at reduction temperatures and (2) to provide carbon monoxide to reduce the zinc oxide to metallic zinc in the vapor phase, and maintain it in the vapor phase above the fuel bed until oxidized. This starting material is placed on a perforated cast iron grate in a furnace and burned by an air blast until the zinc contained is practically all reduced and vaporized. The zinc vapors and reducing gases generated in the furnace are completely oxidized with air, and the hot gases, in which solid zinc oxide particles are suspended, are cooled in a long metal flue and filtered in a baghouse to recover the zinc oxide.

Each of these essential steps will be considered at some length later, but first we should examine the peculiar physical and chemical properties required in zinc oxide to make it acceptable to the paint and rubber industries. These manufacturers originally required a pigment-type of zinc oxide, and while zinc oxide is now used in many other industries, the equipment now in use was designed to produce the peculiar physical and chemical properties required of a pigment zinc oxide.

Pigment Zinc Oxide

In general, only those properties required by the paint and rubber industries and influenced particularly by the processing described below (see also Chapter 12) will be featured as essential to pigment zinc oxide.

Physical Properties. It is important to note that in the selection of a zinc oxide for use in these industries, the choice has always been made on the basis of the physical properties of the oxide and this accounts, in a large measure, for the different methods of making zinc oxide which have developed in the United States during the past forty years.

Particle size and shape and particle-size distribution are the first and most important attributes of a pigment zinc oxide. Only those methods of producing zinc oxide that have been able to control these important properties have survived. Pigment zinc oxide has a very small and a very uniform particle size. Particle shape must be controlled so that uniformly round or uniformly acicular particles, as may be required, are produced. The distribution of shapes and sizes must likewise be controlled to gain an oxide with the proper physical properties. For example, for use in rubber, the oxide must be uniformly fine, as shown in Figure 8-1(a). Such an oxide has an average particle size of 0.25 micron and a specific surface of 3.5 square meters per gram, which is equivalent to an average spherical diameter of 0.35 micron.

On the other hand, the most preferred oxides for use in exterior paints are those with a substantial proportion of the particles acicular or needle-like in shape,[1] such as are illustrated in Figure 8-1(b). Although such oxides have a wide variety of shapes, they should have substantially the same specific surface as the rubber oxides, i.e., 3.5 square meters per gram.

Zinc oxide was used first in the paint industry in the middle of the 19th century. At that time, it was incorporated with white lead and linseed oil in the manufacture of house paint. A high specific surface was required in order to provide an optimum hiding power, the most favorable particle size being in the neighborhood of 0.25 micron. This situation continued until late in 1930 when the more extensive use of titanium dioxide as the whitening and hiding ingredient in paints became general. There remained, however, certain important properties of paint which demanded the chemical reactivity of zinc oxide. These include resistance to mildew, tint retention, durability and rheological properties. High specific surface,

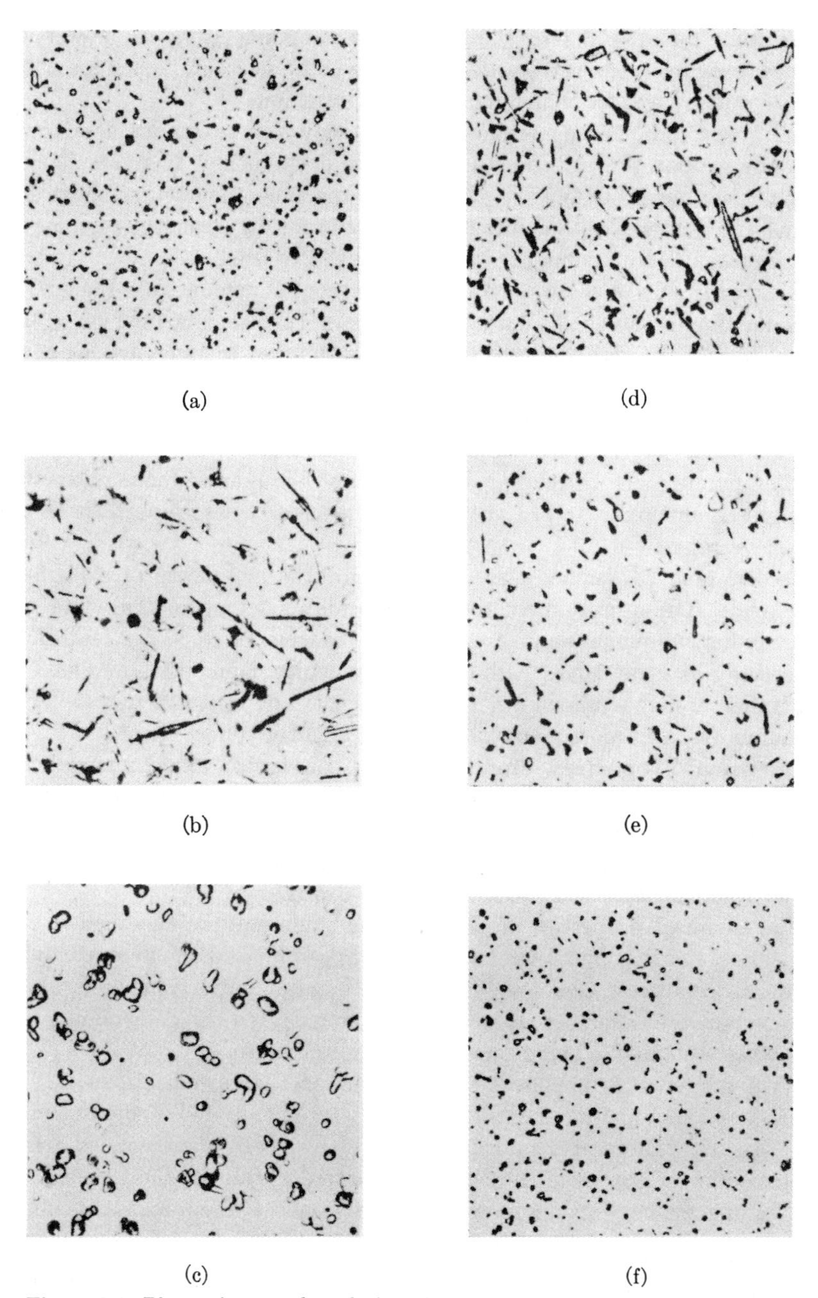

Figure 8-1. Photomicrographs of American process zinc oxide. Magnification 1000×.

or extreme fineness, is required in order that maximum hiding power can be obtained from the zinc oxide in paint which, as a rule, averages about 30 per cent by weight of the pigment concentration.

Oil absorption, tinting strength, consistency and reactivity are largely dependent upon the particle size and shape distribution. In rubber, aciculars are not desirable as they affect tear resistance adversely. A high specific surface improves the curing rate and the tensile strength, even at the low loadings (3 to 5 parts per hundred) currently used.

The occurrence of coarse particles in a pigment is rare because the ASTM Specification D79-44 calls for a maximum of 1.0 per cent held on a 325-mesh sieve, a specification that is easily met. Coarse particles are the cause of poor texture in paints and poor tear resistance in rubber. Such particles include not only the small percentage held on the 325-mesh sieve, but also a substantial percentage ranging from 44 microns (325 mesh) down to the median particle size. Special precautions must be taken in the manufacture of pigment zinc oxide to prevent the development of coarse particles in the sub-sieve range.

Second only to particle size are the color characteristics of a pigment zinc oxide. These must meet the high standards set up in the white pigment industries which require a brightness of from 80 to 85 per cent and a practically neutral tone. Although many white pigments are blued to neutralize a slight yellowness, this practice is not followed in zinc oxide manufacture, and most zinc oxides have a slightly yellow tone.

Chemical Properties. From a chemical standpoint, all commercial zinc oxides have a high degree of chemical purity. Certain industries set up their now special specifications covering the chemical purity required for their processes. However, for the most part, specifications cover the physical rather than the chemical properties of the oxide.

The chemical properties of pigment zinc oxide may be classified in two categories, the first being the chemical composition and the accompanying chemical impurities, and the second the chemical performance characteristics, which are difficult to assay but which are often the determining factors when an oxide is being selected for a specific use.

With reference to chemical composition, ASTM Specifications D79-44 and D80-41 provide simple limitations, as seen in Table 8-3. From this table it is seen that there may be from 1.0 to 2.0 per cent of impurities, which includes almost any metallic compound. Actually, this does not occur with any of the primary oxides now produced in the United States. Primary oxides are produced from carefully selected zinc calcines or sinters that are practically free from chemical impurities. Sulfur, lead, cadmium, copper, iron and insolubles occur in minute quantities in American process oxides and the distribution of these impurities is illustrated in Table 8-4, which

TABLE 8-3. ASTM SPECIFICATIONS FOR ZINC OXIDE AND LEADED ZINC OXIDE, DESIGNATIONS D 79-44 AND D 80-41

	American Process		French Process
	Leaded	Lead-Free	
Zinc oxide, minimum, per cent	62 to 67	98.0	99.0
Moisture and other volatile matter, maximum, per cent	0.5	0.5	0.5
Total impurities, including moisture and other volatile matter, maximum, per cent	2.0	2.0	1.0
Coarse particles (total residue retained on a No. 325 (44-micron) sieve), maximum, per cent	1.0	1.0	1.0
Matter soluble in water, maximum, per cent	1.0	—	—
Normal or basic lead sulfate	remainder	—	—

defines a typical American process lead-free zinc oxide made from Mascot, Tennessee, calcines.

Lead and zinc ores from the Missouri-Kansas-Oklahoma area and zinc ores from Wisconsin, Tennessee, Virginia, New Jersey and New York are preferred to the ores from the western states or from abroad because they are free from objectionable metallic impurities.

All American process zinc oxides contain varying amounts of sulfur compounds, normally zinc sulfate; but, in addition, there are certain sulfur-oxygen-zinc compounds that have a special effect on the curing rate when zinc oxide is used as an activator in the compounding of rubber. This property is usually referred to as "acidity" because it can be analyzed by titrating a slurry of zinc oxide in water with a standard alkali using phenolphthalein as an indicator. However, if zinc sulfate is added to zinc oxide, the curing properties obtained by the natural oxy-sulfur compounds are not attained.

When zinc oxide is reacted with an acid, special care must be exercised to prevent the formation of the basic oxy-zinc salt of the acid. For example, if zinc oxide is dissolved in dilute sulfuric acid, a point is reached where the zinc oxide, if added too rapidly, forms the relatively insoluble basic zinc sulfate, and even though an excess of dilute sulfuric acid remains, the basic salts dissolve very slowly. Normally, the basic sulfate is allowed to settle from the neutral zinc sulfate solutions and it can be redissolved easily when strong acid is added in the preparation of the next batch. A similar situation arises frequently when zinc oxide is reacted with organic acids, in which case it is extremely difficult to eliminate the basic salt by any amount of manipulation. Furthermore, many normal zinc salts react with zinc oxide to form basic or oxy-salts.[2]

TABLE 8-4. CHEMICAL AND PHYSICAL SPECIFICATION FOR A TYPICAL AMERICAN PROCESS LEAD-FREE ZINC OXIDE

Zinc oxide	99.0% minimum
Zinc sulfate	0.20 –0.35%
Lead	0.01 –0.03%
Cadmium	0.05 –0.10%
Sulfur	0.02 –0.04%
Acidity (SO_3)	0.02 –0.10%
Water soluble	0.20 –0.40%
Insoluble in HCl	0.05 –0.20%
Ignition loss (91 hr/760°C)	0.05 –0.20%
Loss at 110°C	0.05 –0.15%
Moisture	0.01 –0.06%
Chlorine	0.01 –0.03%
Copper	0.001–0.003%
Iron	0.01 –0.03%
Specific gravity	5.6
One pound bulk gallons	0.02144
Apparent density (lb/cu ft)	32–40
Spec. surface area (sq m/g)	4.00
*Spec. surface dia. (microns)	0.22–0.32
General particle shape	medium round
Fineness (through 325 mesh)	99.95–99.99%
Rub-out oil absorption	14
Gardner oil	18–22

* Specific surface diameter reported in accordance with ASTM Specification D1366-55T.

When zinc oxide is included in paint formulations, one important function is to form zinc soaps in the paint. Zinc forms both the normal and the basic soaps of fatty acids. The purpose is twofold: firstly, to neutralize any free fatty acids that may be present in the paint oils and provide, thereby, a small quantity of zinc soap which adds certain desirable features to the paint. Secondly, oleo resinous paints, when aged, break down into free fatty acid and the presence of the solid zinc oxide particles in the paint film neutralizes these acids as they are formed in the normal aging processes.

The chemical performance characteristics of pigment zinc oxide have been appreciated only in recent years and are a function of both the chemical and physical properties of the oxide. Important factors to be considered are specific surface, particle shape, the surface characteristics of the particles, and whether the crystal lattice is uniform or distorted. These properties are controlled by reheating or refining.

From a chemical standpoint, the kind and location of the impurities are important, e.g., the sulfur impurities may be present as sulfates, sulfites, polythionates or adsorbed sulfur dioxide and they may constitute a superficial coating on the zinc oxide particles or they may be simply admixed with the particles. These performance characteristics have been the cause for the many recent innovations in zinc oxide processes, including reheating, refining, surface treating and washing, discussed later under Conditioning Steps.

Types of American Process Zinc Oxides

American process zinc oxides are produced lead-free or leaded. The latter type has varying amounts of basic lead sulfate incorporated with the zinc oxide. Originally, all leaded zinc oxides were made by fuming or burning a charge containing lead sulfide ores admixed with the oxidized zinciferous material. This practice is followed today, but, in addition, substantial amounts of leaded zinc oxides are sold that have been prepared by making the two components in separate units and blending in the required proportions. These oxides are known as "blends," whereas the former are referred to as "cofumed." A photomicrograph of a typical cofumed oxide is shown in Figure 8-1(f). Blends and cofumed oxides are available containing 12, 18, 24, 35 and 50 per cent basic lead sulfate.

Except for the lead content, the chemical properties of the leaded zinc oxide are substantially the same as those of the lead-free oxides. The physical properties of both types are substantially the same except that acicular particles are seldom found in cofumed leaded oxide. Equipment for manufacturing the two types is the same, the cofumed leaded oxide being produced by introducing the required amount of plumbiferous material, usually lead sulfide into the charge.

Lead-free zinc oxide is available in a variety of types. Those oxides with nodular or round* particles are available in a range of sizes from 0.15 to 0.9 micron (see a and c of Figure 8-1). Oxides with substantial proportions of acicular particles are available. Typical acicular oxides are shown in Figures 8-1 (b), (d), and (e).

Production

To illustrate the design and workings of a typical Wetherill grate operation, a schematic flowsheet of the Columbus plant of the American Zinc Oxide Company is shown in Figure 8-2. This plant operates as a single unit, continuously and almost automatically. At (2) is the charge mixing equipment; below (2) is the Wetherill fuming furnace; (3) specifies the furnace blocks consisting of two blocks of eleven furnaces each, with flues, combustion chamber and accessories where the zinc oxide fume is produced; (4) is the long cooling pipe known as a "trail"; (5) is a large suction fan; (6) is the filtering unit or baghouse; (7) is the refinery; (8) is the packing house and (9) is the warehouse.

* Appearing round under the light microscope, but actually acicular when examined with an electron microscope.

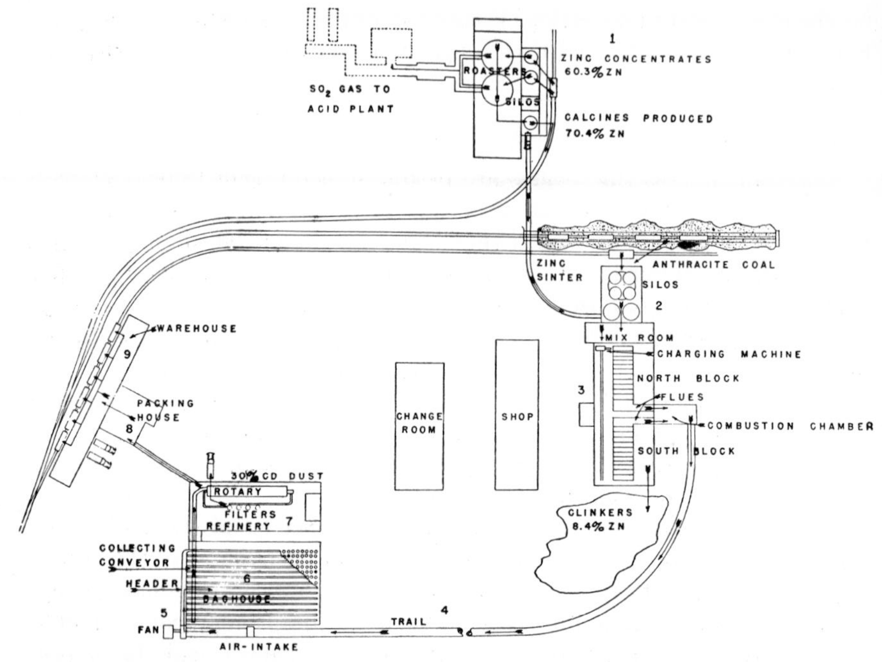

Figure 8-2. Schematic flowsheet of the Columbus plant of the American Zinc Oxide Company.

The Charge. All American process zinc oxides are now made from zinc blend or sphalerite as high-grade concentrates assaying over 60 per cent zinc. These are roasted to remove the sulfur which is used in the manufacture of sulfuric acid. At Columbus, concentrates are delivered by rail from Mascot, Tennessee, and roasted in two Wedge roasters, (1) in Figure 8-2, to produce a calcine having 70.4 per cent zinc and less than 4 per cent sulfur. These calcines are delivered by truck to the mixroom, (2), where they are stored in silos prior to being mixed with coal to form the furnace charge.

As mentioned above, the first essential in production is the preparation of the charge consisting of a zinciferous fraction and a fuel fraction. The former, consisting largely of zinc calcines and sinter, is selected because of its chemical purity. For lead-free oxides, small amounts of cadmium, lead, and sulfur can be tolerated in the primary ores because these impurities may be removed by sintering the calcines prior to charging, or by refining the crude baghouse zinc oxide in a special conditioning process prior to packing. Copper is especially objectionable in an ore because it is not removed in sintering and cannot be removed from the oxide by refining. Copper, cadmium, and other metallic impurities must be low if a product of good color is to be obtained. In addition to the calcines and sinter, there is always a quantity of trail cleanup sweeps, etc., that have to be included in the charge.

The selection of fuels for the charge is determined by cost, availability, and the particular demands of the furnaces used. A low-ash anthracite is to be preferred to coke in most cases. The volatile hydrocarbons are of no value in the grate processes and simply require more secondary air to burn them out. High ash content results in high clinker production and, consequently, low recoveries because of the zinc in the clinker, which ranges from 8 to 15 per cent. Rotary furnaces make higher recoveries because of higher fuel-bed temperatures, but, on the other hand, more metallic impurities, especially copper, are carried over at the higher temperatures.

Several screen sizes of coal or coke are preferred in order to obtain the desired amount of porosity in the fuel bed when the charge is used without briquetting. When briquettes are handled, a fine coal and an organic binder are used. When the charge is to be used without briquetting, it is simply moistened with sufficient water to prevent dusting and classification in handling and to give it a "set" when charged on the grates. It is the practice at Columbus to allow the "green" charge to "bake" on the grate without blast for one hour, during which time the charge sets up and forms a porous rigid bed through which the blast penetrates easily and uniformly without blowing calcines and coal dust into the zinc-bearing gas stream above. This improves color and keeps insolubles low in the product.

In the Columbus mixroom, therefore, a charge is prepared by drawing the proper proportions of coal, calcines and sinter from the silos by an automatic weighing device, adding water and mixing in a large drum mixer.

Furnaces. The charge is stored in overhead bins and delivered to the furnaces by a charging machine. At Columbus there are two blocks of eleven Wetherill grate furnaces located in the furnace room, (3) of Figure 8-2. Figure 8-3 shows one block with the charging machine in operating position. The grate area of each furnace is 12 feet wide and 7.5 feet long, and the perforated cast iron grates are supported by rails above a blast-tight pit. The eleven furnaces of a block are placed side-by-side to form a flat, continuous grate surface 12 by 82 feet. Above the grates, a furnace block consists only of one long tunnel covered with a single silica-brick arch and two firebrick walls in which are located working doors for each side of a furnace area. A single furnace is defined by the area that can be worked through any two opposite doors. One end of the tunnel is walled up and the other opens into a flue of the same cross-sectional area.

Reduction to Zinc Vapor. The function of the furnace is to drive out of the charge the greater part of its zinc content as a fume or vapor and leave behind a clinker or residue low in zinc content and in a condition

Figure 8-3. Furnace block at Columbus. Mechanical charging machine in position.

facilitating easy removal from the furnace. In the burning process, the charge is first placed on the grate by the machine and leveled to a depth of 14 to 18 inches by the workers. One furnace per block is charged hourly. The charge bakes on the grate for one hour before the blast is turned on. Ignition begins on top. The blast of air coming through the grates and up through the charge promotes combustion and raises the temperature to that of the producer-gas reaction, approximately 1100°C (see Eqs. 1 and 2, p. 355). This being an endothermic reaction, the temperature is prevented from going higher to the reduction temperature of reaction (3) because of the excess carbon in the charge.

Although a burning zone is formed in the charge bed, very little zinc is reduced until this zone has migrated to the bottom of the bed just above the grate, which requires about two hours. During this downward migration, quantities of carbon monoxide are generated by the producer-gas reaction in the zone above the burning zone. This carbon monoxide is required to prevent the premature oxidation of the zinc vapors as they travel the length of the tunnel to the flue and combustion chamber. When the burning zone reaches the bottom of the charge, there is a rapid jump in the burning-zone temperature to 1300°C or more because of the higher air-fuel ratio. Reduction of the zinc oxide of the charge now begins. Zinc vapor, accompanied by the carbon dioxide of reaction (3), moves upward through the incandescent overburden where it is reduced to carbon monoxide, and a mixture of gases—Zn, CO, N_2 and H_2O—emerges from the top of the charge.

Three hours more are required until the reduction rate reaches a maximum. The very hot reduction zone remains just above the grate and the overburden of red-hot but unreduced charge slowly migrates downward to supply more fuel and zinc oxide for the reduction reaction, which continues at a uniform rate until the overburden has disappeared and there remains only a hot, porous bed of clinker on the grate. This requires about four hours. At regular hourly intervals the clinker of one furnace on each block is "pulled" and a new charge put in. After the burning period, the furnace lies "dead" for one or two hours until "pulled" and recharged. In the "pulling" step, the clinker on the grate is broken up with slice bars, raked from the furnace into hoppers and taken in trucks to the clinker pile. Clinker may be treated in a Waelz kiln to recover much of the remaining zinc.

The time schedule for one furnace averages one hour for the green furnace to bake without blast, then the blast is turned on and 2 hours are required for the burning zone to reach the grate, 3½ hours to develop maximum reduction, 3⁶⁄₁₀ hours of maximum reduction and 1⁸⁄₁₀ hours dead with no blast, to complete the 12-hour cycle. From time to time, it may be necessary for the workers to level the charge and fill in "blow holes" because the burning and reduction over the entire grate area may not be uniform.

The reduction step in the metallurgy of zinc oxide is substantially the same on all grate furnaces. In those cases where oxidation occurs only above the furnace charge, a saving in time and coal can be made if a supplemental fuel bed of coal is ignited on the grate and the charge placed on top of it.[3] In such cases, the downward movement of the burning zone may not precede the development of the high-temperature reduction zone at the grate level. In rotary furnaces, because of the tumbling of the charge, no stratified zones exist and the charge is rapidly brought to reduction temperature by a gas flame, but reactions (1) and (2) in the charge account for the production of zinc vapors.

Chemical Reactions in the Metallurgy of Zinc Oxide

$$(1)\ C + \frac{\text{Air}}{O_2 + 3.76\ N_2} \rightarrow CO_2 + 3.76N_2$$

$$(2)\ C + CO_2 \rightarrow 2CO$$

$$(3)\ CO + ZnO \rightarrow Zn + CO_2$$

$$(4)\ 2Zn + \frac{\text{Air}}{O_2 + 3.76N_2} \rightarrow 2ZnO + 3.76N_2$$

$$(5)\ 2CO + \frac{\text{Air}}{O_2 + 3.76N_2} \rightarrow 2CO_2 + 3.76N_2$$

$$(6)\ Zn + CO_2 \rightarrow ZnO + CO$$

$$(7)\ ZnO + C \rightarrow Zn + CO$$

Oxidation to Zinc Oxide. The second metallurgical step is the oxidation of the vapors produced by the reduction step. Reactions (4) and (6) illustrate the oxidation step. Although simple, it is the determining factor in the control of particle size, color and type.

As explained above, the vapors issue from the bed of charge as metallic zinc, carbon monoxide and nitrogen. Oxidation can be effected in various way. According to reactions (4), (5) and (6), it may proceed as soon as the vapors escape from the charge bed. This is frequently referred to as simultaneous reduction and oxidation although the former chemical process must always precede the latter and occur in a locale apart from the latter by a definite, albeit very small, distance. Or, oxidation can be delayed by retaining the zinc in the vapor phase until the furnace gases reach the combustion chamber at which point all the oxidation is conducted under controlled conditions. Actually, this latter process is ideal and difficult to maintain for long periods because the necessary charging cycle interrupts, except on chain grate furnaces.

Case one was formerly the practice of all producers. Air was inducted into the furnace by suction over the charge area where oxidation of carbon monoxide and zinc was substantially complete before the fume from one furnace joined the fume from the other furnaces. An excess of air was necessary to prevent excessive furnace temperature which resulted in early failure of arches and furnace walls, but which also affected adversely the quality of the product, causing yellow color, large particle size and low oil absorption. To obtain fine particle size, good color and high oil absorption, it was necessary to combine the fume from only a few furnaces, two or four, in order to keep at a minimum the time that the fume was incandescent. Such arrangements have been referred to as "Eastern-type processes".[4, 5]

At Columbus, eleven furnaces were joined under a single arch (Figure 8-2(3), and Figure 8-3) and the fume from all of them oxidized in a single combustion chamber at one end of the block. This is the arrangement for case two, where an attempt is made to have all the zinc vapors remain fully reduced until they reach the combustion chamber in which all the oxidation is done under controlled conditions. This requires that certain conditions be maintained to prevent premature oxidation over the grates and in the long passage over the furnaces to the combustion chamber. The first condition is that a minimum of air be inducted into the furnace through charging doors, and this is best accomplished by maintaining a slight positive pressure, i.e., "back pressure," on the entire block of furnaces. The second condition is to maintain an excess of carbon monoxide in the gas stream and thereby prevent reaction (6). This excess is easily provided by the carbon monoxide production of the freshly charged furnaces mentioned above. If the carbon monoxide concentration is maintained at the proper level, the formation of zinc oxide particles in the gas stream is prevented and a perfectly clear or transparent gas is delivered to the combustion chamber. This is imperative if particle size control is to be maintained at its best.

Having delivered the clear gas to the combustion chamber, the oxidation procedure depends upon the type of oxide to be made. If the zinc-bearing gas stream is blasted with air from a great many small jets, resulting in the rapid oxidation of both the metallic zinc vapors and the accompanying carbon monoxide under condition of violent turbulence in a small chamber, extremely small and uniform particles of zinc oxide are produced. If, then, this condition is of short duration and the gases are cooled quickly by an excess of air[6] or by a water spray,[7] there is only slight particle growth and a particle size like that shown in Figure 8-1(a) will be obtained. If, however, the zinc-bearing gas stream is allowed to remain at a high temperature, the particles grow rapidly and there results a final product, like that illustrated in Figure 8-1(c). On the other hand, if the zinc and carbon monoxide vapors are burned with a lazy flame[8, 9] in a large chamber and with an excess of air to prevent the development of excessively high temperatures, oxides are produced with varying degrees of acicularity, such as are shown in Figures 8-1(b), (d), and (e). At Columbus, to produce the fine type, the oxidation is conducted in the flues, whereas for the acicular type, most of the burning is conducted in the combustion chamber proper.

From the above, it may be seen that very accurate control of the amounts of secondary air must be maintained. If exactly the theoretical amount of air is used, the maximum temperature is obtained and this is detrimental not only to the quality of the oxide, but also to the walls of the flue and combustion chamber. When water sprays are not in use to maintain the temperatures below limits, an excess[6] or a deficiency of air must be used. In the latter case, the residue of unburned gases, mostly carbon monoxide (because the zinc will be oxidized by the carbon dioxide) will burn further along in the system, causing delayed particle size growth, poor color and cooling difficulties in the trail.

Cooling Tubes and Fan. The hot zinc oxide-bearing fume is drawn into the cooling tubes or trail by a large fan. Two general types are used, a long horizontal flue of steel known as a "trail," or a series of vertical pipes connected by elbows, known as "goosenecks." The cooling is always done by the natural circulation of air around the pipes. Fans, sprays, etc., have not been found effective. At the end of the trail and ahead of the suction fan there will be found an automatic device for admitting diluting air to maintain the temperature at an optimum level before the gases are blown into the baghouse. Baghouse practice is similar to that in zinc extraction processes (Chapter 6).

Baghouses. In the United States, zinc oxide is collected in baghouses or automatic mechanical-filtering systems. It is the preferred practice to collect all of the marketable oxide in these systems and not attempt to collect part of the oxide in settling chambers located ahead of the filter system

Figure 8-4. Collecting hoppers and collecting conveyor in baghouse, Columbus.

as is done in Europe. This system simplifies the collection of the deposited zinc oxide and reduces to a minimum contamination arising from scale and hard deposits of oxide that build up on the walls and hoppers of settling systems. Figure 8-4 illustrates the type of baghouse in use at Columbus, Ohio. In the building are hung 1,008 asbestos bags, 72 inches in circumference and 32 feet long. The bags are attached to rows of hoppers. The hoppers, fourteen in number, act as conductors for the incoming gases and are connected to a header on the discharge end of the fan. In each hopper is a 12-inch helical conveyor which delivers the filtered oxide to a collecting belt after passing through a rotary seal. The belt conveyor passes over a weightometer for controlling the rate at which the oxide is being withdrawn and delivered to a rotary refining furnace. The bags are shaken by hand at 30-minute intervals, two rows at a time. While being shaken, the two rows are shut off by dampers at the header end. The static pressure in the bags ranges from ½-inch water column after shaking to about ¾ inch just prior to shaking. The oxide falls from the bags into the hoppers and is withdrawn by the conveyors as required to maintain a uniform feed to the refinery and, therefore, to a limited extent the hoppers provide storage. A baghouse of this type has a minimum of steel plate in contact with the oxide, thereby keeping scale and hard oxide at a minimum. Rust is prevented by always keeping the interior of the system above the dew point, and in a system using asbestos bags, the temperature of the entering gas is maintained automatically at 400°F. More recent installations consist of filter units using Orlon bags which are automatically shaken at frequent intervals. In all types of filter systems, the temperature must be kept above the dew point to prevent rust and scale developing on the steel surfaces, to prevent caking of the oxide on the bag surfaces and in the conveyors, but more particularly, to reduce hydration of the oxide and sulfation when sulfur gases are present.

Conditioning Steps. All pigment zinc oxide now produced in the United States is given a series of conditioning steps prior to packing, depending on the character of the crude oxide and the uses to which the oxide is to be put.

These conditioning steps include screening, bolting, milling in hammer mills or pulverizers, reheating, refining, densifying, washing, surface treating, deaerating and pelletizing, as described below.

Milling and Screening. Originally, the baghouse product was screened or bolted to remove lumps of hard oxide developed in the baghouse and handling system; these lumps produced poor texture when incorporated in paint. Texture was greatly improved by passing the oxide through high-speed pulverizers prior to packing, and today all zinc oxide is given one or two passes through a high-speed pulverizer before packing.

Reheating and Refining. Many of the American process oxides produced early in the century were of poor color. To improve color, the oxide was heated in a muffle furnace at a low red heat for a short time. If, however, the oxide contained objectionable quantities of water solubles, such as zinc sulfate, a temperature in excess of 720°C was required, and at that temperature, even for a few minutes, a particle growth occurred which altered the physical properties of the product to an objectionable degree. To prevent this particle growth and the accompanying densification of the oxide, carbon black or other carbonaceous material was added prior to reheating, permitting the decomposition of sulfates to take place at lower temperatures; thus densification was prevented, while color was improved and sulfur compounds were eliminated.[10]

The refining process used in Columbus[11] is unique in that in addition to improving color and eliminating sulfur compounds, it removes cadmium, and, by such recovery, a substantial credit against operating expense is obtained.[12] The refining unit is illustrated schematically in Figure 8-5. A sealed rotary kiln, 70 feet long and 5 feet inside diameter, is heated by an automatic burner system that alternates the furnace atmosphere from re-

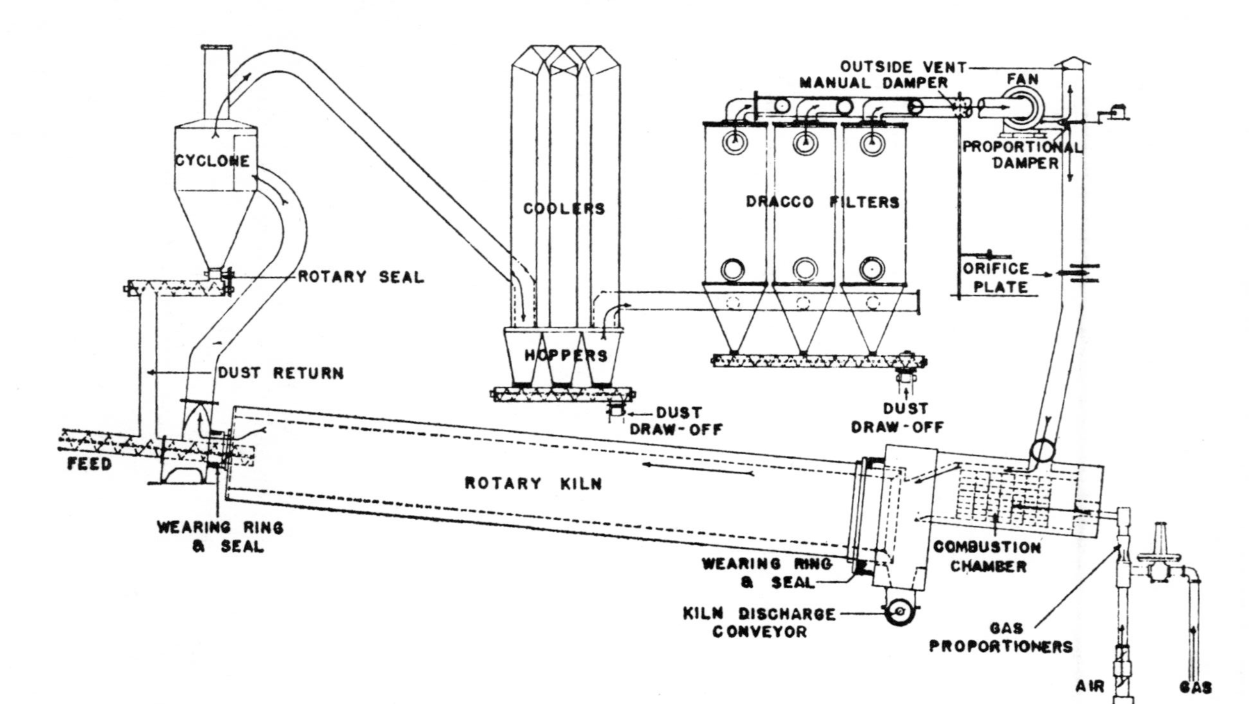

Figure 8-5. Schematic outline of zinc oxide refining unit, Columbus. (*Courtesy American Institute of Mining Metallurgical and Petroleum Engineers*[12])

TABLE 8-5. ROTARY REFINING OF ZINC OXIDE

	Before	After
Zinc oxide	98.4%	99.3%
Cadmium	0.35%	0.06%
Lead	0.003%	0.002%
Total sulfur	0.36%	0.02%
Zinc sulfate	1.40%	0.10%
Acidity	0.35%	0.05%
Brightness	76.0%	83.5%
Apparent density (lb/cu ft)	33	33

ducing to oxidizing at 5-minute intervals. The temperature, composition of the heating gases and rate of flow through the kiln are all controlled automatically.

The crude oxide passes through the kiln countercurrently to the heating gases and is retarded and agitated by a series of lifters not shown in the illustration. A temperature of 770°C is maintained at the discharge end. This temperature in an oxidizing atmosphere would normally result in excessive densification, but the reducing cycle effectively prevents this. The exit gases carry the sulfur and cadmium impurities, together with some zinc oxide (less than 2 per cent), through a cyclone and filter system where the solids are removed and the clean gas from the filters is divided by the proportional damper, permitting a regulated fraction to be recirculated through the kiln so as to maintain the gases at an optimum rate of flow of about 50 feet per second (see Figure 8-5). In Table 8-5 are shown analyses of the zinc oxide before and after refining. Oxides so refined are fast curing when used in rubber because of low acidity and do not retard thiuram accelerators because the low percentage of cadmium remaining is largely in the form of cadmium sulfide.

Washing. The acidity of zinc oxide may also be reduced substantially by washing with dilute ammonium hydroxide. This process[13] depends upon the formation of Warnerian compounds which are much more soluble than the zinc oxysulfur compounds. The latter combine with the free ammonia to form the soluble Warnerian compounds which are easily washed out.

Surface Treating. Pigment zinc oxide is now being sold with a wide variety of additives so processed as to produce a superficial coating on the zinc oxide particles. These additives include stearic,[14] propionic,[15] and other fatty acids; esters, zinc compounds, such as zinc propionate,[2] and oils. The function of these additives is to provide a zinc oxide that disperses better and incorporates faster when used in paint and rubber. For example, an untreated zinc oxide, when incorporated with crude rubber in a master batch of 100 parts zinc oxide to 100 parts rubber in a modern Banbury mixer, would require 20 minutes to mix, whereas when surface-coated with basic zinc propionate, it would require only 10 minutes. Zinc oxide is frequently too active chemically for incorporation in certain paints and enamels. This activity can be substantially reduced by a coating of zinc phosphate on the oxide particles.[16]

Pelletizing, Densification and Deaerating. All zinc oxide for domestic shipment is packed in 50-pound multiwall paper bags, either tied, sewn, or of the valve type. The bulk density of pigment zinc oxide, however, varies from as low as 20 pounds to as high as 90 pounds per cubic foot. This great variety in the sizes of the packing bags complicates packing, handling, and shipping, and many users with limited storage space dislike the large packages. Furthermore, many batch operations are delayed because the bulky zinc oxide requires that two additions be made to a mix, thereby lengthening the processing time.

Then there are certain uses, as in the ceramic industry, where a dense oxide is required because it produces a superior product. All users like the free-flowing and dust-free properties of densified zinc oxide. For those users who do not require that the densified product retain the characteristics of the undensified product, or who prefer that the particle size be large and aggregated, the simple expedient of calcining or reheating the product to build up particle size suffices. On the other hand, there are users who prefer a small particle size but require all the other properties of a densified oxide such as low bulk density and moderately easy dispersion properties. In this case zinc oxide pellets or granules are available. The bulky zinc oxide powder can readily be converted into pellets by passing it through a tumbler or revolving drum such as the one shown in Figure 8-6

Figure 8-6. Pelletizing drum, Columbus.

without the addition of water or binder of any kind. Such zinc oxide is dustless, free-flowing and packs out as high as 90 pounds per cubic foot. However, pellets are not recommended for use in rubber, especially soft stocks, nor for paint, except in the cases where pebble mills are used. They are very satisfactory for ceramics and for chemical uses where complete solution in acids or other reagents can be obtained in a reasonably short time.

Deaeration yields a product which is a compromise between pellets and the normal zinc oxide powder. This is accomplished by the removal of a substantial quantity of the entrained air which reduces the bulk but does not alter the particle size. Deaerated oxide is free-flowing and dust-free and can be incorporated in paint or rubber almost as readily as the original material.

Blending. Blending equipment is a necessity in all zinc oxide packing houses because the many varieties that have to be made cannot possibly be carried in stock, but may be supplied by blending. For example, a high oil-absorption oxide may be blended with low, to produce any intermediate oil absorption that may be demanded without adversely affecting the color or chemical properties of the oxide.

Packing. Pigment zinc oxide is packed in multiwall paper bags, either tied, sewn, or of the valve type. In Figure 8-7 is seen a typical packing bay where the bags are packed with a Mogul packer, automatically weighed, flattened on the machine at the left, and stacked on the leveling device prior to being taken up by a fork-lift truck. Every manufacturer has his own preferred method of packing, but the final product is a stack of 50-pound bags called a unit and containing from 40 to 60 bags. These units are easily handled on a fork-lift truck such as shown in Figure 8-8.

Figure 8-7. Packing bay showing packer, flattener and bags on leveler, Columbus.

Figure 8-8. Units of bags being stacked by fork-lift truck, Columbus.

PREVAILING PRACTICE IN THE UNITED STATES

Although the fundamental metallurgy remains the same, the furnaces and equipment used in the manufacture of pigment zinc oxide have undergone substantial changes over the units of 20 years ago.[4, 5, 17, 18] During this time, the character of the zinc ore has changed. The coarse jig and table concentrates are no longer available and only flotation concentrates are used. The mixed lead and zinc ores are no longer available for leaded-zinc manufacture and the ores low in lead and cadmium, such as franklinite and willemite, are no longer available in quantity. Improvements in sintering practice have made possible the elimination of lead and cadmium. An outstanding development is the chain-grate modification of the conventional Wetherill grate furnace made by The New Jersey Zinc Company.[3, 19] The following is a description of this furnace, prepared by W. M. Peirce* and illustrated in Figure 8-9.

Traveling-Grate Zinc Oxide Furnace

A furnace producing pigment-grade zinc oxide directly from ore, known as the traveling-grate furnace, was developed by the New Jersey Zinc Company in the early twenties and they are the exclusive users of this type of furnace.

The furnace consists of a traveling grate similar to that used in the Coxe stoker. At the feed end there is provision for placing a layer of coal briquettes on the grate, and above this a layer of ore-coal briquettes. As this bed of briquettes travels through the furnace, under-grate air burning the coal briquettes generates the necessary heat to reduce the ore-coal briquettes. The zinc vapor given off is burned either within the furnace itself or in an external combustion chamber by the controlled admission of air. The resulting zinc oxide is collected in bag rooms. The residue, of course, is discharged by the continuous movement of the grate.

* Assistant to the Executive Vice President, The New Jersey Zinc Company.

Figure 8-9. Traveling-grate furnace. (*Courtesy The New Jersey Zinc Company*)

Many grades of oxide, to satisfy the specific requirements of various uses in paint, rubber and other products, are produced by controlling the composition of the ore and the combustion conditions to yield a pigment of the desired particle size and shape and chemical composition. Subsequent treatment may be given the pigment to modify the surface characteristics of the individual particles in a desired manner and the pigment may be pelleted to facilitate handling in certain types of end-use operations.

Rotary-Kiln Processes for Zinc Oxide

Rotary kilns for producing American process zinc oxide were proposed as early as 1912 and other attempts in this direction have been made from time to time,[20, 21] but no successful production of pigment zinc oxide had been recorded until the furnace now being used by the Eagle-Picher Company was developed. The following is a description of this furnace, supplied by the Eagle-Picher Company, as illustrated in Figure 8-10.

For a number of years, Eagle-Picher has been producing zinc oxide directly from high-grade sintered zinc concentrates using a rotary kiln instead of a Wetherill grade. Previously, small rotary kilns had been employed to some extent for this purpose but they had not been used by any of the large producers for making pigment-grade zinc oxide.

Eagle-Picher is now operating rotary kilns for the production of zinc oxide. These furnaces are 8 feet in diameter and 40 to 50 feet long. A 6-inch lining of super-duty firebrick is placed inside the steel shell.

These kilns are operated in a manner somewhat different from conventional rotary-kiln practice. The charge is introduced at the firing end and both the solid residue and zinc vapor issue from the opposite end. The kilns rotate at a very slow speed and are set up with a small slope in order to give sufficient retention time to eliminate zinc from the charge.

Such a kiln produces 15 to 20 tons of bag-room oxide per day. An additional 10 to 15 per cent settles out in the cooling system and is normally recirculated by mixing with the charge.

A typical charge mixture consists of 65 per cent high-grade sinter and 35 per cent

Figure 8-10. Rotary zinc oxide furnace and combustion chamber. (*Courtesy The Eagle-Picher Company*)

Pennsylvania anthracite. The charge is premixed by suitable mixer and elevated to a large feed hopper. From this hopper the feed mixture is fed by a constant-weight device and belt system to the furnace.

Draft on the kiln and the fuel-air ratio are closely controlled to maintain a reducing atmosphere in the kiln. The exit gases from the kiln, therefore, contain unburned zinc vapor. The atmosphere in the kiln, as well as the method of burning the zinc vapor after it leaves the kiln, are varied according to the type of zinc oxide it is desired to produce.

Combusion air for the burner is supplied by a one-pound blower. Gaseous fuel, either propane or natural gas, is used at the rate of 12 million to 18 million Btu per hour.

The zinc vapor and combustion gases leaving the kiln enter a firebrick combustion chamber where additional air is added by means of a low-pressure fan. The air inlets are arranged so that the amount of air admitted can be regulated and the zinc vapor-air burned in a short flame or more gradually, as desired.

The residue from the charge—consisting of a small percentage of zinc, gangue from the ore and some unconsumed coal—drops through a refractory lined chute into a water quench tank. A slow-speed belt drag removes the residue from the quench tank and another belt conveys it to a storage pile.

From the combustion chamber a flue leads to a series of cooling goosenecks and thence to the baghouse. The goosenecks are approximately 40 feet high and may be from 3 to 5 feet in diameter. The total area of the cooling system is approximately 15,000 square feet. The furnace gases enter this system at approximately 1600°F and are cooled to about 400°F. Dilution air dampers are provided immediately before the baghouse in order to further cool the gases to a maximum of 250°F.

The baghouses used with these kilns are of the modern automatic type, with a

TABLE 8-6. PRODUCERS OF AMERICAN PROCESS ZINC OXIDE

Company	Location	Type of Furnace	Type of Oxide Produced
American Zinc Oxide Co.	Columbus, Ohio	Wetherill Grate	Lead-Free
American Zinc Oxide Co. of Ill.	Hillsboro, Ill.	Wetherill Grate	Lead-Free Leaded
Eagle-Picher Co.	Hillsboro, Ill.	Wetherill Grate Rotary	Lead-Free Leaded
Eagle-Picher Co.	Galena, Kansas	Rotary	Lead-Free Leaded
New Jersey Zinc Co.	Palmerton, Pa. Depue, Illinois	Travelling Grate Travelling Grate	Lead-Free Lead-Free Leaded
Ozark Smelting & Mining Co.	Coffeyville, Kansas	Wetherill Grate	Leaded
St. Joseph Lead Co.	Monaca, Pa.	Electrothermic*	Lead-Free

* See p. 382.

filtering area of about 14,000 square feet. A typical baghouse consists of 14 cells of 78 bags each. Wool or synthetic bags, 6 inches in diameter and 8 feet long, are used.

Induced draft for the entire system is provided by one or two high static fans. Total volume handled by these fans is in the order of 40,000 cubic feet per minute, requiring 100 to 150 hp total.

Both the cooling goosenecks and the baghouses have hoppers beneath them from which the zinc oxide drops into screw conveyors. The oxide is transported by screw conveyor and belt to the blending and packing department.

In Table 8-6 are listed the United States' manufacturers of American process pigment zinc oxide, the types of furnaces used and the types of zinc oxides produced. It is interesting to note that in 1919 there were eighteen zinc oxide plants operating in the United States and producing 117,639 tons of oxide,[18] as compared with eight plants producing 154,955 tons in 1956. However, in 1919, selective flotation had not become the general practice and much of this oxide was recovered from mixed lead and zinc ores which could not be separated by the concentrating methods then known, and much of the oxide produced was not of pigment grade.

REFERENCES

1. Calbeck, J. H., Eide, A. C., and Easley, M. K., "Acicular Zinc Oxide," *Offic. Dig. Federation Paint & Varnish Production Clubs*, **No. 208**, 391 (1941).
2. Coulter, Marvin K., U. S. Patent 2,785,990 (March 19, 1957).
3. Singmaster, J. A., Breyer, Frank G., and Bruce, Earl H., U. S. Patent 1,322,142 (Nov. 18, 1919).
4. Faloon, Dalton B., "Zinc Oxide," pp. 49–96, Princeton, D. Van Nostrand Co., 1925.
5. Holley, Clifford Dyer, "The Lead and Zinc Pigments," pp. 152–199, New York, John Wiley and Sons, 1909.

6. Lewis, Warren K., U. S. Patent 1,442,485 (Jan. 16, 1923).
7. Maidens, W. T., U. S. Patent 2,139,196 (Dec. 6, 1939).
8. Maidens, W. T., U. S. Patent 2,150,072 (March 7, 1939).
9. Cyr, Howard M., U. S. Patent 2,331,599 (Oct. 12, 1943).
10. Wemple, Leland E., U. S. Patent 1,292,976 (Jan. 28, 1919).
11. Calbeck, John H., U. S. Patent 2,414,044 (Feb. 18, 1947).
12. Calbeck, John H., Maidens, W. T., and Hassel, Oscar J., "The Calbeck Process for Refining Zinc Oxide," *Trans. Am. Inst. Mining Met. Engrs.*, **188**, 757–63 (1950); *J. Metals*, **2**, 757–63 (1950).
13. Depew, Harlan H., U. S. Patent 2,372,367 (March 27, 1945).
14. Eide, Alwin C., U. S. Patent 1,997,925 (April 16, 1935).
15. Silver, Bruce R., and Bridgewater, Ernest R., U. S. Patent 2,303,330 (Dec. 1, 1942).
16. Gamble, David L., and Haslam, James H., U. S. Patent 2,251,869 (Aug. 5, 1941).
17. Liddell, D. M., "Handbook of Non-ferrous Metallurgy," 2nd ed., Vol. II, pp. 1217–19, New York, McGraw-Hill Book Co., Inc., 1924.
18. Hofman, H. O., "Metallurgy of Zinc and Cadmium," 1st ed., 2nd imp., pp. 285–303, New York, McGraw-Hill Book Co., Inc., 1922.
19. Singmaster, James A., U. S. Patent 1,112,853 (Oct. 6, 1914).
20. Hughes, H. H., U. S. Patent 1,014,062 (Jan. 9, 1912).
21. Kirk, M. P., U. S. Patent 1,566,103 (Dec. 15, 1925).

Burning Zinc Metal

John H. Calbeck

Director of Research, Pigment Division
American Zinc Oxide Company
Columbus, Ohio

In those methods for burning zinc metal currently in use in the United States, the gaseous products of the reduction step do not contaminate the zinc vapors when they are burned in air to produce zinc oxide. These are known as "indirect" processes and the product is known as "French process zinc oxide." Indirect processes using primary zinc metal and processes using secondary metal are discussed here; other secondary zinciferous materials and scrap are treated in the last section of the chapter.

Indirect processes were developed in Europe on a commercial scale in the nineteenth century. Le Clair[1] in 1847 burned the vapors issuing from a horizontal retort, collected the zinc oxide fume in settling chambers and promoted the use of the product in paints.

Properties of French Process Zinc Oxide

The outstanding property of French process zinc oxide is its chemical purity. Metallic zinc vapors of high purity are used in its manufacture and the product is uncontaminated by the products of combustion which are present in direct-process oxides.

The pharmaceutical grade (U.S.P.) of zinc oxide is well known. It is used extensively in medicine, cosmetics and lotions. Calamine lotion, originally prepared from calamine ores, is a mixture of zinc oxide with an oil, and was used for the treatment of skin diseases by the ancients. Recent studies[2, 3] indicate that the effectiveness of this lotion is due to the photogenic synthesis of hydrogen peroxide catalyzed by the zinc oxide even in subdued light. U.S.P. zinc oxide must be packed in special containers and the manufacturing facilities inspected periodically by inspectors of the Food and Drug Administration.

The chemical performance characteristics of French process oxides are different from those of American process oxides. This is largely due to differences in the crystal lattice and especially to the extent and character of the imperfections in the lattice. When the oxidation step is not carefully controlled, indirect-process oxides may contain detectable amounts of metallic zinc, and even if no metallic zinc can be detected by chemical analysis, it has been stated[4] that traces of metal remain that affect the physical-chemical behavior of the oxide. Furthermore, the fact that there is no superficial contamination by sulfur-oxygen compounds (see p. 349) may account for its superiority to American process oxide for many uses. For example, French process oxide is very fast-curing when used in rubber because of the absence of the sulfur-oxygen compounds. French process oxide becomes badly carbonated in the package and when stored in humid climates may develop a zinc carbonate content amounting to as much as 30 per cent. American process zinc oxide carbonates alter only slightly in similar circumstances.

Second to chemical purity is the outstanding color and brightness of French process oxide, which often grades as high as 86 per cent, as compared to 83½ per cent for American process zinc oxide. Although the indirect process lends itself to all the devices for the control of particle size and shape employed in direct processes, as a rule the indirect-process oxides have a finer particle size, with specific surfaces ranging from 5 to 15 square meters per gram. Figure 8-11 is a photomicrograph of a typical French process zinc oxide.

Production

The simple metallurgy of the process has not changed since the time of Le Clair, and, to a limited extent, the horizontal retort is still being used. However, newer methods for vaporizing the metal and burning the vapors have been developed in recent years. Collecting, conditioning and packaging the product are substantially the same as for American process oxide (see p. 357). Prevailing practice in the United States includes use of (1) horizontal retorts, (2) electric arc furnaces, (3) vertical refining columns, and (4) rotary burners.

Figure 8-11. Photomicrograph of French process zinc oxide. Magnification 1000×

Figure 8-12. Molten zinc is poured into horizontal retorts, Hillsboro, Illinois (*Courtesy The American Zinc Company of Illinois*)

Horizontal Retorts. For use in horizontal retorts, the Special High Grade metal is first melted in a remelter and the molten zinc (Figure 8-12) poured into one end of special carborundum retorts mounted in a furnace laboratory in single rows and fired with gas or oil. The metallic zinc vapors issuing from each retort are drawn by suction into a small vertical combustion chamber where they burn to zinc oxide in an excess of air drawn into the combustion chamber with the zinc vapors. The grade of metal used determines the chemical purity of the oxide produced. Special High Grade

Figure 8-13. Combustion chambers, French unit, Hillsboro, Illinois. (*Courtesy The American Zinc Company of Illinois*)

metal is preferred for use in horizontal retorts if oxides of the highest purity are required. The zinc oxide fume is drawn into a cooling and filtering system, collected, conditioned, and packed out, as described in the previous section of this chapter.

A photograph of three combustion chambers in the French unit operated by the American Zinc Company of Illinois at Hillsboro is shown in Figure 8-13. A row of these chambers, one for each horizontal retort, located opposite the clean-out door, is connected to an overhead manifold which delivers the oxide-laden fume to the cooling and filtering system. There is a distinct advantage in having small combustion chambers, in that the heat of combustion is readily dissipated by the time the fume reaches the header, and particle size growth is inhibited, resulting in an oxide of exceptional uniformity of particle size and high specific surface. Maintenance and supervision of a gang of small combustion chambers are, however, more costly than when large, single-combustion chambers are employed.

Electric Arc Vaporizers. Patents[5] were issued January 6, 1925 for a unique process of volatilizing zinc in an electric arc furnace and burning the vapors with a blast of air that chilled the resulting fume almost as rapidly as it formed. The grade of metal used determines the purity of the oxide produced. Oxides made in this manner are extremely fine, having specific surfaces as high as 15 square meters per gram. The particles (less than 0.2 micron in diameter) are much smaller in size than the optimum for color strength, and products using it are only slightly colored by it. For example, almost transparent rubber stocks can be made with substantial loadings of this zinc oxide and enamels using it are not tinted by the oxide used. Because of the high specific surface, this product is very reactive and has found wide acceptance in the chemical industries.

Figure 8-14. Top of refining column. (*Courtesy The New Jersey Zinc Company*)

Vertical Refining Columns. The most recent development in French Process zinc oxide has been made by The New Jersey Zinc Company utilizing their vertical refining columns (see Chapter 7) to provide a pure zinc vapor and thus produce a French Process oxide of exceptional quality. Prime Western metal may be used in the refining column to produce the pure zinc vapors required. Figure 8-14 shows the zinc vapor burning as it issues from the top of the refining column. The following description, supplied by W. M. Peirce,* outlines the method.

French Process Zinc Oxide Process. A method of making zinc oxide, commonly called the "French process," involves first the production of zinc metal from the ore and then the reboiling of this metal and combustion of the vapor. Since the vapor to be burned is pure zinc vapor, undiluted by carbon monoxide, nitrogen, and other products of combustion which are present in the so-called American process of producing zinc oxide directly from the ore, the resulting pigment is different in certain characteristics from American process oxide, particularly with respect to gases adsorbed on the surface of the particles.

The New Jersey Zinc Company produces French process oxide by burning the vapor from their vertical refining columns. These columns furnish a convenient means of reboiling zinc and, at the same time, of refining it to any desired degree of purity. Thus the rigid purity requirements for pharmaceutical zinc oxide can be met by this method.

Control of particle size and shape is achieved by control of the conditions under

* Assistant to the Executive Vice President, The New Jersey Zinc Company.

which the zinc vapor is burned, and, as in the case of American process oxide, treatments to modify the surface condition of the pigment particles may be employed and the product may be pelleted.

Rotary Burners. A simple method used extensively in Canada and Europe and to a limited extent in the United States makes use of a small rotary furnace to melt, vaporize and oxidize simultaneously the slab zinc fed into it mechanically. A revolving drum is heated to a white heat by a gas or oil burner and the feed of slab zinc is started. The zinc melts and begins to volatilize and the vapors are ignited by the excess of air admitted around the burner. The burning zinc vapors provide more heat for melting and vaporizing so that eventually no fuel is required and the burning keeps pace with the feed of slabs. The zinc oxide fume, after passing through a small combustion chamber, is cooled and collected in the usual manner. However, a true French-type of oxide is not produced because there is little or no particle-size control. The zinc vapors are oxidized at a high temperature, and no chilling of the fume is provided until a large particle size and much sub-sieve coarseness has developed.

Conditioning and Packing

French process oxides may be given the same conditioning treatments before packing as American process oxides receive, and the equipment is the same (see p. 359). Because of the high purity, refining or washing is not necessary. Surface treating of French oxides is common.

REFERENCES

1. Holley, Clifford Dyer, "The Lead and Zinc Pigments," p. 152, John Wiley & Sons, Inc., New York, 1909.
2. Reese, Davis R., and Guth, Earl P., "Formation of Hydrogen Peroxide in Calamine Lotions," *J. Am. Pharm. Assoc., Sci. Ed.*, **43**, 491–5 (1954).
3. MacNevin, Wm., Garn, Paul D., Livitsky, Michael, McBride, Harold, Florence, L. K.; and Crabb, Norman, "Experimental Paramaters in the Photochemical Formation of Hydrogen Peroxide Over Zinc Oxide," Read Sept. 10, 1957, American Chemical Society, Division of Inorganic and Physical Chemistry (for subsequent publication).
4. Papazian, Harold A., Flinn, Paul A., and Trivich, Dan, "Influence of Impurities on the Photoconductance of Zinc Oxide," *J. Electrochem. Soc.*, **104**, 84 (1957).
5. Breyer, Frank G., Gaskill, Earl C., and Singmaster, James A., U. S. Patents 1,522,096; 1,552,097; and 1,522,098 (Jan. 6, 1925).

Electrothermic Process

JAMES J. RANKIN

Production Engineer
St. Joseph Lead Company
Monaca, Pa.

Electrothermic methods for smelting zinc-bearing material have been a subject of study for many years. This is not surprising, since electric smelting with its possibilities for building large units with high energy efficiency and clean processing seemed attractive as compared with other procedures. Early attempts at electric smelting date back to at least 1885. They were concerned mainly with the recovery of zinc as the metal. However, in France, about 1907, the Cote and Pierron process was used experimentally to produce pigment zinc oxide. A U. S. Bureau of Mines bulletin[11] contains a discussion of this and other early attempts at electric zinc smelting. Of the early procedures, the most successful was undoubtedly the Swedish Trollhättan process (a metal process);[8] but due to operational difficulties, it was discontinued about 1931. Prior to this, about mid-1926, the St. Joseph Lead Company began to consider the possibilities of electrothermic zinc smelting in connection with its acquisition of zinc mining properties.

Pilot Plant Zinc Oxide Production

The development of a process was based on ideas on electrothermic zinc smelting by E. C. Gaskill, who had been doing pioneer work in the field and had become associated with the St. Joseph Lead Company. The process proposed using sized zinc sinter and coke—preheated for better conductance—as a resistor in a vertical smelting furnace. A series of pilot plant tests was inaugurated, beginning in late 1926.

An 18-inch inside diameter shaft-type retort furnace was built at the Company's Herculaneum, Missouri, lead smelter. The furnace was designed with three independently supported sections mounted on skewback rings. Annular carbon electrodes in the upper and lower sections (spaced 15 feet apart vertically) contacted the charge of mixed sized zinc sinter and coke. The charge was fed from an overhead bin through a rotary gate and was preheated in passage to the furnace by heat transfer contact in an oil fired preheater chamber situated above the top furnace section. Appropriate tuyère openings, or ports, permitted emergence of the distilled zinc vapor plus carbon monoxide gas from the reduction of the zinc oxide in the sinter by the carbon of the coke.

The evolved vapors could be burned to zinc oxide and carbon dioxide and the zinc oxide collected by filtration in a bag room, or ceramic surface condensers could be attached for tests on metal condensation. Furnace residues discharged through a cooling chamber to an intermittently operated discharge pan for further treatment to recover contained values. The charge, which served as a resistor, was supplied with power at approximately 225 kw to bring it to smelting temperature. In operation, no attempt was made to slag the charge unduly. Coke in excess of that theoretically required was carried, and the residues could be handled readily for treatment by screening and jigging to separate discard slag from excess coke and high-zinc residue sinter, both of the latter returning to the main furnace charge. Tests showed furnace recoveries of about 56 per cent per pass from sinter feed assaying 58 per cent zinc and 0.3 per cent sulfur.

In May 1927, the furnace was shut down after producing, from trial runs, about 25 tons of pigment zinc oxide and a little metal. It was realized quickly that the condensation of zinc vapor to metal was a separate problem and best left to later development. Consequently, all efforts were directed toward pigment production.

The 18-inch diameter furnace was followed by a 48-inch inside diameter furnace which started operation in April 1927, and with which experimental campaigns were made until May 1929. For some test runs, this furnace was modified to 57-inch diameter. The furnace was built in four sections, of which the top and lower were electrode sections. The annular electrodes of the 18-inch furnace were supplanted, because of power segregation difficulties, by individually controlled 12 by 12 by 60-inch length carbon electrodes mounted downward through entry ports. The vertical distance between electrode sections was about 22 feet. Three electrodes—top and bottom—120 degrees apart provided power distribution to the feed.

In order to obtain the data required in the design and building of a commercial furnace, many ideas were tested during the period of operation of the 48-inch furnace, with gradual encouraging progress toward a successful outcome. Coordinated with this was the development of a free-flowing charge suitable for furnacing, as well as a procedure for treating the residues whereby slag could be discarded and values of partially smelted sinter and coke recovered for return to the circuit.

As developed, the furnace operated with a charge approximating 30 pounds per minute consisting of ⅔ sinter and ⅓ coke. The sinter size ranged from ¼ to ⅝ inch; and the coke, from 3/16 to 9/16 inch. The power to the charge approximated 850 kw with an average voltage per leg of about 250. With a 59 per cent zinc sinter, 1.25 kwh was required per pound of zinc oxide produced. One-pass recovery was about 72 per cent with production averaging 7.5 tons zinc oxide per day.

Approximately 750 tons of pigment zinc oxide were produced during experimental runs with the 48-inch furnace, with the longest continuous run being of 15 days' duration. Most of this oxide was of good commercial grade.

With the process showing evidence of successful development, plans were made to build a plant designed initially to produce pigment zinc oxide and sulfuric acid. To protect the process, patent applications had been filed in the United States Patent Office in the name of E. C. Gaskill[1] and were granted in the year 1930, with the St. Joseph Lead Company operating under these patents.

The Commercial Plant and Early Practice

The site selected for the plant was on the Ohio River at Josephtown, near Monaca, in western Pennsylvania. The smelter which started operations in December 1930 had the following main divisions: roaster and acid plant; sinter plant with coke preparation and residue-treatment units; furnace plant and zinc oxide plant with collecting, packing and shipping units. The supporting facilities were an office and laboratory building, shops, change rooms, as well as yard and rail facilities.

The roaster plant comprised two 12-hearth Nichols-Herreshoff roasters. The feed consisted at first of zinc concentrate from the Company's Edwards Mine in St. Lawrence County, New York. Shortly afterward, the production of the nearby Balmat Mine was added. The concentrates approximated 57 per cent zinc and 32 per cent sulfur, with iron as sulfide the other major constituent. The roaster gases were converted to commercial grades of sulfuric acid in an adjoining vanadium pentoxide contact-acid plant.

The roaster calcine of about 1.5 per cent sulfur content was treated in a nearby sinter plant which operated with three 42-inch by 44-foot Dwight-Lloyd sintering machines. The feed consisted of the roaster calcine, coke, reclaimed products, silica and a considerable load of return sinter fines. The process was operated to produce hard sinter sized ¼ to ¾ inch, with 59 per cent zinc and 0.2 per cent sulfur. By-product coke was sized in a sinter-plant unit to about sinter size. Both materials were moved to their respective furnace-plant bins. Any lead or cadmium present in the ores was largely eliminated during the roasting and sintering operations.[7] A unit housed in the sinter plant also treated the residues from the electrothermic furnaces, rejecting low-zinc, high-iron products and reclaiming both coke and values with high zinc content.

The furnace plant as erected contained eight electrothermic furnace settings with five of these utilized for oxide furnaces and the remaining three for pilot plant work in connection with metal development. The oxide furnaces were of low-porosity firebrick construction[9] and were 57 inches in inside diameter and 37 feet in shaft height. They were designed with six independently supported sections mounted on steel skewback rings to provide expansion joints sealed with refractory cement. A rotating table at the shaft bottom removed furnace residues.[16]

Stepped-down power was supplied for operation through two sets of three carbon electrodes 12 x 12 x 60 inches in length. The electrodes—in diamond position—contacted the charge at a slope. They were spaced equidistant horizontally and 24 feet apart vertically. Power input per furnace was about 1200 kw at about 260 volts per leg. Charge input per furnace approximated 52 pounds per minute, proportioned 70 per cent sinter and 30 per cent coke. The charge was preheated to redness by passage through a gas-fired rotary drum and entered the furnace through a distributor top which also sealed the furnace. Output per furnace day approximated 10 tons with about one kwh required per pound of zinc oxide produced. Appropriate ports permitted escape of distilled zinc vapor plus carbon monoxide into affixed vertical manifolds where combustion occurred to zinc oxide fume and carbon dioxide. The combustion products passed under suction of fans (30,000 cubic feet per minute capacity) from the manifolds to a duct for delivery to the baghouse filtering system.

A furnace was started primarily on sized coke, gradually changing to normal sinter-coke mix as the furnace reached operating temperature. Extensive instrumentation permitted close operating control. Through regulation of burning conditions, three primary grades of lead-free pigment were produced, of fine, medium and coarse particle size. The oxides, of excellent purity, had particle characteristics of American process manufacture, with nodular, plate, rod and twin forms present. Through calcination of the furnace pigments, further diversification was obtained. The grades met satisfactorily the requirements of major consumers in the rubber, paint, ceramic and chemical fields.

Starting actively in 1936, the Josephtown Smelter began the commercial production of zinc metal, (see Chapter 6), in addition to zinc oxide, by the electrothermic process. Operating experience had indicated that enlargement of the original furnaces with resultant increase in furnace production capacity was now possible. Over the next several years this enlargement program was pressed actively; and with it, there was a general rearrangement of furnaces to permit expansion of metal production since furnace enlargement had provided considerably more oxide capacity than was needed at this time. The technological improvements and developments in the process have been described in detail in the literature (Refs. 12, 17 and 18).

Growth and Present Practice

The Josephtown Smelter had operated until about 1940 with ore concentrate originating from the Company's own mines in New York State.

With the continued growth of the plant, supplementary concentrates were needed; and over the years, sources both at home and abroad have been developed. Also, it was deemed advisable in 1948 to divide the plant operations into two parallel circuits; and in 1956, into three circuits. Two of these circuits are known as "Prime Western" and "Intermediate." They treat concentrates and zinc bearing products suitable for the electrothermic production of lead-bearing and other alloy grades of zinc metal. The other circuit, known as "High Grade," operates so that, through ore-concentrate selection and plant processing, it supplies both the oxide and high-grade-metal furnaces with a zinc sinter relatively free from lead, cadmium or other impurites. The operating steps involved are shown diagrammatically in Chapter 6, p. 271.

In the "high grade" circuit, the concentrates are roasted in two steps. As received, they approximate 57 per cent zinc, 0.6 per cent lead and 32 per cent sulfur; the remainder is mainly iron as sulfide. The preliminary roasting step, which is also a deleading operation, is conducted in five modified 12-hearth Herreshoff roasters, 19 feet 3 inches in inside diameter. The maximum hearth temperature is about 1000°C. The Herreshoff discharge calcine has a lead content of about 0.04 per cent (approximately 95+ per cent lead elimination) and a sulfur content of about 25 per cent. This calcine for further desulfurization moves to either of two fluidized-bed roasters, one 20 feet, and the other 22 feet 9 inches in inside diameter. These roasters operate at a temperature of about 950°C with a bed of approximately 5 feet under air pressure. The roaster gases under suction pass through waste-heat boilers, a cleaning system, and with about 6.0 per cent sulfur dioxide content, move to the acid-plant converter and absorption system for the production of commercial grades of sulfuric acid. The finished calcine, which approximates 67.0 per cent zinc, 2.0 per cent sulfur and 0.03 per cent lead, moves to the High-Grade section of the sinter plant for further treatment.

The sintering process at Josephtown[10] is designed to produce from the calcine a sized, porous sinter sufficiently strong to move through the electrothermic furnaces without undue breakdown. Also, the High Grade sinter circuit is operated to minimize unwanted impurities. These results are accomplished in two steps. In the first step, the calcine, plus an addition of return sinter fines and fine coke, passes over two 60-inch by 44-foot sinter machines to form a soft sinter. The top portion of the soft sinter cake, which contains the least impurities, is removed by a slicer; the bottom portion recirculates. In the second step, the top-portion sinter, after grinding, is mixed with return sinter fines, high zinc residues, coke and sand and passes over two 60-inch by 44-foot sinter machines. The product, after crushing and sizing to $\frac{1}{4}$ to $\frac{3}{4}$-inch range, enters the circuit feeding the High Grade electrothermic furnaces. This finished sinter assays approximately 57 per cent zinc, 10 iron, 8.0 silica, 0.1 sulfur, 0.007 lead and 0.007 per cent cadmium. Sinter gases pass through Cottrell precipitators. The recovered fume is treated in a wet chemical plant to reclaim values of zinc, cadmium and lead. Airborne dusts from sintering operations pass by means of fans to bag filters, the product returning to the sinter circuit.

A section of the sinter plant houses equipment to recover contained values of sinter and coke from furnace residues through a combination of screening, magnetic and air separation. The reclaimed coke obtained moves to the electrothermic furnace feed, the reclaimed sinter moves to the sinter machine feed, and the slag plus high iron residue moves to discard.

Coke required for the furnace charge, as well as coke required in the sinter mix, is processed in a section of the sinter plant. Incoming by-product coke selected for purity is sized to a range of approximately ¼ to ⅞-inch furnace feed. The undersize fine coke plus purchased coke breeze is ground by a rod mill to supply the requirements of the sinter-machine feed.

At present (1957) there are fifteen electrothermic furnaces at the Josephtown Smelter, of which four are zinc oxide furnaces and the remainder are metal furnaces. There are two sizes of oxide furnaces, one of 69-inch diameter and the other of 96-inch diameter, the shaft height of both being 37 feet. They are basically of the same sectionalized construction as the earlier furnaces. The smaller furnace produces daily about 22 tons of zinc oxide and the larger furnace about 34 tons. The feed for these furnaces (from the High Grade section of the sinter plant) is of optimum quality to produce high-purity zinc oxide or premium-quality zinc metal.

The older type 12-inch square carbon electrodes have been displaced by round 12-inch diameter by 60-inch length graphite electrodes because of greater economy and handling ease. The vertical distance between electrodes approximates 27 feet. The 69-inch diameter furnace operates with four equally spaced electrodes, top and bottom; and the 96-inch furnace operates with six electrodes similarly arranged. The upper electrodes protrude into the charge through ports, at a slope, for an average distance of about 13 inches beyond the inside wall and are adjusted at regular intervals to maintain proper charge contact. The lower electrodes are set horizontally, through entry ports, approximately 6 inches into the charge, and, since they suffer little wear during operation, are reset infrequently. Electrical contact is maintained from the power supply buses by means of water-cooled copper contact clamps.

The properly proportioned charge of sized sinter, coke or other sized zinc-bearing material feeds from overhead bins through a rotary, refractory-lined, gas-fired preheater. The charge, heated to about 800°C, moves through a rotary distributor and furnace top seal which delivers the material in a definite pattern to the furnace so that the charge is relatively coke-rich in the center and coke-lean toward the outside wall. This permits operating the furnace under load so that the major path for the applied power is through the hot coke-rich center with the wall charge relatively cool. A temperature gradient of from about 1300 to 950°C has been observed with these conditions. At the bottom of the furnace shaft, the wall is protected and supported on a steel water-cooled ring. A refractory-lined rotary discharge table with adjustable lift removes spent furnace charge at the desired throughput rate. The zinc elimination on one pass through the furnace approximates 75 per cent, with over-all recovery of better than 90 per cent. Cleanup of the discharged material takes place in the adjoining sinter plant as already described.

Plant power is at present supplied by the local public utility company from two 69,000-volt lines and stepped down through transformers to 2300 volts. Further distribution to the electrodes is by furnace transformer banks operating at 2300/290–190 volts with 10 per cent regulation possible by induction regulators. A new 100,000 kw generating station (plant operated) soon will modify the incoming power distribution system. The kilowatt load per electrode is automatically controlled with the load limited to about 600 kw. The 69-inch oxide furnace operates at about 2000 kw with a charge input of about 81 pounds per minute consisting of 70 per cent sinter and 30 per cent coke. The 96-inch furnace operates similarly at about 3000 kw with a charge input of about 115 pounds per minute. Figure 8-15

Figure 8-15. Control panel, St. Joseph Lead Company electrothermic furnace instrumentation.

shows furnace-control instrumentation. Power requirements for smelting approximate 1.05 kwh per pound of zinc oxide produced.

The manpower requirement for operating the small oxide furnace is one man per shift with two men required for the large furnace. In addition, there are operators regulating incoming feed and outgoing discharge residue, who look after the needs of a bank of furnaces. Furnace campaign life varies considerably and is determined mainly by slag build-up inside the furnace shaft. An average life figure for all conditions approximates 7 to 8 months, but it is not uncommon to run for over a year's time. Usual downtime for repairs is from 7 to 10 days.

With normal charge of sinter and coke in the furnace, the carbon of the coke reduces the zinc compounds of the sinter and liberates approximately one volume of zinc vapor to one volume of carbon monoxide. The liberated vapors, passing through the exit ports, burn to zinc oxide fume and carbon dioxide gas in attached manifolds under controlled temperature, rate of oxidation and flow. Usually there are four ports to a manifold; however, in the case of the newer 69-inch furnace there is but one large port per manifold. The combustion manifolds of a small furnace join to a common 42-inch diameter furnace duct. Due to greater output, the combustion manifolds of a large 96-inch furnace deliver to two 42-inch ducts. It is possible to produce a fine and coarse grade of oxide simultaneously with the large furnace by appropriate construction and control of each manifold duct system.

The oxide furnace manifolds and ducts are constructed of sheet steel. They are either unlined or refractory-lined well into the duct, depending upon the grade of oxide produced. In the production of fine-particle grades, the heat of reaction is quickly dissipated through the unlined manifolds with oxidation through air ingress and flow rate relatively fast. Conversely in the case of the production of coarse-particle oxides, the heat of reaction is conserved in the refractory-lined manifolds with oxidation through air ingress and flow at a relatively slow rate so that the microcrystalline particles have time for growth. Operating temperature in the manifolds, depending upon the grade desired for production, may range from about 200 to 1000°C. Figure 8-16 illustrates diagrammatically the oxide furnace producing system. For certain uses, it is possible to modify the type of pigment formed through the addition of metallic zinc to the furnace charge or through the burning of a fuel such as natural gas in the manifolds, in this way varying the ratio of zinc vapor in the combustion manifold gases.

The oxide-laden gases from the furnace manifolds, under fan suction, pass through water-cooled tubular units integrated into the duct system. On the pressure side of the fan, they pass through a cleaning cyclone and, before entry to the baghouse, are further cooled if needed to a temperature of 275°F, by the automatic introduction of auxiliary air.

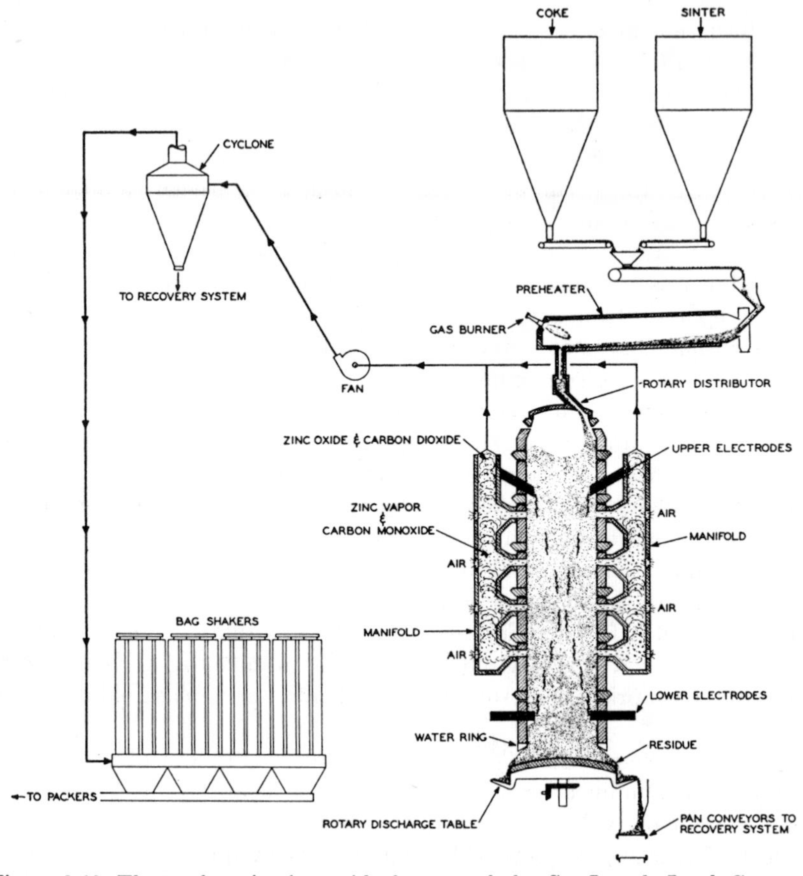

Figure 8-16. Electrothermic zinc oxide furnace of the St. Joseph Lead Company.

The separation of the oxide from the gases is accomplished by filtering through cotton fabric bags, 20 inches in diameter by 45 feet in length, equipped with long-wearing nylon boots. Filter area is about one square foot per pound of zinc oxide per day. The bags are shaken automatically in a timing cycle. The freed oxide falls from the fabric into a sheet-metal collect hopper from which it moves by conveyor and elevator to large overhead storage bins. From here, it moves by conveyor to screens and packers or continues onward for further processing. Normally the pigment's zinc oxide content is 99.5 per cent or higher unless for special uses it is intentionally modified by the introduction of additives. When further processing is required the pigment may be subjected to operations such as calcination for densifying or brightening it, pelletization or granulation to convert it to a compact, free-flowing, dustless form or coating, with appropriate reagents to change its surface activity.

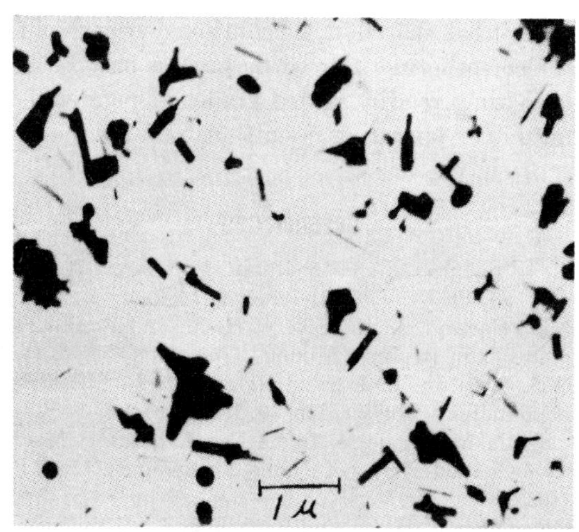

Figure 8-17. Electron microscope photograph of electrothermic zinc oxide pigment. Magnification 10,000×.

St. Joseph Company's electrothermic zinc oxide is a white, smooth-textured pigment of relatively fine particle size, *viz.*, with a particle range of from about 0.01 to 1.0 micron numerical diameter, as observed under the electron microscope. The specific surface, depending upon grade, may range from about 10 or more to 3 or less square meters per gram, determined by the nitrogen absorption procedure. Fine particle size and high specific surface are complementary, indicating in the case of zinc oxide a pigment of a high order of chemical reactivity. Coarse particle size and low specific surface, on the other hand, indicate a relatively low order of activity. Either may be of value depending upon use; in a protective coating, for example, the fine pigment imparts body and color retention, while the coarse pigment imparts stability with low oil demand and permits high pigment loading. Figure 8-17, shows a fine grade of pigment at a high magnification.

As packaged, the oxide grades vary in density from about 20 to 85 pounds per cubic foot; and they are supplied to the market in paper bags, drums or barrels. For economy in handling, the packaged grades may be formed into units of about one ton and shipped as such, or the granular or pelleted form may be shipped in bulk carloads. Data by the Josephtown technical staff concerning electrothermic zinc oxide in application have appeared in the literature.[2-6, 13-15]

SUMMARY

The electrothermic process for producing zinc oxide as practiced by the St. Joseph Lead Company has been in commercial operation for over 25 years. For the first five or six years of smelter operation, oxide was the only zinc product from the plant. Since the successful introduction of electrothermic zinc metal, it has shared in the continued growth of the Josephtown operations. The electrothermic zinc oxide process may be characterized by its flexibility in yielding readily varied grades of pigments, by the precise controls it permits over operating conditions and by the excellent quality of the pigments produced.

REFERENCES

1. Gaskill, E. C., U. S. Patents 1,743,886 (Jan. 14, 1930), 1,743,964 (Jan. 14, 1930), 1,773,779 (Aug. 26, 1930), 1,775,591 (Sept. 9, 1930).
2. Havenhill, R. S., Carlson, L. E., Emery, H. F., and Rankin, J. J., "Electrical Contact Potentials in Banbury Mixing," *Ind. Eng. Chem.*, **45,** 1128 (1953).
3. Havenhill, R. S., Carlson, L. E., and Rankin, J. J., "Electrical Potentials in Rubber Compounding," *Rubber Age*, **79,** 1, 75 (1956).
4. Havenhill, R. S., and MacBride, W. B., "A New Laboratory Machine for Evaluating Breakdown Characteristics of Rubber Compounds," *Ind. Eng. Chem., Anal. Ed.*, **7,** 60 (1935).
5. Havenhill, R. S., O'Brien, H. C., and Rankin, J. J., "Electrostatic Properties of Rubber and GR-S," *J. Appl. Phys.*, **15,** 731 (1944).
6. Havenhill, R. S., and Rankin, J. J., "The Role of Zinc Oxide in Compounding Government Synthetic Rubber," *India Rubber World*, **107,** 365 (1943).
7. Isbell, W. T., and Weaton, G. F., U. S. Patent 1,940,912 (Dec. 26, 1933).
8. Landis, W. S., "The Trollhättan Electrothermic Zinc Process," *Trans. Am. Inst. Mining Met. Engrs.*, **121,** 573 (1936).
9. MacBride, W. B., "Notes on Refractories for the Electrothermic Zinc Industry," *Bull. Am. Ceram. Soc.*, **14,** 389 (1935).
10. Najarian, H. K., Peterson, K. F., and Lund, R. E., "Sintering Practice at Josephtown Smelter," *Trans. Am. Inst. Mining Met. Engrs.* **191,** 116 (1951).
11. O'Harra, B. M., "The Electrothermic Metallurgy of Zinc," *Bull. 208, U. S. Bur. of Mines*, (1923).
12. Rankin, J. J., "Zinc Oxide Production at the Josephtown Smelter," *Mining and Met.*, **28,** 402 (1947).
13. Rankin, J. J., and Behan, L. J., "Lead-Free Zinc Oxide in Modern Protective Coatings," *Paint Ind. Mag.*, **70,** 11, 17 (1955).
14. Rankin, J. J., and Havenhill, R. S., "Zinc Oxide—Properties and Commercial Applications," Technical Bulletin, St. Joseph Lead Co., 1948.
15. Reising, J. A., "Characteristics of Zinc Oxide as a Raw Material Used in the Manufacture of Ferrites," *Ceram. Age*, **62,** 3, 13 (1953).
16. Weaton, G. F., U. S. Patent 1,932,388 (Oct. 31, 1933).
17. Weaton, G. F., "St. Joseph Lead Company's Electrothermic Zinc Smelting Process," *Trans. Am. Inst. Mining Met. Engrs.*, **121,** 599 (1936).
18. Weaton, G. F., Najarian, H. K., and Long, C. C., "Production of Electrothermic Zinc at Josephtown Smelter," *Trans. Am. Inst. Mining Met. Engrs.*, **159,** 141 (1944).

Production of Zinc Oxide from Secondary Sources

T. R. JANES

Superior Zinc Corporation
Philadelphia, Pa.

In the early thirties secondary zinc oxide was developed from a number of sources in various ways. In one method, zinc dross, scrap zinc, zinc stampings, etc., were used as feed for a reverberatory furnace. This proved unsatisfactory however because the recovery was poor and it was impossible to hold back the impurities. From the reverberatory furnace a double-arch muffle type furnace was developed in which any type of zinc die-cast metallics and a certain amount of other metallic zinciferous material could be used. This type of furnace held back the objectionable material such as aluminum, copper, iron, antimony, tin and lead, and a very good secondary French-type zinc oxide was obtained. Also, from this same furnace, during the same operating run, part of the zinc distilled could be condensed and tapped out as zinc metal of a quality comparable to Intermediate and Brass Special grades.

Secondary zinc oxide has also been produced from dross, scrap zinc, stampings, etc. using upright retorts or bottles, or those set at an angle, in which the high-boiling-point impurities such as lead are held back in the base of the retort or bottle. The advantage of using the retorts or bottles is that almost any metallic material, even one fairly high in impurities, can be charged. The above materials produce a very good secondary oxide by this process.

The secondary zinc industry is a large producer of zinc dust, used primarily in the chemical field. As a by-product of the zinc dust used by the chemical industry there is a large quantity of hydrosulfite sludge. In the early thirties one company obtained a patent for cleaning up this sludge to a degree that it could be used in the ceramic and rubber industries. This process consists of calcination, quenching, hydro-classification, thickening, drying and milling. The material obtained in this manner is an exceptionally good zinc oxide for the ceramic industry in that it is very dense and works well in glazes. Practically all secondary zinc oxide is shipped to the rubber and ceramic industries.

There are five companies refining this sludge, four of which work with their own by-product, using sludge obtained from processes in their plants. Zinc oxide made by this form of processing accounts for a major portion of the secondary zinc oxide used east of the Rockies.*

In recent years, some secondary zinc oxide has been produced from the fume from mining and smelting plants.

* See also Chapter 12 for additional information on zinc oxide produced by wet or chemical processes.

CHAPTER 9

THE PROCESSING AND USES OF METALLIC ZINC AND ZINC-BASE ALLOYS

Grades of Zinc and Fields of Use

C. H. Mathewson

Professor Emeritus of Metallurgy and Metallography
Yale University, New Haven, Conn.

With a Supplementary Section by G. H. Turner

The section on Consumption in Chapter 2 should be consulted by way of introduction to the present chapter, which is addressed specifically to the details of processing and use of the metal zinc in its commercially pure or alloyed form. Primary metallic zinc comes into processing and fabricating plants as slab zinc (produced by one of the methods treated in Chapter 6) for remelting, casting and the mechanical treatment that furnishes sand, slush, or die castings, and rolled or otherwise wrought zinc products.

The old established large-scale uses of the metal are (1) for galvanizing (2) for brass-mill and foundry products and (3) for rolling into sheet or strip. However, Table 9-1 shows that a new product—namely, die castings —came so prominently into use in the period beginning about 1925 and continuing to the present, as to constitute an outlet for slab zinc, now practically equivalent to galvanizing in industrial importance.

During the war years, 1939–45, the use of brass for cartridge cases increased enormously, so that in 1942–44 more slab zinc was used for this than for any other purpose. But in 1945, as shown in the table, brass-mill and foundry products reverted to second place, and shortly thereafter (1946) to the third rank of importance now held, as a composite vehicle for the consumption of slab zinc.

Quality Standards

The major uses also determine the course of technical advance in establishing quality standards for the raw material. For example, the large-scale manufacture of cartridge cases during World War II directed attention to the production of metal relatively low in impurities, particularly lead, required for the desired hot-rolling of brass prior to forming. And earlier it had been realized that any sensible contamination of the zinc used for die casting could not be tolerated. It was the discovery of this high

TABLE 9-1. USES OF SLAB ZINC IN THE UNITED STATES (SHORT TONS)*

	1925	1930	1935	1940	1945	1950	1955†
Galvanizing	283,000	217,000	195,000	287,000	337,181	441,686	439,697
Brass making	165,000	120,000	124,000	232,000	259,377	139,373	144,816
Rolled zinc	71,100	51,400	56,500	58,000	97,589	68,444	50,363
Zinc alloys (mostly die castings)	—	21,500	55,500	58,000	130,836	289,527	404,790

The initial figure for die castings was 13,500, in 1926

* Summarized from yearly data in "Metal Statistics," published by the American Metal Market.
† Corresponding figures for 1955 given in Table 2-8 are not significantly different from these. The previous table gives more detailed data on consumption by industries since 1947.

purity requirement that made the zinc die casting industry possible, since a product made with zinc of ordinary purity is subject to damaging intercrystalline corrosion and swelling.[1]

The preferred alloys in the die casting field according to revised (1957) tentative specifications of the American Society for Testing Materials[2] contain (1) 3.5 to 4.3 per cent aluminum, with 0.25 maximum copper and 0.03 to .08 magnesium; and (2) 3.5 to 4.3 per cent aluminum, with 0.75 to 1.25 copper and 0.03 to .08 magnesium; both restricted to maximum tolerances of 0.1 per cent iron, 0.007 lead, 0.005 cadmium and 0.005 tin. A third composition, with copper up to 3.5 per cent,[3] was for many years included in the group of recommended alloys.

These alloys are generally known by the ASTM or SAE numbers, or as "Zamak" 3, 5, and 2, respectively.* The last, or highest-copper alloy, while superior in strength properties and resistance to wear, as-cast, does not match the others in dimensional stability or retention of impact strength on aging, and is no longer in general use. Tensile strength in the range of 40 to 50 thousand psi, with about 10 per cent elongation, is expected with these alloys. They also have an impact strength of 40 to 50 foot-pounds in the Charpy test at room temperature, but become embrittled when chilled. (See section of this chapter on Zinc-Base Alloy Castings for detailed discussion of die castings.)

The effect of the purity requirements noted above is seen in the chemical requirements of the ASTM Standard Specifications for Slab Zinc, B6-58,[4] reproduced in Table 9-2. It is the Special High Grade metal that fulfills the stringent requirements for purity which have sharpened the techniques for producing the zinc that is needed, particularly for die castings. (See the detailed sections of Chapter 6 on Metallurgical Extraction.) Thus, during the year 1955, as reported in "Metal Statistics" from U. S. Bureau of Mines data, 381,000 tons of Special High Grade metal were produced in

* See special publication of The New Jersey Zinc Company, "Zinc Metals and Alloys," for exact composition limits of the "Zamak" alloys.

Table 9-2. ASTM Specifications for Slab Zinc (Spelter). Chemical Requirements (Maximum Percentages)

	Lead	Iron	Cadmium	Aluminum	Sum of Lead, Iron and Cadmium
(1a) Special High Grade	0.006	0.005	0.004	none	0.010
(1) High Grade	0.07	0.02	0.07	"	0.10
(2) Intermediate	0.20	0.03	0.50	"	0.50
(3) Brass Special	0.60	0.03	0.50	"	1.0
(4) Selected	0.80	0.04	0.75	"	1.25
(5) Prime Western	1.60	0.08	—	—	—

Note 1—Analysis shall not regularly be made for tin, but when used for die castings, if found by the purchaser, tin shall not exceed 0.003 per cent.

Note 2—See Ref. 4.

Note 3—Where it is specified by the purchaser at the time of purchase that the special high grade zinc is to be used for the manufacture of zinc-base die casting alloy ingot, the maximum permissable tin content shall be 0.002 per cent and the maximum permissable lead content shall be 0.005 per cent.

Note 4—See Ref. 4.

the United States (compared with 404,790 tons of die castings for the same year) and 124,000 tons of High Grade metal (compared with the 144,000 tons of zinc used in brass products); while about the same total amount of the lowest grades, namely, 502,000 tons, mostly Prime Western, were produced by the domestic industry.

In spite of the general excellence of the zinc die casting alloys, best exemplified by a normal yearly consumption of more than a quarter of a million tons in the automotive industry (for such items as carburetors and fuel pumps, door latches and handles, trim, grilles, etc.), there is still a clear research incentive for further increasing their strength, and particularly, resistance to abrasion. This is also applicable to the die-casting type alloy (high copper), as sand-cast to make the large forming dies used to shape sheet metal into automobile fenders, aircraft wing coverings, etc. Proprietary compositions hardened by special alloying are known in this field.

Modified procedures, such as vacuum casting and casting combined with simultaneous assemblage of the various units in a final product, have also been exploited in this field.

Use of Common Grades

The greatest outlet for the zinc of ordinary natural or unrefined purity—Prime Western, No. 5 in the tabulated list—is for the general run of galvanized steel products, where any of the slab zinc grades are usually specified as acceptable.[5] However, in viewing the galvanizing industry as a whole, certain higher purity standards have been set up, as in the case of galvanized steel core wire for electrical conductors (utilizing High Grade zinc).

There are two fundamental causes for an impaired ductility in hot-dip zinc (galvanized) coatings; namely, the presence of a brittle alloy layer of zinc combined with iron below the visible surface coating, and the specific tendency toward brittleness of impure zinc itself. In the case of strip-galvanizing in continuous lines, in which aluminum added to the bath inhibits this alloy-layer formation, there is the possibility of intercrystalline oxidation which, as noted above, could be eliminated in die castings only by using a special high grade of zinc; but this does not appear as a recognized defect in the tight coatings now available.

In commercial practice, certain advantages, such as the desirable effect of the cadmium present in Prime Western zinc on the appearance of the spangle, operate against the substitution of High Grade metal. Tin or antimony may also be added for further effect. At the other extreme, in the production of electrogalvanized coatings, the purity of the zinc is a valued feature.

The grades of zinc between High Grade and Prime Western (see Table 9-2) are adapted to various uses in which the high purity required for die castings or highest-grade brass is not essential, and the specified amounts of *lead* and cadmium are either desired or tolerated. Thus, there are many leaded brass-mill products in which lead can come from the spelter, cadmium being generally unobjectionable.

Possible Quality Standards of the Future

In this day of superproducts based on theoretical applications, unusual demands, or improved technology, extreme purity is one of the criteria that has invaded many fields. The electrolytic or modern fractionating refining processes can provide selected products well beyond the purity of 99.99 per cent specified for Special High Grade zinc. Even the most refined commonly known analytical methods will not differentiate degrees of purity beyond 99.999 per cent. Chemically pure (C.P.) zinc can be defined no more accurately than 99.999 per cent. Beyond this purity, the comparatively new process of zone-refining,[6] now becoming well known, or other special treatment, would be required and the purity achieved would be indicated by the properties attained in very special uses.

Although there are as yet no clear indications, it is of course possible that this super-pure zinc may find a place of important application in industry as an alloying component, or even a product in its own right. The following supplementary section gives details concerning its preparation, evaluation and general characteristics, including a present estimate of its utility.

Super-Purity Zinc*

G. H. Turner
Senior Research Engineer
The Consolidated Mining and Smelting Company of Canada Limited

Zinc metal of a purity significantly higher than that of the commercial Special High Grade zinc has been available for many years. While this Special High Grade product has a guaranteed zinc content of 99.99 per cent, purer grades are generally restricted to a maximum impurity content of one or two parts per million when included in the super purity classification. As in all pure materials, the effects of specific impurities are distinctive and even within this stringent specification, each impurity known to be present must be considered in terms of the use to be made of the zinc. For example, the tolerance level of cadmium in zinc oxide phosphors is much higher than that of gallium.[7]

Production. Small quantities of super-purity zinc are available commercially. It is commonly produced by vacuum sublimation.[8, 9] Other methods of refining are, however, available. These include reflux distillation, zone-melting and electrolysis. Each method has limitations.

Electrolysis is theoretically capable of producing very pure zinc. By the use of certain addition agents, the refining of solid zinc anodes to a lead content of 0.2 parts per million has been observed.[10] Amalgam electrolysis is capable of producing metal of purity approaching that required for super-purity zinc.[11] The most formidable problem with electrolysis is the inclusion in the product of additional elements arising from the electrolyte such as oxygen, chlorine or sulfur.

Zone-melting is an effective method for removing many heavy metals such as lead, cadmium and bismuth. However, it is relatively ineffective for removing iron and copper because the segregation coefficients of these two elements in zinc are close to unity.

In all the methods discussed above, the choice of materials of construction for the equipment used and the care taken in controlling the process and handling the product will be equally important in establishing a high degree of purity as the basic features of the purification process. It is perhaps worthwhile to note that combinations of the above processes could be very effective (for example, distillation and zone-melting). The separation of iron and copper, which is not achieved by the latter method, is—at least theoretically—well within the capability of a distillation process.

Analysis. In the field of pure metals, the problem of analysis is at least as important and difficult as that of production. The analytical data guide the development of production techniques. The most commonly used ana-

* This section, through p. 392, by Mr. Turner.

lytical method for pure zinc is emission spectroscopy which currently permits the determination of most metal impurities to one part per million. By less frequently used methods, the sensitivity can be increased for many of the familiar impurities to values as low as 0.1 part per million.

Polarography is a technique capable of determining impurities at very low levels. By using a variation of the concentration technique employed by Truesdale and Edmunds,[9] lead, cadmium and copper can be determined to 0.1 part per million. More recently, the development of square wave polarography has promised a very sensitive analytical method. A polarogram taken at one-fourth sensitivity on a "Model 1336A Square Wave Polarograph" produced by Mervyn Instruments is shown in Figure 9-1. This shows the waves of copper, lead and cadmium corresponding to 1.40, 0.50 and 0.30 part per million, respectively, in zinc metal. Barker and

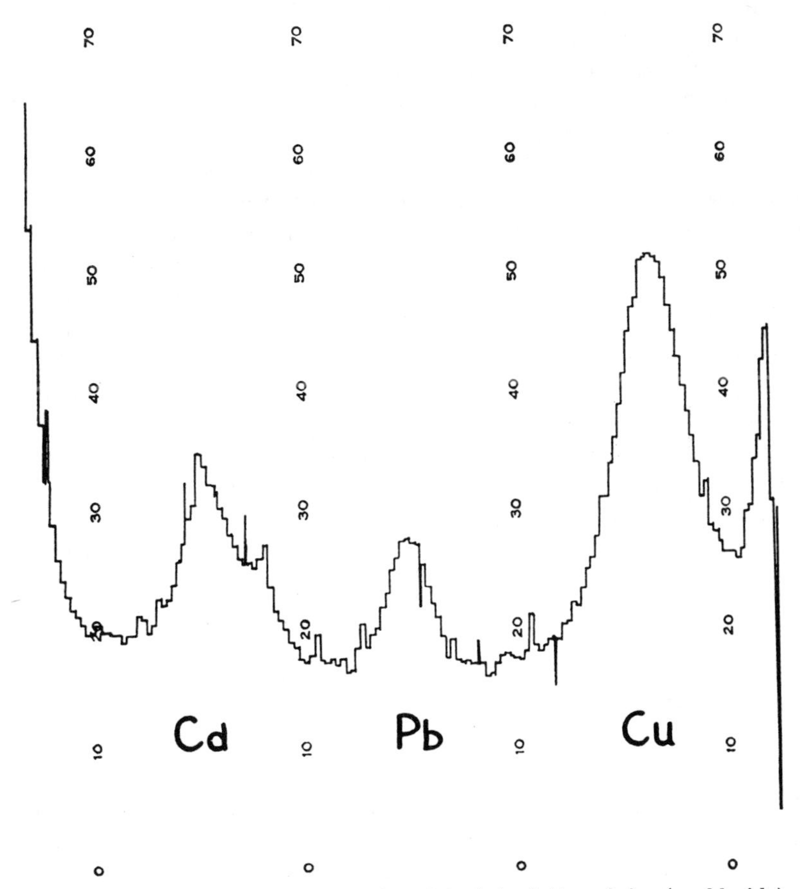

Figure 9-1. Square wave polarogram for a 0.2 g/ml solution of zinc (as chloride).

Jenkins[12] have suggested that this technique can be used to measure impurity concentrations as low as 0.001 part per million in certain cases.

Neutron activation analysis has also been applied to zinc. Measurements have been made at 0.05 part per million of cobalt, and 0.15 of thallium.[13]

The more familiar chemical and spectrophotometric methods may be usefully applied in the case of certain impurities down to 0.1 part per million.

A promising potential method which is still in the research stage is mass spectrometry. It is sensitive to all elements, and concentrations as low as 0.001 part per million can probably be reached with further development.

McLaren[14] has demonstrated that by precision thermometry, it is possible to differentiate between zinc samples of various impurity contents in the range 1 to 100 parts per million by measuring the melting points and melting ranges. Although the technique is not readily adaptable to a general method of analysis, it can be used as an added check for other methods. McLaren's results also emphasize the influence exerted by the presence of a very small amount of foreign elements.

Properties. Truesdale and Edmunds[9] have shown that as little as 0.01 per cent of impurities in zinc can change the mechanical properties sufficiently, so that on compression the impure zinc cracks, while a purer sample (0.001 per cent of impurities) recrystallizes under the same conditions.

It has also been observed that lead[15] and cadmium,[16] when present in zinc in a bulk concentration of only a few parts per million, actually concentrate in clusters in the solid zinc to the extent that they become microscopically visible. One such example is shown in Figure 9-2.

Chemically, impurities in low concentrations (10 parts per million or less) can also exert strong influence. For instance, it is common knowledge that, in pure aqueous sulfuric acid, the rate of solution of zinc is closely related to the zinc purity. In 6*N* sulfuric acid at room temperature, a zinc containing about 10 parts per million of impurities dissolves at a rate approximately 10 times as fast as one containing about 2 parts per million of impurities. It is also known that small concentrations of mercury can retard the rate of solution appreciably while similar quantities of cobalt accelerate the rate of solution manyfold.

Uses. The present uses of very pure zinc are few. It has been used for the production of other high-purity metals such as semiconductor-grade silicon. It can also be used for the production of high-purity chemicals; for example, zinc oxide phosphors. The potential value of pure zinc, as is the case with many other pure metals, does not lie so much in its application in the purest attainable form, but rather in permitting extremely close control of low-level additions. This frontier is being currently opened rapidly, as evidenced by developments in the semiconductor field.

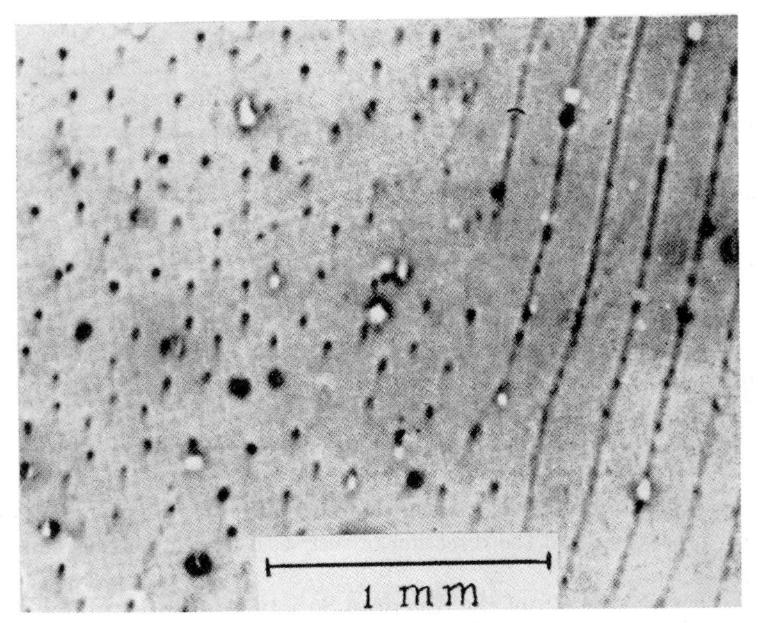

Figure 9-2. Segregation of lead in zinc single crystal.

Casting *vs.* Fabrication of Zinc and Alloys

Zinc can be processed by all of the common mechanical operations of rolling, drawing, forming, extruding, stamping, spinning, etc., although die casting (of suitable compositions) is by far the most prevalent method. However, zinc-base alloys containing more than a small percentage of copper, or fractional percentages of many other metals, such as nickel, iron, magnesium, manganese, chromium, titanium, etc., cannot be worked satisfactorily. This is probably a reflection of the constitutional limitation that there are no zinc-base solid solutions containing more than very small, generally fractional, percentages of the added metal in the low-temperature range of practical operation. Copper dissolves in zinc up to about 2½ per cent at 400°C and 1 per cent at 200°C.[17] Aluminum is undesirable in rolled sheet or strip owing to the danger of intercrystalline oxidation and brittleness, and is extremely sensitive to the simultaneous presence of various impurities in minute quantity. Beneficial hardness, or other specific property attainable with the intermetallic compounds formed by alloying with small fractional percentages of other metals, is overcome by brittleness when larger additions are attempted.

The so-called "Zilloy"[18] alloys containing a maximum of 1 per cent copper with or without a very small addition of magnesium (0.01 per cent) are known in this country as stiffened varieties of rolled zinc. Other special creep-resistant modifications, based on High Grade or Special High Grade zinc, include minor additions of titanium and/or manganese.* The various "Eraydo"[19] alloys consist of a High Grade zinc base (minimum 97 per cent zinc) alloyed with varying amounts of other constituents such as copper, magnesium, manganese and chromium. They are processed through the casting and subsequent rolling operation by a distinctive method for the purpose of improving the engineering properties, both mechanical and physical, over those normally available in the commercial grades of rolled zinc.† Similarly, the "Zinaloy" alloys of the Matthiessen and Hegeler Zinc Company contain copper, with other selected alloying constituents in small amounts, used in connection with specialized processing to improve the rolled product.

In general, the acceptable range of mechanical properties associated with commercial zinc in fabricated form is from about 20,000 to 30,000 psi tensile strength, with elongation in the range of 30 to 65, depending upon purity, processing and the direction of test with respect to the rolling direction.

Specifications (B 69) of the American Society for Testing Materials for Rolled Zinc designate chemical composition only for illustrative purposes, with lead from 0.05 maximum in a high-grade mixture to 0.65 per cent in an ordinary grade; and similarly, iron from 0.010 to 0.020, cadmium from 0.005 to 0.35, and copper from 0.001 maximum to 1.25—together with low lead, cadmium, and magnesium (0.007 to 0.02) in a special composition.

Owing to the low recrystallization temperature of plastically deformed zinc, it is not feasible to cold-work the ordinary commercial metal to specified tempers and then anneal it for softening and grain size control, as it is with the general run of copper, brass and other alloys. The temper of the rolled product therefore depends uniquely on the control of variables such as percentage reduction, speed, frictional characteristics, rate of cooling, etc., and there is abundant opportunity for individual producers to establish a competitive position by developing proprietary compositions and methods. In this connection, small alloying additions of other metals, which generally have a minor effect on the ordinary mechanical properties, may operate disproportionately to raise the recrystallization temperature, inhibit grain growth, or otherwise modify properties. For example, Jevons[20] writes that "bubbling boron trichloride through the molten metal induces a very fine crystal structure in the cast metal and minimizes the natural tendency for zinc to increase in size crystals at temperatures only slightly higher than room temperature."

* Private communication from E. H. Klein, The New Jersey Zinc Company.
† Private communication from W. R. Synnott, Illinois Zinc Company.

The use of rolled zinc appears to be much more prevalent in Europe than in this country: viz., our 50,000 tons (see Table 9-1) serving some 160 million people should be compared with some 200,000 tons for the same population in Europe. This is due to the much greater use of zinc abroad in roofing applications, where long experience with the metal has brought appreciation and acceptance of its peculiarities, which include a high coefficient of expansion, well-defined directional properties and creep phenomena. These characteristics can be altered by alloying[21] so as to bring a closer approach to the familiar cubic metals, aluminum and copper, and require a new evaluation of rolled zinc in some of its competitive lines.

Wire drawing and extrusion are of limited but increasing importance in the zinc field. The largest use of wire is for metallizing, in which the zinc coating is applied by a continuous run of wire through a spray gun.

Wrought zinc and zinc die castings are easily machined by appropriate use of standard methods and tools.

Decorative finishing is commonly required for drawn or cast articles and the familiar electrodeposited or organic coatings are easily applied to zinc. Chromium is commonly plated over primary coats of copper and nickel. A supplementary thin chromate film is sometimes produced on zinc articles by simple dipping in acidulated sodium dichromate solution, to retard or prevent the formation of white corrosion products in special situations.

Outline of Principal Forms and Uses of Zinc

Information concerning the details of these manifold applications is best obtained from the numerous trade publications, or reprints of technical articles in trade journals, distributed by the various producers of basic zinc products; and especially through the American Zinc Institute which is continually alive to the impact of new processes and applications, and can supply reprints or other technical data, as well as the publications of its own staff. A useful summation of wrought and cast zinc alloy properties is available as a reprint from the March 15, 1957 issue of *Design News.*

The 1948 edition of the ASM "Metals Handbook"[22] contains some 15 pages of information on technological aspects of the zinc industry and the properties of zinc and zinc alloys. Other books or scientific papers in this field[23-25] are generally of much earlier date and thus of limited value.

Table 9-3 shows the principal forms and uses of zinc in pure or alloyed form. The use of zinc as a secondary component of other metal and alloy products is discussed in Chapter 10, which also contains reprints of available constitutional diagrams.

For easy reference, beyond the previously outlined chemical requirements associated with the different kinds of slab zinc and zinc die castings, Table 9-4 supplies a full listing of the specifications in common use to govern the marketing of zinc products.

TABLE 9-3. FORMS AND USES OF METALLIC ZINC

Slab zinc (spelter): raw material for all applications
- Castings
  - anodes
    - boiler and hull plates and zinc for protection of underground structures
    - for electroplating
    - for wet batteries
  - die castings—of major importance in the automotive industry and generally seen in wide variety of applications
  - permanent-mold castings
  - sand castings
    - stamping dies (Cu-Al-Mg die-casting type alloys (high copper) or special compositions)
    - miscellaneous
  - slush castings—mainly for ornamental objects
- Coatings—for sheet and strip, wire and wire rope, tube and pipe fittings, and miscellaneous hardware
  - by electroplating
  - by galvanizing
    - hot-dip
    - continuous lines
  - by metallizing (sprayed coatings)
  - by sherardizing
  - by zinc-dust paint (containing oxide pigment) and zinc-rich paint

Rolled zinc: sheet, strip, plate or foil
- anodes
  - boiler and hull plates and zinc for protection of underground structures
  - battery cans (dry cells)
- eyelets and other drawn, pressed, spun or stamped shapes, such as cosmetic cases, flashlight reflectors, etc.
- foil
- frames, mouldings, beading and bindings
- general sheet metal work
- glazier points
- lithographers' and photoengravers' plates or sheets
- mason jar caps
- organ pipes
- ribbon
- roofing
  - leaders and gutters, flashing, weather strip and miscellaneous architectural uses
  - roof covering
  - roofing material in corrugated form (of stiffened zinc alloy as substitute for galvanized steel)
- special nails, washers or gaskets
- stencils, name and address plates, tags, templates, etc.
- terrazzo floor strips
- washing machine parts

Zinc rod and wire: for metallizing and miscellaneous applications
- extruded zinc—now coming into renewed use

TABLE 9-3. FORMS AND USES OF METALLIC ZINC—(*Continued*)

Zinc: for processing in other fields
- desilverization of lead by the Parkes process
- dust, moss, nodules, "feathered zinc," or shavings, as reagent for chemical reduction or precipitation of other metals, and dust for paints
- special alloys for research or laboratory use
- zinc-sodium (2 per cent) for deoxidation and grain refinement of nonferrous alloys
- zinc in various forms for use in the manufacture of zinc compounds and other chemical products

TABLE 9-4. DESIGNATED PRODUCTS WITH ENUMERATION OF SPECIFICATIONS COMMONLY USED IN THEIR COMMERCIAL DISTRIBUTION*

Slab Zinc (Spelter)

Zinc-Coated Steel and Iron Products

HOT-DIPPED GALVANIZED COATINGS

Product	Specification
Hardware	
Products fabricated from rolled, pressed, and forged steel shapes, plates, bars, and strip; zinc (hot-galvanized) coatings on,	A 123
Steel products, assembled; zinc coating (hot-dip) on (tentative),	A 386
Hardware, iron and steel; zinc coating (hot-dip) on,	A 153
Safeguarding against embrittlement of hot galvanized structural steel products and procedure for detecting embrittlement	A 143
Safeguarding against warpage and distortion during hot-dip galvanizing of steel assemblies (tentative)	A 384
Providing high quality zinc coatings (hot-dip) on assembled products (tentative)	A 385
Sheet Products	
Sheets, coils, and cut lengths; zinc-coated (galvanized) iron or steel (tentative)	A 93
Sheets, roofing, 1.25 oz class coating (pot yield); zinc-coated (galvanized) iron or steel (tentative)	A 361
Sheets; zinc-coated (galvanized) wrought iron	A 163
Wire, Strands, and Fencing	
Line wire, telephone and telegraph; zinc-coated (galvanized) "iron"	A 111
Line wire, telephone and telegraph; zinc-coated (galvanized) high tensile steel	A 326
Core wire for aluminum conductors, steel reinforced (ACSR); standard weight zinc-coated (galvanized) steel	B 245
Core wire for aluminum conductors, steel reinforced (ACSR) (with coatings heavier than standard weight); zinc-coated (galvanized) steel	B 261
Tie wires; zinc-coated (galvanized) steel (tentative)	A 112
Overhead ground wire strand: zinc-coated steel (tentative)	A 363
Wire strand; zinc-coated ("galvanized" and Class A "extra galvanized") steel (tentative)	A 122

TABLE 9-4. DESIGNATED PRODUCTS WITH ENUMERATION OF SPECIFICATIONS COMMONLY USED IN THEIR COMMERCIAL DISTRIBUTION*—(*Continued*)

Wire strand; zinc-coated (Class B and Class C coatings) steel (tentative)	A 218
Barbed wire; zinc-coated (galvanized) iron or steel	A 121
Wire fencing, farm-field and railroad right-of-way; zinc-coated (galvanized) iron or steel	A 116
Fence fabric, chain-link; zinc-coated steel (tentative)	A 392
Poultry netting (hexagonal and straight line) and woven poultry fencing; zinc-coated (galvanized) steel (tentative)	A 390
Pipe	
Pipe for ordinary uses, welded and seamless, black and hot-dipped; zinc-coated (galvanized) steel	A 120
ELECTRODEPOSITED ZINC COATINGS	
Zinc on steel; electrodeposited coatings of	A 164
Coating Tests	
Weight of coating on zinc-coated (galvanized) iron or steel articles	A 90
Uniformity of coating by the Preece test (copper sulfate dip) on zinc-coated (galvanized) iron or steel articles	A 239
Zinc Die Castings	
Zinc-base alloys in ingot form for die castings (tentative)	B 240
Zinc-base alloy die casting (tentative)	B 86
Preparation of zinc-base die castings for electroplating	B 252
Electrodeposited coatings of nickel and chromium on zinc and zinc-base alloys	B 142
Zinc-alloy ingot and die-casting compositions Society of Automotive Engineers' Standards	903, 921 and 925
Federal Specification for zinc-base alloy; die castings	QQ-Z-363
Sand Castings	
Military Specification; alloys for forming dies	MIL-Z-7068
Rolled Zinc	
Rolled zinc	B-69
Zinc Anodes	
Military Specification (for) anodes, corrosion preventive, zinc: plate, slab, disc and rod shaped	MIL-A-0018001D
Preparation of Metallographic Specimens and Micrographs	
Section on Zinc and Its Alloys in Tentative Methods of Preparation of Metallographic Specimens	E 3
Tentative methods of preparation of micrographs of metals and alloys	E 2

* Numbers refer to specifications of the American Society for Testing Materials, except in the indicated cases of Government or SAE specifications. Reference should be made in all cases to latest edition of standard listed.

REFERENCES

1. Brauer, H. E., and Peirce, W. M., "The Effect of Impurities on the Oxidation and Swelling of Zinc-Aluminum Alloys," *Trans. Am. Inst. Mining Met. Engrs.*, **68,** 796–839 (1923). See also Figure 9-17 and the text under the heading Die Castings, in the ensuing section of this chapter.
2. ASTM Standards 1958, Part 2, NonFerrous Metals, "Tentative Specifications for Zinc-Base Alloy Die Castings, B 86-57T," pp. 910-2, Philadelphia, Am. Soc. Testing Materials, 1938.
3. 1957 SAE Handbook, "Zinc Die-Casting Alloys," pp. 290–1, New York, *Soc. Automotive Engrs.*, 1957.
4. ASTM Standards 1958. Part 2, NonFerrous Metals, "Standard Specifications for Slab Zinc (Spelter), B 6–58," pp. 724–7, Philadelphia, Am. Soc. Testing Materials, 1955.
5. "ASTM Standards on Zinc-Coated Iron and Steel Products," Separate publication by the Am. Soc. Testing Materials, Philadelphia, 1956.
6. Pfann, W. G., "Principles of Zone-Melting," *Trans. Am. Inst. Mining Met. Engrs.*, **194,** 747–53 (1952); also, "Zone Melting," New York, John Wiley and Sons, Inc., 1958.
7. Leverenz, H. W., "Luminescence of Solids," New York, John Wiley and Sons, Inc., 1950.
8. Cyr, H. M., "Pure Zinc," *Trans. Am. Electrochem. Soc.*, **52,** 349–54 (1927).
9. Truesdale, E. C., and Edmunds, Gerald, "Pure Zinc—Its Preparation and Some Examples of Influence of Minor Constituents," *Trans. Am. Inst. Mining Met. Engrs.*, **133,** 267–79 (1939).
10. Gardiner, W. C., "Zinc, Manganese and Other Metals Recovered by Amalgam Process at Duisburger Kupferhütte," *FIAT*, *Rev. Ger. Science*, Report No. 821 (1946); Hohn, H., "Application of Amalgam Metallurgy," *Research* (London), **3,** 407–17 (1950).
11. Lowe, Sherwin P., Long, Geo. W., Downes, Kenneth W., and Kent, Geo. H., U. S. Patent 2,529,700 (Nov. 14, 1950).
12. Barker, G. C., and Jenkins, I. L., "Square Wave Polarography," *Analyst*, **77,** 685–96 (1952).
13. Jervis, R. E., "Neutron Activation Analysis," *Chemistry in Can.*, **8,** 27–31 (March, 1956).
14. McLaren, E. H., "The Freezing Points of High Purity Metals as Precision Temperature Standards. III, Thermal Analyses on Eight Grades of Zinc with Purity Greater than 99.99%," *Can. J. Phys.*, **36** (May, 1958).
15. Jillson, D. C., "Production and Examination of Zinc Single Crystals," *Trans. Am. Inst. Mining Met. Engrs.*, **188,** 1005–8 (1950).
16. Schilling, Harold K., "Growth of Crystals of Zinc Containing Cadmium by the Czochralski-Gomperz Method," *Physics*, **6,** 111–16 (1935).
17. Anderson, E. A., Fuller, M. L., Wilcox, R. L., and Rodda, J. L., "The High Zinc Region of the Copper-Zinc Phase Equilibrium Diagram," *Trans. Am. Inst. Mining Met. Engrs.*, **111,** 264–92 (1934).
18. "Metals Handbook," 1948 ed. p. 1092, Cleveland, Ohio, American Society for Metals.
19. Woldman, N. E., "Engineering Alloys," p. 182, Cleveland, Ohio, American Society for Metals, 1954.
20. Jevons, J. Dudley, "The Metallurgy of Deep Drawing and Pressing," pp. 285–90, New York, John Wiley and Sons, Inc., 1940.

21. Anderson, E. A., Boyle, E. J., and Ramsey, P. W., "Rolled Zinc-titanium Alloys," *Trans. Am. Inst. Mining Met. Engrs.*, **156**, 278–87 (1944).
22. "Metals Handbook," 1948 ed. pp. 1077–92, Cleveland, Ohio, American Society for Metals.
23. Mathewson, C. H., Trewin, C. S., and Finkeldey, W. H., "Some Properties and Applications of Rolled Zinc Strip and Drawn Zinc Rod," *Trans. Am. Inst. Mining Met. Engrs.*, **64**, 305–77 (1920).
24. "Zinc and Its Alloys," Circular No. 395, Bureau of Standards, U. S. Department of Commerce, Washington, D. C., 1931.
25. Burkhardt, Arthur, "Technologie der Zinklegierungen," Berlin, Julius Springer, 1937.

Physical Metallurgy of Zinc

E. A. Anderson

Chief of Products Application, Research Department
The New Jersey Zinc Company (of Pa.)
Palmerton, Pa.

This and other sections of Chapter 9 deal specifically with the casting, fabrication and other manipulations of zinc into various forms. Chapter 4 is addressed to the fundamental chemistry and physics of zinc technology. In addition to these areas of knowledge there has been developed over the past few decades a substantial amount of information of a basic or explanatory type concerning the physical nature and consequent behavior of zinc when subjected to the various operations of metal technology. Since such information is useful in predicting or explaining metal behavior, the present section has been prepared to make it readily available.

METALLOGRAPHY

In the early period of scientific metal study, the term "metallography" was applied to the entire field of constitution, structure and properties of metals and alloys. Presently, however, it has come to mean only those areas involved in the examination of metal structures with the microscope. Other basic aspects of metal technology now are covered by the term, "physical metallurgy." Both field of study are discussed in this section. Binary and ternary "phase diagrams" or "equilibrium diagrams," of special importance to the zinc metallurgist, will be found in Chapter 10.

Virtually all metallographic examinations of metals are carried out with reflected light from polished surfaces which have been etched to reveal the essential structures. It is necessary in the polishing to avoid producing distortions or alterations of the structure. This is particularly important in the case of zinc in which twinning, rumpling, grain growth and other departures from the original structure can easily be developed. Such artifacts are most difficult to avoid in cast and other zinc objects of large grain size.

Experiments have shown that the distortion produced during the polishing of cast zinc may penetrate to a depth 20 to 100 times that of the deepest abrasive scratch. Each polishing step, accordingly, must remove considerably more metal than that required to eliminate visible scratches from the previous step. A workable procedure to accomplish this is to etch heavily after each polishing step. While modifications may be needed in special cases, the following polishing procedure has been used for many years with good results:

1. Cut specimen from sample with a fine-tooth jeweler's saw using slow strokes and light pressure.
2. Mount specimen. Lump specimens may be mounted in methacrylate plastic at the lowest possible molding temperature. Rolled zinc and galvanized sheet may be mounted in multiple in clamps using spacers of heavy gage, soft zinc between specimens.
3. Grind on belt sander through 30-, 60-, 120-, and 240-mesh grades. Press lightly and cool specimen frequently in water to prevent overheating.
4. Hand polish by rubbing successively with light pressure on 280-, 320- and 400-mesh dry abrasive papers using a plate glass backing.
5. Wet polish with light pressure on rotating wheels using four successively finer abrasive particle sizes prepared according to Rodda's procedure.[1] Using etchant No. 1, described below, etch as follows after each polishing step:

   3½ minutes after final hand polishing
   1½ minutes after No. 1 wheel
   30 seconds after No. 2 wheel
   10 seconds after No. 3 wheel*

6. Give final etch for metallographic examination.

* Omit this intermediate etching when polishing die casting alloys.

Throughout the entire procedure the pressure on the specimen should be as light as possible and the specimen should be rotated 90 degrees between polishing steps. When zinc-coated specimens are involved, the intermediate etching steps should be eliminated to avoid lowering the zinc surface below the plane of the steel. It is helpful in minimizing such differences of level to add a corrosion inhibitor to the wet abrasive suspension.

The most useful etching reagent for zinc materials is pure chromic anhydride, CrO_3, activated by controlled additions of sodium sulfate. The grades used in chromium plating solutions are suitable. The etching solutions most commonly used are:

Solution No. 1

CrO_3	100 g
Na_2SO_4*	15 g
H_2O	1,000 ml
* or $Na_2SO_4 \cdot 10H_2O$	34 g

Solution No. 2

CrO_3	50 g
Na_2SO_4	4 g
H_2O	1,000 ml

Solution No. 3

CrO_3	200 g
Na_2SO_4	7.5 g
H_2O	1,000 ml

Solution No. 4

CrO_3	200 g
H_2O	1,000 ml

Satisfactory results generally are obtained with the etching conditions listed in Table 9-5 but deviations often are made to meet special needs.

Most consistent results are obtained when the etching operation is performed on freshly polished specimens using mild finger-tip rubbing with alcohol to remove loose abrasive and other contamination. To avoid spotty reaction it is desirable to wet the specimen surface with water prior to etching. Staining is avoided if the etched specimen is rinsed in Solution No. 4 before thorough washing in a strong stream of running water. Drying is best done by immersing the piece successively in alcohol and ether followed by exposure to a warm air blast as from a hairdryer.

Microexaminations generally are made with white light illumination except where grain size determinations are to be made. Grain boundaries are defined rather poorly in white light but when illuminated with properly compensated polarized light, each grain assumes a color, the shade of which is probably orientation-dependent. Grain counts can then be made with good accuracy. The polarized light technique further permits the study of grains in flat, smooth surfaces without the need for etching, a fact which allows recrystallization and grain growth to be observed in action.

Illustrative microstructures of zinc and zinc alloys are shown in Figures 9-3 through 9-17. In some cases the impurity contents were increased above normal concentrations to make the structures more evident.

TABLE 9-5. ETCHING SCHEDULES FOR ZINC

Metal	Solution No.	Time-Seconds	
		100×	1000×
Unalloyed Zinc	1	15	1
Rolled Zinc-Copper Alloy	3	15	1
Zinc Die Casting Alloys	2	5	1

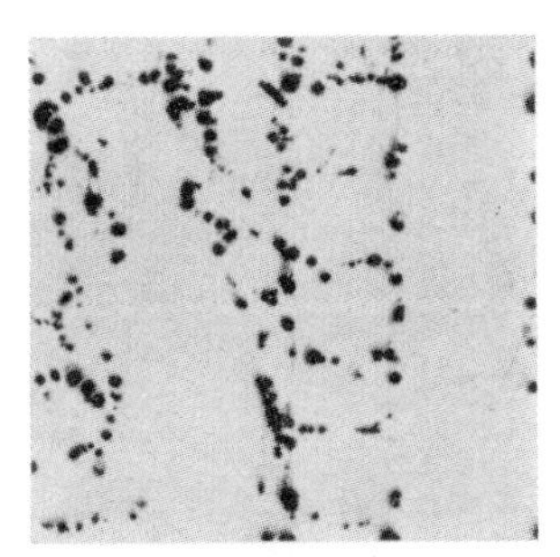

Figure 9-3. Cast Prime Western zinc. Note large grains and abundance of dark lead particles. Magnification 100×.

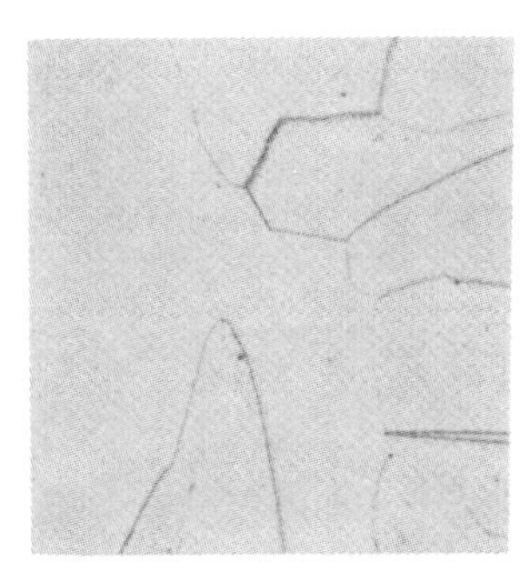

Figure 9-4. Cast Special High Grade zinc. Note nearly complete absence of lead particles. Magnification 100×.

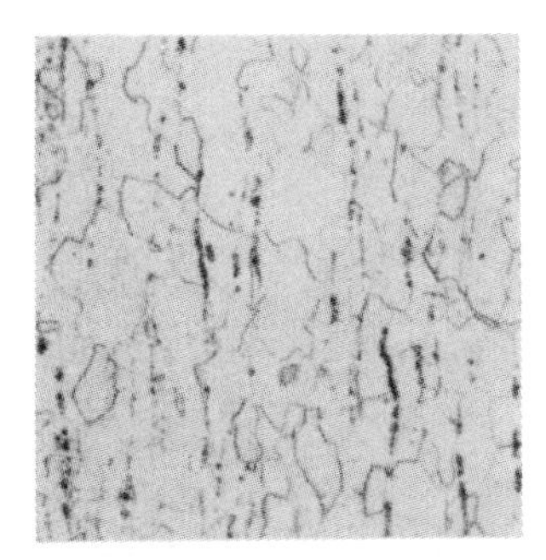

Figure 9-5. Hot-rolled Brass Special zinc—white light illumination. Note poor grain boundary definition. Magnification 250×.

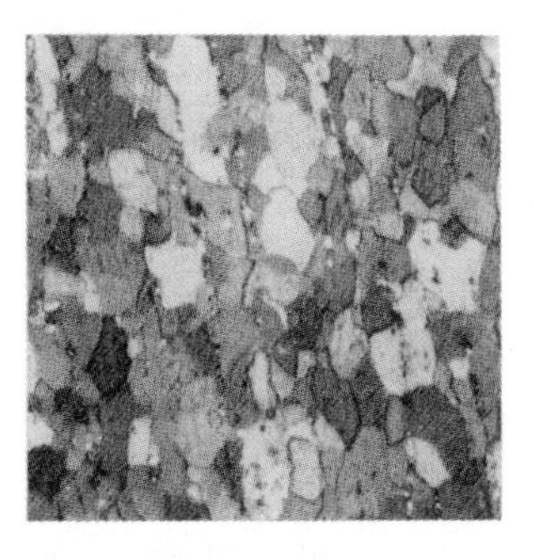

Figure 9-6. Same area as in Figure 9-5 but photographed with polarized light. Note clear definition of grains. Magnification 250×.

Figure 9-7. Cold-rolled Brass Special zinc—polarized light. Note grain distortion. Magnification 250×.

Figure 9-8. Hot-rolled 1 per cent copper alloy—polarized light. Note coring. Magnification 250×.

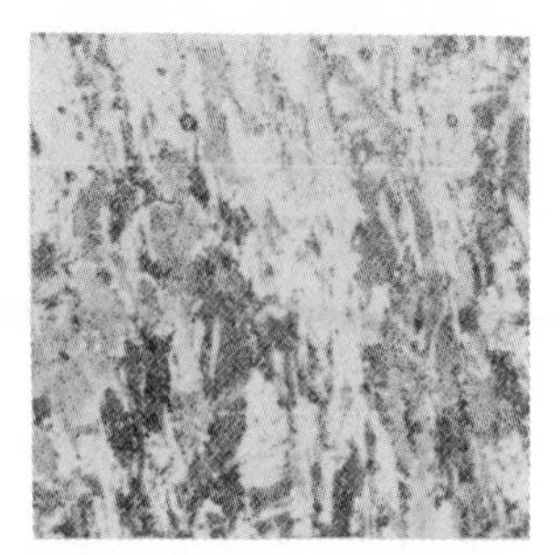

Figure 9-9. Cold-rolled 1 per cent copper alloy—polarized light. Note severe distortion of grains. Magnification 250×.

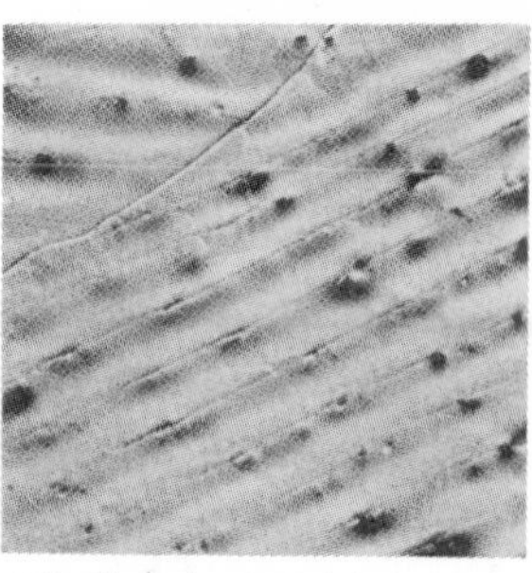

Figure 9-10. Cast zinc containing 0.5 per cent of cadmium. Note coring in grains. Magnification 100×.

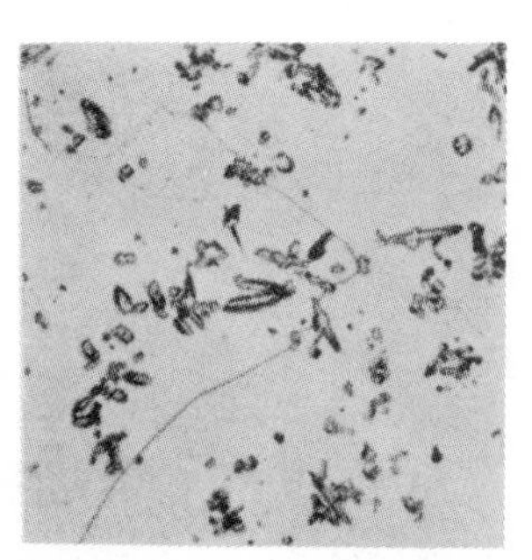

Figure 9-11. Cast zinc containing 0.2 per cent of iron. Note iron-zinc compound particles. Magnification 100×.

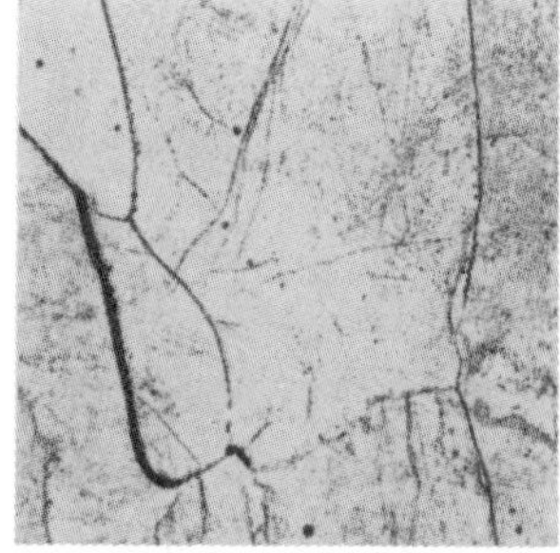

Figure 9-12. Cast zinc containing 0.1 per cent of tin. Note low melting eutectic in grain boundaries. Magnification 100×.

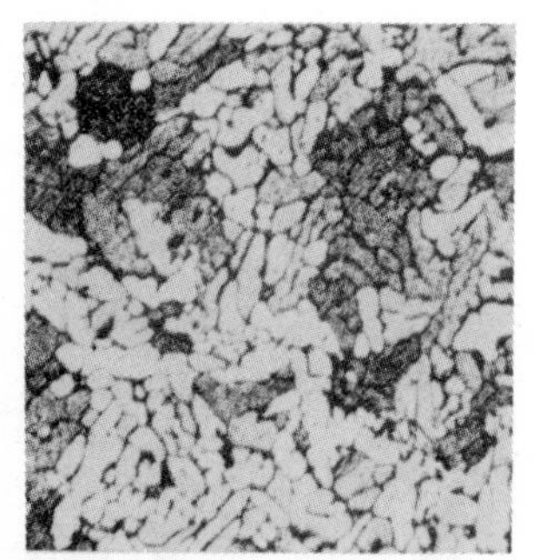

Figure 9-13. Cast zinc containing 0.4 per cent of titanium. Note grain refinement and zinc-titanium compound in boundaries. Magnification 100×.

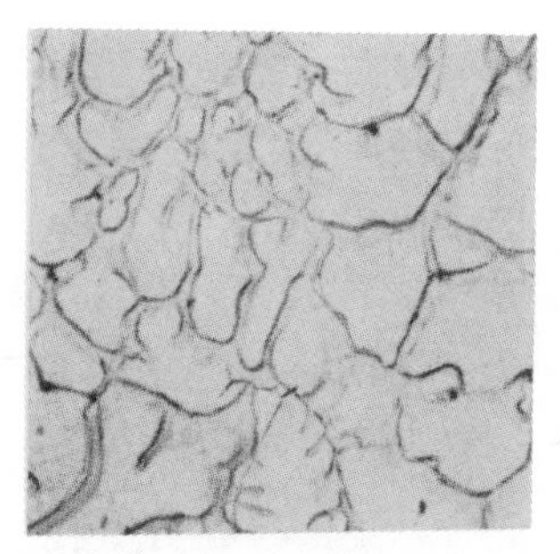

Figure 9-14. Cast zinc containing 1 per cent of copper. Note coring. Magnification 100×.

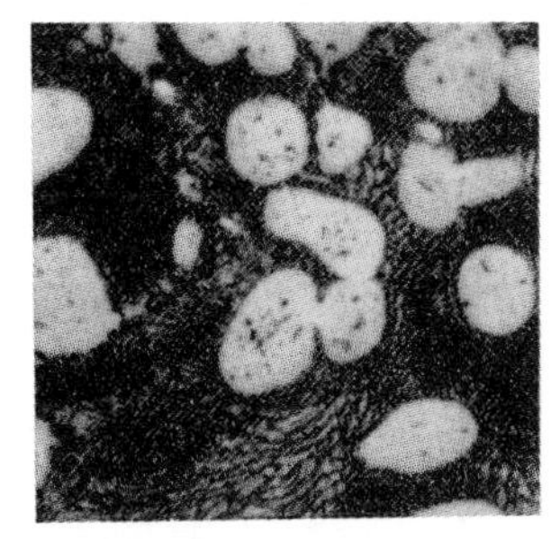

Figure 9-15. Die-cast "Zamak-5" as cast. Note clear primary zinc-rich phase and eutectic. Magnification 1000×.

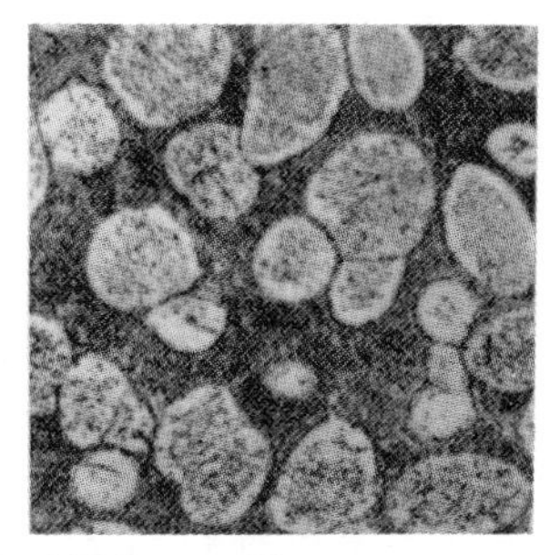

Figure 9-16. Same as Figure 9-15 but after aging for ten days at 100°C. Note precipitate in zinc-rich phase and blurring of eutectic. Magnification 1000×.

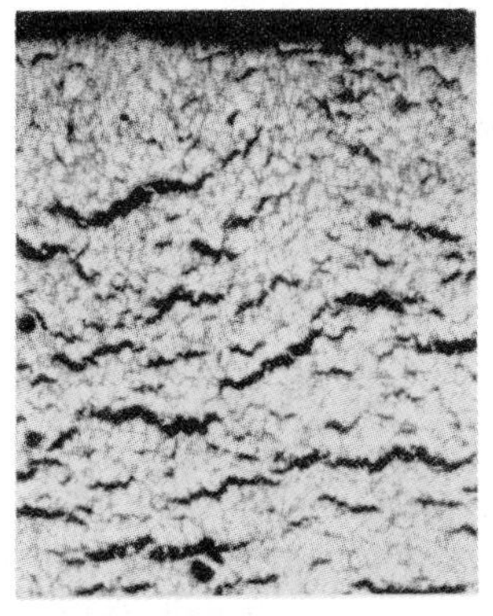

Figure 9-17. Contaminated "Zamak" alloy after ten days in 95°C steam test. Note subsurface corrosion. Magnification 100×.

CAST ZINC

Structural Characteristics

Crystal structure plays an important part in the plastic deformation of zinc and in the development of directional variations in properties. In common with such metals as magnesium, cadmium, beryllium and alpha-titanium, zinc crystallizes in the hexagonal close-packed arrangement of atoms shown schematically in Figure 9-18. The dimensions of the unit cell at 25°C (77°F) are:[2, 3]

a 2.6954 kx
c 4.9370 kx
c/a 1.8564 (axial ratio)

To convert to Ångstroms, multiply kx values by 1.00202.

Since the axial ratio for the hexagonal close-packing of spheres is 1.633, the substantially higher axial ratio of zinc reveals a much wider atomic

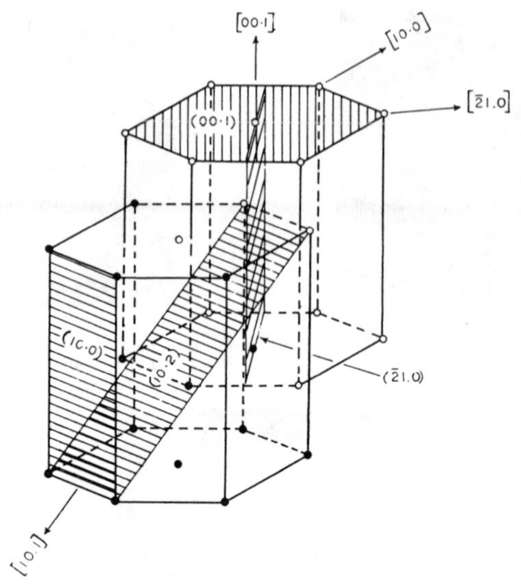

Figure 9-18. Arrangement of atoms in the crystal structure of zinc.

spacing in the *c* direction. A consequence of this is a marked anisotropy in such properties as thermal expansion, thermal conductivity and electrical resistivity. The mechanisms by which zinc may be deformed plastically also reflect the crystal structure and the axial ratio.

Cast, unalloyed zinc solidifies into long columnar grains which extend essentially from the mold surface to the air interface or to a meeting with similar grains growing from another mold face. Increased freezing rates do not result in equiaxed grain structures even under the extreme chill conditions of pressure die casting. Most natural impurities and alloying elements exert little influence on grain shape or size. Notable exceptions are aluminum and titanium.

Except for a thin layer of relatively fine grains at the extreme surface, and a very thin transition zone, the principal orientation in cast zinc places the (00.1), or basal, plane parallel to the long axes of the columnar grains. Crystal growth places the basal plane parallel to the direction of heat flow and hence normal to the principal faces of the casting.[4, 5] No particular direction in the basal plane seems to be favored. The thin layer of finer grains at the mold contact have their basal planes parallel to the surface. In the transition zone, the pyramidal plane (10.1) is disposed perpendicularly to the heat-flow direction.

Effects of Impurities and Added Elements

Virtually all the natural impurities, contaminants and alloyages involved in the production and use of zinc result in readily detected alterations in microstructure and changes in one or more properties. The characteristic effects produced by several elements are described briefly below.

Lead. The solid solubility of lead in zinc is very low. Consequently, the appearance of minute lead droplets is a characteristic of the microstructure of virtually all zinc materials. The example shown in Figure 9-3 is typical of the lower grades of zinc. Lead has little influence on the mechanical properties of zinc but due to the low melting point of this impurity it does limit upper hot-working temperatures. When present in controlled amounts together with suitable cadmium concentrations a desirable influence is noted on the corrosion of zinc in chemical reactions such as occur in dry cells. Lead adds definition to spangle boundaries in hot-dip zinc coatings.

Cadmium. The amounts of cadmium present in most zinc products are in solid solution and produce no change in microstructure other than the coring effect seen in Figure 9-10. Largely because of the coarse grain structure of cast zinc, cadmium has little effect on properties as-cast. In rolled zinc cadmium increases strength, hardness and creep resistance and raises the recrystallization temperature.

Iron. When present in amounts exceeding about 0.001 per cent, iron appears in the microstructure of zinc as an intermetallic compound containing approximately six per cent iron as shown at an uncommercially high concentration in Figure 9-11. The effects of iron on the properties of zinc are controlled by the particle size of the compound which in turn is established by the thermal history of the piece. The largest effects are produced by the smallest particles, which are obtained when the metal is cast from a temperature at which substantially all of the iron is in liquid solution. Prolonged holding at lower melt temperatures will result in coarser particles. Fine compound particles in a casting can be coalesced to coarser form by heating, e.g., for 36 hours at 375°C (700°F).

By providing impediments to grain growth, the addition of about 0.005 per cent iron aids in the control of grain size during the rolling of Special High Grade zinc. When more than about 0.008 per cent iron is present in finely divided compound form in rolled zinc, undesirable increases in hardness and loss of ductility may result. Coarser particles have little or no effect. There is evidence that critical particle fineness is not easily attained when cadmium is present in the zinc.[3]

When present in excess of 0.0014 per cent, iron causes formation of undesirable corrosion product films of high electrical resistivity on zinc galvanic anodes used to protect steel structures in sea water.[6] This effect is reduced substantially when the intermetallic compound is coalesced into a coarser form by heat treatment of the anode. In amounts above 0.0025 per cent, iron greatly stimulates the dissolution of zinc by nonoxidizing mineral acids. Similarly, the open-circuit corrosion of the zinc anodes in alkaline primary cells reaches harmful levels when the iron content of the zinc exceeds about 0.004 per cent.

Tin. Except for certain hand-dip galvanizing operations tin normally appears in zinc only as a contaminant. The solid solubility in zinc is low and tin contents as low as 0.001 per cent can produce a grain boundary eutectic phase which melts at about 200°C (398°F). During the hot rolling of zinc this can result in serious grain boundary ruptures.

Copper. Aside from the natural impurities, copper is the most widely used alloying element in rolled zinc. While capable of precipitating a second phase during long exposure at near room temperature, the added copper (about one per cent) usually is found in solid solution. The cored cast structure shown in Figure 9-14 resembles that produced by cadmium. Like cadmium, the addition of copper results in higher strength, hardness and creep resistance and a higher recrystallization temperature. Copper has undesirable effects on the corrosion of zinc in dry cells but has practically no effect on the rate of corrosion of zinc in the atmosphere.

Aluminum. The effects of aluminum on the microstructure of zinc are evident in the structures of die castings (see later discussion of their characteristics), such as shown in Figures 9-15, 16 and 17, representing "Zamak 5," which contains around 4 per cent aluminum, with about 1 per cent copper. Aluminum is added in amounts of about 0.1 per cent to the zinc melts used in the continuous strip, hot-dip galvanizing of steel.

In the presence of aluminum, molten zinc does not react with iron to form the voluminous, brittle zinc-iron intermetallic compound layer which characterizes the older type of aluminum-free galvanized product. In the absence of this brittle material hot-dip coatings are highly ductile. In similar amounts aluminum improves the performance of zinc galvanic anodes in sea water. The presence of more than 0.005 per cent aluminum in rolled zinc causes a slow embrittlement due to corrosion penetration of the grain boundaries.

Titanium. Titanium has limited solid solubility in zinc[7] and forms a zinc-rich compound with a eutectic at about 0.11 per cent titanium. This compound tends to decrease the grain size of cast zinc and to restrain grain growth in rolled zinc at elevated temperatures. Titanium has little effect on the tensile strength and hardness of rolled zinc but does greatly increase creep resistance, particularly when added with copper.

Die Castings

The effects of adding aluminum to zinc are many, and generally useful. Starting at very low concentrations, such as 0.02 per cent, aluminum sharply reduces the rate of oxidation of zinc at elevated temperatures. At some concentration near 0.1 per cent the reaction of molten zinc with iron begins to decrease as the formation of zinc-iron intermetallic compounds becomes less favored and that of aluminum-iron compounds more favored. At the normal aluminum content (about 4.3 per cent) of the standard zinc die casting alloys, the rate of attack by the melt on iron is sufficiently low to permit die casting in hot-chamber machines in which the iron operating mechanism is immersed continuously in the molten alloy.

Aluminum acts as a grain refiner in cast zinc with the result that the die casting alloys reveal a fairly fine equiaxed grain structure even when chill or sand cast. This desirable structure is a major reason for the strength, ductility and toughness of zinc die castings. It should be pointed out, however, that at and very near the eutectic concentration of 5 per cent aluminum the ductility of the zinc-aluminum alloys is very low.

General reference should be made to the aluminum-zinc phase diagram shown in Chapter 10. At room temperature, the zinc die casting alloys, such as ASTM Nos. XXIII and XXV, consist essentially of the terminal aluminum (alpha) and zinc (beta) solid solutions. During solidification the first material to freeze appears as primary particles of beta. Later, the remaining liquid solidifies as eutectic composed of beta and an unstable high-temperature constituent, alpha prime, producing the characteristic, as-cast, microstructure shown in Figure 9-15.

A number of structural changes take place as a die casting ages, none of which has any marked influence on alloy properties. The beta solid solution may contain about 0.35 per cent of aluminum in a freshly made die casting. During a five-week room-temperature aging period this will decrease to about 0.07 per cent, the excess appearing as minute particles of alpha within the beta structure (see Figure 9-16). This precipitation of aluminum from the beta solid solution causes the very small decrease in dimensions which is known to take place during the five-week interval after casting.[8] Most of the aluminum can be precipitated, and the shrinkage experienced in much shorter time, by aging* the die castings at slightly elevated temperatures.

The alpha prime constituent of the eutectic is stable only at temperatures above 275°C (577°F). At lower temperatures it decomposes eutectoidally into alpha and beta with a substantial liberation of energy as heat and with an increase in alloy density. These effects are at maxima in the eutectoid composition of 78 per cent zinc and 22 per cent aluminum. While additions of such elements as copper and magnesium delay the alpha prime transformation they do not prevent its occurrence. It is now felt that this

* It has been noted in the introductory section of this chapter that the high-copper modification of the zinc-aluminum die castings, formerly included in the ASTM specifications, is no longer current, owing to less favorable aging characteristics. [Ed.]

structural change has little effect on the aging characteristics* of zinc die castings.[9]

Aluminum produces a change in the corrosion characteristics of zinc which must be considered. In the absence of aluminum, zinc suffers no special corrosion in the grain boundaries. When aluminum is present such corrosion may occur with the rate and magnitude stimulated by the presence of such soft, heavy, low melting-point impurities as tin and lead (see Figure 9-17). The mechanism involved is believed to be that of stress corrosion. Additions of magnesium and, to some extent, copper, reduce the severity of attack and when combined with control of the accelerating impurities to low concentration levels result in alloys which suffer no subsurface corrosion in service.

Zinc Single Crystals

There is substantial knowledge of the crystallographic mechanisms by which zinc objects change shape in adjusting to external deformation forces. Since most of this information was obtained through studies made on zinc single crystals, it is proper to mention here the procedures used in making and documenting such specimens.

It is not difficult to grow quite large single crystals from pure zinc, and several procedures are available. The most widely used methods involve progressive freezing of the melt from one end. This has been accomplished by lowering the melt slowly from the furnace,[10] by moving the furnace relative to the melt or by moving only a thermal gradient in the furnace. Single crystals of rather irregular surface configuration can be produced by slowly withdrawing a crystal nucleus from a melt surface.[12] Large grains, and at times single crystals, have been produced by annealing critically-strained rolled zinc specimens. Other known procedures have not been widely used.

Investigations[13,14] have shown that the orientations to be obtained require control of the rate of solidification and the temperature gradient between liquid and solid within rather narrow limits. Departure from ideal conditions may result in changes in orientation along the specimen length, or in the production of internal defects. Basal cleavage fractures of imperfect crystals often reveal a mosaic pattern of small crystal blocks having similar but slightly differing orientations.[15,16] Such crystals have been shown to reorient through as much as 30 degrees during a 48-hour anneal at 400°C (750°F).[11] Attempts to control the orientation of a specimen by means of a seed crystal may result in an imperfect crystal if the growth rate and temperature gradient are not nearly optimum for the chosen orientation.

The production, handling and testing of zinc single crystals require great care to avoid accidental deformation. The use of precision-ground glass

* See footnote p. 409.

tubing as a container minimizes the danger of shrinkage restraints initiating in tube bore irregularities. Continuation of crystal growth to consume the entire melt is advisable. Directional differences in shrinkage in a polycrystalline specimen top can result in serious deformations in the single-crystal portion.

Crystal Documentation. Before they can be used in careful experiments, single-crystal specimens must be examined for degree of crystal perfection and freedom from extraneous small crystals, and their orientations must be determined. Light etching in 50 per cent hydrochloric acid usually suffices to disclose any small superficial grains which may be present. Careful cleaving along the basal plane may reveal mosaic imperfections.[17] Recent studies of etch pits suggest that their presence may be connected with atomic-scale imperfections. Heating at 400°C (750°F) for 48 hours will reveal some imperfections, at least, by a change in orientation. Since small orientation differences will tend to spread the diffracted x-ray beam, an indication of crystal perfection can be had from a comparison of the diffraction spots on the negative with the evidence from the undiffracted beam.

In most instances the orientations of the *a* and *c* axes relative to the specimen axis are determined by standard Laue back-reflection x-ray procedures. The parameters desired are the angle chi (χ) between the basal plane and the specimen axis and the angle lambda (λ) between the [11.0] direction and the specimen axis.

PLASTIC DEFORMATION OF ZINC

Zinc can be fabricated into desired shapes by virtually all of the processes commonly used in metal forming. It can be rolled, extruded, drawn, spun, bent, etc., on a commercial scale. As is the case with many metals the procedures used must be designed to take full advantage of favorable forming characteristics and to minimize the importance of limitations.

Like all metals zinc cannot be changed permanently in shape by rolling, drawing or other fabrication procedures without characteristic plastic deformation of the crystalline grains of which it is composed. Since the mechanisms by which such deformations take place are crystallographic in nature, the behavior of zinc during fabrication is closely associated with its hexagonal close-packed crystal structure. These mechanisms are discussed below.

Strain Hardening-Recrystallization

In most metals a consequence of the crystallographic movements associated with plastic deformation at room temperature is an increase in internal energy which is reflected in increased hardness and strength. The mechanism involved is known as "strain hardening" or "work hardening."

Restoration of the original soft state is accomplished by annealing the hardened metal at a temperature at which recrystallization into new, unstrained grains occurs. This process makes it possible to produce mill products by a sequence of cold reductions and annealings and to subject metals successfully to complex end-use fabrications. Zinc, however, normally cannot be fashioned by these procedures.

Pure zinc recrystallizes at or below room temperature.[18] While solid solution additions of cadmium or copper raise the recrystallization temperature somewhat, the atomic mobility of zinc at room temperature makes it difficult to develop sufficient strain hardening to permit establishing practical procedures of cold rolling and annealing. Pure zinc, for example, commonly reaches a condition of strain during attempted cold working which permits the development of sporadic grains of very large size during the act of rolling or drawing.

The recrystallization and grain growth of zinc of 99.999 per cent purity can be observed in progress on the microscope with compensated polarized-light illumination.[3, 19] To do this, a well-recrystallized rolled specimen is reduced 6 to 7 per cent in thickness on slow-speed, polished rolls; the specimen and the rolls are chilled by dry ice and acetone, using very small (about 0.001 inch) reductions per pass. When transferred immediately to the microscope without etching, the changes in crystal structure can be observed directly.

Since it is impractical to produce rolled zinc in a variety of tempers by cold rolling or to develop specified grain sizes by annealing, the customary practice is to attain the desired properties by manipulation of the final-pass reductions and temperatures. In most cases the choices of composition, pass reduction and temperature are made to produce properties which experience has shown to be best for a given end use.

Characteristic Mechanical Properties. The resistance of zinc to deformation increases as the rate of load application increases. The magnitude of this effect is such that a zinc which will deform appreciably by creep under a dead-weight stress of a few thousand pounds per square inch may have armor-piercing qualities at ballistic velocities. Normal tension-test data have little value for design purposes since polycrystalline zinc shows no yield point and the tensile strength and modulus of elasticity will have different values for every testing speed. The most useful design information is the relation of the steady-stage creep rate to the magnitude of the applied stress.[20, 21, 22]

With the exception of bending there are no tests which truly determine the ability of rolled zinc to withstand end-use fabrication. Routine determinations of hardness, dynamic ductility and temper[23] serve as useful control tests in the rolling mill. Such tests do not predict fabricating properties reliably, however, and in most cases zinc is rolled to mill properties which experience has shown to be satisfactory for the intended use.

Deformation Mechanisms in Zinc Single Crystals

Zinc was one of the first metals to be studied in detail with respect to the manner in which its crystals become adjusted to the change in shape imposed by deformational processes such as rolling, drawing or simple extension or compression.[24] The principal mechanism of deformation is slip on favorably oriented basal planes initially present, or on basal planes reoriented to favorable positions by the ordinary form of twinning. Under some conditions of stress geometry, bending or kinking of the basal plane takes place, producing orientations which involve slip. Finally, zinc crystals can be cleaved along the basal plane by tensile stresses normal to these planes.

The determination of these elements of the deformation process usually is made by microscopic observations of the markings which develop on the surfaces of carefully strained single crystals in predetermined orientations or by analysis of the orientation changes produced during the course of the deformation. The evidence thus obtained defines rigorously the mechanisms involved in producing the observed changes in specimen shape. The substantial ductility displayed by zinc crystals over a wide range of orientations shows that the slip process, supported by the twinning and bending mechanisms, is adequate to take care of most of the changes of form involved in the fabrication of zinc. Under conditions of restraint and in a limited range of very unfavorable orientations other means for change may operate, but such are difficult to identify.

Slip Deformation. Slip takes place by glide on the (00.1), or basal, plane in one of the directions joining closely adjacent atom sites in the basic hexagon of Figure 9-18; for example, in the crystallographically designated direction, $[\bar{2}1.0]$. This is consistent with the general tendency of metals to slip on planes of highest atomic population. It has been hypothesized that the motion in the $[\bar{2}1.0]$ direction in reality is the resolution of a zigzag progression of slip along alternating directions inclined 30 degrees to this direction.[25] It is possible, of course, that under the restraints existing in polycrystalline zinc, and at other temperatures of test, additional slip systems become operative and, indeed, inconclusive evidence of glide on planes such as $(\bar{2}1.0)$ in single crystals at elevated temperatures has been proposed.[26, 27]

A shear stress in the slip direction along the basal plane will result only in reversible elastic deformation until a critical stress level is reached, at which irreversible plastic deformation starts. This critical shear stress is an important parameter of the strength and hardness of a metal. It can be calculated from the attendant specimen stress by resolving this threshold stress along the basal plane in the slip direction by the critical resolved shear stress equation:[28]

$$S = \frac{F}{A} \cdot \sin \chi \cos \lambda$$

where S = critical resolved shear stress, F/A = critical unit stress along specimen axis, χ = angle between slip plane and specimen axis, and λ = angle between slip direction and specimen axis.

The critical resolved shear stress required to initiate slip in zinc is low and varies with temperature and the rate of application of the stress. At room temperature (25 ± .5°C) and with a loading rate of 0.1 g/mm^2/min in the slip direction, the critical stress was found to be 18.4 g/mm^2 regardless of the crystal orientation.[29] Other room temperature tests at higher, and varied, strain rates resulted in an average critical resolved shear stress of 37.0 g/mm^2.[30] The bending of single crystals under their own weight at room temperature indicates that plastic deformation by slip can occur at a critical resolved shear stress as low as 8 g/mm^2 under these dead-weight loading conditions.[31]

The effect of temperature on critical stress has been determined in less rigorous experiments to be as follows:[32]

Temperature (°C)	Critical Resolved Shear Stress (g/mm^2)
20	32.4
−77	48.0
−196	56.1

The smallness of the change with temperature is consistent with the fact that properly oriented zinc single crystals will bend under their own weight at −175°C.[31]

Evidence has been put forward[33, 34] to show that sharp yield points do not occur in nitrogen-free zinc single crystals. The addition of 0.002 per cent of nitrogen produces sharp upper and lower yield points in the resolved shear stress ranges of 80 to 110 g/mm^2 and 70 to 95 g/mm^2, respectively, at an unstated rate of stressing. It is to be expected that other solid solutions, such as those of cadmium and copper in zinc, also will increase the critical resolved shear stress required to initiate slip, but data from controlled tests are lacking.

Slip does not occur simultaneously on all basal planes in the stressed region but initiates on selected planes which are some distance apart.[35] Consequently, the progress of deformation involves small blocks which move relative to each other. Further, it has been noted repeatedly that, at least

.t low stressing rates, the deformation does not proceed smoothly but, ather, takes place in a series of jerky movements.[36, 37] The atomic movenents involved in the slip process have received critical attention.[38]

Shear Strengthening. Stress-strain curves of single crystals charactersticaly show a rise in stress with continued strain beyond the initiation of lip deformation. This strain hardening or shear strengthening plays a most mportant part in the hardening and strengthening of metals during cold vorking. Stress-strain curves for various single crystal orientations may be ransformed into one fundamental curve relating the acting resolved shear tress on the slip plane to the shear or glide strain, to express the essential train-hardening capabilities of a metal. The stress and strain parameters or this resolution may be calculated from the test data by means of the ollowing equations:[39, 40]

$$S = \sigma \sin \chi_0 \sqrt{1 - \frac{\sin^2 \lambda_0}{d^2}}$$

vhere S = resolved shear stress on slip plane, σ = the applied tensile stress, $_0$ = initial angle between slip plane and stress axis, λ_0 = initial angle beween slip direction and stress axis, and d = change in length of specimen l_1/l_0).

$$a = \frac{1}{\sin \chi_0} (\sqrt{d^2 - \sin^2 \lambda_0} - \cos \lambda_0)$$

vhere a = glide strain.

A typical stress-strain curve for a zinc single crystal is seen in Figure -19. It is observed that the resistance to shear increases very little with ncreased shear deformation (in the range of plastic action), a circumstance hich explains the low capacity of polycrystalline zinc for strain harden-

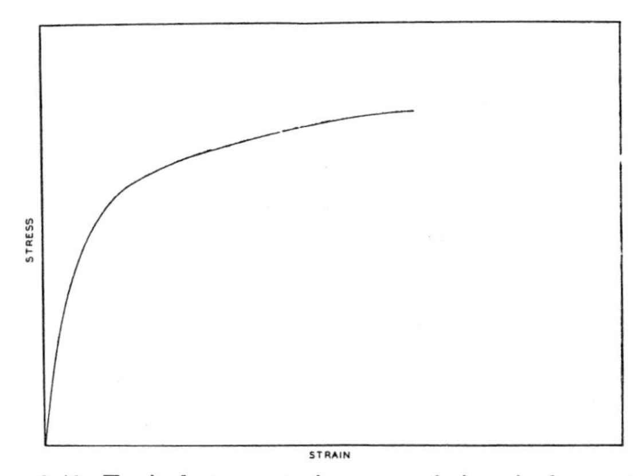

Figure 9-19. Typical stress-strain curve of zinc single crystal.

ing during cold working. One explanation for this behavior is that the high atomic mobility of zinc effects a degree of rapid elimination of the structural hardening changes (recovery) incident to the deformation. Such events are time-sensitive and reflect in the degree of recovery the rate at which the stress is applied.

While creep in polycrystalline zinc also involves grain boundary phenomena there is no doubt that the recovery mechanism and the consequent low rate of strain hardening contribute greatly to the inability of pure zinc to resist continuous deformation under light, dead-weight loads. The resistance of zinc to creep can be increased substantially by alloying, but the mechanism by which this is accomplished is obscure in the case of the very effective titanium addition.[7] The solid-solution addition of copper or cadmium seemingly operates by reducing the recovery rate and, hence, by permitting a higher rate of strain hardening.

Slip Rotation. If it were possible to stress a single crystal in tension without restraint the consequence of simple slip would appear to be an uncomplicated glide along favored basal planes which would continue until the specimen severs by one portion slipping completely off the other. In practice, however, the holding grips restrain any lateral movement and force the specimen to deform along a single stress axis. In order for slip to occur under these conditions it is necessary that the basal plane rotate toward the stress axis as shown in Figure 9-20.[35] The rotation process provides the means by which the crystal is elongated and narrowed in tension or thinned and broadened in compression.

It is apparent that the rotation due to slip can continue until an orientation is reached at which some other mode of deformation, such as twinning, is more favored. The reorientation produced by twinning then permits further slip to take place in the twinned segment.

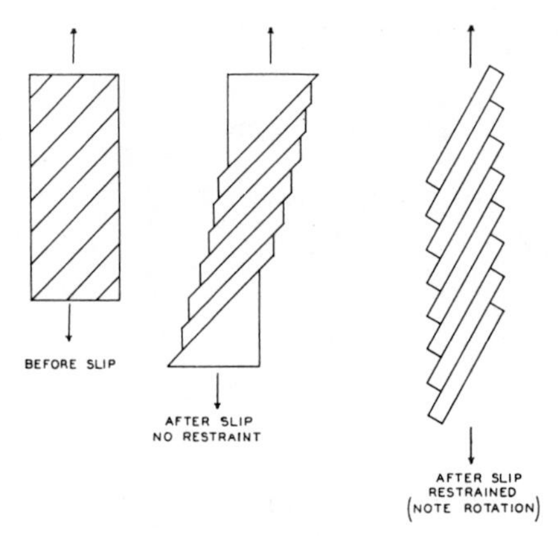

Figure 9-20. Block rotation during slip deformation.

Deformation by Twinning. Twinning in zinc crystals occurs on one of the first order pyramidal planes, such as (10.2), when a compressive stress is exerted normal to the basal plane, or a tensile stress parallel to the basal surface in the corresponding [10.0] direction. (See Figure 9-18.) The mechanism, which has been studied extensively,[41-44] may be considered as a shearing along the (10.2) plane in a [10.1] direction with slight positional adjustments of some atoms to the lowest energy sites. It is possible, however, to take the view that each atom undergoes an individual movement to its new position.

In any event, the result of the twinning movement is a reorientation of the lattice into a position equivalent to a physical rotation of the twin block through an angle of about 94 degrees.[43] In this position the basal plane of the twin is only about 4 degrees removed from the original (10.0) first order prism plane position. The act of twinning reduces the prism height by about 7 per cent with a corresponding increase in width. While this serves to relieve stress and to change dimensions, of greater importance is the fact that the reoriented basal plane is in a position to permit continued slip. This characteristic of twinning from unfavorable to favorable slip orientations contributes substantially to the ease with which zinc can be fabricated.

When twinning occurs a distinct clicking sound is heard, following which continued stress causes the twin to grow larger, but without further sound. Reversal of the stress direction results in a decrease in twin width which continues until all visible evidence of the original presence of the twin has disappeared.[45] The appearance of a typical twin in a zinc single crystal is shown in Figure 9-21.

Efforts to determine the critical resolved stress causing twinning have not been very fruitful. Values ranging from approximately 300 to 600 g/mm^2 have been reported[31] with the observation that the critical stress tends to

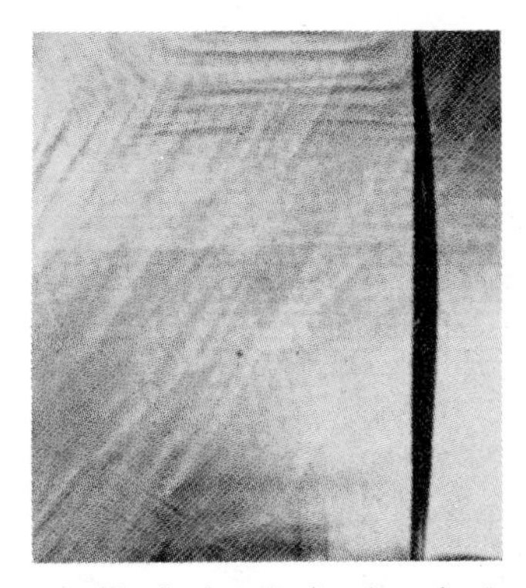

Figure 9-21. Typical twin in zinc single crystal.

decrease with increased prior basal slip and increased speed of deformation. While the critical resolved shear stresses for slip and twinning both increase with decreasing temperature the rate of change is the greater for twinning.[46]

It is evident from the above that at elevated temperatures and with slow rates of load application slip will tend to continue in orientations where twinning would have occurred at lower temperatures and with more rapid loading.

Bending and Kinking. While slip and twinning constitute the principal means by which zinc crystals accommodate to plastic deformation, there exist conditions under which neither can occur directly. Such situations are found when the crystal is compressed parallel to the basal plane[45, 47, 48] or where a portion of the crystal is restrained by grips,[49] by polycrystalline material[49] or by an adjacent grain or twin.[45]

Under each of these circumstances the relief of stress is accomplished by a bending of the basal plane around a close-packed [21.0] direction as an axis. The axes of bend are contained in a bend plane (Figure 9-22) which bisects the included angle of bend and which usually has no specific crystallographic indices. The bending mechanism involves slip since individual basal planes must move relative to one another to accomplish the bending. While not measured, the stress required to produce bends is known to be small.

Although in some cases the volume of metal involved in a series of bends is sufficient to produce macroscopic kinks,[47] bending often occurs around a

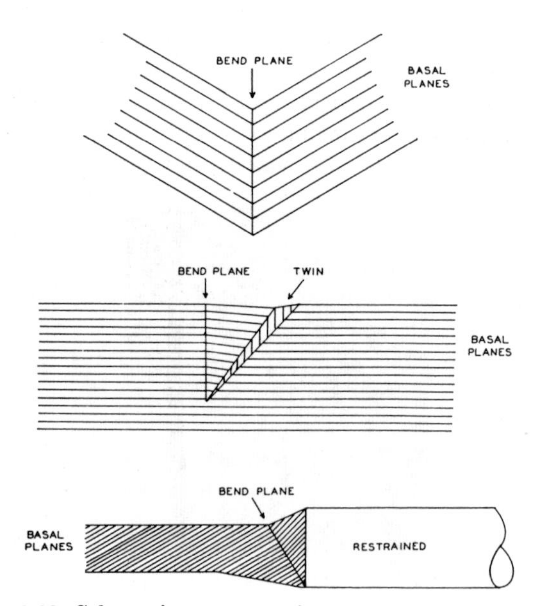

Figure 9-22. Schematic representation of bend plane situations.

very large number of closely-spaced parallel axes giving the effect of a curved basal plane.[45] When such deformed structures are annealed near the melting point, a coalescence of the fine, bend segments into coarser elements occurs.[51]

Cleaving. Zinc crystals can be cleaved along the basal plane by a critical tensile force, normal to this plane, of 180 to 200 g/mm^2.[52, 53] The indications from the data are that the cleavage stress is less influenced by low temperatures than is the shear stress for slip. Accordingly, cleavage is somewhat more likely to occur at the lower temperatures in crystals having borderline orientations.

Regardless of the mechanisms which may operate during the deformation of a crystal, the final rupture occurs either by one segment slipping completely off another or by basal cleavage, with the latter by far the more common. The twinning reorientation can, indeed, provide a means for continued slip but it also can place the basal plane in a favorable position for cleavage. The effects of temperature and crystal orientation on the rupture characteristics of zinc single crystals have received attention.[49]

The Deformation of Polycrystalline Zinc

Rationalization of the events which occur during the deformation of polycrystalline zinc from the knowledge obtained in single-crystal studies is not a simple matter. The restraints imposed by adjacent grains and the special deformations which take place in grain boundaries complicate the analysis. It is possible, however, to obtain an over-all impression of polycrystalline deformation which has value in aiding the interpretation of the behavior of zinc in practical applications.

Studies made with polycrystalline specimens have shown that a subgrain or cell structure develops in individual grains particularly at elevated temperatures and at slow strain rates.[54] Other studies[55, 56] have shown that mere heating and cooling of polycrystalline zinc results in the deformation of individual grains by stresses caused by the anisotropy of thermal expansion in the zinc lattice. These and many other studies reveal the importance of the restraints imposed upon grain movements by adjacent grains.

Since a grain cannot change shape without affecting its boundaries, the deformation characteristics of the boundaries are of importance.[57] The effects on creep strength, particularly at elevated temperatures, have received attention.[58, 59]

Deformation of Cast Zinc. A cast slab or bar of zinc is composed of fairly coarse columnar grains so oriented that the basal plane is parallel to the column axis and, hence, is normal to the slab surface. In this orientation, shortening of the column height by compression must start by basal-plane bending and not by simple slip, which accounts for the high roll forces required in the early stages of the rolling of zinc.

Those grains in which the basal plane is disposed normally to the direction of rolling are subjected to a cleavage stress during rolling. Since the rate of stressing in rolling is high the operation must be carried out at elevated temperatures where the slip mechanism is favored, otherwise cleavage cracks will develop in the slab surface and will report as scale in the rolled product.

Deformation in Rolling. Despite the complications produced by grain boundaries and other restraints there is ample evidence that the deformations involved in the rolling of zinc largely involve the slip and twinning mechanisms. This can be deduced from the typical pole figure, or orientation pattern, of cold-rolled zinc shown in Figure 9-23. It is seen that the majority of the grains are oriented with the basal plane about 20 degrees from the plane of the strip in the direction of rolling, or in positions 94 degrees removed from this orientation.

A simple but apparently valid analysis of the pole-figure evidence suggests that during the rolling operation reduction in thickness and extension in length are accomplished by slip-rotation in the rolling direction within individual grains. As the basal plane approaches the plane of the strip surface the compression component of the rolling force normal to the basal plane increases until that orientation is reached where the twinning mechanism supplants slip. The reorientation due to twinning makes it possible for slip to resume and this sequence of events continues.

When zinc is hot-rolled a population of crystallites in the twin position

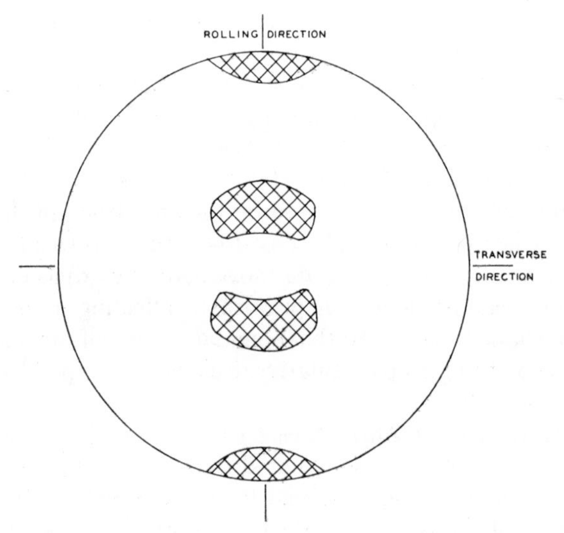

Figure 9-23. Typical pole figure for cold-rolled zinc showing range of positions assumed by the basal planes relative to the rolling plane as plane of projection.

does not develop. The probable explanation for this is that any twinning which may occur is followed immediately by recrystallization to new grains whose orientation is somewhat similar to that of the matrix grain.

There is a striking difference between the deformation by rolling of zinc and hexagonal metals of low axial ratio, such as magnesium. At room temperature magnesium deforms by basal slip but, contrary to zinc, requires a tension stress, not a compression stress, normal to the basal plane to induce twinning. Accordingly, the slip process in magnesium continues until the basal plane is parallel to the surface of the strip at which point further slip becomes impossible and the metal ruptures. Rolling at elevated temperatures, which permits another slip system to come into play, is necessary with magnesium but not with zinc.

The well known differences in with- and across-grain ease of bending in rolled zinc also are associated with orientation, although streaked-out impurities, such as lead and iron-zinc compound, as well as surface scratches and defects, also are involved. It can be seen that during an across-grain bending operation the tension and compression components of the bending stress can be relieved simply by slip on the tilted basal planes. In the with-grain case, however, the planes are not tilted sufficiently, if at all, to permit easy slip, and on the compression side of the bend the requirements for buckling and kinking are met. In a similar manner the metal will flow easily from the with-grain positions in a cupping operation, but with greater difficulty from the lateral areas.

REFERENCES

1. Rodda, J. L., *Trans. Am. Inst. Mining Met. Engrs.*, **99,** 149 (1932).
2. Jette, E. R., and Foote, F., *J. Chem. Phys.*, **3,** 605 (1935).
3. The New Jersey Zinc Co., unpublished data.
4. Edmunds, G., *Trans. Am. Inst. Mining Met. Engrs.*, **143,** 183 (1941).
5. Edmunds, G., *ibid.*, **161,** 114 (1945).
6. Teel, R. B., and Anderson, D. B., *Corrosion*, **12,** 343T (July, 1956).
7. Anderson, E. A., Boyle, E. J., and Ramsey, P. W., *Trans. Am. Inst. Mining Met. Engrs.*, **156,** 278 (1944).
8. Fuller, M. L., and Wilcox, R. L., *Trans. Am. Inst. Mining Met. Engrs.*, **122,** 231 (1936).
9. *Ibid.*, **117,** 338 (1935).
10. Bridgman, P. W., *Proc. Am. Acad. Arts Sci.*, **58,** 165 (1923).
11. Jillson, D. C., *Trans. Am. Inst. Mining Met. Engrs.*, **188,** 1005 (1950).
12. Czochralski, J., *Z. physik. Chem.*, **92,** 219 (1917).
13. Hoyem, A. G., and Tyndall, E. P. T., *Phys. Rev.*, **33,** 81 (1929).
14. Tyndall, E. P. T., *ibid.*, **31,** 313 (1928).
15. Straumanis, M., *Z. anorg. u. allgem. Chem.*, **180,** 1 (1929).
16. Buerger, M. J., *Z. Krist.*, **89,** 195 (1934).
17. Cinnamon, C. A., and Martin, A. B., *J. Appl. Phys.*, **11,** 487 (1940).
18. Masing, G., and Stanau, H., *Metals and Alloys*, **14,** 98 (1941).
19. Brinson, G., and Moore, A. J. W., *J. Inst. Metals*, **79,** 429 (1951),

20. Peirce, W. M., and Anderson, E. A., *Trans. Am. Inst. Mining Met. Engrs.*, **83**, 560 (1929).
21. Anderson, E. A., *ibid.*, **89**, 481 (1930).
22. Ruzicka, J., *ibid.*, **124**, 252 (1937).
23. Am. Soc. Testing Materials, Stand. Spec. B69-39, "ASTM Standards," Part 2, p. 517, 1955.
24. Mark, H., Polanyi, M., and Schmid, E., *Z. Physik*, **12**, 58 (1922).
25. Mathewson, C. H., *Trans. Am. Soc. Metals*, **32**, 38 (1944).
26. Kolesnikov, A. F., *J. Exp. Theoret. Phys.* (U.S.S.R.), **8**, 1031 (1938).
27. Washburn, J., M. S. Thesis, Univ. Cal. (1951).
28. Schmid, E., and Boas, W., "Plasticity of Crystals," p. 105, London, F. A. Hughes and Co., Ltd., 1950.
29. Jillson, D. C., *Trans. Am. Inst. Mining Met. Engrs.*, **188**, 1129 (1950).
30. Gilman, J. J., *ibid.*, **197**, 1217 (1953).
31. Miller, R. F., *ibid.*, **122**, 176 (1936).
32. Deruyttère, A., and Greenough, G. B., *J. Inst. Metals*, **84**, 337 (1955–6).
33. Wain, H. L., and Cottrell, A. H., *Proc. Phys. Soc.*, **63B**, 339 (1950).
34. Wain, H. L., *ibid.*, **65B**, 886 (1952).
35. Reference 28, p. 56.
36. Becker, R., and Orowan, E., *Z. Physik*, **79**, 566 (1932).
37. Reference 28, p. 120.
38. Mathewson, C. H., *Trans. Conn. Acad. Arts Sci.*, **38**, 213 (1951).
39. Reference 28, p. 123.
40. Reference 28, p. 59.
41. Mathewson, C. H., *Trans. Am. Inst. Mining Met. Engrs.*, **78**, 7 (1928).
42. Mathewson, C. H., and Phillips, A. J., *ibid.*, **78**, 445 (1928).
43. Mathewson, C. H., and Phillips, A. J., *Proc. Inst. Metals Div., Am. Inst. Mining Met. Engrs.*, p. 143 (1927).
44. Schmid, E., and Wasserman, G., *Z. Physik*, **48**, 370 (1928).
45. Jillson, D. C., *Trans. Am. Inst. Mining Met. Engrs.*, **188**, 1009 (1950).
46. Davidenkov, N. N., Kolesnikov, A. F., and Fedorov, K. N., *J. Exp. Theoret. Phys.* (U.S.S.R.), **3**, 350 (1953). Also *J. Inst. Metals, Metallurgical Abstr., Series 2*, **1**, 420 (1934).
47. Hess, J. B., and Barrett, C. S., *Trans. Am. Inst. Mining Met. Engrs.*, **185**, 599 (1949).
48. Orowan, E., *Nature*, **149**, 643 (June 1942).
49. Morton, P. H., Treon, R., and Baldwin, W. M., *J. Mech. Phys. Solids*, **2**, 177 (1954).
50. Miller, R. F., *Trans. Am. Inst. Mining Met. Engrs.*, **111**, 135 (1934).
51. Guinier, A., and Tennevin, J., *Compt. rend.*, **226**, 1530 (1948).
52. Boas, W., and Schmid, E., *Z. Physik*, **56**, 516 (1929).
53. Fahrenhorst, W., and Schmid, E., *ibid.*, **64**, 845 (1930).
54. Ramsey, J. A., *J. Inst. Metals*, **80**, 167 (1951).
55. Boas, W., and Honeycombe, R. W. K., *Proc. Roy. Soc. (London)*, **A188**, 427 (1947).
56. Burke, J. E., and Turkalo, A. M., *Trans. Am. Inst. Mining Met. Engrs.*, **194**, 651 (1952).
57. Li, Choh Hsein, Edwards, F. H., Washburn, J., and Parker, E. R., *Acta Metallurgica*, **1**, 223 (1953).
58. Crussard, C., *Rev. mét.*, **43**, 307 (1946).
59. Crussard, C., *Metal Treat.*, **14**, 149 (1947).

Zinc-Base Alloy Castings

J. C. Fox

Die Casting Consultant
Toledo, Ohio

Zinc-base castings, from the common industrial point of view, are almost exclusively die castings. The tendency with all metals to form large crystalline grains when slowly cooled is accompanied in the case of zinc by an enhanced brittleness which makes ordinary casting in sand molds a rather unattractive procedure. However, there are special applications of sand-cast zinc-base alloys which are using several thousand tons of slab zinc annually (see Table 2-8). This section of Chapter 9, addressed to zinc castings generally, deals first with die castings, and then very briefly with other zinc castings.

ZINC-BASE DIE CASTINGS

Die casting, the art of producing accurately finished parts by forcing molten metal into a metallic die or mold under external pressure, has been known, and to a very limited extent, practiced, for the past 100 years. However, the commercial exploitation of the process began about 1905.

From a meagre beginning, the die casting process has grown to such an extent that today it can be considered one of the most important methods of metal fabrication. Of all processes, the die casting process represents the shortest way from ingot metal to finished part.

Modern zinc die casting production really started about 1929 when the high purity elemental zinc (Special High Grade) of 99.99+ per cent purity was first developed. Prior to 1929, zinc-base die castings produced from elemental zinc then available—even from the highest purity of 99.95 per cent—were weak, brittle, and subject to deterioration in use. The development of the Special High Grade (99.99+) zinc has resulted in zinc-base die castings of high strength, mechanical and dimensional stability, and other properties which have been responsible for their expanded usage.

Coupled with the emergence of this Special High Grade zinc was The New Jersey Zinc Company's research and development of the "Zamak"-type alloys which have been, and are still, the standard alloys for zinc-base die casting production throughout the world. In the preparation of these Zamak alloys, the use of the Special High Grade zinc (99.99+) is mandatory, and zinc of any lower grade cannot be tolerated. Further, no other type of zinc alloy composition has as yet been found to displace the Zamak-type alloys. The American Society for Testing Materials, the Society of Automotive Engineers, and other technical societies have adopted these alloys as standard for zinc-base die casting.

Since these developments, there has been a steady increase in the production and consumption of zinc die castings. In the early thirties, the die casting process consumed only about 10 per cent of the total zinc metal produced, whereas in 1956 about 40 per cent of the total zinc was consumed in die casting production. The present rate of production is in the area of 400,000 tons. Only the galvanizing process exceeds die casting in the use of zinc.

The latest available figures indicate that zinc-base alloys represent about 63 per cent of the total die casting production, with aluminum-base alloys next at about 35 per cent, and the remainder divided among magnesium, copper, tin and lead-base castings.

For the successful operation of the die casting process, there are three basic requirements:

1. A properly designed and smoothly operating casting machine to hold and operate a die under pressure.
2. A properly designed and constructed die.
3. A suitable alloy.

In addition, the operation of a die casting plant requires suitable machining facilities for the removal of gates, fins, etc., and other machining operations. Organized facilities for accurate inspection must be provided.

A well organized die casting facility also requires a staff of engineers, metallurgists and technicians to keep all the factors of the process fully coordinated and controlled. Parts must be specifically designed for die casting before they can be produced. Dies must be properly designed and constructed and of the right type of steel; heat treatment of the steel for the die must be specified. The right casting alloy must be selected to meet the service conditions imposed by the part, and the alloy must be carefully prepared and rigidly controlled both physically and chemically during its use.

The Casting Machine

The die casting machine used for zinc-base die casting is the so-called "hot" chamber or submerged plunger type. This type machine is illustrated diagrammatically in Figure 9-24, which shows its various components. Figure 9-25 is an actual photograph of the machine in operation.

In this machine, a metal cylinder or "gooseneck," as it is most generally termed, and a plunger are submerged in a pot of molten alloy with an injection plunger connected to an air or hydraulic-shot cylinder which is located above the furnace as shown. The gooseneck is secured within a cast iron or cast steel pot, which is held in a well insulated furnace. A passage leads from the gooseneck bore through a nozzle into the sprue opening in the die. When the plunger is retracted, molten metal from the pot is admitted through the parts in the gooseneck, thus filling the cylinder up to

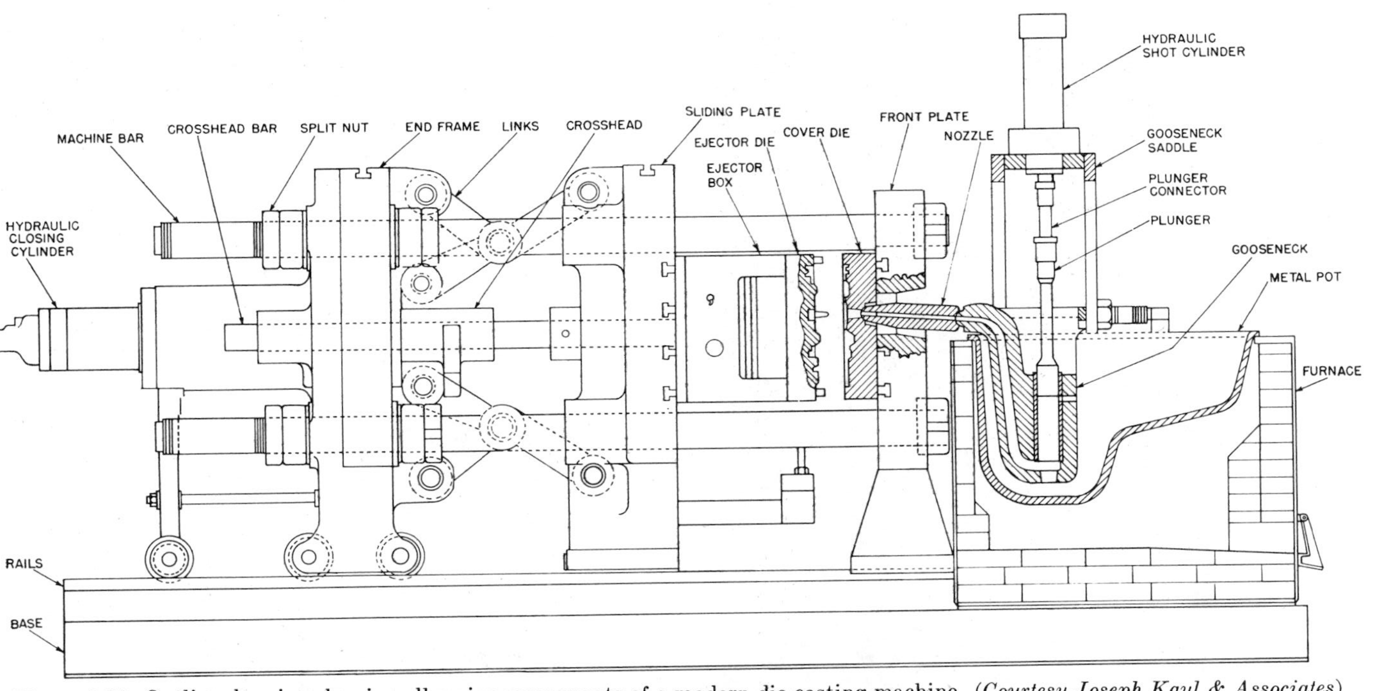

Figure 9-24. Outline drawing showing all major components of a modern die casting machine. (*Courtesy Joseph Kaul & Associates*)

Figure 9-25. Die casting machine for zinc-base alloys. Same machine as shown in Figure 9-24. (*Courtesy Joseph Kaul & Associates*)

the bottom of the plunger, and the passage up to the metal-level in the pot. On the downstroke, the plunger first seals off the part and then pushes the metal through the passage into the die.

The typical sequence of operation of a modern zinc die casting machine is as follows: starting from the opening position of the die, the operator pushes a button which withdraws the ejector pins and actuates the cores and slides into place. The machine then closes the die, bringing both die halves together and locking them. As soon as the die blocks are locked, the plunger moves forward to inject the metal into the die cavity. At the end of a prescribed interval, the plunger retracts, the excess alloy returns to the gooseneck, and the machine opens; the cores are withdrawn and the ejector pins move forward to eject the die casting from the die. The operator then cleans and lubricates the die surfaces and pushes a button to start another cycle. A, B, C, and D of Figure 9-26 represent the theoretical flow of metal into a die.

The modern die casting machine, with the aid of timing devices, electronically controlled valves, relays, and limit switches, makes the casting operation almost automatic. The operator simply pushes a button to close and open the die, and removes the casting. If need be, the operation could be made fully automatic. In fact, there are many fully automatic zinc die casting machines in use, especially for the production of very small parts such as are used in clock, watch and other similar manufacturing.

From a metallurgical viewpoint, and as compared with the "cold" chamber type casting machine used for aluminum and other alloys, the zinc "hot" chamber machine may be said to be as good as can be expected. In the latter machine, there is a greater possibility of forcing only clean liquid metal, free from gases and oxides at a definite temperature, into the die, since the metal is withdrawn from the center of the liquid metal bath. It is only through carelessness, when the level of the metal is allowed to run down below the intake orifice of the gooseneck, that any contamination of metal with dross can get into the die.

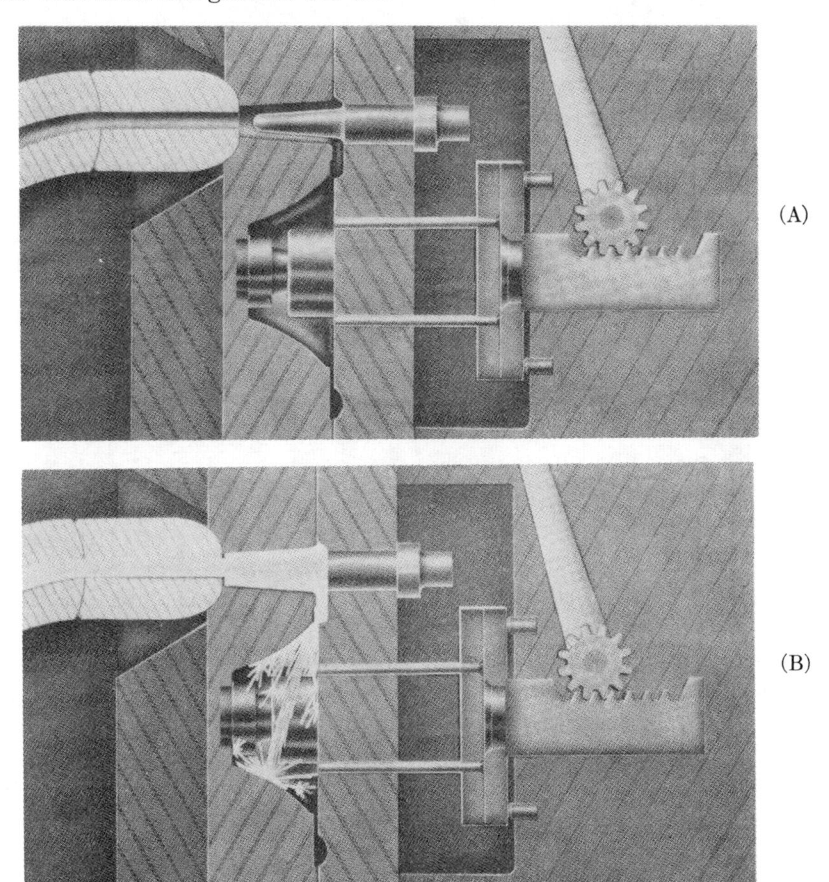

Figure 9-26. Illustrations showing sequence of theoretical metal flow into a die. (*Courtesy The New Jersey Zinc Company*[16])

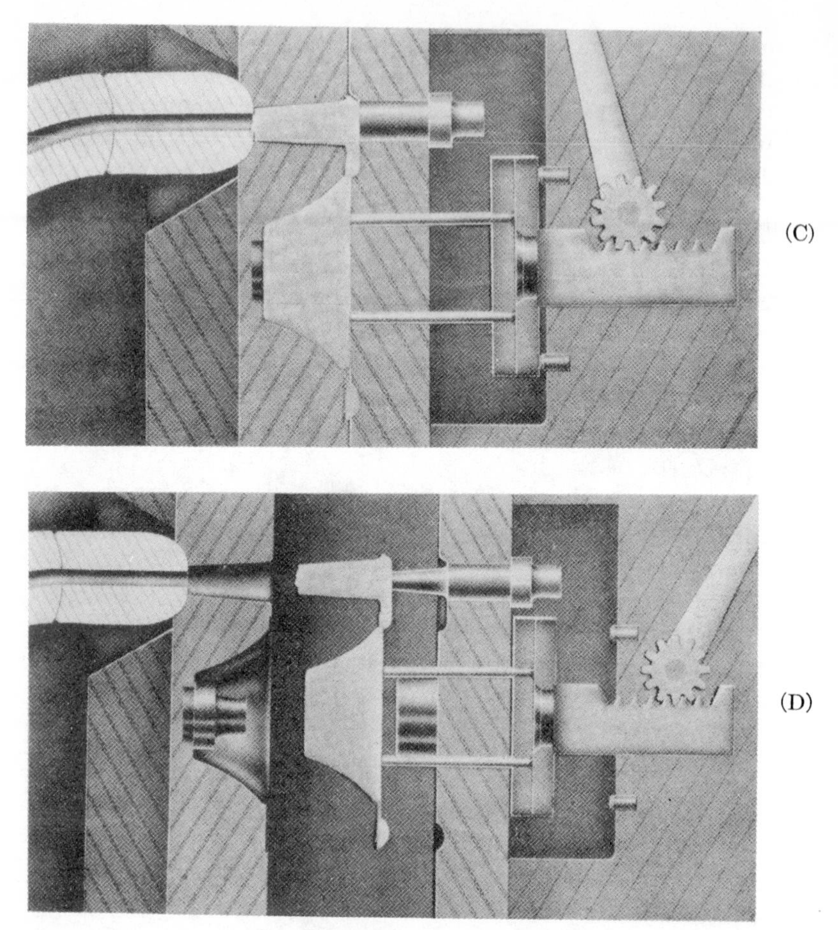

Figure 9-26. Cont.

The metal goosenecks may be made of cast iron, meehanite, cast steel or alloy steel forgings. In cast goosenecks, the practice at times is to shrink a heat-treated and nitrided sleeve in the gooseneck bore to minimize wear by the stroke of the plunger. Goosenecks made of steel forgings may be used without sleeves and simply heat-treated to a given hardness. The plungers may be made of cast iron, cast steel or alloy steel. Some are used with piston rings to retain the pressure on the metal and some are used without rings.

The machine platens may be made from medium carbon cast steel or fabricated from boiler-plate stock.

The toggle links are generally made from cast steel as in the crab and closing cylinder. The link bushings are generally bronze, corresponding to SAE64 bronze.

Hydraulic System. Up to about 25 years ago, most die casting machines were operated by pneumatic power; since then, this has been supplanted by hydraulic power, which is now being used almost universally.

The selection and use of the most suitable hydraulic fluid for die casting is very important. Oil, water and a number of synthetic high-flash organic compounds are the usual materials selected for actuating casting machines and mechanisms.

Where ordinary oil is the medium used, there is a constant danger of fire. Under the high pressures of 1000 psi or more, at which hydraulic systems operate, any break in hydraulic lines can create an explosive condition due to the adjacency of molten metal. It is for this reason that oil is not recommended, nor is it now generally used.

Although most casting machines are self-contained—that is, the hydraulic pumps and accumulators are right at the casting machines—most of the larger die casting companies employ a central system. This eliminates the pumps and motors for each machine, and permits a large pump to handle a given number of machines at a point remote from the casting area. A direct central system, using water containing about 2 per cent of a soluble oil to minimize rusting, is considered the safest and most economical hydraulic system. The use of water-based hydraulic fluids eliminates the fire hazard of oils and other organic materials, and any small leaks in the hydraulic lines, which do invariably happen, can be disregarded until there is time to repair them. With oils and the like, leaks are costly and also create slippery floors which are hazardous. A drawback to the use of water, however, is that the conventional oil valves cannot well be used since closer fits or clearance must be maintained with water systems than with oils or more viscous materials.

Automatic Cycle Control. For maximum speed of operation and casting uniformity, it is essential to have a definite predetermined interlocking cycle control. Pressures and temperatures must be regulated automatically. Dies must remain closed for a given length of time while they are being filled with the molten alloy, and then opened in accordance with a selected time pattern to permit the alloy to solidify before the die is reopened. A regular time cycle of operation also permits the uniform die temperature necessary for good casting production.

Die Casting Dies

A simple die casting die consists of two blocks of steel, each containing a part of the casting impression. These are termed the "cover" and "ejector" die blocks. The cover die block is fastened to a stationary platen on the die casting machine and does not move during the casting cycle. These blocks are so arranged on the casting machine that they are brought together,

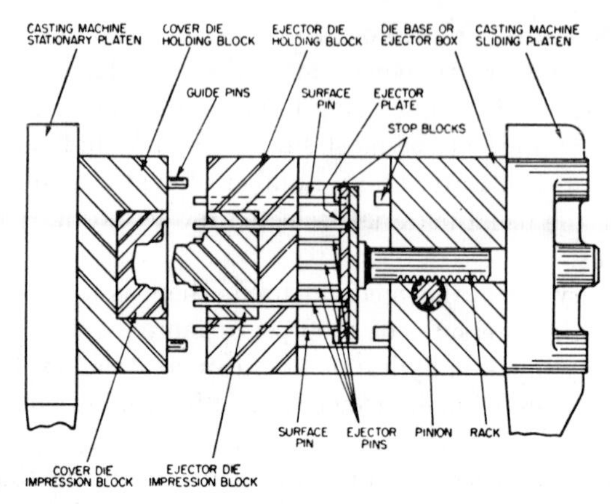

Figure 9-27. Essential components of a simple die casting die. (*By permission of the McGraw-Hill Book Company*[7])

closed tightly and locked into place to make a casting, and opened to remove the finished casting. The "ejector" die block contains, in addition to part of the casting impression, all other movable parts such as the ejector mechanisms, cores and slides. Figure 9-27 shows the elements of a typical die casting die.

The ejector die block is mounted on a die base—cast iron or cast steel—in which there is provided a means for suitably ejecting the casting from the die cavity. Figure 9-28 shows the die base and ejector mechanism.

Die casting dies may be classified into essentially four types: single impression, multiple impression, combination and unit. A *single impression*

Figure 9-28. Die base and ejector mechanism of a die casting die. (*By permission of the McGraw-Hill Book Company*[7])

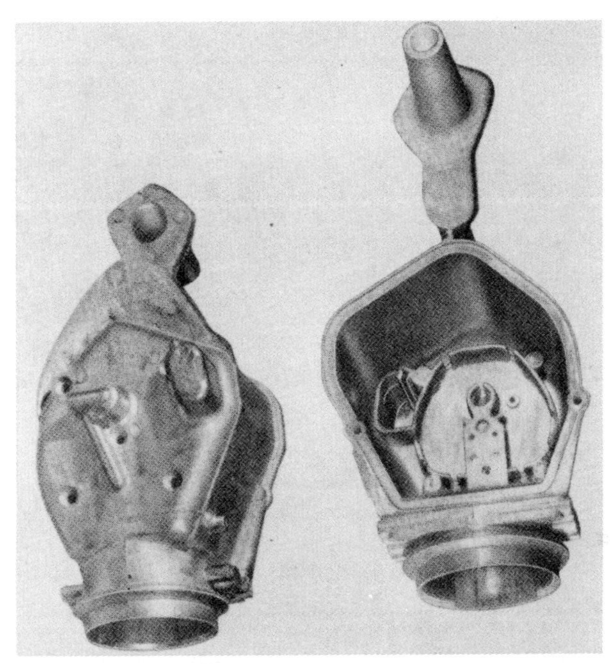

Figure 9-29. Gate and casting from a single-impression die. (*Courtesy The New Jersey Zinc Company*[16])

die (Figure 9-29) is probably most universally used; it is usually the least complicated and easiest to build. It is used for small, simple castings where production requirements are low, and for medium and large-sized parts where it may be the only practical type.

Multiple impression dies (Figure 9-30) are used where large and rapid production of a part is required. The number of impressions in such a die depends upon the size and shape of a part and the rate of production demanded. For some small, simply-shaped parts, the number of impressions may run as high as 32 or more.

Combination dies (Figure 9-31) and multiple dies both have the same purpose; namely, to permit the casting of two or more parts simultaneously. The difference is that the multiple die is used for casting parts of the same shape and size, while the combination die is used for castings which are different. Combination dies are generally used for different parts of one assembly whose production requirements, either large or small, are always similar.

For the die casting of parts of moderate size, especially when the desired production is relatively low, the use of *unit dies* is of considerable advantage. In the unit die arrangement, the cavities are formed in replaceable die

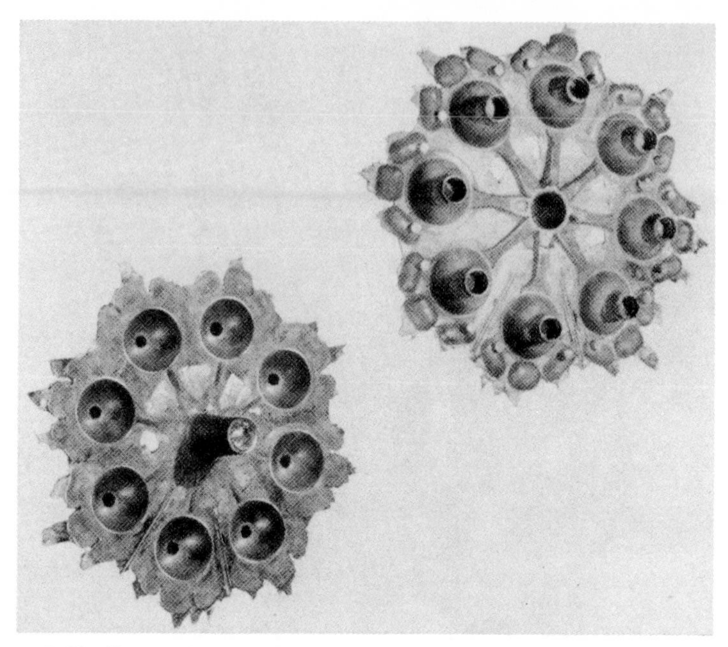

Figure 9-30. Gate and castings from a multiple-impression die. (*Courtesy The New Jersey Zinc Company*[16])

units, which are set into machined locations in a master holding die and to which the charge of metal is simultaneously delivered. Each of the replaceable units constitutes a separate part. Thus, one unit die may be used to die cast a number of different parts for a number of different products and customers. Units are made in various sizes and may be round, square or rectangular in shape. The master holding die may be constructed for multiples of two, four, six or eight units. Figure 9-32 shows a four-impression unit die.

Dies with Separate Block Impressions. Today, many dies are fabricated with the casting impressions made in separate blocks; these blocks are inserted and keyed in a holding block of steel, as an alternative to sinking the impression in a large, single die block. The advantages of this method are threefold:

(1) The smaller impression blocks can be more readily replaced by new inserts when required.

(2) The steel for the impression blocks can be of the highest quality alloy composition, while the holding blocks may be of lower-grade steel, cast steel or cast iron, thus reducing material costs.

(3) When changes in the design of a part are necessary, only the insert block need be replaced or changed.

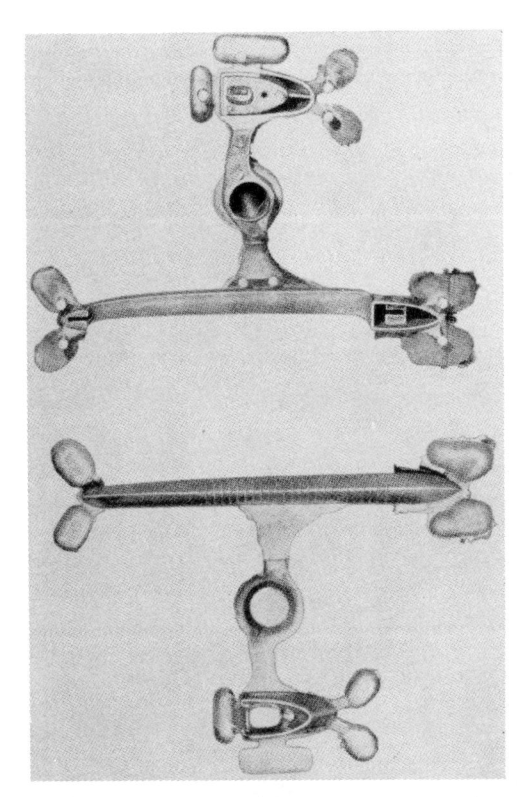

Figure 9-31. Gate and castings from a combination-impression die. (*Courtesy The New Jersey Zinc Company*[16])

Die Components and Operating Features. *Cores.* In die casting, a hole of practically any shape can be included. This is accomplished by the use of cores usually made from a high grade of steel and heat-treated to a definite hardness. Cast holes can be round, square, oblong or of other geometric shapes. Cores may be provided with keyways or other shaped slots, as well as geared or splined.

Cores may be fixed or movable in a die. Fixed cores may be placed in either the cover or ejector die block. If in the cover die, they should be designed with sufficient draft so that the casting is readily freed from the core upon opening the die. Movable cores are those having axes parallel to, or at any angle with, the parting line. Movable cores must be withdrawn from the casting before its removal from the die. The simplest movable core is one having a circular section with axis parallel to the die motion. However, there are other types, such as rotary cores, which are actuated on an arc; collapsible cores, which are only practicable when the part can-

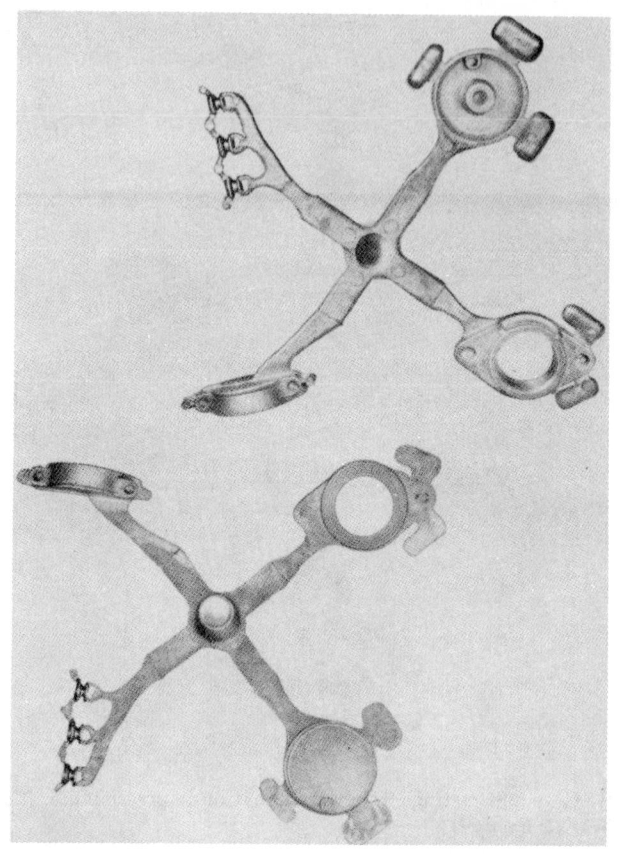

Figure 9-32. Gate and castings from a four-impression unit die. (*Courtesy The New Jersey Zinc Company*[18])

not be produced more economically by using loose cores or by machining the hole; and hinged cores, or spiral cores, which are required in specially designed die castings.

Loose Cores. Loose cores are die members used when it is necessary to form complex shapes, and especially to cast undercuts. Such cores form recesses which cannot be formed by ordinary fixed or movable cores, where it would be impossible to eject the casting from the die because of undercuts.

Loose cores are ejected, along with each casting, from the die and then removed from the casting each time, either manually or by appropriate fixtures. After removal from the die casting, the cores are returned to the die for the next casting cycle. Loose cores must be made to fit closely and accurately in a die, so as to minimize the formation of seams on a casting;

they should be designed to permit their easy placement in a die as well as their easy removal from each cast.

Slides. An undercut in a die casting is a recess in the side wall, or in a cored hole, so disposed that the casting cannot be readily ejected from the die without the use of a type of a core termed a "slide." A slide is a separate block of steel which also may contain cores, inserted in a die and operated independently. Slides and movable cores can form only such undercuts as are accessible from the outside of a casting. Interior undercuts must be formed by loose pieces set in the die before each shot.

Core Actuating and Locking. Cores and slides constructed so they are easy to "pull," or operate, and must be held tightly in their desired position during the casting operation. If a core or slide is difficult to move, the rapidity with which a casting cycle can be made may be affected. And, if a core or slide is pushed out of position by the pressure of the molten alloy as it enters the die, dimensions of the casting may be changed sufficiently to cause rejection of the casting.

There are several types of mechanisms—core pulls—which can be used for actuating cores and slides. These include for example rack and pinion core pulls, pin core pulls and hydraulic core pulls, the latter being the type now most generally used.

Means must be provided for keeping cores and slides in casting position. This is accomplished most effectively by the use of an internal wedge lock. These locks, like the slides, are made with suitable tapers and are accurately fitted to a matching taper in the slide, or core proper.

Parting Line. This refers to a line, or seam, around a die casting where the two die-halves meet. Although it is desirable to have the parting line of a die in one plane for simplicity of design and ease of casting, many dies must be constructed with irregular faces, or parting-line surfaces. This is necessary for casing an irregularly shaped part without producing an undercut. It is the function of the die designer to determine where the parting will be located and to design the casting so that parting can be made as simple as possible and preferably in a single plane, rather than irregularly. Irregular parting-line dies are more costly to fabricate and, in addition, they are apt to cause the formation of flash or fins on the die castings which are more costly to remove than in straight-line parting.

Casting Ejection. An important part of die construction concerns the provision for removing the die casting from the die just as it is formed. Several methods of ejection are in common use; the one most generally employed is by means of ejector pins. These are pins fashioned from finished drill-rod stock which, after a suitable heat treatment, are secured in the ejector plate. The actuation of ejector pins may be manual or automatic, accomplished by either rack and pinion or hydraulic devices.

Gating and Venting. The opening in a die casting die into which the molten metal is injected to reach the die impression is known as the "gate" or metal inlet. When two or more impressions are in the same die, two or more feeder lines must branch out from the main gate to each impression. These lines are usually termed "gate runners." The location and size of a gate in relation to the area of a casting are of considerable importance.

Vents provide means for the escape of air displaced in the dies while filling. They are shallow slots, ranging in depth from 0.006 to 0.015 inch, machined in die blocks at the parting line. Vents are so placed in a die as to allow the air in the cavity to be readily expelled. Air is also removed from the die cavities through clearances in cores and slides.

The type of gating and venting in a die casting die influences the quality of the die casting, especially in relation to its internal soundness and surface finish.

Fillets and Radii. Whenever possible, it is important to design with uniform wall section throughout a casting. When this is not possible, which is frequently the case, the transition from thick to thin sections should be made as gradual as possible and ample fillets should be provided at the juncture of sections. Fillets are necessary in corners where sections meet to avoid stress concentrations. Even the smallest fillet is far better than none at all. Fillets and radii promote stronger sections, improve die life and make for easier finishing of the casting.

Water Cooling of Dies. All die casting dies are provided with channels in the die blocks for water cooling certain sections. This is especially necessary in castings of heavy wall section and at the gating sections where the temperature of the die is raised to a high level. Slides, cores and other moving parts of the die are also water cooled to avoid difficulties of seizure and galling in their operation.

Ribs, Bosses and Projections. Ribs are projections used quite extensively on die castings for the purpose of lending strength and stiffness to a given area; also to decrease the tendency to warpage or distortion. They also aid the flow of metal in filling the die cavity. A heavily-sectioned part may be lightened considerably by reducing the wall section and adding supporting ribs.

Bosses are projections used as fastening or bearing points primarily for fitting a die casting to a mating part.

Inserts. When it is necessary to impart to zinc die castings certain properties which are not inherent in the cast alloy, inserts are used. These are separately fabricated parts that are cast into place in a die casting and are an integral part of that casting. Inserts are provided for superior bearing surfaces; added mechanical strength, hardness, corrosion resistance, spring resiliency, magnetism; and as a base for soldering and welding. Various mate-

TABLE 9-6. AVERAGE WEIGHT AND DIMENSIONAL LIMITS FOR ZINC-BASE DIE CASTINGS

Maximum weight of casting (lb)	35
Minimum wall thickness: large parts (in.)	0.050
Minimum wall thickness: small parts (in.)	0.015
Minimum variation from blueprint dimensions per in. diameter or length	0.001
Cast threads—max. no. per in., external	24
Cast threads—max. no. per in., internal	24
Cored holes—minimum diameter (in.)	0.031
Minimum draft on cores in. per in. of length or diameter	0.001
Minimum draft on side walls in. per in. of length	0.005

rials may be used for inserts, including cast iron, tool steel, spring steel, bronze and other nonferrous alloys, plastics, insulated wire, and even wood or treated paper. The insert may be fabricated as a casting, screw machine part, stamping, punching, forging, sheet metal, bar, rod, or almost any other form.

Average Size and Weight of Zinc Dies Castings. Table 9-6 shows the average dimensional and weight limits for zinc die castings. These are figures for average conditions. Thinner wall sections, closer dimensions and larger and heavier castings are possible when required.

Die Materials. The high cost of die casting dies makes it mandatory to exercise the greatest care in their handling and use in order to obtain the maximum die life, as measured by the number of castings capable of being produced before replacement of a die becomes necessary.

The steel used for the dies should be of the highest quality, produced under the best steel mill practice. It should be uniformly sound and free from internal defects such as forging cracks, seams, slag inclusions, porosity and other similar defects which may be uncovered during machining operations. Besides structural soundness, the steel should possess good machinability, with strength and hardness sufficient to resist deformation in service; sufficient toughness to prevent cleavage cracking; and it should be capable of being heat treated with a minimum of dimensional change, warpage or distortion.

Die materials and die life for zinc base die casting do not present the same problems as are encountered in the die casting of the higher melting-point metals and alloys such as aluminum, magnesium, or brass. With the same steel dies that are used for aluminum casting, zinc will show a life three or four times longer.

Even though it is possible to utilize low-cost unalloyed or low-alloy steels for zinc dies and often obtain a fairly long die life from them, it is expedient to use a high-grade steel, since the initial cost of the steel represents only a minor part in the cost of a die, labor being the major cost.

It is important to select and use a high-grade alloy steel, heat-treated to

TABLE 9-7. PERCENTAGE COMPOSITIONS OF ZINC DIE CASTING DIE STEELS

Steel	Carbon	Manganese	Silicon	Chromium	Molybdenum	Vanadium	Others
1	0.30	0.75	0.50	0.80	0.30	—	—
2	0.40	0.40	1.00	5.25	1.00	1.00	—
3	0.20	0.30	0.30	0.25	—	0.20	Nickel—4.10 Aluminum—1.20
4	0.10	0.30	0.20	5.00	0.90	0.25	—

a minimum hardness of about 340 Brinell, because of the improvements and developments made on the present-day casting machines. The larger and stronger die casting machines, utilizing high pressures and high locking means, have forced the use of die steels of high strength and hardness.

Typical die steel compositions for zinc die casting dies in present use are listed in Table 9-7.

No. 1 of Table 9-7 is a low-cost steel generally supplied in the prehardened condition with a Brinell hardness of 280 to 320; if supplied in the annealed condition, it is hardened after the die is fabricated.

Steel No. 2 is the composition used generally for aluminum-base castings, but it is also used with zinc, particularly for long production runs. This is a high quality alloy steel, air hardening to 400 to 440 Brinell, and is capable of long die life in zinc die casting.

Steel No. 3 is a precipitation-hardening steel which is used at a hardness of about 300 Brinell. It has good resistance to heat checking and erosion in zinc casting. It can be readily machined in the prehardened state and acquires a very high polish.

Steel No. 4 is a hobbing die steel. It is hardened after hobbing by air cooling from 1750 to 1800°F and is tempered at 800°F to a final hardness of about 360 Brinell.

Precautions. The proper selection and handling of steel for die casting dies have an important bearing on the economy of the process. Die casting dies are costly and their cost must be justified by the production of a suitable number of castings.

To obtain an economical die life, the steel used must be of suitable composition and must be prepared under the best possible mill practice. A good die casting steel should also possess good machinability, resistance to early fatigue-cracking and to the attack of the molten casting alloy. High thermal conductivity, low expansivity and dimensional stability in heat treatment are also necessary attributes.

In order to gain maximum life of die casting dies, it is essential to consider factors other than the quality of steel used. Factors such as die design, heat treatment, welding, thermal shock, die-surface attack and general good care of the die in production, influence die life to a greater degree than the quality of steel alone.

Effect of Die Design on Die Life. The design of a die can greatly influence die life. If at all possible, all large dies should be made in segments instead of one-piece blocks. There are a number of advantages to be gained in segmental dies. Large one-piece dies show greater variations in temperature than dies made in sections. Segmental dies have more chance for movement without setting up high stresses. There is a better opportunity for producing higher quality through working the steel in small sections than large blocks. It is easier to replace any part of a segmented die which failed, rather than to replace a whole die-half (in the case of one-piece dies).

The one main objection to segmental dies is that seams occur on the castings where the segments join, and these invariably have to be removed from the casting surfaces. However, it is possible sometimes to minimize their appearance through novel die design. Even if they cannot be fully masked, their cost of removal can at most times be justified by the longer die life thus achieved and the lower cost of die replacement, when such becomes necessary.

The proper placement of water-cooling channels on dies to prevent temperature differentials and resulting stresses is of importance, and much thought should be given to this factor when designing a die. A balanced die temperature is significantly important not only in relation to die life but also in the production of good sound castings. Water lines should be smooth and free from tool marks which can act as stress raisers and which may lead to early cracking of dies. Die impressions should also be free from tool marks and preferably should be given as high a polish as possible.

Ample fillets should be incorporated at re-entrant corners. The smallest fillet is better than a sharp corner. Table 9-8 shows the effect of radius on the notch impact strength of die steel No. 2, and readily indicates the significance of avoiding sharp corners in die design.

Die Temperature. Maintaining a uniform die temperature at all times during the operation of a die is extremely important. With zinc-base alloy,

TABLE 9-8. EFFECT OF RADIUS ON NOTCH IMPACT STRENGTH OF DIE STEEL*

Radius at Base of Notch	Brinell Hardness	Average Impact Strength (ft-lb)
0.000	440	3.5
0.010	450	6.0
0.020	430	10.0
0.040	440	18.0
0.080	450	20.0
0.125	440	22.0

* Steel composition No. 2.

TABLE 9-9. EFFECT OF TEMPERATURE ON IMPACT STRENGTH OF DIE STEEL*

Temperature (°F)	Average Impact Strength (ft-lb)
32	2.6
85	3.0
212	5.0
400	8.0
600	12.5

* Steel composition No. 2.

the range of temperature between 350° and 450°F is considered most satisfactory for good casting surface finish, as well as for maintaining good mechanical properties of the casting. Another important reason for keeping the dies at these temperatures is to prevent cleavage cracking, since the impact strength and toughness of the steel increase with rise in temperature. The operation of dies at uniform temperature results in closer control of casting dimensions along with better surface finish and highest mechanical properties. Table 9-9 gives pertinent data on the change of impact strength with temperature.

Die casting dies differ in their heat treatment from many other tools in that this is readily done after the die is completely finished. Therefore, extreme care must be exercised in heat treatment to prevent changes in the surface chemistry of the steel, to minimize dimensional changes, to minimize warpage and distortion, and to prevent scaling and pitting of the impression surfaces. Lack of attention to these details in the heat-treatment procedure can greatly affect the life of a die.

Other causes for poor die life and early die failures are: (1) the promiscuous use of the welding torch for making die changes and repairs, without due regard for the proper welding procedure and treatment after welding; (2) lack of control of die temperatures and of maintenance of uniform die temperatures during casting; (3) lack of care in keeping the die impression surfaces clean and free from attack by the molten casting alloy by use of suitable protective films.

Alloys

There are two zinc-base alloy compositions in current use which are standard for the die casting industry. These alloys, known under the "Zamak" trademark, were developed by The New Jersey Zinc Company following the introduction of the 99.99+ per cent grade of zinc.* They have been adopted and standardized by the American Society for Testing Materials, the Society of Automotive Engineers and other groups, after considerable study and testing.

* See also the discussion under Quality Standards in the first section of this chapter.

TABLE 9-10. CHEMICAL REQUIREMENTS OF ZINC-BASE ALLOY DIE CASTINGS

Alloy Components	Percentage in Alloy AG 40-A XXIII	Percentage in Alloy AC 41-A XXIV
Aluminum	3.5 to 4.3	3.5 to 4.3
Copper	0.25 maximum	0.75 to 1.25
Magnesium	0.03 to 0.08	0.03 to 0.08
Iron	0.100 maximum	0.100 maximum
Lead	0.007 maximum	0.007 maximum
Cadmium	0.005 maximum	0.005 maximum
Tin	0.005 maximum	0.005 maximum
Zinc	remainder	remainder

Chemical Requirements. Table 9-10 gives the chemical requirements of these alloys in accordance with the latest revision of American Society for Testing Materials' Tentative Specifications B-86.

The limits of the various elements, mentioned in Table 9-10, have been established after many years of exhaustive and thorough testing. While there is a minimum and a maximum content for the several elements, there is also an optimum content for each element which promotes the best results.

The effect of each of these elements* is briefly summarized as follows:

Aluminum. Aluminum is added to zinc for increasing strength, reducing grain size, and to decrease the attack of the zinc alloys on iron and steel parts, thus making it possible to utilize the submerged plunger-type machine. It forms a complex compound with any dissolved iron, causing the resulting dross to go to the surface of the molten alloy from which it can readily be removed. The impact strength of the zinc alloy is seriously affected by aluminum contents greater than 4.50 per cent. At 5.00 per cent aluminum, the alloy can be quite brittle. Therefore, it has been found expedient to set the safe maximum limit at 4.3 per cent. The minimum content is set at 3.5 per cent in the above specifications to allow sufficient range in the preparation of the alloy. Aluminum appreciably lower than 3.5 per cent can adversely affect castability and lower the mechanical properties of the casting.

Copper. Copper increases tensile strength and hardness. However, it tends to reduce impact strength and dimensional stability on aging. Alloy AG 40-A can tolerate a copper content as high as 0.40 per cent without too great a loss of these properties on aging. For the majority of commercial applications, a copper content in the range of 0.25 to 0.75 per cent will not adversely affect the serviceability of die castings and should not serve as a basis for rejection.

Magnesium. Magnesium is purposely added to zinc alloys to aid in pre-

* Additional data on the effects of added elements in zinc are included in the earlier section of this chapter on the *Physical Metallurgy of Zinc.*

venting sub-surface corrosion, since it has been found that magnesium can counteract the ill effect of small quantities of impurities such as tin and lead. The minimum magnesium content of 0.03 per cent has been set as the lowest content which will prevent the formation of sub-surface corrosion. The maximum value of magnesium was arbitrarily set at 0.08 per cent in the ASTM specifications, but this amount is rarely encountered in zinc die casting practice since it has been determined that magnesium above 0.05 per cent can cause serious difficulty in casting, mainly hot shortness. High magnesium content can also adversely affect impact strength and ductility.

Lead. With magnesium at the minimum content of 0.03 per cent, the maximum lead content that can be tolerated is 0.007 per cent. This maximum limit for lead has been established from accelerated aging tests which have indicated that sub-surface corrosion takes place when the lead content is about 0.009 per cent or more, but does not appear with lead at 0.007 per cent.

Cadmium. According to tests conducted by the New Jersey Zinc Company, it was found that cadmium in relatively high amounts can be detrimental to the mechanical properties of the casting, and can affect castability as well. The amount of cadmium normally present in the 99.99+ per cent grade of zinc is too low to have any effect on castability. The maximum limit for cadmium of 0.005 per cent included in the specifications was set more or less arbitrarily.

Work recently conducted by the Canadian Department of Mines and Surveys has indicated that the ASTM maximum allowable cadmium content has been set unnecessarily low. However, it is believed that the present limit of 0.005 per cent is low enough to prevent any ill effects should the alloy be contaminated by cadmium from extraneous sources, such as cadmium-plated parts, accidentally entering the alloy from scrap during remelting.

Tin. Tin promotes sub-surface corrosion, just as lead does, in zinc-aluminum alloys; therefore, it is necessary to hold this element to a maximum limit of 0.005 per cent.

Iron. Iron does not have any detrimental effect on the properties of the zinc alloys in the amounts found in zinc die casting. Only about 0.02 per cent iron is held in solution in the liquid alloy. Any amount in excess of this figure combines with aluminum to form a dross which rises to the top of a melt where it is periodically removed by skimming. It has been determined that an iron content of 0.10 per cent in die casting has no effect on the permanence of properties or dimensions of die castings. A high iron content in die castings can only result through carelessness in allowing high iron dross on the top of a melt to get into the die. In such cases, the high iron would cause difficulty in the machining of castings.

Other Elements. In addition to the impurities of tin, lead, cadmium and iron normally present in zinc die casting, there are other elements which may be detected, even though present only in infinitesimal amounts. It has been determined that such elements as indium, antimony, arsenic, bismuth and mercury are detrimental to the stability of the zinc alloy if present in amounts greater than 0.001 per cent. Other elements such as boron, germanium, gallium, thallium and the rare earth metals have not been investigated. Beryllium appears to have some advantageous effect but it has not been fully investigated. Silver and titanium are considered neutral.

The elements manganese, silicon, chromium and nickel can be found in zinc die casting. Manganese and silicon are present in the aluminum used for alloying, and occur also through contact of the zinc alloy with cast iron and steel parts, which may contain these elements. Nickel and chromium are found in die castings primarily through the remelting of electroplated scrap. After considerable study, the American Society for Testing Materials has decided to include these latter four elements in the latest revision of the B-86 Tentative Specifications in the form of a footnote, as follows:

"Zinc alloy die castings may contain nickel, chromium, silicon and manganese in amounts up to their solubility (0.02, 0.02, 0.035 and about 0.5 per cent, respectively) at the freezing temperature. No harmful effects have ever been noted due to the presence of these elements in these concentrations and, therefore, analyses are not required for them."

Mechanical Properties. Typical mechanical properties of the two zinc alloys in the as-cast state and after indoor and outdoor aging treatments are shown in Table 9-11.

The effects of heating and cooling on the strength-properties of the zinc alloys are shown in Table 9-12.

Physical Constants. Useful physical constants of the two predominant zinc alloys are summarized in Table 9-13.

Melting and Alloying. Zinc-base alloys are melted in cast iron, cast steel or welded steel pots, the capacities of which may range from 2,000 to 8,000 pounds. These pots are set in furnaces which are comprised of a steel cylindrical shell, lined with refractory, and suitable insulation. They are generally fired with gas, although oil or even electric immersion heaters have been used. To some extent, low-frequency induction furnaces are being used, but gas-fired furnaces are generally employed.

There are essentially two methods used in die casting practice for getting the alloy into the liquid state for casting. One of these is to melt the ingot and scrap in a pot, crucible or furnace at the casting machine; the other is to melt all metal, new and scrap, at a central point away from the casting machines and supply only cleaned liquid metal to the machine holding pot.

The first method—direct melting at the casting machines—has definite

TABLE 9-11. MECHANICAL PROPERTIES AS-CAST AND AFTER AGING

	Tensile Strengh (psi)	Elongation (% in 2 in.)	Impact Strength (Ft-lb)	Dimensional Change
Alloy—AG 40-A				
As-Cast	41,000	10	43	—
After 10 yr* indoor aging	33,800	20	41	−0.0001
After 10 yr† outdoor aging	32,200	13	30	0.0000
Alloy—AC 41-A				
As-Cast	47,600	7	48	—
After 10 yr indoor aging	38,400	13	36	−0.0001
After 10 yr outdoor aging	36,800	6	25	0.0000

* Indoor exposures conducted at Coco Coco, Canal Zone; Tuscon, Arizona; and New Kensington, Pa.
† Outdoor exposures conducted at Key West, Fla.; Sandy Hook, N. J.; Altoona, Pa.; State College, Pa.; and New York City.

TABLE 9-12. MECHANICAL PROPERTIES AT ELEVATED AND SUB-NORMAL TEMPERATURES

Property	Temperatures					
	−40°F	−4°F	32°F	70°F	104°F	203°F
Alloy AG 40-A						
Tensile strength (psi)	44,800	43,700	41,300	41,000	35,500	28,300
Elongation (% in 2 in.)	3	4	6	10	16	30
Impact strength (ft-lb)	2	4	23	43	42	40
Brinell hardness	91	87	82	82	68	43
Alloy AC 41-A						
Tensile strength (psi)	48,900	49,400	48,300	47,600	42,900	35,100
Elongation (% in 2 in.)	2	3	6	7	13	23
Impact strength (ft-lb)	2	4	41	48	46	43
Brinell hardness	107	104	99	91	89	62

disadvantages with little or no utility. It is practically impossible to maintain uniformity of metal temperature by this method. There is always a wide range of temperature throughout a melt, due to the periodic addition of cold, solid metal which reduces the temperature of the melt in the area where the addition is made. With this method, more heat is generated in

TABLE 9-13. PHYSICAL CONSTANTS[17]

	Alloy AG 40-A	Alloy AC 41-A
Brinell hardness	82	91
Compression strength (psi)	60,000	87,000
Electrical conductivity (Mhos/cm, 20°C)	157,000	153,000
Melting point (°F)	727.9	727
Modulus of rupture (psi)	95,000	105,000
Shearing strength (psi)	31,000	38,000
Solidification point (°F)	717.1	716.7
Specific gravity	6.6	6.7
Specific heat (cal g/°C)	0.10	0.10
Thermal conductivity (cal/sec/cm²/cm/°C)	0.27	0.26
Thermal expansion (per°F)	0.0000152	0.0000152
Transverse deflection (in.)	0.27	0.16
Weight per cu in. (lb)	0.24	0.24

the machine areas because of the higher temperatures necessary to melt the solid metal.

The second method—the central melting system—offers by far the better procedure, having the following advantages:

1. Precise control of metal temperature.
2. Cleaner metal going into the castings, since it is possible to thoroughly flux and clean the metal before its delivery to the machine pots.
3. Less heat and discomfort to the casting operator.
4. Better control of alloy composition; analyses and composition adjustment can easily be made before delivery.
5. Increased casting efficiency through reduction of metal handling by the casting operators.

In good melting practice, one usually starts with a small "heel" of metal left in the pot from a previous heat. The use of a heel of metal, instead of a completely empty pot, greatly enhances the life of the pot. The required amount of aluminum in the form of shot, which generally contains the required magnesium content for the alloy, is charged to the pot, followed by the addition of the slab zinc, and the whole melted down. Some operators prefer to melt the aluminum—in the form of ingot—separately, and add the liquid aluminum directly to the molten zinc.

As soon as the zinc is melted, the solution of the alloy constituents and the thorough mixing of the melt are assured by the use of a mechanical stirrer. This stirrer has multiple-blade propellers driven by an electric motor on a shaft inserted in the melt. For the zinc alloy containing copper, the aluminum shot may also carry the required copper content.

Fluxing. Periodic fluxing and cleaning of the molten zinc alloy are extremely important. It has been found that uncleaned metal can cause "soldering", "wash-outs" and other defects due to the action of the molten alloy on the die-steel surfaces, which is much more rapid than that of clean metal. Some authorities believe that in the preparation of the zinc alloy with all new virgin metals, fluxing of the alloy melt is not altogether necessary, but it is the author's contention that all metal melted should be fluxed and cleaned before use. When scrap is remelted, it is absolutely necessary to thoroughly flux the melt to get rid of all extraneous matter held in suspension. It is certainly mandatory to flux the zinc alloy whenever plated scrap is remelted.

The fluxes for zinc alloys may be straight zinc chloride, or mixtures of zinc chloride and other halide salts. There are also a number of proprietary mixtures on the market for this purpose. Halide salts have the property of removing some of the magnesium content of the alloy and it is necessary, therefore, to compensate for any losses of this element through fluxing, especially in the remelting of scrap. Thus, it is necessary to check the magnesium content of all zinc heats, especially scrap heats, before their re-use. The results obtained with clean alloys can readily justify this extra effort.

The finished molten zinc alloy may then be pumped, siphoned, or hand-ladled into suitable ladles or containers for delivery to the casting machines. The ladles are usually suspended from an overhead monorail and moved by hand, or contained in a gasoline or battery-operated truck.

Finishing of Zinc-Base Die Castings

Zinc-base die casting alloys have good corrosion resistance.* In fact, they are used in many applications in the as-cast state without any additional finish or treatment. However, the surfaces of zinc die castings do tarnish in moist atmospheres and, therefore, in most cases it is expedient to apply finishes for (1) protection against moisture and other corrosive media, (2) improvement of appearance for sales appeal and (3) keeping parts clean and sanitary.

Because of the smoothness of their as-cast surfaces, zinc die castings can be economically finished with practically every available type of treatment, including:

1. *Mechanical Finishes*—which embrace buffing, polishing, tumbling and other similar treatments of the cast surfaces by abrasion.

* The New Jersey Zinc Company[17] reports corrosion penetration rates with the principal die casting alloys, compared with those of rolled zinc and galvanized iron, as follows:

	At Palmerton, Pa.	At New York City
Zamak-3 (Alloy-type AG 40-A)	0.000078 in/yr	0.00022 in/yr
Zamak-5 (Alloy-type AC 41-A)	0.000063 in/yr	0.00028 in/yr
Rolled zinc	0.000064 in/yr	0.00028 in/yr
Galvanized iron	0.000052 in/yr	0.00027 in/yr

2. *Chemical Treatments*—in which castings are coated by immersion in various chemical solutions for decoration and for inhibiting surface corrosion. These coatings also act as bases for subsequent paint finishes, and include products such as "Cronak," "Iridite," "Bonderite" or other similar phosphate coatings; "Molyblack" and others.

3. *Electrolytic Treatments*—in which various metals such as copper, nickel, chromium, brass, gold and silver are deposited by electroplating for both decoration and protection.

4. *Organic Finishes*—in which various paints, lacquers and enamels are applied by spraying, brushing or dipping, thus producing colored and multi-textured effects for both appearance and protection. Special organic finishes have been developed to produce novel effects such as wrinkle, crystalline, crackle, veiling, hammered, pearl or opalescent surface. Also in this class are flock finishes in which particles of vegetable or animal fibers such as cotton, velvet, felt, fur and the like, suspended in a suitable vehicle, are blown over the cast surfaces, producing coatings which are used for sound-deadening, mar-proofing and insulation.

Applications

Zinc-base die castings are used extensively in practically every industry. One cannot look anywhere—in the home, the office, the factory or in the street—without seeing some zinc die casting application.

The automotive industry has been, and continues to be, the largest single user of zinc die castings. Next in extensive use is the home appliance industry, followed by machinery, tools, plumbing and heating supplies, business machine and office equipment, optical products, sporting goods, toys, novelties and many other industrial outlets. Zinc die castings are used in these industries because of their comparatively low cost, ease of machining and finishing, and excellent stability. The following table (9-14) shows

TABLE 9-14. END-USE DISTRIBUTION OF ZINC-BASE DIE CASTINGS

	Per Cent by Weight
Motor vehicles	58.0
Home appliances	17.5
Commercial machines and tools	6.9
Builders' hardware: plumbing, heating	5.8
Business machines—office equipment	3.4
Optical, photographic instruments	2.5
Timing devices	1.3
Agricultural, mining construction equipment	1.2
Electronic equipment	0.8
Sporting goods, toys, novelties, etc.	0.7
Military	0.7
Transportation (other than motor vehicles)	1.2

the end-use distribution of zinc die castings in accordance with the latest figures.

The automotive industry utilizes zinc-base die castings for parts and assemblies such as carburetors, fuel pumps, radiator grilles, trim molding, ornaments, and a variety of interior and exterior hardware.

For home appliances, zinc die-cast parts are extensively used in such equipment as washing machines, ironers, sewing machines, lighting fixtures, gardening tools, oil burners, kitchen and food-mixing equipment, radios and television sets and many other items.

The hardware industry uses zinc die castings in plumbing fixtures, bathroom accessories, furniture hardware, padlocks and other locking devices, refrigerator hardware, and in waste-food disposal.

In commercial machinery, zinc die castings are found in machine tools, hand tools, scales, shoe machinery, slicing machines, textile machinery, grinders, sanders, vending machines, and packaging machinery.

Office equipment and business machines utilize zinc die castings in adding and accounting devices, typewriters, hardware, cash registers, desk accessories and hardware filing cabinets.

In optical industry applications, one finds zinc die castings in cameras, projectors, light meters, flashlights and other optical instruments.

Practically all other industries utilize zinc die castings in some degree for their economy, appearance and properties.

OTHER ZINC CASTINGS

Of principal importance is the sand-casting of zinc-base alloys for short-run production of steel or aluminum alloy stampings in the automotive and aircraft industries. This stems from the use of zinc alloy dies in the latter industry during World War II for shaping aluminum alloy elements with presses or drop hammers. Cost savings are considerable with these low-melting zinc-base alloys, which are easy to polish, machine, weld and remelt.

Menton[14] has described open mold casting as practiced in Detroit (1947) with the alloy known as "Kirksite A," composed nominally of zinc alloyed with 3.5 to 5 per cent aluminum, 4 per cent copper and less than 1 per cent magnesium. The National Lead Company issues a comprehensive handbook[15] describing the development, foundry practice, properties and uses associated with these Kirksite alloys, which are available as ingot, sheet or rod.

Castings of Kirksite A are expected to show tensile strength of 33 to 35 thousand psi, impact strength (Charpy) of 5 foot-pounds, elongation of 2 per cent and Brinell hardness of 100; while those of an improved alloy, designated "Hi-Phy Kirksite," are somewhat higher in tensile strength, elongation and Brinell hardness, with greatly improved impact strength.

The alloys are recommended for dies and punches to be used in forming, blanking, trimming, and for many other applications, such as chuck jaws, molds, bearings, guides, rollers, etc.

Patterns may be of wood, but plaster is more commonly used. Open-mold casting is advocated, with very rigid flask and platform of wood or steel, and use of fine-grained sand of low moisture content, with little or no binder. The pattern is laid on the platform with the contours of the die or punch facing upward. The flask is placed so as to provide clearance of at least 3 inches around the pattern to prevent metal from running out at pouring. Venting is ordinarily not required as these alloys do not produce gases, but steam from the sand may cause rough spots at the bottom of the deepest dies if not vented.

Castings are usually gated at a single point. Ordinary cores and inserts are frequently used. The alloy is melted in shallow bowl-shaped pots or kettles fired by gas or oil under careful temperature control to prevent high-temperature kettle attack and excessive dross formation, or low-temperature alloy segregation. Limiting temperatures are 850 to 900°F and the proper casting temperature is 800 to 850°F. The inner surfaces of the melting pots are coated with a special compound for protection against the solvent action of molten zinc alloy.

Metal from the tilting-type furnaces is poured into casting ladles handled by overhead crane. Dross is rejected in casting by use of a skimmer at the pouring lip. The solidified casting is lifted by a crane and immersed in water or sprayed to obtain a moderate improvement in hardness.

While usually poured in sand cavities, zinc alloy dies can also be cast in preformed plaster and steel molds. They are easily ground, take a brilliant polish, machine readily, and with proper techniques are easily welded. They also possess good abrasion resistance and are somewhat self-lubricating. A great advantage of the zinc alloy dies is that they can be reclaimed by remelting and casting.

It may be stated that these alloys are of the general type so well known for die casting, also made with 99.99+ zinc to keep at a minimum the impurities lead, tin and cadmium. Copper is in the high range, however, and frequently modifications are made by extra alloying with components calculated to contribute strength, hardness, wear resistance and toughness without undue segregation due to density differences. Holzworth and Boeghold[10] describe the casting, properties and applications of a proprietary zinc-base alloy for drawing, forming, flanging and bending dies. This alloy contains (typically) 0.8 per cent nickel, 0.2 titanium and 0.15 magnesium, in addition to 4 per cent aluminum and 3.25 copper.

Federal Specification MIL-Z-7068 lists two classifications of alloys for forming dies, as follows:

Classification I	Classification II
Al 3.5–4.2	Al 3.9–4.3
Mg 0.02–0.10	Mg 0.02–0.05
Pb 0.007 max	Pb 0.003 max
Sn 0.005 max	Sn 0.001 max
Cu 2.5–3.5	Cu 2.5–2.9
Fe 0.100 max	Fe 0.075 max
Cd 0.005 max	Cd 0.003 max
Zn (99.99 % pure) remainder	Zn (99.99 % pure) remainder

Specified properties (Classification I) are: tensile strength, 37,860 psi; percentage elongation in 2 inches, 3; and Brinell hardness, 100 (500 kg load on 10 mm ball).

Today, zinc alloy sand-cast dies are used for forming, drawing, stretching, blanking and trimming large shapes and bodies from light-gage sheet aluminum, magnesium, stainless steel, low-carbon steel and other metals. These dies vary in size, some weighing several tons, and are capable of producing up to several thousand stampings before replacement or major repair. The process is now so well established that its use has extended far beyond the aircraft industry. Truck, passenger car, and bus manufacturers now use zinc stamping dies for many body and fender parts. The same zinc alloy in the form of rolled plate is used to form blocks and blanking dies.

Zinc alloy dies are in constant use in almost all types of forming and drawing tools including drop hammers, hydraulic presses, double acting presses, stretch presses and brakes. Some of the industries—other than aircraft and automotive—that use zinc alloy dies, and the items formed with these dies, are listed below.

Industry	Parts and Usage
Electrical	Boxes, panels, rings, fixtures, blanks and housings
Vending and business machines	Frames, housings, panels
Toys and novelties	Sheet metal parts
Metal furniture	Backs, seats and tube bending
Cooking utensils	Many aluminum parts, both shallow and deep
Jewelry	Formed parts in copper, gold, silver and sterling
Ceramics	Pressed block tile
Rubber	Special tire molds, small molded parts and novelties
Plastics	Injection molding and many diverse parts from a variety of plastic compositions
Food handling and dairy equipment	Stainless steel trays, pans and evaporating equipment

The net result of this diversity is a growing market embracing thousands of tons of zinc alloy annually.

Slush and sand castings, aside from the stamping dies discussed above, take several thousand tons of slab zinc annually. A brief description of the slush-casting process as applied to zinc-base or lead-base alloys is given in the 1948 edition of the "Metals Handbook."[21]

In essence, this process gives hollow castings of thin wall section by freezing a metal shell on the inner surface of a split mold, and quickly inverting the mold to permit the bulk of metal to flow away from the skull-casting left behind. Zinc alloys used for this purpose are of the zinc-aluminum type, containing around 5.5 per cent aluminum in a base of Special High Grade zinc; or of the zinc-aluminum-copper type, with something less than 5 per cent aluminum and about 0.25 per cent copper, also made with Special High Grade metal. Various elements, in carefully controlled amounts are necessary to avoid cracking, warping and improper flow characteristics. Principal uses for slush castings are in the lighting fixture and novelty fields.

Permanent mold* castings[17] (aside from slush castings) may be used as low-cost and inferior substitutes for die castings where only moderate production of adaptable shapes is indicated and strength requirements are not severe. Such castings leave the die with good luster and finish so that only moderate cost is involved in preparing the casting for any desired surface treatment.

There are special applications involving simple shapes and moderate production requirements in which the relatively cheap process of sand-casting alloys of the "Zamak" type may be considered preferable to die casting. Ordinary foundry methods are applicable to these alloys and in addition to the preferred open-mold casting of heavy sections, closed molds can be used for thinner sections when provided with large gates and risers. The proper casting temperature is 800 to 875°F. Allowance must be made in all patterns for a shrinkage of 0.14 inch per foot. Mechanical properties expected are: tensile strength, 20 to 30 thousand psi; impact strength of ¼ in. square bars, 2 to 4 foot pounds; Brinell hardness, 70 to 100; and compressive strength, 60 to 75 thousand psi.

REFERENCES†

1. Anderson, E. A., "The Finishing of Zinc Base Die Castings," *Proc. Am. Soc. Testing Materials*, **41**, 949–52 (1941).

---

* At the smelter, special grades and shapes of zinc castings (see last section of Chapter 6) such as anodes for cathodic protection, cast, as in the case of ordinary slab zinc, in permanent cast iron molds, are sometimes produced for direct application by the consumer.

† Principally for supplementary reading.

2. Anderson, E. A., and Werley, G. L., "The Effect of Variations in Aluminum Content on Strength and Permanence of the A.S.T.M. No. XXIII Zinc Die-Casting Alloy," *ibid.*, **34,** Part I, 261–9 (1934).
3. Anderson, E. A., and Werley, G. L., "The Impact Strength of Commercial Zinc Alloy Die Castings," *ibid.*, Part II, 176–81.
4. Bell, R. C., Edwards, J. O., and Meier, J. W., "The Effect of Certain Metallic Impurities on the Properties of Zamak 3 Type Zinc-Base Die Casting Alloys," *Trans. Can. Inst. Mining Met.*, **55,** 252–61 (1952).
5. Brauer, H. E., and Peirce, W. M., "The Effect of Impurities on the Oxidation and Swelling of Zinc-Aluminum Alloys," *Trans. Am. Inst. Mining Met. Engrs.*, **68,** 796–832 (1923).
6. Colwell, D. L., "Development of Zinc-Base Die-Casting Alloys," *Proc. Am. Soc. Testing Materials*, **30,** Part II, 473–92 (1930).
7. Doehler, H. H., "Die Castings," New York, McGraw-Hill Book Co., Inc., 1951.
8. Fox, J. C., "Finishing of Die Castings," *Proc. Am. Soc. Testing Materials*, **36,** Part I, 193–203 (1936).
9. Fuller, M. L., and Wilcox, R. L., "Phase Changes During Aging of Zinc Alloy Die Castings," *Trans. Am. Inst. Mining Met. Engrs.*, **117,** 338–56 (1935).
10. Holzworth, J. C., and Boeghold, A. C., "Gmoodie—A Low Cost Die Material," *Metal Progress*, **69,** 49–53 (1956).
11. Kelton, E. H., "Fatigue of Zinc Base Die Castings," *Proc. Am. Soc. Testing Materials*, **42,** 692–707 (1942).
12. Kelton, E. H., and Grissinger, B. D., "Creep Data on Die-Cast Zinc Alloy," *Trans. Am. Inst. Mining Met. Engrs.*, **161,** 466–71 (1945).
13. Maxon, C. R., "Tumble Burnishing of Die Castings," *Metals and Alloys*, **13,** 182 (1941).
14. Menton, S., "Casting Zinc Alloy Dies," *Foundary*, **75,** 76–9; 267–9 (June, 1947).
15. National Lead Company, "Kirksite A Handbook," New York, 1953.
16. New Jersey Zinc Company, "Die Casting for Engineers," New York, 1958.
17. New Jersey Zinc Company, "Zinc Metals and Alloys," New York, 1955.
18. New Jersey Zinc Company, "Practical Considerations in Die Casting Design," New York, 1955.
19. Peirce, W. M., "Metallography of Zinc-Base Die-Casting Alloys," *Proc. Am. Soc. Testing Materials*, **30,** Part I, 334–5 (1930).
20. Werley, G. L., "A Study of Die Design Changes for Improvement of Soundness and Uniformity of Test Bars, *ibid.*, **37,** Part I, 223–54 (1937).
21. Werley, G. L., and Hiers, G. O., "Slush Casting," in "Metals Handbook," 1948 ed., p. 741, Cleveland, Ohio, American Society for Metals.
22. Williams, H. M., "Swelling of Zinc Base Die Castings," *J. Am. Inst. Metals*, **11,** 221–4 (1917).

Zinc Coatings

JOHN L. KIMBERLEY

Executive Vice President, American Zinc Institute
New York, N. Y.

With Sections by Divisional Specialists

During modern times and presently the largest use of zinc in the United States has been and continues as a protective coating for steel. Such coatings are applied by five different methods:

1. Hot Dip Galvanizing—immersion in molten zinc.
2. Electrodeposition.
3. Metallizing—spraying with droplets of molten zinc.
4. Sherardizing—diffusion of zinc powder into steel surfaces at elevated temperatures.
5. Painting.

The discussions of these methods as presented in the following subdivisions have been prepared by individuals who—through direct contact with the processes as inventors, developers, or operators—are eminently qualified to present the basic principles and operating details of their respective topics.

Hot-dip galvanizing is by far the most prevalent method and accounted for the major part of the 451,000 short tons of zinc (see Table 2-8) used for coatings in the peak consumption year of 1955. The most recent development in hot-dip galvanizing is that of coating steel strip from large coils on continuous lines. About 90 per cent of the 2,957,991 tons of galvanized sheet produced in 1956 were so coated. Three of the various continuous-strip galvanizing processes now in use are discussed here.

Hot-Dip Galvanizing of General Products

K. S. FRAZIER

Chief Research Engineer, Fenestra Incorporated
Detroit, Michigan

The popularity of the protection of iron and steel by hot-dip galvanizing stems from three factors: (1) The intrinsic value of the properties of iron and steel in all forms of construction work, combined with its susceptibility to corrosion when exposed to the elements. (2) The sacrificial property of zinc in preferential corrosion to iron giving full protection even though small voids occur in the coating. (3) The economy of application of hot-dip galvanizing as compared to other forms of lesser protection.

Figure 9-33. General galvanizing (hot-dip). (*Courtesy Johan Vis and Company, Rheem Manufacturing Company, Equipment Steel Products, Empire Metal Products, Fenestra Incorporated, H. K. Porter Company, Witt Cornice Com-*

These factors have been instrumental in the growth of this industry where we find large plants with four or five galvanizing kettles with individual capacities of from 20 to 200 tons of molten zinc, and numerous smaller plants with one or two kettles.

The products galvanized are in the thousands and prominent among them are: hot water tanks; all types of containers, such as pails, tubs, wire baskets and racks; pole line hardware; eavestroughs and downspouts; pipe fittings; steel windows; playground equipment; fire escapes; and structural steel. See Figure 9-33.

Each different product must be studied carefully for galvanizing procedure so that runs, sags, warping and flux spots are eliminated and quality work is produced. Special care and experience must be used in the galvanizing of such items as prefabricated pipe railings, complicated assemblies of fin coils for refrigeration, and other items where it is possible for pockets to form, which will trap water and cause explosions. The American Society for Testing Materials has written specifications and guides which are helpful in galvanizing practice.[1]

Each step in the routine of hot-dip galvanizing is important to the finished product, and of prime importance is the cleaning of base metal.[1] It is common practice to use hot alkali washes to remove grease and oils and this method is found adequate for most materials. Stubborn inclusions of oil often found adjacent to welds can be removed by the trichlorethylene vapor degreasing process.

Mill scale and superficial rust are removed in an acid bath. It is interesting to note that by allowing material to attain a slight rust on the surface, the edges of scale are loosened and their removal with acid is more rapid. Throughout Britain—with the exception of a few plants—hydrochloric acid is used for this cleaning process. It has been found in the United States, however, that the slower acting sulfuric acid lends itself to closer controlled pickling and causes less atmospheric contamination, and is therefore in popular use, while the hydrochloric and more caustic acids are used only for special work.

After pickling and thorough rinsing, the base metal is found to be so clean that superficial rusting would start within a few minutes. This must be avoided by dipping into a flux bath or by passing the work through a flux foam blanket into the molten zinc for galvanizing.

Post-treatment after galvanizing is not of paramount importance for long protective life, but the proper treatment with phosphates leaving a somewhat absorbent flat gray finish, or with chromates leaving tinted or clear bright hard finish, not only increases the protective life of the product by eliminating the natural rapid oxidation of a free zinc as a protective oxide film is forming, but also acts as a barrier to the chemical reaction of some paints with free zinc where painting over galvanizing is desired.

Definitions

The term "hot-dip galvanizing" refers to the process of dipping clean steel or iron into molten zinc and letting it remain until it reaches the temperature of the molten zinc. It is then withdrawn slowly, allowing complete retention of the accepted coating.

In *wet galvanizing*, a blanket of flux foam—usually sal ammoniac—is built up on the surface of the zinc bath to act as a wetting and cleaning agent to the work as it passes through to the molten zinc. In *dry galvanizing*, the work is dipped in a flux bath, usually zinc ammonium chloride, and dried in an oven. This forms a thin film over the work, protecting it from surface oxidation and acts as a wetting agent as the work enters the molten zinc.

There are many variations from the normal full coating process, to conserve zinc, attain greater ductility and promote other desirable characteristics. Small amounts of aluminum are added to the bath, up to 0.05 per cent, reducing the rapid oxidation of surface zinc, thereby allowing the applied zinc to retain its natural fluidity and produce a more uniform coating free from sags and runs.

Larger amounts of aluminum are in some cases added, particularly in continuous sheet galvanizing, causing a reduction of zinc-iron alloy growth and a more ductile coating. This allows for more severe forming after galvanizing. Reduction of zinc-iron alloy growth may also be accomplished through preheating of the work by passing it through hot gas chambers or drawing it through molten lead before galvanizing.

In sheet galvanizing it is common practice to withdraw the sheets from the molten zinc through pressure rolls to produce a reduced uniform coating. Where the coating of zinc is specified on sheets as 2 ounces of zinc per square foot, for example, it implies 1 ounce of zinc on each side of the sheet. This is sometimes confusing since on all other products the weight of zinc is specified in ounces per square foot of surface, or double the weight under the similar sheet specification.

How Galvanizing Resists Rust

When two dissimilar metals are placed in an electrolyte, a small battery is formed. One piece of metal becomes the anode or positive pole and the other, the cathode or negative pole. A small electric current is established, which flows ions from the anode, breaking it down and depositing it on the cathode. In plating, this natural phenomenon can be reversed by imposing a more powerful current in the opposite direction.

Zinc in the potential or electromotive series, as shown in the standard books on corrosion (see, for example, Evans[2]), is less noble than iron so

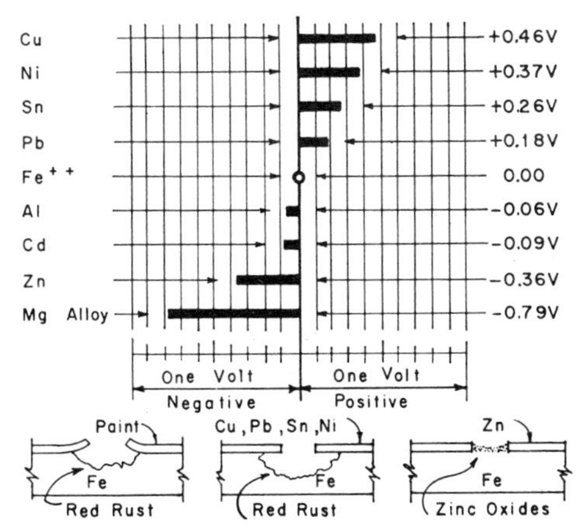

Figure 9-34. Above, sacrificial protection of metals as related to iron. The chart shows the voltage difference between iron, arbitrarily placed at 0, and other metals in salt water and similar solutions. (Transposed from data by LaQue and Cox.[4]) Below, action of corrosion at exposed spots through various coatings on steel. (After Daesen.[6])

that zinc dissolves in the presence of an electrolyte, thus sacrificing itself to protect the iron from corrosion.

The chart shown in Figure 9-34 is based generally on these considerations, but for more direct application in the zinc coating field, it indicates graphically the voltage action between various metals and iron, in the middle position, whether anodic or cathodic toward a neighbor in the presence of a salt water or similar electrolyte. Positive voltage, with metals nobler than iron, tends to draw the iron into solution to protect the nobler metals (red rust). Negative voltage, with metals less noble than iron, tends to draw them into solution to protect the iron. In protective coatings, unless there is a breakthrough to base metal, this electrolytic action does not take place. Zinc is sufficiently anodic toward iron to maintain its protective function under many unfavorable or abnormal conditions. Cadmium is a thin-coat protection.

Care should be taken in the interpretation of the standard potential series, in which the metals are arranged in the order of their dissolution tendencies, or in order of potentials generated when they dissolve as ions. These values are adjusted for equilibrium in solutions containing in effect one atomic weight of the metal expressed in grams of ions per 1000 grams of water at 25°C. The potentials, and sometimes the relative positions (where close to one another), are dependent not only on the specific tendencies of the metals to dissolve, but also on their ionic concentrations in the dissolving environment, and on various circumstances and conditions that can affect their action. Thus, Evans[2] notes with respect to aluminum, that this metal "unless amalgamated gives nobler values owing to the presence of an oxide film." Burns and Bradley[3] note further that "the insulating oxide film on aluminum-coated steel is more noble than aluminum itself, thus restricting electrochemical protection of bare steel at cracks and flaws in the coating."

LaQue and Cox[4] contribute potential measurements of metals and alloys in salt water and similar solutions, with inclusion of a galvanic series in sea water. They remark that the intensity of the effect with metallic combinations will be determined by "the relative areas of the metals forming the couple, by the distance apart in the series, and by the incidental conditions of exposure which affect corrosion reactions." In spite of these divergencies, zinc remains sufficiently anodic toward iron to maintain generally its protective function.

Table 9-15 (American Zinc Institute[5]) may be taken as a good practical guide to the galvanic activity of metals and some common alloys in contact under ordinary conditions.

When a zinc-coated iron surface is completely encased in zinc, no sacrificial action takes place. If, however, there is a pinhole, check or abrasion

TABLE 9-15. TABLE OF METALS IN GALVANIC SERIES[5]

Corroded end (anodic or less noble)

Magnesium
Zinc
Aluminum
Cadmium
Iron or steel
Stainless steels (active)
Soft solders
Tin
Lead
Nickel
Brass
Bronzes
Nickel-copper alloys
Copper
Stainless steels (passive)
Silver solder
Silver
Gold
Platinum

Protected end (cathodic or most noble)

Any one of these metals and alloys will theoretically corrode while offering protection to any other which is lower in the series, as long as both are electrically connected. As a practical matter, however, zinc is by far the most effective in this respect.

in a zinc protection of bare iron, the surrounding zinc will sacrifice itself to protect the bare spot.

In comparison, iron is less noble than tin, lead or nickel so that if a pinhole is present in one of these coatings the bare iron will sacrifice itself to protect the coating, and corrosion of iron at this point will actually be more rapid than if the coating were not present. If zinc-coated iron has a bare spot, the influence of the current is still present, allowing only a superficial rusting but no pit or destructive rusting (see Figure 9-34). Daesen[6] states that there is a force of 0.43 volt urging the zinc into solution around an opening in the protective layer and that as long as zinc remains in the vicinity none of the iron will dissolve. Moreover, if the opening is very small it will fill up with corrosion product and stop the action. Spowers[7] finds that "When iron or steel, which has been coated with zinc, is exposed to the atmosphere, a galvanic action is set up, although extremely slight. The result, therefore, is that with the slight galvanic action set up on galvanized iron or steel when exposed to the atmosphere, a corrosion of the zinc takes place. If this did not follow, there would be no protection. In this case the electropositive metal, zinc, suffers corrosion while protecting the electronegative metal, iron."

English authorities[8] write that "Nickel or tin coatings do of course completely protect iron as long as they are continuous, and prevent the penetration of oxygen and moisture; but if there are pinholes or breaks in these coatings, the iron will be attacked preferentially. This hidden rusting may spread rapidly and ultimately detach the coating altogether. Zinc on the other hand behaves very differently, as it has a greater tendency to lose electrons than iron. The whole position is reversed when iron and zinc are immersed together in a solution."

We have noted that a bare steel area on new coated material will show first a superficial coating of red rust, which disappears as the sacrificial action takes effect, contrary to the continued action permitted by other coatings. Sisko[9] writes in this connection: "For most purposes the best and the cheapest protective coating is paint, thoroughly and frequently applied. This method of protection is so well known that further attention is unnecessary here. The other method of protecting ferrous materials from corrosion is to coat with zinc, tin, lead, chromium, or nickel. In general, a zinc coating is less noble than iron and will protect the steel even if the coating is broken. This is important, as coated steel is rarely free from pinholes and scratches."

It thus appears that many authorities are agreed on the benefits of sacrificial protection of zinc over iron, although they may vary in their explanatory approach to the problem of rust prevention. Hot-dip galvanized steel windows have been examined after 25 and 30 years of service and have been found in good condition; here, the lack of spotty corrosion common with carelessly maintained paint protection may be attributed to this remarkable protective adaptability of zinc.

Testing the Protective Life of Galvanized Coatings

Many technicians have tested hot-dip galvanized coatings over a period of many many years and almost without exception have found them to offer excellent protection for steel. The protective life of a galvanized coating is tremendously influenced by the characteristics of the exposure to which it is subjected. The actual life prediction of this protection depends not only on the type of testing and the intended use but also on the experience of the authors in interpreting their tests.

It has been found in salt-spray testing for instance that similar materials hot-dip galvanized in the same process under identical conditions will not consistently show quite the same results in testing. This indicates that many elements taken singly or in combination, some of which may not as yet have been isolated, affect the corrosion rate of zinc. Salt-spray testing with either the standard 20 per cent concentration or the more recently used 5 per cent concentration is recognized as being indicative of coating value, but most attempts to convert the results of these tests into predictions of years of protective life have fallen short of being conclusive. Humidity testing appears to be more uniform in results but is short of prediction accuracy since it involves only one phase of protective life.

The Preece test, where test pieces are dipped in copper sulfate for periods of one minute and rinsed between dips until a deposit of copper shows on bare steel, is influenced not only by the thickness of coating but also by the crystalline structure of the coating, which varies with molten zinc bath analysis, bath temperature, time of immersion, rate of withdrawal and rate of cooling. This test, then, does not lend itself to a qualified prediction of life under normal atmospheric conditions nor does it define the thickness of coating. The Preece test has merit when used as a comparative check for continuous runs of uniform work, and results can be calibrated to indicate coating thickness within reasonable tolerances for this type of work.

It is recognized that the thickness of a zinc coating produced by hot-dip galvanizing indicates its comparative protection life, but other factors must be considered in ascertaining its true protective value. Thickness-of-coating tests by stripping indicate the life of the protection but even this falls short of actual service prediction since it does not anticipate coating construction, brittleness, ductility, uniformity or adhesion characteristics.

The zinc-iron alloy growth in a coating is generally more corrosion resistant than the free zinc layer but is at the same time more brittle, so that the proper ductility of the complete coating for a given usage is predicated on the proportion of zinc-iron alloy growth to free zinc. Microstructural examination of zinc coatings reveals coating thickness, proportion of zinc-iron alloy to free zinc, ductility and uniformity, as illustrated in Figure 9-35.

The above limitations do not mean that the protective life of hot-dip galvanizing is not known, for results of field testing have been recorded for over 200 years and the basic principles of hot-dip galvanizing have not changed. Only those control factors which create uniformity, appearance, ductility, desired crystalline surface structure and economical production have been improved.

Field inspection has shown that the popular service chart (Figure 9-36) is conservative for general usage and numerous individual cases have shown a protection substantially exceeding the periods shown.

Anticipated Coating

Although it is established that thickness of coating can be controlled by molten zinc temperatures and time of submersion, it is recognized that the best quality of coating is produced when the work is withdrawn at the time or soon after it reaches the bath temperature.

On this basis light materials less than $\frac{3}{16}$ inch in minimum cross section and of small mass will not accept the weight of coating that heavier materials will accept.[7] This factor is considered in ASTM designation A 386 – 55 T, Table 1.

Figure 9-37 shows the coating expectancy with small irregular sections. Heavier work such as structural steel will accept 2 ounces or more of zinc per square foot of surface in a quality coating. The rate of increase in weight of hot dip coatings over 2 ounces decreases rapidly and is not proportional to the time of submersion. There is then, an economic limit to the thickness of coating produced.

The composition of base metal to be galvanized has considerable bearing on both the structure of coating and the weight of coating applied. Steel with silicon of 0.05 to 0.15 per cent or a combination of high phosphorus and high carbon will increase the zinc-iron alloy growth producing a heavy, almost complete alloy coating with a dull gray finish known to the trade as "gray galvanizing" as opposed to "bright galvanizing." Although corrosion resistant, this coating tends to be brittle and is ordinarily not desirable.

In the process of galvannealing where galvanized work is placed in ovens at 1000 to 1200°F for the purpose of promoting alloy growth throughout the coating, the crystalline structure is refined to retain reasonable ductility. This coating then appears as gray galvanizing.

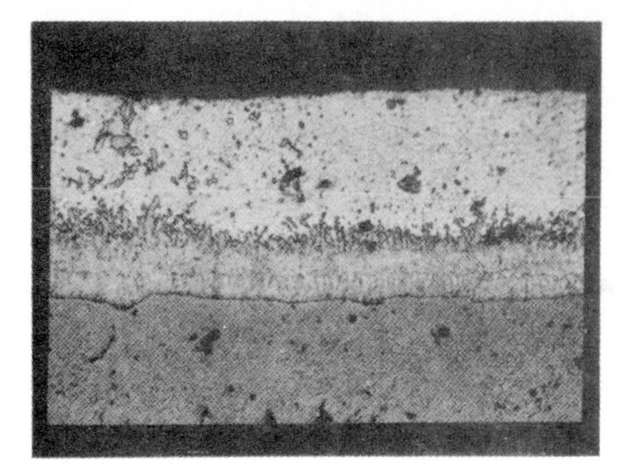

A. Good coating, ductile, adhesive. Magnification 250×.

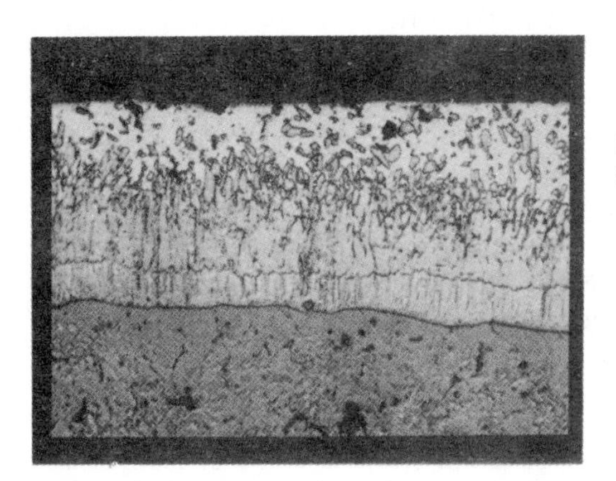

B. Forced zinc-iron alloy brittle, adhesive. Magnification 250×.

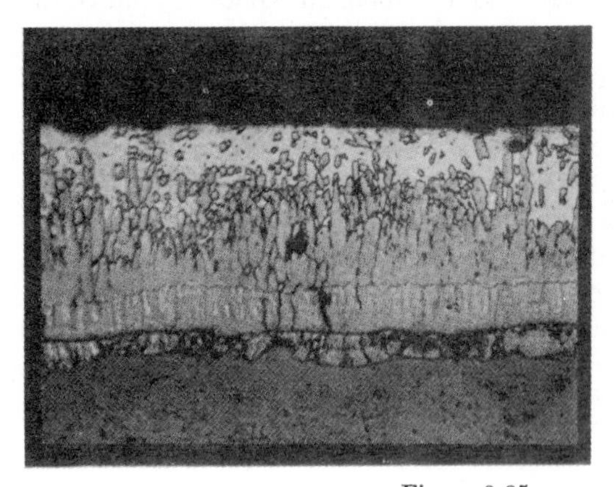

C. Forced zinc-iron alloy, brittle, nonadhesive. Magnification 250×.

Figure 9-35.

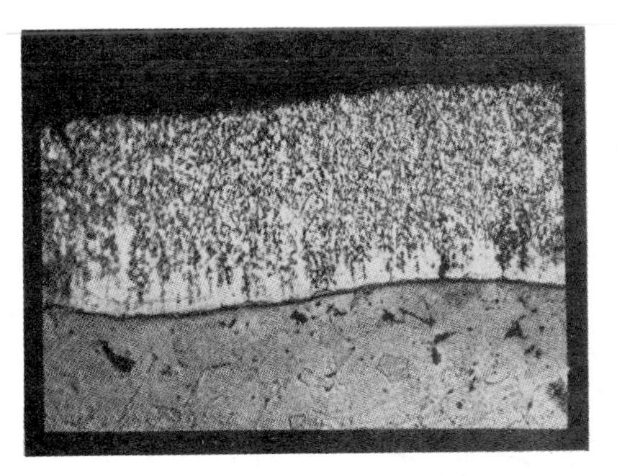

D. All zinc-iron alloy (gray galvanizing), adhesive. Magnification 250×.

E. Centrifuged galvanizing of small parts. Magnification 250×.

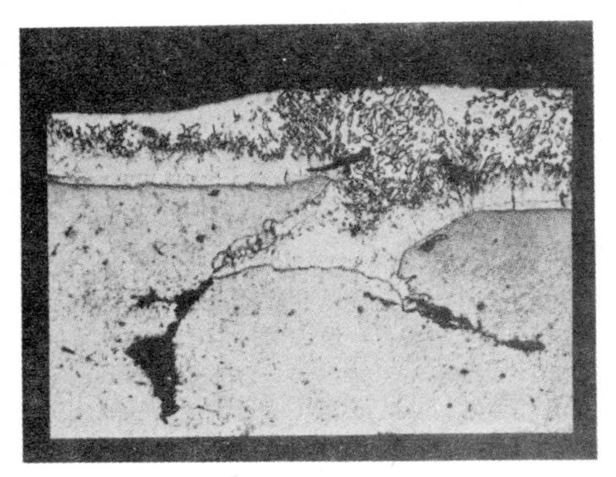

F. Penetration of zinc at surface defects. Magnification 125×.

Figure 9-35, cont.

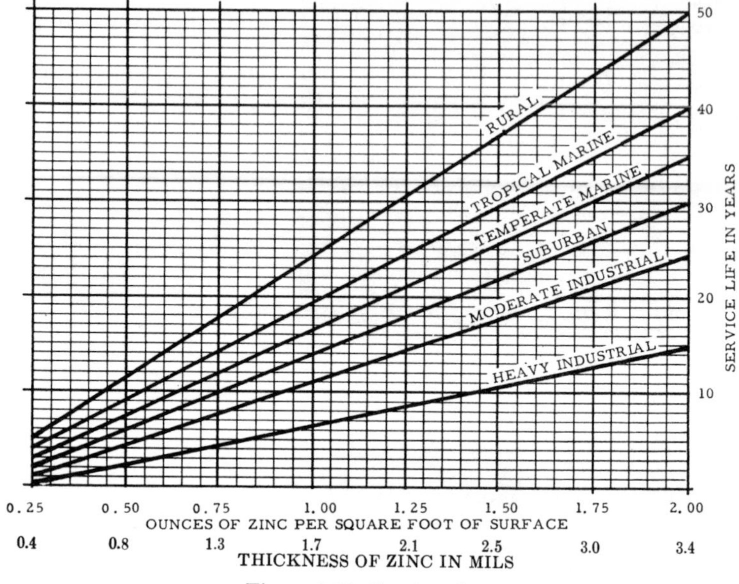

Figure 9-36. Service chart.

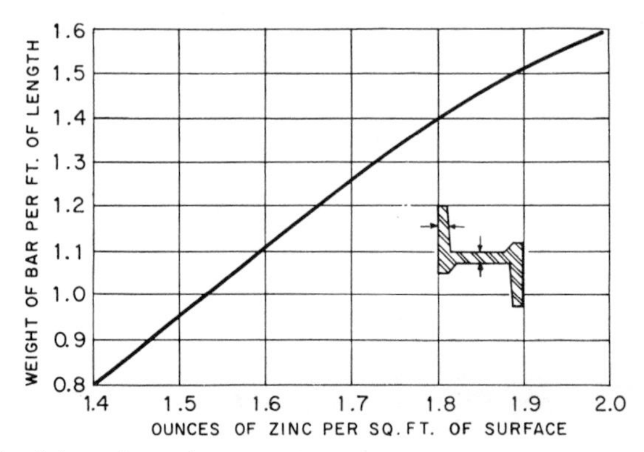

Figure 9-37. Anticipated coating on light irregular sections with minimum cross section less than 3/16 inch.

Conditions and Improvements[10]

For general galvanizing, a rimmed low-carbon, low-silicon, low-phosphorus, aluminum-killed basic open hearth steel is preferred. Prime Western zinc is accepted by the general galvanizer because it is a product which insures an excellent coating with minimum kettle attack and dross formation, tending toward a clean and effectively controlled bath.

Because of the varied nature of the products found in general galvanizing, much of this work must be handled with simple tongs and hoists so that timing depends on the experience of the workmen. Where production is similar and continuous, many plants have installed conveyor systems with dropped sections or automatic hoists, and have found an advantage in production and uniformity of work.

The introduction of improved methods of cleaning material to be galvanized, electronic temperature controls, electrically controlled timing, and the design of kettles to handle work without the necessity of double dipping, all show an advance in the art and bring the assurance of quality work.

Hot-Dip Galvanizing of Sheet by Conventional Pot Methods

NELSON E. COOK

General Superintendent of Galvanizing, Wheeling Steel Corporation
Wheeling, West Virginia

"Conventional pot galvanizing" is a term applied to the method for zinc-coating sheet material in cut lengths. The process was introduced in the early part of the nineteenth century.

The procedures in the early days were very elementary and were accomplished without benefit of modern metallurgical techniques. Cut lengths of sheets, usually rolled from fagot iron or puddled iron, were pickled in acid, washed, fluxed and introduced into a molten zinc bath contained in an iron pot usually heated with coke or coal. The sheet being coated was held in the coating bath until it attained the coating temperature, then it was slowly withdrawn from the bath by hand or pulled directly upward from the bath by some simple contrivance such as a block and tackle. Regulation of the weight of coating depended almost entirely on the skill of the workman in withdrawing the sheet at constant speed. The coated surface was rough and extremely dirty as compared with the modern product. Coatings were very heavy and relatively nonadherent.

Modern conventional galvanizing methods have changed measurably since these early beginnings as a result of gradual and important improvements in the process, particularly since the beginning of the twentieth century. Sheets were greatly improved in quality and uniformity as Bessemer and then open-hearth steel became available. Rolling procedures have been constantly improved, culminating in the development of continuous hot- and cold-rolled steel strip in the late twenties. Annealing practices have been improved to the point where annealing is well controlled and uniform, even to the development of good deoxidized annealing which makes possible the coating of sheets without prior pickling.

Figure 9-38. Side view of an old type conventional galvanizing unit, coal-fired, showing the driving mechanism for three pairs of rolls, all connected to one common drive, with motor directly mounted over the bath. This is a design of early twentieth century, some of which are still in operation.

Perhaps the greatest advances have been made in designing machines for galvanizing. A galvanizing machine consists essentially of a set of rolls to guide the sheet into a small acid tank for final cleaning, and a set of wiping rolls to wipe the excess acid and guide it into a flux bath floating on the metal bath. Then there are guides leading to either one or two pairs of driven rolls operating near the bottom of the coating bath, and above normal dross levels; more guides lead the sheet up through the driven coating-rolls which are half submerged in the metal at the exit end. There are, of course, many variations in the types of rolls, cooling and handling equipment, and other auxiliaries. Machines have been operating at very high speeds for cut-length coating, some as high as 160 feet per minute, with all rolls and equipment in the line synchronized to give perfect control of coating insofar as mechanical and electrical devices can accomplish this.

With the development of better steels there were built better steel containers for the molten metal, better rolls, and better galvanizing machines, which minimized the solution of iron into the zinc, thereby minimizing dross formation in the operation and giving longer life to the equipment.

Slab zinc for coating purposes has gradually improved in quality, making possible greater uniformity and control of the process. Great strides have been made in the heating of the galvanizing kettle and most of the more modern installations utilize gas or electric heat with positive control. Pickling and cleaning procedures have become more varied and more adaptable

Figure 9-39. A view of one of the most modern conventional lines looking from the exit inspection toward the pot and entry section. This line, with all rolls and equipment completely synchronized and with controlled cooling, operates at speeds of 160 feet per minute. The metal bath is heated with infrared gas multiple burners.

to the successful processing of steel for coating. New and better fluxes have been formulated.

Generally there are two types of conventional galvanizing processes for cut-length sheets. One is the zinc-bath process, most widely used, in which the pickled sheet is passed through a flux blanket and through a bath made up entirely of zinc with normal impurities and additions. The other process is the lead-zinc process which utilizes a combination metal bath. The pickled sheet is passed through the flux blanket into the lead. The sheet travels most extensively through the lead bath, where it is preheated without coating. Then it passes through a relatively shallow bath of zinc floating on top of the lead at the exit end of the bath. The coating rolls are positioned in this small zinc bath. The advantage of this process is that it produces thinner coatings, which are more economical. Also, in the conventional process thinner coatings are generally more adherent and therefore this process has gained wide acceptance where the coated material is subjected to forming operations.

Conventional coatings are generally of Prime Western zinc, with alloying additions for better spangle appearance and thinner and more adherent coatings. The metal contains the normal impurities such as iron, lead and cadmium, and addition metals are usually tin or antimony or both. Tin is the favored addition metal for the reason that relatively small percentages make it possible to operate the bath at lower temperatures, to get better flowering or spangle appearance on the product, and to get thinner and more adherent coatings.

The one addition metal which makes for tightly adherent coatings—aluminum—has never been too successfully used for conventional sheet galvanizing. There are some installations, but difficulties with fluxing of material and coating of rolls, and difficulties of operating coating machines, have made the process generally unsatisfactory.

The conventional coating process embraces so many variables that its operation is truly an art. The process is fairly economical, it can produce different weights and types of coatings that can be fairly well controlled, but it has very definite limitations in that the formation of brittle zinc-iron alloys cannot be minimized sufficiently to guarantee against peeling of the coating. Despite all the excellent developments in controlling the many variables, the process has now become practically obsolete for the reason that the demand is for extremely adherent coatings as now produced on continuous lines, where small percentages of aluminum can be used in the zinc bath to virtually eliminate the brittle alloy layers. While conventional galvanizing has all but faded from the scene, it was in this field that most of the researches, discoveries and developments were made during the past century, to make possible the excellent coatings now being produced on continuous-strip lines.

Hot-Dip Galvanizing of Sheet in Continuous Lines by the "Cook-Norteman" Process

NELSON E. COOK

General Superintendent of Galvanizing, Wheeling Steel Corporation Wheeling, West Virginia

The "Cook-Norteman" process, an annealing-out-of-line process utilizing an induction-heated bath and prefluxing and preheating of the fluxed material, is a very recent development in applying tightly adherent zinc coatings to a steel or iron base. Equally adaptable to the zinc coating of wire, pipe, tanks, ware and other steel articles, its principal application is in the tight zinc coating of strip in continuous lines. Basic principles of the process were established from laboratory research many years ago, but means and equipment for practical application of this knowledge were not developed until 1950 when the process was put into pilot operation for coating dipped articles and continuous strip. The first mill application of the process for continuous coating of strip was made in 1953. Since that time the use of the process for all types of galvanizing has grown rapidly throughout the world.

The process, which utilizes basic metallurgical principles of the zinc-iron-aluminum metal system, is applied commercially in a very new and original

Figure 9-40. View from delivery section looking toward pot section of two continuous "Cook-Nortemen" lines at Wheeling Steel Corporation. The longer line processes 48-inch wide strip and the other line, 36-inch wide strip. Both operate at speeds of 320 feet per minute. (*Photography by Cress Studios*)

way. In a molten zinc-iron-aluminum system the aluminum has a greater affinity for iron than for zinc. Therefore, when the zinc-iron-aluminum system is composed of zinc of the coating bath, iron of the strip or article to be coated, and aluminum additions to the bath, the aluminum tends to unite with the iron which would normally form very brittle zinc-iron alloy layers on the article being coated. The aluminum-iron alloys formed are eliminated by floating to the top of the bath, and the article being coated is substantially free of the brittle alloys which, in conventional coating processes, cause the coating to be nonadherent.

Actually, in the tight zinc coating of steel, the metallurgical procedures are somewhat more complicated than stated. There are many other variables in the coating process which affect the balance of the zinc-iron-aluminum system. The machine which guides the strip through the coating bath is made of steel which slowly dissolves in the molten bath and reacts with the zinc and aluminum of the bath. The container for the coating metal, which is usually steel, tends to dissolve slowly and add iron to the system. Different kinds and analyses of steel base which are used for galvanizing purposes react differently, and at different rates, with the zinc and aluminum of the coating bath, and will form different types and quantities of alloy layers on the article being coated.

Unannealed strip is more conducive to the formation of brittle alloy layers than annealed strip. Unclean strip, which retains the oils and iron products from the rolling process, tends to form alloy layers rather rapidly. The same is true of aluminum-killed steel and many other types of steel, each of which reacts in its own peculiar way. Speeds of travel through the coating bath and times of immersion also affect the coating-system balance. Therefore, since the aluminum is added to the bath for the purpose of uniting with any iron which otherwise would form brittle alloys on the coated product, it follows that the percentage of aluminum in the metal system must be varied, and controlled instantaneously, to meet the many variables, if the advantages in the use of aluminum are to be fully employed in the production of tightly adherent coatings.

The Cook-Norteman process simplifies the science of tight zinc coating to the degree that the control of adhesion is positive regardless of any other factors such as the chemical, physical or surface properties of the steel base. All of the control of the process is localized in the coating area which consists of the coating bath itself and the prefluxing region.

In the fluxing area, the Cook-Norteman process applies a flux film of aqueous zinc ammonium chloride to the precleaned strip or article. This flux bath is maintained at 180°F. The flux is so formulated that it can be dried and preheated to a temperature approximating 500°F without breakdown or deterioration of the very thin flux layer. Applying, leveling and controlling of the film thickness of the flux is done automatically regardless of any conditions of speed, gauge or surface.

The flux layer is preheated to a temperature at or above 350°F and the fluxed material is introduced into the coating bath. The reaction of the flux with the coating metal, particularly with the aluminum in the metal, occurs at approximately 350°F and, as long as the prefluxed metal is introduced into the coating metal at or above this temperature, the reaction of the flux with the metal takes place immediately at the point of entrance into the coating metal, and there is no possibility that flux will carry through on the finished article. The reaction of the flux with the metal normally unbalances the metal system, but with thermally circulated metal in the induction bath, the desired metal mixture is maintained in the correct proportions at all times in all parts of the coating bath.

The coating bath in the Cook-Norteman process utilizes a ceramic-lined, low-frequency induction-heated pot which is engineered electrically so that the metal is constantly being stirred by thermal circulation regardless of the input of heat demanded by the variations in gage and speed. The thermal circulation of the bath at all times makes it possible to maintain constant metal mixtures. The container being ceramic, there is no iron to dissolve which would complicate the function of the zinc-iron-aluminum system. The machine to convey the material through the bath is made of very low-carbon steel or pure iron which minimizes the problem of iron

Figure 9-41. View of the coating section and the control desks and panels of a 48-inch wide "Cook-Norteman" continuous-strip line at Wheeling Steel Corporation. The coating bath is heated with low-frequency electric induction heating. (*Photography by Cress Studios*)

additions to the metal system. With constant circulation of the metal in the bath, any percentage of aluminum which is required to minimize the alloy formation on the product can be provided for any set of conditions.

With this positive control, tightness of the coating is directly proportional to the amount of total iron in the coating on the finished article. A simple rapid iron-titration test of the product suffices to control the process. When additional amounts of aluminum are needed to meet any set of conditions, it is added directly into the induction throats and is almost immediately completely mixed into the system. This quick control assures tight coatings at all times and under all conditions.

The Cook-Norteman process has many other advantages. It is applicable to a wide range of gages and types of steel base and to a wide range of weights and types of coating. Because of the fluxing system, and the requirement that only the surface of the material being coated be raised to the temperature of the coating bath, speeds of coating up to 320 feet per minute for strip are a reality, and greater speeds will be attained with further development of mechanical and electrical equipment. The process may be operated continuously or intermittently as required. By-product losses in the process are extremely small. Most important, with simple and positive metallurgical control, tight coatings can be assured at all weights of coating, at all speeds and on all types of steel base.

Hot-Dip Galvanizing of Sheet in Continuous Lines by the "Sendzimir" Process

KASIMIR OGANOWSKI

Associate Director of Research, Armco Steel Corporation
Middletown, Ohio

It has long been recognized by galvanizers in their efforts to produce a better zinc coating that three major factors affect the manufacture of galvanized sheets: (1) surface preparation of the base metal, (2) temperature control during coating operation, and (3) composition of the spelter bath. When all these factors are simultaneously controlled at the optimum of their range of variation, a coated product of high quality can be produced consistently.

The Sendzimir coating process[11,12] was the first to attain commercial success by fulfilling the basic coating requirements in a continuous galvanizing operation. A coating line installed in 1936 at Butler, Pennsylvania, by Armco Steel Corporation proved the soundness of the process and stimulated developments in zinc coatings that revolutionized the galvanizing industry. The Sendzimir process is outlined in the following pages.

Surface Preparation

Surface preparation of the base metal is accomplished in two steps: (1) oxidation of the surface, and (2) subsequent reduction of the oxidized surface. Combination of these two steps assures a high degree of cleanliness regardless of a broad variation in surface contamination of the base metal when it arrives at the coating line. The surface can be oxidized either by heating in an oxidizing medium or by other chemical means.

When oxidation by heat is used, a very uniform oxide layer is formed by bringing the base metal to a predetermined temperature. Rolling lubricants and other combustible materials are removed from the surface as gaseous products of reaction while the base metal is being heated to the oxidizing temperature. Oxide thickness is controlled by regulation of the base-metal temperature. In actual practice the base metal is passed through either a radiation-type furnace containing air atmosphere or an open-flame furnace, and is heated to a temperature in the range of 750 to 850°F.

An alternative means of oxidation employs chemical methods. Alkaline cleaner removes rolling oils and is followed by rinsing and drying to bring about oxidation of the surface. Immediately after oxidation, reduction of the oxidized surface is performed by heating the base metal in the reducing atmosphere. During the reduction of oxides some surface desulfurization and decarburization also occurs. Since products of these reactions are gaseous, any undesirable surface contamination is prevented. The resulting surface consists of freshly formed metal and is fully prepared for the molten zinc bath.

Surface consistency is regulated by simply controlling the composition of the atmosphere and temperature of the reducing furnace.

Temperature Control

For reasons of economy it is desirable to combine surface-preparation steps with the heat treatment of the base metal. Consequently the full-hard, cold-reduced material generally used in the Sendzimir coating process is annealed at the same time the oxidized surface is being reduced. Any heating and cooling cycle practicable for a continuous annealing operation can be used in this process.

Temperatures required for annealing the base metal and reduction of the surface oxides are higher than the temperature at which the base metal should be immersed in the coating bath to obtain the most satisfactory coating bond. Thus, after completion of the reducing and annealing operations the base metal is cooled in a nonoxidizing atmosphere to the proper coating temperature. If the base-metal temperature at the time of immersion is maintained slightly above the coating-bath temperature, a very satisfactory bond is obtained consistently, with minimum losses in zinc dross and maximum life of the zinc kettle.

In practice, reduction of surface oxides and annealing and cooling of the base metal to the immersion temperature are done in one furnace consisting of two parts. In the first part the base metal is heated to a desired annealing temperature in the approximate range of 1350 to 1750°F. In the second part the base metal is cooled to a temperature range of 900 to 950°F. This part of the furnace is generally equipped with both cooling and heating devices so that the base metal can be brought to the bath at the desired temperature, regardless of variation in the amount of heat being dissipated due to variation in the strip width.

The reducing atmosphere of the first part of the furnace and nonoxidizing atmosphere of the second part are generally obtained by passing dry dissociated ammonia in counterflow to the incoming strip.

Zinc Bath Composition

When base metal with a properly prepared surface and at correct temperature is immersed in the zinc, wetting and bonding occur instantaneously. Rapid reaction between zinc and base metal also takes place. Unless suppressed, this reaction would result in the formation of brittle iron-zinc compounds in the coating and ductility of the coating would be impaired.

The rate of formation of iron-zinc alloys in the coating can be suppressed by adding small quantities of aluminum to the zinc bath. Aluminum additions can be readily maintained and controlled because the Sendzimir process does not require fluxes for surface preparation. In ordinary galvanizing processes the desired amount of aluminum cannot be retained in the zinc bath owing to the rapid reaction of aluminum with the flux.

Addition of 0.2 per cent aluminum to the bath in the form of zinc-aluminum alloy, or use of spelter containing this quantity of aluminum, constitutes the basic control of bath composition. Control of other impurities in the bath can be exercised if control of size and flatness of spangle is desired.[12]

Operating Principles

The above described requirements are met by heating and cooling the base metal to predetermined temperatures in predetermined atmospheres and immersing it in a prealloyed zinc bath. Under the continuous type of operation these processing requirements can be met with a high degree of uniformity and consistency.

Basically, the equipment should provide for continuous operation, so that the base metal in a form of strip can flow through the processing components of the coating line at a given rate without interruption. In processing by continuous heating and cooling, the rate of production is governed by maintaining constant tonnage for a given width and changing speed so as to make it inversely proportional to the change in thickness.

Rated capacity, type of equipment, type of fuel, degree of automation, annealing cycles, type of steel and other factors may vary widely between various localities. However, all Sendzimir coating lines have three common

Figure 9-42. Feeding section of a "Sendzimir" coating line showing double uncoiler, two sets of pinch rolls, shear, welder and loop car.

Figure 9-43. Oxidizing furnace preceded by feeding pinch rolls and strip tension device.

principal sections. The first of these is the *feeding section*. This facilitates unwinding of incoming coils and joining of the coils without disrupting the flow of material at uniform speed through the next section (Figure 9-42).

In the *processing section*, the base metal passes at the rated speed through the oxidizing process, preferably through the furnace that preheats it in the oxidizing atmosphere. Then it goes through the reducing furnace, which completes the heating to a desirable annealing temperature in the reducing atmosphere. The next phase is the cooling chamber, where the strip is cooled to the coating temperature in a nonoxidizing atmosphere. Finally the strip passes through the coating bath, which deposits the coating of required thickness on its surface (Figures 9-43 and 9-44).

Third is the *delivery section*, which completes the cooling of the coated material and delivers it in the form of coils or cut-length sheets. Here again the uniform speed is not disrupted (Figures 9-45 and 9-46).

Depending on the specification, the three basic sections can be equipped more or less elaborately and may perform additional functions compatible with the main processing requirements. For example, the feeding section may incorporate strip-tension control and tracking devices. Furnaces of the coating section may be of the recuperative type that utilizes heat dissipated in the cooling chamber for preheating the incoming strip. Such furnaces are used where fuel is costly and the base metal requires high annealing temperatures. The delivery section may be equipped with mechanisms for indicating continuous weight of coating, surface treatment for protection from wet storage stain, stenciling and assorting facilities.

Figure 9-45. Delivery section with shear, roller leveler, inspection table and oiler.

Specifications for coating lines become increasingly more demanding, not only in respect to rated capacities, but also in respect to gages for indicating the range of widths of strip and other engineering features incorporating more and more advanced mechanization and automation.

Versatility of the process is seen in the fact that various Sendzimir coat-

Figure 9-46. Delivery section picturing the double coiler.

ing lines over the world* are processing strip from 0.007 to 0.170 inch in thickness, and up to 60 inches in width, at the rated capacities from one ton per hour or less to 20 tons per hour and more, with weight of coating from 0.2 ounce to 2.50 ounces per square foot. Cut-length rigid conduit also is being continuously galvanized by this process.

The world-wide acceptance of the Sendzimir process* attests to the quality of the product and economy of processing. The product is characterized by (a) absolute adherence† of the coating, (b) attractive lustrous appearance, and (c) excellent workability or forming properties of both coating and base metal.

Metallurgically, this process eliminates the brittle iron-zinc alloys from the coating structure as shown in comparative photographs (Figures 9-47 and 9-48). The economy of the process stems from the fact that cold-rolled strip is converted into a fully finished galvanized product in one continuous mill operation.

* Sendzimir galvanizing lines are now being operated in these countries: Argentina, Belgium, Canada, England, France, Italy, Japan, Union of South Africa and the United States.

† According to the author's definition, adherence of the coating is absolute if it does not peel or flake off on the deformation of coated material severe enough to fail the base metal.

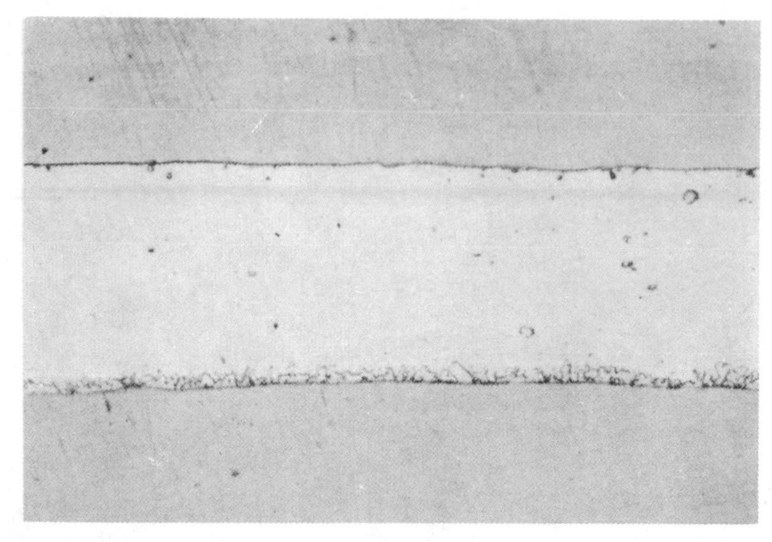

Figure 9-47. Typical structure of zinc coating produced by the Sendzimir process Note the highly suppressed zinc-iron alloy. Magnification 1000×.

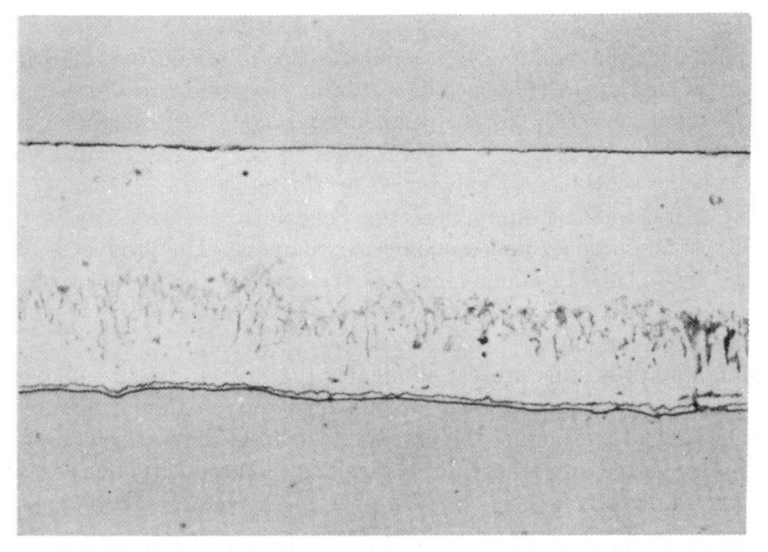

Figure 9-48. Typical structure of a coating produced by conventional means using pickling and fluxing for surface preparation and aluminum-free zinc bath for coating. Note the well-developed multilayer of zinc-iron alloys. Magnification 1000×.

Commercial availability of galvanized sheets with a high degree of workability, produced economically, has greatly broadened applications for these sheets and increased the use of zinc in sheet galvanizing.

Hot-Dip Galvanizing of Sheet by Continuous-Line Operations at the United States Steel Corporation

UNITED STATES STEEL CORPORATION STAFF

Pittsburgh, Pennsylvania

Since the end of World War II, continuous galvanizing has almost completely displaced the galvanizing of individual sheets at the mills of the United States Steel Corporation. Only one sheet-line remains in operation and this is a modern installation at Pittsburg, California, for coating very heavy-gage sheets. The choice between the two types of operation is largely a matter of economics. Continuous processing, however, is inherently more uniform and some problems incident to the coating of individual sheets completely disappear; for example, the continuous process eliminates the gap between successive sheets which results in a variation in the weight of coating metal applied to the leading and trailing ends of the sheets.

Although the processing of sheet steel in the form of a continuous strand permits a number of manufacturing steps to be grouped in a single in-line sequence, several options are presented to the designer of continuous galvanizing equipment. In selecting the steps to be included in a specific installation, attention must be given to factors such as mill product-mix, the capacity of existing facilities for cleaning and annealing and similar considerations which can vary widely from mill to mill. Conditions at United States Steel's mills have favored cleaning, annealing, coating, finishing and inspecting in-line.

Bright annealing is practiced, and the annealed strip is cooled to slightly above pot temperature and introduced into the zinc while still protected by furnace atmosphere. This serves a twofold purpose: heat from the annealing operation is utilized to maintain the coating bath at the proper temperature and the use of liquid fluxing agents is unnecessary. Figure 9-49 is a schematic side elevation showing the arrangement of processing equipment.

Eight lines of this general design are in operation at the mills of the Corporation. Four of these are light-gage lines (18-gage maximum); the others are designed to handle strip up to 10 or 11 gage. As indicated in Figure 9-49, the lines are divided into three operating sections separated by slack accumulators. Details of the operation are discussed below; some difference in construction and practices on individual lines are noted.

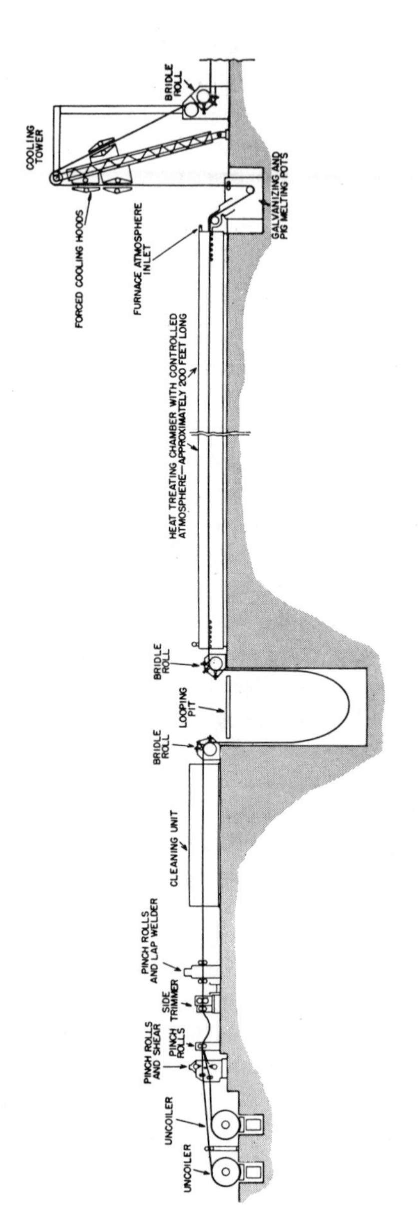

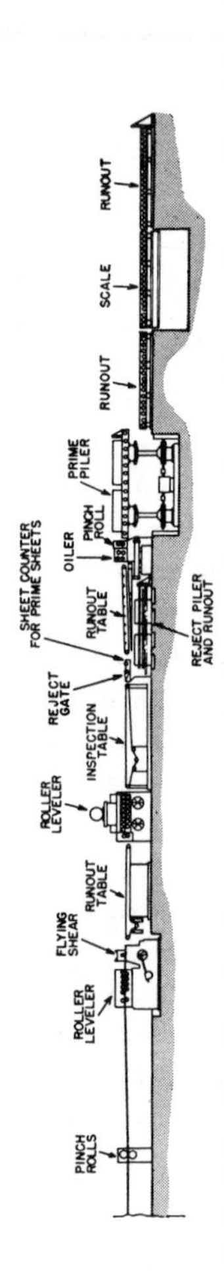

Figure 9-49. Side elevation showing arrangement of processing equipment for continuous line galvanizing operations. United States Steel Corporation.

The first or entry section of the line includes the uncoiling, joining and cleaning equipment. For most products the starting material is a 5 to 50 thousand pound coil of full-hard cold-reduced strip steel; however, for some specialties, cold-reduced, alkaline-cleaned, box-annealed and temper-rolled coils may be used and for certain heavy gage products (16 gage and heavier) hot-rolled, pickled strip is sometimes used. Processing is essentially the same for all material except that line speed is adjusted as required by differences in annealing characteristics.

The coil is mounted on an uncoiler and fed to a squaring shear. The squared head end is then advanced to a welder where it is joined to the squared tail end of the coil being processed. Lap, butt or mash welders are used. During joining operations the coil-ends are manipulated by power-driven pinch rolls. Equipment for continuously side trimming to ordered width is located either before or after the welder.

From the welder the strip enters a cleaning unit where residual cold-mill lubricant and any mill dirt picked up in prior processing or storage are removed. Any of the conventional alkaline cleaning practices are satisfactory for this purpose; dipping and scrubbing with a hot alkaline solution is used on some lines; electrolytic alkaline cleaning is used on others. Polarity of the strip during electrolytic cleaning, the nature and concentration of the cleaning agent and other cleaning conditions vary from line to line. Cleaning is followed by thorough rinsing, and the strip is dried by hot-air blasts.

The strip is moved through the entry section by a drive bridle at the end of the cleaning station, which feeds the clean, dry strip into a slack accumulator. A looping pit, 30 to 90 feet deep, is usually provided for this purpose. Where excavation is impractical, a tower-type or a horizontal car-looper is used. The speed of the entry section drive bridle is automatically controlled to maintain a full loop in the pit and the slack so stored permits the entry section of the line to be stopped for the joining operations without affecting the operation of the balance of the line.

The second section contains the annealing and coating units. The strip is pulled through these units by a drive bridle located at the end of the section. Since steel is relatively plastic at annealing temperatures, excessive tension in the strip in the furnace must be avoided; however, tension if properly regulated will correct minor defects in strip shape and will improve flatness. To facilitate control of tension in annealing, the strip is pulled from the slack accumulator through a tension bridle and passed into the furnace with a minimum of intervening equipment. The cleaning step occurs in the entry section of the line for the latter reason. The tension bridle operates to help or retard the forward movement of the strip, as required, to maintain a uniform tension in the strip entering the furnace. Where required, conveyor rolls in the furnace are driven to avoid build-up of tension.

Annealing is conducted at subcritical temperatures. Electrically heated, multiple-vertical-pass furnaces are used on the light-gage lines; gas-fired radiant-tube, horizontal-pass furnaces on the heavy gage units. The furnaces are divided into several independently controlled zones and are designed to heat the strip to annealing temperature as rapidly as possible, then cool the strip to about 950°F, at which temperature it is introduced into the coating bath. Cooling is accomplished either by recirculating a portion of the furnace atmosphere of the cooling section through heat exchangers, or by means of tempered air circulated through tubes positioned adjacent to the strip in that section. The furnace atmosphere is a mixture of cracked ammonia and NX gas. The discharge end of the furnace extends below the surface of the molten zinc. The gas, which may be introduced at several points along the length of the furnace, flows counter to strip movement to escape at the entry end of the furnace. Gas-flow is regulated to maintain the furnace atmosphere at a positive pressure of one half to one inch of water. A considerable portion of the gas is introduced near the discharge end of the furnace to aid in cooling the strip.

The coating pot contains a sink-roll for submerging the strip and exit rolls for controlling the thickness and distribution of the coating. The exit rolls are driven and are mounted to operate about half submerged in the molten zinc. As in sheet practice, the weight of coating is controlled by the adjustment of the exit-roll pressure, the grooving of these rolls and their relative position with respect to the level of zinc in the pot. As stated earlier, the pot is kept at operating temperature (840 to 900°F) by heat carried into the bath by the strip. Electric or gas-fired means are provided for heating the pot during extended shut-downs and initially to melt the coating metal at the start of operations.

About 0.15 per cent of aluminum is maintained in the zinc bath, added as an aluminum-zinc alloy. Sufficient lead and cadmium for the production of an attractive frosty spangle are present in the spelter added to maintain the bath level. Small amounts of antimony are sometimes added, either as an alloy with zinc or as the metal itself, to assist in spangle-control.

The coated strip passes vertically to a large-diameter roll mounted at the top of a cooling tower. Air blasts are directed against the strip to solidify the coating before it reaches the tower roll. Additional air blasts may be used on the downward pass to further cool the strip.

Galvannealed coatings can be produced by replacing the cooling hoods in the up-pass of the tower with a gas-fired furnace. Space is usually provided between the cooling tower and the drive bridle for any special treating facilities, e.g., bonderizing equipment, provision being made to bypass or withdraw such apparatus when not in use. The drive bridle of this section is provided with automatic controls for maintaining the strip speed constant at a selected value. The latter depends on the gage and width being processed and the mechanical properties specified.

Operations in the third section of the line require occasional interruption. For this reason looping facilities similar to those described for the entry section are provided. Strip is pulled from this loop by a set of pinch rolls, the speed of which is automatically controlled to maintain a minimum loop during normal operation. From the pinch rolls the strip passes to a roller leveler and flying shear where it is cut to ordered length. Sheets are conveyed to a second leveler and discharged onto the inspection table. Automatic devices permit the inspector to divert defective sheets to the reject-piler, and pass the prime material through a sheet counter and run-out table to the prime-piler. Some lines are equipped with a recoiler, located after the loop pinch rolls, for the production of coiled product.

The galvanized product is inspected for appearance, and samples from the line are taken at regular intervals by metallurgical laboratory personnel to be tested for weight of coating, adherence, hardness and formability of the base metal. In addition to ASTM and similar industry-wide specifications, mill standards have been established for the several grades of product produced at each plant. The material is accepted or rejected for a particular application or order on the basis of these tests. The test results are made available immediately to the line operator for his guidance.

Productive capacity is determined primarily by the annealing capacity of the individual line. The furnaces of earlier lines were designed to anneal and cool at a rate of 10 tons per hour; larger furnaces have been used in later installations, the latest being rated at 27 tons per hour when processing 48-inch and wider strip.

Hot-Dip Galvanizing of Wire

J. Paul Tierney

Assistant Superintendent; Rod, Wire, and Conduit Departments
The Youngstown Sheet and Tube Company
Youngstown, Ohio

Observing a wire galvanizing unit in operation gives the impression that it is a very simple operation, and although not the most complex from a technical viewpoint, it does require considerable attention to produce wire within the desired specifications. Even before the wire is delivered to the Galvanizing department for coating with zinc, or galvanizing, preparations for correct end results must be exercised by those involved in its manufacture.

Selection of steel of suitable composition is essential in order to attain the required tensile strength and enhance the response to coating deposition from the various types and grades available. Rimmed, capped, semi-killed and killed steels are used in the manufacture of galvanized wire and the application of each type is limited by the properties required in the finished product. Rimmed and capped steels are used where tight adherence of coating is not mandatory, but even in the low-carbon grades where coating adherence is a primary requisite silicon-killed steels are used. It has been found through experience that a minimum silicon content of 0.07 per cent is sufficient to insure satisfactory adherence providing other factors are controlled. Some consumers specify semi-killed steel for minor applications such as fence wire to obtain the benefit of a slight amount of silicon and its resultant effect on coating adherence.

A large portion of the entire range of wire sizes is galvanized. The wire is loaded on pay-off reels and fed to lead pans for tempering as desired. Commonly, the wire is either "hard" galvanized or "soft" galvanized and the temper is controlled by varying the length of immersion or the temperature of the lead bath. Soft galvanized wire is generally processed through two lead pans approximately 20 to 25 feet long with the lead temperature at about 1200 to 1300°F. The lead serves a dual purpose as it also burns off any substance on the surface of the wire which is volatile at the operating temperatures.

After passing through the lead, the wire is introduced into a muriatic acid bath for further cleaning. One of the "tricks of the trade" is to submerge a pig of zinc in the muriatic, or hydrochloric, acid to improve coating propensities. After the acid cleaning the wire passes through a water rinse usually containing ammonium chloride. This water rinse is applied to wash iron salts and finely divided iron from the surface of the wire since contamination of the spelter by iron or iron salts accelerates the formation of dross.

After passing through the water rinse, the wire enters the "galvanizing pot" through ammonium zinc chloride which acts as a flux to promote fusion of the steel wire surface and the zinc. Flux can be eliminated if the material is free from oxide. This latter practice requires protection by an inert, or reducing, atmosphere. The zinc deposits on the wire surface in layers, and the thickness of the coating may be varied by length of immersion measured in seconds of time or rate of travel, and the temperature of the zinc. This temperature normally is held at 861°F and within a range of 840 to 900°F. A heavier iron-zinc alloy layer is usually associated with high temperatures and slow rate of travel through the zinc or, analogically, a longer immersion time. A heavy iron-zinc layer is conducive to brittleness in the coating and though some authorities contend that a heavy layer of pure zinc contributes greatly to corrosion resistance, experience indicates that the thickness of the combined layers is the determining factor.

The thickness of the coating after deposition is controlled by the type of wipe. Regular galvanized wire generally used for miscellaneous purposes is tight wiped. The tight wipe is effected by pulling the coated wire through asbestos held in place by mechanical fixtures.

Type I (a designation peculiar to the wire industry) wire is also asbestos wiped to remove excess zinc and smooth the surface. Regular galvanized wire is similar in coating characteristics to Type I wire, but the Type I wire is manufactured and supplied according to specific coating requirements as determined by the Preece test for uniformity of coating, and weight-of-coating tests. Types II and III wires are pulled through a charcoal mixture which leaves a heavier coating of zinc on the wire surface than in the case of the tight asbestos wipe.

Brightness of the zinc is commonly attained by quenching with cold clear water, preferably under pressure, immediately after the wire emerges from the wiping apparatus. Other methods of obtaining brightness are used under special conditions and these involve the addition of brighteners in the zinc bath.

Winding of Type III-coated wire on the take-up blocks is delayed to permit complete solidification of the coating. This delay is effected by diverting the wire over rolls and/or sheaves designed for this purpose, after wiping and quenching to effect a longer travel distance before reaching the take-up blocks. A water-soluble wax is used to promote packing on the final take-up blocks. The wire may be either sprayed with the wax or passed through a tank containing the waxy solution. The deposition of wax creates a smooth, slippery finish which packs more compactly, and consequently permits storage of a greater quantity of wire on the cage-like extensions of the finishing blocks. In addition, the wax aids materially to retard the formation of "white rust," which results when moisture contacts zinc without free access to air.

During production, control tests are regularly taken to determine the weight of coating, the uniformity of its distribution and to determine compliance with tensile-strength requirements or other specified properties. Modifications in practice, usually of a minor nature, may be dictated by the results obtained from such control tests.

Hot-Dip Galvanizing of Pipe

J. L. Roemer, Jr.

Superintendent, Butt Weld Tube Mills
The Youngstown Sheet and Tube Company
Youngstown, Ohio

The hot-dip galvanizing of pipe is accomplished by submerging properly prepared pipe in a molten bath of zinc at the correct temperature for a predetermined length of time. The purpose of this process is to produce a pipe coated inside and outside with zinc, thus making it more resistant to corrosion.

The galvanizing of pipe, usually in 21-foot lengths, takes place in a kettle or pot made from low-carbon rimmed steel plates 1½ inch in thickness. These plates are welded together to form the rectangular kettle. Usual dimensions are 5 feet in width, 6 feet in depth, 26 feet in length, and the kettle contains approximately 170 tons of zinc.

Prime Western zinc slabs, containing a minimum of 98.32 per cent zinc, with a maximum of 0.08 per cent iron and 1.60 per cent lead, are used to fill the galvanizing kettle. It generally requires about 40 hours to melt down the zinc in a new kettle installation. Heat is applied to the sides of the kettle from a series of evenly spaced gas-fired burners. The flame from these burners is directed at the base of a brick baffle wall over which the products of combustion pass so as to avoid direct impingement of the flame against the sides of the pot. A firing arrangement of this type forces the heat transfer through the upper section of the side walls of the kettle, which is conducive to increased pot life and minimum dross formation.

Operating temperatures of the molten zinc usually range from 840 to 865°F, and are automatically controlled and recorded by the use of instruments wired to thermocouples set in the molten zinc. Temperatures exceeding the specified maximum result in reduced pot life and increased dross and oxide formation, along with inferior quality of the zinc coating on the pipe.

During the manufacture of pipe, whether by the butt-weld, continuous-weld, seamless or electric-weld processes, scale or iron oxide is formed on the surface of the pipe and must be removed before galvanizing. This also applies to contamination by any foreign substance, such as oil or grease acquired during preparation of the pipe for galvanizing.

Oil and grease are removed by picking up a lift of pipe with an overhead crane and submerging it in a hot alkali solution until clean. It is then removed and rinsed in water (see Figure 9-50). Scale is removed by placing this same lift of pipe in a pickling vat containing about 8 to 10 per cent sulfuric acid and a predetermined amount of an inhibitor. Pickling-solution temperatures generally range between 130 and 150°F. The factors controlling the pickling time necessary to remove scale are the percentages of acid and iron in the solution, temperature, pipe size and type of steel. Immediately after pickling, the lift of pipe is dipped and thoroughly rinsed in two separate water tanks to prevent the carry-over of iron salts into the next tank, which contains the flux solution.

This flux solution consists of zinc ammonium chloride and water heated to 120°F, and has its density maintained between 20° and 22° Baumé.

The pipe is submerged in this solution for the purpose of covering the pickled surface with a protective coating to prevent oxidation prior to galvanizing. Double rinsing of the pickled pipe in water prior to fluxing is necessary to control the iron content in the flux solution. This iron carry-over has a definite effect on the dross build-up in the galvanizing pot.

After preparation the pipe is placed by crane on the charging table at the galvanizing kettle. From this point it is conveyed by chain or rolled manually into the pot through a flux blanket of zinc ammonium chloride. This flux blanket covers the molten zinc on the entering side of the pot and is separated from the exit side by a longitudinal flux shield which prevents the flux from flowing over to the exit side. The flux blanket is necessary to protect the surface of the molten zinc from oxidizing and to remove any slight oxide contamination which may have formed on the pipe while on the charging table.

Upon entering the molten zinc the pipe may be moved to the exit side of the kettle by various methods. Some of these in common use are screw-type conveyors, star wheels, chain mechanisms, overhead plungers actuated by air cylinders, or a series of plungers operated from an overhead rachet. Except with the screw-type conveyor it is necessary to use guides or saddles in the pot for the pipe to rest on as it fills up with zinc and is moved from the entering to the exit side of the pot.

The pipe is removed from the molten zinc on the exit side by a potman who uses a long hook made for this purpose. It is withdrawn from the pot by magnetic rolls or tongs on a drawbench. As a zinc-coated pipe is drawn from the pot it passes through a perforated ring supplied with superheated air which removes the excess zinc from the outside, resulting in a smooth surface. The pipe is removed manually from the magnetic rolls or drawbench by the use of tongs, and is blown internally with superheated steam which removes excess zinc. The galvanized pipe is chain-conveyed into and through a heated water bosh for the purpose of setting the zinc coating. From the water tank it passes over an inspection table before being sent to the finishing floor for further processing. A flowsheet of the process is shown in Figure 9-50.

The zinc coating applied to pipe by the hot-dip process theoretically consists of three distinct layers. The first layer on the steel is known as the *bonding layer*, the second, or middle one, is referred to as the *zinc-iron alloy layer*, and covering this is the one containing more of the pure zinc coating. Three factors are the controlling elements in obtaining a satisfactory coating by the hot-dip method. They are (1) properly prepared surface, (2) correct temperature and (3) time in the pot.

By-products incidental to the galvanizing of pipe are dross, zinc dust, zinc oxide and flux skimmings. Dross has a greater density than the zinc

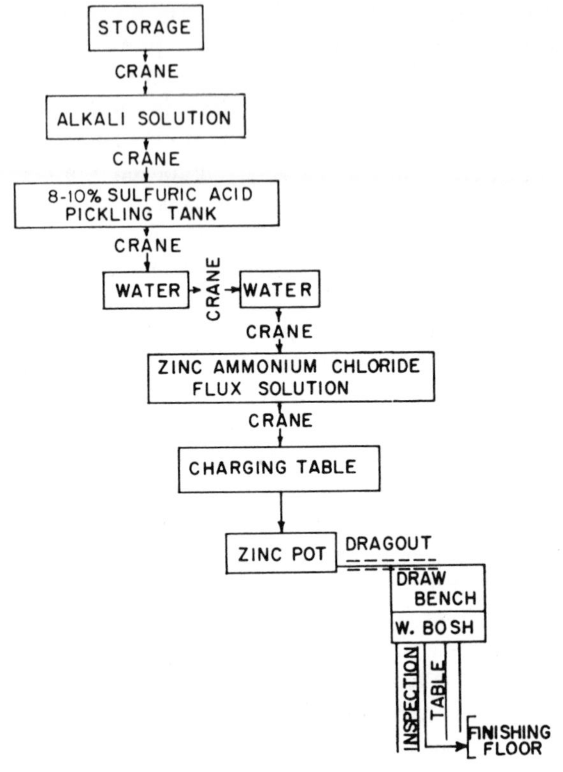

Figure 9-50. Flow sheet of pipe-galvanizing process

and therefore settles out in the bottom of the pot and must be removed periodically. It has an iron content of about 3.50 per cent. Zinc dust is formed from blowing the pipe internally with superheated steam. It has a zinc content ranging from 93 to 95 per cent. Zinc oxide forms on the surface of the molten zinc bath on the exit side of the pot and has a zinc content of about 75 per cent. Flux skimmings form on the entering side of the pot and have a zinc content of about 65 per cent, together with a chloride content of about 13 per cent.

Electrogalvanizing of Sheet

H. H. Greene

Metallurgist, Sheet and Strip Division
Republic Steel Corporation
Cleveland, Ohio

Flat-rolled steel has been electrogalvanized in considerable amount for quite a few years. Originally, the widths produced were under 20 inches, with the majority under 12 inches, and the product was mostly in the specialty class. Widths over 24 inches were not produced until the late 1930's; those up to 60 inches became available in 1940. Two lines built during World War II for electrotin were converted to zinc shortly after the war. Both of these, adding to the available wider capacity, have a maximum width of approximately 36 inches. In 1957 estimated United States capacity of electrogalvanized flat-rolled product in widths over 24 inches was 25,000 tons per month.

The chief reasons for the increased production of electrogalvanized sheets have been:

1. Complete adherence of zinc to the base metal, even to the point of fracture in a draw.
2. A smoother, better surface for painting, free from the spangle characteristic of the hot-dipped product.
3. Better physical properties for deep drawing, and ability to control them in the coating process.
4. Good control of zinc thickness and uniformity.
5. Pure zinc coatings result in better corrosion resistance.

Most of the wide sheet produced carries coatings weighing in the range of about 0.06 to slightly less than 0.2 ounces per square foot (0.000050 to 0.000170 inch thickness, each side). A small amount carries considerably less, i.e., about 0.025 ounce per square foot. Very little is produced in the wide widths, with coatings heavier than 0.2 ounce at the present time. The particular coating thickness supplied by each producer is generally dictated by prevailing economics and equipment.

Approximately 90 per cent of the 0.06 to 0.2-ounce coatings is phosphatized immediately after coating so that it can be painted by the customer without the metal surface being treated in any way other than by cleaning of storage and fabricating soilage. All the very thin zinc coatings (0.025 ounce) are phosphatized for final painting. The portion of production not phosphatized was not designed to compete with hot-dipped coatings, but despite this, has found an economic place for inside uses where a spangled coating may be less desirable, or where severe forming and moderate corrosive conditions make the electrogalvanized coatings adequate for the purpose. Some uses of this nature have been reported in which these nonphosphatized, nonpainted coatings have shown remarkable life. The nonphosphatized coatings are purchased extensively by customers who have phosphatizing equipment available for treatment of the fabricated part immediately before painting.

All producers use as the basis metal a regular cold-rolled sheet which is furnished to the coating lines in coil form as dry and as clean as possible. This takes advantage of the smooth surface of the cold-rolled product which, due to the nature of the coating process, tends to be reflected in the final coating. There is no record of any producer using a hot-rolled pickled product as basis metal.

All the coating lines are continuous, with the strip from the coil fed through suitable entry equipment—welder, side trimmer, looping pit, etc.—into the precleaning units. These units consist at least of an alkali cleaner and an acid bath, either or both of which may be electrolytic. The acids used are either hot sulfuric or cold hydrochloric. Suitable water rinsing and brush scrubbing are introduced after each bath. It should be pointed out that in most cases the strip speed and the resultant time in the precleaning units are such that only very light surface contamination can be successfully removed. It is fairly well established from experience what can be tolerated in this respect.

In the electroplating unit, the solutions are nearly always of the acid type, using either zinc sulfate or zinc chloride, plus other salts to improve conductivity. Solution temperatures are held under 150°F and current densities vary widely from 120 to 350 amperes per square foot. Solution agitation is accomplished by strip travel and pumps, and this, together with constant solution filtering, helps to insure maximum quality of the plate. The anodes used are nearly always 99.99 per cent zinc. Direct current from either generators or rectifiers is generally supplied at up to 12 volts, with at least one line having a total of 120,000 amperes available for plating. Electrical contact with the moving strip is made by either rolls or fingers, depending on the design. Other rolls and guides are used to maintain proper anode-cathode spacings.

Design and construction of the various lines differ considerably. The two newest, designed originally for electrotin, are "three story" types, in which one side of the moving strip is plated at a time. The initial bottom is coated in the first level, followed by a doubling back and turning over of the strip in the second level, where the original top is coated as the bottom. The strip again reverses onto the third level, where rinsing and reclaim of plating solution occurs. Roller contactors for electrical connection to the strip are used between each small plating cell. An older line plates both sides in a straight-line operation, in which case many finger contactors are used to complete the electrical circuit.

After plating, the coated strip is very thoroughly water-rinsed. This is essential, particularly if the material is to be phosphatized later. At this point the coating has a dull, silvery-white appearance which is very desirable for some end uses.

If the stock is not to be phosphatized, it is usually put through a hot, dilute chromic acid-type final rinse which "passivates" the active zinc surface and, to a degree, tends to protect the surface in transit, storage or handling. This rinse does not alter the silvery-white color.

The phosphatizing operation takes place generally on the same continuous line as the coating, immediately after the water rinsing. This process is chemical in nature, whereby a very slight amount of the zinc surface is converted into a layer of insoluble tertiary zinc phosphate crystals. This is accomplished by immersion in, or spraying or brushing the zinc surface with, an aqueous solution of the phosphatizing agent (usually a proprietary product). The action takes place in a range of 6 to 12 seconds. The strip is then given a thorough water-rinse, followed by the hot, dilute chromic acid-type rinse, and finally drying with hot air. The phosphatized surface is somewhat darker than a natural zinc surface.

After drying, the strip is either recoiled or cut to length as desired. Some lines recoil only, with the cutting to length performed on a separate unit.

Use of the phosphatized, electrogalvanized, flat-rolled material has spread into many fields, among which are: sign work, trailers for both truck and home use, lighting fixtures, refrigeration, automotive products, and a variety of cabinet applications. It is particularly valuable for the small manufacturer who has no phosphatizing equipment. The nonphosphatized product has found many applications in the domestic stove, automotive, container, and other industries.

Zinc Coating of Steel Wire by Electrodeposition

J. E. CROWLEY*

Division Head, Chemistry; Research Department, Bethlehem Steel Company Bethlehem, Pa.

History

Electrogalvanizing of steel commodities is by no means a new art. Electrodeposition affords a means of applying zinc coatings for a wide range of requirements which cannot be met by hot-dip galvanizing. It is probable that the first electrogalvanizing process used on a large commercial scale was that developed at the Langbein-Pfanhauser Werke in Germany. Langbein used neutral zinc sulfate solution as an electrolyte, with zinc anodes at moderate current densities—from 50 to 100 amperes per square foot. While there was nothing very new about the actual electrodeposition process at that time, Langbein apparently first developed the necessary wire-handling methods combined with a more satisfactory wire-cleaning process, thus making possible a commercial scale operation in the production of reasonably heavy coating weights.

* Contributions to this discussion from the following sources are gratefully acknowledged: American Steel and Wire Company, Indiana Steel and Wire Company, Republic Steel Corporation, and the MacWhyte Company.

With certain variations in electrolyte composition and methods of wire cleaning, essentially similar processes employing soluble zinc anodes have been used in electrogalvanizing plants installed in the United States. In the early 1930's, processes employing insoluble anodes were developed, by which virgin zinc is electrolytically deposited on the steel.

Properties of Steel Wire Coated with Zinc by Electrodeposition

Zinc is usually electrodeposited on steel wire in the range from 0.3 to 3.0 ounces per square foot. Coatings up to 5.0 ounces per square foot have been applied but there is no demand for such a product. Since the thickness of coating governs the wire-processing speeds and costs, there is an economic limit of acceptance by the customer.

Diameters of plated wire, including wire cold-drawn after plating, are usually in the range from 0.009 to 0.192 inch. Carbon contents are available from 0.08 to 0.85 per cent. Tensile strength ranges from 50,000 to 300,000 psi. The wide selection of properties and sizes available results from application of the zinc to cold-drawn or heat-treated steel wire, and also the ability of the coated wire to be cold-drawn without injury. The heat-treated and coated wire can be cold-drawn to the extent of approximately 95 per cent reduction in area, depending upon the chemical composition, type of heat treatment and coated-wire diameter.

The coated wire can be rolled, wrapped around its own diameter and swaged or bent into an endless number of designs. The electrodeposited coatings are of uniform longitudinal thickness and concentricity, and afford protection against corrosion in proportion to thickness of coating. When necessary, the coated wire is furnished with a medium or high luster provided by a polishing operation. Also, various surface films can be applied, such as light oil, wax or chromate passivation film.

Typical Sequence of Operations in Electrodeposition Lines

Steel wire is fed from rotating reels in coil form. The coil size depends on the billets used and is in the range from 250 to 400 pounds. Spools may also be used. Successive coils of wire are joined by welding, there being no interruption in the wire travel.

Low-carbon wire is annealed during its passage through molten lead contained in gas-fired steel pans. Other wire is passed through the lead and is not heated to the full annealing temperature, but only to a point where wire-drawing compounds are rendered quite soluble in the subsequent pickling step. High-carbon wire is usually heat-treated separately and prior to being passed through the electrodeposition lines, because the characteristic treatment is done at slower linear speeds, which are less than the speeds used for electrodeposition.

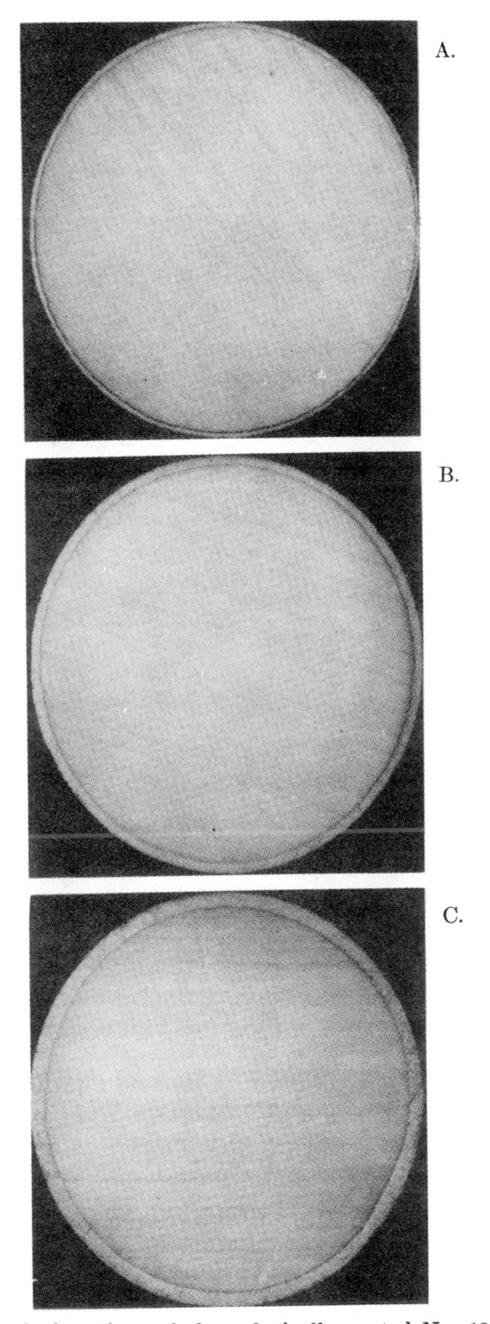

Figure 9-51. Unetched sections of electrolytically coated No. 12 gage wires. A, coating weight, 0.82-ounce. B, coating weight, 1.66-ounces. C, coating weight, 2.51-ounces. Magnification 20×.

Figure 9-52. Etched structure of electrolytically coated wire. No. 6 gage with 3-ounce coating weight. Steel base on top. Magnification 500×.

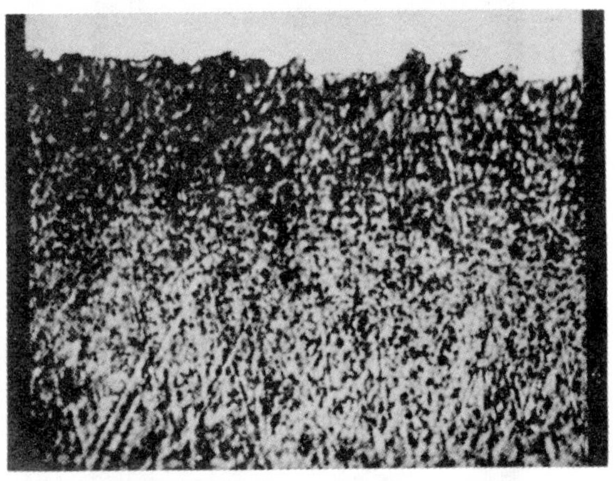

Figure 9-53. Unetched section of electrolytically coated wire. No. 6 gage with 3-ounce coating weight. Steel base on top. Magnification 500×.

An electrolytic alkaline cleaner may be used prior to the pickling treatment in muriatic acid. The wire is then subjected to anodic electrolytic treatment in cool sulfuric acid solution. This treatment not only completes the necessary precleaning but also renders the wire receptive to an efficient initial electrodeposition of zinc.

The electrodeposition cells contain anodes of slab zinc immersed under the steel wires in the soluble anode processes. Insoluble anode processes use grids between which the wires pass.

Direct current from motor-generator sets is introduced into the wire at

Figure 9-54. View of electrolytic cells at Sparrows Point showing method of introducing direct current into wires, and overhead electrolyte circulation launders. Take-up blocks located at extreme right.

intervals by conductors which contact the wire. The heat generated by electrodeposition must be dissipated by coolers. If this is not done, no protective foam blanket can be maintained on the electrolyte to retain acid fumes so that proper working conditions are maintained. In addition, the electrochemical efficiency is depressed by excessive electrolyte temperatures.

Wires pass from the electrolytic cells through a water rinse. Subsequent coatings with light oil or chromates can be applied.

The matte electrodeposit has a light gray color and crystalline appearance. Bright appearance can be developed with rotary burnishers and a bright luster finish can be applied when the wire is given a slight reduction through a polishing die.

Finished wire is taken up on rotating "blocks" and is tied into coils. For some products, it is necessary to remove the welds which were made at the entrance of the line.

Soluble Zinc Anode Processes

There are presently in production several important operations which use soluble zinc anodes with a low-acid, zinc-sulfate type of electrolyte. Relatively lower current densities are usually employed in the soluble anode system, varying from 125 to 875 amperes per square foot, depending upon the wire size and the coating weights being produced. This compares to a range from 500 to 5,000 for the insoluble anode processes. These soluble zinc anode processes embody the essential features of the method of electrodeposition known as the "Meaker Process" but some modifications of this process have been made by the Republic Steel Corporation. The soluble anode process is also used by the MacWhyte Company and the American Steel and Wire Company.

The zinc anodes are placed under the wires in the electrolytic cells and are replenished as they dissolve. No leaching circuit is required but there is a requirement for filtration or purification of the electrolyte. The balance of cathode and anode current efficiencies in the soluble anode processes is carefully controlled so that electrolyte losses by drag-out do not cause the solution composition to change.

Insoluble Magnetite Anode Process

The Indiana Steel and Wire Company produces the heavier weights of zinc coatings, that is, Class B and Class C coatings, by electroplating zinc from a concentrated solution of complex zinc tetrammine salts with an insoluble anode of cast magnetite (Fe_3O_4). The plating solution is made by dissolving zinc oxide or other zinc-bearing material in a solution containing ammonia and ammonium chloride and/or sulfate. The complex salt is assumed to contain four atoms of ammonia and one atom of zinc in the molecule, as is shown in the formulas $Zn(NH_3)_4Cl_2$ or $Zn(NH_3)_4SO_4$.

When zinc is plated out of the solution, ammonia and ammonium salts are liberated, which makes it possible for the solution to again dissolve zinc. Zinc in the plating solution is replenished by continuously withdrawing some of the solution from the plating cell to a leaching plant adjacent to the plating line. In the leaching plant, galvanizer's zinc ashes, calcined zinc ore or other zinc-bearing materials are added to the solution. The free ammonia and ammonium salts will readily dissolve the zinc oxide and other zinc compounds to form the complex zinc tetrammine salts. Zinc metal is very slowly dissolved. The replenished plating solution which contains 1.00 to 1.25 pounds of zinc per gallon is then treated with oxygen gas or other oxidizing agent to oxidize iron and related metallic impurities so that they form hydroxides in the ammoniacal solution, which precipitate and are filtered out of the solution. The solution is finally treated with zinc dust and filtered to remove copper and other elements below zinc in the electromotive series. The replenished and purified solution is then fed into the plating cell at approximately the same rate as depleted solution is withdrawn to the leaching plant.

In the plating cell the wire passes between plates of cast magnetite,[13] the insoluble anodes, which are arranged in rows spaced about two inches apart. The cathodic current is brought to the wire by shielded contactors which are tipped with tantalum[14] where they contact the moving wire.

Insoluble Lead Anode Process

This process was originated by U. C. Tainton and has been developed by the Bethlehem Steel Company from the pilot plant built in 1932. Roasted zinc concentrates are dissolved in the high-acid mechanical leaching process. Manganese dioxide is used for oxidation in leaching so that dissolved iron will precipitate upon neutralization by the zinc raw material. This is followed by filtration in a Burt or Oliver precoat filter to remove insoluble matter and occluded impurities such as germanium and antimony. This type of leaching process is also used at the Bunker Hill and Sullivan Company's electrolytic zinc plant. An outstanding innovation has been developed, perfecting control of the colloidal iron derived in the leaching step, by the addition of a small amount of phosphate, so that uniformly high production rates can be maintained through the filters.[15] A final purification is conducted, at controlled acidity, with zinc dust followed by filtration to remove heavy metals such as copper and cadmium. The procedure is controlled by chemical analysis.

Before admission to the electrolytic cells, a sample of the purified zinc sulfate solution is used to make an electrolyte from which a zinc plate of commercial coating thickness is electrodeposited on aluminum. This zinc plate is easily stripped from the aluminum, and its ductility is measured with an Erichsen testing machine. The ductility is quite remarkable and is of the same order of magnitude as that of rolled zinc.

The electrolytic cells contain grids of insoluble anodes through which the wires pass. The anode material is a silver-lead alloy developed by U. C. Tainton. This alloy has had world-wide acceptance by electrowinning plants which produce Special High Grade zinc. Normal chloride concentrations found in solution make-up water have no effect on lead-anode life. This is the situation, except for one electrowinning plant located abroad, where it is necessary to resort to a chloride removal treatment to insure good anode life. Manganese dioxide added in the leaching plant dissolves as sulfate and is reformed as the dioxide at the cell anodes. It is periodically washed out of the cells.

The presence of a saturation concentration of calcium sulfate in the electrolyte results in high maintenance requirements for efficient heat-exchanger operation in lead coils or pipes which employ well water as a coolant. The Research Department of Bethlehem Steel Company developed air coolers which have solved this problem.

The Tainton process usually employs current densities less than 2,000 amperes per square foot. There is also in operation at Bethlehem an electrodeposition process with current densities up to 5,000 amperes per square foot.[16]

The American Steel and Wire Company has used a process employing insoluble lead anodes with a zinc sulfate electrolyte of medium acid content. A zinc regeneration system is employed and the electrolyte is filtered continuously.

Control of Coating Quality

In the insoluble anode processes, the purified zinc sulfate liquor is fed at controlled rates to the electrolytic cells to maintain uniform electrolyte composition. The electrodeposition of zinc releases an equivalent quantity of sulfuric acid. Spent electrolyte is accumulated and returned to the leaching plant to dissolve zinciferous material thereby making a regenerative cycle with respect to sulfuric acid. There is no recycling of acid in the soluble anode processes.

As the loss of electrolyte through drag-out on the wire is substantially greater than is experienced in the electrowinning of zinc, there is inherently a greater tolerance threshold for cumulative impurities in the wire-coating process. Also, the time of electrodeposition in the wire-coating process is measured in minutes or seconds as contrasted with the many hours required for electrowinning zinc. This time factor also results in a greater tolerance threshold for impurities in the wire coating process as compared to electrowinning. In the processes which employ a leaching step, co-depositable impurities such as copper and cadmium are removed, as previously described, to a level well below their tolerance limit. Excessive concentrations would lower the ductility of the zinc electrodeposit. Nondepositable impurities, such as germanium and antimony, are removed, as previously described, to a level well below their tolerance limit. Excessive concentrations of such impurities would depress the electrochemical efficiency. The use of high-grade slab zinc prevents introduction of impurities from zinciferous raw material in the soluble anode processes.

The contributions of electrolyte purity and cleanliness of steel surface in controlling ductility and bending of the electrodeposit on steel wire are illustrated by the wrap test. This is a routine inspection test in which the wire is wrapped around its own diameter and examined visually. The outer fiber elongation is 50 per cent. Such an elongation value is normally beyond the properties of electrodeposited zinc coatings. Success in the wrap test obviously must be influenced by the tenacious adherence of the zinc to the steel base.

A clean steel wire surface is necessary for the production of uniform and adherent zinc electrodeposits. Control of the steel surface begins in the steel-making operation and reaches the final stages in the electrolytic lines.

COMPANIES FOR COATING STEEL WIRE

Process Component	Bethlehem Steel Co.	Republic Steel Corp.	MacWhyte Co.	Indiana Steel & Wire Co.	American Steel & Wire Co.	
Type	Insoluble Anode	Soluble Anode	Soluble Anode	Insoluble Anode	Soluble Anode	Insoluble Anode
Raw material for leaching	Roasted zinc concentrates	None	None	Zinc skims or roasted zinc concentrates	None	Slab zinc
Leaching solvent	H_2SO_4	None	None	NH_4OH $(NH_4)_2SO_4$ NH_4Cl	None	H_2SO_4
Leach residue filter	Burt or Oliver precoat	None	None	Settling plus filter press	None	None
Leach purification	Zinc dust with filter press	None	None	Zinc dust with filter press	None	None
Heat treatment	Molten lead & electric resistance	Molten lead	None	Molten lead	Molten lead	Molten lead
Alkaline cleaner	Occasional	Yes	Yes	No		No
Muriatic acid pickle	Yes	Yes	Yes	Yes	Yes	Yes
Anodic nitric acid	No	No	Yes	No		No
Anodic sulfuric acid	Yes	Yes	Yes	No		No
Electrodeposition anodes	Silver-lead alloy	High grade zinc	High grade zinc	Cast magnetite	High grade zinc	Lead[17]
Electrolyte filtration	No	Yes	Yes	Yes	Yes	Yes
Electrolyte composition (g/L)	$ZnSO_4$: 200 H_2SO_4 : 200	$ZnSO_4$: 200 to 250 $(NH_4)_2SO_4$: 25 to 30 *p*H: 3 to 4½	$ZnSO_4$: 200 to 250 *p*H: 3 to 5	Zn: 95 to 130 Cl_2 : 100 to 135 NH_3 : 140 to 153 SO_4 : 35 to 50 *p*H: 9 to 10	*p*H: 2 to 4	
Electrolyte temp. (°F)	115 to 160	100 to 130	90 to 110	90 to 100	120	120
Electrodeposition current density (amps/sq ft.)	750 to 5000	325 to 875	125 to 400	500 to 1500	200 to 500	1000 to 2000
Zinc coating types	I, II, A, B, C	I, II, A	I, II, A, some B	B, C	I, II, A, B, C	I, II, A, B, C
Wire diameters coated	0.300 to 0.030	0.192 to 0.0625	0.120 to 0.054	0.238 to 0.060	0.157 to 0.050	0.157 to 0.050

Zinc coating types are described in detail in ASTM Specifications A-122 and A-218.

Table 9-16 contains data on the important features of coating steel wire by electrodeposition of zinc, as practiced by the leading producers.

General Electroplating with Zinc

R. E. Harr

Chemical Engineer, Metal Finishing Development
Western Electric Company, Incorporated
Chicago, Illinois

The term "general electroplating" is used here to distinguish between the method of electroprocessing parts after fabrication and the methods already described, which apply to the processing of continuous lengths of sheet stock and wire. In electroplating parts, two different methods of handling are employed, barrel and rack plating. In barrel, or mechanical, plating, parts are handled in bulk as distinguished from racking methods in which the individual parts are handled separately, to attach them to a conducting support in the form of a wire, hook or clip.

Barrel plating is used for plating all small parts which are not subject to nesting or tangling and are not likely to be damaged by tumbling. It is less expensive than rack plating due to the fact that labor is the largest single item of cost in applying any electroplated coating, with the exception of the precious metals. In zinc plating this is especially true due to the low cost of metallic zinc and the chemicals used in the preparation of plating solutions.

In rack plating, as well as in barrel plating, when the volume of similar parts justifies it, handling of racks or barrels by conveyers can be used to reduce the labor cost. In addition, advantages are gained in greater uniformity of treatments. Manually operated equipment for plating parts in barrels or on racks is provided either for small production requirements or for situations involving large variations in the size or shape of the parts to be plated.

Two types of zinc plating solutions, the acid and the alkaline-cyanide, are used for general-purpose plating. Details of solution composition and operating conditions may be found in the references.[18-20] A comparison of the capabilities and inherent characteristics of each type of bath are summarized in Table 9-17.

Equipment for Electroplating with Zinc

Facilities for electroplating with zinc are designed to meet production requirements based on the thickness of coating and the area to be plated. To meet high production schedules, highly mechanized units are employed for efficiency and economy. Tanks for cyanide solutions are not lined. When

TABLE 9-17. COMPARATIVE CAPABILITIES OF ZINC PLATING SOLUTIONS

Acid	Cyanide
Lesser throwing power.	Excellent throwing power.
High plating rate due to good conductivity and high permissible current density	Good for plating irregular shapes.
Low operating costs.	Poor for plating malleable and cast iron.
Requires more elaborate surface preparation.	Bright zinc deposition can be used for decorative purposes and is less subject to staining.
Solution compositions are simple and less critical.	Surface preparations are simple.
Lined chemical-resistant tanks required.	Unlined steel tanks satisfactory. Special formulation of the solution is not required to use insoluble auxiliary anodes.
May be highly corrosive if chloride used.	More complex in chemical nature and more critical as to the balance of constituents in the solution. Cyanide in rinse water must be destroyed when dilution is not sufficient for local regulations.

an acid solution is used, linings of lead, rubber, "Koroseal" or asphalt may be used.

The acid plating solution, although used widely in the early days of electroplating, is currently used only for special applications in general electroplating. These include the plating of long cylindrical rods which by their nature do not require a solution of good throwing power to produce a coating of uniform thickness. Two other special applications favoring the use of acid zinc solutions are the plating of cast iron and plating in a locality requiring the elimination of cyanide in effluent rinse waters.

By far the largest amount of general-purpose zinc plating is from the alkaline-cyanide solution. The principal reason for its preference is the good throwing-power characteristic.

The bright-plating cyanide baths, introduced about 1935, not only eliminated troublesome variations in the color of the zinc coatings, but also reduced their tendency to stain. The need for anode bag in plating was also eliminated by the concurrent introduction of zinc of 99.99 per cent purity for anodes. The principal differences between the older cyanide plating solutions and the newer bright-plating solutions favor the latter, in the following respects:

1. Maintenance of higher purity
2. Higher concentration of sodium hydroxide
3. Fixed ratio between the zinc content and the total cyanide
4. Lower operating temperature
5. Higher current density
6. Optional use of organic and metallic brighteners or acid bright-dip

Figure 9-55. A barrel-plating unit at the Western Electric Company, Indianapolis Works. At left is the refrigeration unit originally installed to cool the plating solution because of an inadequate supply of cooling water. It is now used to permit the use of higher than normal plating current per unit volume of solution. In the center, the control panel and barrel-plating unit may be seen. On the right is a centrifugal dryer.

Figure 9-56. A close-up view of control panel and a plating barrel.

While these baths operate at lower temperatures with less breakdown of cyanide to carbonate, there is still a considerable accumulation of sediment in the bath, which is best removed on a frequent periodic schedule, such as on a weekly or monthly basis. This sediment, consisting mainly of foreign metal sulfides, ferrocyanides and carbonates, may be conveniently removed by decantation during a week-end period. Decantation is preferred in handling zinc solutions due partly to the difficulties of maintaining the pump, filter and piping system in a leak-free condition, and also in part to the difficulty of maintaining good circulating rates of the solution while removing the filter-clogging slimes normally present in the alkaline cyanide solution.

Thickness of Electroplated Zinc Coatings

Most of the electroplated zinc coatings are applied in thicknesses of from 0.00015 to 0.0005 inch for limited protective qualities and/or for decorative purposes in accordance with types RS and LS in ASTM Specification A164-55. While coatings in this thickness range are not equivalent to hot-dipped galvanized coatings in protective qualities, they do provide adequate protection for many types of mild exposure. Also, they are invaluable for parts having close dimensional tolerances, and are less expensive. Finishes in this thickness range are frequently treated with a chromate conversion coating solution. This treatment usually improves the appearance and reduces staining, as well as the tendency to form bulky white corrosion products under conditions of high humidity.

Zinc coatings, 0.5 to 2 mils thick, are frequently competitive with hot-galvanized finishes for outdoor applications and usually are preferred for threaded parts and other parts requiring close dimensional control. Life extension of thick zinc coatings for outdoor use by the application of chromate conversion coatings appears doubtful according to several investigations.

Metallizing with Zinc

SAM TOUR

President, Sam Tour and Company, Incorporated
New York, N. Y.

"Metallizing" is a general term applicable to the various processes used for applying a metallic surface by means of a spray or blast of molten or semimolten metallic particles. This blast or spray may be obtained by the atomizing of previously molten metal by introducing the metal in powder form into the heart of a blast burner, or by introducing the metal in wire form into the hot zone of a blast torch. As the end of the wire melts, the molten material is converted into a spray of finely divided molten metal particles. This method is known as the "wire gun method."

The wire gun has evolved to be the only extensively used industrial gun for the purpose of applying commercial zinc coatings. The first successful wire type of gun was developed some 45 years ago. While the basic concepts of the original gun have been retained throughout these years, many improvements have been made.

Wire-type spray guns have been improved to such a remarkable degree as to make the process of metallizing a fully practical "putting-on tool" for continuous use in production work. The complete process of metallizing requires more than just a metallizing gun. Proper auxiliary equipment is necessary for operation and maintenance of the gun. There is still more involved in the present-day process of metallizing; surfaces must be prepared to take the coating of sprayed metal. Depending upon the particular wire used in combination with the method of surface preparation and the actual spraying procedure, a large variety of types of coatings may be obtained.

Built into the wire gun itself is a wire feed mechanism which draws the wire from a reel or a coil and feeds it into the gun at a very constant rate. This rate of feed is not affected by small kinks, bends or small surface variations in the wire. The wire feed mechanism is adjustable as to speed of feed to suit the particular metal wire being used. Wire speeds range from as low as 1½ feet per minute to as high as 25 feet per minute. Modern guns maintain this speed of feed constant in spite of air pressure fluctuations, kinky wire or unusual wire drag during operation.

The modern wire gun is a scientifically engineered precision-built unit which will operate continuously, day in and day out, as all modern production tools should operate. The zinc capacity of regular hand-operated guns using ⅛ inch wire is 27 pounds, and using 3/16 inch diameter wire, 61 pounds of metal per hour.

Magnetic thickness gages have been developed for measuring the thickness of nonmagnetic coatings on an iron or steel base. The net result of all of these developments is that now a shop can install the complete process with equipment designed for the specific work intended, and management can expect uniform production from day to day. A complete metallizing installation showing the modern equipment from the compressor to the gun is illustrated in Figure 9-57.

Sprayed metal, or metallized coatings produced by the metallizing process, differ considerably both chemically and physically from the original wire used in the wire gun to produce the metal spray. While the mechanism of the bond between sprayed metal and base metal and the mechanism of

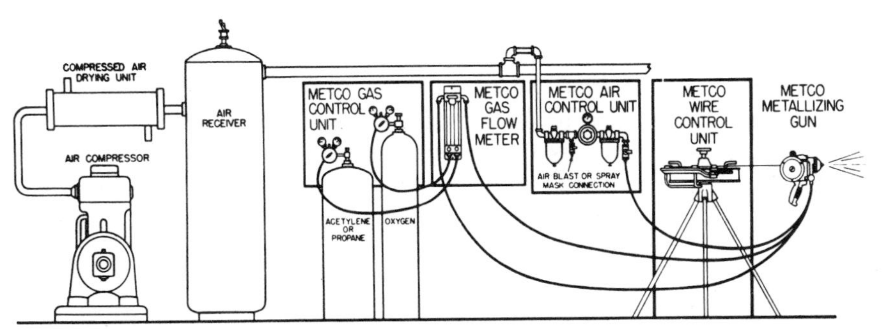

Figure 9-57. A complete metallizing installation. (*Courtesy the Metallizing Engineering Company*)

the bond between the individual particles of sprayed metal are not fully understood, a great deal has been learned about them. A sprayed metal deposit consists of many individual particles with certain mechanical interlocks, a small amount of welding or alloy bonding between minute portions of particles, and attachment through joint oxide layers between particles. The over-all properties of the sprayed metal must, therefore, be a composite of the properties of the metal particles themselves, the degree of attachment between individual particles, the properties of the oxide layers between particles and the degree of porosity within the structure.

Sprayed zinc shows an ultimate tensile strength of 13,000 psi, which is much higher than any tensile strength normally found for unalloyed zinc metal. The elongation under load before fracture of sprayed zinc is of the order of 1.40 per cent.

Spray-Metallized Zinc for Corrosion Protection

Probably the most common use for zinc is in the field of corrosion protection, where its qualities have become well known. Galvanizing is specified on many types of steel and iron structures and on equipment where paint would fail rapidly. However, galvanizing is limited because of tank sizes and the thickness of metal which may be applied by this method. The application of metallized zinc is not limited in such respects. Structures of any size can be so treated before or after fabrication, regardless of location, and any thickness of zinc may be applied.

The electrolytic potential of the sprayed zinc coating has been found to be identical to that of hot-dipped galvanizing and will thus provide electrolytic protection to bare areas. In salt or industrial atmospheres, treating or sealing of the metallized zinc coating with the proper organic materials is usually recommended. Coatings treated in this manner will be dissipated at a rate of from 0.0001 to 0.0006 inch per year, depending on the type of atmosphere.

The relative value of fine versus coarse sprayed coatings of zinc and aluminum for the corrosion protection of steel has long been a controversial subject. Comparative tests of the behavior of such coatings have been made. These tests indicate that coarse coatings of zinc are equal to, if not superior to, fine coatings, for salt and fresh water service.

Spray-metallized zinc coatings have a certain degree of inherent porosity. Experience has shown that this porosity is not detrimental to the corrosion protection quality of the coating as compared to a nonporous coating containing the same quantity of zinc per unit of surface area. During the last few years the scope of the metallizing process has been broadened considerably by the use of sealing treatments designed to fill the pores of sprayed metals. In the corrosion field, vinyls, chlorinated rubbers, phenolics and various other organic materials have proved useful for sealing sprayed zinc.

Some typical applications of spray-metallized pure zinc for corrosion protection include tanks, bridges and trestles, barges, fishing boats, dock piling, "Christmas Tree" control columns, steel work and fabricated steel products, and rubber mixing units. Figures 9-58 to 9-61 show several of these uses. A special application of spray-metallized pure zinc is for the electrical shielding of electronic instruments such as that shown in Figure 9-62.

Spray-Metallized Zinc as a Base for Paint

In the last ten years the emphasis in the metallizing industry has been placed upon combination coatings which consist of a layer of sprayed metal sealed or coated with organic materials such as vinyls, chlorinated rubbers, phenolics and silicones. These coatings generally are referred to as "metallized system coatings." The emphasis in such systems has been on the sprayed metal, with the organic coatings usually being referred to as "sealers." This is only natural inasmuch as the sprayed metal layer in most cases is of considerable thickness and offers the primary protection. When the sprayed metal is 10 mils thick or more, such coatings offer excellent protection against many severe corrosive environments. Metal coatings of these thicknesses, however, are relatively expensive as compared with paint coatings. The extensive publicity given the thicker coatings has tended to push into the background the use of thin sprayed metal coatings as a base for paints.

It is true that the difference between thick and thin metal coatings is quantitative, but there are a number of qualitative differences as well. *Thick* sprayed metal coatings are thought of as the primary protective barrier, whether or not paints and organic sealers are used. The cost is relatively high and service requirements must justify the expense. Where *thin*

Figure 9-58. 10,000 square feet of these gas holders metallized with zinc in 1932. Only 0.003 inch originally applied which resulted in a few thin spots showing up in 1941, nine years later. Spots were touched up with additional zinc at that time. Although these tanks are exposed to sea-coast atmosphere and refinery acid fumes, the surfaces were still in first class condition in 1950, 18 years later. (*Courtesy the Metallizing Engineering Company*)

Figure 9-59. Kansas City; girders of this highway overpass metallized with 0.012 inch zinc to protect the steel from corrosive effects of flue gases from locomotives passing beneath 24 hours per day. (*Courtesy the Metallizing Engineering Company*)

coatings (as thin even as 2 to 4 mils) are used as a base or preparation for paint, the story is quite different. The primary protection in this case is from the paint and the primary purpose of the sprayed metal is to enhance and prolong the properties of the paint film. Coatings of this kind are rela-

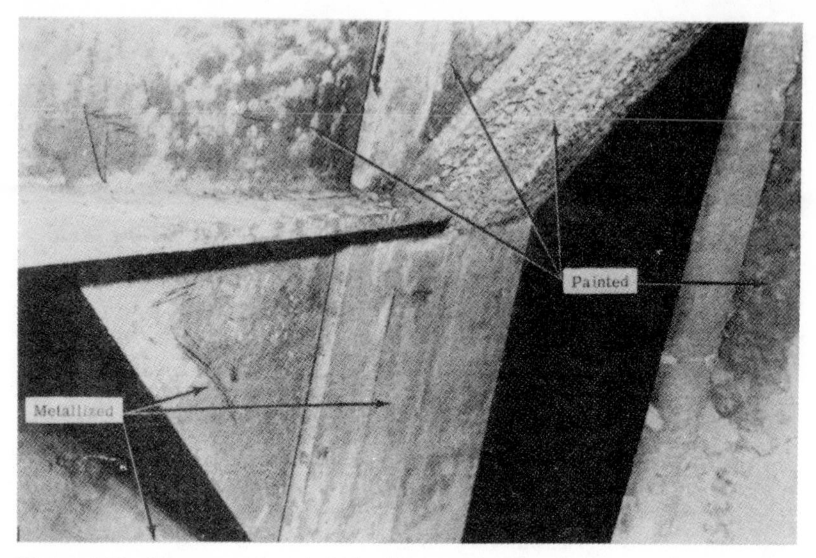

Figure 9-60. Close-up of steel dock piling, showing excellent condition of metallized portion and serious rusting on painted parts. The top 15 feet of each pile had been metallized with 0.010-inch zinc, the remaining portions of the piling having been painted. There has been no corrosion of the metallized areas of the piling. Even where it had been scratched and marred by docking vessels, the zinc around these areas has protected the scratches and pin holes cathodically. Painted portions of the piling have rusted severely. (*Courtesy the Metallizing Engineering Company*)

tively inexpensive and in many cases prove less costly than paint alone, even in a few years of service.

Sprayed zinc coatings are porous, and, unlike galvanized surfaces, offer an excellent anchor for paint. In spite of the porosity and even when no paint is applied, sprayed zinc provides long-term protection to steel. This is true for two reasons:

1. These coatings are anodic to steel and provide cathodic protection even at small voids in the coating.

2. Sprayed metal coatings become less permeable with time when the conditions of the environment favor the formation of insoluble salts and oxides within the pores. The protective life is, however, related to the thickness, and the thickness to the cost. Where these thin coatings are painted over, the pores are filled and very little, if any, bare metal is exposed directly to the corrosive environment.

With a clean steel surface, metal spraying is an intermediate step which may be considered as part of the preparation. At this stage a thin coat of zinc is quite inexpensive. One of the biggest cost factors for a heavy industrial metallizing coating is the grit-blast operation required for preparation.

Figure 9-61. Practically all major manufacturers of high-quality degreasers use metallized zinc coatings to protect this equipment against corrosion. (*Courtesy the Metallizing Engineering Company*)

Figure 9-62. Metallized zinc made this dictaphone time-master recorder lighter, more compact and less expensive to manufacture. (*Courtesy the Metallizing Engineering Company*)

Where thin coatings such as 2 to 4 mils are used, particularly for zinc, the requirements of the blast job are considerably reduced. In fact, a first-class blast job for paint application is often sufficient and the zinc may be applied without any additional blasting costs. In such cases the benefit derived from the sprayed metal base far outweighs cost in extending the life of the paint coating.

A relatively thin coating of metallized zinc provides far stronger adhesion than can be obtained by any other method. Such a coating effectively prevents rusting at the interface, which is the most common cause of paint failure. In the long run, the main consideration is maximum protection per dollar.

Figure 9-63. Before metallizing. Conduits shown in typical condition six months to one year after painting. (*Courtesy the Metallizing Engineering Company*)

Figure 9-64. After metallizing. Conduits shown after coating according to "Metco System 103." (*Courtesy the Metallizing Engineering Company*)

The paints and primers used must be compatible with the sprayed metal, and some materials compounded for steel or galvanized zinc are not satisfactory over a metallized base. In particular, sprayed zinc should never be confused with galvanized zinc. On galvanized surfaces strong acid primers sometimes are used which actually attack the galvanized surface and etch it. Sprayed zinc on the other hand offers a natural bond for paint; primers should be used, not to etch the coating, but as inhibitors. For instance, wash-coat primers containing phosphoric acid are frequently recommended on sprayed zinc for inhibitive purposes, but the acid content must not exceed 4 per cent. Some primers for galvanized surfaces have acid contents up to about 16 per cent and these would be deleterious to sprayed metal.

When applying paints directly to sprayed metal, it frequently is advisable to thin the first one or two coats so as to obtain better penetration into the base. At least one large metallizing contractor advises applying one coat or two with the same paint, unpigmented and thinned. For the best job the base coat is unpigmented and diluted with three parts of thinner to one part of paint, and the second coat unpigmented and diluted one part of thinner to one part of paint. This is followed by full-strength pigmented material in the desired number of coats. The thinned paint vehicle, however, is not a substitute for inhibitive primer, which as mentioned before should always be used where the job is subject to severe moisture conditions.

Examples of spray-metallized zinc as a base for paint are shown in Figures 9-63 and 9-64.

Sherardizing

C. T. Flachbarth, *Engineer*

and

W. H. Taylor, *Advertising Manager*
Walker Brothers
Conshohocken, Pa.

Sherardizing was discovered accidentally in the year 1900 by an Englishman, Sherard Cowper-Coles. It is a process used to apply a zinc coating to metal for the purpose of providing protection against corrosion. Sherardizing is a process of sublimation, alloying, occlusion and adhesion. The process is well suited for zinc-coating articles such as tubing, conduit, bolts, nuts and castings. The sherardizing process is employed in many plants under different conditions for varied purposes. It is generally regarded as a method of galvanizing, along with electrogalvanizing and hot dipping.

Description of Process

Sherardizing requires, first of all, clean metal surfaces. Manufacturers of sherardized steel tubing and rigid steel conduit generally employ a pickling operation, followed by cold and hot water rinses. The cold rinse is performed in continuously running water to remove all traces of the pickling solution. This leaves the metal clean. Since surfaces dry better when they are hot, the hot rinse is performed last to promote rapid drying and so reduce the possibility of oxidation before the sherardizing operation begins.

After cleaning, the articles are placed in a metal drum which may serve as the shell of a furnace if heating is done electrically. Metallic zinc dust is added to the drum. This zinc dust is exceedingly fine in particle size. For large or irregular work, sand is generally added to improve the uniformity of zinc distribution throughout the drum. The drum must be sealed to prevent escape of zinc vapor. The drum is then heated and slowly rotated. The work and zinc dust together should not completely fill the drum, so that a slight tumbling action will occur as the drum rotates. The tumbling helps produce a uniform coating. The heated drum is allowed to soak at an elevated temperature for a period of time. Sherardizing continues during cooling, until the temperature falls below a minimum point. Cooling is done slowly to prevent loss from exposing hot zinc dust to the atmosphere. Time and temperature relationships depend upon the character and quality of zinc dust used and on the physical properties required of the sherardized

Figure 9-65. Drum with head removed showing sherardized ¾-inch rigid steel conduit. The conduit is evenly coated with zinc and is of a solid gray color.

material. The temperature which is best for both zinc dust and product must be determined for each specific condition. Generally, heating is done for several hours at 700 to 800°F.

Structure of Coating

A sherardized coating has excellent resistance to cracking, splitting or flaking when the sherardized article is bent or worked. This is one of the advantages inherent in the process. When an iron or steel article is sherardized, the zinc alloys with the iron at the elevated temperature. Thus the protection gained exceeds that of an equivalent coating of zinc alone. The outer surface is almost pure zinc. The region just under the surface is a zinc-rich, zinc-iron alloy. Further within the workpiece, the composition blends gradually to an iron-rich alloy; at deeper penetration, a few thousandths of an inch within the article, the original composition remains unaffected. If the conditions of sherardizing are kept constant, a uniform coating can be produced at a controlled thickness of from 0.1 ounce of zinc per square foot of surface to about 0.4 ounce per square foot. Lighter coatings do not offer as good protection. Heavier coatings tend to lose their superficial zinc when they are bent or struck sharply, indicating superfluous zinc usage. In such a case, the real sherardized surface under the superfluous zinc gives all the corrosion resistance possible.

Zinc-Rich Paints

G. W. Ashman

Assistant to Manager of Pigment Division, The New Jersey Zinc Company New York, N. Y.

The bulk of this section of Chapter 9 on Zinc Coatings is concerned with coatings of metallic zinc either in commercially pure form or alloyed with minor quantities of other metallic components. They are applied either by thermal processing or by electrodeposition.

Zinc-rich paint is another metallic protective coating of wide acceptance and growing commercial importance. It is formulated with zinc dust* to give dry films containing high concentrations (85 to 95 per cent by weight) of the zinc powder. This coating is rich in metallic zinc that is in intimate contact with the iron or steel, and it appears to give sacrificial electrochemical or cathodic protection to the basis metal, such as occurs on galvanized

* The conventional zinc-dust paints contain nonmetallic pigments such as zinc oxide and iron oxide in considerable quantity and cannot be classified as zinc coatings. For example, the 80 per cent zinc dust-20 per cent zinc oxide pigment mixture known as "Metallic Zinc Paint" is in common use. See also the section of this chapter on Miscellaneous Processes and Applications.

products. Also, like galvanizing, the zinc-rich paint film is electrically conductive. This dual property of direct metal to metallic zinc contact and good electrical conductivity of the dry film distinguishes zinc-rich paint from other rust-inhibitive paints.

Published results of large-scale commercial applications report excellent long-time metal protection in atmospheric or underwater service for a number of types of zinc-rich coatings. Technical papers on laboratory investigations and service evaluations of newer types of zinc-rich formulas indicate the widespread and growing interest in the exceptional qualities of this unique metal-protective paint.

Zinc dust was used as a pigment in organic coatings for the protection of ferrous metals as early as 1840.[21] Since then it has been used with a variety of organic and inorganic vehicles in metal primers and metal-protective paints. For several decades certain organic vehicle zinc-dust formulas, such as the heat-resistant types, have been applied at conventional pigment concentrations(50 to 65 per cent by weight in the air-dried film) but were converted to zinc-rich coatings during use on heated surfaces. This was accomplished when the binder was baked or partially burned out by heat from the metal it was protecting. The resulting shrinkage and loss of weight of the organic binder proportionally increased the weight and volume of the pigment component, thus producing a zinc-rich type coating. The converted film is dense, firmly adherent and gives excellent service, even with repeated heating and cooling of the metal.*

Investigations made by the University of Cambridge gave impetus during World War II to the use of zinc-rich coatings. After extensive research, laboratory and field tests, these investigators developed zinc-rich paints that have received wide commercial acceptance in the protection of iron and steel. Published results[22] of their work show that such coatings provide excellent protection on new or rusty iron and steel surfaces, under atmospheric exposure or in fresh or salt-water service.

The Cambridge investigators, under the sponsorship of the British Iron and Steel Research Association, are still actively engaged in studying and reporting the properties of zinc-rich coatings in a wide variety of applications. The accepted value of their work is attested by the adoption by the British Admiralty of a specification covering the use of zinc-rich paint as a rust-inhibitive primer on the exterior of small vessels. Another example is the successful service still being obtained on a 223-mile pipeline in Australia that was coated in 1942 with a zinc-dust sodium silicate application. Figure

* Present day heat-resistant paints, such as the silicone resin-zinc dust or dibutyl titanate-zinc dust types that are used on rocket launchers, are capable of withstanding considerably higher temperatures than the alkyd resin-oil-zinc dust formulas discussed above.

Figure 9-66. Pipes, structural steel and tanks of this chemical plant are coated with inorganic zinc-rich primer and vinyl top coat for maximum abrasion and corrosion resistance. (*Courtesy Amercoat Corporation*)

9-66 shows an application of inorganic zinc-rich primer to various components of a chemical plant.

To prove the outstanding electrochemical or cathodic protection provided by zinc-rich coatings on iron or steel exposed to corrosive atmospheres, the Cambridge investigators stressed two characteristics: (a) the ability to prevent corrosion of scratch marks and (b) the resistance to "rust creep" under the paint film. Their tests show that when pickled steel was coated with zinc-rich paint, then scratch-marked and partially immersed in sea water (a very severe test), no rust was visible at the scratch after 20 months. Again, if the panel were not scratch-marked until after 15 months of partial immersion, rust formed at the scratch within 24 hours after reimmersion, but rusting soon ceased, owing to a stifling action of the metallic zinc, and it did not creep under the paint in the vicinity of the scratch line.

In another series of tests, they used rusty steel that had been weathered on a roof for over a year. The loose rust was removed by a mechanical wire brush just before applying the zinc-rich paint. After 1 or 2 days' partial immersion in sea water, the panels acquired the same potential of zinc (determined by careful electrical measurements) as in the case of pickled steel coated with zinc-rich paint, and throughout the 15-month test period slight corrosion took place at only one or two places. The investigators explained that this shows the capability of zinc-rich paint to make metallic contact with the underlying metal through a thin layer of rust.

Mayne, one of the early workers in these investigations, offers further results on the subject in his 1954 report.[23] Also active were Pass[24] and Blowes, Meason and Pass.[25] For further information on the history and development of zinc-rich paint, reference may be made to a survey of literature, 1939–1954, on "The Use of Zinc Dust in Paint" that is given in a comprehensive bibliography published in May, 1954 by the Zinc Development Association, Oxford, England.

The use of metal anodes to effect galvanic protection of buried iron or steel pipelines, or of tanks, pilings, ships and other structures subject to rusting is discussed in a subsequent section of this chapter. Evidence of similar cathodic protection of iron or steel surfaces by zinc-rich coatings, as reported in the literature, has been cited above. To be effective, the anodes must be electrically connected with the unit that is to be preserved from corrosion. With zinc anodes, a ratio of about 400 square feet of ferrous surface to one square foot of zinc-anode surface is commonly accepted. In a zinc-rich paint, electrical contact with the basis metal must also be maintained to secure cathodic protection. The ratio, however, of ferrous surface to metallic zinc surface is strongly reversed. Because of the large surface area presented by the extremely finely divided zinc-dust pigment, and because of the thickness of the zinc-rich coating, several hundred square feet of surface area of metallic zinc may be providing cathodic protection for each square foot of ferrous surface that is being protected. Of course, to be a truly cathodic protective system, the zinc-rich paint film must not only be in good electrical contact with the basis metal but must also be an electrically conductive film.

With anodes used for cathodic protection, the highest effective electrical protection is needed; hence, the advantage of the added potential of high-purity zinc is evident. With zinc-rich coatings, no experimental evidence has yet been reported to indicate an advantage for the added potential obtained from high-purity zinc dust. Possibly, one explanation may be the very large surface area of an adequately thick zinc-rich coating. Conversely, no evidence has been published to indicate that metallic impurities in zinc dust are not a disadvantage. Certainly, one might expect localized galvanic action in the area of the metallic impurities, accompanied by a wasteful dissipation of the zinc dust involved; wasteful, first, in respect to the cathodic protection afforded the iron and steel surfaces that are to be protected and, second, with regard to the possibility of creating voids and unnecessary porosity in the paint coating. The need for experimental study on this subject is evident.

Considering the effects of the particle size of the zinc dust, it would appear that an extremely fine pigment would not be needed to provide adequate surface area for cathodic protection. There is a possibility, however, that a critical relationship exists between average particle size of the pigment and the concentration of zinc dust in the dry paint film, or between the average particle size and the optimum thickness of the zinc-rich coating. Laboratory studies reported by Blowes, Meason and Pass[25] indicate decidedly different rates of settling in the liquid paint of several zinc dusts, where the average particle size ranged between 10 and 3.4 microns. Comments were offered also on zinc dust of 2 to 3-micron size. The British investigators have also reported[25] that the settling rates of different particle sizes of zinc dust varied rather widely in three different types of organic vehicles. In this field of effects of particle size, the need for experimental study is evident.

Today, zinc-rich paint is commercially available with either organic or inorganic binders, together with suitable solvents or thinners that allow its application by brush, spray, dipping or other conventional painting methods. Most of the organic-binder formulas are of the very rapid-drying type and the inorganic binders are either self-curing, heat-curing or are separately treated with chemical or other curing agents. Both the organic and the inorganic types may be formulated with such rapid film set-up that additional paint applications may be made in a short time and, if desirable, multiple applications may all be made the same day. Despite the high content of zinc dust, the selected binders have sufficient strength to hold the active pigment portion and yet provide good flexibility and adhesion to iron and steel. This is due to the high weight, but relatively low bulk, of the zinc-dust pigment. Zinc-rich paints customarily range between 85 and 95 per cent by weight in the dry film (depending upon the type of vehicle used and the intended service), but with the low bulking of the zinc dust, the dry films contain 35 to 50 per cent binder by volume. It is the relatively large volume of the binder that provides the necessary film strength and adhesion, and pigment-holding properties.

Among the commonly used organic binders are chlorinated rubber, isomerized rubber, chlorinated paraffin, plasticized polystyrene, polyvinyl acetate; acrylic, epoxy, epoxy ester, alkyd and phenolic synthetic resins. The most commonly used inorganic binder is sodium silicate. Other types of binders have been proposed or are being used. The list of the commonly used types, however, is extensive and many arguments pro and con have been offered by the respective advocates. Because of the complexity of advantages and disadvantages of each, and because many of the newer formulas represent advanced developments that are "stock in trade" of the commercial zinc-rich paint producers, no attempt will be made here to analyze the effectiveness of the respective types.

In preparation of the ferrous surface for painting, the producers of inorganic-type coatings recommend sand blasting of bright metal before painting. With the organic types, a less exacting surface preparation is frequently acceptable, including application over tightly adhering mill scale. Much laboratory study and service evaluation has been devoted to the several types, but because of the many evident advantages indicated for zinc-rich paint, still more study and development are needed. Indeed, an increasing amount of work is now in progress, and as new developments or new commercial products are announced, further investigations are to be expected.

The observation that "paint is a compromise of many properties" is especially true for zinc-rich paint. The wide range in effective types of binders has been indicated above, and from these a variety of satisfactory formulations may be made. As with other paints, it may be necessary to compromise with certain properties or characteristics to obtain maximum effectiveness in other specific conditions or properties. Zinc-rich paint is a specialty and, in general, specialties are expensive. To achieve the required properties, costly paint ingredients and surface preparation are frequently used. The result is that zinc-rich paint, as initially applied, is usually an expensive coating. This, however, means economy in terms of extended service, but it also means that this specialty should be used only when less expensive but normally good metal-protective paints will not provide the desired service or prolonged metal protection.

The determination of the maximum ratio of zinc-dust concentration in the dry film is receiving considerable study. Originally, the consensus favored approximately 95 per cent by weight, and this ratio still finds staunch support, both for organic and inorganic binders. For better application, and because of the complexity of the requirements that paint coatings must meet, the final formulations now usually represent a compromise (previously mentioned) between the maximum amount of binder that the system can tolerate and the maximum zinc dust for good electrical conductivity in the dry film. For these reasons, the optimum zinc-dust concentrations in the zinc-rich systems are of much practical importance. Both Elm[26] and Nantz[27] have reported results of their studies on the subject, and in both cases they have worked with 100 per cent zinc-dust pigmentation.

In investigating cathodic protection by a paint system, Elm[26] has measured the concentration of zinc dust necessary to provide electrical contact with the substrate and to maintain good electrical conductivity within the film. He finds that maximum conductivity (minimum electrical resistance) is obtained at about 85 per cent zinc-dust concentration, as shown in Table 9-18.

At 85 per cent zinc dust by weight in the dry film, the pigment volume

TABLE 9-18. EFFECT OF ZINC-DUST CONCENTRATION ON COATING RESISTIVITY

Pigment Concentration (% by wt)	Resistivity (100 Ohms) per sq in.
95	12.7
90	4.0
85	2.5
80	10.5
75	290
70	1,900
65	11,000
clear vehicle	150,000

concentration in the dry film is about 48 per cent. As a matter of interest, the standard 80-20 zinc dust-zinc oxide, raw linseed oil paint, designed to meet Federal Specification TT-P-641b, Type II, has 35 per cent pvc and the heat resistant zinc-dust paint has about 43 per cent pvc in the dry, unheated film.

Nantz, studying zinc-rich films at the Wright Air Development Center, has given the results of a comprehensive investigation.[27] The studies report tests with several types of organic vehicles on concentrations of zinc dust ranging from 65 per cent upward. Measurements of conductivity are cited for films dried from 2 hours to 20 days, and at dry film thicknesses ranging from 2.5 to 14.5 mils. He suggests a film thickness of 10 mils minimum for optimum conductivity. A major achievement of the investigation was the development of a simple method of making reproducible tests on conductivity of the dry film.

A wide range in average dry-film electrical resistance is reported for specific vehicles with 85 per cent zinc-dust concentration (in dry film). For example, Nantz notes: "(a) nitrocellulose lacquer, too high to read on ohmmeter (1×10^6 plus, ohms); (b) (a proprietary) lacquer, 25,000 ohms; (c) "Epon" vehicle, 80,000 ohms; and (d) alkyd vehicle, 170,000 ohms."

Commenting on the findings reported by Blowes *et al.*[25] that a concentration of 90 to 95 per cent zinc by weight is necessary for proper conductivity through a coating, Nantz states:

> "In our tests on a proprietary material, it was found that the best conductivity was obtained at a zinc concentration of 75 to 85 per cent zinc. It is believed that the big reason for this disagreement lies in the difference in vehicle used. The vehicle in our tests was of the phenolic drying oil type where J. Blowes *et al*. report using styrene, vinyl and rubber vehicles."

In subsequent discussion, Nantz notes that minimum resistance was obtained at a zinc concentration of 83 per cent, which agrees with Elm's finding of about 85 per cent concentration for minimum resistance.

As a result of the United States Air Force studies, a military specification for zinc-rich coatings has been issued. This is MIL-P-26915 (USAF), dated 12 June 1956, and entitled "Primer Coating, Zinc Dust Pigmented, for Steel Surfaces." Two types of formulas are specified, namely: Type I, air-dry cure (organic binder); and Type II, bake cure (inorganic binder). Type I specifies that a spray coating of primer at least 2.0 mils in dry film thickness shall dry dust-free in not more than 10 minutes and shall dry through in not more than 1 hour. For Type II it is specified that a similar sprayed dry film shall dry dust-free in not more than 20 minutes, and shall be fully cured after being air dried for 20 minutes and baked for a maximum of 1 hour at 225°F.

The United States Navy has also developed a military specification for zinc-rich paint for application on galvanized iron, namely MIL-P-21035 (SHIPS), dated 23 August 1957 and entitled "Paint, High Zinc Dust Content, Galvanizing Repair." They find it particularly effective for coating rusted areas of galvanized metal.

In their 1954 report, Blowes, Meason and Pass[25] consider the dilution of zinc dust with other pigments, including zinc oxide, and find that such materials will, as expected, increase the electrical resistance of the film. Nevertheless, they state:

> "In practice, however, provided total pigmentation is maintained at a high level, it would seem that appreciable quantities of zinc oxide can be present and cathodic protection of a steel surface still be produced."

The optimum content of zinc dust in a zinc-rich coating is primarily dependent upon the type of formula. The optimum ratio may vary with different types of binder and with the extent to which the zinc dust is replaced by nonconductive pigments. Such other nonmetallic pigments are added in some instances to improve application properties or to change the film color, e.g., zinc-oxide or mineral pigments. It is not the purpose of this article to discuss the variety of pigment combinations applicable in zinc-rich coatings, or to cite their advantages or disadvantages. It is merely indicated in passing that the pigmentation is not necessarily limited to 100 per cent zinc dust, with certain types of binder.

From the above, it is evident that a wide variety of formulations of good quality zinc-rich paint is possible. Many paint laboratories are intensively studying the subject, including several which are conducting exposure tests in various atmospheres, in cooperation with the American Zinc Institute. A rapidly growing number of commercial applications are in service and are being made (see for example, Figure 9-67). There are, at this writing, at least two dozen paint companies in this country producing zinc-rich formulas, and the number will probably increase rapidly.

The versatility of this so-called cold galvanizing system, both as to ease

Figure 9-67. Interior surfaces of this fresh-water surge tank and differential tube in a hydroelectric installation are protected against corrosion and abrasion by two coats of organic zinc-rich paint and a phenolic-resin barrier coat. Tank is 150 feet high by 50 feet in diameter. (*Courtesy Debevoise Company*)

of application and as to effectiveness of cathodic protection of iron or steel, makes it a most attractive coating. It may be applied as a primer or as a multicoat system. Where a barrier paint coat on the surface is needed, as in protection against excessive abrasion or extremely corrosive atmospheres, the zinc-rich film offers an excellent surface for adhesion of the top coating. The fast-drying properties of the zinc-rich applications offer many advantages. There is some difference of opinion as to the optimum minimum dry-film thickness, and justifiably so, since, in addition to differences in formulation, much depends on the uniformity of application and on the uniformity of the basis metal. In general, a thicker application is needed on sand-blasted areas, pitted surfaces or rusted surfaces than on a relatively smooth surface such as a cold-rolled steel panel. The guiding criterion should be a minimum dry-film thickness of at least 5 to 6 mils on the highest spots of the surface. As to recoating, the zinc-rich applications can be effectively cleaned and spot-primed, provided loose material and oil or grease are first properly removed.

REFERENCES

1. ASTM Standards 1958, Part I, Ferrous Metals, "Recommended Practice for Safeguarding against Warpage and Distortion During Hot-Dip Galvanizing of Steel Assemblies (Tentative)," A384-55T, pp. 1232–4, Philadelphia, Pa., Am. Soc. Testing Materials, 1958; *ibid.*, "Recommended Practice for Providing High Quality Zinc Coatings (Hot-Dip) on Assembled Products (Tentative)," A385-55T, pp. 1235–7; *ibid.*, "Tentative Specifications for Zinc Coating (Hot-Dip) on Assembled Steel Products," A386-55T, pp. 1238–40.
2. Evans, U. R., "An Introduction to Metallic Corrosion," Table I, London, Edward Arnold, Ltd., 1955.
3. Burns, R. M., and Bradley, W. W., "Protective Coatings for Metals," ACS Monograph Series, 2nd ed., New York, Reinhold Publishing Corp., 1955.
4. LaQue, F. L., and Cox, G. L., "Some Observations on the Potentials of Metals and Alloys in Sea Water," *Proc. Am. Soc., Testing Materials*, **40**, 670–87 (1940).
5. "How Zinc Controls Corrosion," New York, American Zinc Institute.
6. Daesen, J. R., "Galvanizing Handbook," New York, Reinhold Publishing Corp., 1946. An elementary guide.
7. Spowers, William H., Jr., "Hot Dip Galvanizing Practice," Cleveland, Ohio, Penton Publishing Co., 1947.
8. "Hot-Dip Galvanizing and Rust Prevention," Hot-Dip Galvanizers Association (Great Britain), Oxford, Alden Press, 1949.
9. Sisko, F. T., "Modern Metallurgy for Engineers," New York, Pitman Publishing Corp., 1941.
10. Frazier, K. S., "Hot-Dip Galvanizing," *Steel*, **134**, 102 (Feb. 22, 1954) and continuations, 98 (Mar. 1), 138 (Mar. 8).
11. Sendzimir, T., U. S. Patents 2,110,893 (Mar. 15, 1938), 2,136,957 (Nov. 15, 1938), and 2,197,622 (Apr. 16, 1940).
12. Ellis, O. B., and Oganowski, K., U. S. Patent 2,703,766 (Mar. 8, 1955).
13. Burns, R. H., (to Indiana Steel and Wire Co., Inc.), U. S. Patent 2,393,517.
14. Holmes, A. W., and Burns, R. H., (to Indiana Steel and Wire Co., Inc.), U. S. Patent 2,708,181.
15. Roberts, C. E., (Research Department, Bethlehem Steel Co.), U. S. Patent 2,754,174.
16. Wick, R. M., (Research Department, Bethlehem Steel Co.), U. S. Patent 2,392,175.
17. DeWitz, A., and Roy, F., (Research Department, Bethlehem Steel Co.), U. S. Patent 2,695,269.
18. Gray, A. G., "Modern Electroplating," New York, John Wiley and Sons, Inc., 1953.
19. Blum, William, and Hogaboom, George B., "Principles of Electroplating and Electroforming," 2d ed., New York, McGraw-Hill Book Co., Inc., 1949.
20. Graham, A. Kenneth, "Electroplating Engineering Handbook," New York, Reinhold Publishing Corp., 1955.
21. Mallet, R., "Second Report Upon the Action of Air and Water, Whether Fresh or Salt, Clear or Foul, and at Various Temperatures, Upon Cast Iron, Wrought Iron and Steel," *Brit. Assoc. Advance. Sci.*, **10**, 221–388 (1840).
22. Mayne, J. E. O., and Evans, U. R., "Protection by Paints Richly Pigmented with Zinc Dust," *Chemistry & Industry*, 109–10 (1944).
23. Mayne, J. E. O., "Effect of Rust and Environment on Inhibition by Zinc-Rich Paints," *J. Iron Steel Inst. (London)*, **176**, 140–3 (1954).

24. Pass, A., "Zinc Dust as a Protective Pigment," *J. Oil & Colour Chemists' Assoc.*, **35**, 241–61 (1952).
25. Blowes, J., Meason, M. J. F., and Pass, A., "Zinc Dust as a Protective Pigment," (Part II), *J. Oil & Colour Chemists' Assoc.*, **37**, 480–507 (1954).
26. Elm, A. C., "Reevaluation of the Function of Pigments in Paints," *Offic. Dig. Federation Paint & Varnish Production Clubs*, **29**, 336–70 (1957).
27. Nantz, Lt. D. S., "Conductivity of Zinc-Dust Pigmented Coatings," Technical Report, Wright Air Development Center, No. 57-185, pp. 1–23 (1957).

Rolled Zinc

R. K. MARTIN

Technical Director
Matthiessen and Hegeler Zinc Company
LaSalle, Illinois

Rolled zinc is produced as sheet, strip, plate, rod and wire in numerous compositions and alloys, depending upon the ultimate requirements of the rolled products. When desired, strip is slit into narrow, or ribbon, widths. Historically, the commercial grades of rolled zinc contain varying amounts of natural impurities; namely, lead, cadmium and iron. In recent years, the trend is to alloys of High Grade and Special High Grade zinc, using copper, magnesium, manganese, aluminum, chromium, and/or titanium as the alloying metals. This advance in alloyed rolled-zinc products has made necessary many changes in melting, casting and rolling practices.

Melting

Commercial mixes of zinc for the various types of rolled products are melted in reverberatory-type melting furnaces. These furnaces can be gas, oil or coal fired. For small volume production, pot-type furnaces with gas or oil over-fired heating can be used. The zinc-alloy type metals are more effectively melted and efficiently mixed in low-frequency induction furnaces, although high-frequency induction furnaces are used for certain alloys requiring critical control of their analysis. All types of furnaces must be refractory lined to eliminate iron pick up by the molten metal. The temperature of metal in furnaces is held to a range of 830 to 950°F depending on metal analysis and casting requirements.

Casting

Rolling slabs are cast in either open or closed book-type molds. Open molds generally have fins on the bottom for water cooling. Closed-type molds are air-cooled. Hot tops are generally used on open molds to allow release of trapped gases and to reduce shrinkage cavities on the top surface of the cast slab. Casting molds are constructed of cast iron and contact surfaces with the metal must be smooth and clean to allow unrestricted shrinkage of the cast slab. Mold lubricant is not necessary, but if used should be held to a minimum.

Casting temperatures vary with types of casting and metal analysis. The normal temperature range is from 830 to 950°F. Mold temperatures vary from 175 to 250°F depending on the type of mold and metal analysis. The pouring of slabs in both type molds must be at such a rate as to hold turbulence to a minimum and provide an even flow of metal across the bottom surface of the mold. Slabs cast in open molds must be skimmed immediately to remove surface oxide.

Rolling slabs are cast from ¾ to 4 inches thick. The thickness, width and length of the slabs are determined by the gage and size of finish-rolled products, and the capacity of the rolling equipment.

Rolling

Sheet Zinc. Temperatures of rolling slabs delivered to the slab roll range from 350 to 500°F, depending for the most part on the analysis of the metal being rolled. Reductions on the slab roll start at 10 per cent and are increased to approximately 30 per cent as the rolling progresses. This rate of reduction is controlled by metal analysis, roll shape and mill capacity. Normal practice is to break down two slabs at a time.

Slab-rolled material is cut into pack sheets. The length of the pack sheet is determined by the width at the finish roll. Packs are made up of from 2 to 40 sheets, depending upon the final gage requirements, and are rolled at 90° to the direction of rolling on the slab roll. The side of the pack sheet that was originally the bottom of the slab is called the "good side." This surface has a finer grain structure and fewer defects. It is necessary to have this face in proper position in the pack.

Packs are finished at starting temperatures of 350 to 450°F in the pack-rolling process. The stacked sheets are rolled as a unit in the finish-rolling operation. In rolling, the center sheets in the pack tend to stretch faster than outside sheets, making it necessary to split the pack. Splitting also helps to keep a uniform temperature throughout the pack. Uniform lengths of pack sheets are necessary for proper gage accuracy in the finished sheet. Packs are rolled at light pressures, with resulting loss in temperature. Temperatures of finished packs will vary from 175 to 250°F depending upon final thickness and metal analysis.

Rolling lubricants used in producing sheet zinc are mixtures of tallow and paraffin. Care must be taken to keep the lubricant at a minimum in order to obtain proper surface finish.

Where extreme flatness of sheet is required, sheets are stretcher-leveled before final inspection and shearing. The majority of zinc sheets produced for use as lithograph plates are surface-ground.

Strip Zinc. Strip slabs are heated to 350 to 450°F prior to slab rolling. Reductions on the slab roll vary from 8 to 30 per cent. Light passes are required at the start of the rolling cycle and reductions are increased as the metal is rolled. Reductions are controlled by the pressure required for proper flatness and also by the temperature build-up due to rolling friction. Temperatures of over 450°F are undesirable due to the possibility of annealing the higher-purity-analysis metals. This is especially critical for material that will be subjected to deep-drawing operations. Alloys with approximately 1 per cent copper content do not present this problem. Strip is rolled flat on the slab roll to a thickness of 0.125 to 0.200 inch depending upon the type of coiling equipment. It is then coiled and handled in coiled form through the balance of the rolling cycle.

Strip is finish rolled at starting temperatures of 125 to 350°F and reductions of 15 to 45 per cent, depending upon metal analysis, width of coil, type of finish rolling equipment, and final surface and grain requirements. Surface requirements of strip zinc range from bright plating quality to a dull satin-grey finish. This range of finish is obtained with highly polished rolls by rolling strip at temperatures in the upper range, and with satin-finished rolls by finishing strip in the low-temperature range. Physical characteristics are controlled primarily by metal analysis, but proper reduction and temperature control are necessary to assure uniform quality in the finished strip.

Rod and Wire. Rod and wire are produced from individually cast bars, as well as by the continuous casting method. The temperature of the bars entering the initial breakdown roll is 300 to 350°F for both types of bar. Cast bars vary from 3 by 3 to 4 by 4 inches in cross section and from six to ten feet in length. The rolls in the breakdown mill are grooved with alternating square and diamond-shaped openings. The bar is fed alternately into the shaped openings, with each pass reducing the cross section by 5 to 20 per cent. The rod is then rolled in a mill with round grooves. Bar stock is rolled to the proper diameter and is finished with shave dies on draw-benches to the proper finished diameter. Wire stock is rolled to the proper diameter and coiled on reels.

The bar from the continuous-casting machine has a modified triangular shape and is fed directly into a multistand rolling mill. This mill has alternate triangular and round-shaped rolls that reduce the rod to the proper diameter for coiling.

Coiled rod stock to be used in the production of wire is then run through drawing dies to the finished diameter. This may be done on single draw-benches using a dry soap lubricant on the die, or with a multihead drawing machine using a soapy water solution as lubricant and coolant.

Terrazzo shapes are rolled from coiled bar of the proper size on shape rolls that roughly form the desired shape. This shape is pulled through dies that produce the exact finished shape.

Properties and Specifications

Coils or sheets cut from strip zinc, pack-rolled zinc and zinc plates are specified with respect to permissable variations in thickness, width and length, and other marketing characteristics by the American Society for Testing Materials, Standard Specifications for Rolled Zinc, B69-39 (Reapproved, 1958), also approved as "American Standard" by the American Standards Association. Methods of testing and typical compositions are included for general reference. Conformable specifications are used by procurement agencies of the United States Government.

The American Zinc Institute supplies tabular data on wrought zinc alloys (reprinted from *Design News*, March 15, 1957), including tensile strength, elongation, Brinell hardness, density, melting point, specific heat, thermal expansion, thermal conductivity, impact strength and fatigue strength. Typical uses and fabricating processes are indicated.

A data sheet on "Typical Properties of Wrought Zinc Alloys" is issued by *Materials and Methods*,[6] and the "Metals Handbook" of the American Society for Metals also contains data (pp. 1090–2) on commercial rolled zinc and copper-hardened rolled zinc alloys (with or without a small addition of magnesium). The special alloys noted at the beginning of this section have not been standardized and are used as proprietary compositions by the various producers, who have developed their own procedures for improving the performance of wrought zinc in special applications.

Uses of Rolled Zinc

Rolled strip with a zinc content of 99.80 per cent or more provides proper drawing characteristics, if the iron content is held to a maximum of 0.014, for forming a wide range of finished products. The proper grade of metal in this range can be drawn into battery cans, mason jar caps, eyelets, flashlight reflectors, grommets, cosmetic cases, etc.

Strip zinc of lower purity (higher lead, cadmium and iron content) is used in the sides and bottoms of dry batteries, address plates, name plates, roof flashings and valleys, cable hangers, counterpoise strip and weather strip.

High Grade zinc with varying copper, manganese, chromium, and/or magnesium content, in strip form, provides a material with a wide range of temper and strength characteristics. Strip with lower alloy content is especially adapted for use as flashings, valleys, corner bead and other architectural uses.* The higher alloy range produces a strip with higher resili-

* The American Zinc Institute supplies a useful pamphlet on "Zinc for Architectural Uses."

ency and tensile strength that is used as weather strip, terrazzo strip, wall ties, moldings and downspouts.

Commercial-quality sheet zinc is used for lithograph plates, organ pipes, dry cell battery sides, bearing plates, paper plating, name plates and weather strip.

Alloys of High Grade zinc in sheet form are used for flashings, valleys, facia strips, gravel stops, gutters, organ pipes and casket shells.

Rolled plates of Special High Grade zinc with a maximum iron content of 0.0014 per cent are produced on the sheet slab roll for use as cathodic protection plates, boiler and hull plates. Zinc rod in various diameters is produced from metal of the same analysis for use in cathodic protection and boiler-water treatment.

Commercial-grade rods are produced in a wide range of diameters for use in making machined parts and in zinc coating (by tumbling at elevated temperature, as in sherardizing). Alloy rod is produced with 0.85 to 1.00 per cent copper content for stone anchors and machined parts.

Zinc wire with relatively high purity (99.95 to 99.98 per cent zinc content) is produced for metallizing, in diameters from 0.057 to 3/16 inch.

Terrazzo shapes are produced in a wide range of sizes for installation as the divider strip in marble terrazzo floors.

Spinning[8] is easily accomplished with rolled zinc and is a preferred method of forming where only a limited number of parts are required at minimum cost, or where certain favorable shapes are desired.

A few important applications of rolled zinc requiring specialized treatment, as in the printing industry, are described below in some detail.

Rolled Zinc for Photoengraving*

Lewis S. Somers, III

Superintendent, Photoengraving Zinc Manufacturing
Imperial Type Metal Company
Philadelphia, Pa.

The manufacture of photoengraving zinc begins with the spelter (virgin zinc) purchased from the refiner. The grades are especially picked for their low impurities. This material must be free from copper, tin, indium and aluminum. Some of these are present in most zinc ores, and careful control of the spelter, with respect to its character and origin, must be exercised. For two of the alloys manufactured the spelter must contain very small controlled amounts of iron to regulate the grain structure under elevated-temperature conditions in the final manufactured product. Each shipment of the spelter, once a supplier is established, is then very carefully analyzed before it goes into production.

* This section, through p. 528, by Mr. Somers.

The first operation in the manufacture of photoengraving zinc is the remelting of the spelter in electric induction furnaces. In these furnaces very accurate control of temperature and alloy is maintained. Here, the metal is alloyed with additions of cadmium, magnesium and manganese. These alloying elements are used in very minute quantities depending on the alloy and do not in general exceed two-tenths of one per cent in any case.

Once the metal is poured from these furnaces, the alloy, the width and the engraving surface are established. The molds are specially equipped to permit controlled cooling to give as perfect a slab, or bar of metal, as possible. These slabs are then inspected for defects. If passable, they are placed in an oven and held at a controlled temperature, depending on size and alloy, for rough rolling.

The slabs are progressively reduced from 2 inches to a thickness depending on the final gage which the finished product is to take. The edges of the resulting sheets are trimmed and the sheets cut into smaller sizes. Gage and quality are checked. After cooling to room temperature the material is cold-rolled on another mill. The rolls on this mill are of a high finish and the mill is built for very precise work. Sheets are accurately gaged as they leave this mill and are trimmed to their final width.

This material is then degreased and prepared for painting. The backs are painted with a special acid-, alkali-resistant material and then the sheets are baked and annealed at a temperature commensurate with the requirement for the particular alloy. After cooling, the sheets are ground in pairs on a wide-belt grinding machine.

These grinders use coated abrasive belts up to 50 inches in width. The face of the sheet is ground, washed, gauged and inspected. Defects found at this point can sometimes be corrected. If so, the material is corrected and reground; if not, the material is scrapped.

The next operation, polishing, is done on a similar machine. The coated abrasive belts used here are manufactured specifically for the producer of photoengraving zinc. The sheets are again cleaned and treated with a preservative to protect the finish from corrosion. They are then gaged and inspected. Sheets with defects at this point must be scrapped. The sheets are then sheared to smaller lengths, depending on customer requirements, inspected again, individually wrapped in tissue and kraft paper and placed in wooden shipping containers.

Because of the gage requirements and the perfection of surface finish necessary to the photoengraving industry, there is a large percentage of reject material throughout the operations. A quality-control laboratory is maintained to continually check all phases of the operation. Physical and chemical checks are taken at all points in the production cycle.

Rolled Zinc in Lithography. The modern commercial lithographic process is almost always referred to as "offset lithography." It is based on the principle that grease and water do not mix. This was the principle used by Alois Senefelder in Bavaria when he invented, in 1798, his third method of stone printing which he called "chemical printing." In this process, only the image to be printed takes the ink, with the non-image areas being protected by water. Prior to 1889, all lithography had been produced using flat press plates of stone, tile or copper.

In 1889, the rotary principle was applied to the lithographic press. The image was produced on a thin zinc sheet which was mounted on a cylinder. This direct-rotary press had its limitations in transferring true reproductions and, in 1904, the offset press was developed.

A third cylinder was incorporated into the rotary press, between the plate and impression cylinders. This cylinder was covered with a rubber-coated fabric blanket. The plate transferred its ink to the blanket which in turn transferred the image to the paper. This allowed the necessary fidelity in transfer of fine tones and color, and it was now possible to produce full-color copy in four colors. From this press has been developed today's modern offset presses for lithographing papers, metals, plastics, etc.

Rolled zinc sheets as offset press plates have played the major part in the development of the present high standard of offset lithography. Zinc lithograph plates are for the most part pack-rolled sheets. These sheets are designed by chemical composition and rolling technique to provide a press plate capable of meeting the precise demands of the modern offset press. The plate must have sufficient tensile strength to allow proper clamping without distortion of the plate cylinder. It must also have toughness to resist fatigue and clamp bend breakage.

Particular care must be taken to assure flatness and surface perfection. In recent years, a large percentage of zinc lithograph plates are being produced with a ground printing surface to assure better surface quality in subsequent processing.

Zinc offset press plates may have either a plain or grained surface. Considerable research has been carried out in recent years on the plain-surface zinc plate, but the grained-surface zinc plate must be considered standard for the industry.

The grained printing surface of a zinc plate is produced in a graining machine in which zinc plates are placed in such a manner as to cover the bottom of the machine, and are securely clamped with the printing face up. The plates are covered with steel balls, water and an abrasive. The graining tub is then oscillated in a horizontal plane to produce a scouring action on the printing surface of the plate, called "graining." The fineness, depth and sharpness of the grained surface is controlled by the size of ball, type and coarseness of abrasive, amount of water and speed of oscillation of the graining tub. After graining, the plates are removed from the graining machine and are thoroughly washed to remove all abrasive from the grained surface. They are then dried, using a minimum of heat in order to prevent recrystallization of the zinc plate. The finished, grained plate must be clean, uniform in grain depth and size, and free from scratches, in order to provide a satisfactory printing surface.

The grained surface of the plate provides a tooth for the printing image and ink, and enables the plate to carry the proper amount of water in the nonprinting areas. A wide range of grain characteristics can be produced on a zinc plate to fit specific requirements in offset lithography.

There are two basic types of zinc offset plates: namely, Surface Plates, and Deep Etch Plates. Copy to be lithographed by either type plate is photomechanically reproduced on film. This copy is divided into two categories: (1) line copy, which includes type proofs, pen and ink drawings or lettering and solid reverses, etc.; and (2) tone copy, which is photographed through the halftone screen to get the halftone dots of varying size to show tones of shading; this includes photos, wash drawings, paintings, etc. The film copy in either type is the exact size defining the image to be lithographed.

Films are positioned on a flat in such a manner as to fit the size of press plate required. This flat is placed in contact with a sensitized press plate in a vacuum frame and burned in with a carbon-arc lamp. The same procedure can be carried out in a step-and-repeat machine where one or several films can be positioned, burned in, then accurately repositioned and burned in for a number of cycles until the plate area is covered.

In making a Surface Plate the film is a "negative" and the exposed portion of the coating becomes the base for the ink-receptive image. The unexposed coating is removed during development and becomes the water-carrying area.

Deep Etch Plates are made from "positive" film where the exposed coating acts as a stencil for the image area. The coating on the image area is removed and this area is etched to a depth of 0.002 to 0.003 inch. Lacquer and ink are applied to the image areas to provide the ink-receptive surface and the non-image areas are cleaned of coatings to provide the water-carrying surface. The Deep Etch Plate will produce more impressions per plate than can be obtained from a Surface Plate.

If the image to be printed is in more than one color, color separations are made in the preparation of the positive or negative films, and plates are made for each color.

Offset presses are made for either single or multicolor. If color work is to be produced on a single-color press, the material being lithographed must be rerun through the press for each of the colors required. The same is true for the multicolor press if the number of colors required exceeds the color capacity of the press.

The completed press plate is carefully positioned on the plate cylinder of the press. This is especially important in color work, where register is most necessary. Adjustments are made to assure proper ink and water pickup as well as the "squeeze" contact with the rubber blanket on the transfer roll. This contact is two to four thousandths of an inch depending upon the type of press plate, blanket and material being lithographed. The image is transferred from the press plate to the rubber blanket and then, in turn, to the material on the impression cylinder using the same amount of "squeeze" as between the blanket and the press plate.

Zinc press plates fit snugly to the plate cylinder and in color lithography have sufficient "give" to assure perfect register. Finished lithographed material is then cut, folded, stamped, etc., depending on the type of finished product.

In many cases, after the completion of the press run, the zinc press plates are removed, cleaned, flattened and regrained. This makes the plate available for further press runs. It is not uncommon for zinc plates to be regrained from five to ten times. It is also common practice to store press plates at the completion of a run where a rerun is anticipated. The plate is cleaned, coated with asphaltum and stored for future use.

Rolled Zinc in Dry Batteries. The first practical dry cell was produced by Gassner in 1888. The term "dry cell" is only relative, as the contents of the cell are in pasty form compared to liquid in other type batteries. The first standard dry cells measured 2¼ by 6 inches. Early uses of this cell were in the telephone and auto-ignition field. Later developments produced the "A" and "B" batteries used in radio receiving sets.

These cells used a cylindrical shell of zinc. The inside surface of the zinc shell is covered with paper board. A carbon rod is placed in the center of the cell and the space between carbon and zinc shell is packed with a depolarizing paste that varies with the ultimate requirements of the finished battery. The specific compositions of this paste are withheld by the various battery manufacturers, but the composition is approximately ten parts each of manganese peroxide and carbon, or graphite, two of salammoniac and one of zinc chloride. The zinc chloride is a wetting agent to prevent loss of moisture in the water paste.

These shell-type batteries are now produced in sizes from the small penlite photo flash and flashlight sizes, to the standard dry cell. The smaller-type shells are in some instances combined in various-sized units and connected in series to produce 50 volts or more.

The other primary dry battery is known as the flat-cell type. This consists of flat zinc of the proper size, gage, surface finish and shape as the electrode. The zinc is coated on one side with pasted pulp-board and on the other side with waterproof, conducting plastic, the latter serving as the cathode terminal. The plastic-coated side is covered with a mix cake consisting of a suitable mixture of manganese dioxide, carbon and electrolyte. These units are stacked one on another until a sufficient number of cells are combined in a block to produce the desired battery characteristics. Electrical contact between cells is obtained through the coated zinc. No separate connections are required. These units are then encased and sealed as individual units, or combined in multiple units, depending upon the type of finished battery.

The original 2¼ by 6 inch dry-cell can was produced using sheet zinc for the battery side and bottom. In recent years most producers of this type battery have used strip zinc for the shell. The battery can of this size is formed and has soldered seams.

The standard flashlight battery was first produced with a soldered can using strip-rolled zinc for side and bottom. In recent years this type of battery and also the smaller types have used drawn zinc cans as the battery shell. The flat-cell type of battery uses strip-rolled zinc. Impact-extruded cans are also being used in and below the flashlight size.

The requirements for, and capacities of, dry cell batteries are ever-changing. Electronic developments constantly demand new types of, and better performance from, dry cell batteries.

Fabrication and Finishing of Zinc Articles

As noted previously, zinc strip of approximate High Grade purity and with low iron content is suitable for a wide range of drawing operations. A temperature somewhat above ordinary room temperature is desirable and lubrication with warm soapy water is generally satisfactory.[8, 10] Self-annealing between drawing operations permits successive operations and separate annealing must in general be avoided owing to the danger of coarse grain growth.

Thickness of the stock should be maintained and the reduction in diameter from blank to cup should be moderate (not more than 40 per cent[8]), with considerable decrease in this reduction during the successive drawing stages (not more than 20 per cent[8]). Sizable die clearances and generous fillets are required.

With zinc of lesser purity, or alloyed varieties, the forming characteristics must be especially evaluated in any desired operation. Some of the alloyed zincs strain-harden considerably and can be annealed between draws without excessive grain growth.

Bending is favored across the grain, i.e., around an axis normal to the strip length, and in all cases sharp bends must be avoided.

Rolled zinc is easily soldered with half-and-half solder and a cut muriatic acid flux. Overheating should be avoided by working rapidly with a moderately hot iron.

Rolled zinc may be buffed, painted, plated, lacquered, chemically colored and enameled to provide a large variety of finishes of diverse characteristics. The American Zinc Institute has a brochure on "Preparing and Painting Zinc Surfaces" and can supply information from many sources concerning current finishing techniques. Standard Specifications for an "Electrodeposited Coating of Nickel and Chromium on Zinc and Zinc-Base Alloys (B 142-58)" are issued by the American Society for Testing Materials. For comparison, the last part of the text on zinc-base die castings in a preceding section of this chapter, designated "Finishing of Zinc Die Castings" may be reviewed.

REFERENCES*

1. Anon., "Copper, Brass and Zinc Rolling," (Brinsdown Works of Enfield Rolling Mills, Ltd.), *Metal Ind.* (London), **84,** 421–2 (1954).
2. Anon., "Rolling of Zinc Strip," *ibid.*, **58,** 107–13 (1941).
3. Arend, A. G., "Characteristics of Zinc Plates for Lithographing," *Ind. Chemist*, **27,** 453–63 (1951).
4. Arend, A. G., "Preparation of Zinc Sheets for the Engraver," (Casting, rolling, grinding, polishing and inspection), *Process Engraver's Monthly*, **58,** 234 (1951).
5. Carlson, E. G., "Vapor Blasted Grain for Lithographic Plates," *National Lithographer*, **55,** 39, 106 (1948).
6. Editorial, "Wrought Zinc Alloys; Typical Properties," *Materials and Methods*, **32,** 91 (Oct., 1950).
7. England, F. E., "Grain of Lithographic Plates," *National Lithographer*, **58,** 28–9 (Dec., 1951).
8. Gent, E. V., "Spun and Drawn Zinc Parts Have Many Applications," *Materials and Methods*, **36,** 102–4 (Dec., 1952).
9. Hofmann, W., and Trautman, B., "Die Walztextur von Zink und ihr Einfluss auf die technologischen Eigenschaften, insbesondere die Tiefziehigkeit," *Z. Metallkunde*, **39,** 293–303 (1948).
10. Jevons, J. Dudley, "The Metallurgy of Deep Drawing and Pressing," pp. 285–90, New York, John Wiley and Sons, Inc., 1940.
11. Kelton, E. H., "Working of Zinc and Zinc Alloys," in "The Metals Handbook," pp. 1082–3, Cleveland, Ohio, American Society for Metals, 1948 edition.
12. Peirce, W. M., "The Rolling of Zinc," in "Nonferrous Rolling Practice," Institute of Metals Division, Symposium Series, No. 2, pp. 117–28, New York, Am. Inst. Mining Met. Engrs., 1948.
13. Roberts, C. W., and Walters, B., "The Rolling of Zinc and Zinc-rich Alloys," *J. Inst. Metals*, **76,** 557–80 (1949–50).
14. Sheridan, S. A., "Plate Metal and Grain Standardization," (Zinc and Aluminum), *Modern Lithography*, **14,** 30–5 (1946).

* There is no extensive bibliography on the rolling of zinc. These references will supply some material for collateral reading, including additional information on a few important applications. (Ed.)

Zinc Anodes for Cathodic Protection

T. J. Lennox, Jr.

Research Investigator, American Smelting and Refining Company
Central Research Laboratories
South Plainfield, N. J.

The corrosion of metallic structures buried in soils, in contact with sea water or other electrolytes, has long been a serious engineering and economic problem. The annual loss to the American pipeline industry alone, from actual destruction by corrosion and the cost of preventing corrosion, is estimated to be in the order of 600 million dollars.[2]

About 25 years ago, the most common method of reducing corrosion of buried structures was by the application of protective coatings. In the interim, the principle of cathodic protection was applied to buried structures and has rapidly come to the forefront as the dominant method for mitigating corrosion on these structures.

Fundamentally, cathodic protection consists of establishing an electromotive force on a submerged structure to make the entire structure cathodic. This may be accomplished by using sacrificial anodes such as zinc, magnesium or aluminum, or by using relatively inert anodes to impress an electromotive force from an outside source. Current will flow from these auxiliary anodes and will oppose galvanic currents on a structure to mitigate the corrosion process.

History

Marine Service. Cathodic protection can be traced back to 1824, when Sir Humphrey Davy[3] advised the Royal Society of his successful use of zinc protectors to prevent corrosion or chemical changes of copper sheeting on ship hulls exposed to sea water. In his early experiments Davy fastened small pieces of zinc to polished plates of copper and iron and placed them in sea water. He observed that not only was the copper preserved, but also the iron remained bright.

Davy even stated that when the metallic protector (zinc) ratio was from 1:40 to 1:150 there was no decay of the copper. Sheets of copper defended by 1:35 to 1:80 of their surface of zinc became coated with a white matter, which on analysis he claimed was principally carbonated lime and carbonate and hydrate of magnesia. His tests were conducted in Portsmouth Harbor both under stagnant conditions and on harbor boats.

Copper was placed on the bottom of wooden naval vessels to protect the wood from destruction by worms and prevent the adhesion of weeds, barnacles and other shell fish. Bimetallic couples were formed between the copper-tin nails and copper sheathing. The nails formed corrosive deposits which protected them, but adjacent copper corrosion was accelerated, the copper being worn into deep and irregular cavities.

In his first active ship experiment in March 1824 on the "Sammarang," cast iron equal in surface to about 1:80 of copper was applied in four masses; two near the stern and two on the bow. Inspection in January 1825, after a trip to Nova Scotia, showed that for a considerable space around protectors, at both stern and bow, the copper was bright. The color became green toward central parts of the ship; yet even here the corrosion was a light powder and evidently there had been little copper lost in the voyage.

Davy also stressed the need for remembering that the larger the ship, the more the experiment is influenced by the imperfect conducting power of the sea water, and consequently the proportion of protecting metal may be larger without being in excess.

With the advent of steel ships the principle of cathodic protection continued to be used. The area of most concern was the aft end where the bronze propeller is in contact with the ship's steel hull. Slabs of zinc were generally attached to ship hulls in a routine fashion. Zinc so attached gave random performance. In some cases the zinc anodes were consumed, while in other instances the zinc would have the same shape and size as when originally installed.

The most significant causes for the random performance in the early use of zinc anodes included (1) disregard for the quality of zinc, (2) poor attachment method, and (3) disregard for the need of not painting the zinc, so that it would work sacrificially. An erroneous criterion that developed in this early phase of cathodic protection was to consider zinc anode material most effective when it lasted longest.

Work done at the Naval Research Laboratory by May, Gordon and Schuldiner[5] indicated that pure zinc (99.99 per cent) developed the least resistance in its corrosion products. The pure metal also suffered least from electrode polarization. In this work, zinc anodes measuring ½ by 6 by 12 inches were mounted on a ship's hull near the stern. Exposures of four and six months showed that the pure zinc shed its corrosion products quite readily and was very clean upon removal. The impure zinc anodes were coated with hard, tenacious layers of corrosion products which reduced the extent of corrosion and lowered the anode current output.

Pipeline Service. The possibility of using zinc anodes on a large scale for the cathodic protection of pipelines was discussed by Rhodes.[9] One of the earliest successful installations of zinc on a pipeline was reported in a paper by Smith and Marshall.[10] This paper describes zinc installations beginning in 1935 on 6- and 18-inch welded lines.

A paper by Mudd[6] presents data on the performance of zinc anodes used to protect a pipeline cathodically. Of particular interest in this paper is the observation that zinc anodes installed in soils containing gypsum showed less tendency to become inactive than anodes installed in other soils of the region. Brockschmidt[1] described an installation of zinc anodes made in 1939 on an 18-inch pipeline located in severely corrosive soil.

Wahlquist[15] has stated that pure zinc continues to produce current over a period of several years when used to protect buried pipelines. His studies also led to the conclusion that backfills high in sulfate reduced the tendency of zinc anodes to form electrically resistant corrosion-product films.

The results reported in these papers provide encouraging evidence as to the practicability of using zinc under a wide range of conditions. They have also illustrated, however, the need for more intensive study of long-term behavior of zinc when used a as galvanic anode.

Recent Developments

Through the efforts of many investigators, the zinc industry produces zinc anodes that remain active for their entire life. Performance of the best zinc anode composition for marine use has been reported in a paper by Reichard and Lennox.[8]

Galvanic Series

The potential difference between metals in a given electrolyte will determine the direction of current flow. The magnitude of current is, however, the only true measure of rate of galvanic corrosion. If two metals listed in Table 9-15, p. 458, are connected electrically and immersed in a common electrolyte, the metal closer to the top is expected to be the anode (corroded) and the other, the cathode (protected). This and the other tables commonly found in corrosion literature should be used only as a general guide, however, since the metals will have characteristic potentials in a given electrolyte. The chart of Figure 9-34 p. 457 gives a potential series referred to sea water and similar solutions.

The corrosion current and hence the rate of metal loss will depend on the driving voltage, but will be limited by the resistance of the electrolyte and by the reaction products which accumulate on the anode and cathode surfaces as the result of current flow.

Current flowing from one metal to another in a common electrolyte will cause corrosion of the most active metal. This same principle, however, is used to mitigate corrosion.

Cathodic protection is the physical act of reversing the electrochemical force and the subsequent mitigation of the process which destroys metals. The theory behind the application of cathodic protection is relatively simple.

It means the application of a current of sufficient magnitude from an auxiliary anode which is immersed or buried in the same electrolyte. Current for protection is obtained from the most active metals, i.e., zinc, magnesium and aluminum. This current flow from the auxiliary anode opposes the galvanic currents on the structure to eliminate the loss of metal on the structure. In the process, the auxiliary anodes are consumed or sacrificed to protect the structure.

Soils and Waters

Electrochemical cells resulting from dissimilarities in the environemnt are regarded as a common cause of pitting on immersed or buried metals. The cells which cause potential differences on a structure may be caused by differences in electrolyte concentration, electrolyte composition and oxygen concentration.

In addition, differences in metal character caused by weld stresses or bimetallic couples such as new pipe connected to an old pipe, cause accelerated corrosion.

Failures as a result of corrosion may occur sooner on a well-coated structure than on a bare structure because corrosion current will be concentrated at holidays or when there are damaged areas in the coating.

It is usually relatively easy and less expensive to apply cathodic protection to a coated structure as compared to a bare structure. The combination of a good coating plus cathodic protection has proved successful on numerous structures.

On existing bare pipelines it may be more economical to use galvanic anodes for "hot spot protection." Hot spot protection is the placement of galvanic anodes in areas where more corrosive conditions exist or where anodic pipe sections have been located.

While corrosion by soils is a well-known fact, means of identifying corrosive soil are usually regarded as inadequate. The most common methods of identification are measurement of soil resistivity, observation of wetness or dampness, measurement of soil *p*H and presence of dense soil. None of these methods, however, give quantitative information as to the corrosiveness of a soil.

Corrosion and pitting of buried structures have also been identified with poorly aerated environments in which anaerobic bacteria flourish. The conditions which favor anaerobic bacterial corrosion are absence of oxygen and presence of moisture, sulfates and organic matter. Sulfate-reducing bacteria thriving in such environments reduce sulfates to sulfides.

The exact effect this type of corrosion plays in the current requirements for cathodic protection has yet to be determined. It has been stated[4] that the current requirement in polluted water in the presence of oxygen-producing and iron-reducing bacteria is double the normal.

The corrosion products can be tested for sulfide by placing them in dilute hydrochloric acid. The presence of hydrogen sulfide gas as indicated by the darkening of moistened lead acetate paper is an indication that bacterial corrosion has occurred. This test can only be conducted after the corrosion has occurred, however, and cannot be used to predict the corrosiveness of soils before laying a pipe.

Considerable work has been done by the American Gas Association in developing methods to detect corrosive soils. Starkey and Wright[11] have designed a "redox" probe for this purpose. Oxidation-reduction, or redox, is a measure of the tendency for a substance to give up or take electrons. The redox potential of an electrolyte is the potential difference between an inert metal electrode (such as platinum) and a reference electrode (such as calomel).

The potential difference measured depends only on the nature of the substances in the electrolyte and their concentration. The redox potential, however, expresses only the intensity of the reduction or oxidation and gives no evidence of the amounts of oxidized or reduced ions in the electrolyte.

To establish the feasibility of the use of the soil redox potential as an index of corrosivity of a soil, Starkey and Wright[11] measured the redox potentials in many soils along pipeline distribution systems. These results were correlated with severity of corrosion on the pipes and the resulting qualitative corrosion criteria were derived:

Range of Soil Redox Potential	Corrosiveness Classification
Below 100 mv	Severe
100 to 200 mv	Moderate
200 to 400 mv	Slight
Above 400 mv	Noncorrosive

The apparatus for making such measurements is not readily portable, however, and requires considerable revision to make it suitable for rapid field measurements.

In any event, redox measurements will only be an aid for the corrosion engineer and will furnish him information on one soil property. This measurement will supplement, though not supersede, existing measurements of soil resistivity, potentials of structures to environment, potential gradients and other measurements now necessary in corrosion control work. The interpretation of measured redox values should be left to the corrosion engineer who is familiar with field conditions where the measurements are made.

Stray Currents

Most stray-current problems originate in leaking rails of street railways or mines and in grounded d.c. circuits of manufacturing plants. The source of stray current in railway lines is the d.c. generator in the substation. The electric current flows from the positive terminal of the generator to the overhead trolley wire, through the electric motors of the cars to the rails and back by way of the rails to the negative terminal of the generator.

Since the rails are usually poorly insulated from the earth, a portion of the rail current leaks off into the earth in areas distant from the substation. This leakage current in its path toward the railway substation flows onto foreign buried metallic structures. It is carried by these structures to the vicinity of the substation where it is discharged to the earth to return to the negative terminal of the generator.

All current that flows on a buried structure does not produce corrosion. Only that portion of the current that flows from the structure into the environment causes corrosion. In the case of ferrous pipe, electrochemical corrosion occurs at the rate of 20 pounds of iron lost per year per ampere discharged.

Similarly, leakage currents from gounded d.c. circuits in manufacturing plants may flow along buried metallic pipes and cause electrochemical corrosion where they are discharged into the earth to return to the source of the current.

Stray-current corrosion can usually be controlled by use of mitigation bonds. That is, the structure which is suffering the stray-current corrosion should be bonded to the current source through a metallic conductor.

A few of the methods used to determine the existence of stray currents are listed below:

1. Correlation of pipe to soil potential with railroad to soil potentials.
2. Measurement of current flow on a pipeline or other buried structure.
3. Millivolt drop in earth with cathodic protection system off.

Galvanic Anodes *vs.* Impressed-Current Systems

Galvanic anodes as a class possess definite advantages over impressed-current cathodic protection systems where current requirements are small. Properly designed galvanic-anode systems require less maintenance and inspection, are free from power failures, and stray-current interference on foreign unprotected structures is largely eliminated. It is also impossible to hook up a galvanic-anode system improperly. All one has to do is electrically secure the galvanic anodes to a structure in a common electrolyte. The current can only flow from the anode to the cathode, protecting the latter. An impressed-current anode system connected in reversed polarity would cause accelerated corrosion of the structure to be protected.

Generally, zinc anodes can be used effectively for cathodic protection in electrolytes of low resistivity. Zinc is not ordinarily applicable for use in locations where high soil resistivity limits the current output to smaller values. As a general rule, it might be mentioned that for bare uncoated lines, zinc anodes should not be considered when the soil resistivity is above 3000 ohm-centimeter.

Zinc anodes have been used, however, to protect cathodically well-coated structures in soil of relatively high resistivity. On well-coated pipelines the resistivity of the pipe coating may be high enough that only small currents are required for protection. In addition, the high resistance of the coating, rather than the soil resistivity, would then control the current distribution.

Magnesium anodes are being effectively used to protect cathodically bare structures in environments with resistivities up to 10,000 ohm-centimeter. As in the case of zinc, the useful range of magnesium anodes is extended when the coating on the structure is of high enough resistance to be the controlling factor in current distribution.

Aluminum anodes of the 5 per cent-zinc variety may still be considered in the experimental stage. Data for zinc, magnesium and aluminum anodes are given in Table 9-19.

Usually, the choice of what type of cathodic protection system to use depends on economic factors. Where power is relatively cheap and available, impressed-current cathodic protection systems may be called for. Other important considerations are the effects a cathodic protection system will impose on a foreign structure. Interference problems are usually at a minimum when galvanic anodes are used, for the reason that galvanic anodes can be located much closer to a pipeline than impressed-current anodes. In addition, higher voltages are used with impressed-current cathodic protection systems than are obtained from a galvanic-anode system. These higher voltages can cause an increase in current requirements if paint is stripped from the structure.

For better current distribution, anodes should be spaced at a distance

TABLE 9-19. CHARACTERISTIC DATA FOR ZINC, MAGNESIUM AND ALUMINUM GALVANIC ANODES

	Zinc	Magnesium*	Aluminum
Theoretical amp hr per pound	372	1000	1352
Current efficiency (%)	90	50	56†
Actual amp hr per pound	335	500	750†
Pounds per amp per year	26	17	12
Solution potential sea water (sat $Cu\text{-}CuSO_4$) (volts)	1.10 (anodic)	1.62	1.02
Driving potential (volts)‡	0.2	0.72	0.12
Specific gravity	7.14	1.94	2.92
Pounds per cu ft	445	121	182

* Nominal composition 6% Al, 3% Zn, 0.2% Mn-Mg alloy.

† This value has been reported by some investigators, but insufficient data are available on the 5 per cent zinc-aluminum alloy.

‡ Working against a structure having a polarized potential of 0.9 volt to a saturated $Cu\text{-}CuSO_4$ half cell

from the structure to be protected. Economics may at times, however, justify installing anodes in the trench right along a pipeline when it is being constructed.

Not all pipelines lend themselves to cathodic protection by such simple straightforward methods. There are some urban and surburban areas where stray currents make it desirable to conduct field tests to determine the best corrosion mitigation method.

Zinc Anode Development

Effect of Composition. Teel and Anderson[12] reported the effect of iron on zinc anode performance in tests conducted at "INCO'S" Harbor Island Test Station under the auspices of the American Zinc Institute. Their work showed that zinc anodes performed well when the iron content was kept below 0.0015 per cent. The Navy has since adopted a 0.0014 per cent iron-content zinc anode (MIL-A-0018001D) for their use.

Reichard and Lennox[8] have shown that merely controlling the iron content of zinc anodes at 0.0014 per cent is not sufficient to obtain maximum performance. They report best anode performance when aluminum and cadmium are added to the zinc. Their data were obtained on commercial-size zinc anodes operating in sea water, under quiescent conditions as well as on active ship hulls.

Insulating Barrier. In the tests conducted by Reichard and Lennox[8] it was found that an insulating material on the anode's faying surface is not required when zinc anodes are mounted on a painted steel hull.

Mounting Technique. Anodes should not be mounted by drilling through the zinc. Accelerated local anode attrition will then occur with premature anode failure, usually resulting in the anode becoming detached from the ship's hull. Brackets should be cast into the anode and should be either welded directly to the hull or mounted over studs welded to the hull. When anodes are mounted over studs, lock washers should be used.

Economics again play a deciding role with respect to the mounting technique used. Some marine operators feel that it is less expensive to chip off straps from "welded expended anodes" and reweld the new anode brackets directly to the hull.

Application and Design

Soil. One of the most outstanding uses of zinc has been in soil installations in the Houston, Texas, area. The cumulative leak record on a zinc anode, protected old line showed that five corrosion leaks occurred in the 7-year period from 1944 to 1951, compared with 140 leaks in the period from 1932 to 1944, prior to the zinc installation. No additional leaks have occurred to date.

The American Zinc Institute has recently published "Zinc as a Galvanic Anode—Underground." This brochure covers uses of zinc anodes on transmission pipelines, gas distribution systems, transmission line towers, ground rods, grounding cells and domestic oil tanks.

Where cinders are used for backfill on a pipeline, they tend to penetrate pipe coatings and produce severe galvanic corrosion. This type of attack cannot be completely stopped by cathodic protection since the cinders electrically shield the pipe from cathodic protection currents. It is important, therefore, to make sure that cinders are not used for backfilling around any metallic structure.

Various methods can be used to determine the number of anodes required for protection. One involves obtaining representative soil-resistivity data at pipe depth and at depths of 10 to 20 feet. Current requirements for the structure can be estimated when the total surface area of the structure is known. The total number of anodes required can be calculated when the current output of an anode of particular size and shape is known for a given electrolyte.

An alternate and better method is to obtain current requirements from field tests made after the pipe has been backfilled. Current requirements for soil application depend on the quality of coating, soil resisitvity, oxygen permeability of the soil, amount of rainfall and climatic conditions. The current required for protection may be determined by tests of isolated sections. When the average current requirements of representative sections have been determined, these values usually may be used for preliminary design of the remainder of the system. With modern coatings of good quality, a test current that quickly produces a change in potential of 0.3 volt usually is sufficient for protection. When current requirements are determined from field tests, it is not necessary to measure soil resistivity except for the purpose of selecting locations for the anodes.

When installing zinc anodes in soil, the backfill material is normally installed dry around the anode, in an augered hole dug for the purpose. The backfill is thoroughly tamped during installation in order to avoid the possibility of any unfilled cavities at points surrounding the zinc anodes. It is installed dry so that, as it takes up moisture from the surrounding soil, the resulting slight swelling will completely fill the space between zinc anode and augered-hole walls. If the backfill material were mixed with water prior to installing in the hole, immediate current output would be obtained from the anode; however, there would be a possibility of slight shrinkage of the backfill with time which would result in somewhat poorer long-term performance than if it were installed dry and allowed to take up moisture from the soil. A commonly used backfill, in which zinc anodes are buried to insure good performance in a variety of soils, consists of 50 per cent ground gypsum and 50 per cent bentonite clay.

Zinc anodes are particularly well suited for use in applying cathodic protection in congested areas. Interference effects on piping or cable systems will be at a minimum when zinc is used.

One of the most important features when employing cathodic protection systems in congested areas is to insulate the system to be protected from other underground metallic structures. Otherwise, excessive current drain may occur on the anodes and the structure to be protected will receive insufficient current to achieve protection.

Chemical spillage at industrial plants or highly acidic mine effluents can greatly increase local cathodic-protection requirements. Cathodic-protection designs for these areas should, therefore, be based on field tests.

Regardless of the method employed to provide cathodic protection, it is usually advantageous to locate anodes in the lowest resistivity soil that can be found within a reasonable distance from the structure. Anodes should be placed in locations favorable to moisture retention by the backfill. Points of low elevation in the proximity of streams or ditches should be selected in preference to high ground, steep slopes or well drained areas.

It is desirable to have augered holes as deep as possible so that anodes will be less affected by seasonal variation in moisture. In general, anodes in augered holes should be installed at a depth that will provide 4 feet of earth above the backfill.

After installation of the anodes, a period of several weeks should be allowed for conditions to stabilize. A pipe-to-soil potential survey may be indicated, to determine adequacy of protection. Since there always exists the possibility of mechanical coating damage or accidental contact with unprotected piping, periodic checks of pipe potentials are usually required to assure success of a cathodic protection system.

Marine Uses. Zinc is an ideal metal for cathodic protection of ships in sea water because painted surfaces in the immediate area of the anodes are not subject to damage from high potentials. These high potentials, which are most common with impressed-current cathodic protection systems, may be injurious to commonly used paints. Nevertheless, good marine paints, sound application practices and reasonable maintenance are, of course, necessary.

As mentioned earlier, the use of zinc anodes for cathodic protection of stern-frame areas is not a new subject. Numerous designs have been proposed to protect cathodically either stern-frame areas or the entire ship, and are in wide use.

In designing cathodic protection systems for the exterior hulls of active ships, the total underwater area should be calculated. A cathodic-protection design system can be based on this total underwater area for a certain type of coating system or can be based on this underwater area after assuming that a certain percentage of the steel is bare of coating. Either of these design methods can be used and the approximate number of anodes determined, when the current output of a galvanic anode and the cathode current density required for protection are known. Anodes should be located to obtain maximum current distribution in line with good operating practices.

The areas where the hull will be bared of coating cannot be determined in advance. It has been the author's experience that with the varied conditions that exist among operating ships, a design which would appear adequate in one instance would be entirely inadequate in another. Therefore, it can be said that any initial design may have to be altered after experience has been obtained on a particular vessel or structure. One observation that deserves mention is that service and commercial ships may well be separated into different classes for cathodic protection purposes.[4, 13]

Naval vessels usually have heavier applied coatings compared to those observed on commercial vessels. This is not meant to infer that paints used by commercial operators are inferior to those used by the Navy, or vice versa. It is recognized that economics play a big part in determining the length of time a vessel can be dry-docked for repairs and maintenance.

As an example, the author has observed a number of steel tugs on which approximately thirty 24-pound 6 by 12 by 1¼-inch zinc anodes were attached to each vessel. On three such tugs, corrosion mitigation was highly successful for periods up to sixteen months. In other instances, with the same design, the tugs were not in as good condition. One of the obvious reasons for this difference in corrosion-mitigation appearance was the difference in the coating condition.

Although differences exist between the various types of paints used on a hull, care in application and operating conditions could more than offset differences which might exist between two paint systems. Paint will generally be removed on underwater hulls that pass over sand bars or are used to break ice during their operations. Normally, general deterioration of a coating system is severe. This would naturally increase the bare steel area and the current requirements for cathodic protection.

Steel H-piles and sheet pile structures in sea water usually show the most severe corrosion in the splash zone, i.e., the area just above mean high tide. Metal cladding and premium-quality coatings may be used to control the excessive corrosion which occurs in this zone.

The rate of corrosion on the continuously submerged areas of H-piles is dependent upon a number of factors among which are water temperature, velocity of water currents, amount and distribution of dissolved oxygen in the water and underlying mud, salinity of the water, pollution and the presence of marine life such as slime.

The corrosion occurring below the water line can be effectively controlled by the use of cathodic protection.

In the case of oil-well drilling platforms and petroleum loading piers, there is a possibility that some of the cathodic-protection current will be lost to the drill casing or tank form. This condition can be eliminated by the proper use of insulating fittings which serve to isolate the piling from other extensive grounded structures.

The American Zinc Institute has recently published "Zinc as a Galvanic Anode—Underwater." This brochure covers uses of zinc anodes on marine craft, H-piling in sea water, traveling water screens, spillway and lock gates and water storage tanks.

Zinc Anode Life

The life of zinc anodes will depend on the current discharged from them and the efficiency at which they operate. The current output will depend on the circuit resistance and driving potential, which are both characteristics of the environment in which the anodes may be installed.

Zinc anodes can be expected to have an operating efficiency of 90 per cent.This means that during its life 90 per cent of the zinc consumed will be used in supplying useful current, while the remaining 10 per cent will be used in the self-corrosion process. The relationship between zinc-anode weight, current discharge and life expectancy is given by the formula:

$$\text{Life in years} = \frac{(0.0425)\ (\text{anode weight-pounds})\ (0.9)}{\text{current output-amperes}}$$

Effect of Current Density on Galvanic-Anode Performance

It can be affirmed at the outset that there is probably a current density and type of electrolyte in which the performance of any anode material used for cathodic protection will be best. Current density and electrolyte composition may affect the potential, current-output characteristics, efficiency or corrosion characteristics of the anode.

Zinc Anodes. Current density at the zinc-anode surface will have an important effect on the long-time behavior of the anode. In low-resistance soils, with the anode operating at a high current density, the film formation may be too rapid to be leached by soil waters. On the other hand, the general reaction at the anode surface would tend to be an increase in the degree of acidity (lowering of *p*H). This in turn affects the character and solubility of the corrosion products.

Another factor related to current density is the possibility of depletion of the salts in the soil adjacent to the zinc anode. Such an effect might tend to be most pronounced in installations designed to give a high anode current density.

The use of chemical materials in the soil adjacent to the zinc anodes reduces the tendency of the anodes to become passive where soil conditions are unfavorable. Gypsum-bentonite packed around the anode has been used satisfactorily in practical installations. The low solubility of gypsum prevents the material from leaching away. This is an important consideration where backfills are used in porous soils.

The potential of zinc anodes in soil has been observed to be 1.1 volt measured against a saturated $Cu/CuSO_4$ half-cell. As corrosion products develop on the zinc, the potential has been noticed to change by approximately 50 millivolts.[15] This reduction is observed usually soon after the anodes have been in operation, and is a stable value which will continue at approximately this level during the life of the anodes.

Some observers have noted that zinc plates buried in soil of high calcium carbonate content developed a hard, impervious carbonate scale which resulted in the anode becoming passive. Gypsum tends to counteract the tendency for carbonate films to form on zinc-anode surfaces.

In sea water, anode composition is a controlling factor on anode performance. As discussed previously,[8] zinc anodes that contain aluminum plus cadmium are recommended by the author as the most desirable material for sea-water applications.

Magnesium Anodes. Satisfactory performance of magnesium anodes in both soil and sea water has been obtained over the last ten years. The 6 per cent aluminum-3 per cent zinc-magnesium alloy anode gives current efficiencies in the neighborhood of 500 to 600 ampere hours per pound at anode current density levels under which magnesium generally operates in these environments. As a general design figure in estimating the life expectancy of a magnesium anode cathodic protection system, the lower value of 500 ampere hours per pound of metal is used.

Anode attrition is usually very uniform in sea water where the alloy anode operates at a high current density. When the anode current density is lower, as in soil applications, anode attrition is usually less uniform.

Aluminum Anodes. Little information is available on the performance of aluminum anodes.

Protective Criteria

There is no one method to ascertain the effectiveness of cathodic protection; many have been used and found effective. A few of the methods used in either underground or sea-water cathodic-protection work are discussed briefly below.

Visual Observations. This procedure has been used most extensively to evaluate cathodic protection systems on both the interior and exterior of ship hulls. The presence or absence of rust, calcareous deposits, retarding of pitting and condition of the paint are characteristics noted when this method of evaluation is used.

Potential Measurements. The use of potential measurements to deter-

TABLE 9-20. STANDARD REFERENCE ELECTRODES[13]

$$E = E_{25°} + \frac{de}{dt} \cdot t$$

Name	Cell	$E_{25°}$ (volts)	$E_{25°}$ (strong acids)	$E_{25°}$ (buffers)	de/dt (volts/°C)
0.1*N* Calomel	$Hg/Hg_2Cl_2/KCl$ (0.1*N*)	0.3337*	0.3386†	0.336†	-0.7×10^{-4}
1.0*N* Calomel	$Hg/Hg_2Cl_2/KCl$ (1.0*N*)	0.2800*	0.2848†	0.2828†	-2.4×10^{-4}
Saturated calomel‡	$Hg/Hg_2Cl_2/KCl$ (sat.)	0.2415*	0.2457†	0.2434	-7.6×10^{-4}
Silver chloride	$Ag/AgCl/KCl$ (0.1*N*)	0.2881*	—	—	-6.5×10^{-4}
Copper sulfate**	$Cu/CuSO_4/CuSO_4$ (sat.)	0.316	—	—	9.0×10^{-4}

* The *E* values so marked are true values in that they do not include liquid-junction potentials.

† The *E* values so marked include an approximate value for liquid-junction potentials. See W. J. Hamer *Trans. Electrochem. Soc.*, **72**, 62 (1937).

‡ The *E* of this electrode is subject to pronounced hysteresis effects and should be used only at constant temperature.

** Ewing, S., "The Copper-Copper Sulfate Half Cells for Measuring Potentials in Earth," p. 10, Committee Report, American Gas Association, 1939.

mine the effectiveness of a cathodic protection system has been used extensively by corrosion engineers. In this phase of corrosion-control work, the potential method was adopted for economic reasons. Even with this criterion for cathodic-protection effectiveness, varied adaptations of the method are used. It is not uncommon to find some engineers using 0.85 volt potential to a saturated copper/copper sulfate reference electrode as a protective criterion, while others use a change in potential of 0.3 volt from the value measured before the cathodic protection system was installed.

The characteristics of standard reference electrodes that are used in corrosion work either in the laboratory or field are shown in Table 9-20.

A few words of caution when measuring potentials should be mentioned here. Potentials measured when metals are coupled do not give the actual potential of a particular electrode because the measurement includes an *IR* drop. To obtain actual potentials of each electrode, readings can be satisfactorily obtained by using a current-interrupter method. If this method is used, readings should be observed immediately after breaking the circuit.

Another alternative method in measuring electrode potentials is to use a bridge-type circuit which eliminates the *IR* drop. Pearson[7] has made extensive use of this bridge circuit to eliminate the *IR* error.

Drilled Test Holes. Drilling of metal structures and measuring wall thickness with a micrometer has been used in determining corrosion values. A few of the drawbacks of this method of testing are:

1. Limitations imposed by number of holes that can be tolerated by a structure without causing reduction in structural strength.

2. Micrometer measurements obtained are representative for only the limited area actually measured.

3. Composite steel thickness values are obtained.

Steel Test Panels. In this method, steel panels are placed at selected locations. Some panels should be electrically connected to the structure while others should be isolated to serve as controls. Evaluation may be made by pit-depth measurements and metal weight loss.

In some instances test coupons have been sand-blasted before exposure to obtain relatively uniform surfaces on all panels. Regardless of the surface preparation method used, surfaces of test panels are never identical with the surface of the structure. The possibility of test panels being anodic or cathodic to the structure always exists.

Actual positioning of test panels will also affect results obtained. In any case it is usually impossible to duplicate exactly the critical conditions of moisture, temperature, flexing etc., encountered on the actual structure.

Audigage Measurements. Audigage measurements to determine changes in metal thickness are being used to evaluate corrosion losses and effectiveness of corrosion-control systems in the marine field. This method utilizes high-frequency sound waves. The surface being measured must be ground smooth and polished with emery to a mirror finish. A 6-inch diameter test station is prepared in the above manner at each location of interest on the structure's surface.

In order to eliminate errors caused by regrinding the surface each time measurements are taken, plastic films have been used to eliminate corrosion at the test-station areas between readings. If a particular tank's bulkhead surfaces are to be evaluated, a test station is therefore prepared in an adjacent tank. The cover material in some instances is not removed to obtain subsequent readings. A correction factor is used to obtain metal thickness: i.e., the plastic-film thickness (usually 1 mil) is deducted from the average thickness value observed.

Five measurements are usually taken at each test station and the average used as the reading of record. Measurements are normally taken at six-month or one-year intervals over a period of two to four years.

Accuracies from ½ to 1 per cent are claimed for audigage measurements when made by a highly trained operator. This would be equivalent to a variation in thickness measurements of from 2.5 to 5.0 mils on a ½ inch thick steel plate.

Signal strength is reduced when reflecting surfaces opposite the crystal are rough or pitted. Therefore, only a qualitative estimate of surface conditions is obtained on rough or pitted surfaces, even when scale conditions on the opposite surface are comparable. It is also possible to reach a point at which thickness variation under the crystal is so great that no signals can be obtained. In general, when the peak to valley depth of pitting exceeds 20 per cent of the total thickness, it will be impossible to measure thickness. This condition normally is not encountered in bulkhead and deck plating measurements, but sometimes is present on hull plates.

In spite of the fact that average metal thickness is observed with audigage techniques, they usually are accepted by regulatory bodies as indicative of plate deterioration. Regulatory-body surveyors must first approve of the engineer conducting such tests. In addition, although the audigage technique should be supplemented with pit-depth measurements, it is generally considered cheaper and faster than drilling holes, calipering and rewelding.

Plaster of Paris Molds. These have been used by some to determine changes in surface contours. This practice has not, however, been widely adopted.

***p*H on Surface of Protected Structure.** An alkaline reaction of the protected structure to *p*H paper has been used by some as an indication that corrosion is being controlled. This method is based on the fact that the tendency of steel to corrode is reduced when the solution is in the alkaline *p*H range.

When reviewing this or any other criterion it should be kept in mind that the application of cathodic protection is not an exact science. Each job must be considered as an individual one. Experience gained in one installation is the basis for future adjustment and the design of other installations.

SUMMARY

Factors affecting underground or underwater corrosion include electrolyte resistivity, moisture content of soils, soil and water characteristics, dissolved oxygen content, temperature, seasonal variations of environment, protective coatings, dissimilar metals, location of other metallic structures and stray currents.

Cathodic protection can be used effectively to mitigate underground or underwater corrosion. The theory behind the application of cathodic protection is relatively simple. It is the physical act of reversing the electrochemical force and subsequent mitigation of the process which normally destroys metals. Cathodic protection is obtained through the application of a current of sufficient magnitude from an auxiliary anode immersed or buried in the same electrolyte as the structure to be protected. This current flow opposes the galvanic current to eliminate the loss of metal on the structure.

Usually the choice of what type cathodic protection system to use depends on economic factors. Where power is relatively cheap and available, impressed-current cathodic protection systems may be indicated. Another important consideration is the effect a cathodic protection system will impose on a foreign structure. Interference problems usually are at a minimum when galvanic anodes are used.

Zinc anodes with dependable long-term galvanic properties are being marketed by the zinc industry. Special attention should be given to the composition of zinc galvanic anodes used for cathodic protection. Reichard and Lennox[8] have reported on the best composition for sea-water applications.

Zinc anodes with cast-in brackets are preferred for mounting on a ship's hull. Anodes can be mounted over studs or welded directly to the hull. Insulating barriers are not required on the faying surface of zinc anodes installed on a painted ship's hull.

Zinc anodes have been used effectively for cathodic protection in soils for the past fifteen years. They are most effective for the protection of bare structures in electrolytes of low resistivity, and in relatively high-resistivity electrolytes when the structure is coated with a high-resistant material. A gypsum-bentonite mixture has been used as backfill material in a variety of soils to insure good anode performance.

The application of cathodic protection is not a patterned-type process. A combination of electrical, mechanical and chemical measurements is usually required to locate and account for corrosion, and to permit adjustments of a cathodic protection system. Each job should, therefore, be considered an individual one and experiences gained in one installation are used for others.

REFERENCES

1. Brockschmidt, C. L., "A Practical Application of Zinc Anode Protection to an 18-inch Pipe Line," *Petroleum Ind. Elec. News*, **11,** 31 (1942).
2. Correlating Committee on Cathodic Protection, "Management Information on Cathodic Protection of Buried Metallic Structures Against Corrosion," *Corrosion*, **4,** (9), 3–6 (1948).
3. Davy, Sir Humphrey, "On the Corrosion of Copper Sheathing by Sea Water, and on Methods of Preventing this Effect; and on their Applications to Ships of War and Other Ships," *Phil. Trans. Roy. Soc.* (London), **114,** 151–158 (824); "Additional Experiments and Observations on the Application of Electrical Combinations to the Preservation of the Copper Sheathing of Ships, and to Other Purposes," *ibid.*, **114,** 242–6 (1824); "Further Researches on the Preservation of Metals by Electrochemical Means," *ibid.*, **115,** 328–46 (1825).
4. Graham, D. P., Cook, F. E., and Preiser, H. S., "Cathodic Protection in the U. S. Navy, Research-Development-Design," *Trans. Soc. Naval Architects and Marine Engrs.*, **64,** 241–317 (1956.)
5. May, T. P., Gordon, G. S., and Schuldiner, S., "Anodic Behavior of Zinc and Aluminum-Zinc Alloys in Sea Water," Cathodic Protection—A Symposium by Electrochem. Soc. and Nat. Assoc. Corrosion Engrs., p. 158–69, Houston, Texas, Nat. Assoc. Corrosion Engrs., 1949.
6. Mudd, O. C., "Experiences with Zinc Anodes," *Petroleum Ind. Elec. News*, **13,** 11 (1943).

7. Pearson, J. M., "Electrical Instruments and Measurements in Cathodic Protection," *Corrosion*, **3**, 549–66 (1947).
8. Reichard, E. C., and Lennox, T. J., Jr., "Shipboard Evaluation of Zinc Galvanic Anodes Showing the Effect of Iron, Aluminum and Cadmium on Anode Performance," *Corrosion*, **13**, 410–16 (1957).
9. Rhodes, G. I., "Cathodic Protection or Electrical Drainage of Bare Pipe Lines," Monograph, Am. Gas Assoc., New York, 1935.
10. Smith, W. T., and Marshall, T. C., "Zinc for Cathodic Protection of Pipe," *Gas Age*, **84**, 15–18; 35–6 (1939).
11. Starkey, R. L., and Wright, K. M., "Anaerobic Corrosion of Iron in Soils," Final Report, Am. Gas Assoc. Corrosion Research Fellowship, Monograph, Am. Gas Assoc., New York, 1945.
12. Teel, R. B., and Anderson, D. B., "The Effect of Iron in Galvanic Zinc Anodes in Sea Water," *Corrosion*, **12**, 343–9 (1956).
13. Tytell, B. H., and Preiser, H. S., "Cathodic Protection of an Active Ship Using Zinc Anodes," *J. Am. Soc. Naval Engrs.*, **68**, 701–4 (1956).
14. Uhlig, H. H., Ed., "Corrosion Handbook," New York, John Wiley and Sons, Inc., 1948.
15. Wahlquist, H. W., "Use of Zinc for Cathodic Protection," *Corrosion*, **1**, 119–47 (1945).

Miscellaneous Processes and Applications

Ernest W. Horvick

American Zinc Institute
New York, N. Y.

Aside from the principal topics featured in this and other chapters, there are a number of minor processes and applications based on zinc, alone or in alloyed form, that will be considered briefly here. There is also included a summary of the analytical operations used to determine the zinc content of raw materials and products common to the zinc industry. Additional notes on the evaluation of purity in super-purity zinc have been included on pp. 390–2.

ZINC DUST

Zinc dust is derived as a by-product in the distillation of zinc from ore, by condensing the vapor; in general, it is returned for redistilling. Domestic production is obtained mainly from redistillation of dross and other secondary material in the form of scrap zinc, by thermal or electrothermal processes.

When the zinc dust does not have to be absolutely pure, the usual source of supply is the blue powder formed during the ordinary thermal and electrothermal processes for zinc recovery. The electric furnace gives less pure zinc powder because of the higher temperature at which it operates.

Before 1920, zinc dust had been comparatively crude because known processes of manufacture had resulted in products comprising particles too large in size or generally varying in size and shape. Consequently, efforts were devoted to vaporizing molten metal and condensing the vapor so as to obtain products composed of particles of extreme fineness and uniformity, and not oxidized to an extent incompatible with its efficient utilization.

In the past, when the demand for zinc dust increased in the direction of higher-purity material, dross was melted in a bottleneck crucible and the zinc vapors collected in closed steel cannisters. The resulting product was high in metallic zinc, low in zinc oxide, lead, iron and other impurities. Improvements were made in types of furnaces for melting dross, including electric induction furnaces. The form of vessel holding dross for charging to produce zinc dust varied in the different operations. Large-diameter carborundum retorts up to 6 feet long, with a suitable connector between retort and cannister, have been used to produce the largest percentage of fine zinc dust and maximum recovery from galvanizers' dross.

Several of the the thermal processes for zinc dust manufacture described in the literature make use of methods in which condensation of zinc takes place in an inert atmosphere. The oxygen in the air originally present is usually removed at the beginning of the operation by reaction with the zinc to form zinc oxide. Subsequent operation is thus carried out in an inert atmosphere, care being taken to prevent the entry of air into the system or loss of the inert gas.

At this point, it should be explained that "extreme fineness" in zinc dust generally means a product practically 100 per cent minus 325 mesh, with particle size gradation measured in terms of microns. Coarser dusts may show 100 per cent minus 100 mesh, comprising size gradations up through 200 mesh and into the 325 mesh range, with 25 to 30 per cent of the product acquiring this degree of fineness. Still coarser dusts composed of 50 per cent plus 100 mesh and 50 per cent minus 100 mesh may be found to have some specific applications.

Uses of Zinc Dust

The largest present use of zinc dust in the United States is in the manufacture of chemicals consumed in the processes of printing and dyeing textiles. These chemicals for the most part act as reducing and bleaching agents. Zinc dust has long been used as an effective reducing agent on organic compounds, particularly where a strong acid or alkali reaction from the reducing agent could not be tolerated.

Production of metals in whose manufacture zinc dust is used increasingly includes dye-stuff intermediates and hydrosulfite, which is itself used as a reducing agent in the textile industry. Zinc dust is the starting chemical in the manufacture of zinc hydrosulfite, sodium hydrosulfite and sodium sulfoxylate formaldehyde. These chemicals are used in the bleaching of textiles, clay and paper.

Zinc dust is used in the reduction of nitrobenzene. This is a process material important in the manufacture of many other chemicals and drugs. Important in the latter category is the sulfa-family.

Zinc dust is used in the manufacture of "Rongalite" ($NaSHO_2 \cdot CH_2O \cdot 2H_2O$), a textile-printing reducing agent. It is also used in making adrenalin and dichlorethylene.*

Metallurgical Applications. Zinc dust is used as a precipitant for copper, cadmium, lead, silver, gold and other metals which are less electropositive, or less oxidizable, than zinc. The preparation of the emulsion of zinc dust and the addition of this to the solution has been a matter of prime importance in the precipitation of precious metals from cyanide solution. For efficient operation of the process it is important to eliminate oxidizing conditions by the use of zinc dust.

Treatment of zinc plating solutions with zinc dust has been found the most effective means of removing metallic impurities. One such use is the purification of zinc sulfate solutions preparatory to electrolysis. Zinc dust is also used for precipitating silver from photographic solutions and as a deoxidizing agent for bronze and nonferrous metals.

Explosives and Matches. Zinc dust is used as an ingredient of an explosive primer composition that is stable to shock and friction. It finds use in match-head and tear-gas compositions. Zinc dust and hexachlorethane are the chief ingredients of a smoke-producing composition; zinc chloride (the smoke) is produced from the hot flame which accompanies the reaction.

Paints. Zinc dust is a natural enemy of rust. When the dust is mixed with the proper vehicle, it makes a paint which protects against the harmful effects of corrosion to a greater degree than do ordinary paints, particularly on iron and steel. It is extremely effective in salt-air atmospheres and where the painted surface touches brine. Zinc powder has for some time been recognized as having a specific use as a pigment in the antifouling and anticorrosive ship-bottom paints. For this it is necessary to produce a paint made of finely divided zinc powder suspended in a heavy-bodied drying oil, a spar varnish, or a lacquer. Extreme fineness has been a chief requisite for successful application in this field.

Zinc-rich paint, a new type of paint (developed in England during World War II) and compounded so that the dry film contains 85 to 95 per cent zinc dust, with no zinc oxide, is arousing considerable interest in the United

* For other chemical uses of zinc dust, see "Uses and Applications of Chemicals and Related Materials" by Thomas C. Gregory.

States and abroad. (See the section of this chapter on Zinc-Rich Paints.) It is formulated with the zinc dust in suitable vehicles such as plasticized polystyrene, chlorinated rubber and some inorganic materials. These vehicles must have sufficient strength to carry the high pigment concentration involved, and still provide adherence and flexibility. Evidence based on tests is strong that when properly applied, zinc-rich paint has the ability to protect steel cathodically. This is due to the high zinc content which results in electrical contact between the zinc particles in the paint and the base metal. Before application of zinc-rich paints, thorough surface cleaning is essential.

Paints formulated with zinc dust and zinc oxide provide excellent rust-inhibitive properties, adhesion, film distensibility and abrasion-resistance when used for top coats. They act electrochemically in controlling corrosion.

Zinc dust-zinc oxide paints are well adapted for use on galvanized steel because of their unusually good adherence. They are excellent as a priming first coat on partially rusted surfaces. These paints are highly satisfactory for priming steel for atmospheric and underwater exposure. They are used on many outdoor structures such as bridges, water tanks and dams where rusting must be prevented. They are applied in the natural gray color or tinted red, black, blue or brown. The 80 per cent zinc dust and 20 per cent zinc oxide pigment mixture known as "Metallic Zinc Paint" (MZP) has won wide acceptance.

Other Uses. There are many other uses involving smaller quantities of zinc dust. Aluminum powder and mixtures of zinc and aluminum powders soften hard water when shaken with it, due to the liberation of calcium sulfate, magnesium sulfate and hydrogen. A lubricant suitable for the threads of well casings consists of cup grease and zinc dust to which is added an excess of free fatty substances. Zinc dust finds use as a gas-producing agent in a concrete mix, resulting in the porous structure sometimes desired.

The petroleum industry makes use of zinc dust in the following ways: (1) as a catalyst in making benzene and gasoline, (2) as a condensing agent in making certain condensation products, (3) as a promotor of hydrogen evolution in making soaps from paraffin wax oxidation products and (4) as a starting point in making catalysts for the hydrogenation of hydrocarbons (lamp oil and petroleum).

Zinc dust is also used in the manufacture of fireworks and flashlight powder, in powdered alloys, in the preparation of hydrogen gas, in the purification of sugar and for the treatment of paper surfaces.

ZINC ALLOY SOLDERS

Zinc is an important constituent in various hard solders and brazing metals. Zinc solders exhibit good intergranular penetration making for a strong bond. Cadmium additions increase this penetration and also impart corrosion resistance. Lead, and more particularly, silver, also add corrosion resistance. Silver increases the strength, as does copper.

Zinc Alloys for Soldering Aluminum*

In soldering aluminum, the use of solders containing high percentages of zinc is recommended. The resistance of soldered aluminum joints to corrosion is much more dependent upon the composition of the solder than is the case with similar joints in most commonly used metals. Soldered joints made with conventional low-melting-point solders must be protected by a suitable paint in any but indoor exposure. Unprotected assemblies made with zinc-base solders, on the other hand, exhibit long service life even in marine atmosphere.

Aluminum-magnesium alloys such as ASTM-G1A containing 1 to 1.8 per cent magnesium are subject to intergranular penetration by solders containing zinc. Stresses induced by cold working these alloys aggravate intergranular penetration, but heating to 700°F prior to soldering to relieve such stresses significantly reduces the penetration. When zinc-base solders which melt above 700°F are used, stress relief is accomplished before soldering occurs.

Aluminum solders can be grouped conveniently into low, intermediate and high-temperature solders (See Table 9-21). The low-temperature solders for aluminum are generally tin-zinc alloys, in which tin is present in the larger amount. The tin-zinc eutectic metal, containing 91 per cent tin and 9 zinc, melts at 400°F, wets aluminum readily, flows easily and—of the low-temperature solders—has the highest resistance to corrosion on aluminum. The addition of even a small percentage of zinc to ordinary lead-tin solders improves their soldering characteristics and resistance to corrosion.

Intermediate-temperature solders are usually zinc-tin or zinc-cadmium alloys containing from 30 to 90 per cent zinc. They may also contain other metals such as lead, bismuth, silver, nickel, copper and aluminum. Among the more common of these solders are the 70 tin—30 zinc, 70 zinc—30 tin, and 60 zinc—40 cadmium solders. Because of their higher zinc content, these intermediate-temperature solders generally wet aluminum more readily, form larger fillets, and give stronger and more corrosion resistant joints than the low-temperature solders.

* The limitation of Chapter 9 to zinc and zinc-base materials is disregarded here, owing to the wide range of composition of these solders.

TABLE 9-21. TYPICAL RECOMMENDED SOLDERS FOR ALUMINUM*

Composition Percentage	Type	Melting Range	Density (lb/in.)	Wetting of Aluminum	Fluxes	Corrosion Rating
		(°F)				
91 Sn-9 Zn	low-temp.	398	0.26	fair	chemical and reaction	fair
70 Sn-30 Zn	intermed-temp.	390–592	0.26	"	reaction	"
40 Cd-60 Zn	"	509–635	0.28	good	"	good
30 Sn-70 Zn	"	390–708	0.26	"	"	"
10 Cd-90 Zn	"	509–750	0.26	"	"	"
5 Al-95 Zn	high-temp.	720	0.24	"	"	excellent
100 Zn	" "	787	0.26	"	"	"

* From a "Soldering Manual" published by the American Welding Society.

The high-temperature solders contain from 90 to 100 per cent zinc along with small amounts of silver, aluminum, copper and nickel. These are the strongest of all of the solders for aluminum. They produce joints having the highest resistance to corrosion and are markedly superior to the low and intermediate-temperature solders. To assure high resistance to corrosion, the high-temperature solders should be practically free from lead, tin, cadmium, bismuth and other low-melting-point metals. This necessitates the use of 99.99+ per cent purity zinc in their preparation.

Electroplating aluminum with a metal such as copper produces a surface that can be soldered by the solders and fluxes used with copper. The deposition of copper is generally preceded by a "zincating" treatment in which aluminum oxide is removed from the surface and zinc is deposited by galvanic displacement.

Aluminum can be roll-clad with zinc to produce a surface solderable with most of the fluxes and low-melting-point solders generally used on copper and steel. This product, sometimes referred to as "soldering sheet," can also be used at higher soldering temperatures where the zinc cladding melts to form the joint.

Tin-Zinc Solders. A large number of solders based on tin and zinc have come into use for the joining of aluminum. Alloys containing 70 to 80 per cent tin, with the balance zinc, are used extensively. These alloys have a liquidus between 500 and 590°F. In recent years, the tendency is to add 1 to 2 per cent aluminum, or to raise the zinc content to 40 per cent. These solders are more corrosion resistant, but they have a high liquidus temperature ana are a little more difficult to apply. Compositions and freezing characteristics of the tin-zinc mixtures follow (American Welding Society):

Composition (%)		Temperature (°F)		
Sn	Zn	Solidus	Liquidus	Pasty Range
92	8	388	388	0
80	20	388	578	130
70	30	388	585	197
60	40	388	626	238
30	70	388	698	310

Cadmium-Zinc Solders. Cadmium-base solders containing substantial amounts of zinc have been developed for use under conditions of temperature higher than ordinary lead-tin solders will withstand.

The cadmium-zinc solders are useful for soldering aluminum. They develop joints with intermediate strength and corrosion resistance when used with the proper flux. The 40 Cd—60 Zn solder has found considerable use in the spot-welding of aluminum lamp bases. Compositions and freezing characteristics are as follows, according to data extracted from various sources:

Composition (%)				Temperature (°F)		
Cd	Zn	Ag	Other	Solidus	Liquidus	Pasty Range
90	10	—	—	510	536	26
82.5	17.5	—	—	510	510	0
82.5	15	2	0.5 Sn	—	—	—
78.4	16.6	5	—	—	—	—
60	30	—	10 Sn	—	—	—
58	42	—	—	510	608	98
40	60	—	—	510	599	89
10	90	—	—	510	734	224

Zinc-Aluminum Solder. This zinc-base solder is specifically intended for use on aluminum and develops joints with the highest strength and good corrosion resistance. The solidus of the solder is high, as shown below (American Welding Society), which limits its use to applications where temperatures in excess of 700°F can be tolerated. It also limits the flux that can be used to the reaction type.

Composition (%)		Temperature (°F)		
Zn	Al	Solidus	Liquidus	Pasty Range
95	5	720	720	0

Aluminum-joining techniques generally require flux to dissolve the oxide film which otherwise prevents a consistently sound bond. Most, if not all, aluminum fluxes contain corrosive chlorides and fluorides. Residual and entrapped fluxes attack and weaken the aluminum joint in the presence of moisture. Methods of removing the oxide film without the use of flux are:

(a) Application of ultrasonic vibration to the molten solder while it is in contact with aluminum. This causes cavitation and rupturing of the oxide, permitting the solder to "wet" the aluminum.

(b) Vigorously scratching the aluminum oxide and applying molten solder to the abraded area.

(c) Inert gas metal arc welding.

(d) Plating with copper, nickel or other metal. Then follows the normal soldering procedure for plated metals.

These methods require special equipment. They are directed particularly to the removal, not only of foreign matter, but also of the tenacious aluminum oxide surface film.

Solder does not flow into certain types of joint as well in the case of aluminum as with other metals. Corrosion resistance depends more in this case on the composition of the solder. Wetting, fillet size, joint strength and corrosion resistance generally increase with rising zinc content. Solders high in zinc are usually preferred for marine atmospheres.

At the Bell Laboratories, a simple effective process has been developed for soldering aluminum without the use of any of the above-mentioned methods. In this process, the aluminum-oxide layer is punctured by the solder stick. It is only necessary to heat the aluminum to a temperature that will melt the end of the solder stick. A very slight motion of the stick against the hot surface is then sufficient to wet the aluminum. The solder is made of zinc alloyed with small amounts of aluminum and magnesium. Lead, tin, bismuth and cadmium are kept out of the zinc to prevent intergranular corrosion.

The technique of applying this solder has also been used successfully on galvanized surfaces. Strong flux had normally been used when soldering galvanized steel with soft solder. With this new procedure no flux is required and joints can be made that are much stronger than soft-soldered joints.

Another means of joining aluminum without resorting to the special methods mentioned above, but relying strictly on the solder composition, has recently been described by Freedman.[1] It utilizes a zinc-lead blend that in one form (alloyed with aluminum, copper and tin) develops tensile strength in excess of 50,000 psi, yet melts at 800°F or below. It was found that while a zinc-lead combination solders aluminum satisfactorily, the low solubility of lead in zinc ordinarily causes separation of the heavier metal with resulting embrittlement and powdering. To overcome this, taking advantage of the fact that hydrochloric acid promotes blending of the two metals, a porous vehicle, or pasty carrier material, was used for safe and gradual introduction of the acid, permitting up to 18 per cent lead to be blended with zinc. The material is described as a self-fluxing alloy for joining and repairing a wide range of alloys, including aluminum and zinc-base compositions, galvanized sheet and aluminized steel.

Typical compositions of these self-fluxing aluminum solders are shown below, with indication of their melting points and approximate strength:

Melting Point (°F)	Zinc (%)	Lead (%)	Tin (%)	Aluminum (%)	Copper (%)	Tensile Strength (psi, approx.)
500	90.0	7.0	3.0	—	—	10,000
600	87.0	5.6	2.4	5.0	—	25,000
800	85.0	1.4	0.6	8.0	5.0	50,000

Solderable Zinc-Alloy Plating Process

A new plating process developed in the Sylvania chemistry laboratory provides an inexpensive solderable rust-resistant coating. It combines the corrosion protection of zinc and the solderability of tin or cadmium. The zinc alloy process utilizes both conventional tin and zinc plating solutions in an unconventional manner.

In 1946, a tin-zinc alloy process was introduced in England to provide a substitute for cadmium which was then in short supply. While meeting with some industry acceptance, this process presented certain difficulties because the high tin content—about 75 per cent—rendered the coating somewhat less anodic than zinc coatings to steel and served to increase the actual plating metal.

In the development of this solderable zinc-alloy plating process, an investigation was made of the possibilities of adding a small amount of tin to the zinc to make it more solderable. It was found that a deposit containing 10 per cent tin and 90 zinc could be readily soldered by conventional techniques using rosin flux solders.

EXTRUSION OF ZINC

Extrusion is the process in which a billet or slug of solid metal is forced by high pressure through an orifice shaped to impart the required form to the product. The application of heat to the metal prior to extrusion depends on the method of extruding and the optimum temperature at which extrusion is possible.

Two methods of extruding zinc are used. One is known simply as "extrusion" or hot extrusion, and the other as "impact extrusion." In the first method the general operating conditions are as follows:

Temperature Range	Extrusion Speed	Average Pressure
500–650°F	6–75 ft/min.	50–62 tons/sq. in.

Temperature control is essential because a billet that is too cold will require extreme pressures while an overheated billet will be hot short. The speed of extrusion is limited largely by the fact that increases in speed may raise the temperature of the extrusion beyond the critical temperature and cause breaks.

Vertical and horizontal machines of two types are used. In direct extrusion, the ram acts on a heated billet in a chamber, at the opposite end of which is the die through which the metal flows. In indirect extrusion, a hollow ram carries the die while the other end of the chamber is closed by a plate.

The preheated billets are placed in the container of the extrusion press, and the pressure on the hydraulic ram forces the metal through the die orifice as a strip, tube or shape of continuous cross section and variable length, depending on the size of the billet and the relative bulk of the extrusion per unit length. These extrusions have extremely dense grain structures.

Practically all commercial grades of zinc, and some alloys devised specifically for this process, have been used for extrusions. In American practice, the conventional die casting alloys have been used to some extent.

Use of zinc hot extrusion in the United States has been principally to supply wartime substitutes for brass and copper. Rods, wires, bars, angles, coiled strip, tubes and special shapes have been extruded. Rods for welding and soldering alloys can be extruded in various shapes or further reduced to wire so essential for zinc metallizing.

Impact extrusion differs from hot extrusion in that the processing temperature can range from room temperature to 500°F, depending upon the characteristics required. Essentially, the process consists of subjecting the slug of zinc, which is placed in a die, to a heavy blow by a punch which causes the metal to flow into the cavities of the die and the annular space between the die and punch. Each slug is of sufficient volume to form the single open-end tube resulting from the blow, in contrast to hot-extrusion which forms a product of greater length, but without a closed or formed end.

High-purity zinc (99.99+ per cent pure) is the base material for most work, although alloys containing, typically, 4 per cent aluminum and 0.04 magnesium, or small amounts of copper, 0.20 lead, 0.01 iron and 0.06 cadmium, are also employed. Mechanical presses of crank, toggle, or eccentric type are used and pressures up to 20 tons are applied.

Among parts produced, in addition to flashlight cans, are various other round, square and rectangular cans, usually without ribs. In general, the parts are limited to the simplest shapes. They are used for automobile radio parts because their resonance is low and they can be soldered readily. Similarly, they are used for other types of electrical equipment, soft drink dispensers, noise filters and advertising signs. Small gear blanks have also been produced.

WET BATTERIES

Zinc performs a significant function as the anode of various "wet" batteries. The principal types are considered below.

The Lalande Cell

This is an alkaline primary cell of considerable importance. The negative electrode is cast of zinc containing 2½ per cent mercury, added to prevent local action at the electrode. The positive pole is copper oxide which acts as a depolarizer by removing the hydrogen to form water and metallic copper. The electrolyte is sodium hydroxide. The operating cell electromotive force generally is 0.6 to 0.7 volt.

Three commercial types which vary in mechanical construction are the Edison cell, the Waterbury cell and the Columbia cell. The Edison Lalande cell is a single-fluid cell consisting of plates of amalgamated zinc and compressed copper oxide in a strong solution of potassium or sodium hydroxide. This cell has an electromotive force of only about 0.84 volt, but a low internal resistance so that it may give a current of 15 amperes on short circuit. The cell deteriorates slowly and is extremely reliable for either continuous or intermittent work. It is sometimes used where reliability is essential, as in railway signalling or fire-alarm systems.

Over a million Lalande cells are produced annually in the United States with capacities of 300 to 1000 ampere hours and delivering 1 to 5 amperes.

Eveready Air Cell (see also **National Carbon Air-Cell**)

This cell uses an amalgamated zinc anode, a 20 per cent sodium hydroxide solution as electrolyte and a porous carbon cathode through which air circulates and takes part in the reaction. It has an operating voltage for ordinary loads of 1 to 2, or almost double that of the Lalande cells, with capacities of 300 to 600 ampere hours per cell. It is of the reserve type, that is, it is shipped dry and water is added when required for use. It was designed to supply a radio "A" current, but is also used for telephone and railway-track currents.

Fery Cell

The Fery battery has a zinc anode, a porous carbon cathode and an open circuit electromotive force of 1 to 2 volts. It employs oxygen from the air as a depolarizer. While not used extensively in the United States, the cell is popular in France.

National Carbon Air-Cell Battery

This air-cell battery is designed to supply steady voltage to the filament of 2-volt radio tubes in battery-operated radio sets. It gives a more constant voltage and has longer life than the ordinary dry cell. The positive electrode of the air cell is made of a special grade of carbon which is highly pervious to oxygen. Zinc is used for the negative electrode. The carbon electrode, or "lung," as it is called, has the property of extracting oxygen from the air tract as a depolarizer inside the cell. The electrolyte solution is composed of sodium hydroxide and a little calcium hydroxide. If 0.65 ampere, or less, of current is taken from the cell, the porous carbon is able to extract oxygen from the air and replenish the oxygen within the battery as fast as it is consumed. As long as the carbon electrode contains oxygen it repels water and remains dry. However, if the oxygen is exhausted, water from the electrolyte clogs the pores of the carbon; it is no longer able to draw oxygen from the air and soon the battery ceases to function. Once dead, the battery is not rechargeable. Air cells, when properly used, have an exceedingly long life.

The Daniel Cell

The Daniel cell is one of the simplest practical cells. It consists of a copper cathode immersed in a saturated copper sulfate solution and a zinc anode in a solution of zinc sulfate. The two solutions have different specific gravities and mix only by diffusion. In some Daniel cells the two solutions are separated by a cup of porous ceramic material. In a freshly prepared cell the zinc sulfate solution is dilute, and the electromotive force is about 1.06 volts. With the flow of current the zinc cathode dissolves gradually and the solution becomes more concentrated. The electromotive force therefore gradually decreases with age.

The Grove Cell

The Grove cell has an amalgamated zinc plate placed in dilute sulfuric acid and a platinum plate placed in a pot of porous clay standing in a vessel containing the dilute acid. This porous pot is filled with strong nitric acid which is a good oxidizing agent.

The Bunsen Cell

The Bunsen cell is similar to the Grove cell except that a plate of hard graphitic carbon takes the place of the platinum.

Silver-Zinc Batteries

A comparatively new type of wet battery, in which silver and zinc are the active metals in an alkaline electrolyte, is finding important uses for torpedoes, missiles and other special applications. This battery is able to provide instant electrical power for missiles fired at even sub-zero temperatures.

Silver-zinc batteries offer weight and volume efficiencies greater than those of any other battery system. While suitable for low- or high-rate discharge, they are notable for their ability to be discharged at extremely high rates. These batteries are easily charged with conventional charging equipment. During charge and discharge, silver-zinc batteries are characterized by the absence of corrosive or poisonous fumes. Operation is based on the principle of alternate silver-zinc plate construction and ion exchange.

The chemical reactions occurring during operation of the battery are entirely reversible. They are:

$$Ag + Zn(OH)_2 \rightleftarrows AgO + Zn + H_2O$$

discharged charged

During discharge, silver oxide is reduced at the positive plate and zinc is oxidized at the negative plate. The electrolyte is a solution of potassium hydroxide.

ZINC-BASE BEARING METALS[2-8]

Zinc may be cast in the form of sticks or small ingots for experimental purposes, or for the preparation of various alloys. General compilations such as those given in Woldman's "Engineering Alloys" list a large number of zinc alloys that have been described or marketed at one time or another as special products. Prominent among these are the zinc-base bearing metals.

Zinc-base bearing metals are those bearing metals whose principal alloy constituent is zinc. The large number of compositions listed in Table 9-22 reflects their use as substitutes for the tin-base babbitts, or white metals, and bearing bronzes in war-period emergencies (especially, in Germany). Wartime research by the Germans, who resorted to zinc alloys due to acute copper shortages, seems to indicate that they may be satisfactory as bearing materials, though only in a narrow field; it does not signify any considerable use of zinc-base bearing metals under normal conditions. Generally, they may be considered for limited use in situations where light loads, precise fit and unfailing lubrication minimize the seizing tendency of zinc in contact with ferrous metals. Ideal applications are spindle bearings for high-speed lathes, other machine tools, and electric motors where loads are steady and free from shock, and speeds are high.

German research has shown that in comparison with a standard bronze bearing metal, the centrifugally cast zinc-base bearing containing 4 per cent aluminum, 0.7 to 1.0 copper, 0.03 magnesium, and the balance zinc, has a higher affinity for oil, a lower coefficient of friction and superior mechanical properties. Other tests on this alloy and several other zinc alloys proved that they were very satisfactory as sleeve valves, not to be considered as substitutes, but as new, high-quality materials with excellent properties. One investigator recognized as superior for bearings zinc-base mixtures of the following percentage compositions:

(1) 4 Al, 0 to 0.5 Cu, 0 to 0.04 Mg, balance Zn

(2) 1.2 Cu, 0.2 Mn, balance Zn

Table 9-22. Zinc Alloys that have been Used as Bearing Metals

Name or Designation	Composition (% Alloying Metals, Balance Zn)							Remarks
	Cu	Al	Sb	Sn	Pb	Fe	Mg	
GZn-Al4-Cu1	0.7–1.0	4.0	—	—	—	—	0.03	
Zn-Al7-Cu4	3.9	7.0	—	—	—	—	—	
Zn-Al4	0–0.5	4.0	—	—	—	—	0–0.04	
Zn-Cu1	1.2	—	—	—	—	—	—	Mn, 0.2
GZn-Al10-Cu1	0.7–1.0	10	—	—	—	—	0.03	
	1	10–14	—	—	—	—	—	
	1	30	—	—	—	—	—	
Zamac	1–3	4	—	—	—	—	0.03–0.05	Plus hardener of 0.1% Si, Mn, Li or Ni.
ZA-4	—	—	—	28–29	—	—	—	Underwater bearing (Brit)
French Navy anti-friction	7.5	—	—	7.5	—	—	—	
Lumen bronze	10	4	—	—	—	—	0.10	Very hard and brittle.
Dunlevic and Jones anti-friction metal	—	—	10	5–8	—	—	—	
" "	8.0	—	10					
" "	1.6	—	0.4	46.0	—	—	—	Also railway axles.
" "	—	—	20	60	—	—	—	
Dunlevic and Jones (Russian)	8.0	—	12.0	—	—	—	—	
Ehrhards B. M.	10.9	2.5	—	0.2	1.2	—	—	
" " "	4.0	—	—	4.0	3.0	—	—	
" " "	3.0	—	—	6.0	2.0	—	—	
Fenton's B. M.	6.0	—	—	14	—	—	—	
" " " (Schutt)	5.5	—	—	14.5	—	—	—	
Fenton's B. M. II	5.5	—	14.5	—	—	—	—	
Glievor Bearing	4.3	—	8.9	6.9	4.9	—	—	Cd, 1.4
Heavy Axle Bearing	1.0	—	6.0	38.0	4.0	—	—	
Ledebur's Bearing (a)	5.0	—	10.0	—	—	—	—	
Ledebur's B.M. (b)	5.5	—	—	17.5	—	—	—	
Leddell Bearing	6.25	6.25	—	—	—	—	—	
" "	5.0	5.0	—	—	—	—	—	
Propeller bushing	5.0	—	7.0	19.0	—	—	—	
Salge antifriction	4.0	—	—	9.9	1.1	—	—	
Schomberg's B.M.	0.38	—	0.2	39.9	0.21	0.15	—	
Sorel's Alloy	1.0 or 10.0	—	—	—	—	1.0	—	
Vaucher's Alloy	—	—	2.5	18.0	4.5	—	—	
Zinc Babbit	5.0	—	3.0	26.0	—	—	—	
Heine's Biddery	11.4	—	—	1.4	2.9	—	—	
Aalener Zinc Alloy	3.5	—	3.0	20.5	21.6	—	—	

TABLE 9-22.—(*Continued*)

Name or Designation	Composition (% Alloying Metals, Balance Zn)							Remarks
	Cu	Al	Sb	Sn	Pb	Fe	Mg	
Bearings for common work (Hiorns)	8.0	—	2.0	2.0	—	—	—	Also "Cheap Krupp Mill"
" "	5	—	—	10.0	—	—	—	
" "	4.2	—	—	29.3	—	—	—	
" "	8.2	—	—	16.6	—	—	—	
Iridium II	1.12	—	Trace	21.63	—	—	—	
Iridium B.M.	1.25	—	Trace	15.75	—	—	—	
Kneiss B.M.	—	—	—	25.0	25.0	—	—	Also "German" B.M.
" "	3.0	—	15.0	—	42.0	—	—	Also for "ordinary mills"
Pierrot's B.M.	2.3	—	3.8	7.6	3.0	—	—	
Sampson Metal	4.0	8.0	—	—	—	—	—	
" "	10.0	2.0	—	—	—	—	—	Good for high speeds and light loads.
"Aluminum bearing Metal"	0.55	20.20	—	27.7	1.25	—	—	
Becker's alloy for rapidly revolving shafts	5.6	17.47	—	—	—	—	—	Also "Russian" and "English for bearing loads."
Cheap Krupp mill	7.0	—	1.5	1.5	—	—	—	Also "for heavy work" and "Russian state railroad"
English bearing metal	5.5	—	—	14.5	—	—	—	
For propeller bushings	5.0	—	—	26.0	—	—	—	
Germania bearing bronze	4.4	—	—	9.6	4.7	0.8	—	
Glyko metal	2.4	2.0	4.7	5.0	—	—	—	
Hamilton metal	3.5	—	1.5	—	3.1	—	—	
Krupp Mills II	—	—	7.70	—	15.38	—	—	
Russian Packing	—	—	—	0.98	0.32	0.16	—	
Saxonia metal	6.0	0.2	—	5.3	3.0	—	—	
Siemens & Halske	—	—	5.0	—	—	—	—	Cd, 47.5, Also "Automobile" B.M.
Sulzer's antifriction	4.0	—	—	9.9	1.2	—	—	
Tandem	7.0	0.4	—	21.5	4.8	—	—	
Babbit Metal	62.1	—	3.0	19.0	5.0	—	—	4.0, As
English Babbitt	4.0	—	3.0	19.0	5.0	—	—	
Krupp Mills I	—	—	9.0	—	45.5	—	—	
For Propeller Boyes	57.0	—	—	14.0	—	—	—	
Russian B.M.	1.3	—	—	72.2	—	—	—	
Stephenson's Alloy	19.0	—	—	35.0	—	31.0	—	

Zinc alloys of the following percentage compositions have been used in Switzerland:

(1) 4 Al, 0.7 to 1.0 Cu, 0.03 Mg, balance Zn

(2) 10 Al, 0.08 Cu, 0.03 Mg, balance Zn

The first was said to be as good as aluminum in worm wheels and as bearings in machine tools.

The Germans also experimented with zinc alloys in two other ranges of aluminum content, each also containing about 1 per cent copper. One, containing 10 to 14 per cent aluminum, was adopted for bearings on quite a wide scale in general and locomotive engineering. This alloy has properties which lie between those of white metals and bronzes. Bearings could be cast more accurately from these than from competing alloys, with a corresponding saving of machining. The second alloy, with about 30 per cent aluminum, tested in tractor service, seemed destined to outlast the emergency for which they were designed.

The table (9-22) compiled from various sources, is a summary mainly of zinc-base alloys that have been used as bearing metals. It is intended merely to indicate the many compositions that have entered this field at one time or another, often under economic pressure, and not to reflect current engineering practice, except as noted above.

TIN-ZINC ALLOY PLATING

Tin-zinc alloys for plating are notable for their unique combinations of properties which include:

1. Favorable operating characteristics, excellent covering and throwing power, easy control.
2. Economy.
3. Good corrosion resistance.
4. Excellent solderability and retention of solderability on storage, surpassing tin.

The whole range of alloys of tin and zinc can be electroplated; however, principal emphasis has been placed on the alloy of 80 per cent tin and 20 zinc. This, on the basis of experience, appears to be the most useful.

The bath used for producing the tin-zinc deposit is basically the potassium stannate tin bath plus a small amount of zinc cyanide together with a large proprotion of potassium cyanide. Two formulations are in vogue, one for still tanks and automatics, and the other for barrels. Anodes are of the same composition as the deposit, i.e., 80 per cent tin, 20 zinc.

"PEEN" PLATING

"Peen" plating is a new and cheap method of plating small hardware with such metals as zinc and brass without the use of electricity. Typical objects which can be treated by this method are nails, hose clamps, lock washers, chains, powder-metallurgy parts, hinges, coat hooks, springs and small stampings of all types. In the process, the articles are given a suitable pretreatment and then put into a hexagonal tumbling barrel with the impacting material. A promoter solution and the coating metal in the form of dust and water are added. For zinc plating, the barrel is usually tumbled for about 45 minutes. Particles of metal dust are welded onto the objects by the hammering action of the impacting material. The plating efficiency of zinc, for example, can be as high as 98 per cent.

Pretreatment

The pretreatment required is essentially comparable to that required for electroplating. Objects for plating must always be free from grease and, in certain cases where a very thick coating is to be applied, an acid pickle or a shot-blasting pretreatment may be advisable. In some cases, particularly with zinc plating, a very thin copper coating on the objects aids the strike of the plating metal. With ferrous articles this may easily and quickly be accomplished by immersing the article in a copper sulfate solution.

One of the advantages of the method is that plating is done in bulk with very little manual handling of the objects. On a commercial scale, a batch of objects is loaded into a basket, a batch being anything up to 600 pounds. The basket is moved by means of a hoist or crane through whatever immersion pretreatments are required. The articles are then loaded directly into the Peen plating barrel.

Post-treatment

On completion of the plating run, the entire contents of the barrel are emptied into a vibrating sieve which separates the impacting material from the treated objects. The plated objects are removed for subsequent post-treatment (for example, coating with chromate) and thence dried. The liquid used in the plating process is of no further use and is discarded. The disposal of this commodity presents no effluent problem. The impacting material is returned to the barrel ready for the next plating run. Commercial experience in the United States over a number of years has shown that the life of the impacting material would appear to be very long indeed, and that it is only necessary to add small quantities of new impacting material from time to time to account for inadvertent losses.

Suitable Metals

Peen-plated coatings of zinc can be applied at any normal thickness. Metals which can be plated by this method include zinc, brass, cadmium, tin and aluminum. Any of these metals can be plated onto highly tempered objects, such as springs and lock washers, without the danger of hydrogen embrittlement. Powder-metallurgy parts can be plated without subsequent corrosion due to the absorbed electrolyte.

VARIOUS USES OF ZINC AS A METAL

Zinc Coinage

Zinc is not used extensively for this purpose. Under an executive decree in the 40's it was authorized as a substitute for nickel in Ecuadorian fractional coins.

Zinc Wire

Aside from its principal use in the metallizing process, previously described, zinc wire finds some commercial usage in brake linings. For this purpose it provides maximum resistance to wear, makes for less wear on brake drums and gives long service life. Brake linings woven with zinc wire have superior heat conductivity, thus dissipating the heat generated at the frictional surfaces. The coefficient of friction of brake linings woven with zinc wire is retained at the highest temperature developed in service, so that brake fade is eliminated. Also, there is freedom from any tendency to "score" the drum and there is no "squeal" or "judder."

Miscellaneous

1. Paper is metallized by exposing it *in vacuo* to action of vapors of molten zinc.
2. In feathered form, zinc finds use as a reagent in analytical processes involving control and research work; a reagent in making photographic chemicals; and a stripping agent for making photographic solutions.
3. In mossy form zinc is used as a catalyst in organic synthesis; an ingredient of coloring composition for face brick; and an ingredient of chimney soot-removing compositions.
4. Zinc is used for socketing connections of steel cable in the "high wire" method for logging. High Grade zinc is recommended for this application.
5. Zinc in the amalgamated form is used for covering the cushions of frictional electric machines. Here, it is alloyed with 25 per cent each of mercury and tin. The zinc and tin, as powder, are added to mercury held close to its boiling point.

BY-PRODUCTS

Much of the sulfuric acid needed annually in the United States is produced as the major by-product of the zinc industry. It is one of the most important base chemicals and is a dense, oily liquid, colorless to light yellow, depending upon purity. It is made from the sulfur dioxide in the gases generated by the roasting and sintering of sulfide concentrates. Essentially, the process of manufacture consists of oxidizing purified sulfur dioxide to sulfur trioxide in the presence of a catalyst. The sulfur trioxide is then combined with water to form sulfuric acid. A detailed treatment of this important industrial process is beyond the scope of this book.

Sulfuric acid is probably the most widely used chemical and it contributes indispensably to the chemical and oil industries. It serves for the chemical cleaning (pickling) of steel and other metals during their fabrication. It is used in the manufacture of cellulose film, paints, pigments and rayon; in the solution, oxidation, dehydration or sulfonation of most organic compounds; also in the processing of inorganic compounds, dye intermediates and medicinals. It is vital as an electrolyte in conventional lead storage batteries. The "oleum" form is used with other acids in the manufacture of high explosives. Probably the greatest tonnage is consumed in making superphosphates for fertilizers. An interesting modern use is as a leaching agent in uranium mills.

Other by-products of zinc refining are the metals which occur with zinc in the ore bodies. These include lead, silver, gold, copper, indium, gallium, germanium and cadmium—the last of which is used as a bearing metal in special alloy compositions as well as for solders and plating.

CHEMICAL ANALYSIS

Special methods employed in the analysis of super purity zinc are discussed in the first section of this chapter. Ordinary analytical methods for determining zinc are indicated here. The latest version of a method* used by 90 per cent of the zinc industry to determine the zinc content of ores, residues, calcines and other raw zinc materials is described below, followed by ASTM Methods for Analysis of Slab Zinc.

"Titration in Acid Solution with an Outside Indicator"

Solutions Required

Ammonia Wash Solution—Dissolve 100 g of NH_4Cl (C.P.) in 1000 ml of distilled water and add 100 ml of NH_4OH (Sp Gr 0.90).

Sulfuric Acid (1-1)—Mix cautiously 500 ml of H_2SO_4 (sp gr 1.84) with 500 ml of distilled water.

---

* The New Jersey Zinc Co.

Potassium Ferrocyanide—Dissolve 31.4 g in 1000 ml of distilled water and allow to stand for three months before using.

Uranium Nitrate—Dissolve 5 grams in 100 ml of distilled water.

Method

Weigh the sample[1] into a 400-ml beaker, moisten with water and add 30 ml of HCl (sp gr 1.19). Boil gently for 10 minutes with the beaker covered, add 20 ml of HNO_3 (sp gr 1.42) and continue the boiling for 10 minutes more. Wash the cover and the sides of the beaker, add 15 ml of H_2SO_4 (1–1) and evaporate to dryness.[2] Cool, add 5 ml of HCl (sp gr 1.19), 10 g of NH_4Cl and dilute to 50 ml with water. Warm until all salts are in solution. Cool, add 1 ml of liquid bromine,[3] neutralize with NH_4OH (sp gr 0.90) and add 20 ml in excess. Heat to boiling and boil for 1 minute.

Filter[4] the solution into a 600-ml beaker and wash the precipitate three times with ammonia wash solution.[5] Wash the precipitate from the paper into the original beaker, add 5 ml of HCl (sp gr 1.19) and 5 ml H_2O_2 (3%) to dissolve the manganese. Cool, add 1 ml of liquid bromine,[3] neutralize with NH_4OH (sp gr 0.90) and add 20 ml in excess. Heat to boiling and boil for 1 minute. Filter and wash as before, catching the filtrate and washings in the beaker containing the first filtrate and washings.

Add 1 g of test lead to the combined filtrates and boil for 20 minutes, acidify[6] with HCl (sp gr 1.19) and boil for 10 minutes more. Add 10 ml of HCl (sp gr 1.19), dilute to 450 ml with water, heat to boiling and titrate with standard $K_4Fe(CN)_6$ as described under Standardization.

Calculate the percentage of zinc as follows:

$$\frac{(A - B) \times F \times 100}{W} = \% \text{ Zn}$$

where A = ml of $K_4Fe(CN)_6$ solution required for the sample, B = ml of $K_4Fe(CN)_6$ solution required for the blank, F = g of Zn per ml of $K_4Fe(CN)_6$ solution, and W = weight of the sample in grams.

Notes

[1] The size of the sample depends on the zinc content. Weight 0.5 g of zinc concentrates to 5.0 g of tails. High-grade materials that contain metallic shot should have larger samples taken and the solutions aliquoted to 0.4 g or 0.5 g.

[2] Evaporation should be carried out on a warm plate until the solution is ready to emit SO_3 and then continued on a hot plate. Samples high in lead should be diluted at this point with 100 ml of water, boiled, cooled and, after standing for about 1 hour, the $PbSO_4$ should be filtered off and, after adding 10 g of NH_4Cl to the filtrate, the method should be carried out as written.

[3] Sufficient bromine should be added to precipitate all the manganese. If no manganese is present the bromine may be omitted. If the manganese is high the filtrate should be tested with additional bromine to be sure all the manganese has been removed.

[4] Use a No. 2 Whatman 15 cm filter or similar paper. With large precipitates a larger paper may be used.

[5] A precipitate of $Fe(OH)_3$ holds zinc by occulsion. It is desirable to remove as much zinc as possible, especially in samples where the zinc is high, so that the amount of zinc retained in the second precipitate will be reduced to a negligible quantity.

[6] Use litmus paper as an indicator. If necessary add more test lead to remove large amounts of copper. Most of the NH_4OH should be removed during the boiling.

Standardization of Ferrocyanide Solution

Weigh 0.3 to 0.4 g of zinc (C.P.)[1] into a 600-ml beaker and dissolve in 50 ml of distilled water and 15 ml HCl (sp gr 1.19). Add 25 g NH_4Cl, dilute to 300 ml neutralize[2] with NH_4OH. Acidify with HCl, add 1.0 g of granulated test lead and boil for 30 minutes. Add 10 ml HCl (sp gr 1.19) and heat to boiling.

Pour about 100 ml into a small beaker, and add $K_4Fe(CN)_6$ solution to remaining portion until a drop added to uranium nitrate test solution on a spot plate gives a brown color. Add most of zinc solution in the small beaker to the main portion and continue addition of $K_4Fe(CN)_6$ solution until the endpoint is just passed. Now wash all remaining solution from the small beaker into the 600 ml beaker and finish the titration 2 drops at a time.

Blank Determination. Dissolve 25 g NH_4Cl in 300 ml of distilled water in a 600-ml beaker and acidify[2] with HCl (sp gr 1.19). Add 1 g of granulated test lead and boil for 30 minutes. Add 10 ml HCl (sp gr 1.19), titrate to the end point.[3]

Calculate the factor as follows:

$$F = \frac{W}{A - B}$$

where A = ml of $K_4Fe(CN)_6$ solution required for the standard zinc solution, B = ml of $K_4Fe(CN)_6$ solution required for the blank, W = weight of zinc in the standard solution, and F = grams of zinc per ml of $K_4Fe(CN)_6$ solution.

Notes

[1] This is satisfactory for standardizing a $K_4Fe(CN)_6$ solution for concentrates. In case of tails, the standardization should follow the method of analysis adding an equivalent amount of iron and lime present in the tails, and making the same number of separations in the prescribed manner. Another suitable method is to standardize the $K_4Fe(CN)_6$ solution by means of similar tails of known zinc content. The point of all this is that the retention of zinc by the ferric hydroxide precipitate will give low results if the factor obtained with pure zinc is used. The higher factor obtained for tails must, of course, only be used for titrations of solutions from that type of materials.

[2] Use litmus paper as the indicator.

[3] Titrate all samples and standards to the same coloration as the blank. A white porcelain plate covered with a smooth layer of clean paraffin makes a better spot plate than the usual plates provided with hollows.

Methods of Chemical Analysis of Slab-Zinc

According to *ASTM Standard SPECIFICATIONS FOR* SLAB ZINC (Spelter) (B6-58)

Section 7—

"The chemical compositions enumerated in these specifications shall be determined by the following methods of the American Society for Testing Materials (in cases of dispute, Methods E 40 shall be used, except for the Special High Grade and the High Grade, where Methods E 27 may be used as an alternate):

	ASTM Designation*
Chemical analysis	E 40
Spectrochemical analysis	E 27
Polarographic analysis	E 68

* These designations refer to the following methods of testing:

Methods for Chemical Analysis of Slab Zinc (Spelter) (ASTM Designation: E 40),

Methods for Spectrochemical Analysis of Zinc-Base Alloys and High-Grade Zinc by the Solution D-C Arc Technique (ASTM Designation: E 27),

Method for Polarographic Determination of Lead and Cadmium in Zinc (ASTM Designation: E 68)."

For details concerning the above prescribed methods of analysis consult the "1956 Book of ASTM Methods of Chemical Analysis of Metals."

REFERENCES

1. Freedman, Samuel, "Fluxless Aluminum Joining Avoids Joint Corrosion," *Iron Age*, **176,** 71–3 (March, 1956).
2. Gillet, H. W., Russell, H. W., and Dayton, R. W., "Bearing Metals from the Point of View of Strategic Materials," *Metals and Alloys*, **12,** 749–58 (1940).
3. Holder, G. C., "Antifriction Metal," *Metal Ind.* (*London*), **15,** 153 (1917).
4. Jones, J. L., "Making Thin Linings," *Metal Ind.* (*London*), **17,** 232 (1919).
5. Kühnel, R., "Werkstoffe fur Gleitlager," pp. 309–21, Berlin, Julius Springer, 1939.
6. Samans, Carl H., "Engineering Metals and Alloys," Chap. 19, New York, The Macmillan Company, 1949.
7. Schmid, E., and Weber, R., "Lagermetalle auf Feinzinkbasis," *Metallwirtschaft*, **18,** 1005–10 (1939).
8. Vought, A., "Erfahrungen mit Lagermetalle von Zinklegierungenabfallen," *ibid.*, p. 749.

Note: Much of the information on solders is of commercial origin which has not been confirmed by the Editor or American Zinc Institute.

CHAPTER 10

ZINC AS AN ALLOYING AGENT

C. H. Mathewson

Professor Emeritus of Metallurgy and Metallography
Yale University
New Haven, Conn.

Zinc does not form a continuous series of solid solutions with any other metal, but dissolves rather freely in solid aluminum, copper, silver and gold. Maximum solid solubilities here range from more than 80 per cent in aluminum to less than 15 per cent in gold. Also, among high-melting metals, manganese and the metals of the iron group (nickel, cobalt and iron), as well as palladium and platinum, are capable of dissolving large percentages of zinc in the solid state. In these cases, special techniques such as melting under pressure, some form of sintering, or other special processing would have to be conducted in order to prepare the alloys, since in general their melting points lie far above the boiling point of zinc. Solid lithium and magnesium are reported to dissolve about 16 and 8 per cent of zinc, respectively, at corresponding eutectic temperatures.

With other metals, the maximum solid solubilities for zinc are either in the range 1 to 3 per cent, or restricted to small fractional percentages, and decrease with falling temperature.

The maximum solid solubilities of other metals in zinc are always small, amounting to 0.5 to 1 per cent with manganese and aluminum; around 2 to 3 per cent with cadmium, palladium and copper; about 8 per cent with silver; and probably 10 to 15 per cent with gold. In all cases, the solubility is smaller at room temperature. Only fractional percentages, or minute amounts, of other metals enter the structure of the zinc crystal.

Zinc forms many intermediate phases in its association with other metals, and participates in many of the so-called electron compounds, which are characterized by significant ratios of the number of valency electrons to the number of atoms specified by the chemical formula. Thus, the 3:2 ratio is represented by AgZn, AuZn, CuZn, and $CoZn_3$ (O-valency ascribed to Co); the 21:13 ratio by Ag_5Zn_8, Cu_5Zn_8, and also, with O-valency ascribed to the combining metal, Co_5Zn_{21}, Fe_5Zn_{21}, Mn_5Zn_{21}, Pd_5Zn_{21}, Pt_5Zn_{21}, and Rh_5Zn_{21}; while the 7:4 ratio is seen in $AgZn_3$, $AuZn_3$, and $CuZn_3$.

Copper-Base Alloys

Brass products constitute the largest use of commercial zinc in which the zinc is an alloying component, but not the base metal (primary constituent) or a surface coating. About 14 per cent of the 1955 slab zinc consumption in the United States was thus apportioned, roughly 40 per cent as much as for zinc die castings.

Zinc alloys readily with copper in all proportions, but the interval from 40 to 50 per cent by weight of zinc designates a change-over from usefully workable and sensibly tough alloys to brittle and generally undesirable compositions. The scientific basis for this behavior is best appreciated by reference to the well-known phase diagram, shown here as Figure 10-1, in the form presented in the Copper Monograph,[1] but with some added documentation.

Thus, it is seen that the solidified and cooled alloys, which may be regarded as approximately in the state of equilibrium designated below 400°C in the diagram, change from a state of mixed alpha plus beta phases, beginning close to 40 per cent zinc, to the beta phase alone, as their composition approaches 50 per cent zinc (at about 46 to 47 per cent zinc). Below the near-40 per cent boundary these alloys are composed entirely of the alpha phase which is especially characterized by an unusual adaptability to cold processing by the usual operations or rolling, drawing, pressing, spinning, etc. The beta phase is freely plastic at red heat but on cooling becomes too brittle for most uses. However, in combinations with alpha, in the approximate range of composition 39 to 46 per cent zinc, where at suitably elevated temperatures the alloys revert entirely to the beta condition, the hot plasticity is followed on cooling by fair cold-working properties, due to the reappearance of the alpha phase in substantial amount. Alloys of this group are called "Muntz metals," from an early English patent, and are typically represented by the 60–40 composition. It is observed that the melting points decrease with increasing zinc content, which favors the use of the normally brittle alloys containing 50 per cent, or even more, zinc for brazing the common brasses. Here, brittleness is alleviated by diffusion and a more favorable adjustment of composition in the brazed region.

The intermediate phases beyond beta—viz., gamma, delta and epsilon—are compound-like associations of zinc and copper atoms into sufficiently unsymmetrical crystal types to prevent the coherent slippage and strain patterns that are essential for plasticity. The alpha phase, face-centered cubic, is ideal for plastic deformation (cold); and the beta phase, above the dotted line in Figure 10-1, is body-centered cubic with a random arrangement of the zinc and copper atoms, which also favors plasticity (hot). Below the line the two kinds of atoms take up fixed positions in an ordered crystalline structure which does not permit far-reaching deformation without fracture.

It is evident from the sloping boundaries that define the limiting com-

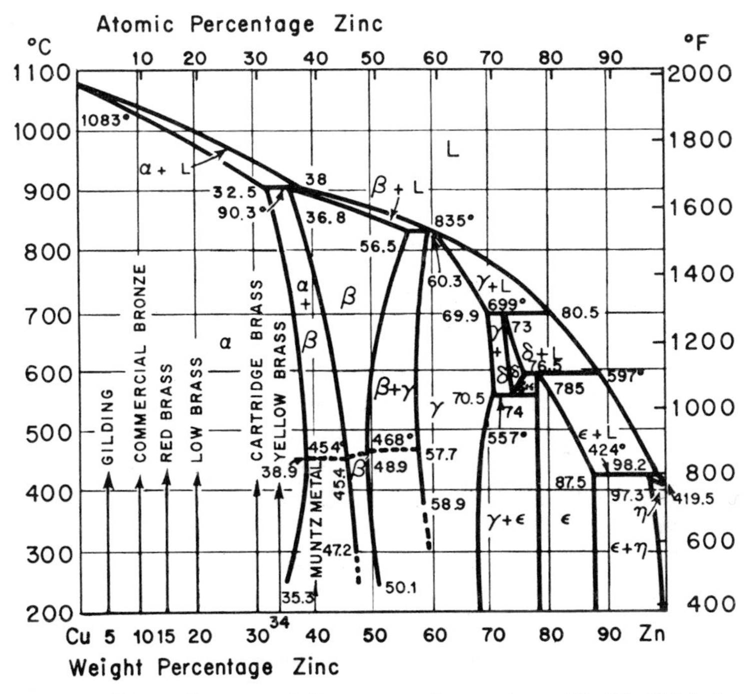

Figure 10-1. Phase diagram of the copper-zinc system. (R. M. Brick in ACS Monograph "Copper".[1])

positions of the alpha, alpha plus beta and beta fields at different temperatures that the usual operations of heat treatment—quenching, reheating at various temperatures, varying the rate of cooling, etc.—would affect the constitution and properties of commercial alloys, particularly in the composition region around 40 per cent zinc.

Wrought Alloys. Not much use is made of such procedures because it is found that the simple operations of cold working to various so-called tempers, and annealing to produce variations in grain size, develop a wide range of strength properties all the way from the soft fully annealed condition of low strength and high ductility to the hard, springy and nonductile condition produced by severe cold working.

Annealing conditions are standardized in terms of grain size (0.010 millimeter average grain diameter, to 0.20 millimeter, at a magnification of 75×), as exhibited in the ASTM publication, E79-49T; and the condition of temper produced by cold working is designated fractionally "hard" all the way from "quarter," or even "eighth hard," up to "extra hard," or even "spring" temper. These designations are usually defined by specifications of tensile strength and hardness in the particular type of material under consideration. They are often interpreted as so many "numbers hard," meaning generally the Brown and Sharp gage numbers traversed to reach the end reduction, or more explicitly, the actual percentage reduction in area of section thus effected. For example, soft cartridge brass strip rolled from B & S No. 17 (0.04526 inch thick) to No. 18 (0.0403 inch) would be one number hard, having a cold reduction of about 11 per cent in thickness, or area of section, and an increase of tensile strength, say, from 48,000 to 56,000 psi, making it about "quarter hard." Characteristic data for the plain brasses are given here in the first part of Table 10-1, conveniently abstracted from the more complete data presented in tabular form by Weaver.[2]

Some special copper alloys are strengthened by the form of heat treatment commonly known as "age hardening," but these are more generally in the bronze field and do not contain zinc. Outstanding is the well-known beryllium copper, or beryllium bronze, probably the strongest in heat-treated form of all copper-base alloys. Copper alloys hardened by nickel phosphide (copper-nickel-phosphorus) have attained some importance in recent years and zinc may appear as a constituent of some compositions.[3]

For easy perception and classification, the names of 7 simple or "plain" brasses published as "standard commercial" by the Copper and Brass Research Association[4] have been written into the phase diagram seen in Figure 10-1, and listed in Table 10-1, along with their tensile properties, hardness and electrical conductivity in the form of strip. It is seen that all of these except Muntz metal are alpha brasses. Crampton[5] writes concerning the nature and use of these important alloys:

"The alpha brasses as a group are endowed with one fundamental property of great importance; namely, a degree of plasticity in cold-working operations unattained by any other alloy that is at the same time cheap enough and strong enough to serve as a generally useful engineering material of construction. Obviously, since zinc is cheaper than copper, the composition ordinarily used for the many familiar objects rolled, drawn, stamped and spun from the yellow brasses contains something approaching the maximum allowable percentage of zinc. Some years ago, the alloy known as 'common high brass,' now preferably designated 'yellow brass' and containing 65 per cent copper and 35 per cent zinc, was used in greater quantities than any other in this series. With a strong trend in recent years toward hot-rolling as a basic procedure for the making of strip, that alloy has been largely supplanted by 70–30, or cartridge brass. Further, producers and fabricators both have learned that the many minor variations of alloy sometimes previously made, including 65, 66⅔, 67, 68, 70, 71, 72 and 75 per cent copper, have no practical significance, so that it seems probable that in years to come the 70–30 alloy will be used substantially to the exclusion of all others in this range.

"The color of brasses varies widely with the composition. When increasing amounts of zinc are added to copper, the resulting brasses show a range of color all the way from copper red, which persists for about 5 per cent zinc through a bronze color at about 10 per cent and increasing dilution of color to the typical brass yellow at around 30 per cent zinc. With more than 38 per cent zinc (alloys obtainable only in rod form) the alloy again takes on a buff red cast. This question of color is tremendously important in some applications, particularly costume jewelry, fasteners, fuse boxes, architectural trim, and so forth. In the range from about 80 to 90 per cent copper, the color changes very rapidly, and for such uses a given nominal alloy must be held within rather narrow limits in order that different parts of an assembly may not show variations in color.

"Red brass, containing 85 per cent copper, is, next to cartridge brass, the most important of the nonleaded brasses, being produced in very large quantities in all the common forms but particularly favored for pipe for use in plumbing because of its high resistance to aqueous corrosion, and for a host of manufactured parts for which a combination of ductility, good fabricating properties, golden color, and ease of polishing render it especially well suited.

"A 90–10 mixture, commonly known as 'commercial bronze' because of its color, finds a wide field of application for such things as angles, channels, costume jewelry, etching bronze, grille work, hardware, projectile rotating bands, screen cloth, screws, rivets, and many others.

"The group of brasses containing both alpha and beta solid solutions is typically represented by the 60-40 mixture, and the name Muntz metal originally applied to it has come to mean anything in the entire group covering the interval 61 to 54 per cent copper. The essential features that distinguish the Muntz metal brasses or mixed alpha-beta brasses from the plain alpha brasses are extreme plasticity at red heat followed by conditions tending in the opposite direction when cooled to room temperature: viz., these alloys can be worked hot easily and drastically by rolling, extrusion, and other methods, but are not suitable for severe cold-working operations. The former condition makes for cheapness of production because heated billets may be almost instantly extruded through dies into rod or other shapes at a minimum expenditure of energy, and the long extruded rods may be finished by a cold-drawing operation to give them the exact dimensions and mechanical properties required by a given specification."

The entire range of compositions from gilding to Muntz metal are commonly modified for ease of machining by adding lead in amounts up to 3 per cent for the free-cutting brass, but usually around 0.5 per cent for low-leaded mixtures and around 2 per cent for the high-leaded products. Specific compositions listed as standard by the Copper and Brass Research Association[4] are: Leaded commercial bronze, Cu, 89–Pb, 1.75–Zn, 9.25; Low-leaded brass (tube), 66–0.5–33.5; Low-leaded brass, 65–0.5–34.5; High-leaded brass (tube), 66–1.6–32.4; High-leaded brass, 65–2–33; Extra high-leaded brass, 63–2.5–34.5; Free-cutting brass, 61.5–3–35.5; Leaded Muntz metal, 60–0.6–39.4; Free-cutting Muntz metal, 60–1–39; Forging brass, 60–2–38; and Architectural bronze, 57–3–40. These do not differ significantly in tensile strength, yield strength and hardness from comparable nonleaded mixtures, but are a little lower in ductility (per cent elongation) depending upon the lead content. This wide range of compositions gives ample opportunity for the selection of material to meet varied mill and fabrication requirements, as well as final properties ranging from the softness and pliability of alpha-brass mixtures to the stiffness and strength of

TABLE 10-1. TENSILE PROPERTIES, HARDNESS AND ELECTRICAL CONDUCTIVITY OF COPPER BASE ALLOYS

Alloy[a]	Tensile Strength (psi)	Yield Strength 0.5% extension under load (psi)	Elongation in 2 in. (%)	Modulus of Elast. E (psi)	Rockwell Hardness		Electrical Conductivity (% IACS STD.)
					F	B	
Gilding, 95%				17,000,000			56
0.050 Anneal	34,000	10,000	45		46	38	
Quarter Hard	42,000	32,000	25			38	
Half Hard	48,000	40,000	12			52	
Hard	56,000	50,000	5			64	
Spring	64,000	58,000	4			73	
Commercial Bronze, 90%				17,000,000			44
0.050 Anneal	37,000	10,000	45		53		
Quarter Hard	45,000	35,000	25			42	
Half Hard	52,000	45,000	11			58	
Hard	61,000	54,000	5			70	
Spring	72,000	62,000	3			78	
Red Brass, 85%				17,000,000			37
0.050 Anneal	40,000	12,000	47		59		
Quarter Head	50,000	39,000	25			55	
Half Hard	57,000	49,000	12			65	
Hard	70,000	57,000	5			77	
Spring	84,000	63,000	3			86	
Low Brass, 80%				16,000,000			32
0.050 Anneal	44,000	14,000	550		61		
Quarter Hard	53,000	40,000	30			55	
Half Hard	61,000	50,000	18			70	
Hard	74,000	59,000	7			82	
Spring	91,000	65,000	3			91	

Cartridge Brass, 70%				16,000,000			28
0.050 Anneal	47,000	15,000	62		64		
Quarter Hard	54,000	40,000	43			55	
Half Hard	62,000	52,000	23			70	
Hard	76,000	63,000	8			82	
Spring	94,000	65,000	3			91	
Yellow Brass, 66%				15,000,000			27
0.050 Anneal	47,000	15,000	62		64		
Quarter Hard	54,000	40,000	43			55	
Half Hard	61,000	50,000	23			70	
Hard	74,000	60,000	8			80	
Spring	91,000	62,000	3			90	
Muntz Metal, 60%				15,000,000			28
Soft Anneal	54,000	21,000	45		80		
Eighth Hard	60,000	35,000	30			55	
Half Hard	70,000	50,000	10			75	
Free-Cutting Phosphor Bronze, Cu88-Pb4-Zn4-Sn4-P 0.10 (Rod, 1 in section)				15,000,000			19
Half Hard	60,000	45,000	20			75	
Inhibited Admiralty,* Cu71-Zn28-Sn1				15,000,000			25
1200°F Anneal	45–48,000	13–14,000	62–69		59–60	9	
Hard-6B & S Nos. to 0.040 in. strip	88–114,000	72–74,000	1–4		109–113	90–97	
Naval Brass, Cu60-Zn39.25-Sn0.75 (Rod, 1 in section)				15,000,000			26
Soft Anneal	57,000	25,000	47			55	
Quarter Hard	69,000	46,000	27			78	
Half Hard	75,000	53,000	20			82	

TABLE 10-1. (*Continued*)

Alloy	Tensile Strength (psi)	Yield Strength 0.5% extension under load (psi)	Elongation in 2 in. (%)	Modulus of Elast. E (psi)	Rockwell Hardness		Electrical Conductivity (% IACS STD.)
					F	B	
Leaded Naval Brass, Cu60-Zn37.5-Sn0.75-Pb1.75 (Rod, 1 in section)				15,000,000			26
Soft Anneal	57,000	25,000	40			55	
Quarter Hard	69,000	46,000	20			78	
Half Hard	75,000	53,000	15			82	
Manganese Bronze A, Cu58.5-Sn1-Fe1-Mn0.3-Zn39.2 (Rod, 1 in section)				15,000,000			24
Soft Anneal	65,000	30,000	33			65	
Quarter Hard	77,000	45,000	23			83	
Half Hard	84,000	60,000	19			90	
Aluminum Brass,* Cu76-Al2-Zn22 (general data, tubing)				15,000,000			23
1050°F Anneal	62,000		52		77		
Hard-extruded, reduced and cold drawn to $\frac{3}{4}$ by 0.049 in.	83,000		17		106		
Nickel Silver,[a] Cu65-Ni18-Zn17				18,000,000			6
0.035 Anneal	58,000	25,000	40		85	40	
Quarter Hard	65,000	50,000	20			73	
Half Hard	74,000	62,000	8			83	
Hard	85,000	74,000	3			87	
Nickel Silver,[a] Cu55-Ni18-Zn27				18,000,000			5.5
0.035 Anneal	60,000	27,000	40		90	55	
Hard	100,000	85,000	3			91	
Extra Hard	108,000	90,000	2½			96	
Spring	115,000		2½			99	

Nickel Silver,[a] Cu65-Ni12-Zn20				18,000,000			8
0.035 Anneal	56,000	21,000	42		78	37	
Quarter Hard	65,000	45,000	23			70	
Half Hard	73,000	60,000	11			80	
Hard	85,000	75,000	4			89	
Extra Hard	93,000	79,000	2			92	
Nickel Silver,[a] Cu65-Ni10-Zn25				17,500,000			9
0.035 Anneal	53,000	20,000	43		76	35	
Quarter Hard	65,000	45,000	25			70	
Half Hard	73,000	60,000	12			80	
Hard	86,000	75,000	4			89	
Extra Hard	95,000	76,000	3			92	

[a] In form of 0.040 in strip metal. Mechanical properties of rod and wire will differ somewhat from above values, particularly in the harder tempers, i.e., the tensile strength of yellow brass wire, 0.08 in. in diameter is given as 128,000 psi in the Weaver tables.

* Data from Wilkins and Bunn. "Copper and Copper Base Alloys," New York, McGraw-Hill Book Company, Inc., (1943).

Muntz metal, which in this leaded form can be cut rapidly on high-speed automatic machines into screws and other shapes.

Crampton states:[5]

"The alloy that is used in greater quantity than all others in this group combined is free-cutting brass, containing 61.5 per cent of copper, 3 of lead and 35.5 of zinc. It can be cut at high speeds with low tool pressure, causing minimum rate of tool wear and with very short chips that clear the tool well. However, since free-cutting brass does not lend itself particularly well to cold-working procedures often required in addition to machining operations, a medium-leaded or high-leaded brass, or perhaps even Muntz metal, might be used in preference: viz., a compromise is necessarily made between optimum machining properties obtainable only with the higher lead content and good cold-fabricating properties obtainable with lower lead contents (and usually somewhat higher copper content).

"Where hot pressing or hot forging is to be used as the principal shaping procedure, forging brass containing nominally 60 per cent of copper, 2 of lead and 38 of zinc is by all odds the favored material. This alloy, which is extruded easily, also can be forged into intricate shapes over a wide range of temperature, and the 2 per cent of lead facilitates extensive and ready machining thereafter.

"In fabricating various articles from strips involving drawing or forming operations and where some machining operations are also necessary, various lead-bearing alloys are available, including low-leaded brass, medium-leaded brass, high-leaded brass and extra-high-leaded brass, the total annual consumption of which is substantial. The choice between cartridge brass containing no lead and one of these four lead-bearing brasses will depend on the relative importance of the cold-fabricating operations necessary, on the one hand, and the degree of machining required, on the other, it being impossible to attain the maximum of both these qualities in any one material."

There are many modified or special brasses in which metals such as tin, aluminum, manganese and iron are added or combined in low percentages, usually to strengthen or improve the corrosion resistance of the plain brasses. Five compositions of this chraacter are included in the Copper and Brass Research Association's list and shown here in Table 10-1, along with mechanical properties.

The admiralty alloy and the aluminum brass are used primarily for heat-exchanger tubes, the former well known in this field for many years and now generally modified by adding 0.02 to 0.10 per cent arsenic, antimony or phosphorus to inhibit dezincification. The latter is especially suited to resist the impingement attack of water at high velocities, especially in marine condensers, and is also made with an inhibitor.

The naval brasses have good strength and resistance to salt water corrosion, while the so-called manganese bronze, containing only a fractional percentage of manganese together with larger amounts of tin and iron, is a higher-strength structural material also resistant to salt water corrosion.

Among the Copper and Brass Research Association's standard phosphor bronzes (tin bronzes with a trace of phosphorus) only a free-cutting alloy contains zinc, in addition to copper, lead and tin. This is included in Table 10-1. Phosphor bronzes are preferred alloys for springs and other uses where high strength, elasticity and corrosion resistance are sought.

A very important group of copper-base wrought alloys containing zinc as an essential component is the copper-nickel-zinc series, called "nickel silvers," or originally "German silvers." Four of these compositions are listed in Table 10-1, together with mechanical properties of strip material. These alloys are silver white in various tints and superior to the brasses in strength with hardly inferior cold-working properties. The nickel silvers are used indispensably as base alloys for silver-plated flat or hollow tableware and for many miscellaneous items such as rivets, screws, zippers, optical goods, costume jewelry, dials, etc.

Detailed information on the chemical, physical and mechanical properties of the brasses and other copper-base alloys can be found in the data sheets issued by the Copper and Brass Research Association,[4] and the ASM "Metals Handbook,"[6] in addition to "Copper" (ACS Monograph),[1] from which most of the data given in Table 10-1 have been abstracted.

Casting Alloys. The principal copper-base casting alloys containing zinc that are used today in the United States are listed in Table 10-2.[7] Most of these compositions are mixed brass and bronze, usually in the alpha range of composition, with various admixtures of lead and minor components. Only the manganese bronzes and the 60–40 yellow brass are outside this range. All of these alloys except the first two and No. 8 contain lead, primarily for easy machining, but also where bearings are concerned to promote "wearing in" and minimize seizing with inadequate lubrication. The first two, known as "Government bronze" and "modified Government bronze," are modern adaptations of the original gun metal used for ordnance before the age of steel. Numbers 2 to 6 inclusive define a wide range of machinable foundry alloys in which the principal items of cost, strength and corrosion resistance would be balanced against the copper, zinc and tin contents. For higher strength combined with low cost, the manganese bronzes are indicated. The last three numbers represent leaded tin-bearing nickel silvers used for hardware fittings, valves, plumbing fixtures, dairy and soda-fountain equipment, marine castings, furniture equipment, trim, ornamental castings, etc.

Besides the conventional sand castings, selected copper-base alloys are used for permanent mold castings, die castings, centrifugal castings, precision-investment castings and plaster-molded castings. Details on these applications can be found in the 1948 edition of "The Metals Handbook".[6] One modern alloy of some interest is a copper-zinc-silicon type, 80–16.5–3.5 to 81.5–14–4.5, especially adapted to die castings.

TABLE 10-2. COPPER BASE ALLOYS IN INGOT FORM FOR SAND CASTINGS (From ASTM B 30 – 54)[7]

Classification	Alloy Number	Commercial Designation	Nominal Composition (%)								Tensile Requirements	
			Cu	Sn	Pb	Zn	Ni	Fe	Al	Mn	Tensile Strength (minimum psi)	Elongation (%)
Tin bronze	1A	88–10–0–2 or "G" bronze	88	10		2					40,000	20
	1B	88–8–0–4 or modified "G" bronze	88	8		4					40,000	20
Leaded tin bronze	2A	Steam or valve bronze	88	6	1.5	4.5					34,000	22
	2B	87–8–1–4 (Navy P-c)	87	8.5	0.5	4					36,000	18
High-leaded tin bronze	3B	83–7–7–3	83	7	7	3					30,000	12
	3C	85–5–9–1	85	5	9	1					25,000	8
Leaded red brass	4A	85–5–5–5 or No. 1 composition	85	5	5	5					30,000	20
	4B	Commercial red brass, 83–4–6–7	83	4	6	7					29,000	15
Leaded semi-red brass	5A	Valve composition, 81–3–7–9	81	3	7	9					29,000	18
	5B	Semi-red brass, 76–3–6–15	76	3	6	15					25,000	15
Leaded yellow brass	6A	High copper yellow brass	71	1	3	25					35,000	25
	6B	Commercial No. 1 yellow brass	67	1	2	30					30,000	20
	6C	60–40 yellow (Naval) brass	61	1	1	36.7			0.3		40,000	15
Leaded high-strength yellow brass (leaded manganese bronze)	7A	Leaded manganese bronze	59	0.75	0.75	37		1.25	0.75	0.5	60,000	15

High-strength yellow brass (manganese bronze)	8A	No. 1 manganese bronze	58			38.5		1.25	1.25	1.0	65,000	20
	8B	High strength manganese bronze†									90,000	18
	8C										110,000	12
Leaded nickel brass (leaded nickel silver)	10A	12% leaded nickel silver	57	2	9	20	12				30,000	8
Leaded nickel bronze (leaded nickel silver)	11A	20% leaded nickel silver	64	4	4	8	20				30,000	8
	11B	25% leaded nickel silver	66.5	5	1.5	2	25				45,000	15

† Several nominal compositions are available for alloy numbers 8B and 8C which meet the physical requirements for these alloys.

Sintered Brass Powders. Brasses of varied composition are found to be readily adaptable to powder-metallurgical processing by compacting under pressures ranging from 20 to 50 tons per square inch (using a small amount of lubricant such as zinc stearate) and sintering at temperatures of 850 to 950°C. A composition which has attained prominence in the manufacture of structural parts by this specialized method is as follows: copper, 77 to 80 per cent; lead, 1 to 2 per cent; and zinc, remainder. The tensile properties that can be achieved in these compositions depend upon the density, which under optimum conditions can reach about 90 per cent of the theoretical value, with strength and ductility considerably inferior to the values obtained in comparable cast or wrought and soft-annealed alloys, but nevertheless in a useful range. Typical properties for the above alloy are: tensile strength, around 30,000 psi; percentage elongation, approximately 20; and Brinell hardness (500 kilogram load), 43.

However, this is a field susceptible to active development, and special brass powders[8] containing small additions of nickel and phosphorus or nickel, phosphorus and iron are already available for the attainment of properties approximating 45,000 psi tensile strength, 30 per cent elongation and 65 Brinell hardness. Shrinkage during sintering, which usually amounts to 1 or 2 per cent, can also be reduced practically to zero, or raised substantially, according to specific requirements.

Light Metal Alloys

Next in importance among alloys containing zinc as a secondary or minor ingredient are the light metal alloys, which in recent years have consumed close to one-half per cent of the slab zinc in the United States. Zinc is used both in cast and wrought aluminum alloys, but not as extensively as copper, magnesium and silicon, which must be regarded as essential in the practice of our extensive aluminum-alloy technology. Similarly, it is a common constituent of both cast and wrought magnesium alloys.

Alloys Based on Aluminum. The aluminum-zinc phase diagram (Figure 10-2), according to Anderson's evaluation,[9] indicates at both extremities a constitutional condition favorable for the production of useful alloys: i.e., there is far-reaching solid solubility of zinc in aluminum and limited solid solubility of aluminum in zinc. The use of aluminum as an essential constituent of zinc-base die castings is discussed in Chapter 9.

The "Alcoa Aluminum Handbook,"[10] shows chemical composition limits of 37 wrought alloys utilizing various amounts of silicon, iron, copper, manganese, magnesium, chromium, nickel, titanium and zinc as alloying constituents, among which 27 contain only 0.10 to 0.30 per cent of zinc, but four are so-called zinc-type alloys. These are: No. 7072, with 0.8 to 1.3 per cent zinc, 0.7 silicon plus iron, 0.10 copper, 0.10 manganese, 0.10

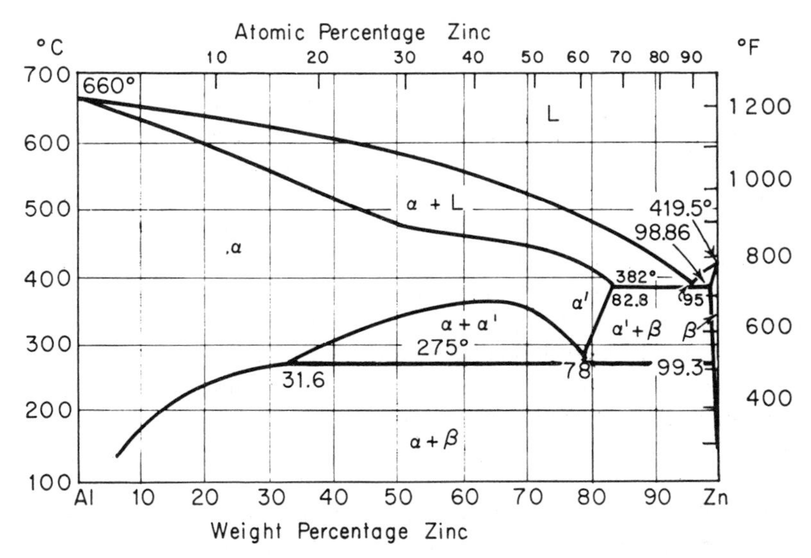

Figure 10-2. Phase diagram of the aluminum-zinc system. From E. A. Anderson, "Metals Handbook," 1948 ed. (*Courtesy American Society for Metals.*)

magnesium, and 0.15 total impurities (an alloy for certain forms of cladding); No. 7075, with 5.1 to 6.1 per cent zinc, 0.5 silicon, 0.7 iron, 1.2 to 2.0 copper, 0.3 manganese, 2.1 to 2.9 magnesium, 0.18 to 0.40 chromium, 0.2 titanium and 0.15 others; No. 7076, with 7.0 to 8.0 per cent zinc, 0.4 silicon, 0.6 iron, 0.3 to 1.0 copper, 0.3 to 0.8 manganese, 1.2 to 2 magnesium, 0.2 titanium and 0.15 others; and No. 7277, with 3.7 to 4.3 per cent zinc, 0.5 silicon, 0.7 iron, 0.18 to 1.7 copper, 1.7 to 2.3 magnesium, 0.18 to 0.35 chromium, 0.1 titanium and 0.15 others.

Of these, the first two types are also found in Reynolds' "Aluminum Data Book"[11] and in "Tentative Specifications of ASTM for Aluminum-Alloy Sheet and Plate."[12] The "Tentative Specifications for Aluminum-Alloy Bars, Rods and Wire,"[13] "Aluminum-Alloy Extruded Bars, Rods and Shapes,"[14] "Aluminum-Alloy Extruded Tubes,"[15] and "Aluminum-Alloy Die Forgings,"[16] also include the alloy with 5.1 to 6.1 per cent zinc, which was originally known as "75s" and is superior to all other wrought aluminum alloys in strength. Typically, it can be heat-treated to a tensile strength of approximately 80,000 psi with yield point around 70,000 psi and percentage elongation close to ten. Thus, zinc is an essential component of the highest-strength wrought aluminum alloy now in common use.

In the field of cast aluminum alloys, zinc appears as a principal alloying component in a rather small proportion of the recognized commercial compositions. Thus, five out of the twenty alloys listed in the "ASTM Tentative Specifications for Aluminum-Base Alloy Sand Castings"[17] are of the zinc type, with from 2.7 to 8 per cent zinc combined with smaller amounts of copper, magnesium, iron, silicon, manganese, chromium and titanium. Most of the others have zinc in minor quantities (usually below 1 per cent) as part of their composition. As in the case of wrought alloys, in general the highest strengths are realized with these zinc-type alloys in heat-treated condition, although other types are competitive in this respect, particularly for strength combined with moderate elongation. The attainable tensile strength, yield point (0.2 per cent offset) and percentage elongation are approximately 37,000 psi, 30,000 psi and 1, respectively—in material that may be described as essentially a 4½ per cent Zn, 2 per cent Mg alloy with some Cr and Mn. The particular disadvantage of the zinc alloys as a group is diminished strength at elevated temperatures. They are especially suitable for brazing.

Similar conditions apply in the ASTM Specifications of alloys for permanent mold casting,[18] with four zinc-type alloys out of 21 tabulated. Most alloys for soldering aluminum are said to contain from 50 to 75 per cent tin, with the remainder usually zinc.[19] Brazing alloys, however, are aluminum-base, commonly of the silicon type, but sometimes sharing zinc with silicon.

Alloys Based on Magnesium. Zinc alloys freely with magnesium, with a favorable condition of limited and temperature-dependent solid solubility on the magnesium side and the unfavorably prompt appearance of an extremely brittle intermetallic phase on the zinc side, without appreciable solid solubility. Figure 10-3 shows the binary phase diagram.[20] Taken in connection with the useful function of zinc in aluminum, it is not surprising to find that zinc is used in both cast and wrought magnesium alloys, generally in combination with a larger amount of aluminum, as for example: Mg-Al, 2.5 to 3.5 per cent; Zn, 0.6 to 1.4 per cent; plus minor quantities of other metals. This is an alloy specified by ASTM[21] for sheet and is expected to have a tensile strength around 40,000 psi and a yield point close to 30,000 psi, with percentage elongation about 4 to 5, in the hardened condition.

An interesting alloy specified for bars, rods and shapes[22] contains 4.8 to 6.2 per cent zinc and 0.45 per cent minimum of zirconium, with tensile strength requirements of 45 to 6 thousand psi, 36 to 8 thousand psi yield and 4 per cent elongation in the prescribed condition.

Magnesium-base alloys for sand castings[23] containing 5 to 10 per cent aluminum, with 0.3 to 3.5 zinc, plus a small amount of manganese, run from 18,000 psi tensile strength and 10,000 psi yield point, as-cast, to approximately 35,000 psi strength, 18,000 psi yield point with small elongation, in heat-treated form. For permanent-mold castings,[24] magnesium with approximately 10 per cent aluminum and 2½ per cent zinc (and 0.10 man-

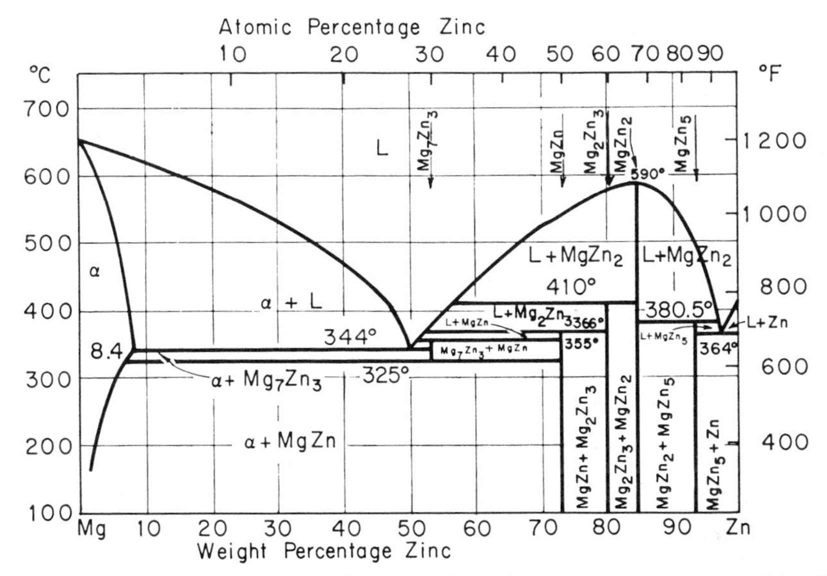

Figure 10-3. Phase diagram of the magnesium-zinc system, as constructed by J. B. Clark and F. N. Rhines up to 85% zinc,[20] and J. B. Hess, beyond 85% zinc. From "Metals Handbook," 1948 ed.[20] *(Courtesy American Institute of Mining, Metallurgical and Petroleum Engineers and American Society for Metals.)*

ganese) is one of the five alloys listed and another contains 3 per cent zinc without aluminum. For die castings aluminum is in approximately the same amount with zinc down to one per cent, or even less, including the customary small amount of manganese. A high-aluminum (8.3 to 9.7 per cent), low-zinc (1.7 to 2.3 per cent) alloy is also supplied as filler metal for brazing magnesium.[25]

Miscellaneous Alloys Containing Zinc as Second or Minor Constituent

Next to copper, aluminum and magnesium, the metal most frequently utilizing zinc in its varied combinations is tin. From a list of about 150 tin-base alloys presented in the second edition of the Tin Monograph,[26] those shown in Table 10-3 contain zinc in significant amount.

It is seen that the principal plain tin-zinc alloys are (1) the fusible metal corresponding to the eutectic containing about 9 per cent zinc and melting at 199°C, as shown in the phase diagram, Figure 10-25, and (2) the simple aluminum solders, composed of various amounts of the zinc phase associated with the eutectic mixture. These alloys are, however, greatly modified and supplemented by other compositions, as described on pp. 555–6.

Britannia metal, pewter and such especially designated metals as

TABLE 10-3. TIN-BASE ALLOYS CONTAINING ZINC*

Name	Percentage Composition									
	Sn	Zn	Al	Pb	Ag	Sb	Cu	Fe	Bi	Others
Fusible Metal	91	9								
Aluminum Solder	85–50	15–50								
Burgess	76	21	3							
Grimm	69.1	1.4		28.8	0.7					
Grimm	50	25		25						
Wegner & Guhr	80	20								
	87–73	8–15	5–12							
Various	33.4	44.4		22.2						
	49.5	20.3		26.1		3.4	1.1			
Ashberry Metal	82.5	1.0		14.4		14.4	2.1			
Ashberry Metal	79.8	2.0				15.2	3.0			
Blatt Silver	91.1	8.3		0.4				0.2		
Britannia	77.9	2.8				19.4				
Birmingham II	85.5	3.0				10.5	1.0			
Cast	85.5	1.0				10.5	1.0			
English	85.5	3.0				9.7	1.8			
German	84.0	5.0				9.0	2.0			
Minofor	68.5	10.0				18.2	3.3			
Tutania Plate	90.0	1.3		6.0			2.7			
Tutania Engl.	80.0	1.3				16.0	2.7			
Spoons	84.7	1.0				5.0	3.7		4.9	
Spoons	84.5	1.5				5.6	3.7		4.9	
Hammonia Metal	64.5	32.3					3.3			
Kamarsh (bearing)	87.9	1.5				5.2	3.7		1.7	
Minofor (bearing)	66.7	9.1				20.2	4.0			
Nonoxidizable (Lemarquand)	9.0	37.0					39.0			Co, 8; Ni, 7.
Nonoxidizable (Marties)	10.0	18.0					17.0	10.0		Ni, 35; Cr, 10.
Parson's White Brass (Bearings)	60.0	35.0					5.0			
Pewter	88.4	0.9				7.2	3.5			
Platinum-Gold (Cooper's Mirror)	27.5	3.5					58.0			
Queen's Metal	88.5	1.0				7.0	3.5			
Queen's Metal	52.6	13.0		17.2		17.2				
Queen's Metal	87.0	1.0				8.5	3.5			
Queen's Metal	73.4	8.9		8.8		8.9				
Queen's Metal	88.5	0.9				7.1	3.5			
Silver Foil	90.0	10.0								
Silver Leaf	91.4	8.3								

* Including several alloys containing less than 50 per cent of tin.

"Queen's metal" are typically tin, hardened and otherwise modified by additions of antimony and copper, tending toward the lower alloyages in pewter, but sometimes modified by additions of zinc or other metals, and not clearly differentiated. The old-fashioned pewters, essentially tin-lead compositions, are no longer important. Tin-base bearing metals, or Babbitts, are also tin-antimony-copper alloys, with occasional modifications, as in the listed compositions containing zinc. Lead Babbitts and the ordinary type metal are lead-antimony-tin alloys, no longer tin-base but very rich in lead and not well adapted to the inclusion of zinc as an alloying addition; the phase diagram (Figure 10-22) shows limited liquid miscibility between lead and zinc. However, a few special bearing alloys and type metals containing zinc are included in Table 10-3.

Silver solders are composed of silver, copper and zinc, with cadmium, tin, nickel and manganese also used in some of the alloys. Woldman[27] lists the compositions shown in Table 10-4. The variable property of special importance in these silver solders is melting point. This may be surveyed in the basic ternary system, Ag-Cu-Zn, by reference to the liquidus surfaces of the diagram shown in Figure 10-32.

A proprietary silver brazing alloy containing 18 per cent cadmium, in

TABLE 10-4. SILVER SOLDERS*

Name	Percentage Composition			
	Copper	Silver	Zinc	Other
Sterling	2.5	80	18	
Hard	13	80	6.8	
Bur. Standards	14	40	6	
General Electric	20	65	15	
	45	20	35	
	4.5	55.5		40 Cadmium
Medium	20–23	75–70	5–7.5	
Quick	21–25	63–57	10–12	3.8–6.2 Tin
French	23	32	17	18 Cadmium
Handy IT	16	80	4	
Handy RT	25	60	15	
Handy ETX	34	50	16	
Handy ET	28	50	22	
Handy DE	30	45	25	
Handy SS	30	40	28	
Handy DT	36	40	24	
Handy NT	38	30	32	
Handy ATT	45	20	30	
Handy NE	52.5	25	22.5	
Handy TL	53	9	38	

* From data by N. E. Woldman in "Engineering Alloys," (American Society for Metals, 1954).

addition to silver-copper-zinc, has broad commercial application. Binary silver-zinc alloys containing about 25 per cent zinc are used for brazing purposes requiring a tarnish-resistant, corrosion-resistant alloy of about the same color as silver. Silver brazing alloys are extensively used on practically all nonferrous metals and alloys and steel because of their high strength, low melting points, free-flowing properties and resistance to corrosion.

Useful tabulations of widely accepted brazing compositions are shown in Tables 10-5, 6 and 7, as extracted from various sources.

Zinc is important in another precious metal field, namely, as an almost indispensable component of the white golds and white-gold solders, although

TABLE 10-5. BRAZING FILLER METAL ASTM B-260-52T, SILVER

AWS—ASTM Classification	Ag (%)	Cu (%)	Zn (%)	Cd (%)	Ni (%)	Sn (%)	Other Elements (total %)
BAg-1	44–46	14–16	14–18	23–25	—	—	0.15
BAg-2	34–36	25–27	19–23	17–19	—	—	0.15
BAg-3	49–51	14.5–16.5	13.5–17.5	15–17	2.5–3.5	—	0.15
BAg-4	39–41	29–31	26–30	—	1.5–2.5	—	0.15
BAg-5	44–46	29–31	23–27	—	—	—	0.15
BAg-6	49–51	33–35	14–18	—	—	—	0.15
BAg-7	55–57	21–23	15–19	—	—	4.5–5.5	0.15
BAg-8	71–73	27–29	—	—	—	—	0.15
BAg-9	64–66	19–21	13–17	—	—	—	0.15
BAg-10	69–71	19–21	8–12	—	—	—	0.15
BAg-11	74–76	21–23	2.5–3.5	—	—	—	0.15

TABLE 10-6.

ALLOYS FOR BRAZING BRASS AND BRONZE

Composition (%)					Remarks
Zn	Cu	Sn	Al	Other	
50	50	—	—	—	For brazing low-melting-point brasses
50	44	4	—	Pb 2	
48	43	—	—	Ag 9	
52	—	30	18	—	For brazing aluminum bronzes
52	—	30	17.5	Ag 0.5	

ALLOYS FOR BRAZING NICKEL-SILVERS

Composition (%)			Remarks
Zn	Cu	Ni	
57	35	8	The joint is strong and the color most closely matches that of the parent metal.
50	37.5	12.5	

TABLE 10-7. PERCENTAGE COMPOSITION OF BRAZING FILLER MATERIAL (ASTM B260-52T)

AWS-ASTM Classification	Cu	Sn	Fe	Mn	Ni	P	Pb	Al	Si	Ag	Other	Zn
Copper-Zinc												
B Cu-Zn-1	58.0–62.0	—	—	—	—	—	0.05	0.01	—	—	0.50	Rem
B Cu-Zn-2	57.0 (min)	0.25–1.0	—	—	—	—	0.05	0.01	—	—	0.50	"
B Cu-Zn-3	56.0 (min)	1.1	0.25–1.25	1.0	1.0	—	0.05	0.01	0.25	—	0.50	"
B Cu-Zn-4	50.0–55.0	—	0.10	—	—	—	0.05	—	—	—	0.50	"
B Cu-Zn-5	50.0–53.0	3.0–4.5	0.10	—	—	—	0.05	—	—	—	0.50	"
B Cu-Zn-6	46.0–50.0	—	0.25	—	9.0–11.0	—	0.05	0.005	0.15	—	0.50	"
B Cu-Zn-7	46.0–48.0	—	—	—	10.0–11.0	0.20–0.50	—	—	0.15	0.30–1.0	0.10	"
Magnesium												
B Mg	0.05			0.10 Min	0.01			8.3–9.7	0.3	Mg balance	0.03	1.7–2.3

TABLE 10-8. WHITE GOLD AND WHITE-GOLD SOLDERS*

Name	Percentage Composition					
	Gold	Nickel	Zinc	Silver	Copper	Cad-mium
White Gold	85–75	10–8	2–9			
Solder, 10 kt.	30		12	55	1	2
White Gold, 10 kt.	41.6	25	8.3		25	
White Gold, 14 kt. (A)	58.3	17	7.6		17	
(B)	59	12.3	3.2		25.5	
Solder, 14 kt.	50		11	36	1	2
White Gold, 18 kt.	75		5.5	18.5	1	
Solder, 18 kt. (A)	60.8	6	5.2	14	4	
(B)	61	7	17	13.5	1.5	
(C)	82	10	6			2

* From data by N. E. Woldman in "Engineering Alloys" (American Society for Metals, 1954).

nickel is generally used as the principal whitening agent. A rather large number of compositions are given by Woldman,[28] as shown in Table 10-8.

Alloys of zinc with the common high-melting point metals—iron, cobalt, nickel, manganese, molybdenum, tungsten and platinum—cannot be made in low zinc concentrations, owing to the volatility of zinc. In some cases, as well-known with iron, intermetallic compounds rich in zinc are formed and utilized in special applications. Thus, hot-dipped zinc coatings contain recognizable layers of such compounds between the base metal, iron, and the exterior zinc coating. Zinc coatings are discussed in Chapter 9. Various compounds of this character may be located in the phase diagrams of the ensuing section.

Phase Diagrams

The phase diagrams presented here are intended to serve simply as small-scale concentration-temperature maps of the systems represented, from which a general idea of the phases present under equilibrium conditions can be read. Discussion of the individual diagrams or of the underlying phase theory is not considered appropriate for this book. For the convenience of readers wishing to pursue these subjects further, a supplementary list of references is included.[a-h]

The binary and ternary phase diagrams have been taken generally from the complete collection published by Interscience Publishers, Inc., in the first volume of Colin J. Smithells' "Metals Reference Book,"[29] containing also acknowledgments to authors, publishers and journals covering original sources.

In addition, CeZn and LaZn diagrams, not shown here, were published in Germany during the war[30] and abstracted in the "American FIAT Review of German Science,"

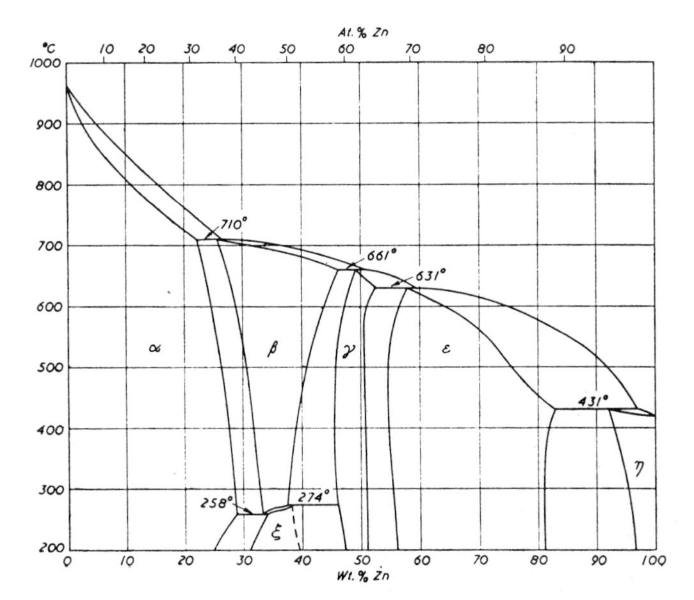

Figure 10-4. Ag-Zn.

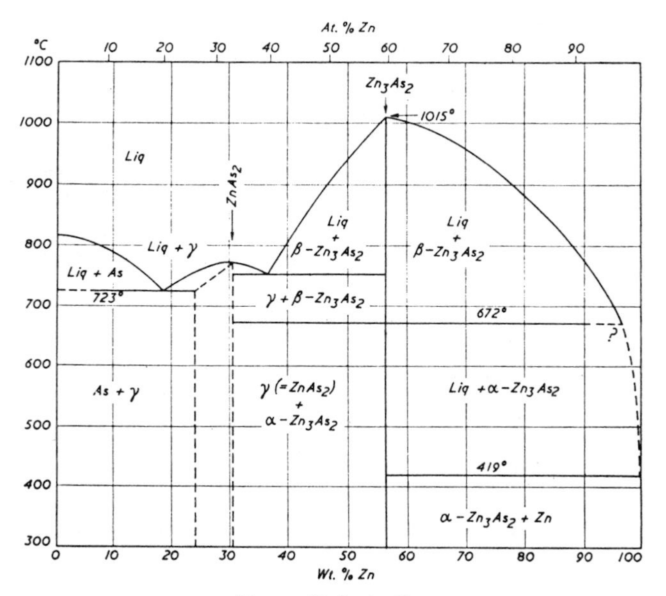

Figure 10-5. As-Zn.

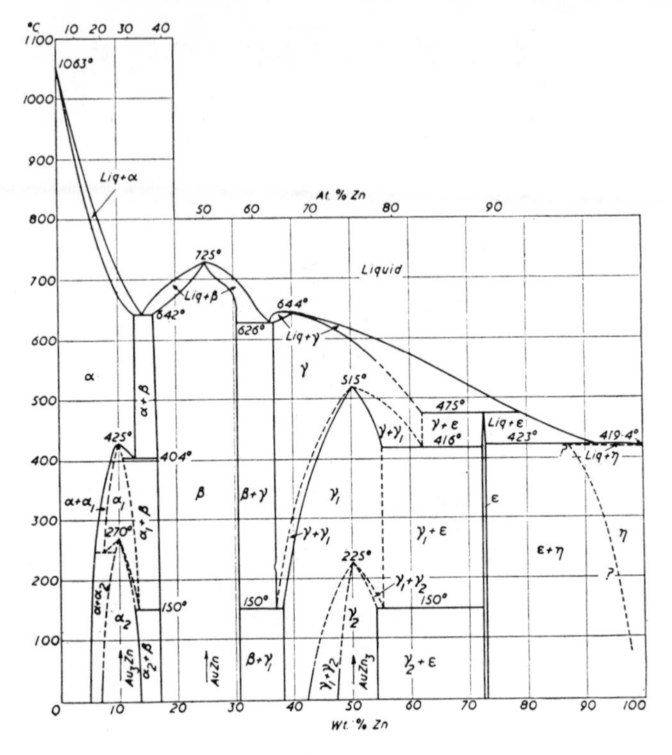

Figure 10-6. Au-Zn.

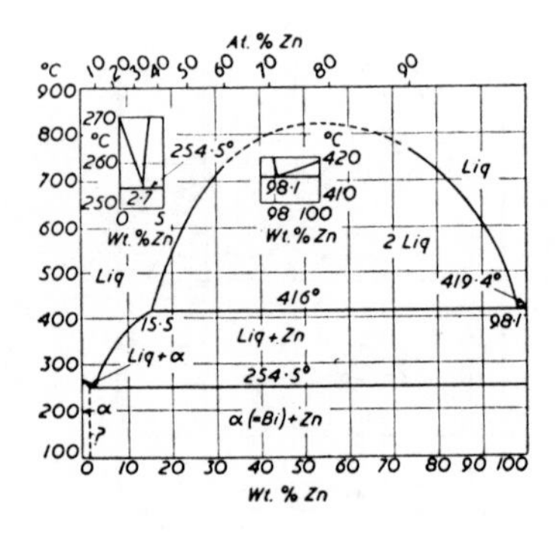

Figure 10-7. Bi-Zn.

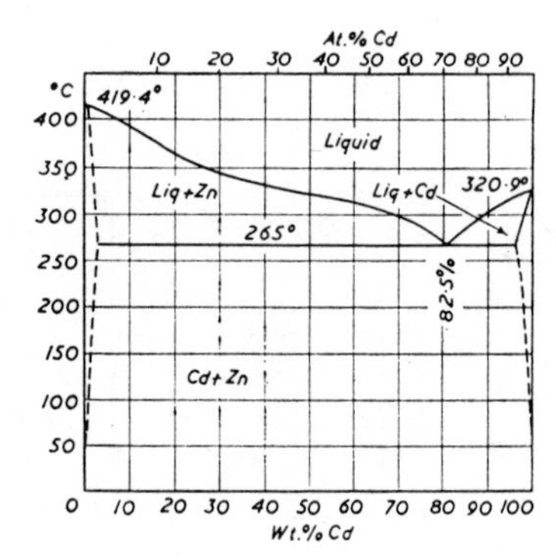

Figure 10-8. Cd-Zn.

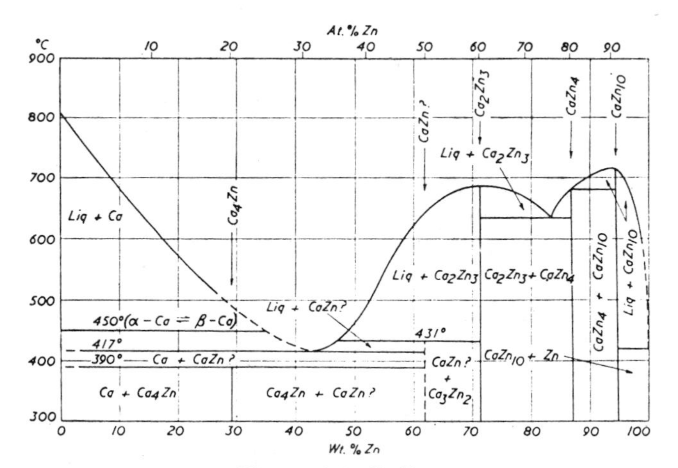

Figure 10-9. Ca-Zn.

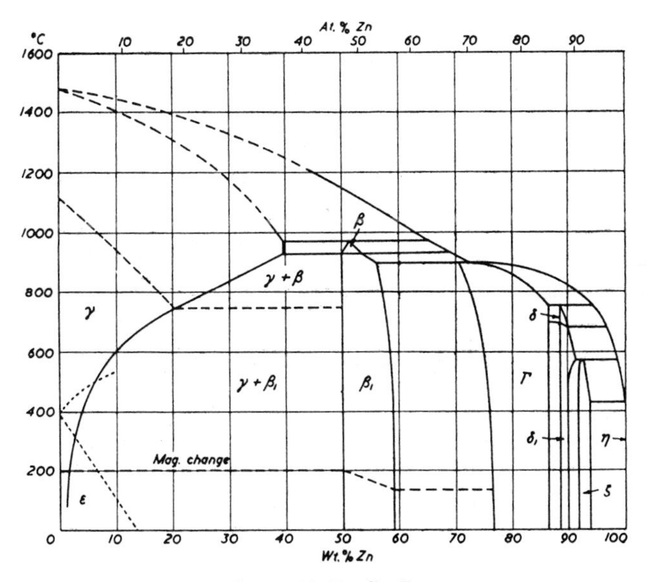

Figure 10-10. Co-Zn.

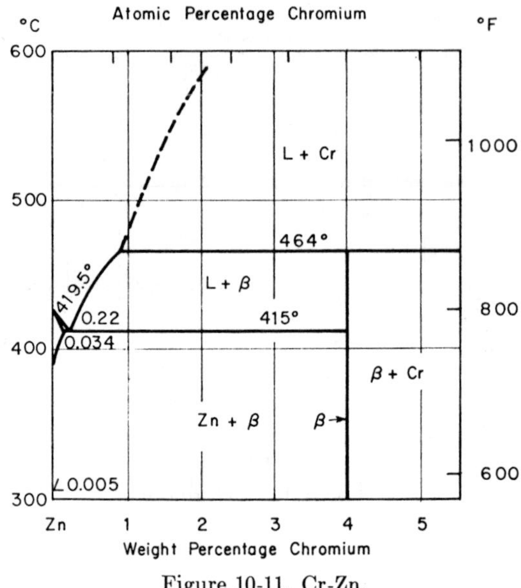

Figure 10-11. Cr-Zn.

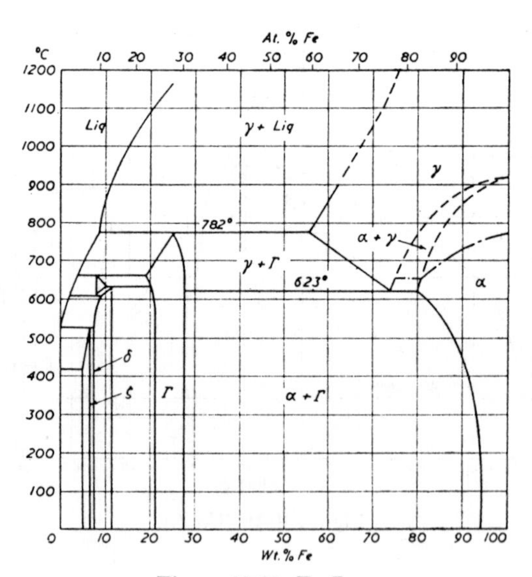

Figure 10-12. Fe-Zn.

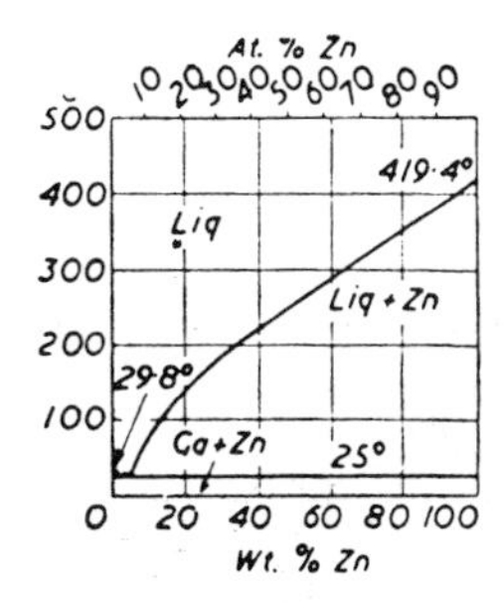

Figure 10-13. Ga-Zn.

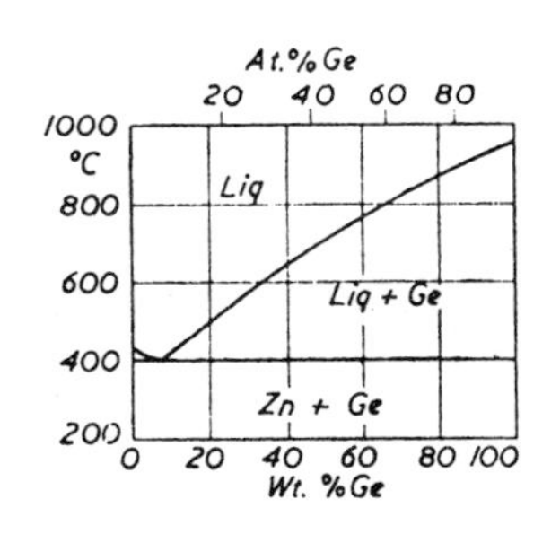

Figure 10-14. Ge-Zn.

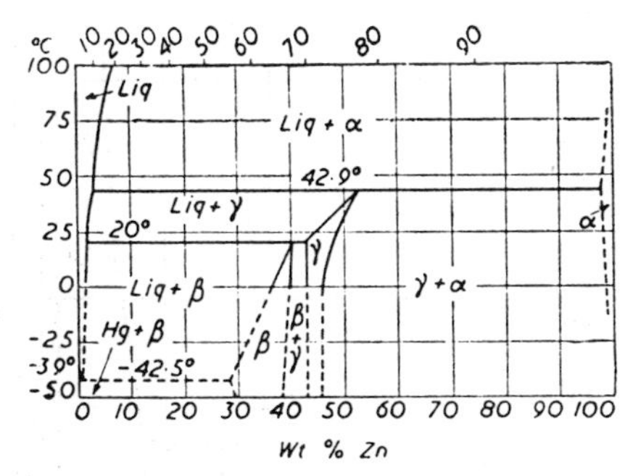

Figure 10-15. Hg-Zn.

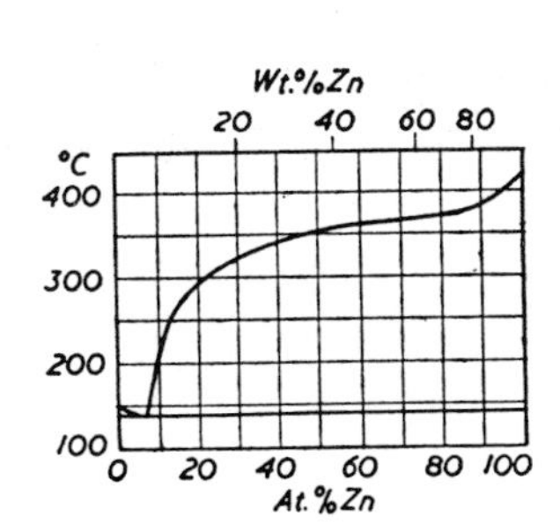

Figure 10-16. In-Zn.

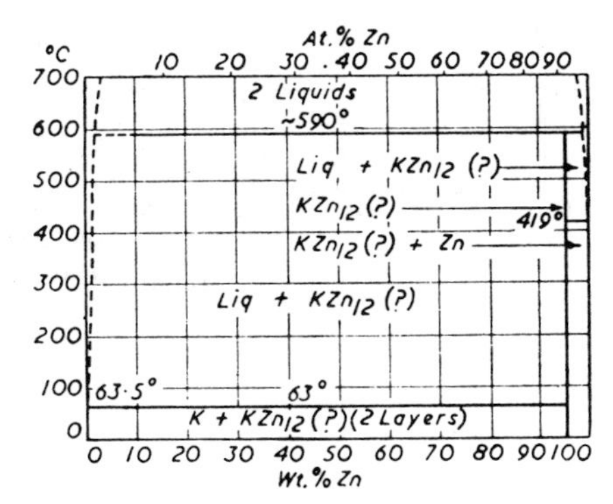

Figure 10-17. K-Zn.

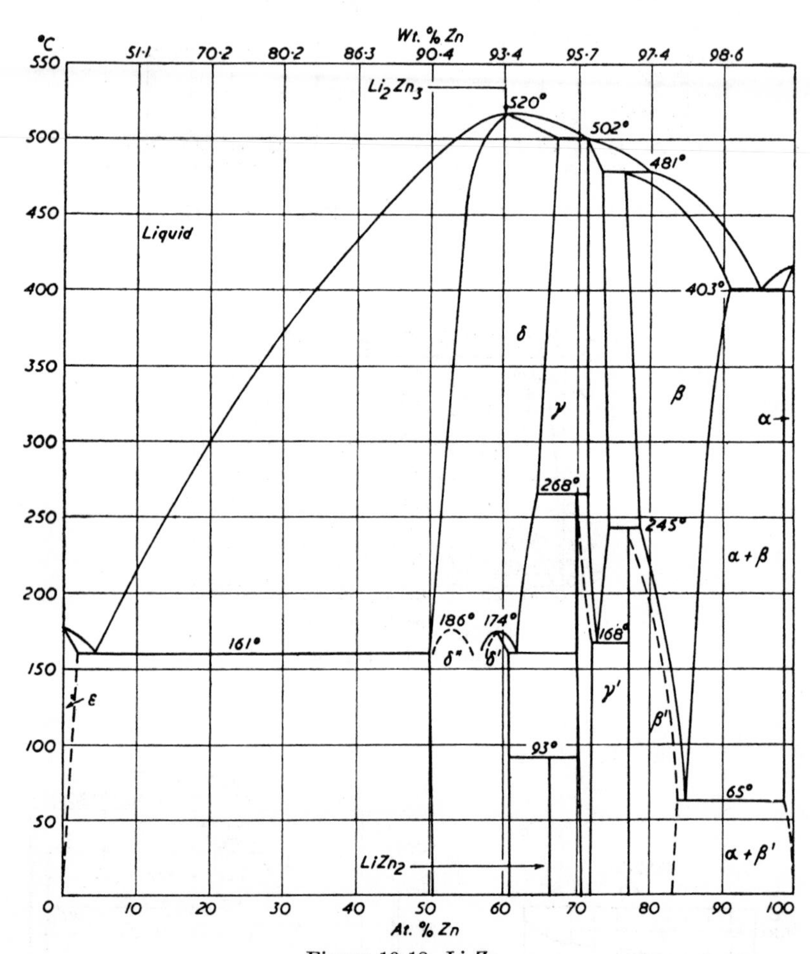

Figure 10-18. Li-Zn.

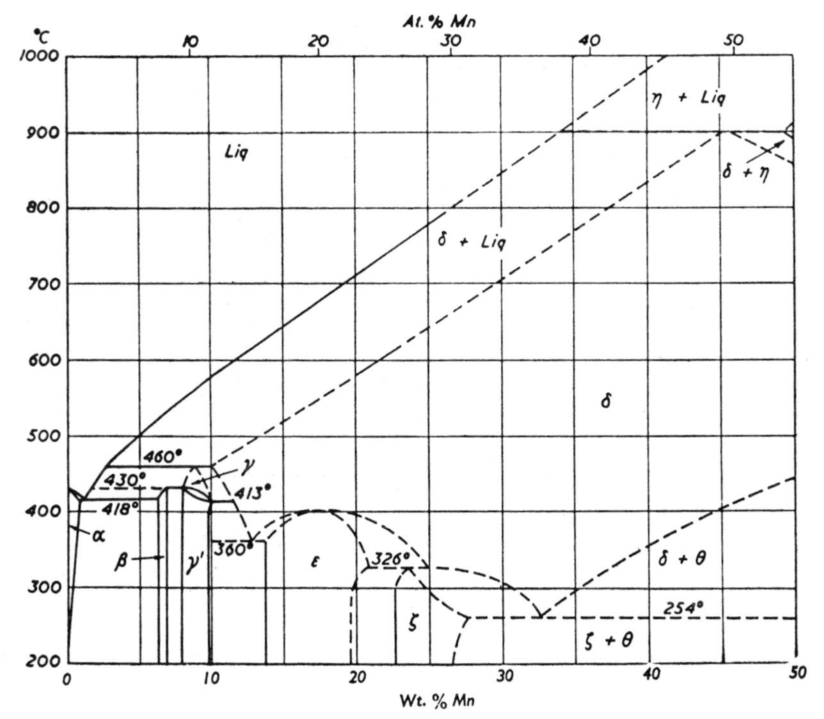

Figure 10-19. Mn-Zn.

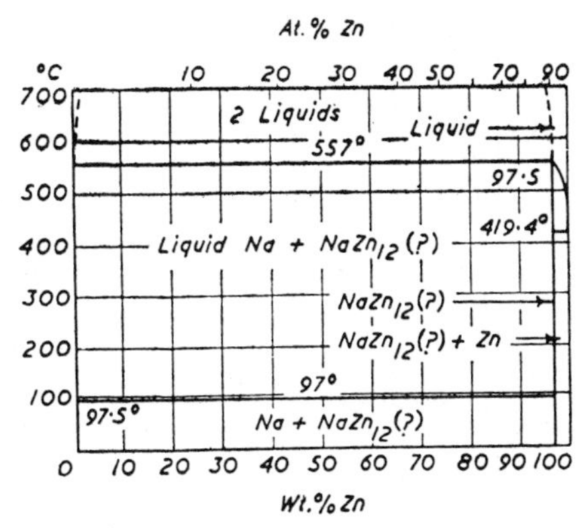

Figure 10-20. Na-Zn.

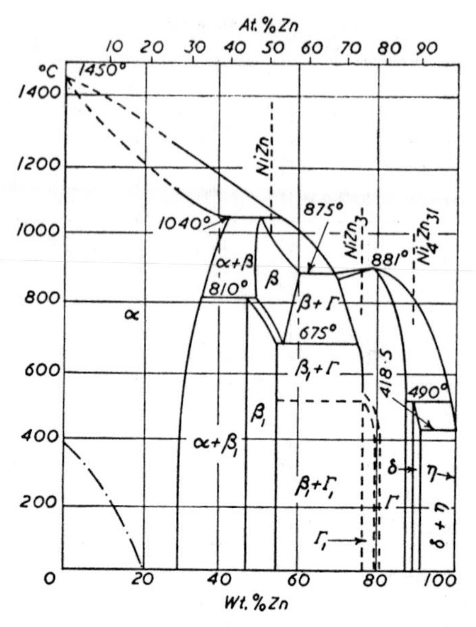

Figure 10-21. Ni-Zn.

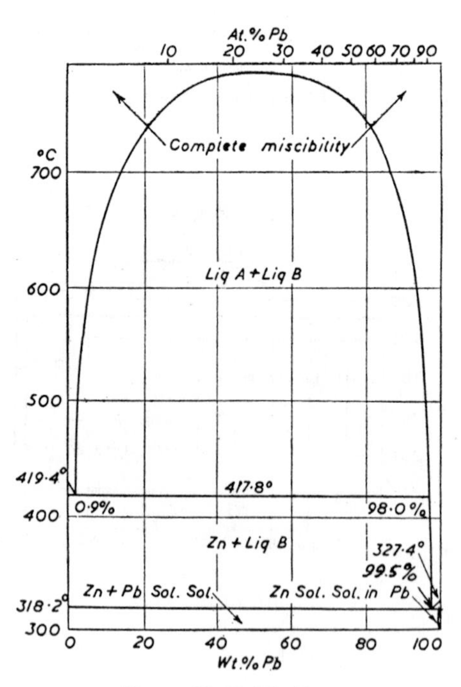

Figure 10-22. Pb-Zn.

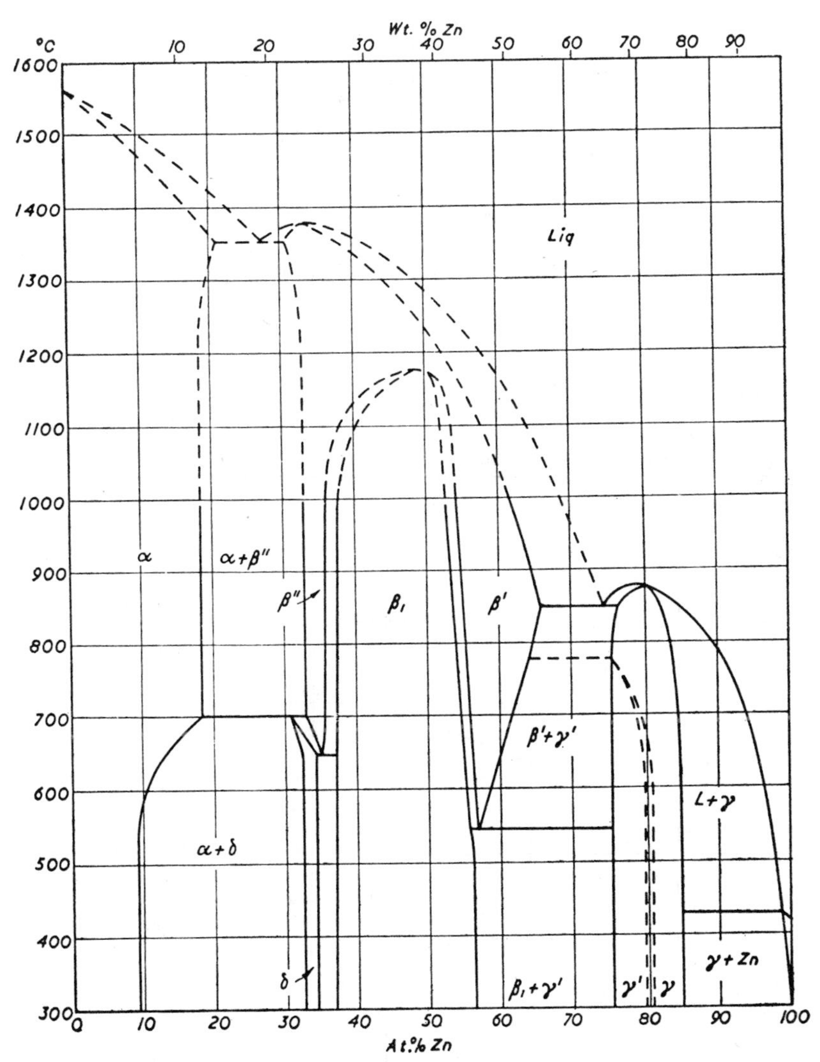

Figure 10-23. Pd-Zn.

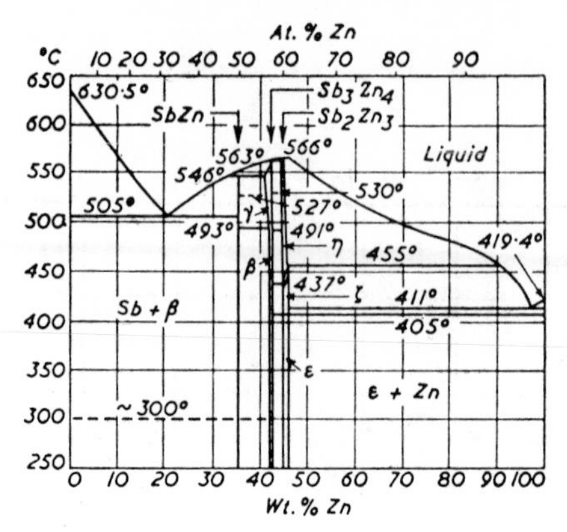

Figure 10-24. Sb-Zn.

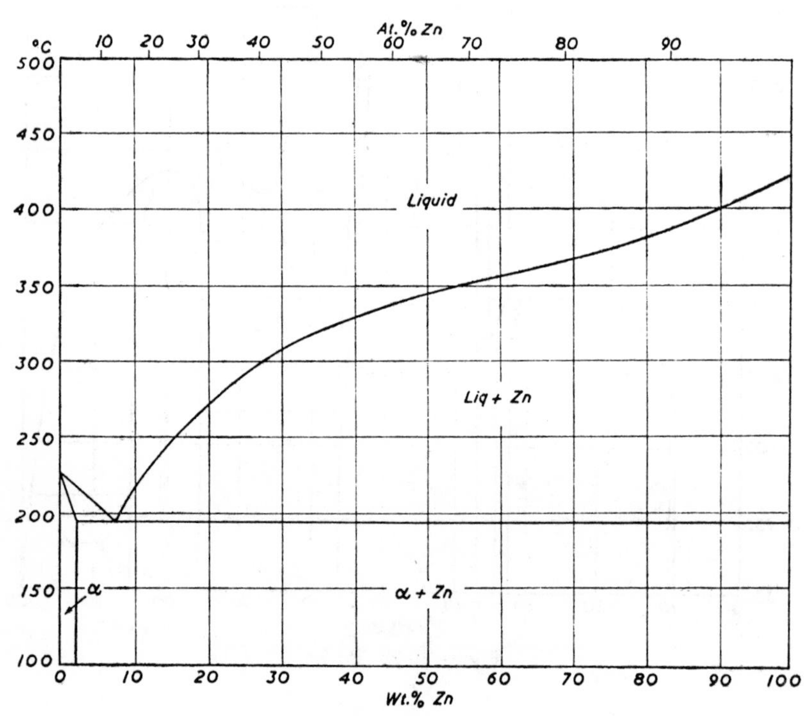

Figure 10-25. Sn-Zn.

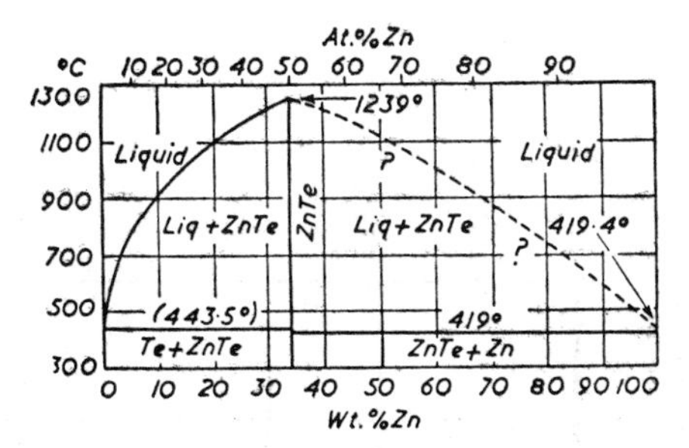

Figure 10-26. Te-Zn.

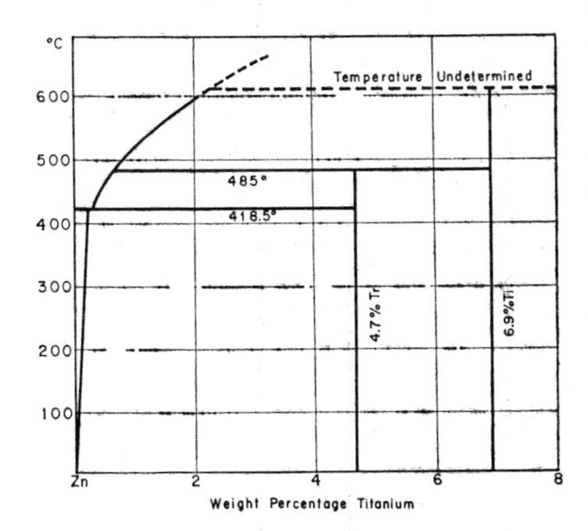

Figure 10-27. Ti-Zn.

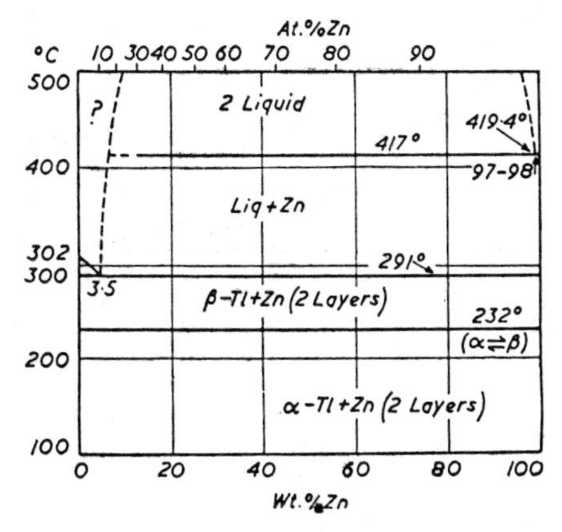

Figure 10-28. Tl-Zn.

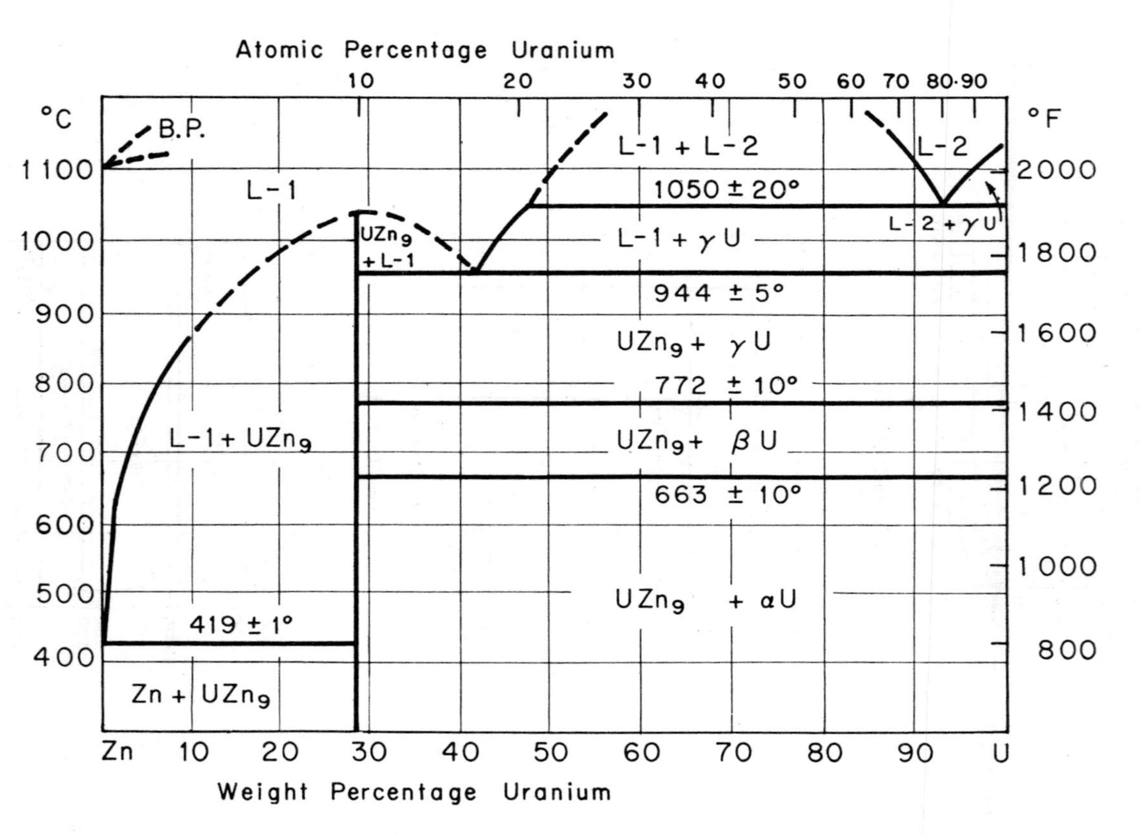

Figure 10-29. U-Zn at 5 atmospheres.

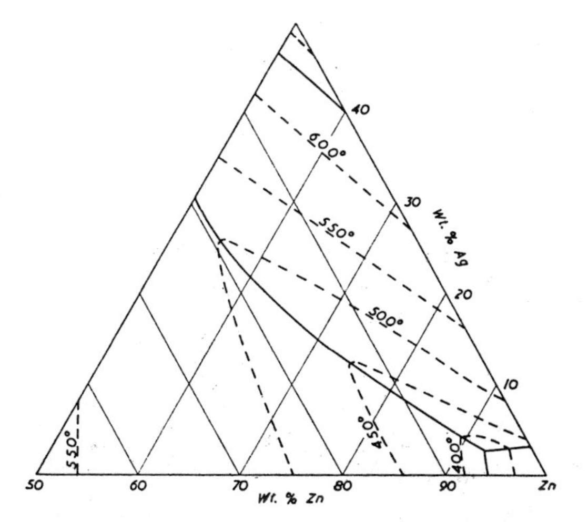

Figure 10-30. Ag-Al-Zn liquidus zinc corner.

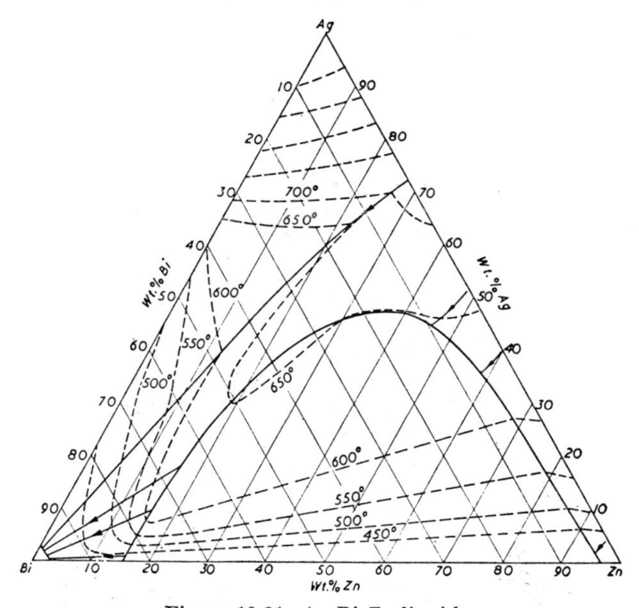

Figure 10-31. Ag-Bi-Zn liquidus.

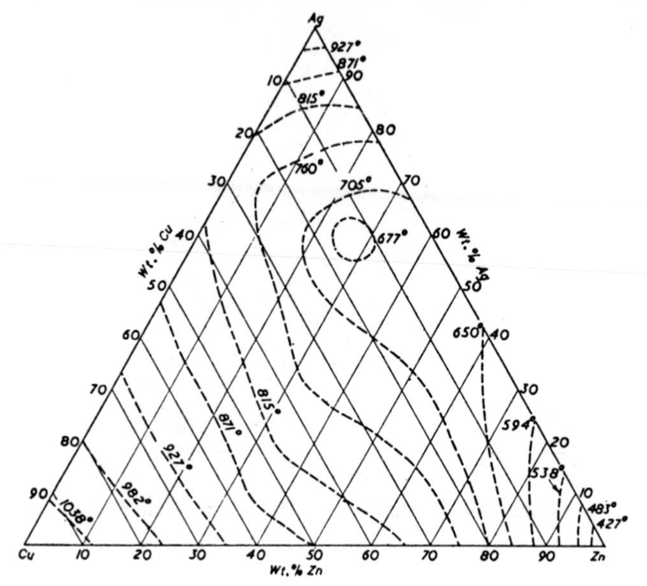

Figure 10-32. Ag-Cu-Zn liquidus.

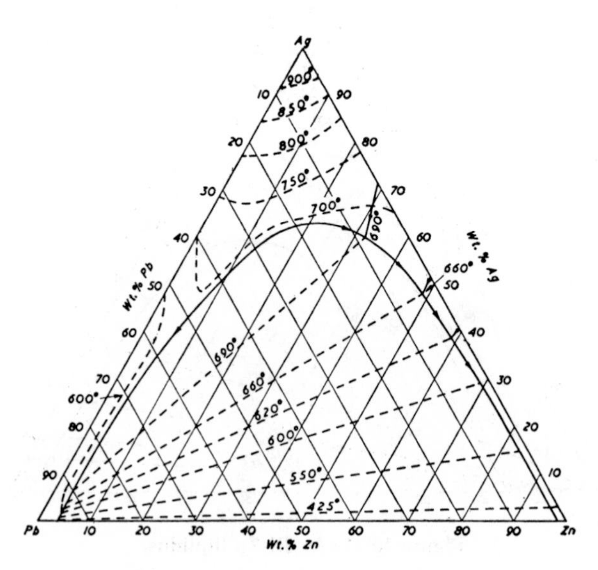

Figure 10-33. Ag-Pb-Zn liquidus.

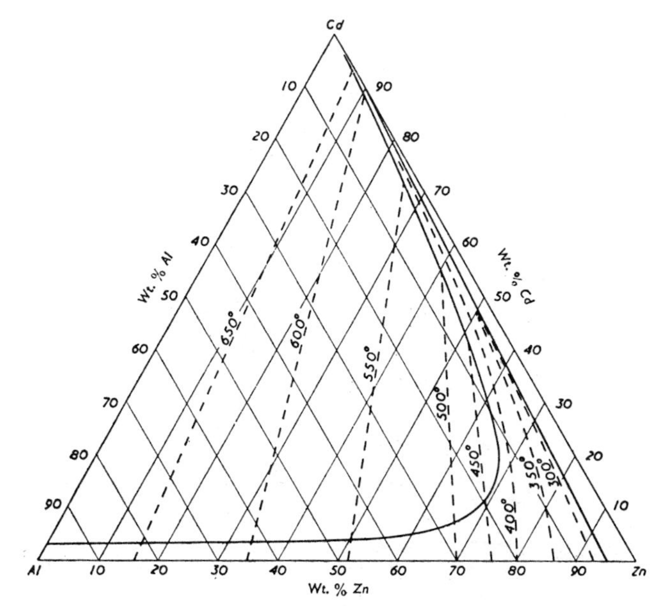

Figure 10-34. Al-Cd-Zn liquidus

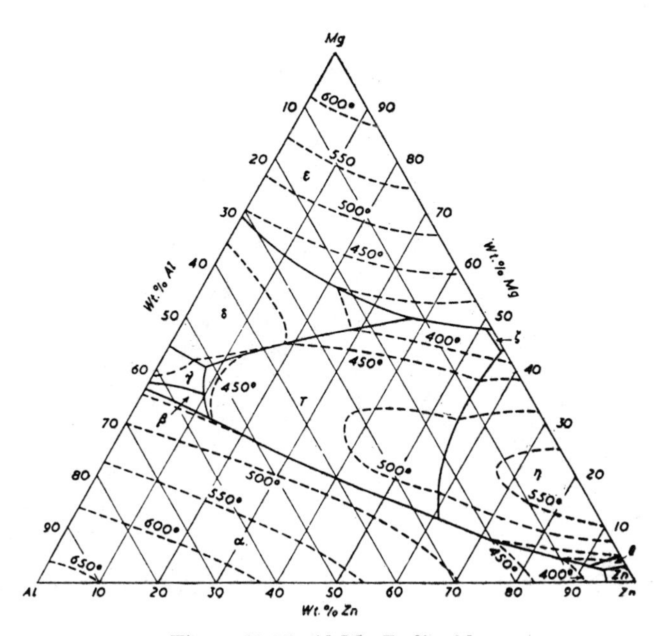

Figure 10-35. Al-Mg-Zn liquidus

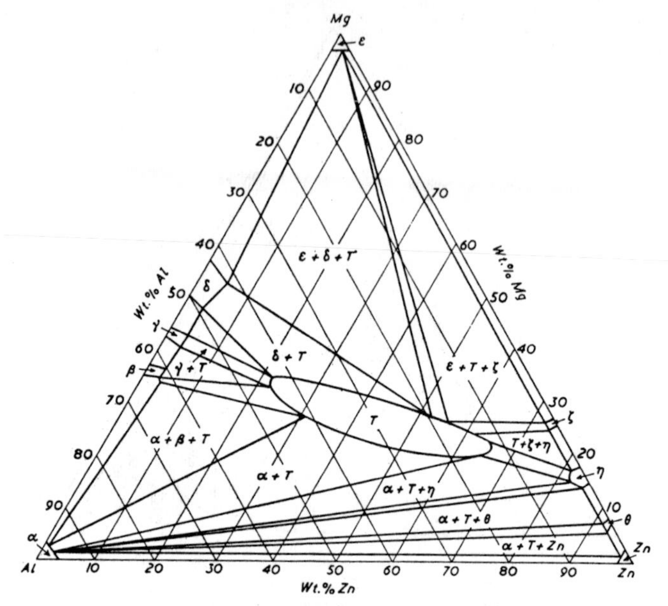

Figure 10-36. Al-Mg-Zn phases in solid alloys.

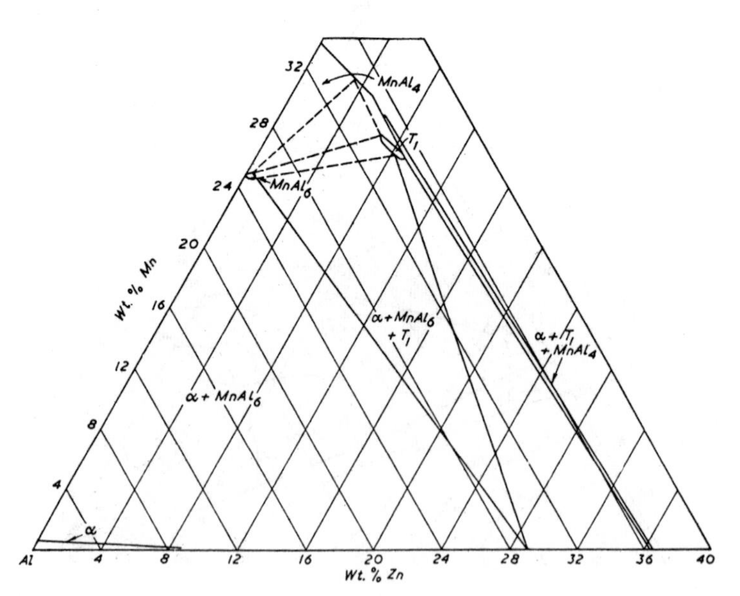

Figure 10-37. Al-Mn-Zn 600°C isotherm.

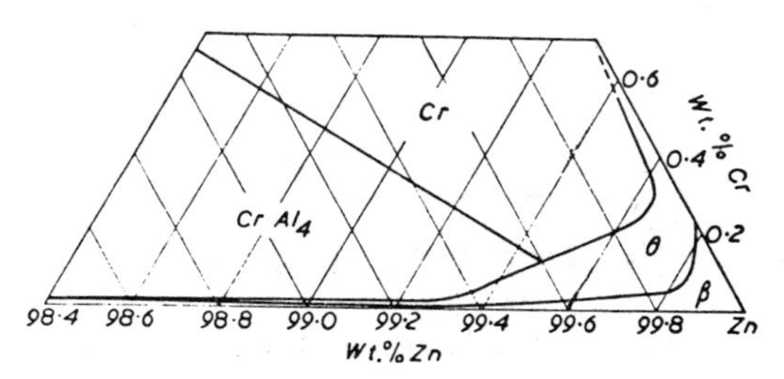

Figure 10-38. Al-Cr-Zn liquidus, Zn corner.

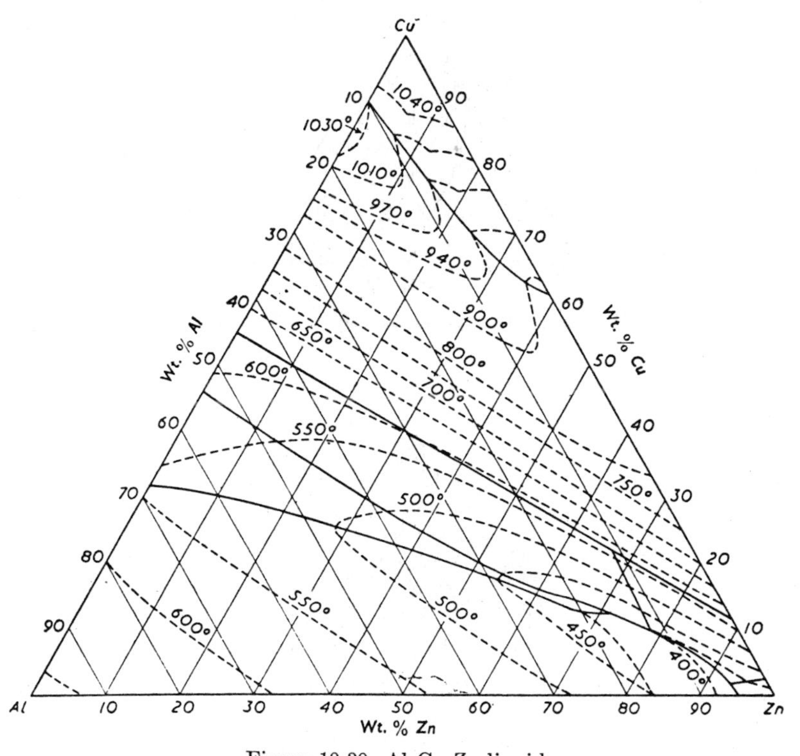

Figure 10-39. Al-Cu-Zn liquidus.

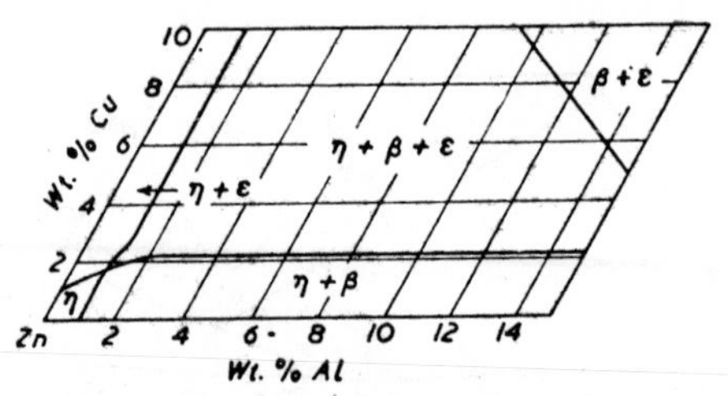

Figure 10-40. Al-Cu-Zn 300°C isotherm. Zn-rich alloys.

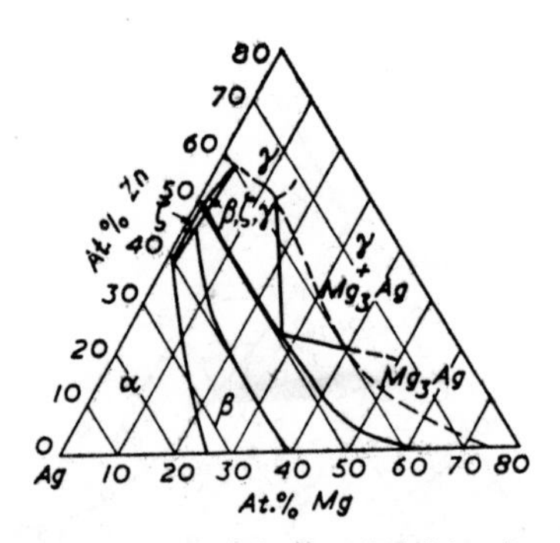

Figure 10-41. Ag-Mg-Zn 250°C isotherm.

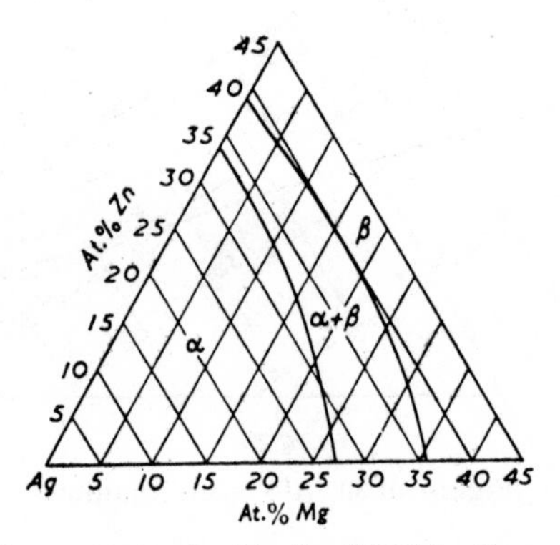

Figure 10-42. Ag-Mg-Zn 650°C isotherm.

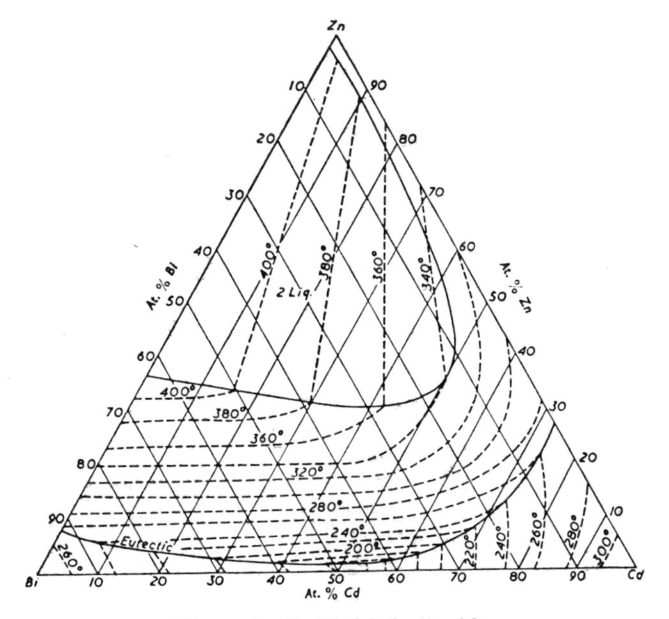

Figure 10-43. Bi-Cd-Zn liquidus.

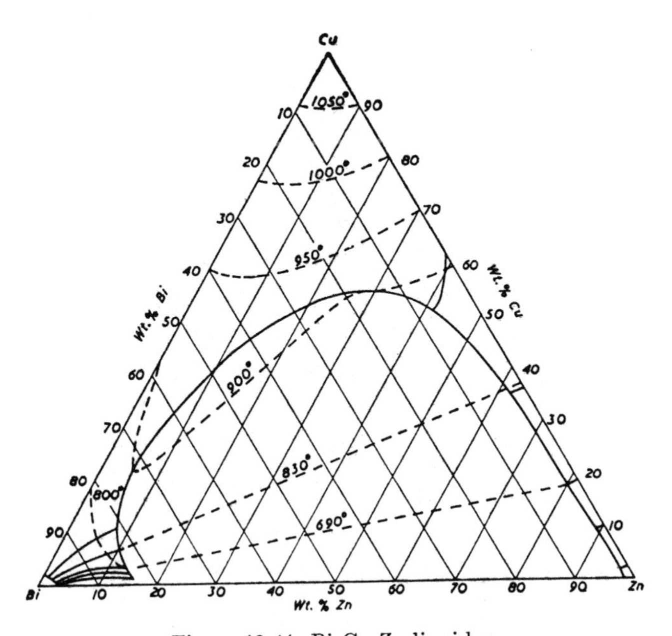

Figure 10-44. Bi-Cu-Zn liquidus.

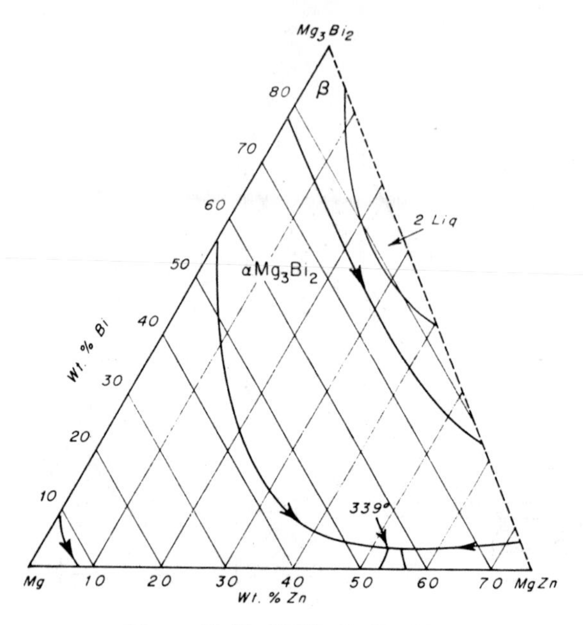

Figure 10-45. Bi-Mg-Zn liquidus.

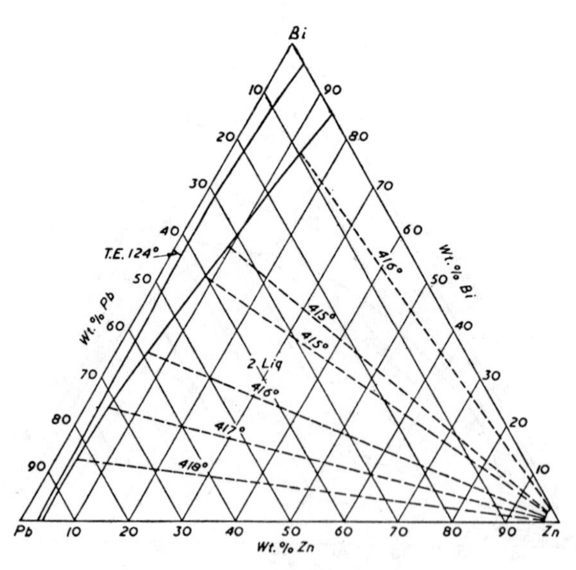

Figure 10-46. Bi-Pb-Zn liquidus.

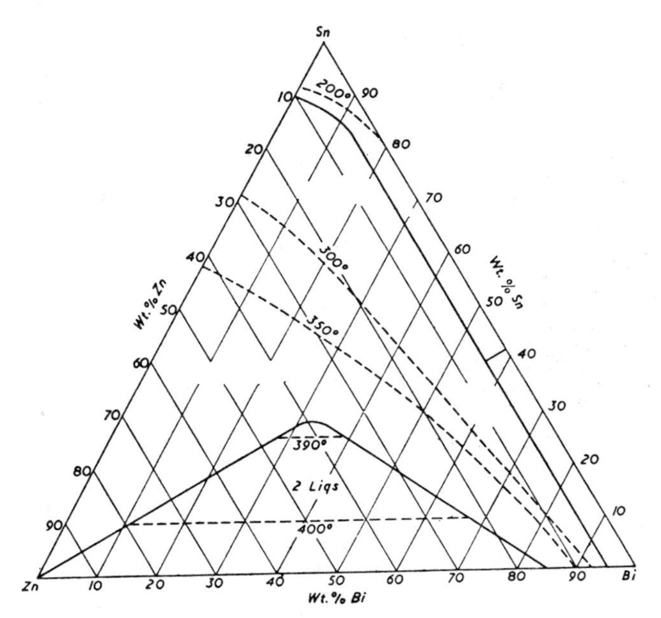

Figure 10-47. Bi-Sn-Zn liquidus.

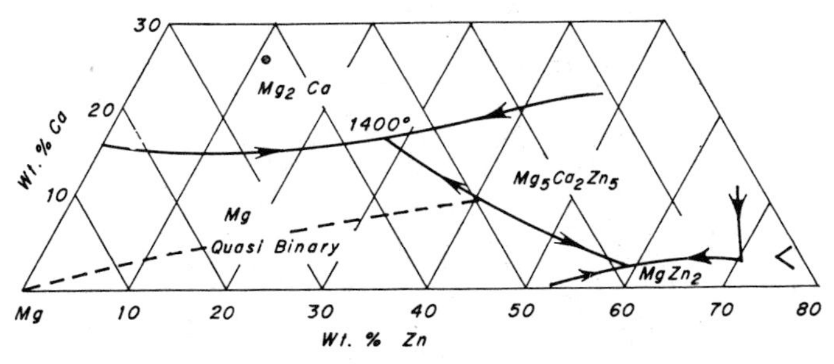

Figure 10-48. Ca-Mg-Zn liquidus.

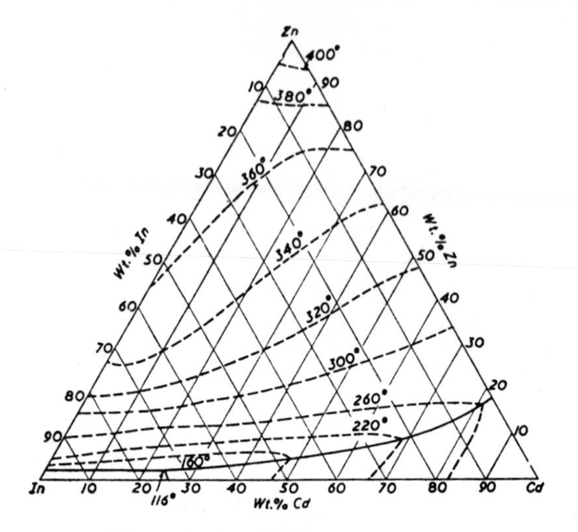

Figure 10-49. Cd-In-Zn liquidus.

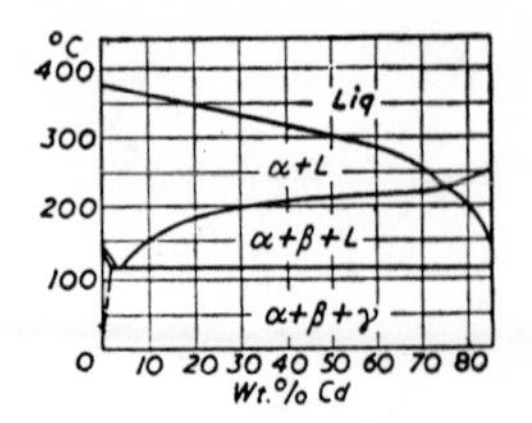

Figure 10-50. Cd-In-Zn, 15% In section.

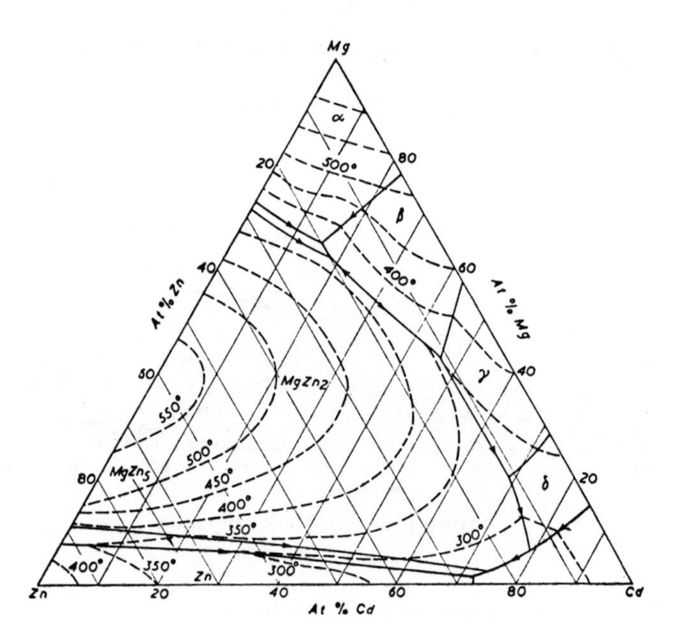

Figure 10-51. Cd-Mg-Zn liquidus.

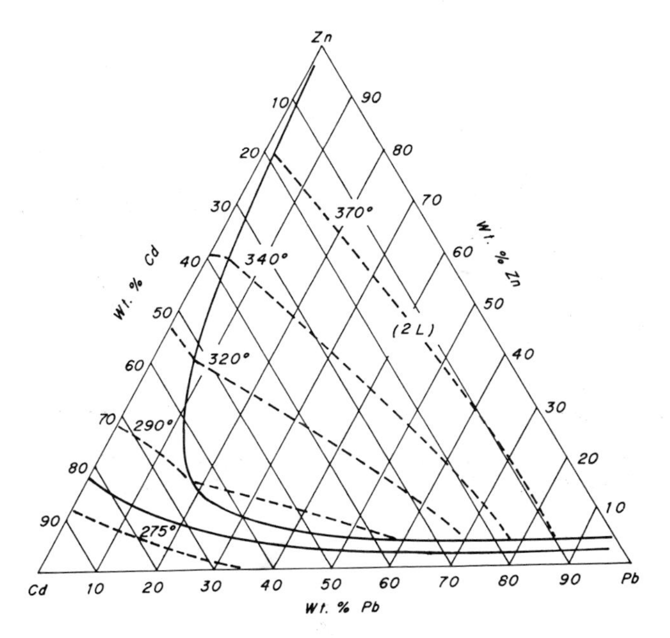

Figure 10-52. Cd-Pb-Zn liquidus.

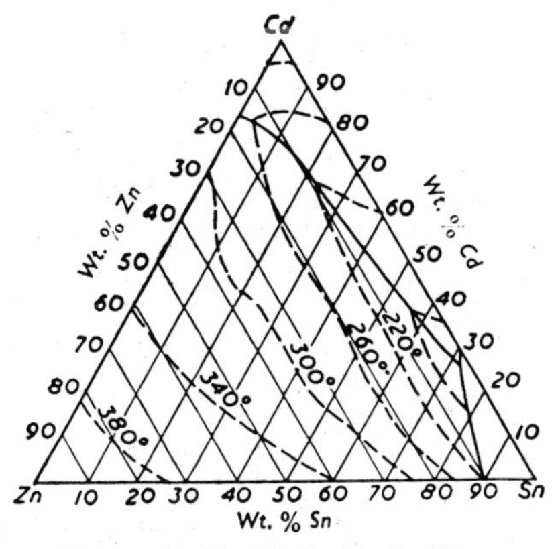

Figure 10-53. Cd-Sn-Zn liquidus.

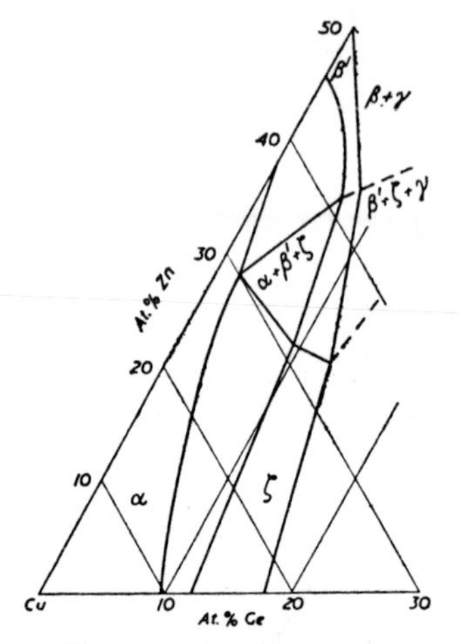

Figure 10-54. Cu-Ge-Zn, 400°C isotherm.

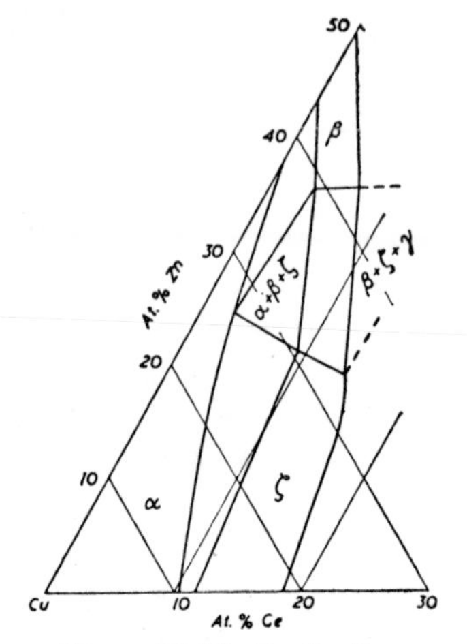

Figure 10-55. Cu-Ge-Zn, 550°C isotherm.

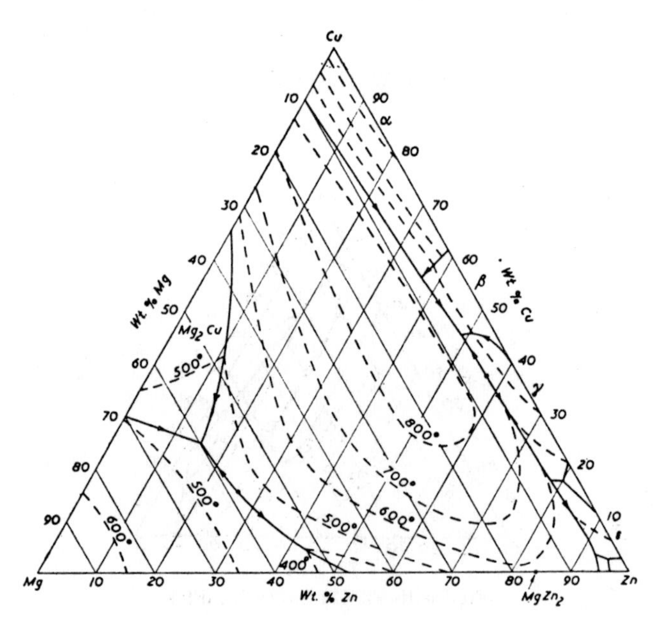

Figure 10-56. Cu-Mg-Zn liquidus.

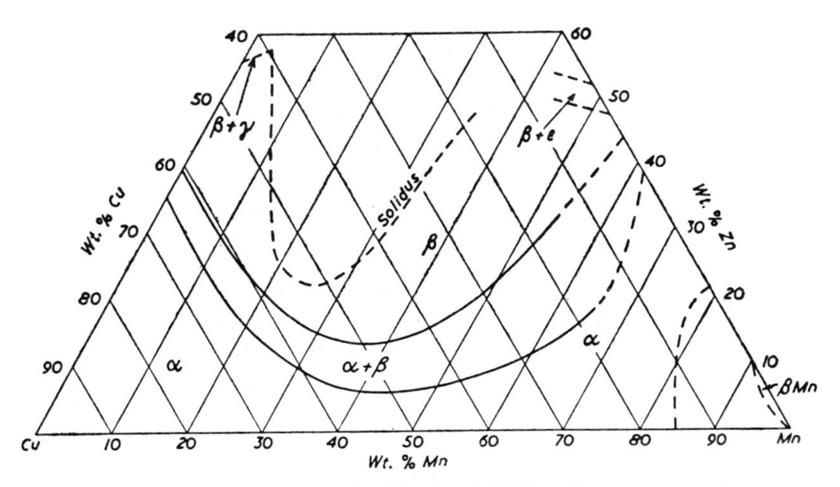

Figure 10-57. Cu-Mn-Zn, 815°C isotherm.

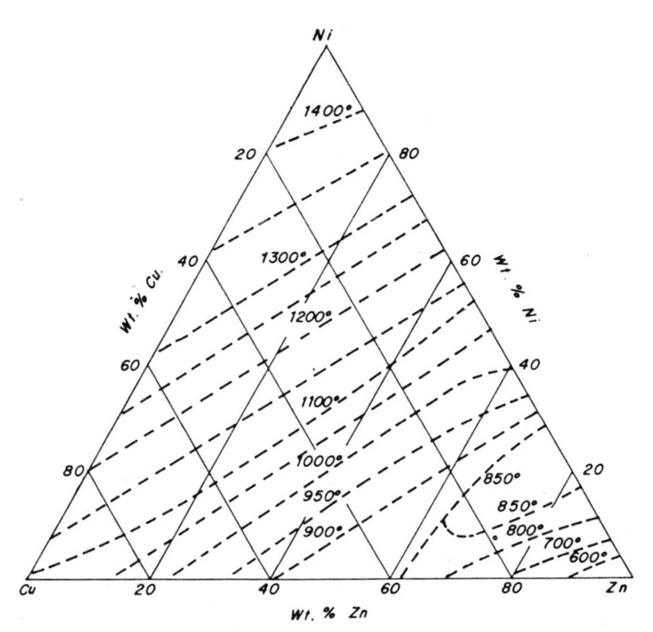

Figure 10-58. Cu-Ni-Zn liquidus.

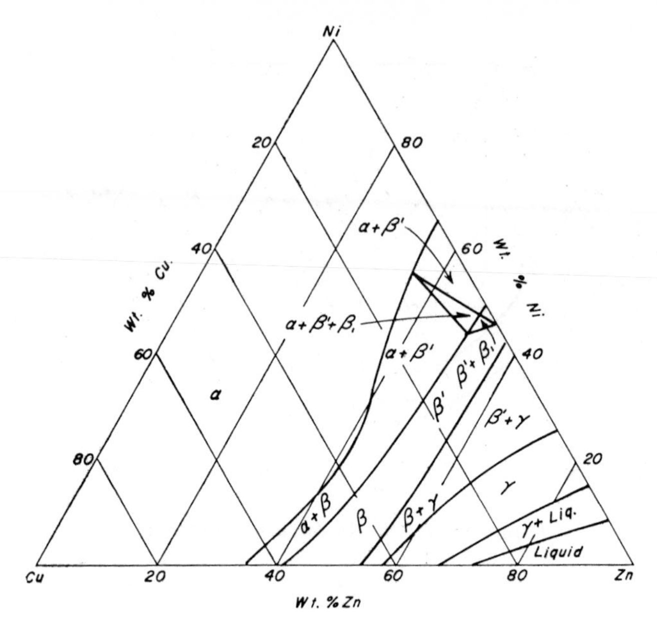

Figure 10-59. Cu-Ni-Zn, 775°C isotherm.

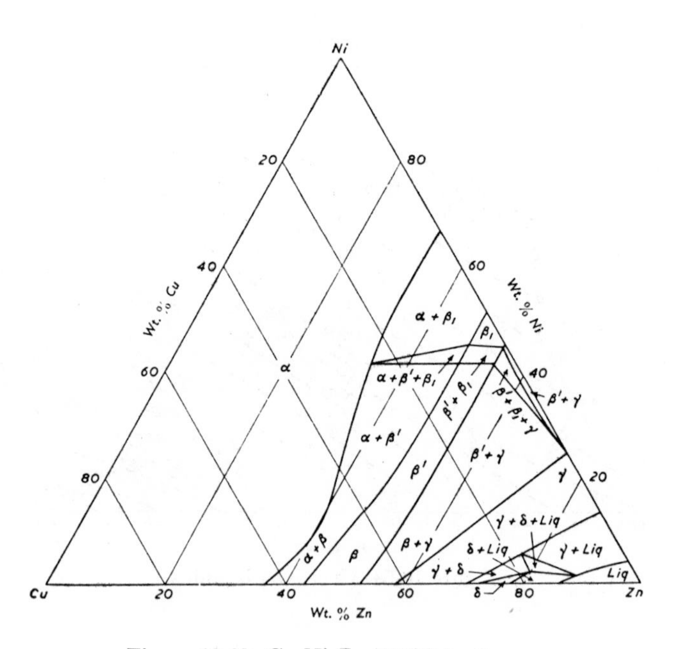

Figure 10-60. Cu-Ni-Zn, 650°C isotherm.

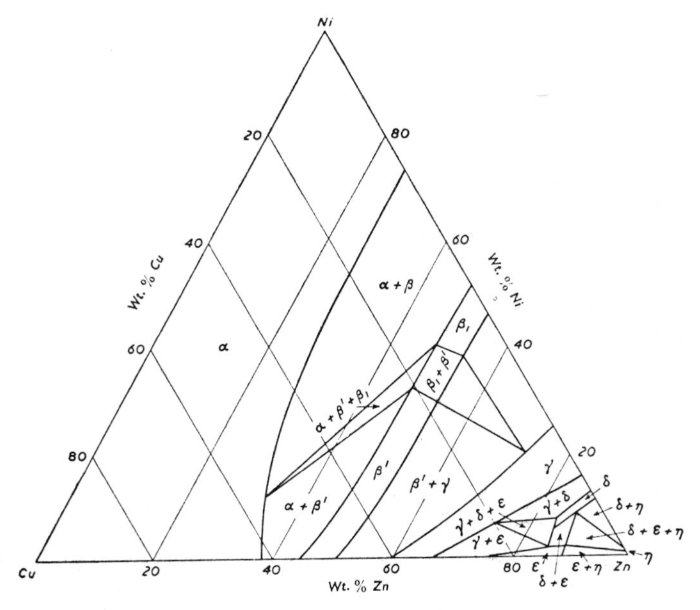

Figure 10-61. Cu-Ni-Zn, 25°C isotherm.

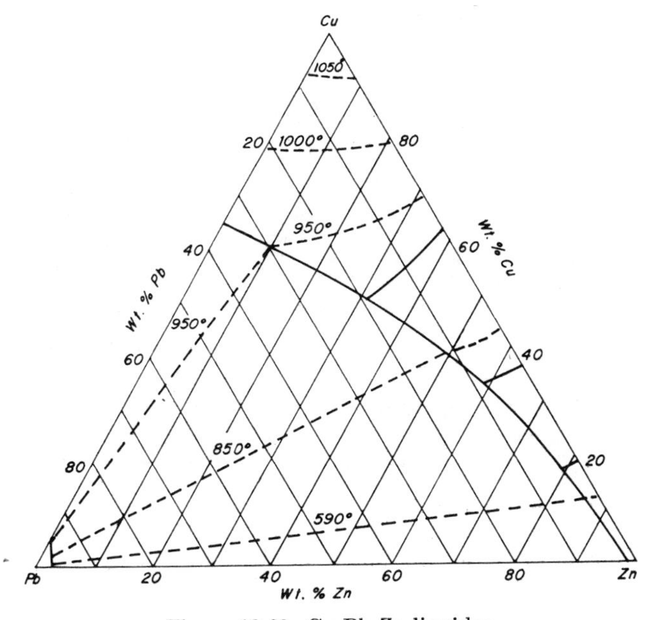

Figure 10-62. Cu-Pb-Zn liquidus.

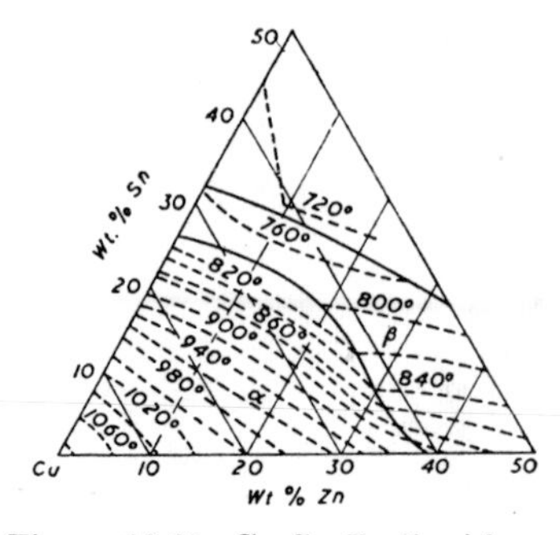

Figure 10-63. Cu-Sn-Zn liquidus.

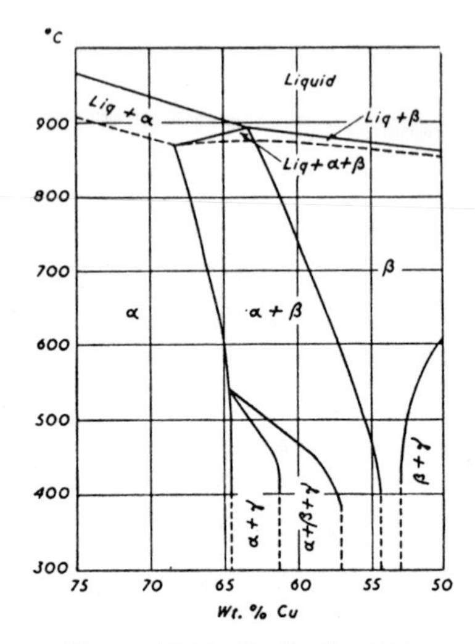

Figure 10-64. Cu-Sn-Zn, 1% Sn sectional diagram.

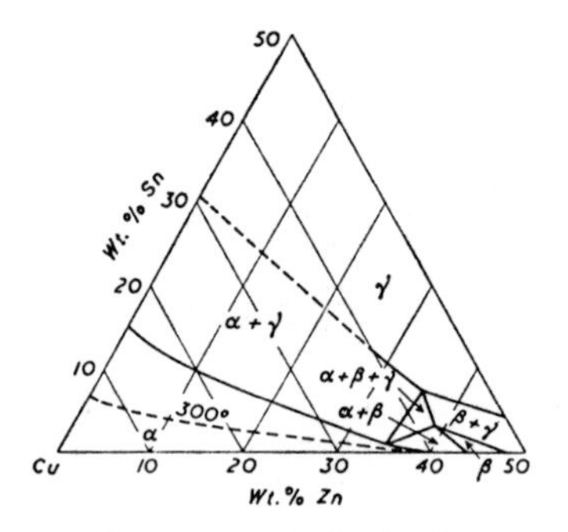

Figure 10-65. Cu-Sn-Zn, 500°C isotherm.

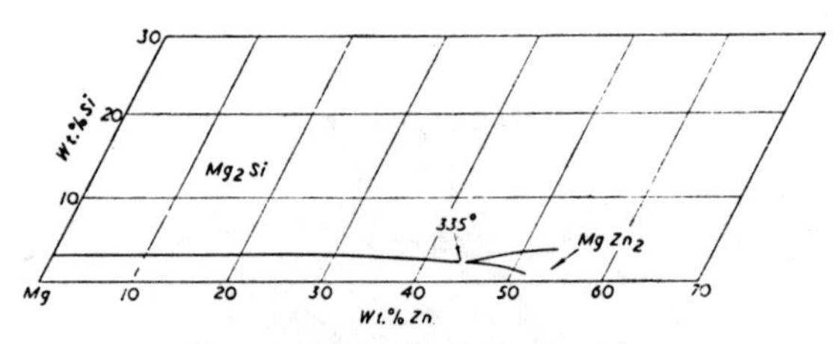

Figure 10-66. Mg-Si-Zn liquidus.

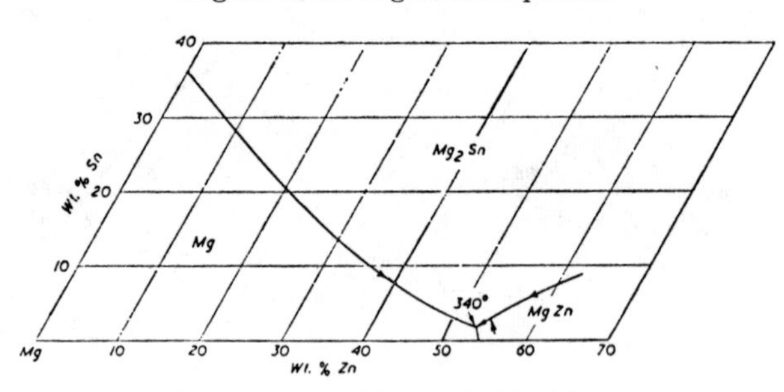

Figure 10-67. Mg-Sn-Zn liquidus.

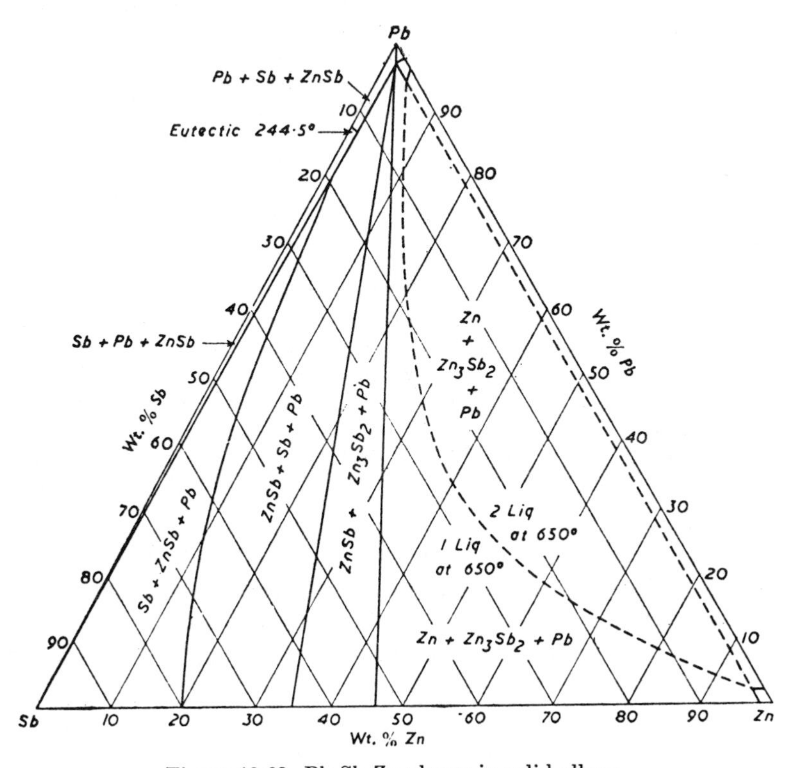

Figure 10-68. Pb-Sb-Zn phases in solid alloys.

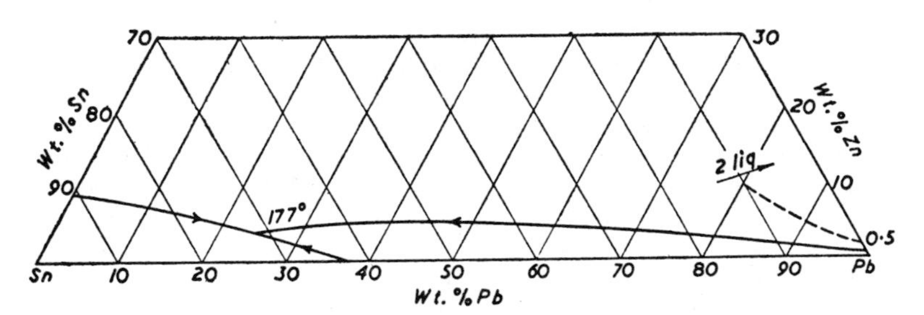

Figure 10-69. Pb-Sn-Zn liquidus.

issued by the O. W. Leiburger Research Laboratories, Inc.[31] A description of both systems is more readily available in abstracted form.[32]

The Leiburger translation also contains a description of the ZnZr diagram between 0 and 2.6 per cent zirconium,[33] from Gebhardt.[34, 35] This may be compared with the partial diagram, ZnTi,[36] shown as Figure 10-27, with a eutectic near 0.1 per cent zirconium and 416°C, instead of 0.12 per cent titanium, and 418.5°C; two zinc-rich phases formed peritectically as in the ZnTi system; a much steeper rise of the primary freezing curve in the case of zirconium (1000°C at 3 per cent zirconium) than titanium (1000°C at 5 per cent titanium); and maximum solubility of 0.02 per cent zirconium, compared with 0.12 per cent titanium, in zinc.

The diagram, CrZn,[37] is shown as Figure 10-11 in the form presented in Volume 2 of the "Chromium" Monograph, No. 132 of this series.

Two UZn diagrams have appeared in a recent publication,[38] one for atmospheric pressure and one for a pressure of 5 atmospheres. The latter is reproduced in Figure 10-29 and also represents equilibrium conditions under atmospheric pressure at temperatures below 910°C, where the zinc-rich liquid has dissociated into zinc vapor and compound, UZn_9.

Data on ternary systems, in addition to those represented on pages 607–23, originally found in the German literature during World War II, may be consulted in abstracted form, as follows:

Mg-Tl-Zn,[39] Al-Sb-Zn,[40] Cu-Si-Zn,[41] Al-Mn-Zn, zinc corner.[42]

Data on five quaternary systems containing zinc and listed by Smithells may be consulted, as follows:

Ag-Cu-Mn-Zn,[43] Al-Cr-Mg-Zn,[44] Al-Cu-Mg-Zn,[45] Al-Mg-Mn-Zn,[46] Al-Mg-Si-Zn.[47]

A second edition of M. Hansen's well-known volume on "Constitution of Binary Alloys," recently issued (1958) by the McGraw-Hill Book Company, contains diagrams of all binary systems exhibited here and, in addition, diagrams of the following systems: Ba-Zn (below 15 per cent Ba), Ce-Zn (to approx. 40 per cent Ce), La-Zn, Pt-Zn (partial), Se-Zn, and Zn-Zr (partial).

REFERENCES

1. "Copper, The Science and Technology of the Metal, Its Alloys and Compounds," American Chemical Society Monograph, No. 122, edited by Allison Butts, Chapter 22, Brick, R. M., p. 487, New York, Reinhold Publishing Corp., 1954.
2. *Ibid.*, Chapter 25, Weaver, V. P., pp. 535–572.
3. (a) Crampton, D. K., Burghoff, H. L., and Stacy, J. T., "The Copper Rich Alloys of the Copper-Nickel-Phosphorus System," *Trans. Am. Inst. Mining Engrs.*, **137**, 354–372 (1940); also, (b) "Modern Uses of Nonferrous Metals," 2d ed., edited by C. H. Mathewson; Chapter 8, Crampton, D. K., p. 142, New York, Am. Inst. Mining. Met. Engrs., 1953.
4. "Manual of Standards, Standard Commercial Wrought Copper and Copper-Base Alloys, Alloy Data Sheets," New York, Copper and Brass Research Association, 1954.
5. Ref. 3(b), pp. 132–134 and 134–135.
6. "The Metals Handbook," pp. 903–942, Cleveland, Ohio, American Society for Metals, 1948.
7. "1958 Book of ASTM Standards, Part 2, Non-Ferrous Metals," pp. 71–6, Philadelphia, Pa., American Society for Testing Materials.

8. Anderson, E. A., and Rennhack, E. H., "Age Hardenable Brass Alloys Through Powder Metallurgy," *Proc. Metal Powder Assoc., 12th Meeting*, Cleveland, Ohio, 1956; also,
"Facts about Pressed Brass and Other Nonferrous Powder Parts," New York, New Jersey Zinc Company, 1952.
9. "The Metals Handbook," p. 1167, Cleveland, Ohio, American Society for Metals, 1948.
10. "Alcoa Aluminum Handbook," pp. 35–36, Pittsburgh, Pa., Aluminum Company of America, 1956.
11. "The Aluminum Data Book," p. 48, Louisville, Ky., Reynolds Metals Company, 1954.
12. "1958 Book of ASTM Standards," Part 2, "Non-Ferrous Metals," pp. 1115–1132.
13. *Ibid.*, pp. 1142–8.
14. *Ibid.*, pp. 1156–64.
15. *Ibid.*, pp. 1194–1202.
16. *Ibid.*, pp. 1221–6.
17. *Ibid.*, pp. 1051–7.
18. *Ibid.*, pp. 1094–9.
19. "Welding Aluminum (Welding, Brazing, Soldering)," p. 152, Louisville, Ky., Reynolds Metals Company, 1953.
20. Clark, J. B., and Rhines, F. N., "Central Region of the Mg-Zn Phase Diagram," *J. Metals*, **9**, 425–30 (1957); "The Metals Handbook," p. 1227, Cleveland, Ohio, American Society for Metals, 1948.
21. "1958 Book of ASTM Standards, Part 2, Non-Ferrous Metals," pp. 1070–5.
22. *Ibid.*, pp. 1086–93.
23. *Ibid.*, pp. 1061–9.
24. *Ibid.*, pp. 1109–14.
25. *Ibid.*, p. 461.
26. "Tin, Its Mining, Production, Technology and Applications," American Chemical Society Monograph, No. 51, 2d ed., edited by C. L. Mantell, pp. 376–81, New York, Reinhold Publishing Corp., 1949.
27. Woldman, N. E., "Engineering Alloys," pp. 131–2, Cleveland, Ohio, American Society for Metals, 1954.
28. *Ibid.*, p. 156.
29. Smithells, Colin, J., "Metals Reference Book," Vol. 1, pp. 301–520, New York, Interscience Publishers, Inc., 1955.
30. Schramm, J., *J. Inst. Metals*, "The Systems Zinc-Cerium and Zinc-Lanthanum," *Metallkunde*, **33**, 358–60 (1941).
31. Gebhardt, E., "Structure of Binary Alloys," *FIAT, Rev. Ger. Sci.*, **31**, 96 and 110 (1950).
32. Schramm, J., *J. Inst. Metals, Metallurgical Abstr., Series 2*, **9**, 312 (1942).
33. Gebhardt, E., "Structure of Binary Alloys," *FIAT, Rev. Ger. Sci.*, **31**, 119 (1950).
34. Gebhardt, E., "On the Partial Systems of Zinc with Titanium and Zirconium," *Metallkunde*, **33**, 355–7 (1941).
35. Gebhardt, E., *J. Inst. Metals, Metallurgical Abstr., Series 2*, **9**, 312 (1942).
36. Anderson, E. A., Boyle, E. J., and Ramsey, P. W., "Rolled Zinc-titanium Alloys," *Trans. Am. Inst. Mining Met. Engrs.*, **156**, 278–87 (1944).
37. Heumann, T., "Beitrag zur Kenntnis des Systems Zink-Chrom," *Metallkunde*, **39**, 45–52 (1948).
38. Chiotti, P., Klepfer, H. H., and Gill, K. J., "Uranium Zinc System," *J. Metals*, **9**, 51–57 (1957).

39. Koster, W., and Kam, K., *J. Inst. Metals, Metallurgical Abstr., Series 2*, **6**, 172 (1939).
40. Koster, W., *J. Inst. Metals, Metallurgical Abstr., Series 2*, **10**, 244 (1943).
41. Masing, G., and Wallbaum, H. J., *J. Inst. Metals, Metallurgical Abstr. Series 2*, **10**, 6 (1943).
42. Gebhardt, E., *J. Inst. Metals, Metallurgical Abstr., Series 2*, **10**, 244 (1943).
43. Moeller, K., "X-Ray and Microscopic Investigations of the Quaternary System, Zinc, Manganese, Copper, Silver," *Metallkunde*, **35**, 27–8 (1943).
44. Little, K., Axon, H. J., and Hume-Rothery, W., "The Constitution of Aluminum-Magnesium-Zinc-Chromium Alloys at 460°C," *J. Inst. Metals*, **75**, 39–50 (1948–49).
45. Strawbridge, D. J., Hume-Rothery, W., and Little, A. T., "The Constitution of Aluminum-Copper-Magnesium-Zinc Alloys at 460°C," *J. Inst. Metals*, **74**, 191–225 (1947–48).
    Saulnier, A., and Cabane, G., "Recent Research on the Alloys, Al-Zn-Mg-Cu," *Rev. mét.*, **46** (1), 13–23 (1949).
46. Butchers, E., Raynor, G. V., and Hume-Rothery, W., "The Constitution of Magnesium-Manganese-Zinc-Aluminum Alloys in the Range 0–5 Per Cent Magnesium, 0–2 Per Cent Manganese, 0–8 Per Cent Zinc. I—The Liquidus," *J. Inst. Metals*, **69**, 209–228 (1943).
    Raynor, G. V., and Hume-Rothery, W., "II—The Composition of the $MnAl_6$ Phase," *ibid.*, **69**, 415–21 (1943).
    Little, A. T., Raynor, G. V., and Hume-Rothery, W., "III—The 500°C and 400°C Isothermals," *ibid.*, **69**, 423–40 (1943).
    Little, A. T., Raynor, G. V., and Hume-Rothery, W., "IV—The Equilibrium Diagram Below 400°C," *ibid.*, **69**, 467–84 (1943).
47. Sauder, W., and Meissner, K. L., "Equilibrium Studies in the Quaternary System, Aluminum-Magnesium-Silicon-Zinc," *Metallkunde*, **15**, 180–3 (1923).
    "The Aluminum Rich Mixed Crystal Area in the System," *ibid.*, **16**, 12–17 (1924).
    Bollenrath, F., and Grober, H., "Aluminum Alloys with Magnesium, Silicon and Zinc," *Metallforschung*, **1**, 116–22 (1946).

SUPPLEMENTARY REFERENCES

a. Bakhuis Roozeboom, H. W., "Die heterogenen Gleichgewichte," Vol. 1, 1901, Vol. 2–1, 1904, Vol. 2–2 (Buchner, E. H.), 1918, Vol. 2–3 (Aten, A. H. W.), 1918, Vol. 3–1 (Schreinemakers, F. A. H.), 1911, Braunschweig, Friedrich Vieweg und Sohn.
b. For interpretation of phase diagrams. Brick, R. M., and Phillips, Arthur, "Structure and Properties of Metals," 2d ed., New York, McGraw-Hill Book Co., Inc., 1949.
c. Cottrell, A. H., "Theoretical Structural Metallurgy," 2d ed., (especially Chapter 11, "The Equilibrium Diagram" and 9, "The Structure of Alloys"), New York, St. Martin's Press Inc., 1955.
d. Hume-Rothery, William, and Raynor, G. V., "The Structure of Metals and Alloys," 3d ed., London, Institute of Metals, 1954.
e. Hume-Rothery, William, "Atomic Theory for Students of Metallurgy," 3d revised reprint, London, Institute of Metals, 1955.
f. Raynor, G. V., "An Introduction to the Electron Theory of Metals," London, Institute of Metals, 1947.
g. Seitz, Frederick, "Modern Theory of Solids" New York, McGraw-Hill Book Co., Inc., 1940.
h. Seitz, Frederick, "Physics of Metals," New York, McGraw-Hill Book Co., Inc., 1943.

CHAPTER 11

USE OF ZINC IN VARIOUS FORMS FOR THE METALLURGICAL EXTRACTION OF OTHER METALS (INCLUDING THE PARKES PROCESS)

T. D. Jones

Chief Lead Refinery Metallurgist
American Smelting and Refining Company
Barber, New Jersey

PARKES PROCESS

History

The phenomenon that lead bullion gives up its silver content to zinc was discovered by Karsten of Germany in 1842. The actual development of the early desilverizing process was described by Alexander Parkes of Birmingham, England, in his patents granted in 1850, 1851 and 1852. The first trial operations on a commercial basis were conducted in 1859 at the plant of Sims, Willyams, Nevill and Company of Llanelly, Wales, Great Britain.[11]

The early efforts in England to develop a commercial process were not too successful, and due to the very small amounts of lead bullion produced in the United States prior to the Civil War, there was no field for the process in this country.

In 1864, Edward Balbach of Newark, New Jersey, patented a process for desilverizing by means of zinc, and in 1867, Edward Balbach, Jr. patented a "movable black lead retort with a neck, placed in a furnace" for the purpose of distilling an alloy of zinc, lead and the precious metals.[5]

The essential difference in the two processes was that in the Parkes process the required zinc (1 to 2 per cent) was stirred into a molten bath of silver-bearing lead bullion, while in the Balbach process the molten softened silver-lead bullion flowed into a kettle containing the required amount of molten zinc and the mixture was stirred. The resultant silver-zinc dross in both processes was skimmed off, liquated, and the residual rich alloy retorted in the black lead, or early clay-graphite, retorts and the distilled zinc returned to process.

By 1872, the Balbach process was so modified that it closely resembled the Parkes process and eventually the process acquired the name "Parkes Process." In fact, the Parkes desilverizing process as generally practiced today differs little (except perhaps in engineering practice) from the original; namely, stirring-in of zinc, cooling, skimming silver-zinc crusts, liquation if necessary, followed by distillation of the crusts and the recovery of zinc which is returned to process.

Chemistry

The lead-zinc phase diagram reveals the unique position of zinc with relation to the commercial removal of values from lead bullion. While other metals can be used, the economy of using zinc is easily understood by referring to this diagram (see Figure 10-22)* which shows that lead at the melting point of zinc (787°F) will dissolve approximately 2 per cent zinc, while at the melting point of lead (621°F) only about 0.5 per cent zinc is soluble.

Thus, in the Parkes process at the elevated temperature of 875°F or higher, more than 2 per cent zinc may be dissolved in the lead, as required to extract the values from bullion containing approximately 100 ounces doré per ton. The limited solubility of zinc in lead at the freezing point of lead is of prime importance in the Parkes process, as excess zinc or repeated zinc additions can be made and silver removed to any desired figure, since the lead-zinc eutectic is approached during each cooling-down cycle and it is impossible to leave more than approximately 0.5 per cent zinc in lead provided proper temperature control is exercised.

The density of zinc (7.13) as compared to lead (11.34) also is of importance in the process, as the silver-zinc crystals are much lighter than lead and concentrate toward the surface of the bath where they can be removed as silver-zinc dross or crust.

The Parkes process is based on the fact that when 1 to 2 per cent zinc is stirred into a molten lead-bullion bath at controlled temperatures, the zinc forms alloys, or intermetallic compounds, with gold, silver and residual copper. Zinc becomes alloyed with gold and copper, preferentially to silver, forming the intermetallic compounds gold-zinc and copper-zinc.

Several refineries, by limiting zinc additions, have adopted the degolding practice; this is possible as gold and copper can be separated out first and with only about one-eighth of the silver being removed. It is claimed that this presents certain economic advantages; however, in general, most refineries remove the gold, copper and silver, together with any traces of platinum, palladium, etc., as so-called silver-zinc crust or dross.

Many published works have contributed to the present-day knowledge of the ternary system Ag-Pb-Zn; however, from a practical standpoint, the field is not clearly understood.[2] One interpretation is that desilverization is

* The scale of this diagram does not permit accurate placement of the solubility values written into the drawing. (Ed.)

caused by the separation of the compounds $AgZn_4$, Ag_2Zn_5, Ag_5Zn_8, $AgZn$, Ag_3Zn_2, as solid solutions which form as the liquid bath cools down through the various temperature zones.[7] In actual desilverizing practice, the silver-zinc compounds are removed from the lead bath by removing the crusts formed and the main object is to remove as high-grade crust as possible with the least amount of entrained lead.

Present-Day Parkes Processes

There are two adaptations of the Parkes desilverizing process in vogue at the present time, known as "batch desilverizing" and "continuous desilverizing" practices.

Batch Desilverizing. Batch desilverizing is carried out in open-type steel kettles which have a capacity of 210 to 270 tons of lead. The doré content usually varies on each individual charge and the amount of zinc required is calculated accordingly. In order to obtain silver-zinc skims which are enriched, the previous day's blocks, and only those blocks which were removed just after the silver-zinc crust was skimmed, are added to the kettle of new, clean, softened lead pumped over from the softener furnace. By following this plan, enrichment takes place, as the zinc content of the blocks at a temperature of 875°F is allowed to alloy with the high silver content of the new charge. A minimum amount of stirring is allowed so as to avoid impregnating the dross with lead. The dross is skimmed with a suitable skimmer into a Howard press and by alternately raising and lowering the plunger, operated by 100 psi compressed air, much of the entrained lead is squeezed out of the dross. (See Figure 11-1.) During the pressing operation, steel separator plates are placed in the basket of the press and thus the dross when discharged is in thin layers and easily broken up. When desilverizing 100 ounces doré in the bullion, the silver-zinc skim discharged from the press contains approximately 3560 to 4000 ounces doré per ton.

Several refineries using the batch desilverizing process employ the well-known silver-zinc-dross sweating or liquation operation wherein a high-grade alloy of the following approximate analysis is obtained: silver, 25 per cent; zinc, 65 per cent; lead, 8 to 10 per cent; copper, 1 to 1.5 per cent. Thus, a large percentage of the lead content of the silver-zinc dross is returned directly to the desilverizing kettle.

After the silver-zinc skim is removed, a hot sample of the bath is taken and assayed to determine the silver content. The correct amount of slab zinc is then added and stirred into the bath. The temperature is gradually lowered in the kettle from the pressing temperature of 875°F, and silver-zinc dross begins to appear on the surface of the bath. This skim is removed to suitable pots, and as each pot freezes, the block is removed and numbered. As the temperature drops, successive blocks are produced and num-

Figure 11-1. Batch desilverizing with Howard press in place.

bered in sequence. From a large kettle, approximately 15 two-ton blocks are removed, and only the first 5 or 6 blocks are added back on the next new charge of lead in order to obtain the highest enrichment possible. Several samples are taken as the bath temperature drops in order to determine whether or not sufficient zinc has been added to arrive at a low silver content in the final lead. During the cooling-down period, the kettles are well scraped in order to prevent any dross from freezing to the sides of the kettles, which if not removed can seriously affect the final silver content.

The bath is cooled to the freezing temperature of lead and at this point a heavy rim will form next to the kettle. Usually, a heavy chain or hooks are embedded in the freezing mass in order that the rim may be easily removed by an overhead crane. The last blocks removed, together with the rim, are very low in silver and are used as cooling blocks on the next charge in line, and aid in lowering the temperature.

After the desilverizing operation has been completed, the temperature of the bath is raised to 700°F and the desilverized charge is pumped to the dezincing kettle where the eutectic zinc content of approximately 0.5 per cent is removed (1) as zinc chloride, by the chlorine process; (2) as zinc oxide, by the Harris process; or (3) as metallic zinc, by the vacuum distillation process.[8]

Continuous Desilverizing Process. The continuous process was developed at the Port Pirie refinery of the Broken Hill Associated Smelters Proprietary, Limited, and forms an integral link in their all-continuous refining process. Essential elements abstracted from a detailed description of the process by Green[6] follow.

The continuous unit is of the kettle type, composed of four cast-iron sections bolted together and also fitted with special joints of welded steel plates, which not only make the connections lead tight, but also give them a degree of flexibility to meet the effects of fluctuating temperatures in the castings. The softened bullion which overflows from the continuous softening furnace enters the top of the desilverizing kettle at 1200°F through a submerged inlet, which prevents any surface dross from being carried into the kettle. The total depth of the Port Pirie unit is 23 feet and the maximum diameter 10 feet; it holds approximately 350 tons of lead and the unit weighs approximately 25 tons. (See Figure 11-2.) The kettle is placed in a brickwork setting suitably arranged so that there are six independent heating chambers, and any portion of the kettle may be heated as required. A siphon, or overflow pipe, is fitted centrally in the kettle and reaches to

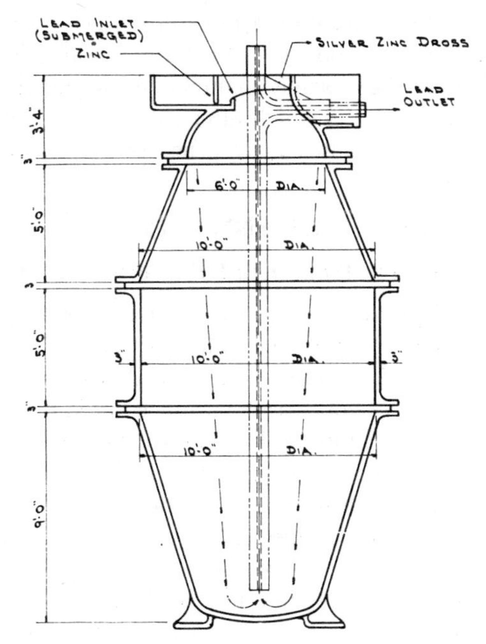

Figure 11-2. Continuous desilverizing kettle; Port Pirie, South Australia plant, The Broken Hill Associated Smelters, Proprietary, Limited. From F. A. Green in "The Refining of NonFerrous Metals," London, 1950. (*Courtesy of The Institution of Mining and Metallurgy*)

within 18 inches of the bottom. It is fitted with a horizontal branch which is fastened to a launder in the top casting, the level of the outlet pipe controlling the working level of the kettle.

When operating, the kettle is filled with lead covered by a layer of molten zinc 3 feet in depth. These metals form a conjugate solution system in the upper section of the kettle; as lead bullion flows in, its contact, both on passing through the zinc layer and at the interface between lead and alloy layers, causes progressive enrichment of the zinc alloy in silver and also in gold and residual copper. When the zinc alloy has reached an enrichment of approximately 6000 ounces of silver per ton, it is hand-ladled to bars of suitable shape for charging to the retorts and recovery of the zinc. The zinc layer is immediately built up again with slab zinc to a depth of 3 feet, and the process of enrichment commences anew. The lead moves downward in the kettle; in passing down the temperature falls, causing zinc-silver crusts to separate from solution and rise to the upper section. The lead actually reaches its freezing point at the bottom of the kettle and a solid lead-zinc eutectic deposits, the amount of which is gaged by the free movement of an iron rod placed in the siphon pipe. The desilverized lead in passing up the siphon is heated in the upper zones and flows freely to the continuous vacuum-dezincing unit where the eutectic zinc content is recovered as metallic zinc. The operation of the continuous desilverizing kettle is easily controlled, as the heating chambers enclosing the kettle castings are connected to exhaust fans which offer the means of controlled heating or cooling around any section to give the correct temperature gradient throughout the bath.

OTHER PROCESSES

Extraction of Cadmium from Lead Baghouse Dusts

In a lead smelting operation, when cadmium is present in the charge, the baghouse fume is circulated until the cadmium content has built up to 3 or 4 per cent, at which point the fume is fused with a siliceous flux in a reverberatory furnace. From this operation a fume containing about 27 per cent is collected in a separate baghouse. In order to operate the leaching process at capacity it is advantageous to process a high-cadmium-content fume, which is made possible by returning a portion of the 27 per cent cadmium fume with the original fume containing 3 to 4 per cent cadmium, so that the charge to the reverberatory furnace contains from 11 to 15 per cent cadmium. The fume resulting from smelting this charge averages 44.5 per cent cadmium and is pulped in a ball mill with wash water to which small amounts of sodium chlorate and potassium permanganate are added. The pulp is then transferred to a lead-lined agitating tank where the proper amount of copperas and sulfuric acid is added, and the *p*H is controlled to

5.5 per cent. The slurry is then filtered and the pregnant filtrate will contain about 140 grams per liter of cadmium.

Precipitation of cadmium with zinc dust is a batch process and for this a suitable batch of the filtrate is run into a conical-bottomed, lead-lined precipitation tank equipped with a turbo-type agitator. The usual procedure is to add one pound of zinc dust for each 1.6 pounds of contained cadmium, maintaining the acidity at about 2.5 grams per liter to avoid precipitation of basic sulfates. The cadmium is precipitated as so-called "cadmium sponge," and by further washing, drying and pressing is finally made into briquettes which are charged with a reducer such as coke into clay-graphite retorts, where the cadmium is distilled and recovered as metallic cadmium in the condensers which are a part of the furnace equipment.

Still another process for treating zinc and lead-smelter baghouse dust is to mix the fume with a sufficient amount of water to absorb all the cadmium into solution. The thin slurry is treated with chlorine gas and during the operation the acidity is maintained at a low point by soda ash additions. The solution containing the cadmium is filtered off, made slightly acid by sulfuric acid additions and treated with zinc dust. The cadmium sponge may be separated by filtration or centrifuging, and is later treated for recovery of metallic cadmium.[3]

Extraction of Gold and Silver from Cyanide Solutions by Precipitation with Zinc Dust

At the present time in the cyanidation process, sodium cyanide (NaCN) or so-called white cyanide has entirely superseded the use of potassium cyanide (KCN). The essential reactions of the cyanide process are as follows:

$$2Au + 4NaCN + O + H_2O = 2NaAu(CN)_2 + 2NaOH$$

$$2Ag + 4NaCN + O + H_2O = 2NaAg(CN)_2 + 2NaOH$$

The reaction with potassium cyanide is exactly the same, substituting KCN for NaCN.

Five methods of precipitating gold and silver from cyanide solutions have been used: by aluminum, charcoal, sodium sulfide, zinc, and electrolytic processing. The first and third were especially developed for silver ores of Cobalt, Ontario. Charcoal has been used in Australia. However, zinc as dust or shavings has been used from the beginning of the cyanide process and continues to be the standard method throughout the world.[4]

The essential reaction with zinc is a replacement of gold and silver, the zinc going into solution as the double cyanide:

$$Zn + 2NaAg(CN)_2 = Na_2Zn(CN)_4 + 2Ag$$

Precipitation of both gold and silver are complicated by side reactions, especially the evolution of hydrogen by reaction of zinc with sodium cyanide, and the entire effect of the precipitation may be expressed by the equation:[1]

$$NaAg(CN)_2 + 2NaCN + Zn + H_2O = Na_2Zn(CN)_4 + Ag + H + NaOH$$

The precipitation of gold follows the same reaction, substituting gold for silver.

Zinc dust is generally preferred to zinc shavings as a precipitant due to the fact that for perfect precipitation each molecule of metal-bearing solution must be brought into contact with a particle of precipitant.

The best zinc dust was formerly imported; however, a more efficient dust is now made in the United States. It is practically free from foreign materials and extremely uniform in size, the finer grades nearly all passing a 300-mesh sieve.

In the precipitation process, zinc dust is sprinkled dry or as an emulsion into large charges into vats of solution under agitation, which are then pumped through square filter presses. One of the best methods for adding zinc dust is by means of a slow-moving horizontal belt on which the zinc dust is spread in a uniform layer, the dust discharging into a small mixing cone through which a trickle of auxiliary solution conveys it to the pump suction pipe. Another method utilizes a hopper and at the outlet a revolving roller or pulley for removing a small ribbon of dust. An auger-like horizontal screw may also be used. The main object of this careful control of zinc-dust additions is to provide for contact between the particles of zinc dust and the value-bearing solutions in order to effect complete precipitation.[9]

Extraction of Thallium by Reduction with Zinc

Thallium is chiefly found in acid-plant flue dust and in residues from the purification of cadmium. The dust and/or residues are boiled with dilute sulfuric acid and filtered. This filtrate is treated with salt or hydrochloric acid and the thallium chloride precipitated. Thallium metal can be obtained by reducing the chloride by means of zinc and melting in an inert gas.

Extraction of the Rare Metals; Iridium, Rhodium, Osmium and Ruthenium, by Use of Zinc as the Precipitant

In general, the rare metals iridium, rhodium, osmium and ruthenium, if present, are recovered as platinum-group metals and are obtained from the residues and solutions available after separation of the platinum.

Zinc plays an important part and acts as a precipitant in the final separation of the rare metals during the various intricate chemical processes required in this work.[10]

Production of Tin-Lead Solder Metal from Tin Tetrachloride and Lead-Zinc Alloy

As explained under "Batch Desilverizing," the lead after desilverizing contains approximately 0.5 per cent zinc which can be removed by several processes, one of which is called "chlorine dezincing." In this process the molten metal containing 0.5 per cent zinc is pumped through a vertical cylinder flooded with chlorine gas, the pump and cylinder forming the compact unit which rests on the kettle rim during the dezincing process, and the zinc is removed as zinc chloride which floats on the surface of the bath.

In the production of solder metal by reaction with tin tetrachloride ($SnCl_4$), this same equipment is used. The fully refined lead bath is saturated with zinc and the liquid tetrachloride is admitted through a port into the reaction cylinder, at which time it gasifies and reacts with the zinc content of the lead, according to the following equation:

$$SnCl_4 + 2Zn = Sn + 2ZnCl_2$$

The released tin enters the lead bath and the operation is continued until solder metal of the desired tin content is obtained.

The zinc chloride formed during the operation floats on the surface of the bath and is skimmed off and transferred to the zinc chloride plant for production of either fused zinc chloride or zinc chloride solution of suitable grade.

The refined solder metal is cast into appropriate shapes for sale to solder manufacturing plants for the production of bar or wire solders.

REFERENCES

1. Clennell, J. E., "The Cyanide Handbook," 2nd ed., p. 123, New York, McGraw-Hill Book Co., Inc., 1910.
2. Davy, T. R. A., "Desilverizing of Lead Bullion," *Trans. Am. Inst. Mining Met. Engrs.*, **200**, 838–48 (1954).
3. Dickinson, S. J., "Cadmium and Cadmium Alloys," in "Encyclopedia of Chemical Technology," Vol. 2, pp. 724–5, New York, The Interscience Encyclopedia, Inc., 1948.
4. Dorr, J. V. N., "Cyanidation and Concentration of Gold and Silver Ores," 1st ed., pp. 190–203, New York, McGraw-Hill Book Co., Inc., 1936.
5. Eurich, E. F., "Development of the Parkes Process in the United States," *Trans. Am. Inst. Mining Met. Engrs.*, **44**, 741–50 (1912).
6. Green, F. A., in "The Refining of Non-Ferrous Metals," pp. 288–300, London, The Institution of Mining and Metallurgy, 1950.
7. Jollivet, L., "Equilibria of the System, Pb-Ag-Zn," *Compt. rend.*, **224**, 1822–34 (1947).

8. Jones, T. D., "Lead," in "Encyclopedia of Chemical Technology," Vol. 8, pp. 237–47, New York, The Interscience Encyclopedia, Inc., 1952.
9. Liddell, D. M., "Hydrometallurgy of Gold and Silver," in "Handbook of Non-Ferrous Metallurgy," 2nd ed., Vol. 2, pp. 325–8, New York, McGraw-Hill Book Co., Inc., 1945.
10. Liddell, D. M., "Minor and Rare Metals," in "Handbook of Non-Ferrous Metallurgy," 2nd. ed, Vol. 2, pp. 696, 702–5, New York, McGraw-Hill Book Co., Inc., 1945.
11. Percy, John, "Metallurgy of Lead," pp. 148–153, London, John Murray, 1870.

CHAPTER 12

INDUSTRIAL ZINC OXIDE, ZINC SULFIDE AND OTHER ZINC COMPOUNDS

A. Paul Thompson

Director of Research
The Eagle-Picher Company
Joplin, Missouri

Physical properties of zinc (Table 4-4) and some of its compounds (Table 4-5), and basic structural considerations affecting the behavior of the metal, alone or in combination, are included in Chapter 4.

Brief comment on the chemical behavior of zinc will contribute to an understanding of the properties of its chief compounds and their applications. The outstanding property of zinc and the basis for many of its most important applications is its strong electropositive character; as a matter of fact, of the reasonably common metals, only aluminum and magnesium are more electropositive. This property of zinc is the basis for its chief commercial use, galvanizing, wherein zinc in contact with iron or steel undergoes sacrificial corrosion. This same property is utilized in the purification step of the electrolytic process for zinc extraction (removal of metals lower in the electromotive series such as copper, cadmium, etc.) and in the metallurgy of cadmium (precipitation of cadmium from impure zinc solutions).

Pure zinc is highly resistant to attack by dry air at atmospheric temperature but the rate of attack increases rapidly above 225°C. In the presence of moist air, attack will occur at room temperature, corrosion being accelerated by carbon dioxide and sulfur dioxide. A hydrated basic carbonate is the end product of normal atmospheric corrosion. Dry halogen gases will not attack zinc at atmospheric temperature but the addition of moisture will induce zinc foil actually to ignite.

Zinc is attacked by mineral acids, reaction being most vigorous in the case of sulfuric, followed by hydrochloric and then nitric. The presence of impurities in the zinc expedites attack. The reaction with nitric acid does not result in the evolution of hydrogen as in the case of the other mineral acids but rather reduction in part of the pentavalent nitrogen to lower valences. This reaction well shows the vigorous reducing power of zinc which, of course, can be intensified by increase of clean metallic surface per unit weight.

The amphoteric character of zinc is demonstrated by the fact that the metal will also dissolve in hot caustic solutions evolving hydrogen and forming zincates. Likewise, zinc oxide dissolves easily in acids giving zinc ions (Zn^{++}), and in caustic solutions, forming zincates. Similarly, zinc hydroxide can be precipitated by the addition of sodium or potassium hydroxide to zinc ions in an acid solution, the precipitate readily dissolving in an excess of the reagent forming zincates (similar to aluminum but distinct from iron and manganese). Likewise, the precipitated zinc hydroxide will dissolve upon acidifying.

Most zinc salts are white in color, the exceptions being largely confined to those compounds of zinc with colored anions. Most of the common zinc salts are soluble in water although such compounds as the carbonates, oxalates, sulfides, phosphates and silicates are either insoluble or only slightly soluble in water.

Zinc is not regarded as inherently toxic. It should be noted that zinc oxide when inhaled after being freshly formed as in the manufacture of brass and other alloys, can cause a mild malady known as "brass founders' ague," "zinc chills" or "oxide shakes." The brief duration of the attack and the absence of any cumulative or chronic aftereffects should be emphasized. Under conditions of continuous exposure, the human system can develop a high degree of resistance. When not freshly formed, zinc oxide dust is practically harmless. Soluble salts of zinc have a harsh metallic taste and taken internally can cause nausea and purging. Small repetitive doses may affect the digestion and cause constipation. Inhalation of zinc chloride fumes can injure the lungs and respiratory tract, the damage very likely being due chiefly to hydrolysis and consequent release of hydrochloric acid.*

The reactions of zinc which are of significance in the manufacture of the industrially important compound, zinc oxide, are:

1. Reduction of the zinc content of calcined concentrates or oxidized ores to zinc vapor by carbon (coal or coke) at 1000 to 1200°C, thus allowing separation from the bulk of impurities in the material being reduced.
2. Oxidation of the zinc vapor to zinc oxide, and
3. Reduction of carbon dioxide to carbon monoxide by zinc vapor (at 900 to 1200°C).

The main reactions are:

1. $2C + O_2 \rightleftharpoons 2CO$
2. $ZnO + CO \rightleftharpoons Zn\ (vapor) + CO_2$ (1000 to 1200°C)
3. $CO_2 + C \rightleftharpoons 2CO$ (1100 to 1200°C)
4. $2Zn\ (vapor) + O_2 \rightarrow 2ZnO$

* See Chapter 13 for further information on physiological, pathological and toxicological aspects of zinc. It is especially important to note that the physiological value of zinc as a trace element normally required in animal nutrition is now becoming well established. (Ed.)

Mention should be made of the fact that the reduction of zinc oxide (Reaction 2) is a reversible reaction, the equilibrium being a function of temperature and of the relative concentrations of carbon dioxide and carbon monoxide; in other words, under proper conditions carbon dioxide will function as an oxidizing agent for zinc vapor. In the reactions noted above, it is desirable to keep the temperature of the furnace at about 1100°C and to provide extra carbon so that Reaction 2 will always proceed from left to right. Selecting two temperatures by way of illustration, zinc vapor reduces carbon dioxide at 900°C if the ratio of carbon dioxide to carbon monoxide exceeds 0.09 by volume, while at 1200°C, reduction of zinc vapor will occur only if the ratio of carbon dioxide to carbon monoxide exceeds 0.94 by volume. Clearly, proper control is necessary for optimum recovery of the zinc and its conversion to zinc oxide. In Chapter 4 the equilibrium constant for this reaction is calculated and its use illustrated with the aid of appropriate graphical constructions. See especially Figure 4–8 showing the percentage of zinc in the gas at various temperatures and pressures.

ECONOMIC ASPECTS

Zinc oxide is the most important zinc chemical with respect to both tonnage and total value. It should be remembered that zinc oxide is or can be the starting point for all zinc chemicals. In recent years, shipments of zinc oxide have amounted to around 150,000 short tons. In 1957, the figure was 151,267 tons with a value of $41,007,424; off about 2 per cent from the previous year in both tonnage and value, since the average price each year was $271 per ton. American-process oxide, lead-free, carload lots in bags with freight allowed is currently quoted at 14½ cents a pound while the corresponding price for White Seal French-process oxide, the brightest and least dense of the "Seals," is 16¾ cents, for Green Seal 16¼ cents, and for Red Seal (the least bright) 15¼ cents. The imports and exports of zinc oxide are relatively minor, the figure for imports in 1957 being 5,245 tons and for exports 3,144.

The next most important zinc chemical from the standpoints of tonnage and value is zinc chloride (50° Baumé), followed by leaded zinc oxide, zinc sulfate and lithopone. Comparative figures for the major zinc chemicals for both 1956 and 1957 are given in Table 12-1. It will be noted that shipments of zinc sulfate and zinc chloride increased in amount 4 and 10 per cent, respectively, with corresponding increases in value of $7 and $3 per ton. On the other hand, leaded zinc oxide was off 11 per cent in tonnage and $33 per ton in value.

As has been the case for years, the rubber industry is the chief consumer of zinc oxide, followed by the paint and ceramic industries. The figures as to shipments of zinc oxide by uses for 1956 and 1957 are given in Table 12-2.

Table 12-1. Zinc Chemicals Shipped by U. S. Manufacturers
U.S. Bureau of Mines

	1956			1957		
	Short Tons	Value		Short Tons	Value	
		Total	Per Ton		Total	Per Ton
Zinc oxide	154,955	41,966,858	$271	151,267	41,007,424	$271
Leaded zinc oxide	27,164	7,647,169	282	24,203	6,028,233	249
Zinc chloride 50° Baumé	53,201	6,590,815	124	58,569	7,455,739	127
Zinc sulfate	32,200	4,917,073	153	33,620	5,368,063	160
Lithopone	38,434	5,630,991	147	No data available		

Table 12-2. Shipments of Zinc Oxide by Uses (Short Tons)
(U.S. Bureau of Mines)

	1956	1957
Rubber	80,459	81,745
Paints	32,485	32,605
Ceramics	10,160	8,459
Coated fabrics and textiles (including rayon*)	8,447	3,263
Floor coverings	1,436	1,249
Other	21,968	23,586
Total	154,955	151,267

* These figures include 7721 tons in 1956 and 2838 tons in 1957 for rayon.

ZINC OXIDE

Properties

Zinc oxide, ZnO, formula weight 81.38, crystallizes in the hexagonal system. The zinc oxide of commerce is white but impurities can affect the degree of whiteness. As a matter of fact, laboratory workers have been able to effect through heat treatment or by other methods a whole series of colors running through yellow, green, brown, red and even other shades. The red-colored oxide has probably received the most study; its color is generally considered to be due to 0.02 or 0.03 per cent of excess zinc atoms in the crystal structure. It should be noted that the mineral zincite (ZnO) is usually a deep red in color, rarely an orange yellow. The color in this case is generally regarded as caused by a small amount of impurities such as iron or manganese. The indices of refraction of pure artificially produced zinc oxide are given as 2.004 and 2.020 while the corresponding figures for the mineral are stated as 2.013 and 2.029. The high refractivity values of the oxide coupled with its availability in fine particle size explain its high hiding power, and are the major reasons for its use as a paint pigment. It also absorbs ultraviolet light, this opacity contributing to its value in such widely differing products as paint, face powder and sunburn ointments.

The density calculated from the lattice constants by the National Bureau of Standards is 5.680 at 26°C. Pycnometric values given in the literature show considerable variation depending upon such factors as starting material, process used to produce the oxide and the exact procedure and liquid employed. The specific gravity values given for the mineral vary from 5.43 to 5.70. Zinc oxide is a refractory material, melting at about 1975°C ± 25°.

Vaporization is apparent below 1300°C; at 1450°C the vapor pressure curve increases rapidly, reaching 12 millimeters at 1500°C. Extrapolation of this curve indicates that it reaches 760 millimeters at about 1700°C which is almost 300°C below the melting point given above. A further interesting fact is that when zinc oxide is heated, the zinc content increases because zinc vaporizes less readily than oxygen, particularly under such conditions as reduced pressure, presence of reducing agents, etc. The resulting product shows many characteristics different from those of the starting material.

Zinc oxide exhibits interesting luminescent, electrical and light-sensitive properties, many of these having become known only in the last two or three decades; and they are still neither fully understood nor utilized. These properties form the basis of the use of the oxide in phosphors and ferrites, as well as in a number of other applications, the most important of which are discussed later in this chapter.

Types and Grades

Zinc oxide of commerce may be divided roughly into three generic types based upon the method of manufacture, two being made by furnace processes and the third by chemical processes. At the present time, all three types are being produced and sold in the United States. However, the trade designations do not necessarily give any indication of the type of oxide nor the method of manufacture; in fact at the present time much less consideration is given to the type of oxide than to the peculiar properties of a particular grade.*

The oldest type of commercial zinc oxide is the "French process" oxide, or indirect type, made by burning zinc vapor produced by heating zinc metal in a retort. French process zinc oxide is characterized by a high degree of chemical purity and extreme whiteness and brightness. The impurity content of French process oxide is a direct function of the type of metal used. Oxide made from Special High Grade zinc or its equivalent will be extremely low in metallic impurities. Since the raw material contains no

* Manufacturing processes are described in Chapter 8.

acid-forming substances such as sulfur or chlorine, the acidity of French process oxide is at a minimum. Due to the lack of contaminants, French process oxide is the brightest and whitest oxide of commerce. As generally produced the particle size is relatively fine and spherical when observed by the light microscope (see Figure 8-11).

Two commercial grades of zinc oxide, pharmaceutical and the "Seals," are normally of the French process type. Pharmaceutical oxide, as its name implies, is used for salves, lotions and cosmetics and must be of the highest purity. The Seal grades are used throughout the chemical industry where a high degree of purity is required and also in some types of paint formulations. Seal grades are differentiated primarily by brightness and apparent density, White Seal is the brightest and the least dense, Green is less bright and Red the least. All these grades command a premium price over the standard lead-free oxide.

"American process" zinc oxide, or the direct type, results from the reduction by a carbonaceous fuel of zinciferous materials, usually a sintered calcine, to produce zinc vapor, and the subsequent oxidation of this vapor. The furnaces in which these reactions take place are of several different kinds, each yielding an oxide having its own peculiar properties and characteristics. One modification, termed the "electrothermic process," is unique in that it depends on heat generated by the electrical resistance of the furnace charge, rather than on heat by combustion processes, for the reduction step; the consequent absence of gases of combustion contributes to low sulfur content of the resultant zinc oxide. American process zinc oxide has become most versatile and therefore supplies the greater part of the United States market.

As indicated above, American process zinc oxide varies widely in its chemical properties and physical characteristics. From a physical standpoint, there are two major variations available, the acicular and the nodular types, as shown in the photomicrographs, Figure 8-1. The acicular (or needle-like) type of American process oxide has found great favor as a paint pigment in the United States. Acicular oxide has demonstrated good durability when incorporated into paint and also offers excellent properties of suspension for the paint system. Nodular (or spherical) oxide is generally preferred for rubber compounding.

Due to wide variations in the material processed and in the process itself, it is virtually impossible to characterize the properties of American zinc oxide except in a general way. It will have a zinc oxide content of about 98 to 99.5 per cent; it will contain a higher level of impurities than French process oxide and will be somewhat less bright. Both the particle size and shape may vary greatly. Clearly, its analysis will depend upon the purity of the raw material, the type of reducing agent and the processing procedure employed. It is characterized by having a higher acidity than the French process oxide but improved manufacturing techniques in recent years have resulted in products approaching the purity of the latter.

American process zinc oxide is sometimes heat-treated or calcined in order to further modify its properties. Application of heat may improve the brightness by removing carbonaceous material. The acidic elements may also be reduced in amount and some of the more volatile metallic compounds partially removed. Heat treatment will also serve to increase the particle size and to reduce the bulk or apparent density.

Lead-free oxide produced by the American process is the standard grade of zinc oxide sold in the United States. There is no further generalized commercial designation but each manufacturer designates a particular type of zinc oxide by an appropriate grade or code number. There are possibly three types of lead-free zinc oxide, all of which sell for the same price. These include (1) pigment oxides designed primarily for use in exterior house paint formulations, (2) rubber oxides designed especially for compounding purposes, and (3) ceramic oxides intended primarily for use in frits and glazes. Zinc oxide which is surface-treated with an organic acid is also being produced for the rubber industry. Pelletized free-flowing zinc oxide has likewise been made available, as well as densified grades which conserve storage and shipping space.

The third major type of zinc oxide is that produced by *wet* or *chemical processes*. The oxides resulting from these processes are frequently designated "secondary zinc oxides" and in recent years have become increasingly abundant. They may be a by-product of a chemical reaction in which zinc dust is employed, or a product of a wet process made by dissolving impure zinc oxides such as dross from galvanizing operations or fume from brass melting, and purifying the resulting solution before precipitating the zinc, usually as either the normal or basic carbonate. The zinc-bearing precipitate is calcined to drive off the water and carbon dioxide. It is then dry-milled in order to reduce it to a desirable particle size. While usually possessing high purity, these zinc oxides are somewhat less bright than the normal types of American process zinc oxide. The individual particles are usually nodular in shape and somewhat coarser than either American or French process zinc oxide. These latter two properties (brightness and particle size) are primarily a function of the necessary calcination of the material subsequent to its precipitation. Wet process zinc oxide is used primarily in the production of other zinc salts or in rubber compounding and is sold as a lead-free grade.

Zinc Oxide in Rubber

Until the discovery of vulcanization a little over a century ago, in 1839, by Charles Goodyear, the uses of rubber were largely restricted to waterproofing of fabrics. The articles so produced were very limited as to tonnage and utility. Although the very first vulcanized compositions did not contain zinc oxide, it must have been recognized almost immediately that the reactions of vulcanization could be assisted by many inorganic materials. The use of zinc oxide as a compounding ingredient for rubber is mentioned in United States patent 6,066, Reissue No. 141, issued to Henry G. Tyer and John Helm in 1849. The changes in properties brought about by vulcanization were so remarkable that a sort of chain reaction was started, and an industry was born which soon mushroomed to enormous proportions—and is still growing.

Over the years, three main processes have evolved for the production of zinc oxide. These method are described above under "Types and Grades." The American rubber industry consumes zinc oxide of all three types, though in tonnage American process oxide is no doubt dominant.

The substances that speed up the reaction of sulfur with rubber are now called "accelerators." In addition to the inorganic bases such as zinc oxide, there are dozens of organic compositions available commercially, and many are known that have not been made commercially. As an inorganic accelerator, in the absence of any organic accelerator, zinc oxide is relatively gentle, much less effective than such bases as magnesia or litharge. Therefore, before the discovery of the organic accelerators, zinc oxide was usually employed for its coloring or toughening effect, rather than as an accelerator. The early tires, made before carbon black become abundant, were toughened by a high loading of zinc oxide. The word "toughened" is used advisedly because the zinc oxide not only improved the tensile strength and resistance to abrasion of the rubber composition, but also protected the rubber by its opaqueness to ultraviolet light and by its high thermal conductivity. Its pleasing white color and freedom from toxicity were also factors in favor of zinc oxide, both in the processing operation and in the finished article. These properties still contribute to the popularity of zinc oxide as a compounding ingredient for rubber.

Although some of the organic substances function as accelerators in the absence of an inorganic base, most of them are much more effective accelerators when used in conjunction with an inorganic base, commonly termed an "activator." Zinc oxide is the activator most commonly employed, and this use now accounts for most of the tonnage of zinc oxide in rubber; indeed, for much of the total consumption of zinc oxide.

To function as an activator for an organic accelerator the zinc oxide must be converted to a rubber-soluble form by reaction with organic acids. These acids are usually present in the rubber, but additions of such materials as stearic acid, palm oil, pine tar, rosin or rosin acids are commonly made, not only to modify the properties of the unvulcanized rubber in order to facilitate processing, but also to insure an adequate supply of acid for the vulcanization reaction. The zinc salt of the organic acid may then react with the accelerator to form a salt or complex, regenerating the original acid which then repeats the cycle; or perhaps a double salt is formed with the accelerator. This zinc-containing substance is the true accelerator. The chemistry of vulcanization is very complex and much has been written about it. Most authors accept the theory that by some mechanism, the details of which may vary from one accelerator to another, sulfur is converted to an active form such as nascent sulfur or metastable polysulfides, and that this active sulfur reacts with the rubber to generate vulcanized rubber. Zinc sulfide is another end product of this reaction.

In the vulcanization of some synthetic elastomers, such as "Thiokol" and "Neoprene," the reaction results in release of acids. Zinc oxide is utilized in such compositions for its acid-accepting characteristics and results in improved properties of the vulcanized article.

Acids are also sometimes generated in vulcanized products by processes associated with loss of serviceable life, called "aging" by rubber technologists. Here again, zinc oxide extends the serviceable life of such articles by accepting these acids of decomposition.

The usual dosage of zinc oxide required for activation is in the range of 3 to 5 parts per 100 parts of rubber, though the optimum concentration may be as high as 10 to 15 parts per 100 of rubber, depending on the type of accelerator. In some formulations, such as those for transparent articles, the loading of zinc oxide may be 1 to 2 parts per 100 of rubber. It might be remarked also that for this application, zinc oxide of fine particle size and relatively low hiding power is preferred.

Zinc oxides for rubber vary considerably as to impurity content, depending on the process and raw material from which they originate. Impurities—aside from coarse particles of foreign substances which are frequently of concern to the rubber compounder—are sulfur, lead, cadmium, copper, manganese, acid-insoluble and water-soluble contents. The relative importance of these impurities is related to the type and intended application of the rubber composition. Organic accelerators are frequently sensitive to certain impurities, e.g., sulfur or lead. Obviously, the concentration of impurity with respect to the accelerator increases with increase in loading of the zinc oxide in the rubber composition. Usually, accelerators are affected by impurities only when the zinc oxide is employed at a dosage well above that required to activate the accelerator; for example, when the zinc oxide is employed for reinforcing purposes.

Commenting in greater detail, it may be said that the effects of lead and cadmium are specific with respect to the accelerator; for example, accelerators of the mercaptobenzothiazole type are activated by these impurities, especially if they are in the oxide form; whereas thiuram-type accelerators are retarded, though the retarding effect decreases with increase in curing temperature. Lead and cadmium tend to affect color of the vulcanizates because they form colored sulfides. However, lead in American process zinc oxide is sulfated and has much less effect on color and cure than it does in the oxide form marking its presence in French process zinc oxide.

Sulfur may be present as sulfur dioxide or sulfites and as sulfates. Although sulfur present as lead sulfate is relatively innocuous, other forms of sulfur tend to cause acidity. The general effect of acidity is to retard curing rate, notably of aldehyde-amine accelerators; however, with accelerators of the mercaptobenzothiazole type the effect is frequently to give a "flat" curing curve—a desirable condition. Acidity may also result from chlorides. Water-soluble salts (sulfates and chlorides) tend to cause coagulation of latex dispersions—an undesirable effect. Low water-soluble content is usually demanded in zinc oxides for the compounding of electrical insulation.

Copper and manganese tend to impair the aging of natural rubber, especially when both are present in significant concentrations. However, they are of less importance than formerly because of the use of age-resisters and because of increasing use of synthetic rubbers, which are more resistant to oxidation than are natural rubbers.

Freshly made zinc oxide is rather fluffy and voluminous. By the time it is packed some densification has occurred, to perhaps 25 to 35 pounds per cubic foot. To conserve shipping and warehouse space, and to aid in high-speed mixing procedures, zinc oxides are being offered in densified and pelleted forms, with bulk densities in the order of 60 pounds per cubic foot and higher. Surface treatments with organic additives, usually a fatty acid and sometimes light mineral oil in addition to the acid, are applied to improve the dispersion of the zinc oxide in rubber and to speed up the mixing process. Rapid and complete dispersion is required of pigments for modern high-speed mixing. The rate of mixing of a zinc oxide is related to several other factors, among which are temperature and plasticity of the batch, type of rubber, particle size and shape of the zinc oxide and surface condition of the particles. Usually, calcined and surface-treated zinc oxides, having no adsorbed or occluded gases on the surface of the particles, are faster mixing than before treatment.

The particle shape of zinc oxides for rubber compounding is generally of the spherical or nodular type, rather than the acicular or "threeling" type preferred for some types of paints. The latter type, at loadings above the minimum for activation of the accelerator, tends to give a composition of greater stiffness than the nodular type.

Zinc oxides for rubber are available in a relatively wide range of particle sizes, as judged by average values. Within a given sample, of course, particle sizes may vary over a considerable range. The actual values reported for particle size of a given sample are dependent on the method of evaluation. Because of the importance of zinc oxide in rubber and paint technology, considerable work has been done on methods of determining particle size. Roughly, the average particle diameter may range from about 0.1 to 0.3 micron, depending on the type of zinc oxide. For latex compounding, zinc oxides of very fine particle size are usually employed to minimize the settling from dispersions. As pointed out above, zinc oxides of very fine particle size—with average particle size below the range giving maximum hiding power—also find utility in transparent rubber goods.

The use of rubber, both natural and synthetic, in latex form continues to increase. Zinc oxide is essential for activation of the organic accelerators. In some types of formulation it also functions as a secondary gelling agent. The zinc oxide is dispersed in water by ball-milling or by a colloid mill; in either case an organic dispersing agent is employed. The dosage required is about the same as in dry-milled rubber, namely, two to five parts per one hundred of rubber. Fine particle size is usually desirable to avoid settling of the zinc oxide from the dispersion, but coarse grades are some times employed to decrease thickening on storage and to permit the preparation of dispersions containing high concentrations of zinc oxide.

Because it is well-wetted by rubber, zinc oxide can be added in relatively large-volume loadings (10 to 20 volumes per 100 volumes of rubber) without producing a "dry" stock. This characteristic makes zinc oxide valuable in surgical tape, in which its antiseptic properties are also a desirable contribution. When vulcanized, stocks containing a high loading of zinc oxide are valuable for service in which rapid flexing occurs because of low heat generation. Zinc oxide having particles of the nodular type rather than the acicular type is particularly preferred for this kind of service. Conduction and dissipation of heat from such articles are encouraged by zinc oxide due to its relatively high thermal conductivity. The thermal conductivity of zinc oxide is severalfold that of the rubber hydrocarbon and of the more common fillers added to rubber.

Relative tonnages of zinc oxides sold to the rubber industry, and of rubber processed, are shown below for the last three years currently available:

	1955	1956 short tons	1957
Zinc oxide for rubber	86,677	80,459	81,745
Rubber, natural and synthetic	1,712,000	1,608,000	1,645,000
Zinc oxide as a percentage of rubber tonnage	5.06	5.00	4.97

Zinc Oxide in Protective and Decorative Coatings

History. Zinc oxide was first used in protective and decorative coatings toward the end of the eighteenth century. Prior to that time it had been used chiefly in artists' colors where it had found wide acceptance because it was the first white hiding pigment which retained its whiteness on aging and resisted darkening when exposed to fumes containing sulfur. It was for these same reasons, plus its nontoxic nature, that zinc oxide was first used in the protective coating field. Pigmentation with zinc oxide in the organic binders then in use produced architectural enamels which were hard and durable and retained their color and gloss much better than any previous coatings. Even today, some enamels are pigmented largely with zinc oxide since there is still some demand for enamels of the "French" or "Dutch" type in certain areas of the United States.

Present-Day Coatings. The largest use of zinc oxide in the protective coating field today is in paints for exterior wood surfaces. Exterior house paint formulations are dependent on zinc oxide for a number of reasons. In addition to being a white hiding pigment, zinc oxide is chemically reactive. Zinc oxide reacts with the organic acids present in the vehicle initially, forming soap; this soap formation aids in mixing the other pigments with the oil and at the same time induces a thixotropic consistency in the finished paint. After the paint has been applied and is exposed to the weather, the zinc oxide continues to react with the acids resulting from the oxidation of the organic binder.

Zinc oxide is the most opaque of any of the white pigments to ultraviolet light. This is important, since the oil-type vehicles currently in use are very sensitive to ultraviolet radiation which readily catalyzes their oxidation. The life of a drying oil exposed to the weather is very short unless it is protected from ultraviolet. In addition, many of the color pigments employed to produce the popular pastels and tints fade prematurely unless the formulation contains an adequate amount of zinc oxide.

Another very important reason zinc oxide is necessary in house paints is its fungistatic action. Although zinc oxide is nontoxic, it exerts an inhibiting action on the growth of mold or mildew. This fungistatic action of zinc oxide in properly formulated house paints cannot be duplicated by any other pigment nor can the use of phenol or mercury compounds effect equal inhibition of mold growth.

Zinc oxide is also helpful in controlling the rate of chalking of a paint film exposed to the weather. Zinc oxide is classed as a nonchalking pigment and adds toughness to the film. Its use in conjunction with free-chalking pigments results in a formulation with the desired chalk rate without becoming soft. In addition, the use of zinc oxide improves the "clean-up" of a paint film, since some of the soaps formed by zinc oxide are water-soluble and impart a detergent action resulting in less dirt retention. These two properties produce a film which stays clean with minimum erosion.

In the last few years another property of zinc oxide has been discovered. The new low-gloss, flat or "breather" type of house paint produces relatively porous films as compared to those resulting from the use of conventional house paints. When these paints were made without zinc oxide, they frequently showed objectionable tannin staining when applied over redwood or red cedar substrate. The inclusion of zinc oxide in these formulations, because it precipitates the tannins, effectively minimizes this staining.

In addition to being a major constituent in exterior house paints, zinc oxide is used in white gloss enamels of the architectural or industrial type. Here a small amount of zinc oxide in the pigmentation improves the enamel in several ways. Zinc oxide imparts hardness and greater abrasion resistance to the film; it also prevents after-yellowing and reduces hazing or loss of gloss.

Types of Zinc Oxide Used. Two types of zinc oxide are used in protective coatings today—one is made by the French process and the other by the American process. The French process type has been in longer use; its starting material is metallic zinc. Oxides made by this process have extremely fine particle size, and as a result, greater reactivity; ordinarily they are also more nearly chemically pure. The French process oxides are used exclusively in enamels and industrial finishes, not in exterior house paints.

American process zinc oxides are made directly from zinc ore and are classed as either lead-free or leaded. The lead-free oxides contain at least 98 per cent zinc oxide and the leaded types will contain a specified amount of basic lead sulfate. Leaded zinc oxides may be made by either cofuming the pigment from a mixture of lead and zinc ores or by blending basic lead sulfate with lead-free zinc oxide. Both leaded types can be formulated into equally durable paints.

Leaded zinc oxides are sold in various combinations, the more common being 5, 20, 35 and 50 per cent (the percentage refers to the amount of basic lead sulfate in the product) although other percentages are supplied as demanded. In effect, they provide the paint manufacturer with his requirements for two important exterior pigments in one package and thus reduce his plant work. The paint industry consumes 99 per cent of the leaded zinc oxide manufactured. Current prices for leaded zinc oxide, 35 per cent grade carload lots in bags with freight allowed are 15⅛ cents and for the 50 per cent grade 15⅜ cents. The consumption of leaded zinc oxide has declined in recent years.

American process lead-free zinc oxides can be produced in a variety of particle shapes and sizes, although usually they are much larger in particle size than the French process oxides. By varying the processing techniques, particle shapes varying from large-nodular to extremely acicular can be produced. The range of particle size can also be controlled. All of these various shapes and sizes can be used to produce satisfactory house paints by formulating with respect to the oil demand of the specific pigment. In general, the highly acicular zinc oxides require more organic binder than do the oxides with relatively larger, nodular particles. However, no matter what type of zinc oxide is used, its particle size will be much finer than that of the white inert pigments.

Types of Zinc Oxide-Containing Formulations. The type of formulation will determine largely the type of zinc oxide used. If the paint is designed for an area subjected to industrial fumes containing sulfur, the zinc oxide used will be lead-free. Such paints are termed "fumeproof" or "fume-resistant."

Paints designed for use in areas of heavy mildew infestation are best formulated with a significant amount of leaded zinc oxide because this type of paint has proved to be the most resistant to spore growth. Although the lead pigment by itself is not fungistatic, its use with zinc oxide imparts mildew resistance greater than that of the zinc oxide alone. In addition, such paints are better formulated with the low-oil-demand pigments since the resultant paint will contain less organic binder to serve as food for the fungi. House-paint primers for use in these areas are frequently based on leaded zinc oxide because the use of zinc oxide in the primer as well as in the topcoat materially improves the total mildew resistance of the system.

Exterior house paints formulated for quality service will contain as a minimum 2½ pounds of zinc oxide per gallon. This may be introduced either as leaded or lead-free zinc oxide. The balance of the pigmentation usually consists of about 1½ pounds of titanium dioxide and from 2½ to 4 pounds of extender, the exact amount dependent on the type. For white paints, the magnesium silicate extenders are usually considered to be superior, and for colored house paints, the use of some calcium carbonate is advantageous. The linseed-oil content of such paints will be in the range of 3¾ to 4½ pounds per gallon, depending on the oil demand of the pigmentation and the film properties desired.

Emulsion-type coatings for use on masonry and cement surfaces also require pigmentation with zinc oxide for optimum performance. These finishes benefit from the inclusion of zinc oxide for the same reasons as do the more conventional oil-base paints. Although the synthetic latices themselves are usually not susceptible to attack by mildew, the various additives and plasticizers are. In addition, these paints are subject to contamination by wind-borne organic dirt on the surface as in the case of any paint. This surface contamination will readily support the growth of mildew if zinc oxide is not present in the pigmentation. Zinc oxide also serves to protect the vehicle and color pigments from ultraviolet and to control chalking. Other benefits from the use of zinc oxide specific to latex paints are the reduction of can-corrosion and *p*H-drift. This is due to the alkaline nature of zinc oxide in this type of paint. Zinc oxide also aids in stabilizing the color in the can on storage. The amount of zinc oxide necessary in these synthetic latices appears to be in the order of one pound per gallon.

Zinc oxide also finds use in metal-protective coatings. Paint pigmented with zinc dust and zinc oxide is one of the most successful coatings for use on galvanized surfaces. Zinc oxide is also used in conjunction with red lead and/or zinc yellow for general-purpose primers for ferrous surfaces. In these priming paints, the zinc oxide aids in producing a tough, adherent film, resistant to abrasion and chalking; for these reasons such paints are specified in many metal-protective coating systems.

Zinc Oxide in Ceramics

While the rubber and paint industries are by far the major users of zinc oxide, the ceramic industry is also an important consumer, having ranked third for many years—just ahead of coated fabrics and textiles. Zinc oxide finds considerable use in such ceramic products as glass, glazes, enamels, frits and like articles.

The reasons for its use here are rather interesting:

1. Zinc oxide is, of course, a white substance and therefore by itself contributes no color to any ceramic composition to which it is added; nevertheless, it will increase the brilliance of clear glasses. It will also intensify the brilliance of various colored glasses, particularly the more delicate shades. Curiously, it is essential to proper color development in some cases such as selenium ruby glass and the yellow color of nickel glass when this powerful colorant is used in a suitably small amount. As would also be expected, based upon their similarity to glass, zinc oxide enhances the brilliance and gloss of glazes and enamels.

2. Zinc oxide is a rather refractory oxide, melting at about 1975°C, and yet when used in moderate amounts in ceramic products, it acts as a flux, decreasing the softening temperature of the product, lowering its viscosity and increasing the maturing range.

3. Interestingly too, it tends to impart a low coefficient of expansion, thus improving the elasticity and giving a product of increased resistance to thermal and mechanical shock.

4. A very striking property of zinc oxide has been noted in recent years—namely, that when this nonmagnetic material is incorporated by standard ceramic procedures in ferrites (ferro-spinels), products of improved magnetic and electrical characteristics result, which are finding rapidly expanding uses in the electrical and electronic industries.

Glasses. The improvement in brilliance and purity of color when zinc oxide is added in small amounts to many glass compositions has already been mentioned. It is commonly a constituent of opal and alabaster glasses, particularly when improved chemical resistance or lighting qualities are required, the significant contribution in the latter case being increased opalescence.

Zinc oxide, up to 10 per cent and more in amount, is a constituent of many types of technical glassware. Such glasses show good resistance to chemical attack and to thermal shock and, therefore, are used in laboratory ware, thermometer and boiler-gauge tubing, ovenware and top-of-stove ware. Also, beginning with the classic work of Schott and Abbe in the late nineteenth century on optical glasses, it has been found that zinc oxide imparts desirable properties such as improved optical characteristics, stability and durability to many types of these glasses, particularly to barium crown and barium flint. Zinc oxide is frequently employed also in spectacle and signal glasses.

Glazes and Enamels. The inclusion of zinc oxide in the formulations for glazes and enamels is helpful for many of the same reasons as in the case of glass. It contributes to the fusibility. It helps to control expansion, thus increasing elasticity and resistance to thermal and mechanical shock. It tends to prevent crazing and similar imperfections from developing. It improves luster and brilliance of color. It enhances opacity in many cases and increases the efficiency of such opacifiers as the oxides of zirconium, tin and antimony.

A preliminary step in the manufacture of many enamels and glazes is the production of a frit made by fusing a mixture of some or all of the constituents. The inclusion of zinc oxide in the raw batch is quite common because it lowers the melting point and increases the fusibility, at the same time improving the stability and reducing the water-solubility of the product.

Because, then, of the desirable properties contributed by zinc oxide, it is a frequent constituent of the surface coating, glaze or enamel used in such varied products as sanitary ware (sinks, toilet and washbowls, bathtubs), earthenware, crockery, refrigerators, ranges, washing machines, wall tile, terra cotta, enamels for sheet iron and aluminum, etc. Because it contributes to electrical resistivity, it is also a constituent of enamels used for coating certain electrical equipment.

It is difficult to generalize as to the zinc-oxide content of these various ceramic products but it can be said that a quantity of 10 per cent plus or minus in these fused compositions will ordinarily produce the desired effects. However, for specific uses and to attain certain characteristics, considerably larger amounts may be used such as in solder glass, hydrofluoric acid-resistant glass and in certain optical glasses.

Ferrites. One of the many recent developments in the electrical field has been the discovery that when zinc oxide is a constituent of ferrites, improved magnetic and electrical characteristics result. Ferrites are ferro-spinels, that is, compounds of trivalent iron oxide (Fe_2O_3) with the oxide or oxides of a divalent metal such as magnesium, cobalt, nickel, zinc, manganese, etc. Rather surprisingly, it has been found that improved magnetic properties result if nonmagnetic zinc oxide is used in part as the divalent oxide constituent of the ferrite, nickel oxide very frequently being the other favored divalent oxide. Ferrites as used in the electrical industry are the product of a sintering operation applied in accordance with standard ceramic procedures to a mixture of the desired oxides in proper physical form and proportion. They find extensive use where high frequencies are involved. An interesting application of ferrites is in the so-called "memory devices" of the newer computers where this high-permeability material can be easily magnetized and demagnetized. Nickel-zinc and manganese-zinc ferrites find use in such devices as transformer cores, cup cores and radio antennas. Numerous developments in the rapidly expanding field of ferrites, both as to manufacture and application, are constantly occurring.

Zinc Oxide in Coated Fabrics and Textiles

In late years, the consumption of zinc oxide in the manufacture of coated fabrics and textiles has been approaching the quantity used in ceramics; as a matter of fact, in 1955 (the last year for which statistics are presently available) the tonnage was actually greater, being 11,263 against 10,617 for ceramic applications. The single most extensive use of zinc oxide in this field is in connection with the production of viscose rayon, particularly high-tenacity cord for use in automotive tires, rubber hose and belting, and other reinforcing applications. Apparently, zinc sulfate monohydrate is being increasingly supplied rather than the oxide. This is not surprising since a sulfate spinning bath is employed. Into this bath at a temperature of about 50°C, a thin stream of viscose solution is injected which coagulates, forming a fiber. The composition of the bath varies but it usually contains about 6 to 13 per cent sulfuric acid, 12 to 25 sodium sulfate, 0.5 to 10 zinc sulfate and commonly 2 to 10 per cent glucose. By careful control of the manufacturing procedure, rayon of high strength, good elongation and satisfactory fatigue-resistance, all so necessary for automotive tires and like applications, can be achieved.

The whole art of textile manufacture and treatment (including fabric coating) has become so complex in the last few decades that it is difficult here to more than generalize with respect to the use of zinc oxide and derivatives thereof in this advancing field of technology. Zinc oxide has long been used in the textile industry because of its inherent properties mentioned earlier in this discussion. It is natural that its whiteness and compatibility with the basic fiber, whether natural or synthetic, and with the various reagents employed in processing operations such as spinning, dyeing and finishing, should render it of general usefulness. Frequently, the zinc oxide is converted to some other compound such as the sulfate, chloride or zincate before or during use. The following are typical of the applications of, and the properties contributed by, zinc oxide in the field of textile technology:

1. Whiteness and compatibility.

2. In combination with surface-active agents, zinc oxide, usually converted to the acetate or formate, will serve to fix dye colors to fabrics.

3. Derivatives of zinc oxide, usually the chloride, can be applied to many fabrics to combat mildew and rot, both induced by fungi and bacteria.

4. Zinc oxide is frequently employed as a weighting material or filler. Used in this capacity, it will provide water-resistance, low hygroscopicity and softness to the fabric, and at the same time will not disturb the dye or finish.

5. Zinc oxide and derivatives thereof are frequently used as constituents of fire-resistant reagents applied to textile fabrics. Occasionally, zinc borate is formed in place, imparting fire-retardant properties. Zinc oxide and zinc borate are also used as "glow-proofers," a term used to signify prevention of the incandescent flameless combustion of a fabric.

6. Further, somewhat as in the case of rayon, solutions of zinc-oxide derivatives can be used as the spinning bath for other textiles; for example, a 10 to 50 per cent water solution of zinc chloride heated to 90° may be employed for acrylic fibers such as "Orlon."

Zinc Oxide in Floor Coverings

The floor-covering industry also uses a considerable quantity of zinc oxide, the tonnage being 2281 in 1955, the latest year for which figures are currently available. This represents a decline from the amount consumed a few years previously.

As in other applications which have been discussed, zinc oxide finds use in the linoleum and tile floor-covering industry because of its white color and in general the properties that render it useful as a paint pigment. The finer and more reactive grades of zinc oxide are usually employed thus providing the proper flow and leveling characteristics. It also contributes toughness and durability. It might be remarked that lithopone has been extensively used in the linoleum industry, but its use is decreasing.

Miscellaneous Uses of Zinc Oxide

Zinc Oxide in Lubricants. Zinc oxide finds use in lubricating oil and greases. In any application it can perform one or more of such functions as the following: a filler, a pigment, or a chemical additive, either by itself or in the form of zinc soaps or salts. Many patents have been granted covering the uses of zinc oxide and zinc soaps in this general field, indicating the extensive work which has been undertaken.

The use of zinc oxide in lubricants goes back to the last century. As a matter of fact, zinc oxide appears to be a preferred metallic-oxide additive to lubricating greases, a complete line up to 10 per cent and more having been produced for years. Zinc oxide acts not only as a filler but also as a pigment, improving color. At times, it may serve as a mild abrasive. It has been further stated that such zinc-oxide-containing greases produce a wear-resistant film on working surfaces, thus reducing friction and power costs. Zinc oxide also functions as an antiacid and seems to inhibit oxidation of the grease.

Zinc soaps, frequently preformed, find considerable use in lubricating greases. Apparently they do not function as bodying agents, although zinc stearate at times seems to thicken fluids by its bulking value. Zinc soaps in lubricating greases are said in varying applications to impart such properties as resistance to hyrdocarbon solvents, fluidity and useableness at low temperatures, and water-repellency in certain hydraulic operations, even preventing the rusting of wire cables that operate in water.

Zinc soaps are also constituents of many mixed-base lubricating greases where they are said to contribute to such properties as retardance of oil separation, water-resistance, and inhibition of oxidation.

Zinc salts of a number of organic acids other than fatty acids are employed as additives to lubricating oils. Such acids include the dithiocarbamates and the alkyl dithiophosphates. These zinc salts are said to function as antioxidants, inhibiting the degradation of the oil by oxidation and heat, and also as detergents, preventing the sticking of piston rings and valve assemblies by removal of gums and carbonaceous deposits.

Zinc Oxide in Chemical Smokes. An unusual application of zinc oxide has been its employment in chemical smokes for military purposes (screening and signaling), during both World Wars I and II. The World War I mixture consisted of about 1 part zinc dust and 2 parts carbon tetrachloride along with zinc oxide and diatomite. Upon ignition, a vigorous reaction occurs, forming the volatile and hygroscopic zinc chloride vapor which condenses to give a dense white smoke. During World War II, there were several developments, resulting in an improved "HC smoke mixture" consisting of aluminum powder, zinc oxide and a chlorinating agent, hexachlorethane being the favored one. Again, the primary constituent of the smoke is hygroscopic zinc chloride. The major chemical reactions are considered to be the following:

$$2Al + C_2Cl_6 \rightarrow 2AlCl_3 + 2C$$

$$2AlCl_3 + 3ZnO \rightarrow 3ZnCl_2 + Al_2O_3$$

Zinc oxide also contributes whiteness to the smoke and retards the rate of burning. To provide stability, the moisture content must be maintained below 0.6 per cent. These smoke compositions are used chiefly in grenades or in smoke pots, if rapid generation of smoke is desired.

ZINC SULFIDE

Zinc sulfide, ZnS, formula weight 97.44 crystallizes in two forms: one, hexagonal, is called wurtzite, with indices of refraction of 2.356 and 2.378, sp. gr. 4.087; and the other, cubic, is called sphalerite, with a refractive index of 2.368 and sp. gr. 4.102. The hexagonal form is the stable phase above 1020°C. Zinc sulfide is white in color. Its high index of refraction is the primary reason for its high hiding properties, thus rendering it useful as a paint pigment. Zinc oxide is the chief impurity in the pigment.

In making zinc sulfide, reaction between the two constituent elements in the vapor phase is sometimes employed. Precipitation methods are also used, the most common procedure being precipitation from a zinc salt solution with hydrogen sulfide, although barium or sodium sulfide may also be used. The dried precipitate is heated at about 725°C in the absence of air to effect substantial conversion to wurtzite, the form preferred for use as a pigment.

Lithopone

Advantage is taken of the high refractive index (and high hiding power) of zinc sulfide by co-precipitating it with barium sulfate to make the pigment *lithopone*. This product combines the good qualities of zinc sulfide with an inexpensive and satisfactory inert material to produce a relatively low-price pigment.

In the manufacture of lithopone, carefully purified solutions of zinc sulfate and barium sulfide are mixed, precipitating zinc sulfide and barium sulfate. The resultant product is filtered, dried, calcined and milled; this material has some qualities not found in a mixture of the two chemical compounds. For example, the thixotropy of lithopone paints is an important advantage in many applications. The usual product is almost 30 per cent zinc sulfide and 70 per cent barium sulfate, although other proportions are made for some purposes, particularly a 50 per cent zinc-sulfide product.

The largest use of lithopone is in paints, primarily interior flats, glosses and enamels. Other uses are as a pigment in linoleum, floor coverings, coated fabrics and textiles, paper and rubber. Although the use of lithopone has declined in recent years, it is still one of the important white pigments.

Another product is *titanated lithopone*, which consists of 15 per cent titanium dioxide and 85 per cent lithopone and has much the same uses as lithopone; its chief utility is based on somewhat greater hiding power than the standard lithopone.

Zinc Sulfide in Phosphors

Zinc-sulfide phosphors constitute a striking and important use of zinc sulfide. Phosphors can best be defined as artificially prepared materials which exhibit luminescence upon irradiation by light. If the luminescence appears only during the irradiation, it is termed "fluorescence"; when the luminescence persists after extinction of the light, it is called "phosphorescence." While there are many natural materials which exhibit fluorescence, they are not generally acceptable because of their variability.

The zinc sulfide which is used in phosphors must be exceedingly pure, for even traces of some materials such as iron, nickel and cobalt exert a harmful effect upon the luminescent qualities. To obtain the necessary purity, the reagents used must be of very high quality and meticulous care must be exercised. In the final step, hydrogen sulfide is added under carefully controlled conditions to the highly-purified zinc solution (usually the sulfate), precipitating zinc sulfide.

To prepare the phosphors, the pure zinc sulfide is mixed with about 2 per cent of sodium chloride and about 0.005 per cent of activator—copper, silver and manganese being the most important—and is fired in a nonoxidizing atmosphere. Zinc sulfide has two crystalline forms, cubic and hexagonal; the transition temperature is 1020°C, the hexagonal form being stable above this temperature. Both crystalline forms, in combination with one of the various activators, have their own desirable properties and the type of phosphor used depends on the final application.

An extension of the properties of zinc-sulfide phosphors can be secured by replacing varying amounts of the zinc sulfide with cadmium sulfide. The same activators are used for this combination as for zinc sulfide. Cadmium sulfide exists only in the hexagonal form but the addition of as little as 10 per cent of this compound to zinc sulfide and a firing temperature as low as 900°C insures a solid solution of the two sulfides, having a hexagonal crystal structure. The type of luminescence is dependent on the ratio of the two sulfides in the phosphor. Because of their wide range of desirable characteristics and dependability, the zinc sulfide and zinc sulfide-cadmium sulfide phosphors constitute a major class of luminescent materials.

It might also be mentioned here that in addition to the sulfide phosphors, there is a second main group known as "oxygen-dominated" phosphors which also finds extensive use. In this group fall such compounds as zinc oxide, zinc silicate, zinc beryllium silicate, zinc aluminate and zinc germanate phosphors.

These phosphors find chief usage in cathode-ray tubes employed in both radar and television, and in fluorescent lamps. Luminescent paints, lacquers and fabrics are applications of increasing importance.

ZINC SULFATE

Zinc sulfate, $ZnSO_4$, formula weight 161.44, sp. gr. 3.74, forms colorless rhombohedral crystals with indices of refraction of 1.658, 1.669, 1.670. It forms three hydrated compounds, $ZnSO_4 \cdot 7H_2O$ (white vitriol), $ZnSO_4 \cdot 6H_2O$, and the monohydrate $ZnSO_4 \cdot H_2O$. The monohydrate loses its water of crystallization on heating above 238°C. On further heating at 740°C, the sulfate decomposes into zinc oxide and sulfur trioxide.

Zinc sulfate is usually prepared by leaching roasted zinc ore concentrates with sulfuric acid solution and filtering out the residue. Other raw materials such as galvanizers' skimmings are used in some cases. After purification to remove metallic impurities, the solution may be evaporated and the $ZnSO_4 \cdot 7H_2O$ crystals separated. To prepare the monohydrate, the solution is evaporated to dryness in a rotary kiln, in a spray dryer, or on a drum dryer depending on the physical characteristics desired in the product.

Zinc sulfate is sold both as the 22 per cent zinc grade ($7H_2O$) and as the monohydrate containing about 37 per cent zinc. It is used in numerous industries, the major one being viscose rayon fiber manufacture (see also "Zinc Oxide in Coated Fabrics and Textiles" above). As the viscose filaments are ejected from the spinnerets, they enter a complex precipitating bath. Here, sulfuric acid and sodium sulfate are the major constituents, along with some glucose to prevent crystallization and some zinc sulfate to promote crenulation of the fiber.

The second largest use is as a trace element in agriculture and in animal nutrition. Some areas of the country are deficient in zinc. (See Figure 14-1.) For proper plant growth zinc sulfate is applied to the soil at rates of 2 to 10 pounds per acre, usually admixed with a nitrogenous fertilizer. Better results on trees—citrus and tung for example—are often obtained by incorporating the zinc sulfate in sprays for foliage application.

Among the smaller uses are in chemical manufacture, textiles and dyeing, wood preservation, froth flotation of minerals, clarification of glue, electrogalvanizing solutions, and in paint and varnish processing.

OTHER ZINC COMPOUNDS

Zinc Acetate

Zinc acetate, $Zn(C_2H_3O_2)_2$, formula weight 183.43, is a white solid, crystallizing in the monoclinic system with a sp. gr. of 1.84. It is moderately soluble in water (30 grams in 100 milliliters at 20°C, 44.6 grams at

100°C); it is decidedly less soluble in alcohol at room temperature (2.8 grams in 100 milliliters at 25°C) but its solubility increases rapidly with rise in temperature (166 grams at 79°C). When heated at 200° at either atmospheric or reduced pressure, decomposition occurs with the formation of acetic acid and a basic salt; at still higher temperatures, zinc oxide results.

The dihydrate, $Zn(C_2H_3O_2)_2 \cdot 2H_2O$, is the stable solid phase in aqueous systems below 100°C. It too crystallizes in the monoclinic system forming colorless plates or granules. As would be expected, its density (1.735) is somewhat lower than that of the anhydrous salt and its solubility slightly higher (31.1 grams at 20°C).

Zinc acetate is manufactured in the conventional manner by heating a slurry of zinc oxide with acetic acid and filtering. A slight excess of acetic acid is added to the filtrate which is then evaporated until crystallization begins. On cooling, crystallization continues. The crystals are centrifuged or filtered.

Zinc acetate may be used as a mordant in dyeing, as a wood preservative and in the manufacture of glazes for painting on porcelain. A weak solution (0.1 to 4 per cent) exhibits mild antiseptic properties and is used on mucous membrane and skin. The technical grade currently sells for about 30 cents a pound and the reagent grade for 53 cents.

Zinc Borate

Zinc borate can be made by two processes: (1) fusion of ZnO and B_2O_3 which gives an anhydrous material, and (2) wet precipitation which produces a hydrated compound.

The phase diagram of ZnO-B_2O_3 shows two possible compounds $5ZnO \cdot 2B_2O_3$ and $ZnO \cdot B_2O_3$. As far as is known, no zinc borate prepared by the fusion process is commercially available at the present time in the United States.

Various methods of precipitating zinc borate from aqueous solution have been described; the ratio of ZnO to B_2O_3 in the final product depends on the specific method. A typical procedure is to react suitable ratios of borax, zinc sulfate and zinc oxide in water to obtain the desired product. Alternately, borax, zinc sulfate and caustic soda may be used. The commercial zinc borate presently available is prepared by wet methods.

Zinc borate is used in fireproofing of fabrics and as a fire-retardant ingredient in paints. It has fungistatic properties. It is used in pharmaceuticals and also in the ceramic field.

Zinc Carbonate

Zinc carbonate, $ZnCO_3$, formula weight 125.39, may exist in either white or colorless rhombohedral crystals, depending upon circumstances of formation. Its indices of refraction are 1.85 and 1.62. It is highly insoluble in water, only about 1 part dissolving in 100,000 parts at room temperature.

Zinc carbonate occurs naturally as the mineral smithsonite. It can be prepared by adding soda ash to a zinc sulfate solution. It can also be made by passing carbon dioxide into zinc-oxide slurry. The product, which in both cases contains some basic zinc carbonate, is filtered, washed and dried. Upon heating zinc carbonate, evolution of carbon dioxide begins at about 300°C and is complete at about 500°C, the precise temperature depending upon duration of heating.

Zinc carbonate finds use as a paint pigment, in pharmaceutical preparations, in rubber products and in ceramics. Technical grade of zinc carbonate is currently quoted at 13 cents a pound.

Basic zinc carbonates can be prepared; their composition depends on the method of formation. One of these compounds, $2ZnO \cdot ZnCO_3 \cdot 2H_2O$, is said to be used in the manufacture of rayon.

Zinc Chromate

Zinc chromate, $ZnCrO_4$, formula weight 181.39, forms yellow prisms. As a commercial pigment it is called "zinc yellow." It is a complex compound, whose composition is somewhat variable depending upon the method of manufacture. The following empirical formula is typical:

$$4ZnO \cdot K_2O \cdot 4CrO_3 \cdot 3H_2O.$$

This would correspond to a composition of about 37 per cent ZnO, 11 per cent K_2O, 46 per cent CrO_3 and 6 per cent H_2O. Actually, the zinc oxide content is usually a little more, and the chromic oxide less, than these figures. The pigment has an oil absorption of about 24 pounds of linseed oil per 100 pounds of pigment and a sp. gr. of 3.40. It may be made by heating a slurry of zinc oxide with potassium dichromate and chromium trioxide, the ratios of the three depending upon the composition desired for the pigment. Sulfuric acid may be added to "catalyze" the reaction. It is sparingly soluble in water. Freedom from sulfates and chlorides in the dried product is desirable since their presence in a paint pigment would encourage corrosion. It is widely used as a corrosion-inhibiting primer, either with or without an iron-oxide pigment, for aluminum, magnesium and iron. It does not bleed in solvents and provides a pigment less subject to staining and discoloration than the chrome yellows, though possessing less hiding power and tinting strength. While the most important use of zinc yellow is in primers, it is frequently used also in decorative finishes, usually in combination with other colored pigments.

Zinc tetroxychromate (ZTO), a basic chromate with the formula

$$4Zn(OH)_2 \cdot ZnCrO_4,$$

is less soluble than the usual zinc yellows and has found use in marine primers as well as wash-primers. It contains about 70 per cent ZnO, 18 per cent CrO_3 and 12 per cent H_2O; it has a sp. gr. slightly under 4.

The usual grade of zinc chromate currently sells for 29 cents a pound, and the basic pigment for 34 cents.

Zinc Cyanide

Zinc cyanide, $Zn(CN)_2$, formula weight 117.42, is a white salt, insoluble in water but soluble in alkali cyanides or hydroxides to form double cyanides. Its primary use is in electroplating. Its price in drums in 1000-pound lots or more is 55 cents a pound.

Zinc Halides

Zinc fluoride, ZnF_2, formula weight 103.38, like all halides of zinc, is a somewhat hygroscopic, white solid. It crystallizes in the tetragonal system and has a sp. gr. of 4.95. Its melting point is stated to be 872°C and its boiling point about 1500°C. As is frequently true, the properties of this halide of zinc are in many cases decidedly different from those described subsequently. For example, the fluoride is relatively insoluble in water—only 1.62 grams per 100 milliliters of solution at 20°C. It forms no acid fluorides. It is very stable thermally. Only one hydrate is recorded in the literature, this being the tetrahydrate. It is formed by adding hydrofluoric acid to a slurry of the oxide. The water of hydration can be expelled with the application of heat, the tendency to form basic fluorides being far less marked than in the case of the heavier zinc halides. Zinc fluoride finds use in making phosphors, in galvanizing baths, as a wood preservative and in ceramic products. Zinc fluoride currently sells for 50 cents a pound.

Zinc chloride, $ZnCl_2$, formula weight 136.29, has a sp. gr. of 2.91 and a melting point of about 300°C. It boils without decomposition at 732°C. It is one of the more highly soluble inorganic compounds, 432 grams dissolving in 100 milliliters of water at 25°C, and 615 grams in 100 milliliters at 100°C. It is also quite soluble in many organic solvents containing oxygen or nitrogen, such as alcohols, ethers, amines and nitriles. It is difficult to remove the last traces of water from the solid for it is about as hygroscopic as phosphorus pentoxide. Hydrolysis of aqueous solutions at moderate dilutions is surprisingly small; however, evaporation of aqueous solutions to dryness, except under suitably controlled conditions, will result in partial formation of basic or oxy-chlorides. The literature mentions several hydrates, the one with the highest proportion of water and undoubtedly the most stable being $ZnCl_2 \cdot 4H_2O$. In this connection, it may be said that the basic system $ZnCl_2 \cdot ZnO \cdot H_2O$ is rather an involved one of limited solubility. What might be termed the converse of this system—that is, $ZnCl_2 \cdot$

$HCl \cdot H_2O$—is far simpler and possesses high water solubility, decreasing however with increasing HCl content.

Several different methods may be used for the preparation of zinc chloride. According to one procedure, zinc sulfide concentrates (frequently containing lead) may be given a chloridizing roast with common salt, thus fuming off zinc and lead chlorides, which are caught in a suitable bagroom. The fume is then treated with either dilute sulfuric acid or zinc sulfate, the resulting lead sulfate precipitate being filtered off and the filtrate evaporated to the desired zinc chloride product. Actually, most, and probably all, of the zinc chloride produced in the United States is derived from secondary zinc sources such as zinc dross, sal skimmings, etc. According to the generally accepted procedure, the source of secondary zinc units is treated with somewhat less than the chemically equivalent quantity of muriatic acid, and the clear liquid is filtered off. The nearly-dry residue is then chlorinated and the resultant zinc chloride leached with water and added to the first filtrate. The clear liquid is then evaporated to the desired stage. Zinc chloride can also be made by the action of hydrochloric acid on the metal or the oxide and evaporation of the resultant product.

Zinc chloride is one of the more important industrial compounds of zinc. In aqueous solution, it may be used alone or in combination with phenol or sodium chromate as a wood preservative (railway ties, poles for communication lines) and for fireproofing lumber. It is used as a deodorant, in disinfecting and embalming fluids, and as a fungicide; in the manufacture of vulcanized fiber and parchment paper; as a mordant in printing and dyeing textiles; in mercerizing cotton, sizing and weighting textiles and in chemical synthesis as a dehydrating agent; for oil refining, galvanizing and in the manufacture of dry batteries.

Zinc chloride is sold in the form of a 50° Baumé solution in drums or tanks (about 6 cents a pound) or as fused in stick form (11 cents a pound) or granulated (12 cents a pound).

Zinc ammonium chloride, $ZnCl_2 \cdot 2NH_4Cl$, formula weight 243.29, crystallizes in thin white plates with a sp. gr. of 1.8. It is readily soluble in water with absorption of heat. It is used as a flux for etching, welding and soldering, and in galvanizing. Its current price in carload lots at the works is 10.25 cents a pound.

Zinc bromide, $ZnBr_2$, formula weight 225.21, is, like zinc chloride, a highly hygroscopic material being of the same order of effectiveness as a drying agent as $CaCl_2$. It is either a white or colorless salt depending upon manner of production and size of crystals. It crystallizes in the rhombohedral system.

It is somewhat more soluble in water than even zinc chloride, the solubility being 471 grams per 100 milliliters of water at 25°C and 675 grams at 100°C. It is readily soluble in such organic solvents as ethyl alcohol, ether and acetone. Values given in the literature for the melting and boiling points are not in agreement but it can be stated that its melting point is approximately 390°C and its boiling point of the order of 660°C. Its sp. gr. is 4.20 to 4.22. It may be prepared by dropwise addition of moist bromine to heated zinc, also by conducting bromine in a nitrogen stream over molten zinc at 600°C. Sometimes an atmosphere of carbon dioxide is employed. If the pure salt is desired, suitable precautions must be taken to prevent the formation of basic bromides.

It forms di- and trihydrates, the former probably being the more common one. Similarly to zinc chloride, the system $ZnBr_2 \cdot ZnO \cdot H_2O$ is fairly complicated; there are a number of basic zinc bromides. Zinc bromide also forms addition compounds with ammonia and hydrazine, with amines and with other organic compounds.

Zinc iodide, ZnI_2, formula weight 319.22, has about the same order of solubility in water at 25°C as the chloride and bromide, but it is somewhat less soluble in boiling water. Like the chloride and bromide it is readily soluble in alcohol but less so in ether. It has a sp. gr. of about 4.70, a melting point of 446°C, and a boiling point of about 620°C.

Zinc Nitrate

Zinc nitrate, $Zn(NO_3)_2$, has a formula weight of 189.4. The pure anhydrous salt is unstable and can only be obtained under carefully controlled conditions. The system $Zn(NO_3)_2 \cdot H_2O$ is rather a complicated one because of the large number of hydrates. Hydrates with mole contents of water of 9, 6, 4, 2 and 1 to one mole of $Zn(NO_3)_2$ are recognized.

Organozinc Compounds

Of historical interest is the fact that the first organometallic compound prepared was diethyl zinc; E. Frankland made it in 1849 by reacting zinc and ethyl iodide at 150°C in an autoclave. Organozinc compounds found extensive employment in organic synthesis until Grignard announced in 1900 his more reactive magnesium-containing reagent, "RMgX" (X being a halogen), when the use of the zinc compounds declined. However, organozinc compounds are still widely used where the greater activity of the Grignard reagent is not desired. Contrary to the situation with magnesium, compounds of the R_2Zn type are better known than the RZnX type.

The lower-molecular-weight zinc dialkyl compounds are liquid at ordinary temperatures while the diaryl compounds are white crystalline solids. The low-molecular-weight zinc alkyl compounds are unstable in air, igniting spontaneously and emitting dense white fumes of zinc oxide. The zinc aryls and higher-molecular-weight alkyls are more stable but will undergo oxidation unless properly protected.

Zinc Phosphates

The $ZnO \cdot P_2O_5 \cdot H_2O$ system is rather complex; besides the expected mono-, di-, and tertiary zinc compounds, there is also a compound $Zn(H_2PO_4)_2 \cdot 2H_3PO_4$. All of them contain some water of hydration; upon heating to 100 to 150°C, they revert to the anhydrous $Zn_3(PO_4)_2$.

The major use of zinc phosphate is in phosphate coatings of ferrous, zinc and aluminum surfaces for corrosion protection and improved paint bonding. Because of their porosity, another important use of zinc phosphate coatings on ferrous surfaces is to provide an oil reservoir for reducing friction between moving metal parts and during drawing operations.

For phosphate coating of iron and steel prior to painting, a solution of zinc phosphate and phosphoric acid is used; appropriate accelerators and oxidizing agents are usually added. The metal is thoroughly cleaned and degreased and then either immersed in, or sprayed with, the solution. The crystalline layer so obtained is essentially zinc phosphate combined with a small amount of iron phosphate. Conditions are controlled to obtain small grain size which gives improved corrosion resistance. The porous phosphate film thus formed provides an excellent anchorage for paints applied over it.

For zinc and aluminum surfaces to be painted, the phosphating solution is essentially zinc phosphate; it is applied by the same techniques as used for ferrous surfaces. Since paint adhesion to these nonferrous metals is usually poor, the bonding action supplied by the phosphate coating markedly increases the durability of the protective-coating system.

The porosity of phosphate coatings, which is of such value in paint bonding, can also be utilized for the absorption of lubricants. Oil films produced in this manner are valuable for lubricating moving metal parts, thus reducing friction. The automotive industry, particularly, uses such coatings.

Heavy zinc phosphate coatings are often applied to articles which must be drawn or formed, particularly where the operations are severe. When properly lubricated, these coatings, because of the attendant reduced friction, improve the speed and efficiency of the operation as well as minimize the wear on the forming dies.

Zinc phosphates are not normally sold as such but are prepared by reacting zinc oxide and phosphoric acid to give solutions of the desired compositions.

Zinc Silicates

The system $ZnO\text{-}SiO_2$ has been the subject of extensive investigation largely due perhaps to the frequent use of ZnO in glasses, glazes and enamels; the equilibrium diagram is available.* The only zinc silicate compound which appears in the diagram is the orthosilicate, Zn_2SiO_4. References to zinc metasilicate, $ZnSiO_3$, listing some of its physical properties, appear in the literature but its existence is quite doubtful and certainly is not presently substantiated by x-ray diffraction data.

Zinc orthosilicate, Zn_2SiO_4, formula weight 222.82, occurs naturally as the mineral willemite. It has a melting point of 1512°C, ±3°, indices of refraction of 1.692 and 1.720, and a sp. gr. of about 3.9. It is, of course, quite insoluble in water. It forms a eutectic with tridymite at 49 mole per cent ZnO content which melts at 1432°C, and another with 77.5 mole per cent ZnO melting at 1507°C. It forms hexagonal crystals, the exact crystal form varying with the circumstances of production and subsequent environment, both trigonal and rhombohedral varieties being mentioned in the literature. Probably the best procedure to secure the orthosilicate is to react the proper molar proportions of fine-particle-size ZnO and SiO_2 at temperatures approaching the melting point of the silicate. The reaction may be carried out under pressure in the presence of H_2O.

The pure silicate exhibits fluorescent properties and finds use in this field. Zinc silicate has at times been added as such to glass batches but ordinarily its two components are used.

A monohydrate of the orthosilicate also exists. It is known as "hemimorphite" ($Zn_2SiO_4 \cdot H_2O$). It occurs as rhombohedral crystals and has a sp. gr. of about 3.45.

Zinc Silicofluoride

Zinc silicofluoride, $ZnSiF_6 \cdot 6H_2O$, formula weight 315.52, is frequently called "zinc fluosilicate." It forms hexagonal crystals of low sp. gr. (about 2.1). It is usually made by the addition of zinc oxide or zinc carbonate to fluosilicic acid (H_2SiF_6), the latter in slight excess. Since the silicofluoride (or fluosilicate) decomposes in the presence of water and excess basic ions, giving the fluoride and silicic acid (or the silicate), suitable precautions as to temperature and concentration must be exercised. It is used as a wood preservative and to precipitate potash from sugar liquors and other solutions. It also finds application as a laundry-sour and as an additive to plaster in order to increase strength. It currently sells in drums for about 13 to 14 cents a pound.

Zinc Soaps

Zinc forms metallic soaps of the general formula $(RCOO)_2\,Zn$, R being an organic radical (usually aliphatic) containing at least 6 to 7 carbon

* E. N. Bunting, *Bur. Standards J. Research*, **4**, 134 (1930); also *J. Am. Ceram. Soc.*, **13**, 8 (1930).

atoms. Zinc soaps are insoluble in water but exhibit solubility in nonpolar liquids such as benzene. In general, they have two types of uses: (a) they are employed in the protective-coatings industry to hasten the drying of oils and of varnish and enamel films, and (b) they are used in fungicides, lubricating oils and greases, rubber, plastics, cosmetics and pharmaceuticals, paints and waterproofing compounds—where their properties are valuable for reasons not directly connected with their catalytic action on drying oils. It is interesting to note that they can be used for waterproofing such diverse material as textiles, paper, fiberboard, concrete, bricks and paint.

Two chief methods of manufacture are used: (a) by precipitation from the reaction between a water-soluble zinc salt, such as the sulfate, and soluble alkali soaps, and (b) by fusion of zinc oxide or other zinc compound with organic acids. In case the zinc soap is to be used as a drier, an organic solvent may replace the water. The manufacture of metal soaps and their correct employment are industrial processes of considerable importance and technological involvement.

The zinc soaps commonly used as paint driers are the naphthenates, resinates and octoates, with zinc contents as high as 10 to 12 per cent. It should be stated that while zinc soaps have an important place in the paint-drier field, they are ineffective when used alone but function catalytically when used with a drier such as cobalt. Very likely the superior action resulting from the inclusion of zinc soaps is due to their marked dispersion characteristics.

Zinc soaps, as previously indicated, find highly diversified nondrier uses. For these applications, in addition to the zinc soaps mentioned above, the zinc salts of saturated fatty acids containing more than ten carbon atoms, particularly lauric, $C_{11}H_{23}COOH$; palmitic, $C_{15}H_{31}COOH$; stearic, $C_{17}H_{35}COOH$; and the unsaturated oleic acid, $C_{17}H_{33}COOH$, are also employed. As a rule, the zinc soap used is a mixture rather than the salt of a single fatty acid; also the degree of neutralization may vary, that is, whether the soap is a mono- or di-type.

In addition to serving as driers for protective coatings, zinc soaps—particularly the stearates, naphthenates, oleates and resinates—also find use in this field as bodying and flatting agents, for grinding aids and for waterproofing; for example, zinc stearate is a superior flatting agent, the naphthenate and oleate are useful as pigment-grinding aids, while zinc resinate is used as a relatively high-melting resin in varnish manufacture.

Zinc soaps find use as fungicides, their pale color often contributing to their desirability; as an example, a solution of zinc naphthenate in an organic solvent is being so used for application to cellulosic articles. Zinc stearate and palmitate, which are not wetted by water, are used for waterproofing a wide variety of materials including textiles, paper, fiberboard, and even concrete and bricks. In some cases, a waterproofing coat is applied by conventional methods; in others, the dry powdered soap is incorporated directly into the product (e.g., concrete).

Zinc laurate and stearate are used in rubber as softeners and vulcanization activators; they not only facilitate the processing of the rubber compounds but also shorten the curing time and improve the properties of the products. Dusting of the mold surface with zinc soaps before vulcanizing the rubber product will also prevent sticking of the rubber to the mold and facilitate removal of the product.

In the plastics industry, zinc soaps are used as plasticizers, internal lubricants and stabilizers. In vinyl resins, they perform the important function of neutralizing and absorbing acid decomposition products formed by heat degradation of the resin, thereby preventing discoloration.

Zinc stearate is used by the cosmetic industry as the base for many face and body powders. It exerts a mild antiseptic action and finds utility in preparations for skin disorders.

Zinc soaps are used sometimes in combination with zinc oxide, in lubricants, as described earlier under the heading "Zinc Oxide in Lubricants."

Prices of zinc soaps vary with the price of the fatty-acid constituent and the zinc content; for example, zinc naphthenate in drums with an 8 per cent zinc content is currently quoted at 27 cents a pound; the 10 per cent grade at 33 cents. Zinc resinate, 7.5 per cent zinc, carries a price of about 36 cents a pound, and zinc stearate, technical grade in cartons, 41 cents.

Zincates

Zinc is amphoteric and can form zincates with alkalies. The only zincates which have been investigated to any extent are the sodium, potassium and ammonium compounds. Zincates are soluble only in alkaline solutions, the solubility increasing with the strength of the solution. Monobasic zincates $(HZnO_2)^-$ are formed in dilute solutions of potassium and sodium hydroxides, dibasic zincates $(ZnO_2)^=$ in more concentrated solutions. Both mono- and dibasic zincates have H_2O molecules bound to them. In strong ammonia solutions, the dibasic zincate is formed; in weaker solutions, the zinc is held in the well-known zinc ammonium complex

$$Zn(NH_3)_4^{++}.$$

Solutions of sodium zincate in alkaline sodium silicate are used to coat asbestos-cement shingles; the shingles are then baked and the final product is quite resistant to moisture penetration and dirt collection. A similar solution is used as a binder for roofing granules, bonding wheels and analogous products.

In the manufacture of foam rubber, ammonium zincate is added to the latex to provide zinc oxide, which aids in the gelling of the foam and in the subsequent vulcanization of the rubber.

Zincates are usually prepared by dissolving zinc oxide in the required alkaline solution. Generally, they are not marketed, but made as needed by the user.

REFERENCES*

1. Brown, Harvey, E., "Zinc Oxide Rediscovered," New York, New Jersey Zinc Company, 1957.
2. Gmelin's "Handbuch der Anorganischen Chemie. Zink." Supplement Volume. System-Number 32, 8th ed., E. H. Erich Pietsch (editor), Weinheim/Bergstrasse, West Germany, Verlag Chemie, GmbH, 1956.
3. Kirk, R. E. and Othmer, Donald F. (editors), "Encyclopedia of Chemical Technology," New York, The Interscience Encyclopedia, Inc., 1947–56.
4. Mattiello, Joseph J. (editor), "Protective and Decorative Coatings," Vol. 2, pp. 369–388, New York, John Wiley and Sons, Inc., 1942.
5. Mellor, J. W., "A Comprehensive Treatise on Inorganic and Theoretical Chemistry," Vol. 4, pp. 398–694, London, Longmans Green & Company, Ltd., 1929.
6. "Minerals Yearbooks—Metals and Minerals (except Fuels)," U. S. Bureau of Mines, Washington, D. C.
7. Rankama, K. and Sahama, Th. G., "Geochemistry," Chicago, University of Chicago Press, 1949.
8. Remy, H., "Treatise on Inorganic Chemistry," Vol. 2, pp. 423–445, Amsterdam, Elsevier Publishing Company, 1956.
9. "St. Joe Zinc Oxides," New York, St. Joseph Lead Company, 1957.
10. Sidgwick, N. V., "The Chemical Elements and Their Compounds," Vol. 1, pp. 262–285, London, Oxford University Press, 1950.

* The author's method of treatment in this chapter has been to select only data that in his experience were considered reliable. The above publications are included for further consulation. Ed.

CHAPTER 13

THE BIOLOGICAL SIGNIFICANCE OF ZINC

BERT L. VALLEE

Assistant Professor of Medicine, Harvard Medical School
and
Scientific Director, The Biophysics Research Laboratory of the Department of Medicine, Harvard Medical School and Peter Bent Brigham Hospital, Boston, Massachusetts

Zinc has been studied in biology for many years.[1] Its physiological importance as a trace element normally required in animal nutrition is now becoming well established. Keilin and Mann offered the first concrete explanation of its mode of action in 1940.[2] They showed that the enzyme carbonic anhydrase contains 0.33 per cent zinc as a part of its molecule and that zinc is essential to the elimination of carbon dioxide. This work foreshadowed the identification of zinc as a participant in other important enzymatic reactions.[3, 4] The present chapter will cover some of the salient facets of the physiological, biochemical, clinical and toxicological implications of zinc metabolism, emphasizing recent observations.

PHYSIOLOGY

The usual human intake of zinc is 10 to 15 milligrams per day. The element is excreted mainly through the intestines; the feces contain about 10 milligrams and the urine about 0.4 milligram per day. The zinc content of human organs varies from 10 to 200 micrograms per gram of wet weight of tissue. Most organs, including the human pancreas, contain about 20 to 30 micrograms per gram.[3] Contrary to some statements, the pancreas does not have unusually large concentrations of zinc although it has been shown that the islet tissue of fish is rich in the element, containing 100 to 1000 micrograms per gram of fresh tissue. No significant difference has been detected in the zinc content of the pancreas in normal and diabetic patients. The liver, voluntary muscles, bladder and bones of man hold 60 to 180 micrograms of zinc per gram of fresh tissue and the normal prostate 859 $\pm$ 96 micrograms per gram of dry weight. Outstanding in their content of zinc are the tissues of the eye. From 500 to 1000 micrograms per gram has been found in the retina and as much as 5000 micrograms of zinc per gram in the iris of fish and mammals other than man.

The total amount of zinc in the human body has been estimated to be 2 grams. Apparently no tissue stores zinc preferentially. Human whole blood contains about 900 micrograms of zinc per 100 milliliters. Normal serum-zinc levels probably range from 80 to 160 micrograms per 100 milliliters. Normal erythrocytes contain 1.44 milligrams of zinc per 100 milliliters of packed red blood cells. Three per cent of all zinc in human blood is found in leucocytes, which contain 0.032 micrograms of zinc per million cells. Blood-zinc concentrations do not exhibit seasonal or diurnal variations, nor do they differ between the sexes. They are normal in diabetics.[5]

The distribution of radioactive zinc (Zn^{65}) was studied in the mouse and dog. In both animals, the liver contained the largest fraction of the injected dose of Zn^{65} early in the experiment. The most rapid uptake was observed in the liver, kidney, pancreas and pituitary; least activity was found in erythrocytes, brain, skeletal muscle and skin. Red blood cells and bone were the only tissue components to accumulate zinc gradually. This isotope has been detected in dogs and humans for as long as eight months to a year after its administration. With the injection of radioactive zinc into a pregnant dog, the passage of the isotope across the placenta into all of seven puppies has been demonstrated.

Zinc deficiency in animals was studied as early as 1922, but the first successful experiments were reported in 1934. Todd, Elvehjem, and Hart[6] found that the growth of rats deficient in zinc was only one-third that of normal ones; the zinc-deficient animals required 52 per cent more ration to gain a gram in weight. Their fur softened and turned black to gray in six to seven weeks. Administration of adequate amounts of zinc subsequent to the termination of the deprivation experiment caused two animals to return to a normal appearance.

Studies of the carbohydrate metabolism in these animals suggested delayed absorption of glucose from the gastrointestinal tract. These rats had a low plasma-protein level and a definite reduction of pancreatic amylase and proteolytic activity, which could not be restored to normal by the addition of zinc, in vitro or in vivo. In all the animals the zinc content of the soft tissues was normal, but that of the bone was lowered.

Hove, Elvehjem, and Hart studied carbonic anhydrase activity in animals deficient in zinc. Although a slight anemia was found, there was no significant lowering of the enzyme activity in erythrocytes. Plasma-uric acid levels of the rats deficient in zinc were found to be doubled. The uric acid blood findings were not reversed until five weeks after zinc had been added to the diet. This increased uric acid content seemed to be caused by increased uric acid formation, rather than by lack of destruction, since uricase activity was completely normal.[7]

Zinc and Hormones. The crystallization of insulin as a zinc salt by Fisher and Scott led to the belief, now widely held, that zinc is an integral part of the insulin molecule. This belief persists in spite of the fact that nickel, cobalt, and cadmium can be used instead of zinc in the crystallization.[8] Actually, Fisher and Scott never stated that zinc was an integral part of the insulin molecule; they reported that these metals and insulin combined in a constant ratio. For zinc, the percentage in combination was 0.52; for cadmium, 0.77; for cobalt, 0.44; these percentages were not changed by recrystallization. Moreover, amorphous insulin is as active physiologically as is crystallized insulin, and there is little convincing evidence at present that zinc and insulin must combine in vivo to form an active compound. There is far more zinc in the pancreas than would be necessary for insulin activation and some of it is now accounted for by carboxypeptidase.[9]

Understanding of the interrelation between zinc and insulin has been complicated further by the finding that 0.02 per cent zinc, added to insulin as the chloride, reduces the immediate activity of the hormone by 40 per cent, though the over-all potency of the hormone is not affected if enough time is allowed for it to act.

The subject of the relationship of zinc to the physiology of the pancreas in general and to that of insulin in particular has been reopened recently. Employing a nonspecific technique based on the use of diphenylthiocarbazone (dithizone), Okamoto[10] found variable amounts of zinc in the islands of Langerhans. With this agent, Kadota[11] produced diabetes in rats; he presumed that dithizone combined with the zinc of insulin to cause this experimental diabetes. Other workers found that the experimentally induced diabetic pancreas contained less Zn^{65} than did its normal counterpart.[12] McIsaac has employed autoradiography to distinguish between the uptake of Zn^{65} by acinar and islet cell tissue of the pancreas.[13] He found that the Zn^{65} concentration of the islet cells remained stable throughout the experimental period, whereas the concentration of acinar tissue, initially high, decreased with redistribution and excretion of Zn^{65}. No conclusions were drawn concerning the postulated association of zinc with insulin in vivo. Thus, the debate on zinc-insulin interaction continues without the achievement of a conclusive settlement.

Zinc and Porphyrins. It seems certain that a zinc uroporphyrin exists. Some investigators have found the zinc complex of the Waldenstrom uroporphyrin to be a constant constituent of the urine and feces in cases of intermittent, acute porphyria.[14] Urinary excretion alone seemed to be the rule in the acute, intermittent cases. In fact, the occurrence of the zinc porphyrin complex is characteristic of this type of porphyrinuria, whereas in the congenital type the porphyrin is usually excreted in a free state. In the latter cases, a zinc uroporphyrin is found in the liver. A zinc coproporphyrin, which has also been described, has been found in cases of lead poisoning, and its presence in the urine of victims of acute rheumatic fever has been confirmed.

BIOCHEMISTRY AND PHYSICAL CHEMISTRY

As with other trace elements, an explanation of the physiological role of zinc has been sought in its relationship to proteins, specifically those having enzymatic activity. A review of some basic considerations is fundamental to the understanding of the problem.[4] For operational purposes, the enzymes that are affected by metals can be considered in two groups: metalloenzymes and metal-enzyme complexes.

Metalloenzymes contain a metal as an integral part of the molecule. There is a fixed amount of metal per molecule of protein. The metal cannot be separated from the protein without complete loss of enzymatic activity, nor can it be removed by dialysis. When the element is split from the protein residue by vigorous manipulation, all measurable biological activity is lost and is not readily restored by the addition of this or any other metal. Some metals are part of porphyrins, as in hemoglobin, catalase, or peroxidase. Metalloenzymes may be considered as a class of biochemical substances. Metalloproteins are found in a variety of phyla; their general metabolic importance is indicated by the fact that they are incorporated into the tissues of species, the evolutionary histories of which are widely diversified.

The study of the enzymatic properties of these proteins has usually preceded the demonstration of their metal contents. This is probably due to the fact that until very recently the methods for the determination of enzyme activities have been more sensitive than the analytical methods for the small quantities of metals involved. In certain cases this course of events has been reversed.

Metal-enzyme complexes compose a far larger group of proteins and enzymes that are very loosely associated with a metal, the criterion of association being the activation of catalysis. In this group, the specificity of the association is lacking. The metals are dialyzable and may substitute for one another in many instances. The metal ion is not an integral part of the molecule, and in most instances the enzyme is apparently independent of the metal ion. These facts increase the difficulties of assessing the biological significance of these in-vitro findings. Some have considered the difference between these two groups of metal-enzyme systems to be quantitative only, representing different types of bond structure but having similar biochemical meaning. Others have thought the metal to be involved in the formation of the intermediate enzyme substrate complex. Whatever the hypothesis, operationally, metalloenzymes lend themselves more readily to a definitive assessment of the physiological role of metal-enzyme systems at this time.

Experimentally, the element and the enzyme may be studied jointly and separately, in vitro and in vivo.

In contrast to the highly colored iron and copper proteins and enzymes, zinc compounds in general and zinc proteins in particular are colorless and do not attract attention to themselves. The red and blue colors of iron and copper proteins seem to have been responsible for their early recognition and study. The identification of zinc proteins has been more or less fortuitous, owing to their lack of color. The investigation of the role of zinc in enzymatic catalysis has not been as systematic as that of iron and copper.

Carbonic Anhydrase. Until recently, the sole accepted physiological role of zinc was related to its presence, as an active component, in carbonic anhydrase. Keilin and Mann[2] first noted that the carbonic anhydrase of the red blood cells of oxen contains 0.33 per cent zinc that does not exchange freely with ionic zinc. This metalloenzyme catalyzes the reaction

$$CO_2 + H_2O \rightleftarrows H_2CO_3 .$$

Without carbonic anhydrase, carbon dioxide exchange could not take place with sufficient rapidity to sustain life. Zinc is an integral part of the molecule of carbonic anhydrase, and the removal of the metal results in irreversible inactivation. Reversible inactivation is effected by enzyme inhibitors that combine with zinc, such as "BAL,"* cyanide, sulfide, azide, and sulfonamides. Carbonic anhydrase is widely distributed in almost all tissues, but its assay has been difficult because of the presence in tissues of various activators and inhibitors of both the enzyme-catalyzed and the uncatalyzed reaction. Efforts have been made to account for the presence of all the zinc in tissues by their carbonic anhydrase content. These attempts have been unsuccessful except in selected instances, such as in erythrocytes of man, where carbonic anhydrase may account for all zinc.

Zinc-Containing Protein of Human Leucocytes. A protein that contains 0.3 per cent zinc per gram of dry weight can be obtained from white blood cells in humans. This zinc protein is responsible for 80 per cent of all the zinc found in human leucocytes and has been differentiated clearly from carbonic anhydrase. While the protein is presumably an enzyme, the nature of its enzymatic activity is not known.[15]

Carboxypeptidase. Carboxypeptidase of bovine pancreatic juice has recently been shown to be a zinc enzyme.[9] It is an exopeptidase that splits terminal amino acids from peptides having a free alpha carboxyl group adjacent to the peptide bond. This enzyme was first crystallized in 1937 and has been the subject of much attention since that time. The enzyme contains one atom of zinc per molecule of protein. Zinc is firmly bound to the protein and is apparently indispensable for its enzymatic activity, thus

* "British Anti-Lewisite," or 2,3-dimercaptopropanol.

constituting a prosthetic group. Like carbonic anhydrase, carboxypeptidase is inhibited by cyanide, sulfide, and BAL, as well as by 1,10-phenanthroline, 8-hydroxyquinoline-5-sulfonic acid, and other zinc-binding agents. It is not inhibited by sulfonamides. This finding seems to explain previous observations in which it was demonstrated that about 6.5 per cent of a dose of Zn^{65} administered to dogs was eliminated in their pancreatic juice within five days. Zinc, at least in part, is eliminated as carboxypeptidase; it is thus implicated in proteolysis.

Zinc-Containing Dehydrogenases. Four enzymes responsible for dehydrogenation reactions have recently been shown to contain zinc that is essential for their action. All four enzymes are dependent upon diphosphopyridine nucleotide (DPN) for their activity; alcohol dehydrogenase of yeast (YADH),[16] alcohol dehydrogenase of horse liver (LADH),[17, 18] glutamic dehydrogenase of beef liver (LGDH),[19] and lactic dehydrogenase of rabbit skeletal muscle (SLDH).[20] YADH contains four atoms of zinc per molecule of protein, LADH contains two, and LGDH contains four to five zinc atoms. The molecular weight of SLDH is not known; thus, no figure for the molar ratio of zinc to protein can be given. All four enzymes have been crystallized. The zinc is firmly bound and is essential to the action of these enzymes, as all of them are inhibited by zinc-binding agents.

The detection of zinc in these enzymes may explain the high concentrations of this element in the liver and the retina, since it has been shown that LADH oxidizes vitamin A_1 and reduces retinene, probably being identical with retinene reductase. Thus far, no other metals have been found to occur in similar dehydrogenases. These findings were the first indication that zinc is in any way associated with their catalytic action.

Zinc Protein Complexes. In blood serum, zinc exists in at least two fractions; firmly bound zinc amounting to about 34 per cent, and loosely bound zinc amounting to 66 per cent of the total zinc content.[5] The firmly bound zinc protein, a globulin, satisfies the criteria of metalloproteins, whereas loosely bound zinc protein should be defined as a metal-protein complex. The firmly bound zinc protein has not been purified adequately to reveal the percentage of zinc that it contains. Neither substance has been shown to exhibit enzymatic properties. The loosely bound complex appears to be concerned primarily with zinc transport. The ubiquitous occurrence of zinc and its participation in cellular and gaseous respiration as well as in proteolysis point to its cardinal role in metabolism. These discoveries may well foreshadow a recognition of its significance in pathological processes.

CHEMICAL PATHOLOGY

Changes in Serum Concentrations. In pneumonia, bronchitis, erysipelas and pyelonephritis, the serum-zinc level is decreased significantly. This correlation has been found to be statistically significant. In untreated pernicious anemia, serum-zinc levels are decreased to about 80 to 90 micrograms per 100 milliliters. Twenty-four hours after institution of therapy, the zinc level rises, reaching a maximal value during the first week.

In all other anemias studied thus far, serum-zinc levels are normal or the data are so sketchy that no definite conclusions may be drawn.[5] Decreased serum-zinc levels have been reported in patients with a variety of primary cancers, but no uniform patterns exist. Normal serum-zinc values have been observed in acute rheumatic fever, ulcerative colitis, diabetes, acute nephritis and peptic or duodenal ulcers. Recent observations demonstrate that the serum-zinc concentrations in Laennec's cirrhosis and myocardial infarction are markedly lowered.[21, 22] The concentrations in cirrhosis were reported as 66.7 ± 19.2 micrograms per 100 milliliters, whereas in myocardial infarction concentrations of 73 ± 20 micrograms of zinc per 100 milliliters were found. The normal control group had a serum-zinc level of 120 ± 19 micrograms per 100 milliliters. These findings were interpreted in the light of the newly detected zinc dehydrogenases. Variations in serum levels are attributed to fluctuations in the loosely bound fraction, the firmly bound zinc remaining constant. The physiology of control of zinc in serum is not understood. No plausible explanations for pathological changes have been offered.[5]

Changes in Erythrocyte Content. The zinc content and carbonic anhydrase activity of red blood cells parallel each other and are significantly correlated. In normal persons as well as those afflicted with anemias, polycythemia vera, secondary polycythemia, leukemia and congestive heart failure, both parameters vary directly with the hematocrit level and the hemoglobin concentration. In untreated pernicious anemia, however, the erythrocyte zinc concentration and carbonic anhydrase activity are nearly normal, although the hematocrit value, erythrocyte count and hemoglobin levels are markedly decreased. In remission, zinc concentration and carbonic anhydrase activity return to normal values. Erythrocyte zinc concentration and carbonic anhydrase activity are increased in sickle cell anemia and acute intermittent porphyria.

The content of zinc and carbonic anhydrase of erythrocytes of infants at birth is only 25 per cent of the adult value and is even lower in premature infants. Both values gradually rise to normal levels, doubling at the age of one year and reaching adult values at the age of 10 to 12 years.[23] Certain intractable types of cyanosis and dyspnea of infants at birth have been attributed to very low carbonic anhydrase activity and may be benefited by transfusion of adult red blood cells. A decrease of 50 per cent has been reported in the concentration of zinc in the whole blood of Chinese suffering from protein deficiency, acute and chronic beriberi, and pellagra. No separate values for erythrocytes and serum were obtained, and no hematological data were given.

Changes in Leucocyte Content. Marked increases of leucocyte zinc have been seen in patients with anemias, refractory to all therapy and accompanied by leucocyte counts below 2000 per cubic millimeter. The zinc concentration of leucocytes in acute and chronic lymphatic and myelogenous leukemia is decreased to 10 per cent of the normal value. With successful therapy, the zinc concentration returns to normal. Administration of zinc salts intravenously fails to raise the leucocyte zinc concentration or to influence the course of the disease. The phenomenon is apparently independent of the maturity of the cells.

Changes in Urine Content. The normal zinc content of urine, 0.3 to 0.4 milligram per 24-hour volume, has been found to rise sevenfold to 2.1 milligrams per 24-hour volume in patients with albuminuria.

TOXICOLOGY AND INDUSTRIAL HAZARDS

Zinc poisoning may occur as a result of three distinct processes: (1) the ingestion of toxic amounts of zinc with food or drink, (2) direct contact with zinc or zinc salts, or (3) the inhalation of fairly high concentrations of freshly formed zinc oxide fumes.[24]

The ingestion of toxic amounts of excess zinc in food results in an acute and transitory illness within a few minutes after ingestion. The symptoms may include malaise, dizziness, tightness of the throat, emesis, colic and diarrhea. Stool cultures are negative and a history of "food poisoning" cannot be elicited.[25] The excess zinc can usually be attributed to the preparation of acid foods (fruits, juices, stewed fruits) in galvanized containers resulting in solubilization of zinc. The treatment is symptomatic and supportive; preventive measures will obviate a recurrence.

In general, zinc and zinc salts are well tolerated by the human skin. Poisoning due to direct contact with zinc or zinc salts has been reported repeatedly though it seems to be a relatively rare occurrence. A contact dermatitis attributable to exposure to zinc chromate has been observed. It is troublesome, persists long after the exposure and is relatively disfiguring. The chromate is, however, presumed to be the prime offender.[26]

The problem of "metal fume fever" ("galvo", "brass-founders' ague", "brass chills", "zinc shakes", "spelter shakes") occupies by far the most prominent role among the possible toxic effects of zinc. Increasing attention to the subject has produced little further insight into the mechanisms of action of the metal fumes in the production of the syndrome.

The inhalation of zinc oxide fumes results in fever, malaise and depression, cough which may become violent enough to induce vomiting, excessive salivation and headache. The occurrence of gastrointestinal symptoms is disputed by most authors, though some claim to have observed colic.[27] A sweetish taste has usually been noted to accompany the attack. A high white blood cell count is typical. Increases in urinary urobilinogen and zinc excretion have been observed.

About eight hours after exposure, while the temperature and white blood cell count have been rising, a chill may occur when the temperature reaches 103 to 104°F followed by a drop of the temperature to normal, while the leucocytosis may persist for another 24 hours. The effect of zinc is not cumulative and does not predispose to other disorders of the respiratory tract. Zinc oxide is not the only metal oxide resulting in this syndrome. Inhalation of cadmium oxide may also produce severe reactions which may be fatal, while similar but less severe reactions may be observed due to the inhalation of magnesium oxide fumes. No reaction is observed following the inhalation of ferric or aluminum oxide.

Metal fume fever is potentially encountered wherever zinc oxide is an industrial product or by-product, as in zinc smelting, the manufacture of zinc oxide and zinc powder, the manufacture of brass or the welding of galvanized iron.[28] The disease is self-limited and its treatment is symptomatic, consisting of general supportive measures.

Obviously, the prevention of metal fume fever is more important than its treatment. Preventive measures are easily instituted and aim at concentration of less than 15 milligrams of zinc oxide per cubic meter for eight hours. When exposures for short periods cannot be avoided, respirators should be used.

SUMMARY

Disturbances of zinc metabolism have been noted in a variety of conditions; their detection, while supplementary to established diagnostic procedures, does not replace these procedures. A great deal more experience is necessary before studies of zinc can be assigned a decisive diagnostic role. At present, there is no known therapeutic role for zinc in systemic disease.

REFERENCES

1. Lechartier, G., and Bellammy, F., "Sur la presence du zinc dans le corps des animaux et dans les végétaux," *Compt. rend.*, **84**, 687 (1877).
2. Keilin, D., and Mann, T., "Carbonic Anhydrase: Purification and Nature of Enzyme," *Biochem. J.*, **34**, 1163–1176 (1940).
3. Vallee, B. L., and Altschule, M. D., "Zinc in Mammalian Organism, with Particular Reference to Carbonic Anhydrase," *Physiol. Rev.*, **29**, 370–388 (1949).
4. Vallee, B. L., "Zinc and Metalloenzymes," in "Advances in Protein Chemistry," edited by M. L. Anson, K. Bailey, and J. T. Edsall, Vol. 10, p. 317, New York, Academic Press, Inc., 1955.
5. Vikbladh, I., "Studies on Zinc in Blood," *Scand. J. Clin. and Lab. Invest.* (*supp. 2*) **3**, 1–74 (1951).

6. Todd, W. R., Elvehjem, C. A., and Hart, E. B., "Zinc in Nutrition of Rat," *Am. J. Physiol.*, **107,** 146–156 (1934).
7. Wachtel, L. W., Hove, E., Elvehjem, C. A., and Hart, E. B., "Blood Uric Acid and Liver Uricase of Zinc-Deficient Rats on Various Diets," *J. Biol. Chem.*, **138,** 361–8 (1941).
8. Fisher, A. M., and Scott, D. A., "Zinc Content of Bovine Pancreas," *Biochem. J.*, **29,** 1055–8 (1935).
9. Vallee, B. L., and Neurath, H., "Carboxypeptidase, a Zinc Metalloenzyme," *J. Biol. Chem.*, **217,** 253–261 (1955).
10. Okamoto, K., "Biologische Untersuchung der Metalle, VII. Uber das Gewebseisen der Malarialeber und-milz, die Zinkverteilung im Tierreich und den Zinkstoffwechsel," *Trans. Jap. Soc. Path.*, **33,** 247–252 (1943).
11. Kadota, I., and Midorikawa, N. M., "Diabetogenic Action of Organic Reagents: Destructive Lesions of Islets of Langerhans Caused by Sodium Diethyldithiocarbamate and Potassium Ethylxanthate," *J. Lab. Clin. Med.*, **38,** 671–688 (1951).
12. Lowry, J. R., Baldwin, R. R., and Harrington, R. V., "Uptake of Radiozinc by Normal and Diabetic Rat Pancreas," *Science* **119,** 219–220 (1954).
13. McIsaac, R. J., "Distribution of Zinc^{65} in Rat Pancreas," *Endocrinology,* **57,** 571–9 (1955).
14. Waldenstrom, J., "Studien uber Porphyrie," *Acta Med. Scand., suppl.*, **82,** 1–254, (1937).
15. Vallee, B. L., Hoch, F. L., and Hughes, W. L., Jr., "Studies on Metalloproteins: Soluble Zinc-Containing Protein Extracted from Human Leucocytes," *Arch. Biochem.* **48,** 347–360 (1954).
16. Vallee, B. L., and Hoch, F. L., "Zinc, a Component of Yeast Alcohol Dehydrogenase," *Proc. Nat. Acad. Sc.*, **41,** 327–338 (1955).
17. Theorell, H., Nygaard, A. P., and Bonnichsen, R. K., "Studies on Liver Alcohol Dehydrogenase: III. Influence of *p*H and Some Anions on Reaction Velocity Constants," *Acta Chem. Scandinav.*, **9,** 1148–1165 (1955).
18. Vallee, B. L., and Hoch, F. L., "Zinc, A Component of Alcohol Dehydrogenase of Horse Liver," *Federation Proc.*, **15,** 619 (1956).
19. Valle, B. L., Adelstein, S. J., and Olson, J. A., "Glutamic Dehydrogenase of Beef Liver, A Zinc Metalloenzyme," *J. Am. Chem. Soc.*, **77,** 5196 (1955).
20. Vallee, B. L. and Wacker, W., "Zinc, a Component of Rabbit Muscle Lactic Dehydrogenase," *J. Am. Chem. Soc.*, **78,** 1771 (1956).
21. Vallee, B. L., Wacker, W. E. C., Bartholomay, A. F., and Robin, E. D., "Zinc Metabolism in Hepatic Dysfunction: I. Serum Zinc Concentrations in Laennec's Cirrhosis and Their Validation by Sequential Analysis," *New Engl. J. Med.*, **255,** 403–8 (1956).
22. Wacker, W. E. C., Ulmer, D. D. and Vallee, B. L., "Metalloenzymes and Myocardial Infarction. II. Malic and Lactic Dehydrogenase Activities and Zinc Concentration in Serum," *New Engl. J. Med.*, **255,** 449–456 (1956).
23. Berfenstam, R., "Studies on Blood Zinc: Clinical and Experimental Investigation into Zinc Content of Plasma and Blood Corpuscles with Special Reference to Infancy, "*Acta paediat.*, (*supp. 87*), **41,** 3–97 (1952).
24. Hegstead, D. M., McKibbin, J. M. and Drinker, C. K., "The Biological, Hygienic, and Medical Properties of Zinc and Zinc Compounds," (suppl. 179 to the Public Health Reports) 1945.
25. Callender, G. R. and Gentzkow, C. J., "Acute Poisoning By the Zinc and Anti-

mony Content of Limeade Prepared in a Galvanized Iron Can," *Military Surgeon*, **80**, 67 (1937).

26. Hall, A. F., "Occupational Contact Dermatitis Among Aircraft Workers," *J. Am. Med. Assoc.*, **125**, 179 (1944).
27. DeBalsac, F. H. H., "Reproduction experimentale de la fievre des fondeurs. Sa forme attenuee; febricule zincique professionnelle des souders," Bull. acad. méd. (Paris), **115**, 555 (1936).
28. Clinton, M., Jr. and Drinker, P., "Industrial Poisoning by Metals," "Metals Handbook," p. 758, American Society for Metals, Cleveland, Ohio, 1948.

CHAPTER 14

USE OF ZINC IN WOOD PRESERVATION AND IN AGRICULTURE

WALTER L. LATSHAW

*Director of Agricultural Department**
United States Smelting, Refining and Mining Company
Salt Lake City, Utah

The use of zinc chloride as a wood preservative dates back over one hundred years. The treatment of timber with chemicals as a practical means of extending its period of usefulness did not become an established practice in this country until the start of the century.

The earliest records of the American Wood-Preservers' Association (organized in 1904) indicate that zinc chloride had an important part in wood preservation from the start, and that by 1908 approximately 15 million pounds of this salt were being used annually. Its use was expanded until a peak was established in 1921, when over 50 million pounds of zinc chloride were used to treat over 93 million cubic feet of timber.

The use of plain zinc chloride as a wood preservative steadily decreased after 1921. About two million pounds were used in 1934 and little if any is used today. Zinc chloride as such is no longer listed by the American Wood-Preservers' Association as a principal preservative. In 1935–6 chromated zinc chloride was introduced to the market and later, copperized, chromated zinc chloride was included. These preservatives, introduced for the purpose of correcting the leaching difficulty experienced with zinc chloride alone, have now taken over and are used in quantities reported in Tables 14-1 and 14-2.[1]

Table 14-1, covering the record of use for the period 1951 through 1955, shows that the consumption of copperized, chromated zinc chloride has steadily decreased, while the use of chromated zinc chloride is holding steady. The present use of these chemicals is for the treatment of prepared timber and lumber, in contrast to former uses of zinc chloride in treating crossties, piles, etc.

Having thus briefly considered the use of zinc for preserving a product of the soil, we may now review a new field for the use of zinc and its compounds, that of agriculture. About the time zinc chloride was attaining its greatest importance as a wood preservative, and along with advancements

* Retired

TABLE 14.1 CONSUMPTION OF PRINCIPAL PRESERVATIVES AND FIRE RETARDANTS AND NUMBER OF TREATING PLANTS IN THE UNITED STATES 1951–55

(Courtesy of American Wood-Preservers' Association)

Year	No. of Plants Reporting	Straight Creosote (gal.)	Creosote-Coal Tar (gal.)	Creosote-Petroleum (gal.)	Total Creosote[1] (gal.)	Total Petroleum (gal.)	Miscellaneous Liquids (gal.)	Penta-chlorophenol (lb)	Chromated Zinc Chloride[2] (lb)
1951....	282	131,531,738,	67,325,385	43,223,134	196,353,792	31,006,507	668,757	2,983,781	1,940,125
1952....	290	123,259,576	80,072,930	50,482,827	200,828,992	38,739,036	1,515,154	4,323,005	2,429,389
1953....	288	104,668,172	70,344,179	56,122,313	178,212,580	45,522,072	42,091	5,397,836	2,924,694
1954....	308	86,862,493	58,223,132	50,378,429	149,415,552	47,693,577	102,413	8,340,997	2,409,857
1955....	317	97,279,719	45,683,874	47,012,922	150,621,518	50,578,878	109,856	10,502,897	2,583,835

Year	Wolman Salts* (Tanalith) (lb)	Celcure* (lb)	Copperized Chromated Zinc Chloride (lb)	Protexol*[3] (lb)	Minalith*[4] (lb)	Chemonite* (lb)	Greensalt* (Erdalith*) (lb)	Other Solids (lb)
1951..........	1,544,181	930,977	1,622,041	932,189	578,946	231,518	228,883	392,546
1952..........	1,658,426	831,028	983,591	598,000	600,566	307,570	222,376	471,205
1953..........	1,900,692	862,740	609,655	422,080	366,023	314,468	45,875	658,976
1954..........	1,966,790	1,088,948	408,638	721,570	307,799	279,766	[5]	490,325
1955..........	2,133,215	1,431,780	333,118	682,709	883,947	359,051	[5]	762,063

* Reg. U. S. Pat. Off.
[1] Does not include creosote used in creosote-coal tar solutions for which proportions were not specified by the reporting plants.
[2] Includes chromated zinc chloride fire retardant.
[3] Used for fire retardant and combined preservative and fire retardant treatment.
[4] Fire retardant.
[5] Included in "Other Solids".

TABLE 14-2. MATERIAL TREATED (IN CUBIC FEET) BY ITEM AND PRESERVATIVE, IN THE UNITED STATES, 1951–55[1]
(*Courtesy of American Wood-Preservers' Association*)

Preservative	Year	Crossties	Switch Ties	Piles	Poles	Wood Blocks	Crossarms	Lumber and Timbers	Fence Posts	All Other	Total
Creosote and creosote-coal tar	1951	76,223,588	7,602,087	16,835,337	71,959,006	3,702,793	3,038,900	22,648,932	7,506,347	2,983,595	212,500,585
	1952	94,011,774	7,924,539	16,147,655	67,599,934	4,185,000	1,293,258	25,435,249	9,948,360	1,686,331	228,232,100
	1953	81,784,078	7,544,369	13,246,777	60,379,500	2,811,105	2,159,939	22,385,931	7,649,179	1,476,529	199,437,407
	1954	64,454,459	5,202,471	11,166,509	45,441,147	2,101,533	1,728,585	18,160,015	7,591,697	999,222	156,845,638
	1955	49,938,586	4,963,438	13,366,852	53,814,684	2,061,671	1,781,175	18,628,467	7,454,527	1,455,600	153,465,000
Creosote-petroleum	1951	37,355,183	2,844,294	200,194	832,378	0	0	2,208,511	2,623,686	314,494	46,378,740
	1952	43,340,547	2,806,560	954,089	825,491	0	104,741	2,519,529	3,378,667	426,605	54,356,229
	1953	44,910,415	3,533,429	1,037,876	1,264,053	832,447	197,434	3,940,031	6,502,077	552,015	62,769,777
	1954	40,660,896	2,206,688	921,327	1,496,659	665,131	15,011	3,325,571	6,445,393	460,548	56,197,224
	1955	35,763,109	2,279,355	502,244	904,095	754,725	2,457	3,138,570	6,443,385	431,278	50,219,218
Petroleum-pentachlorophenol	1951	247,145	80,379	110,919	5,749,743	0	1,127,654	2,026,290	404,283	146,533	9,892,946
	1952	287,838	13,284	41,032	6,877,410	0	1,626,668	1,755,580	456,529	0	11,058,341
	1953	179,880	8,200	19,874	10,501,397	0	1,907,405	2,581,082	251,011	300,110	15,748,959
	1954	392,429	11,136	132,581	15,600,654	0	1,981,315	3,071,787	884,838	58,531	22,133,271
	1955	103,621	8,089	63,786	18,578,486	0	2,534,484	3,425,795	1,908,235	564,711	27,187,207
Wolman Salts* (Tanalith)	1951	316	760	0	3,165	0	1,696	4,801,082	106,556	123,747	5,037,322
	1952	4,409	744	0	39,566	0	0	4,836,412	111,956	387,636	5,380,723
	1953	84	1,260	2,214	0	0	2,289	5,010,853	120,582	629,310	5,766,592
	1954	4,272	409	449	9,034	0	0	5,555,653	170,180	578,699	6,318,696
	1955	282	0	0	622	0	0	6,221,317	166,292	613,729	7,002,242
Chromated zinc chloride	1951	6,005	1,678	1,068	27,561	0	0	2,159,360	514	174,018	2,370,204
	1952	656	0	0	242	0	0	2,674,752	0	183,632	2,859,282
	1953	0	958	0	0	0	422	2,757,577	1,594	191,732	2,952,283
	1954	0	420	0	281	0	0	2,848,979	870	137,915	2,988,465
	1955	0	836	0	1,223	0	0	3,130,857	378	103,826	3,237,120
Celcure*	1951	0	0	0	0	0	0	1,309,356	0	0	1,309,356
	1952	0	0	0	171	0	0	1,218,333	0	0	1,218,504
	1953	0	0	0	0	0	0	1,297,240	282	754	1,298,276
	1954	0	0	0	388	0	0	1,716,070	1,983	7,591	1,726,032
	1955	0	0	0	1,280	0	0	2,313,863	2,181	1,347	2,318,671
All other[2]	1951	89,473	9,469	64,350	2,457,786	6,326	7,918	3,303,208	57,911	1,297,469	7,293,910
	1952	22,672	94	10,680	2,876,483	0	4,716	2,818,306	211,356	1,513,209	7,457,516
	1953	17,731	3,174	21,270	953,992	0	43,030	2,517,196	157,838	881,633	4,595,864
	1954	17,247	1,131	96,902	1,325,125	1,808	18,281	2,328,199	136,739	527,939	4,453,371
	1955	113,379	4,966	1,942	1,465,373	906	11,928	2,574,343	251,744	590,736	5,015,317
All preservatives	1951	113,921,710	10,538,667	17,211,868	81,029,639	3,709,119	4,176,168	38,456,739	10,699,297	5,039,856	284,783,063
	1952	137,667,896	10,745,221	17,153,456	78,219,297	4,185,000	3,029 383	41,258 161	14,106,868	4,197,413	310,562,695
	1953	126,892,188	11,091,390	14,328,011	73,098,942	3,643,552	4,310,519	40,489,910	14,682,563	4,032,083	292,569,158
	1954	105,529,303	7,422,255	12,317,768	63,873,288	2,768,472	3,743,192	37,006,274	15,231,700	2,770,445	250,662,697
	1955	85,918,977	7,256,684	13,934,824	74,765,763	2,817,302	4,330,044	39,433,212	16,226,742	3,761,227	248,444,775

* Reg. U. S. Pat. Off.

[1] Cubic footage for all years shown is based on volume reported each year. These data will not agree, therefore, with data shown in Wood Preservation Statistics reports for 1952 and earlier years where cubic footage was based on a standard set of conversion fact

in the fundamental science of chemistry and analytical techniques, it became fairly well established that many elements—formerly considered of little if any importance in agriculture—were definitely essential to the adequate nutrition of plants and animals.* The list of these essential elements now includes copper, zinc, manganese, fluorine, iodine, boron, molybdenum and possibly others, yet undiscovered.

As knowledge concerning the presence of zinc in plant and animal tissue became general and the essential nature of the element established, it was not long before plant physiologists and agronomists started to trace unknown plant disorders and peculiarities of growth to deficiencies of zinc in the affected tissues of the plants. In many cases, a deficiency of zinc in the plant was traced directly to a scarcity in the soil. This condition of soil scarcity prevails throughout the more humid sections of our country, where no doubt leaching has extracted readily soluble zinc from the soils. Along with progressive research work in the field of zinc deficiency in plants, it was discovered that in basic and sometimes neutral soils plants were found to be suffering from a deficiency of zinc, while growing in a soil medium containing a considerable supply of the element. The answer here was found to be one of solubility. The zinc was unavailable to the plant because it was insoluble.

The following extract from an article by J. H. Hunter appearing in *Country Gentleman* several years ago will help to explain the practical problem as related to the correction of zinc deficiency in pecans. "Fertilizing pecans starts with zinc, since a deficiency of this element is the most important condition hampering the production of these nuts. Also, the pecan requires zinc in larger amounts than other crops growing on the same soils. Zinc sulfate, applied either to the soil or in solution as a spray to the trees, has been found especially beneficial in eliminating the trouble commonly known as 'rosette,' which is caused by a zinc deficiency. Soil applications are effective only in areas where the soils are acid. Where they are neutral or alkaline, spraying is more practical."

The quotation from Mr. Hunter emphasizes the need for zinc to supply a deficiency in a particular plant. A pertinent question here would be, how widespread is zinc deficiency in the plants of our country? In this connection, information furnished in 1955 by George P. Teel, Jr., editor of *Farm Chemicals*, is of great interest. He notes that "areas of known and suspected zinc-deficient soils have been reported from 28 states, while some 41 crops are reported affected by the deficiency. The tier of states running across the southern boundary of our country appears to be affected most seriously. Annual application of zinc salts is placed on crops such as citrus, tung, pecans, deciduous fruits, peanuts, cotton, corn and vegetables."

* See also Chapter 13 for the role of zinc in biology.

Figure 14-1. States where zinc deficiency has been proved, shaded area; area of suspected deficiency, cross-hatched (1955).

Figure 14-1 shows the states where zinc deficiency has been proved (indicated by shaded areas), and the area of suspected deficiency (indicated by cross-hatching). It may be observed that in much of the United States, zinc-deficient plants have been found, and new areas are being found constantly. The zinc-deficient area of this map now covers five states not included in 1955, and two states where deficiency was suspected but not proved at that time. It may be anticipated in the near future, as research and experimental work progress, that all of our states will have areas where economic plants are found to be suffering from a deficiency of this important nutritional element.

The form of zinc used as a nutritive supplement, from a tonnage point of view, as reported by the United States Department of Agriculture and the State of California, is largely confined to zinc sulfate and zinc oxide. Other forms of zinc used for this purpose are zinc-manganese (used extensively in the Northwest), zinc chelate, zinc EDTA (disodium zinc ethylenediaminetetraacetate), zinc chloride, zinc carbonate, stripping acid residue, and no doubt many other zinc compounds. Two of these, which presently come to mind, are the fungicides zineb and ziram. The use of these zinc-containing fungicides has accidentally brought to light a number of instances of deficiency and plant-growth response that could be attributed only to the presence of the zinc ion.

The work of Brown, Viets and Crawford[2] is of particular interest in connection with plant utilization of zinc from various types of zinc compounds. Their results, after working with a zinc-deficient, noncalcareous neutral soil (*p*H 7.2) and limited to two plant subjects, beans and grain sorghum, showed that "zinc utilization was greater from the chelate than any other material, including $ZnSO_4$. Uptake from stripping-acid residue, ZnO, $Zn_3(PO_4)_2$, and $ZnCO_3$ was comparable and of about the same magnitude as from $ZnSO_4$." Blast-furnace slag (17 per cent zinc) and several forms of zinc frit materials showed little or no zinc utilized, although zinc granules (10-mesh) compared favorably with zinc sulfate. A duplication of this experiment using calcareous soils with a higher *p*H and other plant subjects should be of great interest.

As with iron, the amount of zinc required for the balanced nutrition of a plant is very small and comes within the range of a fraction of a part per million in the soil solution. The amount applied to the soil in the case of zinc sulfate and zinc oxide is usually not less than 40 pounds per acre, which may be adequate over a period of several years depending upon conditions. The limbs and twigs of deciduous fruit trees are sprayed while dormant with a 4 to 6 per cent solution of zinc sulfate, while some growing plants have been supplied by dusting with zinc oxide. However, the usual application to the plant in leaf is a spray solution with chelate of zinc, or a 0.5 per cent solution of zinc sulfate that has been partially neutralized to form a basic sulfate.

The data available for estimating the quantity of zinc being used annually for plant nutritional and fungicidal purposes is, in many cases, incomplete and unreliable. The Food and Materials Division of the United States Department of Agriculture supplied some data several years ago from which it was estimated that, in 1953, about 10,000 tons of zinc compounds were used for plant nutritional purposes, and 3,600 tons applied in insecticide and fungicide sprays. This source of information, reporting as of May 1, 1957, states that 8,187 tons of zinc sulfate alone were shipped to agriculture in 1955. In correspondence with the United States Department of Agriculture and with various state authorities where zinc is used for nutritional and fungicidal purposes, it developed that only California could supply reliable statistical data. Their figures for the tonnage of zinc oxide and zinc sulfate used in agriculture[3] are shown in Table 14-3.

TABLE 14-3. SEGREGATED AGRICULTURAL MINERAL TONNAGE

Year	Zinc Oxide	Zinc Sulfate	Total
1950	2	165	167
1951	19	293	312
1952	103	557	660
1953	225	779	1004
1954	145	927	1072
1955	171	1098	1167
1956	150	1602	1752

Aside from specific zinc compounds used directly for nutritional purposes, it is believed that considerable zinc may be added to mixed fertilizers sold in California and elsewhere.

The following note from a recent issue of the *Journal of Agricultural and Food Chemistry*[4] presents some ideas concerning the rapidly expanding use of zinc as a fungicide:

"Zineb (zinc ethylenebisdithiocarbamate) appears to be headed for a big boost in Florida, for control of russet on citrus fruit. Francine E. Fisher of University of Florida's Citrus Experiment Station at Lake Alfred discovered in 1955 that the compound, applied as a postbloom spray, gives virtually complete control of the citrus pest. Although she recommends caution until more detailed tests are run, many growers are plunging into its use on the basis of her limited summer spraying over the past two years. Minute Maid, Florida's largest citrus grower, has just ordered 186,000 pounds of zineb to treat some 60 per cent of its 21,000 acres in July. Independents are reported to have begun using zineb almost indiscriminately; some even started full-scale spraying as early as February."

In the state of Washington, where zinc-deficiency troubles are known to be prevalent over large areas, soil scientist J. L. Wilson of the Irrigation

Experiment Station at Prosser estimates that "1400 tons of ZnMnS and 400 tons of $ZnSO_4$ were sold in his state in 1954." The Washington State Department of Agriculture has some data covering zinc sales to agriculture, showing somewhat less, however, since they have only recently attempted to collect such data. The author therefore is inclined to accept Mr. Wilson's figures. It is of interest to note that the mixture of zinc and manganese is reported to come from Canadian sources.

The response from most states to inquiries concerning the status of zinc in this field is somewhat discouraging, since most of them frankly admit that they have no record covering zinc salts or compounds sold for fertilizer or directly for plant nutritional purposes. It would appear that a qualified agency could perform an outstanding service for the zinc industry were it to furnish statistical data, collected from trade sources, covering the annual movement of zinc by states for agricultural purposes, with particular reference to nutritional and fungicidal applications.

From all indications, the future use of zinc in the agricultural field is most promising. It is probable that we have barely scratched the surface in supplying the potential demand for zinc salts or compounds to be used for nutritional and fungicidal purposes. The data from the progressive state of California, where the use of zinc oxide and zinc sulfate alone for these purposes shows an increase from 167 tons in 1950 to 1752 tons in 1956, gives some idea of what can be expected. With adequate public enlightenment on the influence of zinc deficiency on plant production, and the greater use of zinc salts in the manufacture of adequate commercial fertilizers, it is reasonable to believe that, in the course of 15 or 20 years, an annual consumption of 100,000 tons of zinc salts in the agricultural field could be a conservative figure.

To those who may have an academic interest in zinc deficiency in plants, and for the purpose of showing the widespread interest among research workers in this field, attention is directed to a recent paper by Thorne[5] of Utah State University, which includes an excellent bibliography of approximately two hundred titles, and not only reviews the literature but also brings the subject of zinc deficiency up to date.

REFERENCES

1. *Proc. Am. Wood-Preservers' Assoc.*, **52**, 305–6 (1956).
2. Brown, L. C., Viets, Frank Jr., and Crawford, Earl L., *Soil Sci.*, **83**, 219–27 (1957).
3. California, Department of Agriculture, Bureau of Chemistry, Special Publication No. 255, "Fertilizing Materials," p. 13, 1954; Special Publication No. 260, p. 7, 1955; Bureau of Chemistry Announcement No. FM-293, p. 2, 1957.
4. Research News Letter, "Zineb as Cure for Citrus Russet," *J. Agri. Food Chem.*, **15**, 317 (1957).
5. Thorne, Wynne, "Zinc Deficiency and Its Control," *Advances in Agron.*, **9**, 31–65 (1957).

NAME INDEX

SUBJECT INDEX